VOLUME	EDITOR-IN-CI		
38	JOHN C. SHEEHAN		rint
39	THE LATE MAX TISHLER		print
Collective Vol. IV	A revised edition of Annual Volumes 30–39 NORMAN RABJOHN, *Editor-in-Chief*	1036	
40	MELVIN S. NEWMAN	114	*Out of print*
41	JOHN D. ROBERTS	118	*Out of print*
42	VIRGIL BOEKELHEIDE	118	*Out of print*
43	B. C. MCKUSICK	124	*Out of print*
44	THE LATE WILLIAM E. PARHAM	131	*Out of print*
45	WILLIAM G. DAUBEN	118	*Out of print*
46	E. J. COREY	146	*Out of print*
47	WILLIAM D. EMMONS	140	*Out of print*
48	PETER YATES	164	*Out of print*
49	KENNETH B. WIBERG	124	*Out of print*
Collective Vol. V	A revised edition of Annual Volumes 40–49 HENRY E. BAUMGARTEN, *Editor-in-Chief*	1234	
Cumulative Indices to Collective Volumes I, II, III, IV, V	RALPH L. AND THE LATE RACHEL H. SHRINER, *Editors*		
50	RONALD BRESLOW	136	*Out of print*
51	RICHARD E. BENSON	209	*Out of print*
52	HERBERT O. HOUSE	192	*Out of print*
53	ARNOLD BROSSI	193	*Out of print*
54	ROBERT E. IRELAND	155	*Out of print*
55	SATORU MASAMUNE	150	*Out of print*
56	GEORGE H. BÜCHI	144	*Out of print*
57	CARL R. JOHNSON	135	*Out of print*
58	THE LATE WILLIAM A. SHEPPARD	216	*Out of print*
59	ROBERT M. COATES	267	*Out of print*
Collective Vol. VI	A revised edition of Annual Volumes 50–59 WAYLAND E. NOLAND, *Editor-in-Chief*	1208	
60	ORVILLE L. CHAPMAN	140	*Out of print*
61	THE LATE ROBERT V. STEVENS	165	*Out of print*
62	MARTIN F. SEMMELHACK	269	*Out of print*
63	GABRIEL SAUCY	291	*Out of print*
64	ANDREW S. KENDE	308	
Collective Vol. VII	A revised edition of Annual Volumes 60–64 JEREMIAH P. FREEMAN, *Editor-in-Chief*	602	
65	EDWIN VEDEJS	278	
66	CLAYTON H. HEATHCOCK	265	
67	BRUCE E. SMART	289	
68	JAMES D. WHITE	318	
69	LEO A. PAQUETTE		
Reaction Guide to Collective Volumes I–VII and Annual Volumes 65–68	DENNIS C. LIOTTA AND MARK VOLMER, *Editors*	854	

ORGANIC SYNTHESES

ORGANIC SYNTHESES

Reaction Guide

INCORPORATING COLLECTIVE VOLUMES 1-7
AND ANNUAL VOLUMES 65-68

DENNIS C. LIOTTA
MARK VOLMER
Department of Chemistry, Emory University, Atlanta, GA 30322

John Wiley & Sons, Inc.
NEW YORK / CHICHESTER / BRISBANE / TORONTO / SINGAPORE

Published by John Wiley & Sons, Inc.

"John Wiley & Sons, Inc. is pleased to publish this volume of Organic Syntheses on behalf of Organic Syntheses, Inc. Although Organic Syntheses, Inc. has assured us that each preparation contained in this volume has been checked in an independent laboratory and that any hazards that were uncovered are clearly set forth in the write-up of each preparation, John Wiley & Sons, Inc. does not warrant the preparations against any safety hazards and assumes no liability with respect to the use of the preparations."

Library of Congress Catalog Card Number: 21-17747
ISBN 0-471-54261-X

Printed in the United States of America

10 9 8 7 6 5 4 3 2 1

FOREWORD

The action pioneered by Roger Adams -- to make available to the chemical community at large synthetic procedures that have proven reliability -- has been enthusiastically received for many decades. In typical fashion, the 69th volume of *Organic Syntheses* to issue in 1990 will continue the tradition of featuring a selection of exemplary transformations that are extensively detailed, practically designed, and well suited to laboratory adoption. However, the richness of possibilities covered by this series has created an information retrieval problem not unlike that affecting the entire field of synthetic methodology. Few of us are totally aware of the riches contained therein.

This problem was viewed by me with growing concern. Some three years ago, therefore, a suggestion was offered that was designed to offset this trend by taking advantage of the organic chemist's innate ability to assimilate structural formulas with ease and striking rapidity. The concept of publishing a visual index to *Organic Syntheses* was enthusiastically received by my fellow members of the Editorial Board. Following encouragement by the Board of Directors, the team of Dennis Liotta and Mark Volmer was commissioned to bring the *Reaction Guide* to reality. Their expert accomplishment, fashioned with a great deal of attention to subject organization and layout, has resulted in the availability of a unique work, the usefulness of which will persist for many years to come. It is our hope that you will partake in the scanning of its contents frequently and that the *Reaction Guide* serves as a valuable sourcebook for locating detailed procedural information and practical hints of immediate value to your experimental undertakings.

LEO A. PAQUETTE

Columbus, Ohio
September, 1990

PREFACE

Despite the ever-increasing number of useful compendia of synthetic methodology, *Organic Syntheses* remains a unique reference source to organic chemists. Where else can one find detailed and independently verified experimental procedures for effecting virtually every important organic reaction? Indeed, no other compendium (or, for that matter, primary literature source) provides as reliable and in-depth discussions about the experimental details of reactions as does *Organic Syntheses.*

If there is a problem with *Organic Syntheses*, it is that many (perhaps most) organic chemists really are not aware of the many kinds of procedures it covers. One solution to this problem has already been realized -- a searchable graphics database of *Organic Syntheses* (available from Molecular Design Limited™). While the advantages of this approach are fairly obvious (*e.g.*, the ability to search using substructures, keywords, Boolean operators, etc.), the hardware and software costs associated with this searching protocol place it out of reach of many individuals. Moreover, the computer format, while excellent for specific searches, discourages simple browsing through the entries. As a consequence, we have prepared this *Organic Syntheses Reaction Guide* to provide information about the contents of *Organic Syntheses* in an easy-to-use book format.

The *Reaction Guide* summarizes, in structural format, all procedures published to date in *Organic Syntheses* (*i.e.*, Collective Volumes 1 through 7, as well as Volumes 65 through 68). The only exceptions to this include: (a) selected reagents used in a reported transformation (*e.g.*, the preparation of lithium dialkyl cuprates used in conjugate addition reactions); (b) reactions deemed trivial (*e.g.*, some simple hydrolyses); (c) transformations which appear only in the discussion section of an entry. However, none of these exceptions ever resulted in the complete omission of an entry.

In attempting to classify the entries, three different approaches were considered: (a) the use of keywords; (b) the use of the *Organic Syntheses* Reaction Types; and (c) the use of a small number of general reaction categories *(vide infra).* The disadvantages of the keyword approach quickly became obvious. Since the choice of keywords is somewhat subjective, many entries would require multiple keywords to ensure that each reaction was categorized in a way which could be readily understood and utilized by readers. In order to handle this, an elaborate cross-referencing system would be required. When it became apparent that this approach was far too complicated to be useful, it was abandoned.

When this volume was initiated, it was assumed that it would be based on the Reaction Type categories already used in the Cumulative Indices. However, the Reaction Indices often proved too detailed for our purposes. While detailed indices are highly desirable for specific searches, in our opinion too many sub-categories tend to focus readers unnecessarily and to detract from their ability to obtain an overview.

The chosen method uses an indexing system based on eleven broad classes of reactions which would be readily recognized by the vast majority of readers. The categories, along with their definitions, are as follows:

ADDITION	Any reaction whose product contains all of the key elements of the reactants.
ANNULATION	Any ring-forming reaction.
C-C BOND FORMATION	Any carbon-carbon bond-forming reaction.
CLEAVAGE	Any carbon-carbon sigma bond scission.
ELIMINATION	Any reaction which results in the loss of a molecular fragment or fragments and concomitant formation of a multiple bond or bonds.
MISCELLANEOUS	Any process which does not fit any of the other categories listed here.
OXIDATION	Any reaction which increases the degree of unsaturation and/or the number of heteroatoms at a center.
PROTECTION / DEPROTECTION	Any functional group masking and/or unmasking.
REARRANGEMENT	Any reaction involving either a connectivity change or a π-π or π-σ interconversion.
REDUCTION	Any reaction which decreases the degree of unsaturation and/or the number of heteroatoms at a center.
SUBSTITUTION	Any reaction which involves the replacement of one functional group for another.

In an effort to minimize "subjectivity problems," *i.e.*, classification of reactions in categories which might not be universally recognized, we decided to permit a small amount of redundancy. Thus, many of the entries are listed in multiple categories. In this *Reaction Guide*, the classification of reactions is based primarily on structural changes which occur in individual reactions, rather than in the overall transformation reported in a multiple-step entry. However, since chemists often think about reactions from a mechanistic perspective, we have tried to accommodate this whenever possible. For example, the well-known Claisen rearrangement involves the conversion of an allylic alcohol to a rearranged γ,δ-unsaturated carbonyl compound *(vide infra)*. Since the primary structural change occurring here is the formation of a carbon-carbon bond, these reactions are listed under C-C BOND FORMATION. However, since in a mechanistic sense the reaction involves a [3,3]-sigmatropic rearrangement of an intermediate generated *in situ*, it is also listed as a REARRANGEMENT. (This example represents the exception, rather than the rule. Most individual reactions are classified under a single category.)

OH CH$_3$ $\xrightarrow[\text{CH}_3\text{CH}_2\text{CO}_2\text{H, 138 - 140 °C}]{\text{CH}_3\text{C(OEt)}_3}$ [H OEt O CH$_3$] → H CH$_2$CH$_2$CO$_2$Et CH$_3$

In cases such as the one described above, some redundancy can prove helpful to the reader. In most situations, however, redundancy was avoided since it would only increase the overall bulk of the *Reaction Guide*, thereby making it more difficult to use. In accord with this principle, the following qualifications have been imposed on our classification scheme:

(a) Since all carbon-carbon bond-forming reactions are either substitutions or additions, these reactions are not included in the SUBSTITUTION or ADDITION sections.

(b) Since all annulations are either substitutions or additions, annulation reactions are also not included in the SUBSTITUTION or ADDITION sections. Entries in ANNULATION contain all ring-forming reactions, irrespective of the number of bonds which are made in the process in question, and thus include both one-bond cyclizations and multiple-bond cycloadditions.

(c) Since any annulation which involves the formation of a carbon-carbon bond is also a carbon-carbon bond-forming reaction, these reactions are not included under C-C BOND FORMATION.

(d) The substitution of a heteroatom for a hydrogen formally represents an oxidation. In order to keep the overall size of the volume viable, aromatic substitution reactions are not included in the OXIDATION section.

(e) Hydrolyses of carbonyl derivatives are listed in SUBSTITUTION (usually in the TRIGONAL sub-category, *vide infra*). In certain cases, hydrolyses are also listed in the ADDITION section (*e.g.*, the hydrolysis of a nitrile to an amide).

(f) All protections or deprotections can be classified in at least one of the other categories. Entries included in PROTECTION / DEPROTECTION are those which employ masking functionality specifically to enhance reaction selectivity. In this section, entries have been arranged according to the functional group(s) involved (instead of the chronological approach employed in the other sections) using the following order: Aldehydes (CHO); ketones (CO); hydroxy compounds (OH); nitrogen compounds (NH_X); all others (GENERAL).

(g) Several of the major categories have been further segmented into subsections whose definitions are, by and large, self-explanatory. For example, SUBSTITUTION - TRIGONAL includes substitution reactions which occur at trigonal centers. Perhaps the only exception to this is the C-C BOND FORMATION - COUPLING section, which is defined here as any carbon-carbon bond-forming reaction which results in the formation of a symmetrical product.

(h) For convenience, ANNULATION - CARBOCYCLIC and ANNULATION - HETEROCYCLIC have been listed in order of the size of the ring or rings formed in a particular transformation.

(i) Unless specified (*e.g.*, *R* or *S*), materials containing stereogenic centers are racemic.

Rather than trying to explain how each entry was categorized, it may be more useful to examine specific cases which illustrate both the strengths and the weaknesses of our classification scheme. The first example involves the conversion of a substituted *o*-nitrotoluene to a substituted indole (*Organic Syntheses, CV 7,* 34 (1990); *63,* 214 (1985)). The first step involves the protection of a phenolic hydroxyl group as its benzyl ether. This reaction is listed in PROTECTION / DEPROTECTION. This reaction could have also been classified under a variety of sub-categories of SUBSTITUTION. However, since protection of the hydroxyl group is critical to the success of subsequent reactions, classification under PROTECTION / DEPROTECTION seemed most appropriate. The second reaction is a condensation between an ortho amide and the acidic methyl group of the *o*-nitrotoluene. This is again listed under C-C BOND FORMATION - OLEFINATION. The final step of the sequence is a reductive cyclization and, therefore, this reaction is included under both REDUCTION - HETEROATOM and ANNULATION - HETEROCYCLIC - [5].

OH
NO_2
$PhCH_2Cl$
K_2CO_3, DMF
OCH_2Ph
NO_2
NH
$Me_2NCH(OMe)_2$
OCH_2Ph
N
NO_2
Ra(Ni)
H_2N-NH_2
OCH_2Ph
N
H

PREFACE

Most chemists would probably consider the above classifications to be reasonable. However, since the reductive cyclization step is actually an addition - elimination process, one might question whether this entry should also be included in both ADDITION and ELIMINATION. This was not done, primarily because these descriptors were judged to be too general. In fact, if this concept were taken to the extreme, many simple reactions (*e.g.*, hydrolyses of carboxylic acid derivatives) would also necessarily be included under both of these categories. The effect of this would be an enormous increase in the size of the *Reaction Guide* without any comparable increase in its usefulness.

Sometimes the simplest reactions are the most difficult to classify. Consider the conversion of a secondary alcohol to its corresponding acetate. Although mechanistically this reaction involves an addition - elimination at a carbonyl carbon, on the basis of the structural change the reaction can be classified as either a substitution at a trigonal center or as the replacement of a hydroxyl group with an acetoxy group at a methylene carbon. In this case the latter approach was chosen because the alcohol in question was deemed to be the "more important" of the two reactants. In situations where the choice was less clear-cut, the reaction was listed in both possible categories. For example, the formation of a quaternary ammonium salt from a trialkylamine and an alkyl halide was listed in both SUBSTITUTION - TETRAHEDRAL and SUBSTITUTION - HETEROATOM.

Regarding the format of entries, we have presented each unit transformation with the same information content as in the original publication; unimportant by-products have been omitted. In many cases, especially in the earlier volume entries, some additional information (solvent, pressure, etc.) has been provided for clarity. On occasion, two- or three-step reaction sequences have been included *in toto* in each of their respective categories, rather than depicting them separately. We chose this approach, in cases where the extra information required no significant additional space, to provide a broader context for the individual reactions of a given entry.

Before closing, it is important to re-emphasize a point made at the beginning of this discussion. The reason for compiling this *Reaction Guide* was to provide individuals who utilize *Organic Syntheses* procedures with an inexpensive tool for doing simple structural searches. Additionally, it was hoped that the book format would also facilitate "browsing" through the contents of this resource. It was never our intention to devise a system which renders all of the searching indices obsolete. Quite the contrary: the *Reaction Guide* should complement, and not compete with, these excellent alternatives.

DENNIS C. LIOTTA
MARK VOLMER

Atlanta, Georgia
August, 1990

CONTENTS

CONTENTS

ORGANIC SYNTHESES

ANNULATION

***OS*, *CV* 3, 221 (1955)**

Cl–(CH$_2$)$_3$–CN →(NaOH, Δ) cyclopropyl–CN →(H_2SO_4, H_2O) cyclopropyl–CO_2H

***OS*, *CV* 3, 223 (1955)**

Cl–(CH$_2$)$_3$–CN →(Na, NH_3) cyclopropyl–CN

***OS*, *CV* 4, 597 (1963)**

CH_3–CO–(CH$_2$)$_3$–Cl →(NaOH, H_2O, Δ) CH_3–CO–cyclopropyl

***OS*, *CV* 5, 306 (1973)**

cyclooctene →($CHBr_3$, *t*-BuOK, *t*-BuOH) 9,9-dibromobicyclo[6.1.0]nonane (Br, Br)

***OS*, *CV* 5, 328 (1973)**

Ph–CH(Br)–CH$_2$–CH$_2$–Br →(Zn-Cu, DMF, 7 - 9 °C) Ph–cyclopropyl

OS, CV 5, **509 (1973)**

Ph, Ph C=CH—CO_2H → (LAH; THF, Δ) 1,1-diphenylcyclopropane (Ph, Ph)

OS, CV 5, **514 (1973)**

PhCH(Br)–CO–CH(Br)Ph → (Et_3N; CH_2Cl_2) diphenylcyclopropenone (O; Ph, Ph)

OS, CV 5, **855 (1973)**

cyclohexene → (CH_2I_2, Zn (Cu); Et_2O, Δ) norcarane

OS, CV 5, **859 (1973)**

cyclohexene → ($ClCH_2$—OCH_2CH_2Cl; CH_3Li, Et_2O) 7-H, 7-OCH_2CH_2Cl norcarane

ca. **6 : 1** ***exo : endo***

OS, CV 5, 874 (1973)

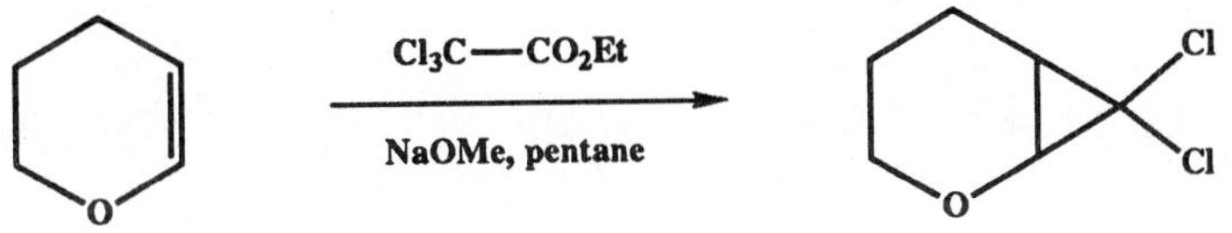

$Cl_3C—CO_2Et$

NaOMe, pentane

Cl

Cl

O

OS, CV 5, 929 (1973)

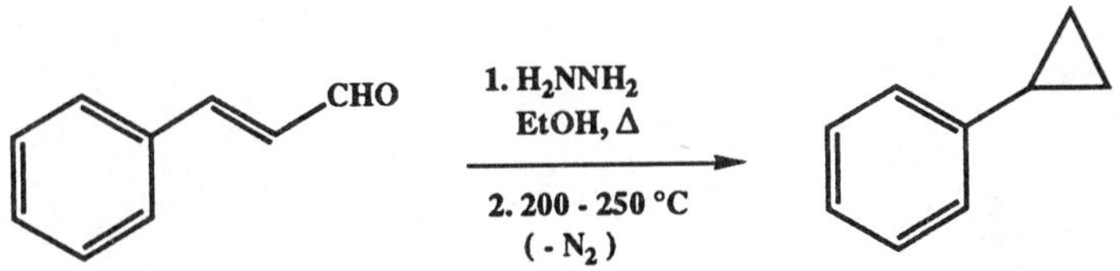

CHO

1. H_2NNH_2
EtOH, Δ

2. 200 - 250 °C
(- N_2)

OS, CV 5, 1058 (1973)

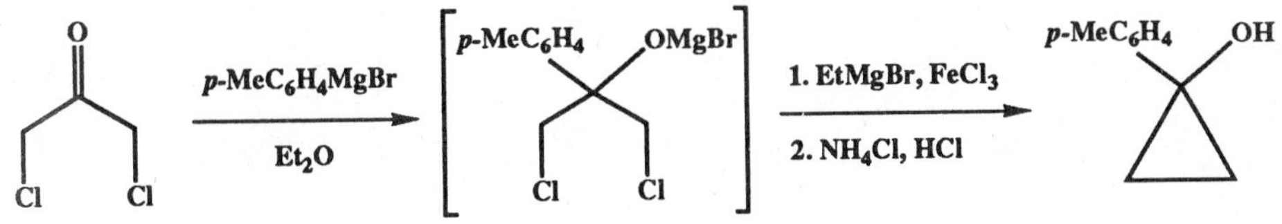

O

Cl Cl

p-MeC_6H_4MgBr

Et_2O

p-MeC_6H_4 OMgBr

Cl Cl

1. EtMgBr, $FeCl_3$

2. NH_4Cl, HCl

p-MeC_6H_4 OH

OS, CV 6, 87 (1988)

t-BuOK

$CHCl_3$, 0 - 5 °C

Cl

Cl

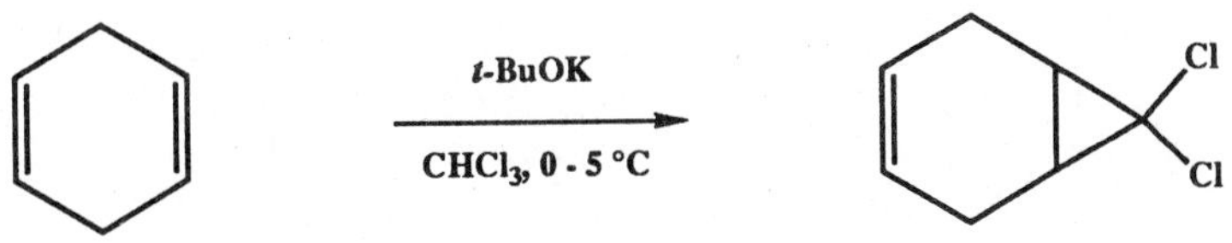

***OS*, *CV* 6, 153 (1988)**

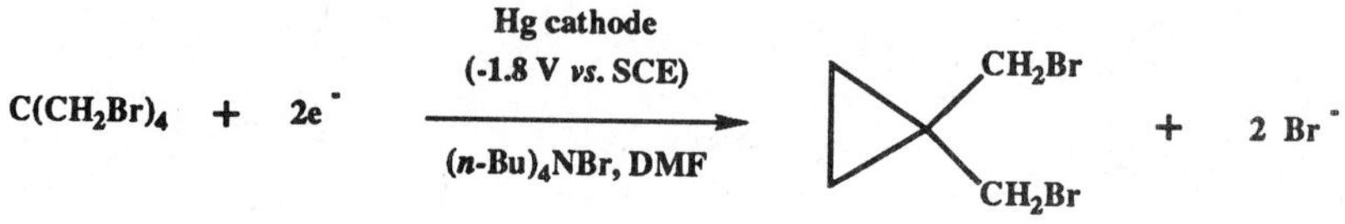

$C(CH_2Br)_4$ + $2e^-$ → Hg cathode (-1.8 V *vs.* SCE), $(n\text{-Bu})_4NBr$, DMF → CH_2Br / CH_2Br + 2 Br^-

***OS*, *CV* 6, 187 (1988)**

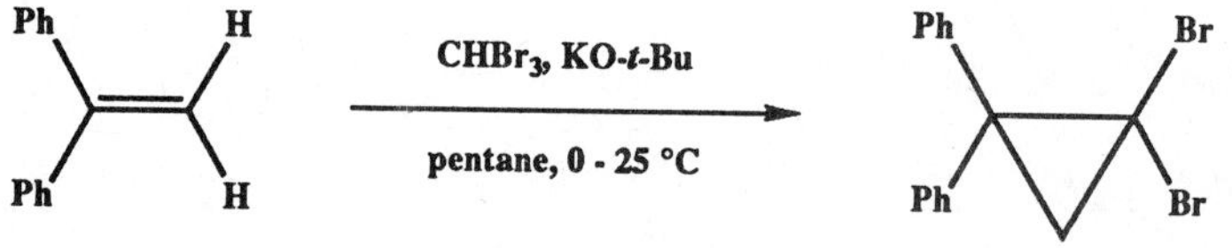

Ph, Ph, H, H → $CHBr_3$, KO-*t*-Bu, pentane, 0 - 25 °C → Ph, Ph, Br, Br

***OS*, *CV* 6, 226 (1988)**

O, N, OEt → hν, *t*-BuOH → [N=C=O, OEt] → Δ, *t*-BuOH → NH—CO_2-*t*-Bu, OEt

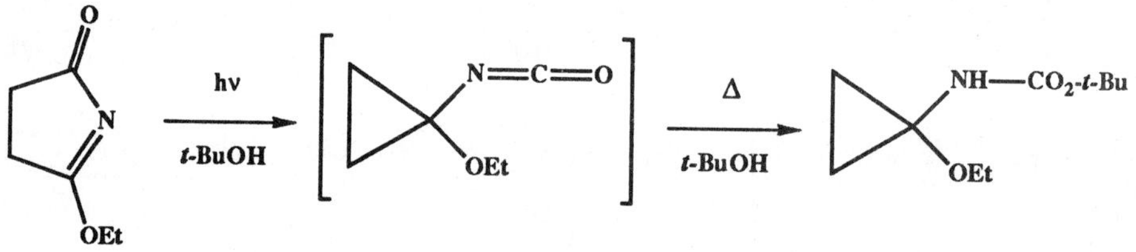

***OS*, *CV* 6, 320 (1988)**

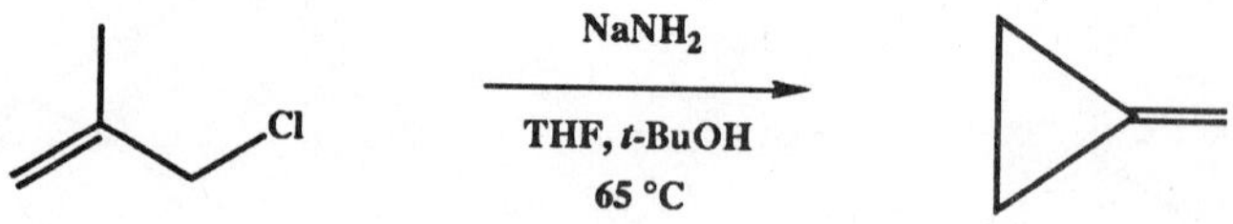

Cl → $NaNH_2$, THF, *t*-BuOH, 65 °C →

OS, CV 6, 327 (1988)

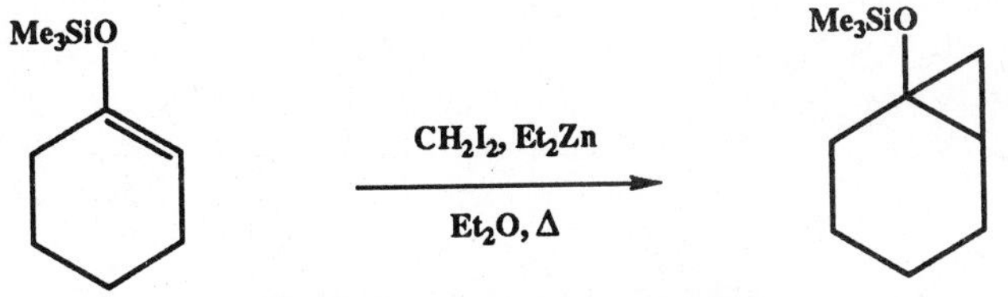

Me_3SiO

CH_2I_2, Et_2Zn

Et_2O, Δ

Me_3SiO

OS, CV 6, 361 (1988)

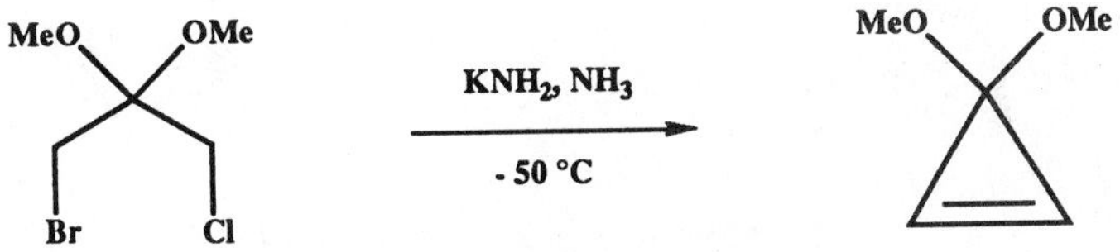

MeO OMe

Br Cl

KNH_2, NH_3

- 50 °C

MeO OMe

OS, CV 6, 364 (1988)

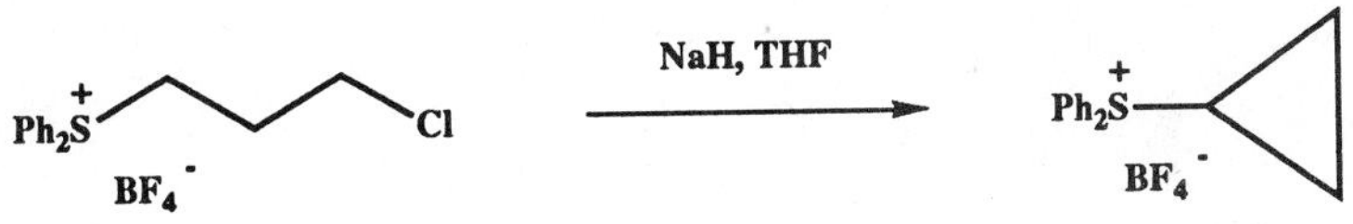

$Ph_2\overset{+}{S}$ Cl

BF_4^-

NaH, THF

$Ph_2\overset{+}{S}$

BF_4^-

OS, CV 6, 401 (1988)

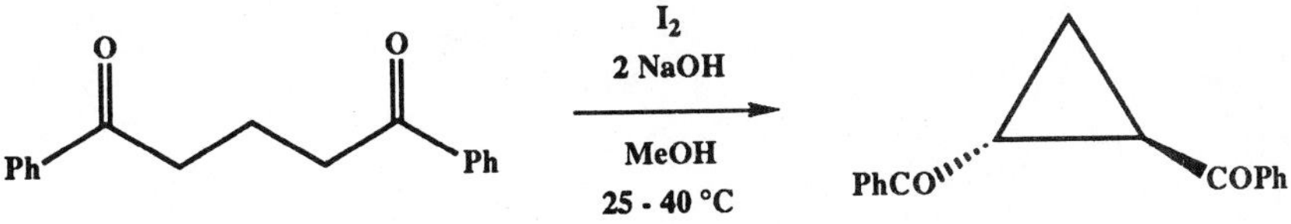

O O

Ph Ph

I_2

2 NaOH

MeOH

25 - 40 °C

PhCO COPh

OS, CV 6, 571 (1988)

CH_3 Cl + OEt → (2,2,6,6-tetramethylpiperidide, N Li; Et_2O, Δ) → CH_3 OEt

OS, CV 6, 731 (1988)

→ ($CHCl_3$, *t*-BuOK; - 30 to 0 °C) → Cl Cl

OS, CV 6, 913 (1988)

Ph + N_2 CO_2Et → (Δ; xylene) → Ph CO_2Et

trans > *cis*

OS, CV 6, 974 (1988)

CH_3 CH_3 CH_3 CH_3 + CHI_3 → (hν; aq. NaOH, CH_2Cl_2) → CH_3 CH_3 CH_3 CH_3 I

OS, *CV 6*, 991 (1988)

2 *t*-BuOK

THF, - 70 °C
then aq. HCl

OS, *CV 7*, 12 (1990); *60*, 6 (1981)

$HCCl_3$, NaOH
40 - 60 °C

phase-transfer
conditions

OS, *CV 7*, 203 (1990); *60*, 53 (1981)

NaOCl
DMSO

H_2O
hexane

benzene
reflux

OS, *CV 7*, 200 (1990); *61*, 39 (1983)

$CHBr_3$

t-BuOK

MeLi

Et_2O

OS, *CV 7*, 131 (1990); *63*, 147 (1985)

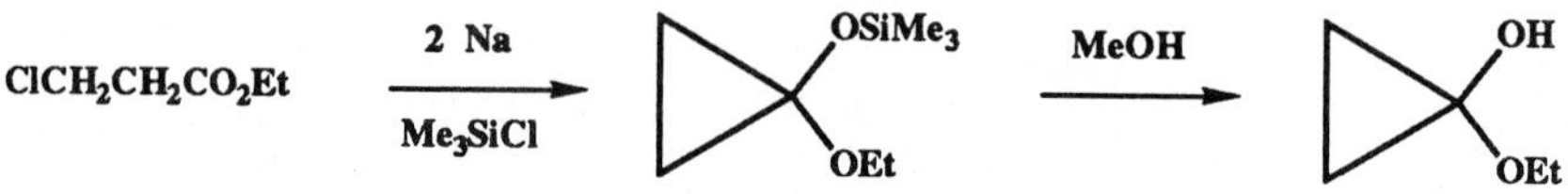

$ClCH_2CH_2CO_2Et$ → (2 Na, Me_3SiCl) → OSiMe$_3$ / OEt → (MeOH) → OH / OEt

OS, *67*, 76 (1988)

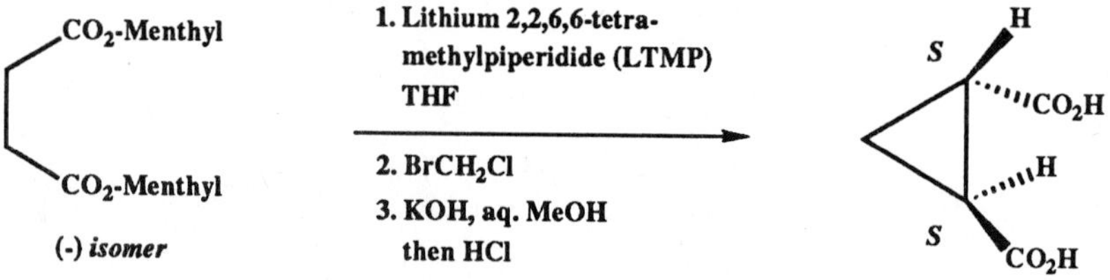

CO_2-Menthyl / CO_2-Menthyl

(-) *isomer*

1. Lithium 2,2,6,6-tetramethylpiperidide (LTMP) THF
2. $BrCH_2Cl$
3. KOH, aq. MeOH then HCl

H, S, CO_2H, H, S, CO_2H

OS, *67*, 176 (1988)

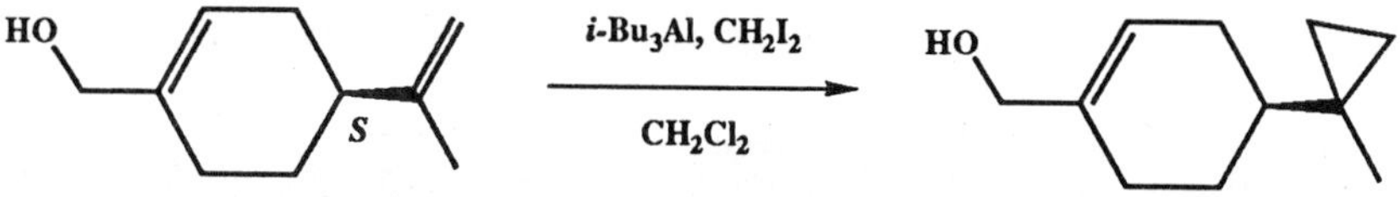

HO, S → (i-Bu_3Al, CH_2I_2, CH_2Cl_2) → HO

OS, *68*, 220 (1989)

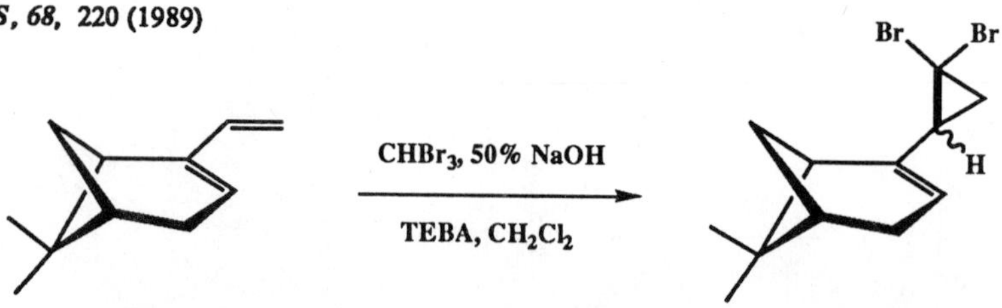

(*1R*) isomer → ($CHBr_3$, 50% NaOH, TEBA, CH_2Cl_2) → Br, Br, H

4 : 1 mixture

OS, CV 3, 213 (1955)

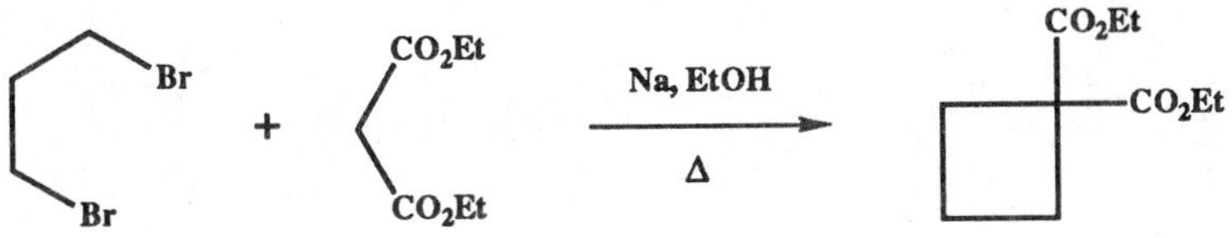

Br, Br + CO_2Et, CO_2Et → (Na, EtOH; Δ) CO_2Et, CO_2Et

OS, CV 4, 288 (1963)

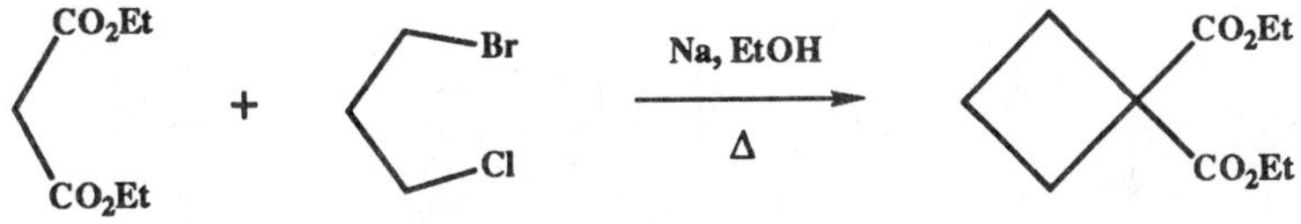

CO_2Et, CO_2Et + Br, Cl → (Na, EtOH; Δ) CO_2Et, CO_2Et

OS, CV 5, 54 (1973)

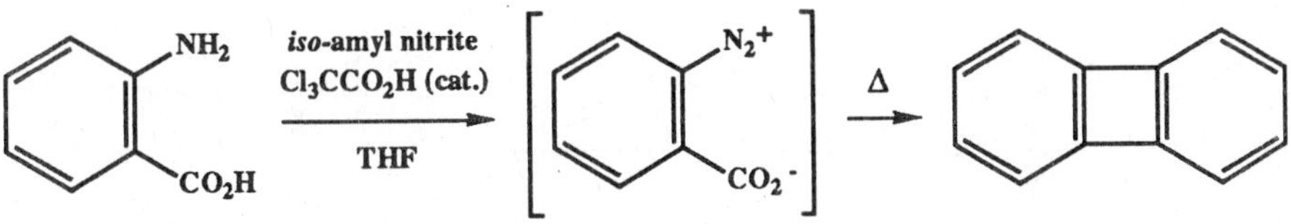

NH_2, CO_2H → (*iso*-amyl nitrite, Cl_3CCO_2H (cat.); THF) [N_2^+, CO_2^-] → (Δ)

OS, CV 5, 235 (1973)

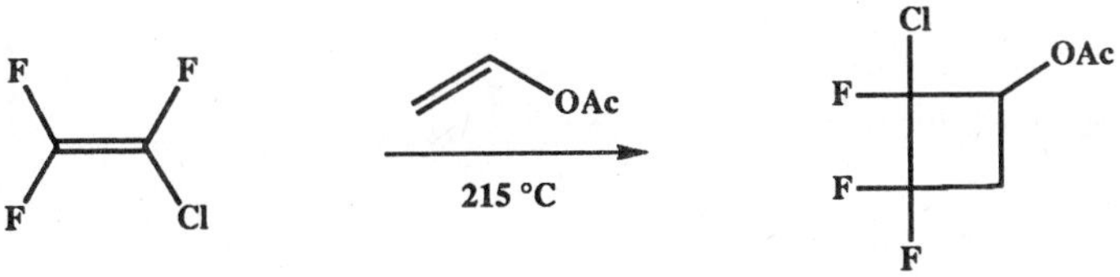

F, F, F, Cl → (OAc; 215 °C) Cl, OAc, F, F, F

***OS*, *CV 5*, 263 (1973)**

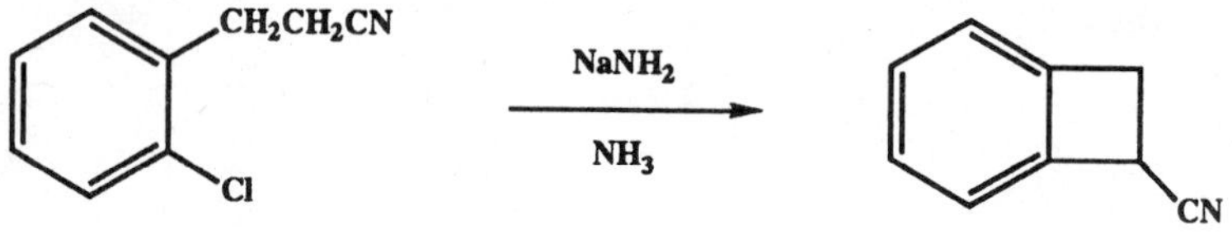

CH_2CH_2CN

Cl

$NaNH_2$

NH_3

CN

***OS*, *CV 5*, 297 (1973)**

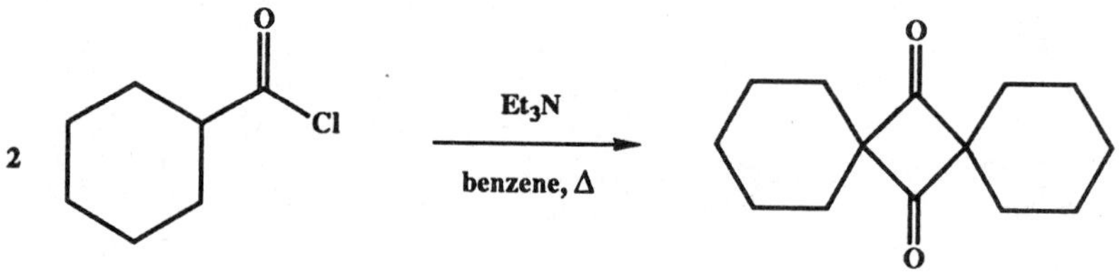

2

O

Cl

Et_3N

benzene, Δ

O

O

***OS*, *CV 5*, 370 (1973)**

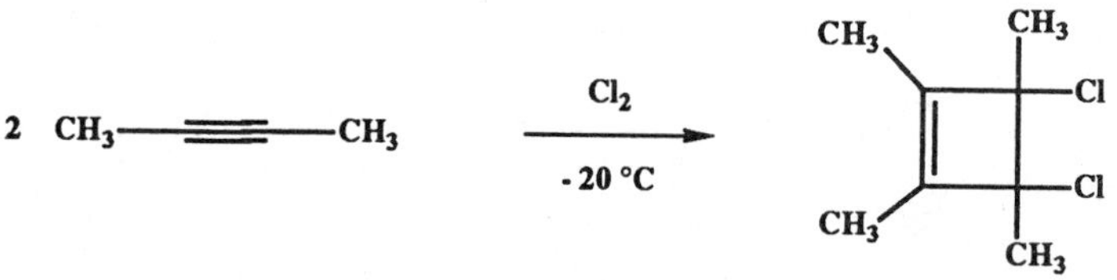

2 CH_3—≡—CH_3

Cl_2

- 20 °C

CH_3 CH_3

Cl

Cl

CH_3 CH_3

***OS*, *CV 5*, 393 (1973)**

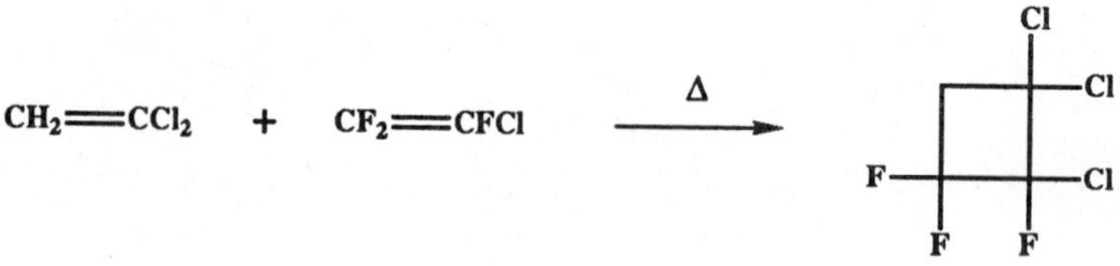

CH_2═CCl_2 + CF_2═CFCl

Δ

Cl

Cl

F

Cl

F F

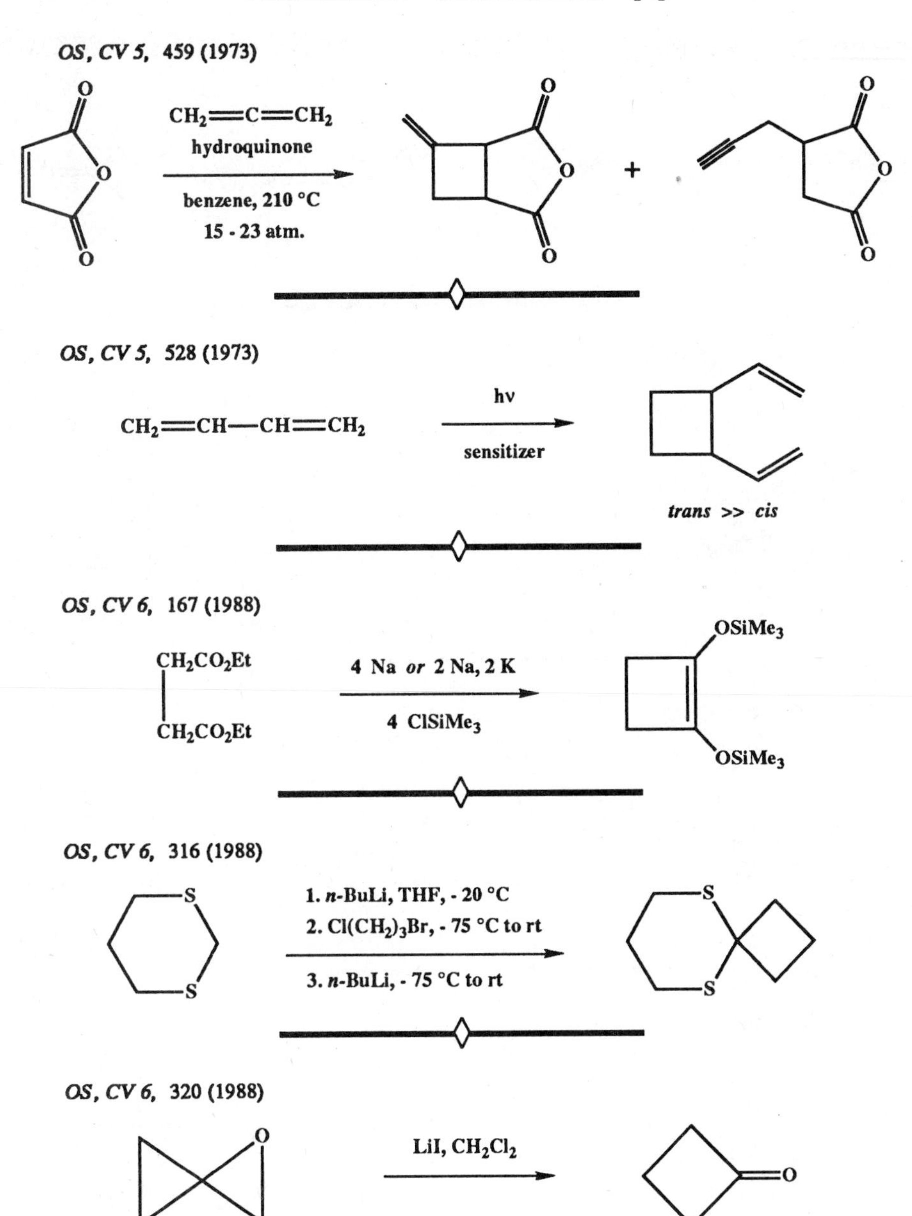

OS, CV 5, 459 (1973)
CH2=C=CH2
hydroquinone
benzene, 210 °C
15 - 23 atm.
OS, CV 5, 528 (1973)
CH2=CH—CH=CH2
hν
sensitizer
trans >> cis
OS, CV 6, 167 (1988)
CH2CO2Et
CH2CO2Et
4 Na or 2 Na, 2 K
4 ClSiMe3
OSiMe3
OSiMe3
OS, CV 6, 316 (1988)
1. n-BuLi, THF, - 20 °C
2. Cl(CH2)3Br, - 75 °C to rt
3. n-BuLi, - 75 °C to rt
OS, CV 6, 320 (1988)
LiI, CH2Cl2

OS, *CV 6*, 324 (1988)

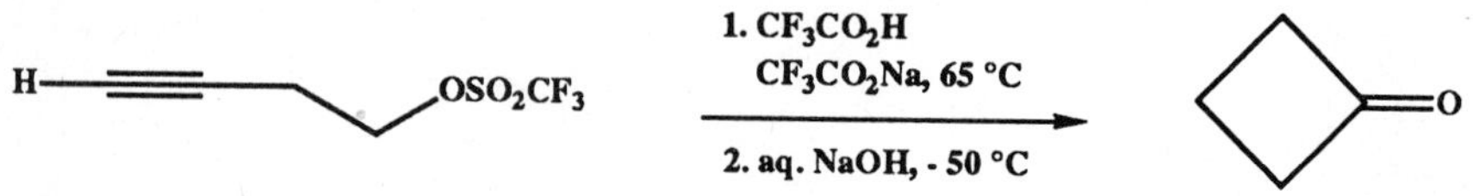
H
OSO2CF3
1. CF3CO2H
CF3CO2Na, 65 °C
2. aq. NaOH, - 50 °C
O

OS, *CV 6*, 482 (1988)

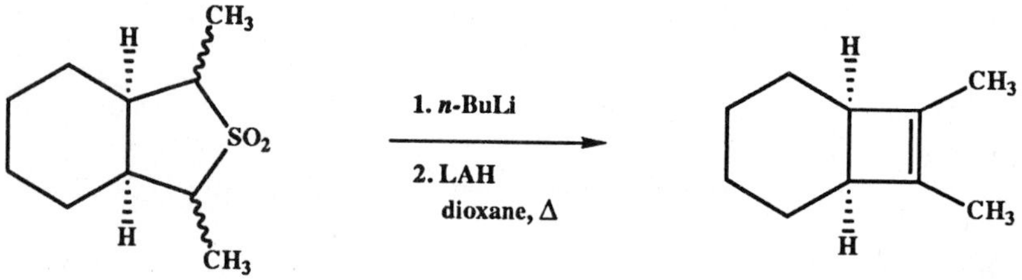
CH3
H
SO2
H
CH3
1. n-BuLi
2. LAH
dioxane, Δ
H
CH3
CH3
H

OS, *CV 6*, 1024 (1988)

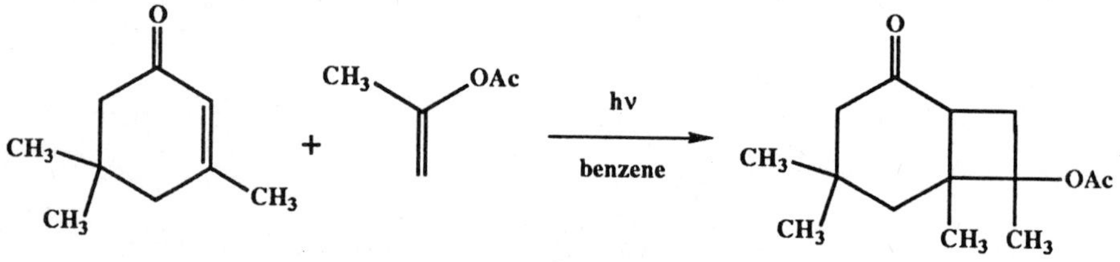
O
CH3
CH3
CH3
+
CH3
OAc
hν
benzene
O
CH3
CH3
OAc
CH3 CH3

OS, *CV 6*, 1037 (1988)

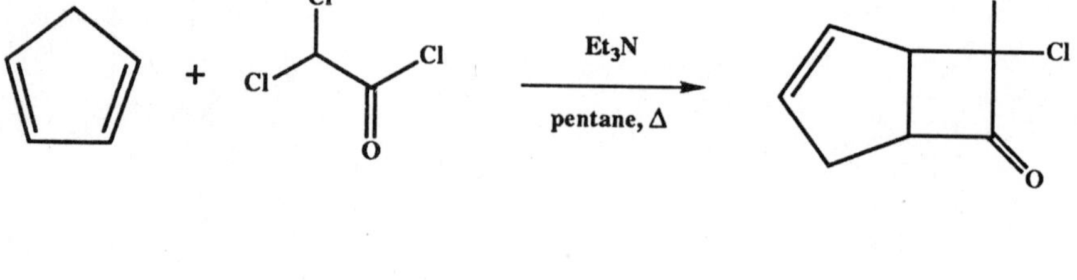
+
Cl
Cl
Cl
O
Et3N
pentane, Δ
Cl
Cl
O

OS, CV 7, 326 (1990); *62,* 74 (1984)

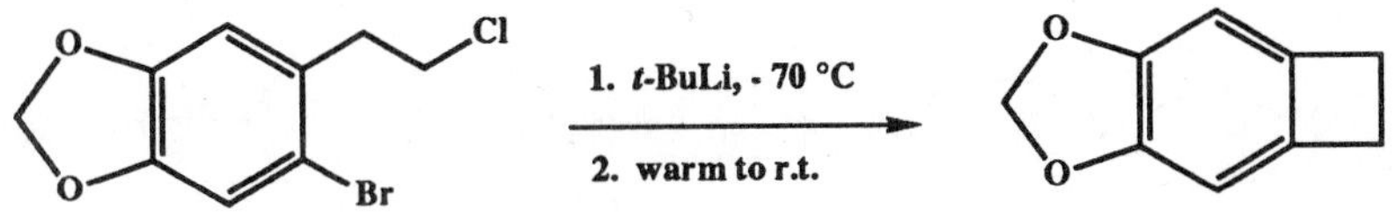
O
O
Cl
Br
1. t-BuLi, - 70 °C
2. warm to r.t.
O
O

OS, CV 7, 315 (1990); *62,* 118 (1984)

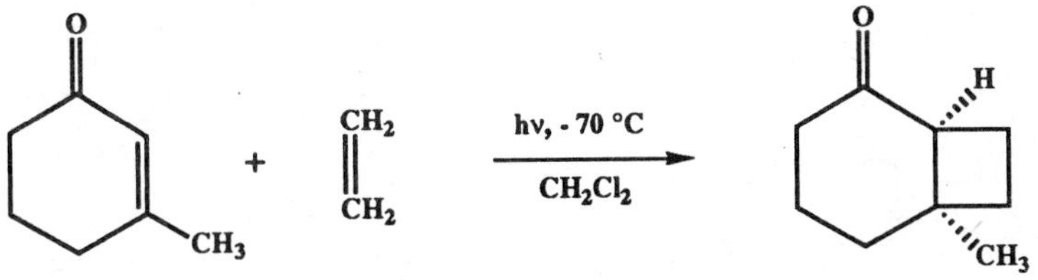
O
CH3
+
CH2
CH2
hν, - 70 °C
CH2Cl2
O
H
CH3

OS, 68, 32 (1989)

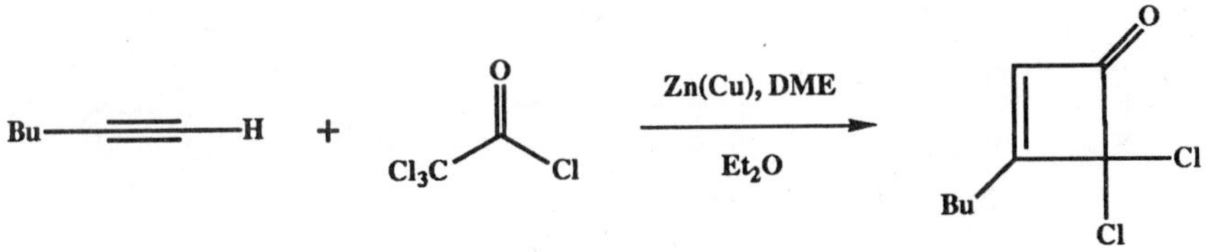
Bu
H
+
O
Cl3C
Cl
Zn(Cu), DME
Et2O
O
Cl
Bu
Cl

OS, 68, 41 (1989)

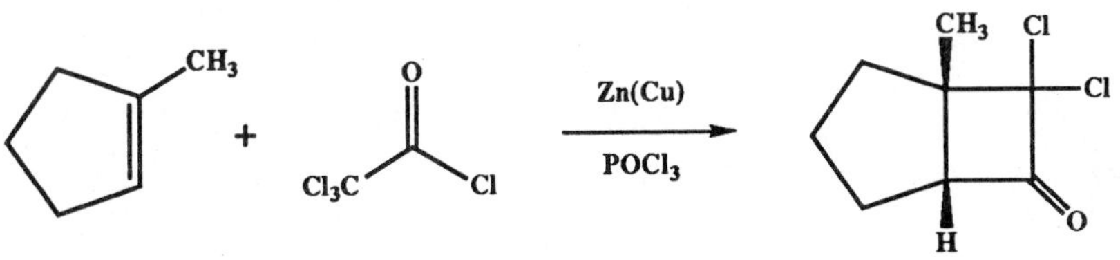
CH3
+
O
Cl3C
Cl
Zn(Cu)
POCl3
CH3
Cl
Cl
H
O

OS, CV 1, 192 (1941)

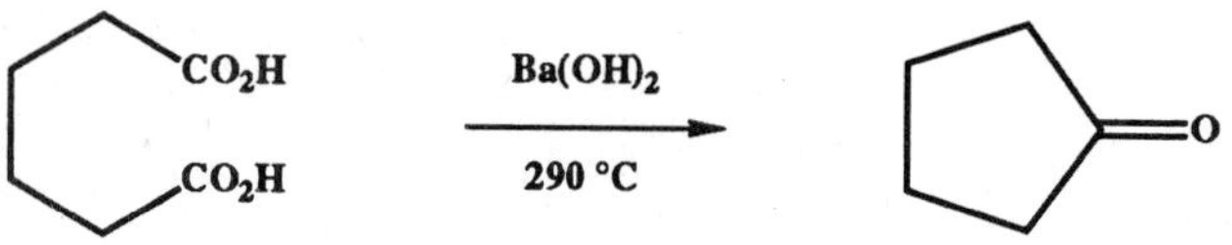

CO_2H / CO_2H → Ba(OH)$_2$, 290 °C → O

OS, CV 2, 116 (1943)

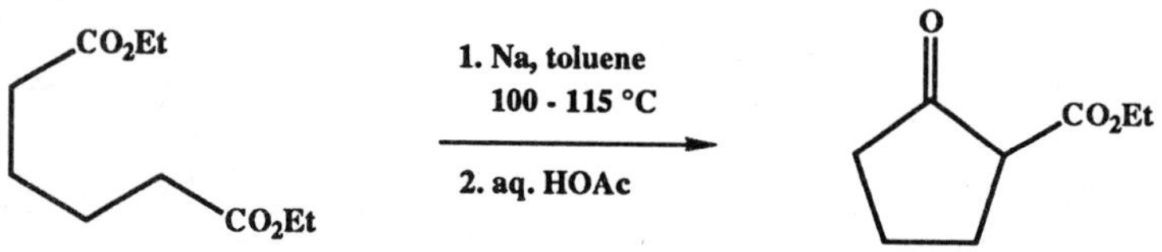

CO_2Et / CO_2Et → 1. Na, toluene 100 - 115 °C; 2. aq. HOAc → O, CO_2Et

OS, CV 3, 353 (1955)

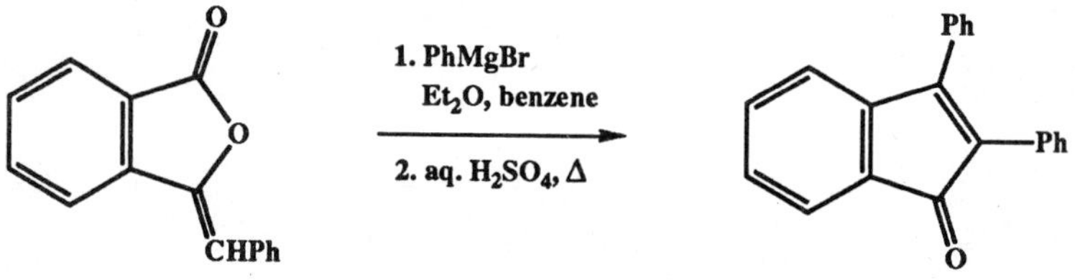

O, O, CHPh → 1. PhMgBr Et_2O, benzene; 2. aq. H_2SO_4, Δ → Ph, Ph, O

OS, CV 3, 806 (1955)

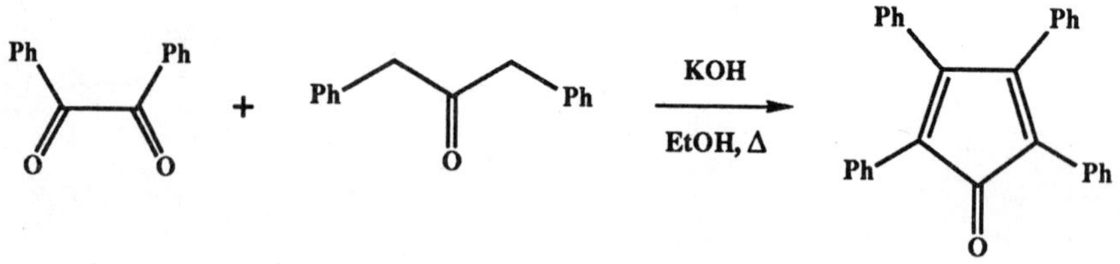

Ph, Ph, O, O + Ph, Ph, O → KOH, EtOH, Δ → Ph, Ph, Ph, Ph, O

OS, CV 4, 482 (1963)

OS, CV 4, 665 (1963)

OS, CV 5, 550 (1973)

OS, CV 5, 747 (1973)

OS, *CV 6*, 28 (1988)

O, O, O + CH$_2$=C(OAc)CH$_3$

AlCl$_3$, Δ
ClCH$_2$CH$_2$Cl

OH, COCH$_3$, O

OS, *CV 6*, 520 (1988)

O, Br, Br + O, N, CH$_2$, Ph

Fe$_2$(CO)$_9$
(- morpholine)

O, Ph

OS, *CV 6*, 774 (1988)

O, CH$_3$, EtO, O

NaOMe, MeOH
xylene, DMSO, Δ

O, CH$_3$, O

OS, *CV 6*, 905 (1988)

CH$_3$, O, CH$_3$, CH$_3$, N, Ph

1. *n*-BuLi
2. Br(CH$_2$)$_4$Br

CH$_3$, O, CH$_3$, CH$_3$, N, Ph, (CH$_2$)$_4$, Br

n-BuLi

CH$_3$, O, CH$_3$, CH$_3$, N, Ph

OS, *CV 6*, 958 (1988)

I, O, *endo*

KOH, MeOH, Δ

O

OS, *65*, 26 (1987)

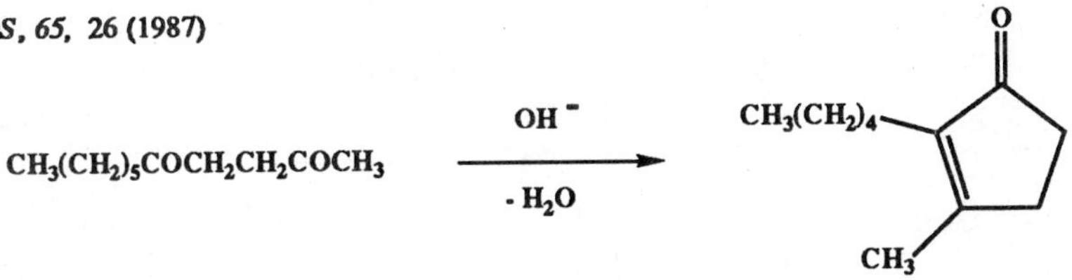

$CH_3(CH_2)_5COCH_2CH_2COCH_3$ — OH^- / $- H_2O$ → $CH_3(CH_2)_4$, O, CH_3

OS, *65*, 32 (1987)

O, O

NC, CN, Ph

NC, NC, Ph, O, O

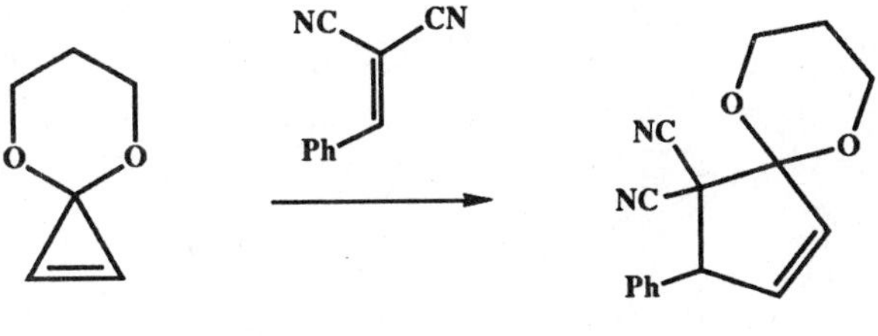

OS, *65*, 42 (1987)

OH

p-TsOH / $- H_2O$

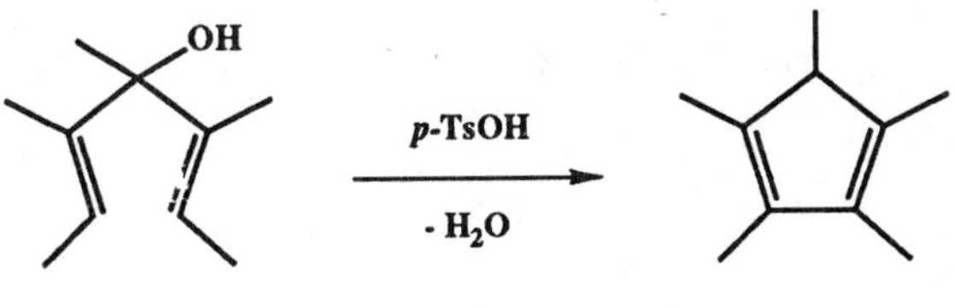

OS, *66*, 8 (1987)

R - (-)

$H_2C{=}C{=}C(SiMe_3)Me$ + (R)-(-)-carvone → ($TiCl_4$) → SiMe₃-substituted bicyclic enone

---

OS, *66*, 52 (1987)

MeO_2C ... CO_2Me → 1. $Li_2Cu(CN)(CH{=}CH_2)_2$ 2. H_3O^+ → CO_2Me

---

OS, *66*, 75 (1987)

$\equiv$—CH_2—$C(CO_2Me)_2$—CH_2—$CH{=}CMe_2$ ($Me_2C{=}CH{-}CH_2$)

Bu_3SnH, AIBN, 80 °C → MeO_2C, CO_2Me, Bu_3Sn → SiO_2, CH_2Cl_2 → MeO_2C, CO_2Me

---

OS, *66*, 87 (1987)

CO_2H, $SiMe_3$ → 1. $(COCl)_2$ 2. $AlCl_3$ → O

***OS*, *67*, 121 (1988)**

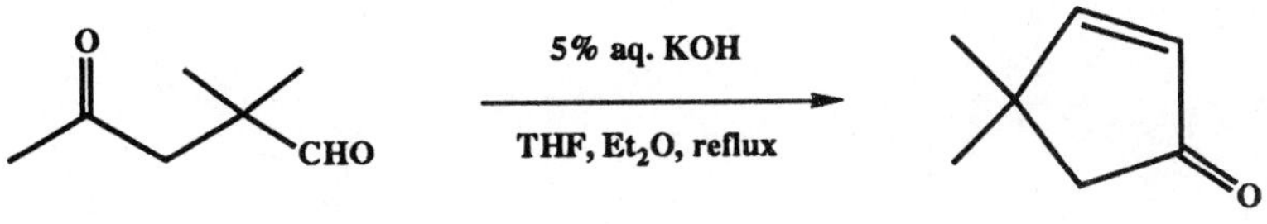

5% aq. KOH

THF, Et_2O, reflux

CHO

***OS*, *68*, 220 (1989)**

Br Br H

MeLi, Et_2O

4 : 1 mixture

(*1R*) isomer

OS, CV 1, 341 (1941)

3 CH_3COCH_3 —(H_2SO_4; (- 3 H_2O))→ 1,3,5-trimethylbenzene

OS, CV 1, 353 (1941)

2-(4-methylbenzoyl)benzoic acid (CO_2H, CH_3) —(fuming H_2SO_4, Δ)→ 2-methylanthraquinone (CH_3)

OS, CV 1, 476 (1941)

phthalic anhydride + 4-chlorophenol (OH, Cl) —(H_2SO_4, H_3BO_3, 200 °C)→ 1,4-dihydroxyanthraquinone (OH, OH)

OS, CV 2, 62 (1943)

anthrone —(glycerol; aq. H_2SO_4, 120 °C)→ benzanthrone

OS, CV 2, 194 (1943)

H_2SO_4

H_2O

CO$_2$Et

CO$_2$Et

OS, CV 2, 200 (1943)

CH$_3$ CH$_3$ CH$_3$

EtO$_2$C

EtO$_2$C

NaOEt

EtOH

ONa

EtO$_2$C

CH$_3$

CH$_3$

1. KOH
2. HCl

CH$_3$

CH$_3$

OS, CV 2, 569 (1943)

HO

1. SOCl$_2$, Δ

2. AlCl$_3$
CS$_2$, Δ

OS, CV 3, 300 (1955)

MeO

MeO

EtO$_2$C

1. HCO$_2$Et
NaOEt, - 10 °C

2. H_3PO_4, H_2SO_4
3. NaOH, then HCl

MeO

MeO

CO$_2$H

OS, *CV* 3, 310 (1955)

CH$_3$ CH$_3$

EtOH, Δ

O O CH$_3$ CH$_3$

OS, *CV* 3, 317 (1955)

2 CH$_3$ O CO_2Et

CH_3CHO

piperidine

EtOH, 0 °C

HOAc

H_2SO_4

Δ

OS, *CV* 3, 637 (1955)

+ EtOMg CO_2Et CO_2Et

Et_2O

EtOH, Δ

H_2SO_4

$Ba(OH)_2$, Δ

aq. dioxane

then H_2SO_4

OS, CV 3, 798 (1955)

HO_2C

polyphosphoric acid, Δ

O

OS, CV 3, 807 (1955)

Ph Ph Ph Ph O

O O O

C_6H_5Br, Δ (- CO)

Ph Ph Ph Ph O O O

Br_2 Δ

Ph Ph Ph Ph O O O

OS, CV 3, 829 (1955)

3 CH_3COCH_3

3 HCO_2Et, Na, EtOH, Et_2O

3 CH_3COCH=CHONa

3 HOAc, 50 °C

$COCH_3$ CH_3CO $COCH_3$

OS, CV 4, 890 (1963)

+ O O O

benzene

O O O

OS, *CV* 4, 898 (1963)

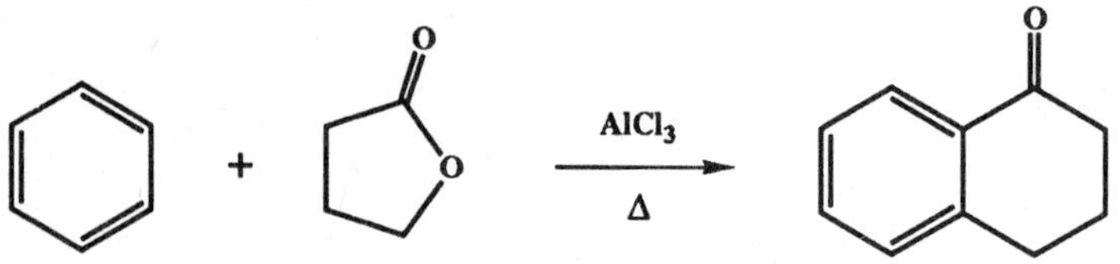

$AlCl_3$, Δ

OS, CV 4, 900 (1963)

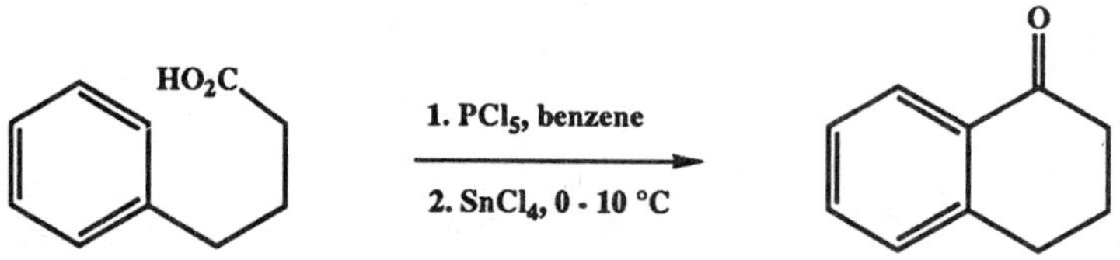

HO_2C

1. PCl_5, benzene
2. $SnCl_4$, 0 - 10 °C

OS, CV 5, 288 (1973)

2 CO_2Et CO_2Et

1. NaOEt EtOH, Δ
2. H_2SO_4, H_2O

CO_2Et EtO_2C

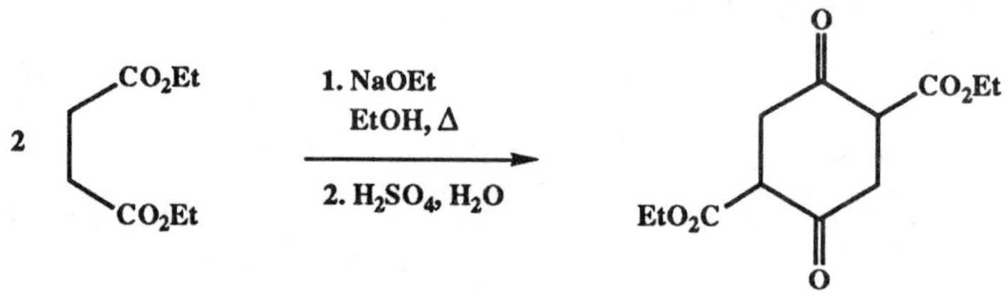

OS, CV 5, 486 (1973)

CH_3

methyl vinyl ketone
KOH MeOH, Δ

CH_3 CH_3

NH
Δ

CH_3

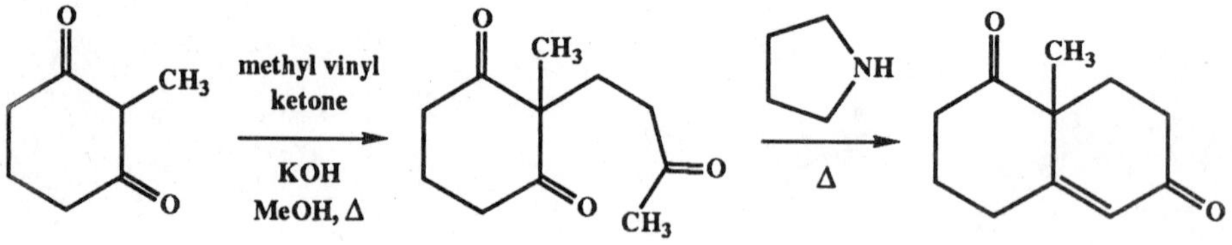

OS, CV 5, 604 (1973)

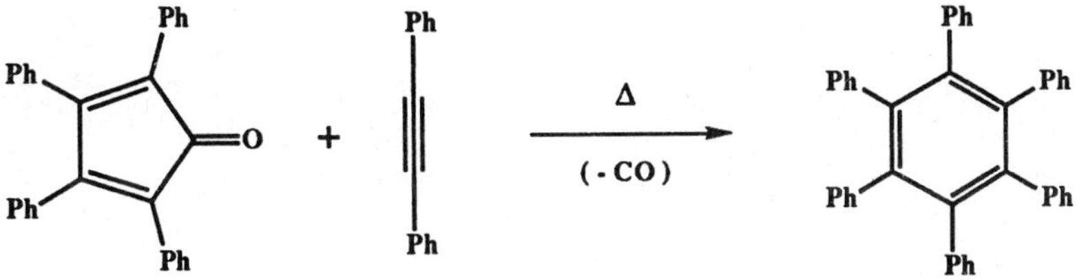

Ph Ph Ph Ph O + Ph Ph Δ (- CO) Ph Ph Ph Ph Ph Ph

OS, CV 5, 869 (1973)

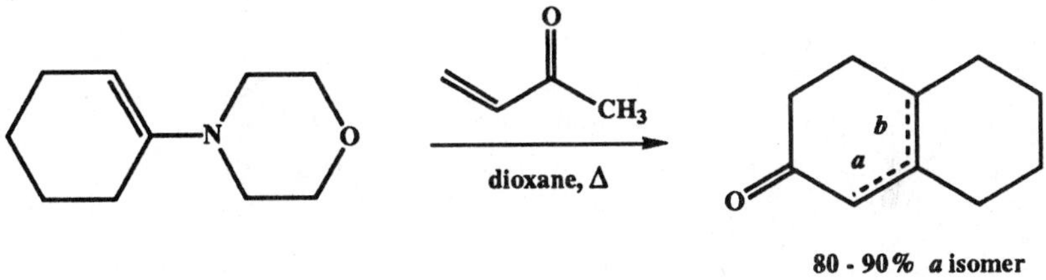

N O O CH_3 dioxane, Δ b a O

80 - 90% *a* isomer

OS, CV 5, 952 (1973)

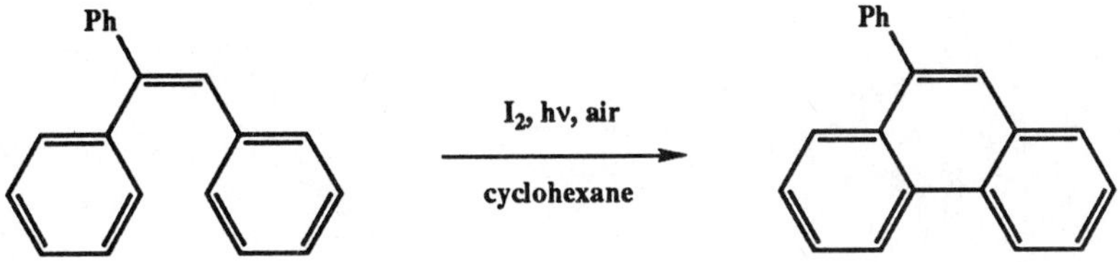

Ph I_2, hν, air cyclohexane Ph

OS, CV 5, 1011 (1973)

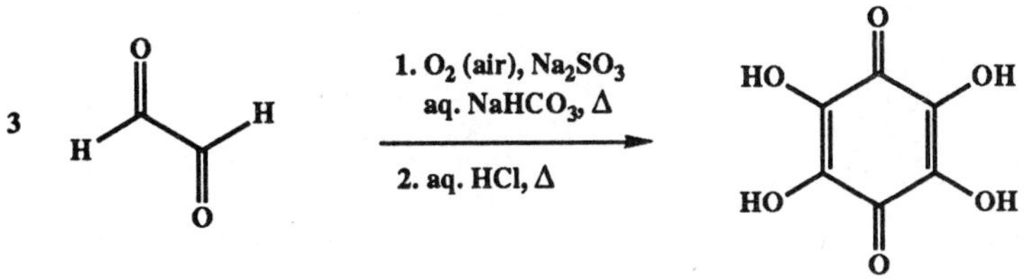

3 O H H O 1. O_2 (air), Na_2SO_3 aq. $NaHCO_3$, Δ 2. aq. HCl, Δ O HO OH HO OH O

OS, CV 5, 1037 (1973)

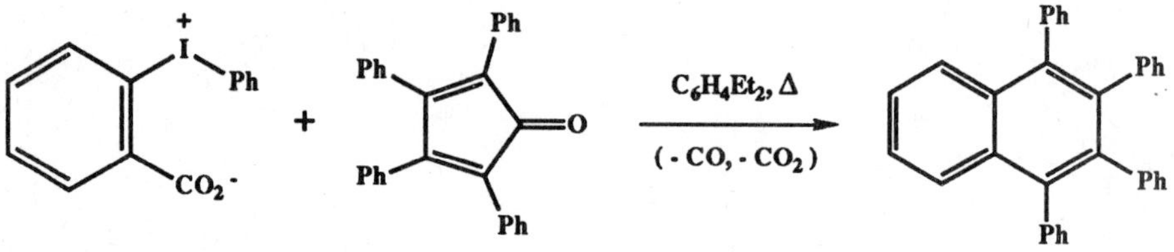

OS, CV 5, 1120 (1973)

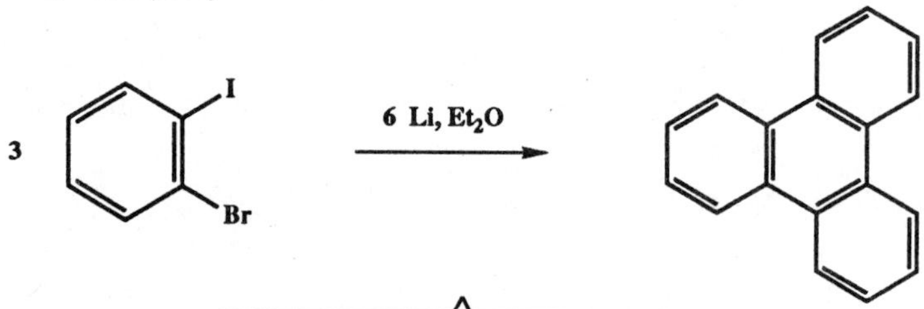

OS, CV 5, 1128 (1973)

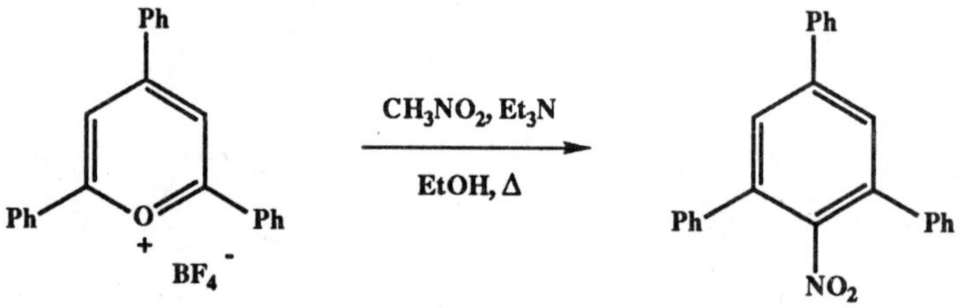

OS, CV 6, 427 (1988)

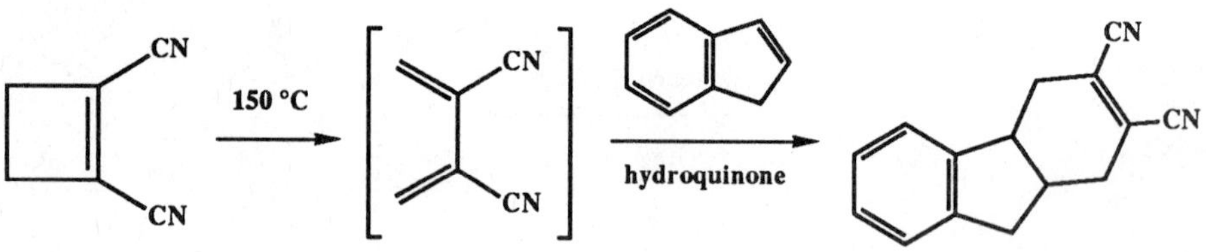

***OS*, *CV 6*, 445 (1988)**

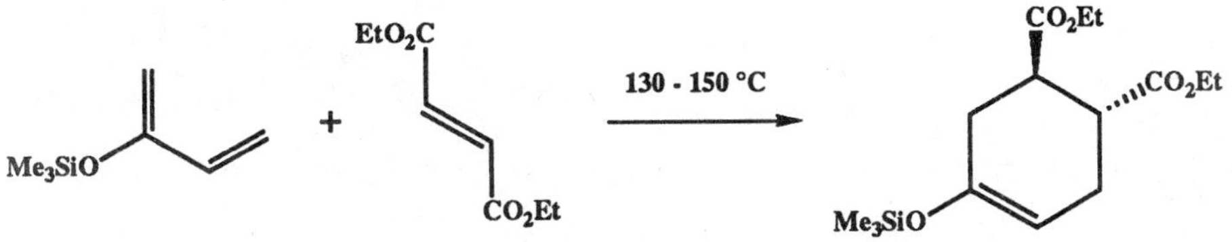

Me$_3$SiO (2-trimethylsilyloxy-1,3-butadiene) + EtO$_2$C–CH=CH–CO$_2$Et —(130 - 150 °C)→ Me$_3$SiO-cyclohexene with CO$_2$Et, CO$_2$Et

***OS*, *CV 6*, 454 (1988)**

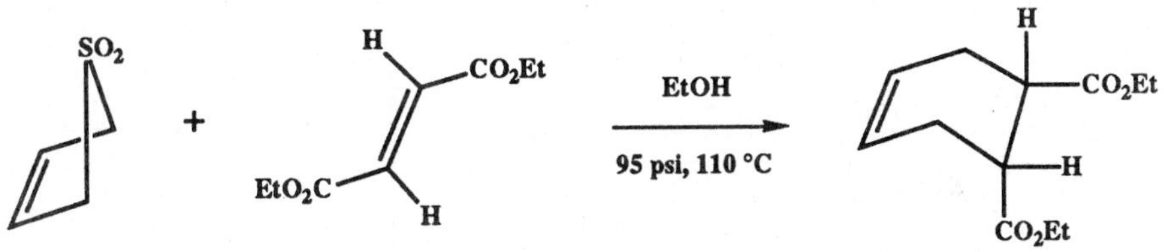

SO$_2$ + H, CO$_2$Et, EtO$_2$C, H —(EtOH; 95 psi, 110 °C)→ H, CO$_2$Et, H, CO$_2$Et

***OS*, *CV 6*, 496 (1988)**

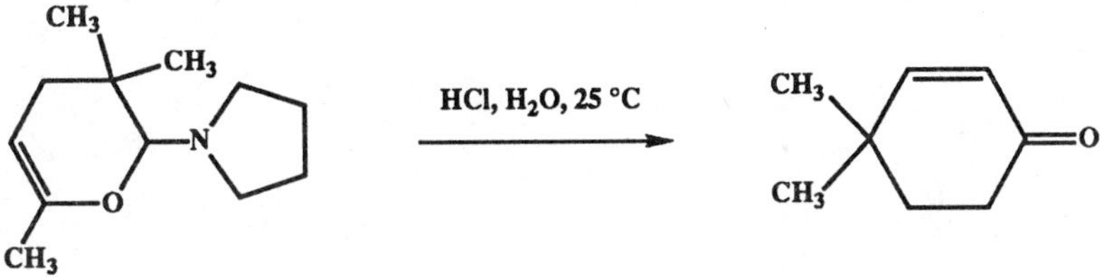

CH$_3$, CH$_3$, N, O, CH$_3$ —(HCl, H$_2$O, 25 °C)→ CH$_3$, CH$_3$, O

***OS*, *CV 6*, 586 (1988)**

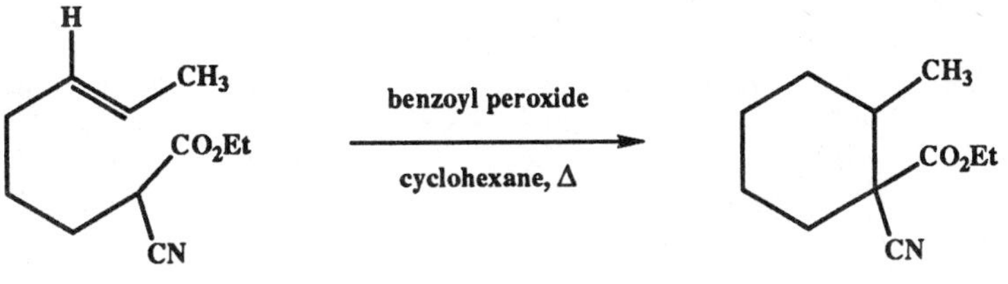

H, CH$_3$, CO$_2$Et, CN —(benzoyl peroxide; cyclohexane, Δ)→ CH$_3$, CO$_2$Et, CN

OS, CV 6, 666 (1988)

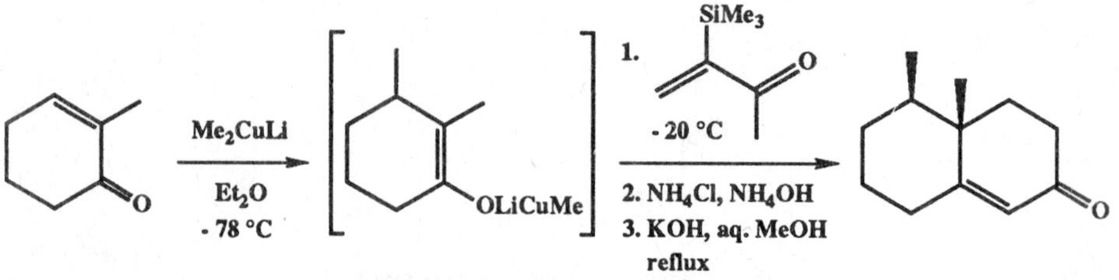

Me_2CuLi, Et_2O, - 78 °C; OLiCuMe; 1. SiMe$_3$ (methyl vinyl ketone), - 20 °C; 2. NH_4Cl, NH_4OH; 3. KOH, aq. MeOH reflux

OS, CV 6, 670 (1988)

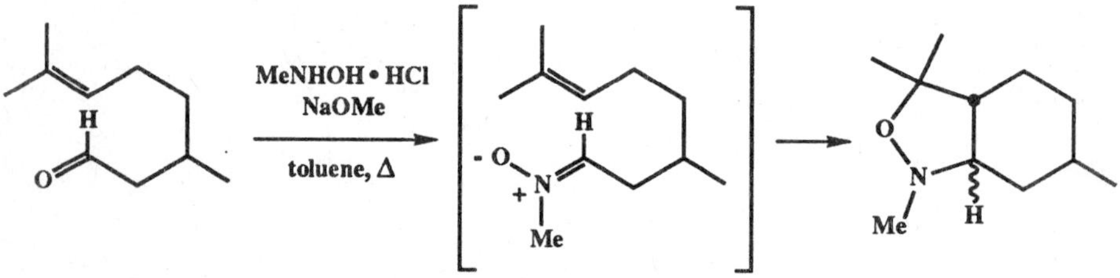

MeNHOH • HCl, NaOMe, toluene, Δ; H; - O; + N; Me; H; O; N; Me; H

OS, CV 6, 744 (1988)

O; Cl; MeO; CH_2═CH_2; $AlCl_3$, CH_2Cl_2; O; MeO

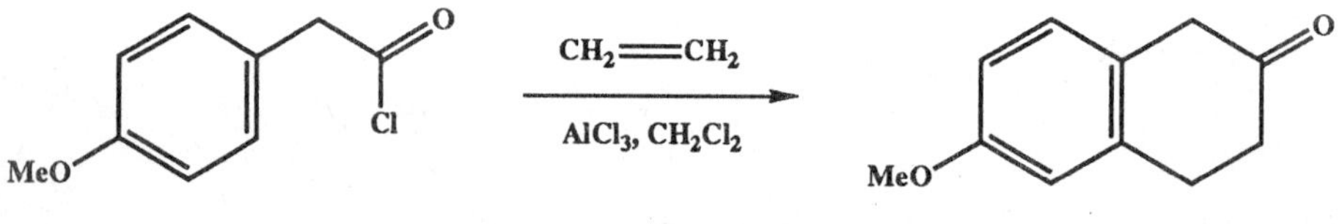

OS, CV 6, 781 (1988)

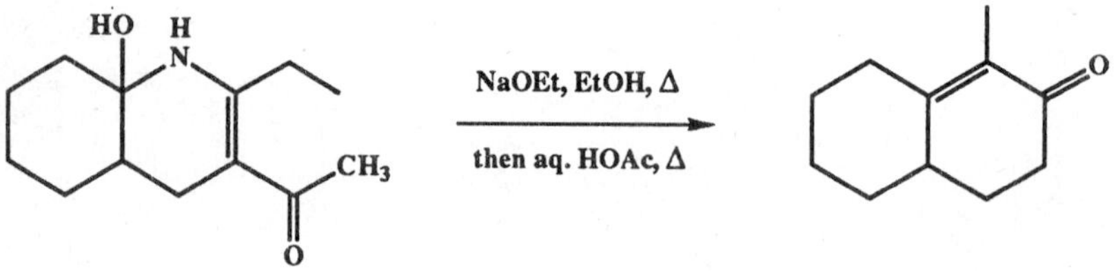

HO; H; N; CH_3; O; NaOEt, EtOH, Δ; then aq. HOAc, Δ; O

OS, *CV* 7, 473 (1990); *61*, 129 (1983)

CHO

1. Piperidine, benzene, reflux
2. CH_2=CHCOMe, EtOH, reflux
3. NaOAc, HOAc, H_2O, reflux
4. 20% NaOH to pH 9-10, reflux

O

OS, *CV* 7, 312 (1990); *61*, 147 (1983)

OMe

Me_3SiO

+

O O O

MeO H O O H O O

OS, *CV* 7, 4 (1990); *62*, 149 (1984)

Ac N O O

160 °C, benzene
Hydroquinone

H Ac N O O H

OS, *CV* 7, 363 (1990); *63*, 26 (1985)

O O O

S - (-) - proline
DMF, 16 °C

O O OH

H_2SO_4
DMF
95 °C

S O O

OS, CV 7, 368 (1990); *63,* 37 (1985)

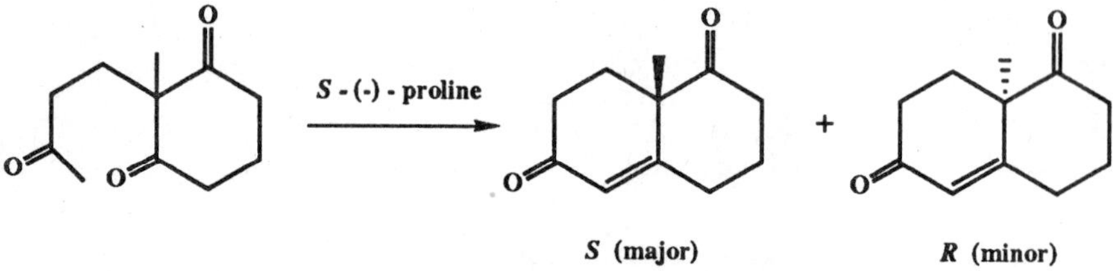

S - (-) - proline

S (major)

R (minor)

OS, 65, 98 (1987)

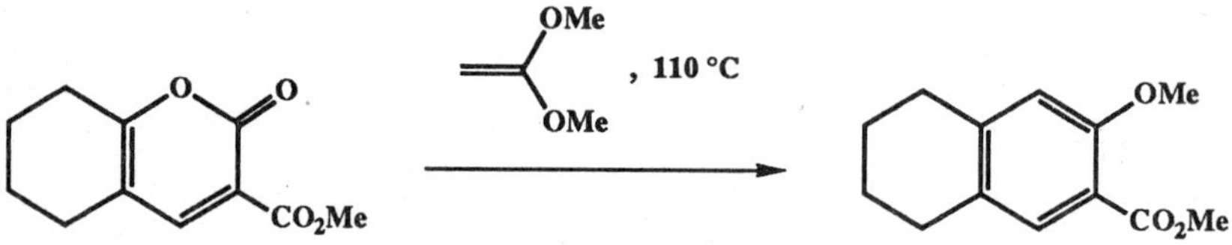

OMe

OMe

, 110 °C

CO_2Me

OMe

CO_2Me

OS, 67, 163 (1988)

SO_2Ph

MeO

$OSiMe_3$

1. xylene, 125 °C

2. H_3O^+

3. Zn, HOAc

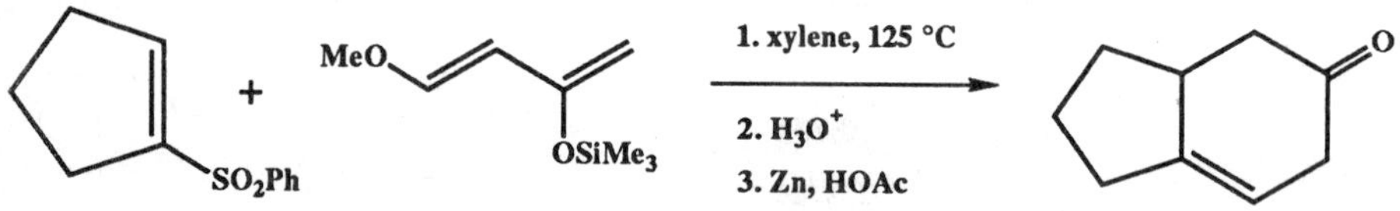

OS, *CV* 4, 840 (1963)

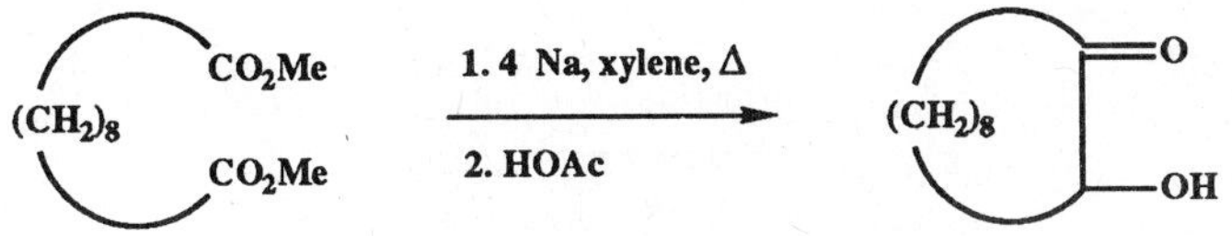

$(CH_2)_8$ CO_2Me CO_2Me — 1. 4 Na, xylene, Δ; 2. HOAc → $(CH_2)_8$ =O, —OH

OS, *CV* 5, 277 (1973)

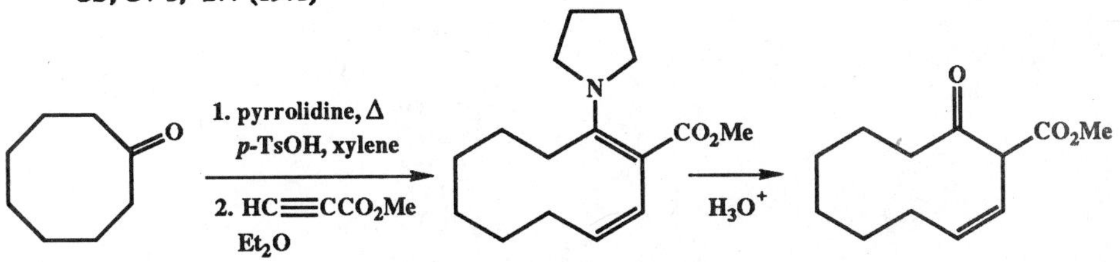

1. pyrrolidine, Δ, *p*-TsOH, xylene; 2. HC≡CCO₂Me, Et₂O → N, CO_2Me — H_3O^+ → O, CO_2Me

OS, *CV* 5, 883 (1973)

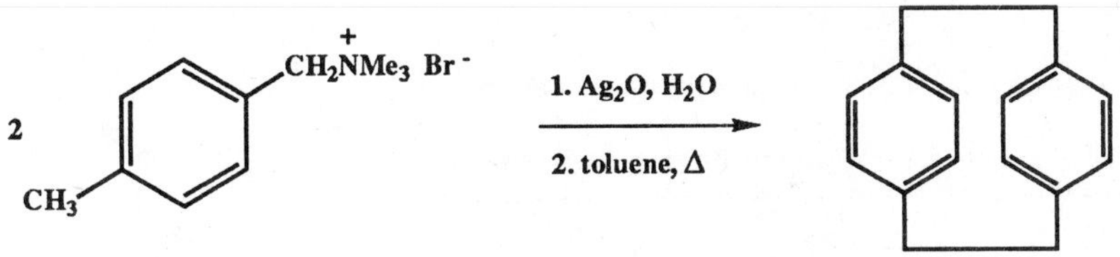

2 CH_3 $\overset{+}{C}H_2NMe_3$ Br^- — 1. Ag_2O, H_2O; 2. toluene, Δ →

OS, *CV* 5, 1088 (1973)

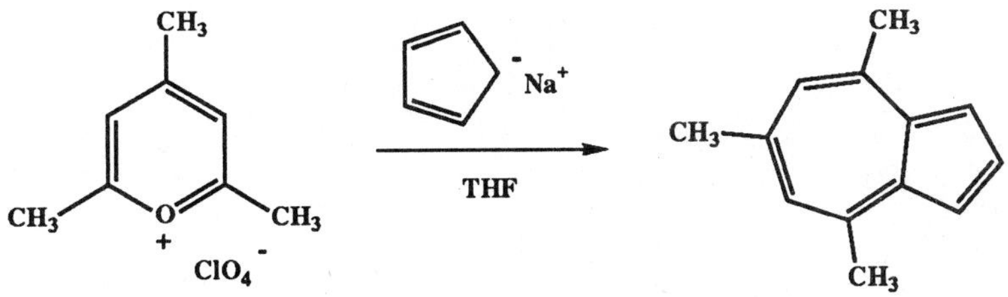

CH_3, CH_3, CH_3, O^+, ClO_4^- — Na^+ (cyclopentadienide), THF → CH_3, CH_3, CH_3

OS, *CV 6*, 68 (1988)

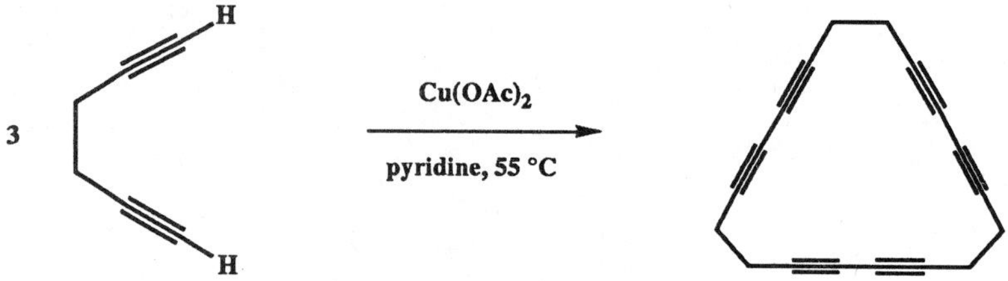

3 [diyne] $\xrightarrow[\text{pyridine, 55 °C}]{Cu(OAc)_2}$ [cyclic hexayne]

OS, *CV 7*, 485 (1990); *60*, 41 (1981)

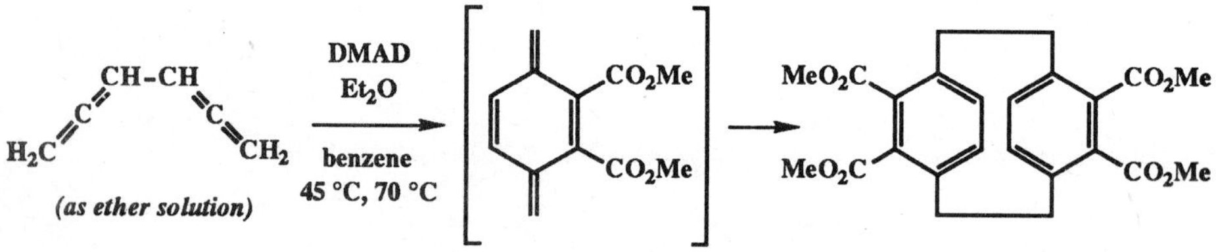

$H_2C{=}C{=}CH{-}CH{=}C{=}CH_2$ (*as ether solution*) $\xrightarrow[\text{benzene, 45 °C, 70 °C}]{\text{DMAD, } Et_2O}$ [CO_2Me, CO_2Me] → MeO_2C, MeO_2C, CO_2Me, CO_2Me

OS, *CV 7*, 15 (1990); *62*, 134 (1984)

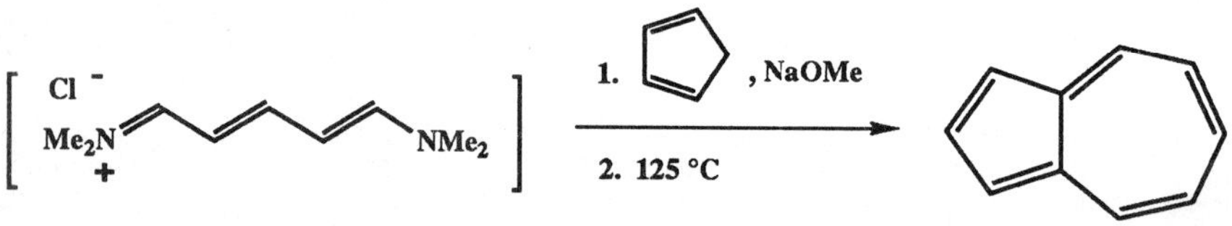

[Cl^- Me_2N^+ ... NMe_2] → 1. [cyclopentadiene], NaOMe; 2. 125 °C

OS, *68*, 234; 238; 243 (1989)

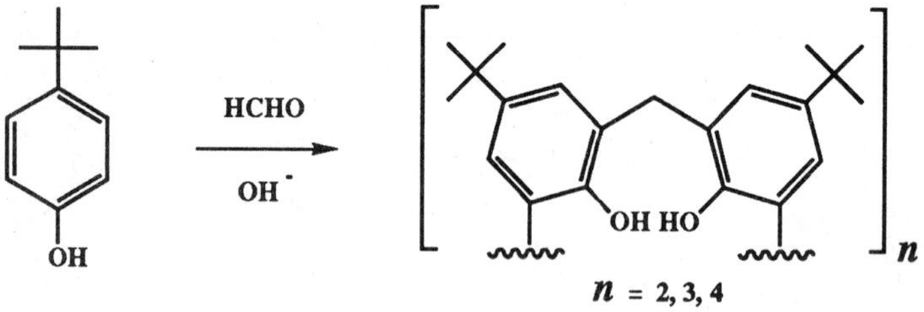

OH → HCHO, OH^- → [OH HO]$_n$

n = 2, 3, 4

OS, CV 4, 278 (1963)

2 (γ-butyrolactone) — 1. NaOMe, MeOH, Δ; 2. HCl, Δ → [$Cl(CH_2)_3$–CO–$(CH_2)_3Cl$] — aq. NaOH, Δ → dicyclopropyl ketone

OS, CV 4, 738 (1963)

dicyclopentadiene — $CH_2{=}CH_2$ (2350 psi), 190 - 200 °C → 2 norbornene

OS, CV 4, 964 (1963)

2-bromofluorobenzene (F, Br) — 1. Mg, THF; 2. Anthracene, 60 °C → [benzyne] — Δ → triptycene

OS, CV 5, 93 (1973)

1,5-cyclooctadiene — benzoyl peroxide, $CHCl_3$, Δ → CCl_3-substituted bicyclo[3.3.0]octane

***OS, CV 5,* 96 (1973)**

N—CO_2Et N CO_2Et → KOH, Δ (then HCl); ethylene glycol → [NH NH] → 1. $CuCl_2$ 2. NaOH 3. Δ →

***OS, CV 5,* 424 (1973)**

Cl Cl Cl Cl MeO OMe → CH_2=CH_2; 180 - 190 °C → MeO OMe Cl Cl Cl Cl

***OS, CV 5,* 863 (1973)**

→ $BF_3 \cdot Et_2O$; HOAc, Δ → OAc + OAc

ca. 90 : 10

***OS, CV 5,* 985 (1973)**

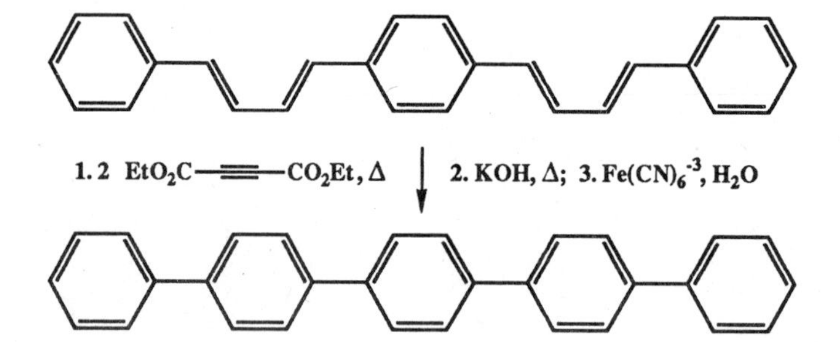

OS, CV 6, 82 (1988)

Cl Cl Cl Cl Cl Cl

n-BuLi

Et_2O

- 78 to - 60 °C

Cl Cl Li Cl Cl Cl

Et_2O

- 60 to 25 °C

Cl Cl Cl Cl

OS, CV 6, 133 (1988)

Cl Br

2 Na

dioxane, Δ

OS, CV 6, 137 (1988)

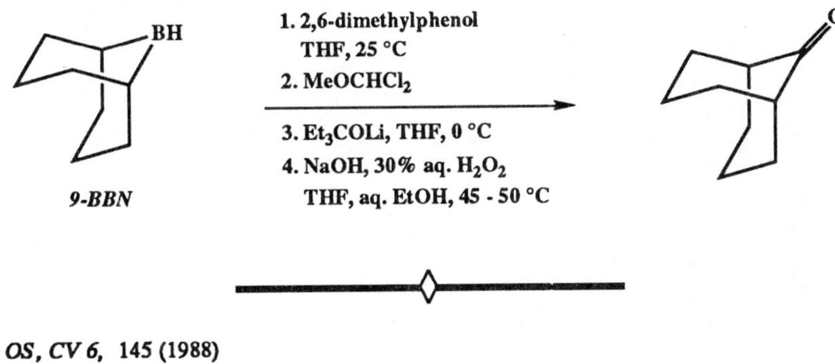

OS, CV 6, 145 (1988)

hν

THF

***OS, CV 6*, 196 (1988)**

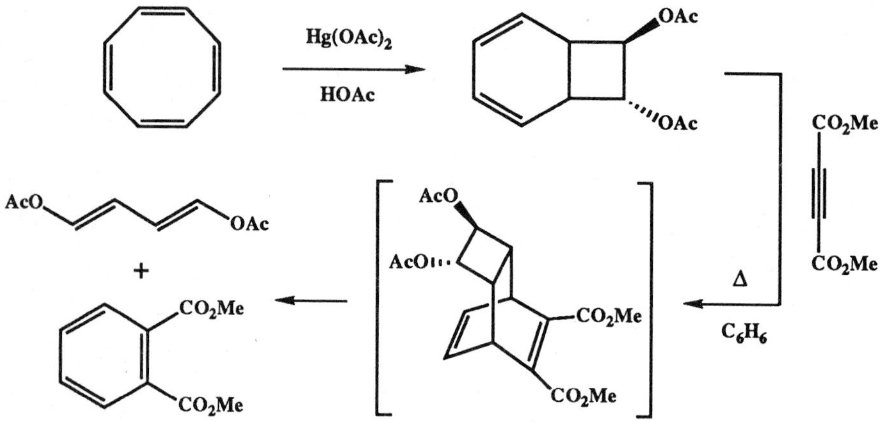

Hg(OAc)$_2$
HOAc

OAc
OAc

CO$_2$Me
CO$_2$Me

Δ
C_6H_6

AcO
AcO
CO$_2$Me
CO$_2$Me

AcO
OAc
+
CO$_2$Me
CO$_2$Me

***OS, CV 6*, 378 (1988)**

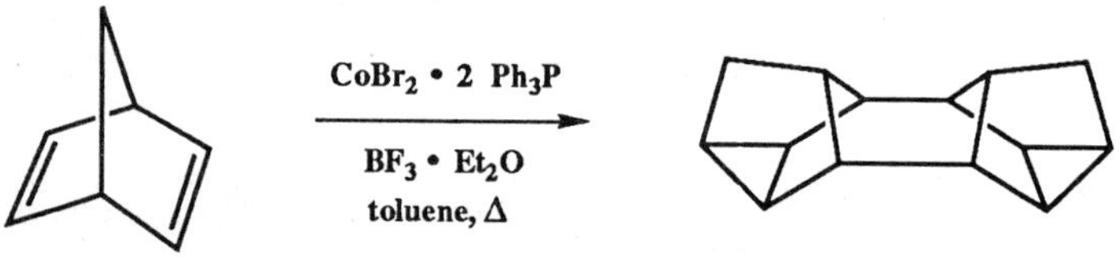

CoBr$_2$ • 2 Ph$_3$P
BF$_3$ • Et$_2$O
toluene, Δ

***OS, CV 6*, 422 (1988)**

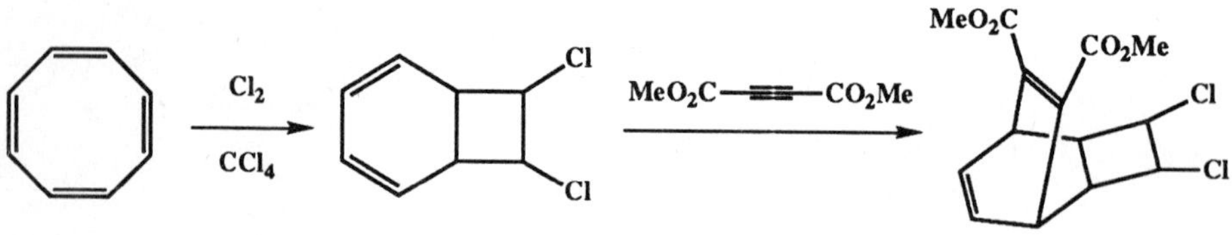

Cl$_2$
CCl$_4$

Cl
Cl

MeO$_2$C—≡—CO$_2$Me

MeO$_2$C
CO$_2$Me
Cl
Cl

OS, *CV 6*, 962 (1988)

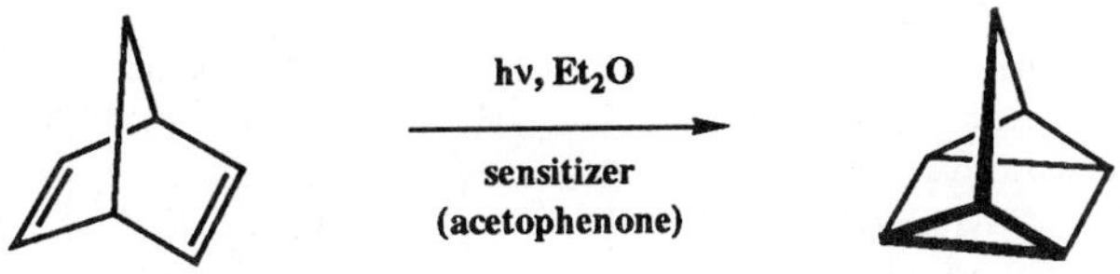

hν, Et_2O

sensitizer (acetophenone)

OS, *CV 6*, 1002 (1988)

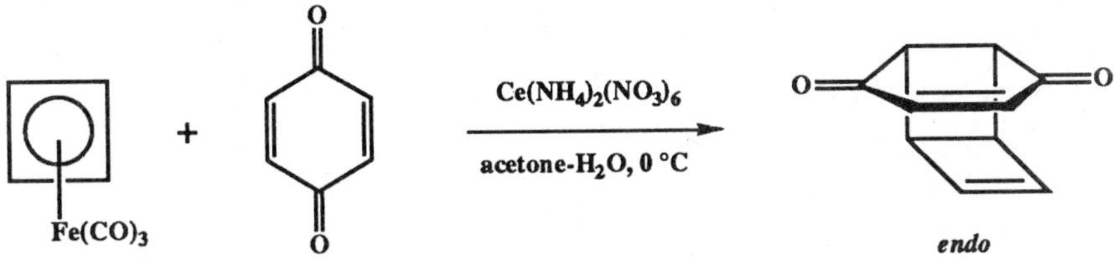

O

$Fe(CO)_3$ + O O

$Ce(NH_4)_2(NO_3)_6$

acetone-H_2O, 0 °C

O O

endo

OS, *CV 7*, 200 (1990); *61*, 39 (1983)

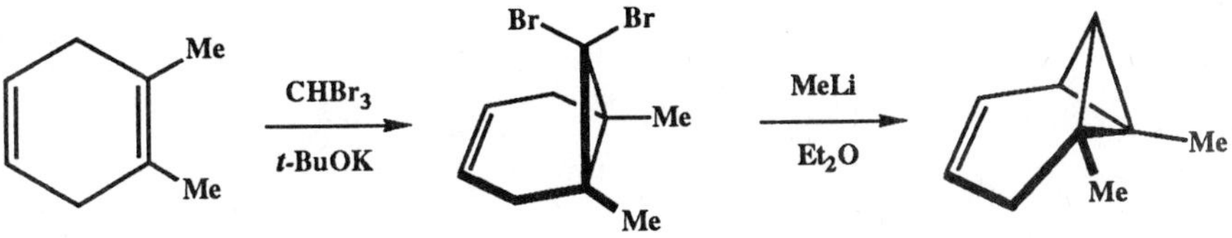

Me Me

$CHBr_3$

t-BuOK

Br Br Me Me

MeLi

Et_2O

Me Me

OS, *CV 7*, 256 (1990); *61*, 62 (1983)

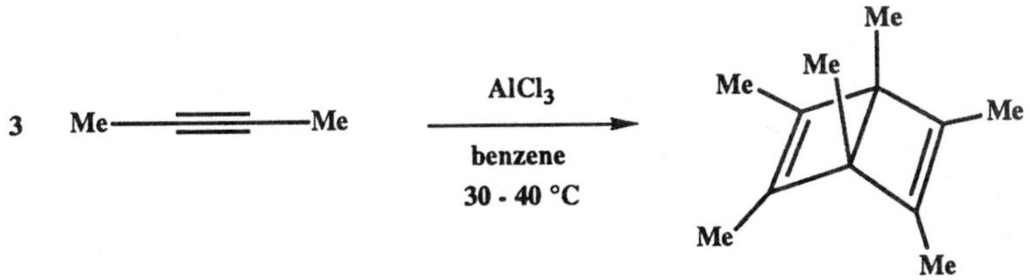

3 Me—≡—Me

$AlCl_3$

benzene

30 - 40 °C

Me Me Me Me Me Me

OS, *CV 7*, 427 (1990); *61*, 103 (1983)

$BH_3 \cdot Et_3N$
diglyme
130-140 and
200 °C

1. CO, $(CH_2OH)_2$
2. H_2O_2, NaOH
95% EtOH

H H B H
H OH H H

OS, *CV 7*, 177 (1990); *62*, 125 (1984)

hν, CuOTf

OH
OH

OS, *CV 7*, 50 (1990); *64*, 27 (1986)

R = H, Me

CO_2Me
CO_2Me

1. NaOH, MeOH, reflux
(H_2O, pH 8.6, 25 °C)
2. aq. HCl, HOAc, reflux

cis

OS, *66*, 37 (1987)

1. LDA
2. MeCH=CHCO₂Me

CO_2Me

OS, 68, 175 (1989)

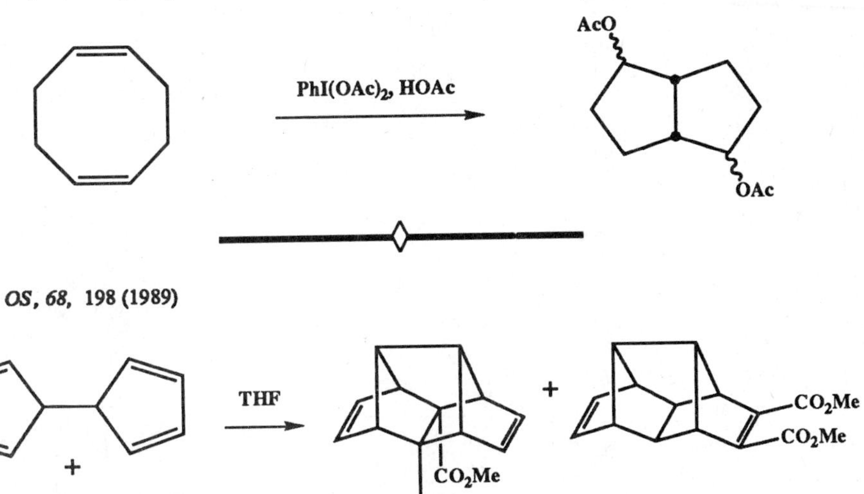
AcO
PhI(OAc)2, HOAc
OAc
OS, 68, 198 (1989)
THF
+
MeO2C
CO2Me
CO2Me
CO2Me
+
CO2Me
CO2Me

OS, CV 1, 185 (1941)

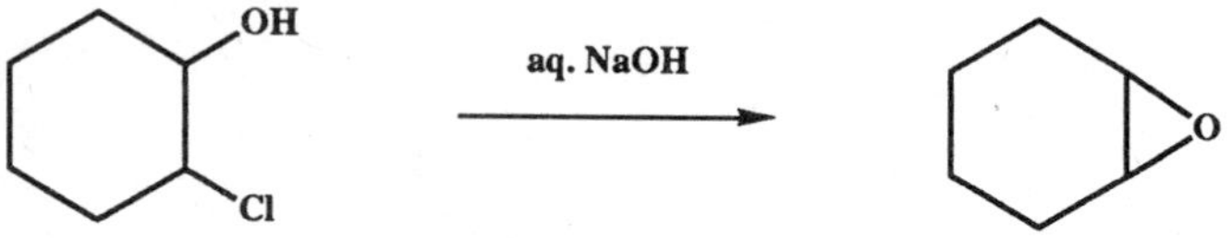

aq. NaOH

OH, Cl

O

OS, CV 1, 233 (1941)

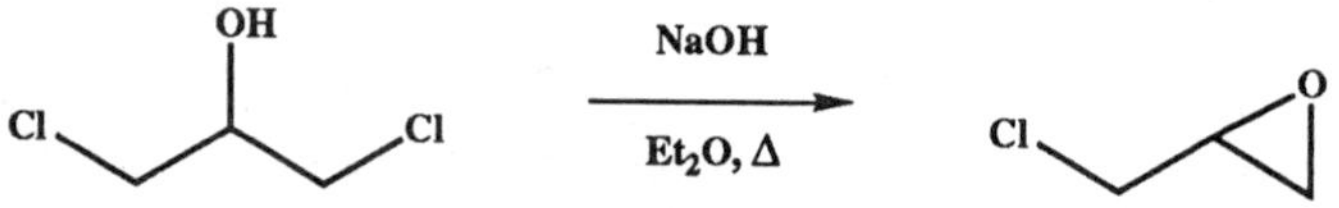

OH, Cl, Cl

NaOH

Et_2O, Δ

Cl, O

OS, CV 1, 494 (1941)

O, Ph, O—OH

$CHCl_3$, 0 °C

O

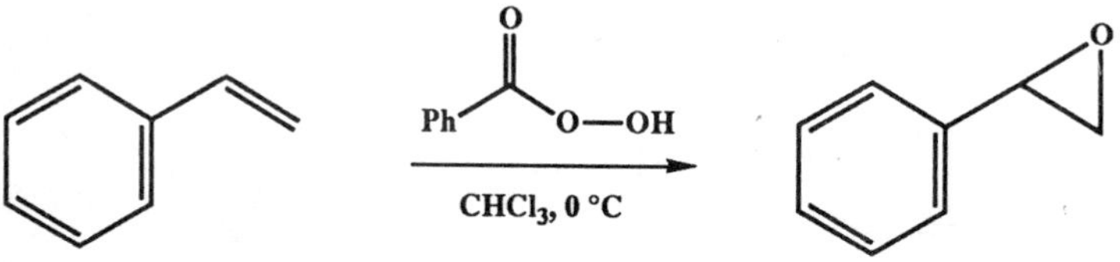

OS, CV 2, 256 (1943)

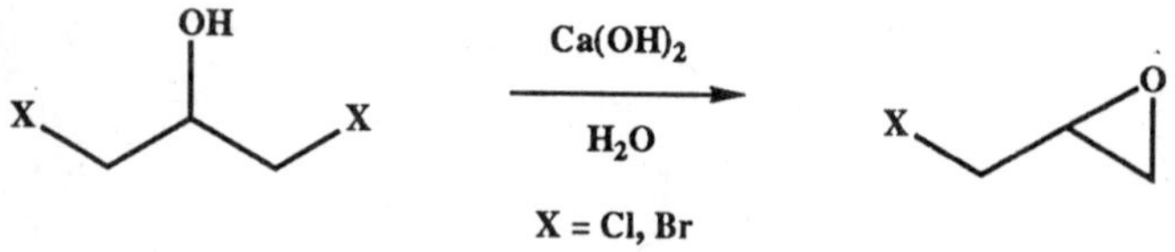

OH, X, X

$Ca(OH)_2$

H_2O

X = Cl, Br

X, O

OS, *CV* 3, 148 (1955)

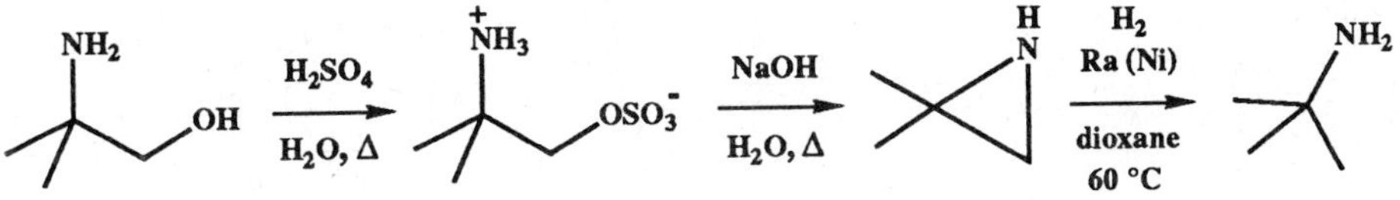

NH_2 OH — H_2SO_4 / H_2O, Δ → $\overset{+}{N}H_3$ OSO_3^- — NaOH / H_2O, Δ → H N — H_2 Ra (Ni) / dioxane 60 °C → NH_2

OS, *CV* 3, 727 (1955)

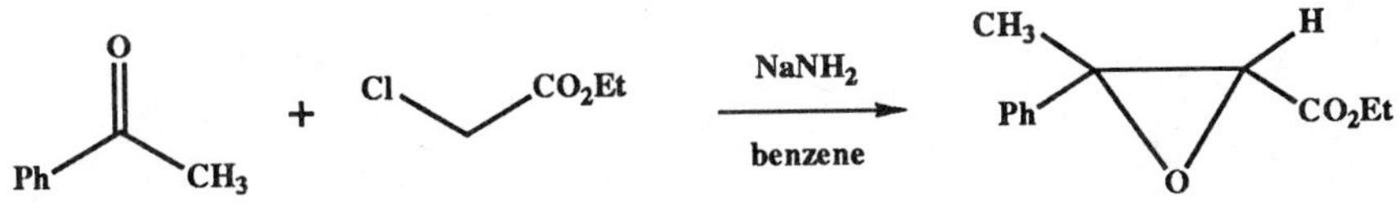

O Ph CH_3 + Cl CO_2Et — $NaNH_2$ / benzene → CH_3 Ph O H CO_2Et

OS, *CV* 4, 232 (1963)

O — KSCN / H_2O, EtOH → S

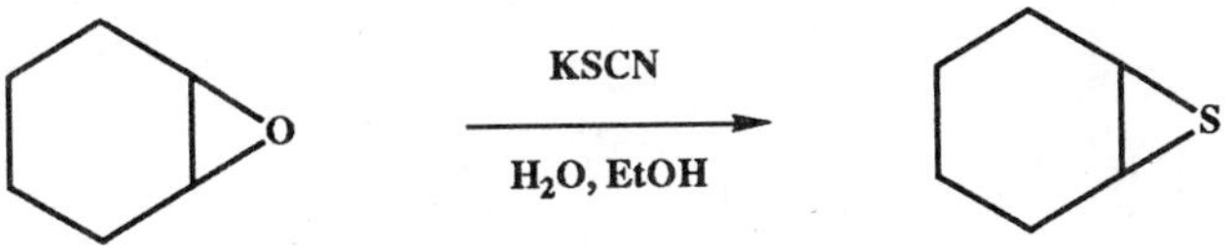

OS, *CV* 4, 433 (1963)

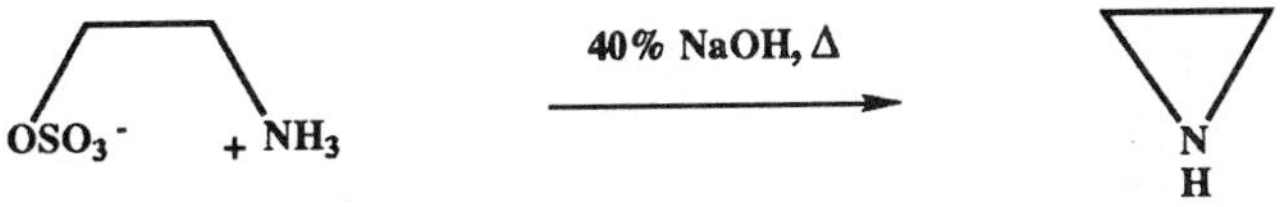

OSO_3^- + NH_3 — 40% NaOH, Δ → N H

***OS*, *CV* 4, 459 (1963)**

Cl CO$_2$Et; KO-*t*-Bu; *t*-BuOH; O; CO$_2$Et

***OS*, *CV* 4, 552 (1963)**

O; CH$_3$; CH$_3$; CH$_3$; H_2O_2, NaOH; MeOH, H_2O; O; O; CH$_3$; CH$_3$; CH$_3$

***OS*, *CV* 4, 649 (1963)**

MeO; O; MeO; CH$_3$; ClCH$_2$CO$_2$Me; NaOMe, Et$_2$O; MeO; CH$_3$; CO$_2$Me; MeO; O

***OS*, *CV* 4, 860 (1963)**

Ph; H; H; Ph; *trans*; peracetic acid; NaOAc; HOAc, CH$_2$Cl$_2$; Ph; O; H; H; Ph; *trans*

***OS*, *CV 5*, 83 (1973)**

$PhCH_2C(=NOH)CH_2Ph$ → (1. LAH, THF, Δ; 2. H_2O) → cis-2-benzyl-3-phenylaziridine + $PhCH_2CH(NH_2)CH_2Ph$

***OS*, *CV 5*, 191 (1973)**

Ph—CH=N—*t*-Bu → ($C_6H_5CO_3H$, benzene) → 2-*t*-butyl-3-phenyloxaziridine

***OS*, *CV 5*, 231 (1973)**

$ClCH_2SO_2Cl$ → (CH_2N_2, Et_3N, Et_2O) → 2-chlorothiirane 1,1-dioxide (SO_2)

***OS*, *CV 5*, 358 (1973)**

2 2-chlorobenzaldehyde (Cl, CHO) → ($(Me_2N)_3P$, Et_2O, C_6H_6) → 2,3-bis(2-chlorophenyl)oxirane

***OS*, *CV 5*, 414 (1973)**

CH_3CO_3H, Na_2CO_3, CH_2Cl_2

H_2O

OH OH HO OH

+

***OS*, *CV 5*, 467 (1973)**

CH_3 CH_3

***m*-CPBA**

$CHCl_3$, Δ

CH_3 O CH_3

***OS*, *CV 5*, 541 (1973)**

Br H N Et

$NaNH_2$, liq. NH_3

Et N

***OS*, *CV 5*, 562 (1973)**

O O O

KSCN, Δ

S

OS, *CV* 5, 755 (1973)

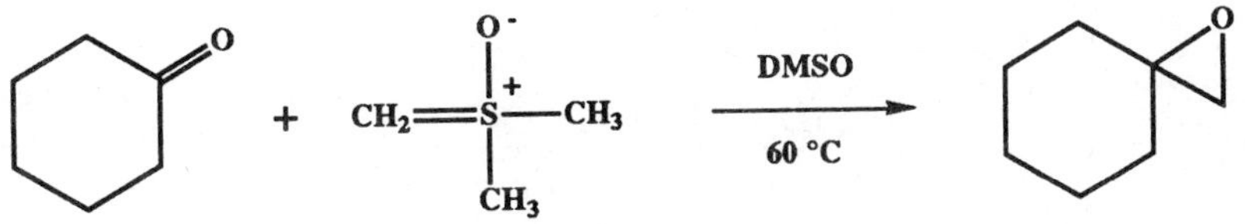

$CH_2{=}S^{+}(O^{-})(CH_3)CH_3$

DMSO

60 °C

OS, *CV* 5, 877 (1973)

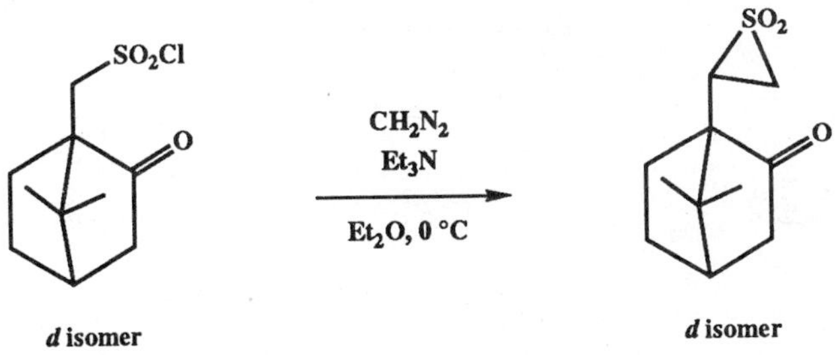

SO_2Cl

CH_2N_2

Et_3N

Et_2O, 0 °C

SO_2

d isomer

d isomer

OS, *CV* 5, 897 (1973)

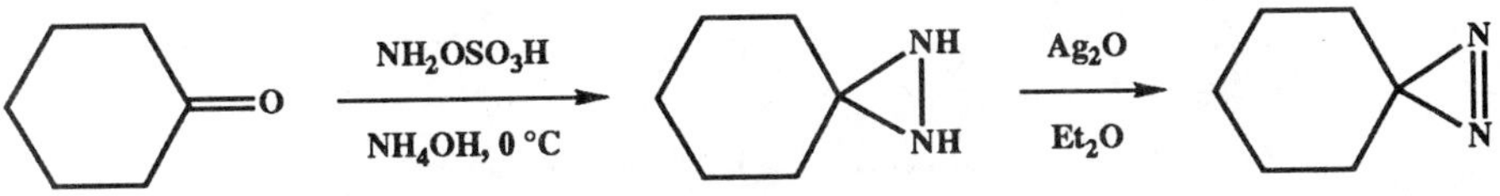

NH_2OSO_3H

NH_4OH, 0 °C

NH

NH

Ag_2O

Et_2O

N

N

OS, *CV* 5, 1007 (1973)

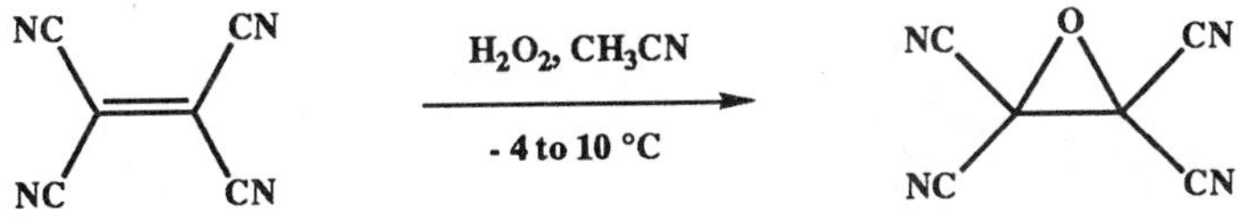

NC CN

NC CN

H_2O_2, CH_3CN

- 4 to 10 °C

NC O CN

NC CN

***OS*, *CV 6*, 56 (1988)**

$Pb(OAc)_4$

CH_2Cl_2, 25 °C

$H_2NNH_2 \cdot H_2O$

pentane, 25 °C

***OS*, *CV 6*, 320 (1988)**

p-NO_2-C_6H_4-CO_3H

CH_2Cl_2

***OS*, *CV 6*, 555 (1988)**

Et_3N

CH_2Cl_2, Δ

***OS*, *CV 6*, 560 (1988)**

K_2CO_3

MeOH, 25 °C

OS, *CV 6*, 679 (1988)

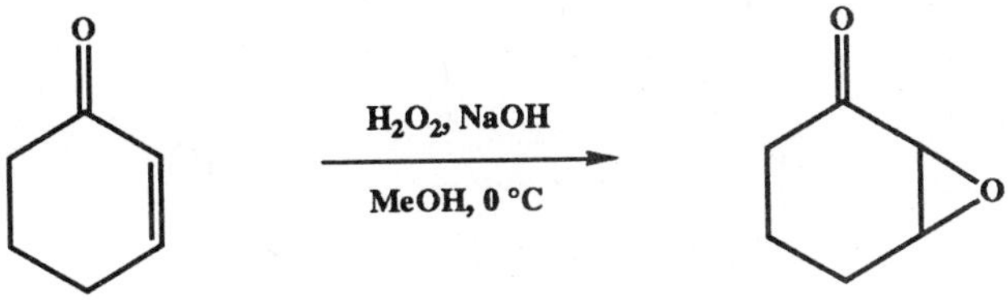

H_2O_2, NaOH

MeOH, 0 °C

OS, *CV 6*, 862 (1988)

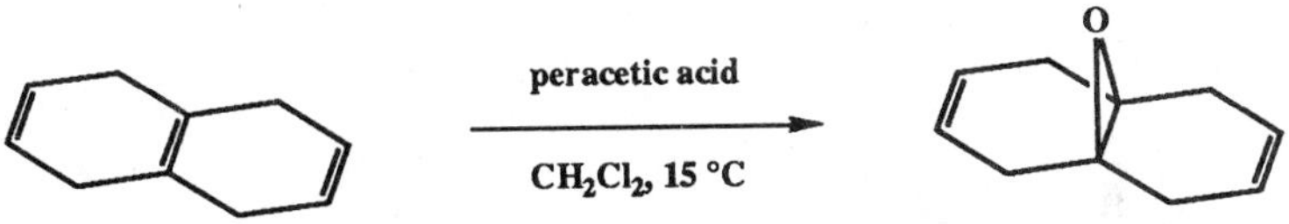

peracetic acid

CH_2Cl_2, 15 °C

OS, *CV 6*, 887 (1988)

$Me_2NCH(OMe)_2$

DMF, THF, Δ

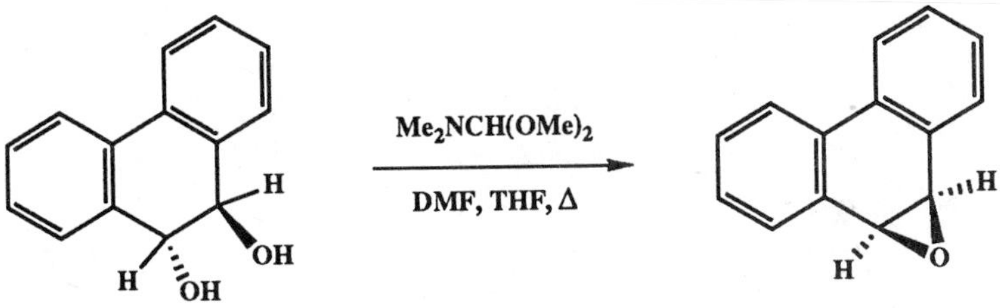

OS, *CV 6*, 893 (1988)

Ph H N₃ CH(OMe)₂

$CHCl_3$, Δ

Ph N CH(OMe)₂

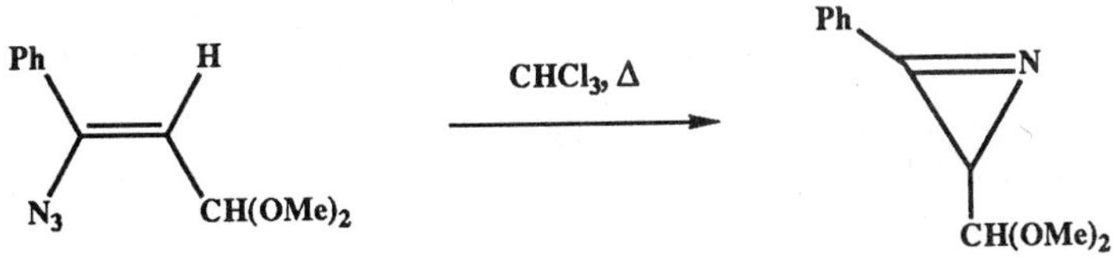

OS, *CV 6*, 967 (1988)

$NHCO_2Me$ I

KOH, EtOH

Δ

H N

OS, *CV 7*, 203 (1990); *60*, 53 (1981)

NH · HCl

Ph NH_2

NaOCl
DMSO

H_2O
hexane

N=N

Ph Cl

Me

Ph

benzene
reflux

Ph Me

Cl H

Ph H

OS, *CV 7*, 461 (1990); *63*, 66 (1985)

HO

$Ti(O\text{-}i\text{-}Pr)_4$
Diethyl (2*R*, 3*R*)-tartrate

t-BuOOH
toluene, CH_2Cl_2

H

HO

O

H

(2*S*, 3*S*)

OS, *CV 7*, 356 (1990); *63*, 140 (1985)

OH

CH_3 CH_2OH

H

S - (+)

HBr
HOAc

OAc

CH_3 CH_2Br

H

S - (-)

n-$C_5H_{11}OK$
$C_5H_{11}OH$

O

CH_3

H

S - (-)

OS, *CV 7*, 164 (1990); *64*, 175 (1986)

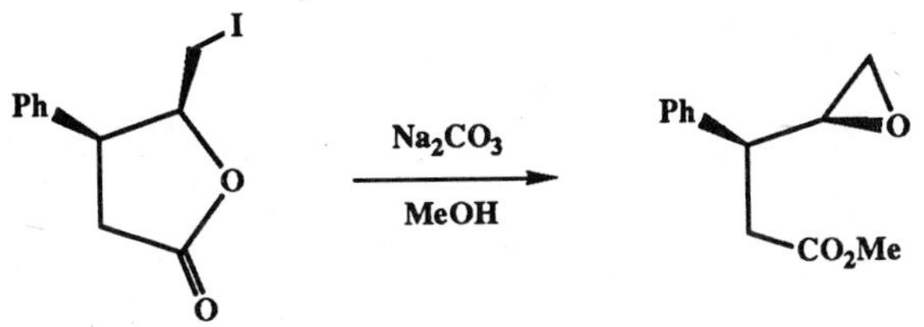

Ph, I, O, O → (Na_2CO_3 / MeOH) → Ph, O, CO_2Me

OS, *66*, 160 (1987)

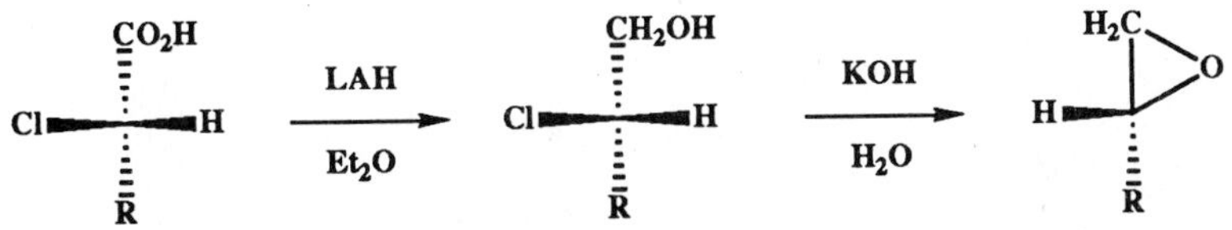

CO_2H, Cl, H, R → (LAH / Et_2O) → CH_2OH, Cl, H, R → (KOH / H_2O) → H_2C, O, H, R

S isomers R = Me, *i*-Pr, *i*-Bu, (*S*) - *sec*-Bu *R* isomers

OS, *66*, 203 (1987)

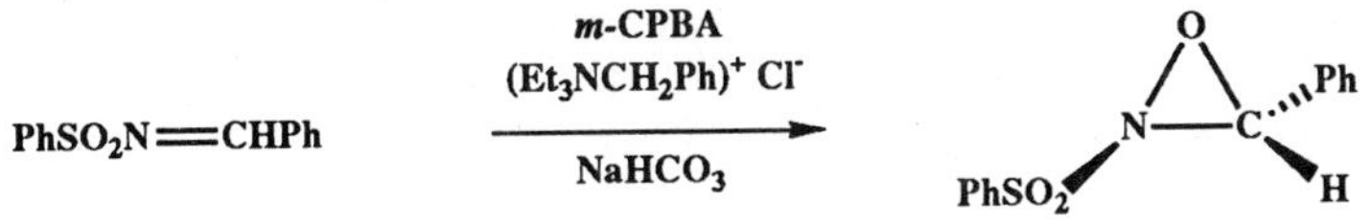

$PhSO_2N{=}CHPh$ → (*m*-CPBA, $(Et_3NCH_2Ph)^+ Cl^-$ / $NaHCO_3$) → O, N, C, Ph, $PhSO_2$, H

OS, CV 3, 508 (1955)

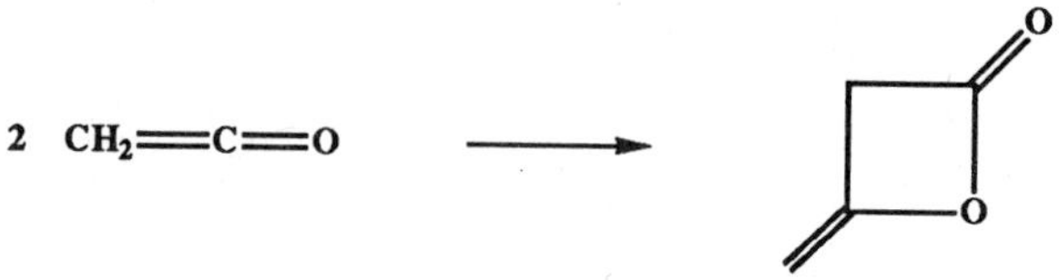

2 $CH_2=C=O$ ⟶

OS, CV 3, 835 (1955)

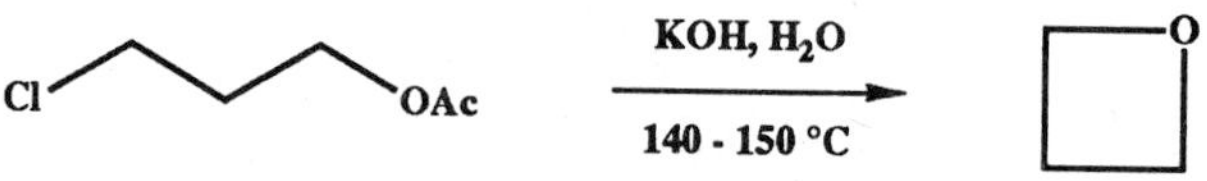

Cl ... OAc

KOH, H_2O

140 - 150 °C

OS, CV 5, 456 (1973)

$AlCl_3$, 135 °C

chlorobenzene

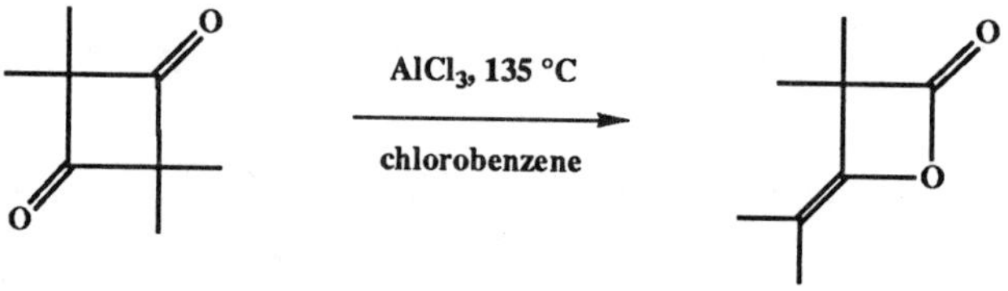

OS, CV 5, 673 (1973)

CH_3 / CH_3 + $N=C=O$ / SO_2Cl

liq. SO_2

- 45 to - 6 °C

CH_3, CH_3, N, ClO_2S, O + O, NH, SO_2Cl

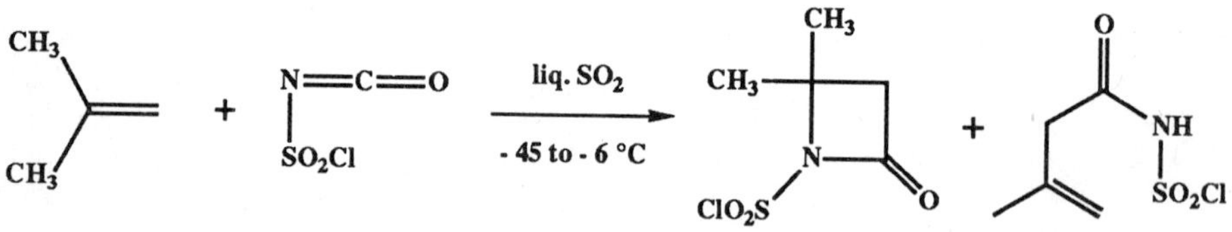

OS, CV 6, 75 (1988)

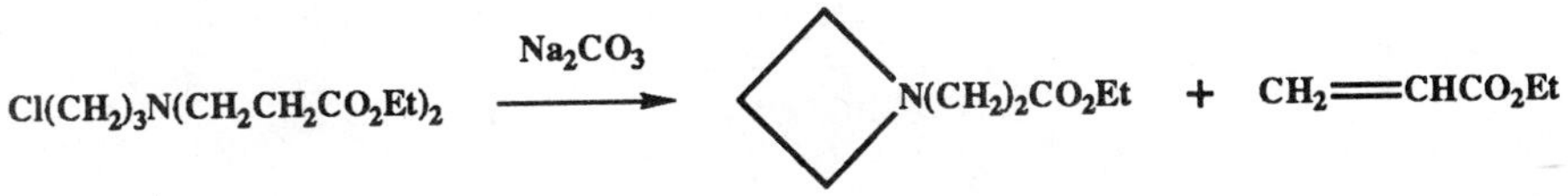

$Cl(CH_2)_3N(CH_2CH_2CO_2Et)_2$ —(Na_2CO_3)→ [azetidine]$N(CH_2)_2CO_2Et$ + $CH_2{=}CHCO_2Et$

OS, 65, 135 (1987)

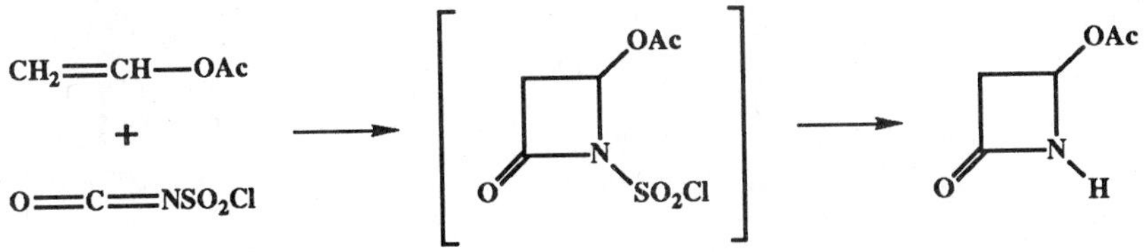

$CH_2{=}CH{-}OAc$ + $O{=}C{=}NSO_2Cl$ → [OAc, O, N, SO_2Cl] → OAc, O, N, H

OS, 65, 140 (1987)

$(CO)_5Cr{=}C(OMe)(Me)$ + $MeN{=}CHPh$ —(hν, pet. ether)→ —(hν, air)→ Me, H, MeO, Ph, O, N, Me

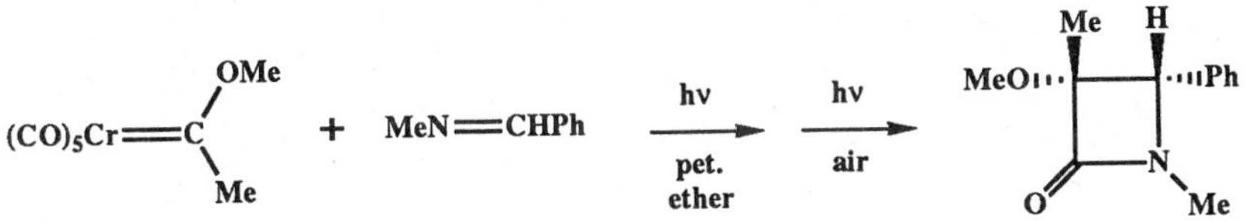

OS, CV 1, 56 (1941)

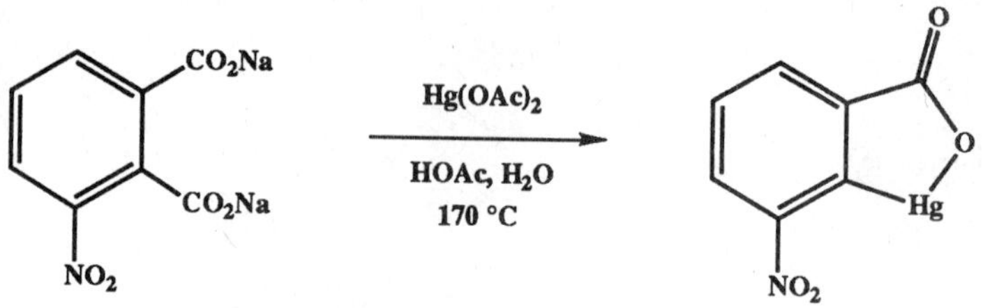

CO_2Na CO_2Na NO_2

$Hg(OAc)_2$

$HOAc, H_2O$

170 °C

O O Hg NO_2

OS, CV 1, 172 (1941)

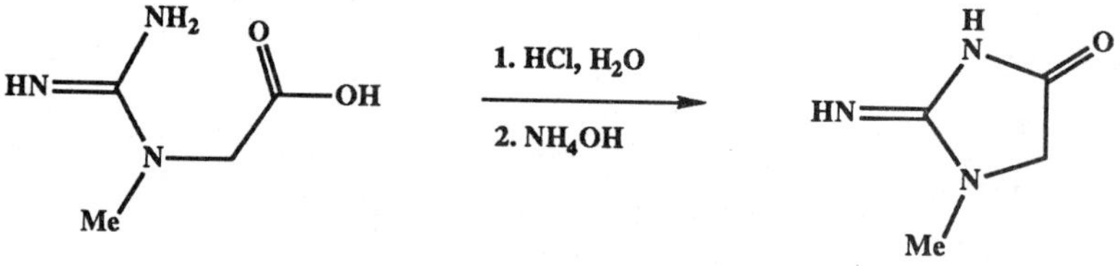

NH_2 O HN= OH N Me

1. HCl, H_2O

2. NH_4OH

H N O HN= N Me

OS, CV 1, 327 (1941)

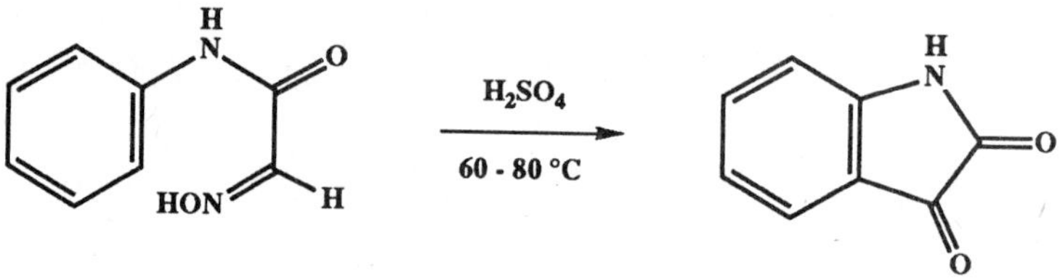

H N O HON H

H_2SO_4

60 - 80 °C

H N O O

OS, CV 1, 410 (1941)

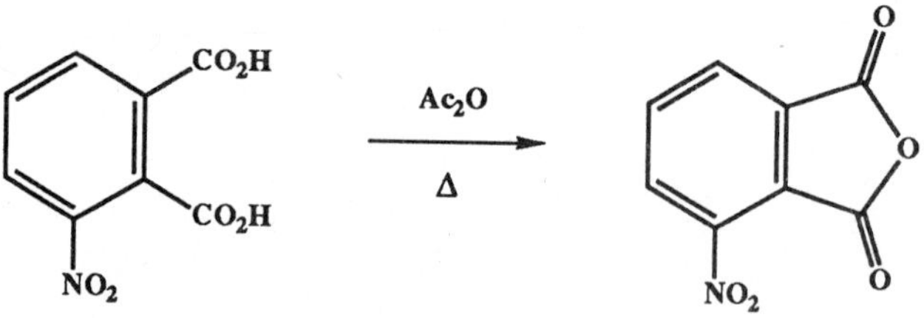

CO_2H CO_2H NO_2

Ac_2O

Δ

O O O NO_2

OS, CV 1, 457 (1941)

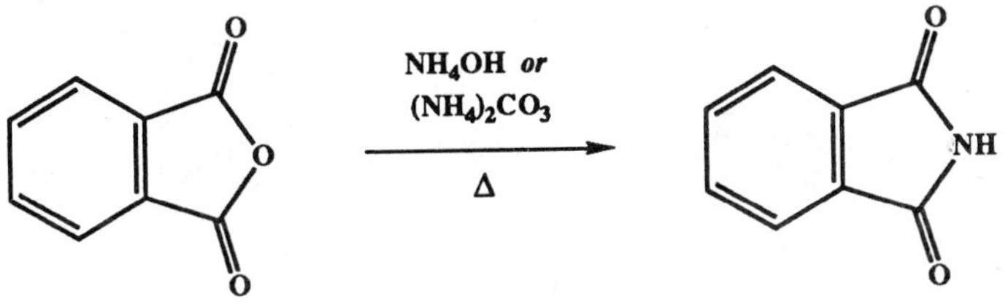
O
O
O
NH₄OH *or*
(NH₄)₂CO₃
Δ
O
NH
O

OS, CV 1, 473 (1941)

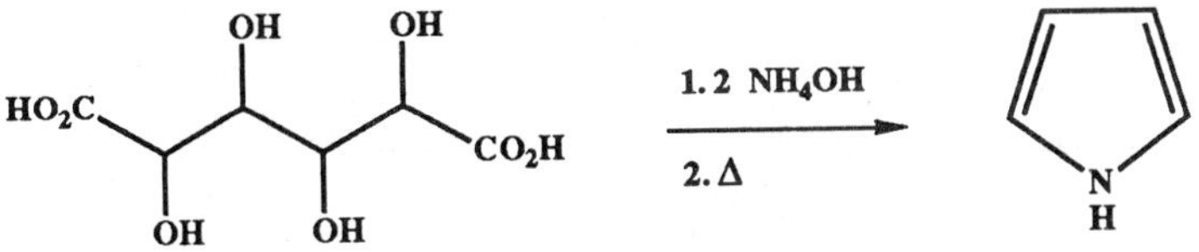
OH
OH
HO₂C
CO₂H
OH
OH
1. 2 NH₄OH
2. Δ
N
H

OS, CV 1, 495 (1941)

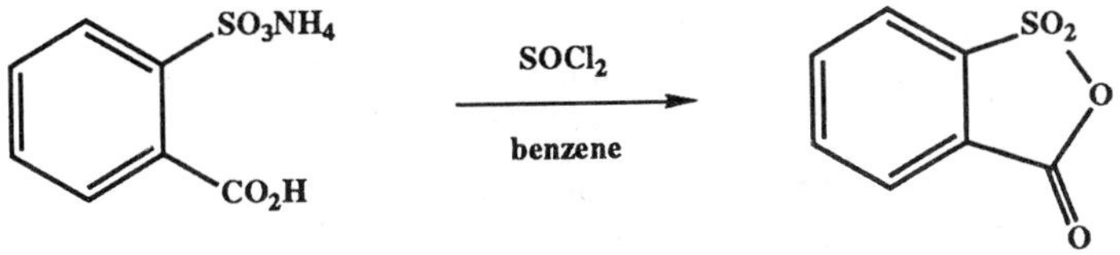
SO₃NH₄
CO₂H
SOCl₂
benzene
SO₂
O
O

OS, CV 2, 1 (1943)

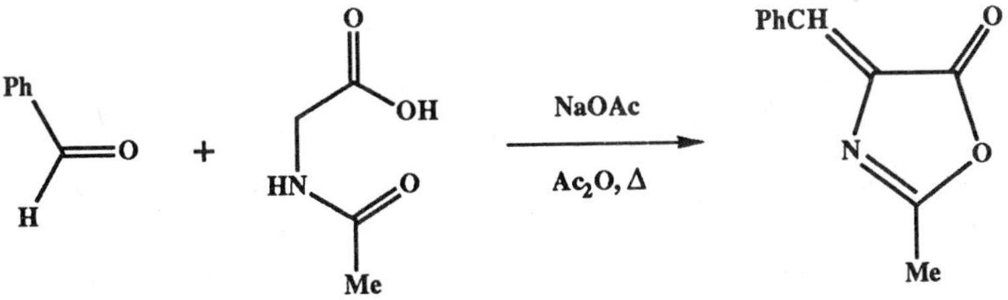
Ph
O
H
+
O
OH
HN
O
Me
NaOAc
Ac₂O, Δ
PhCH
O
N
O
Me

OS, CV 2, 31 (1943)

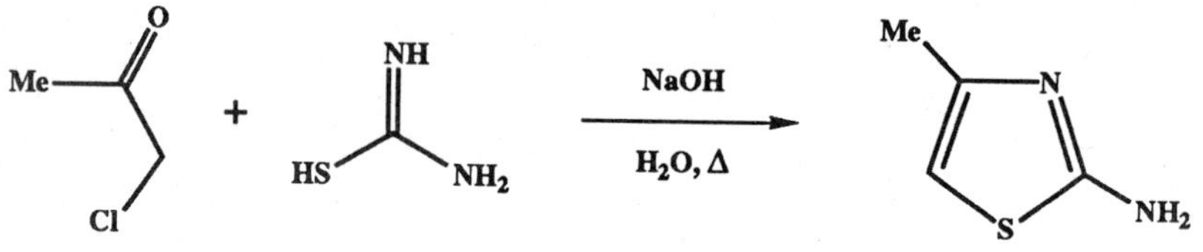

Me O Cl + NH HS NH_2 —NaOH / H_2O, Δ→ Me N S NH_2

OS, CV 2, 55 (1943)

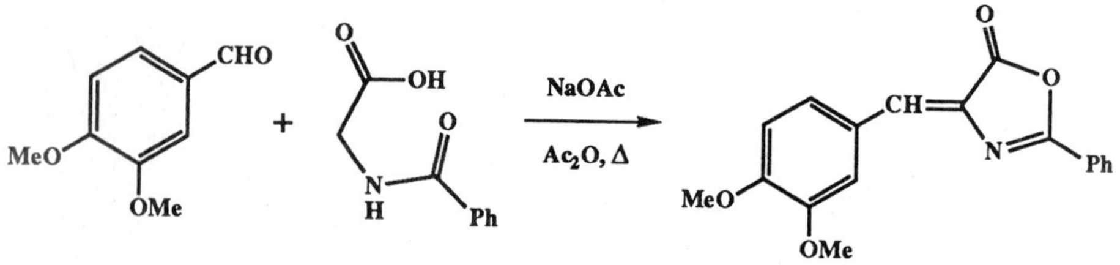

CHO MeO OMe + O OH O N H Ph —NaOAc / Ac_2O, Δ→ O O CH= N Ph MeO OMe

OS, CV 2, 65 (1943)

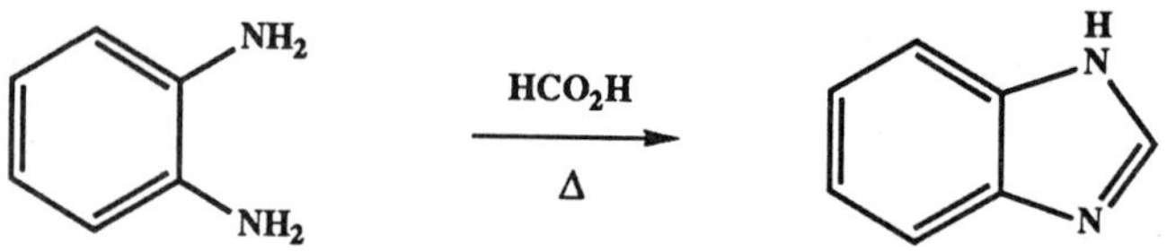

NH_2 NH_2 —HCO_2H / Δ→ H N N

OS, CV 2, 194 (1943)

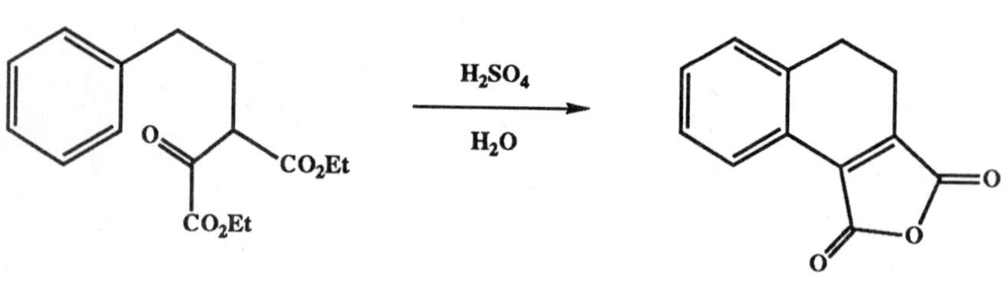

O CO_2Et CO_2Et —H_2SO_4 / H_2O→ O O O

OS, *CV 2*, 202 (1943)

Me, O, EtO_2C, NH_2 + CO_2Et, O, Me → (Δ, HOAc) Me, CO_2Et, EtO_2C, N H, Me

OS, *CV 2*, 219 (1943)

Me, Me, O, O → ($(NH_4)_2CO_3$, Δ) Me, Me, N H

OS, *CV 2*, 231 (1943)

OH, Ph, Ph, O + O, H_2N, NH_2 → (HOAc, Δ) Ph, H N, O, N, Ph

OS, *CV 2*, 368 (1943)

OH, HO_2C, O, O, OH, HO → (Δ) O, O, O

***OS*, *CV* 2, 526 (1943)**

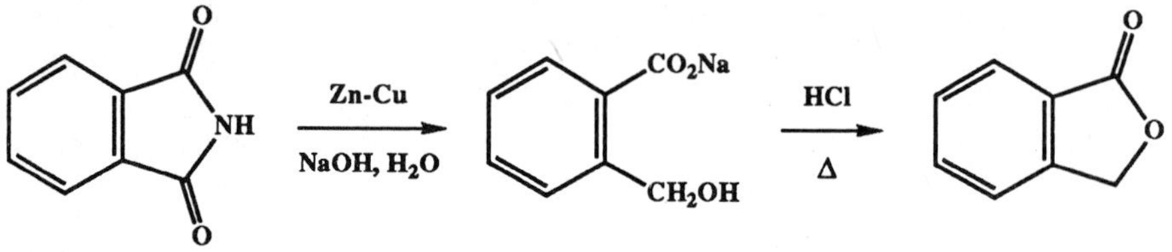

Zn-Cu / $NaOH, H_2O$ → CO_2Na, CH_2OH → HCl / Δ

NH, O, O, O

***OS*, *CV* 2, 560 (1943)**

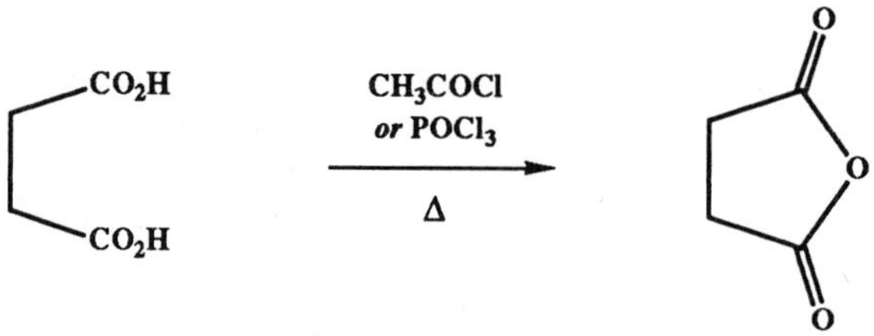

CO_2H, CO_2H → CH_3COCl *or* $POCl_3$ / Δ

O, O, O

***OS*, *CV* 2, 562 (1943)**

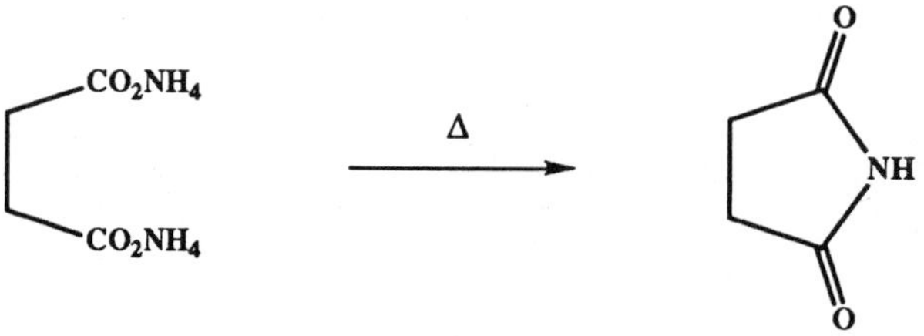

CO_2NH_4, CO_2NH_4 → Δ

O, NH, O

***OS*, *CV* 2, 578 (1943)**

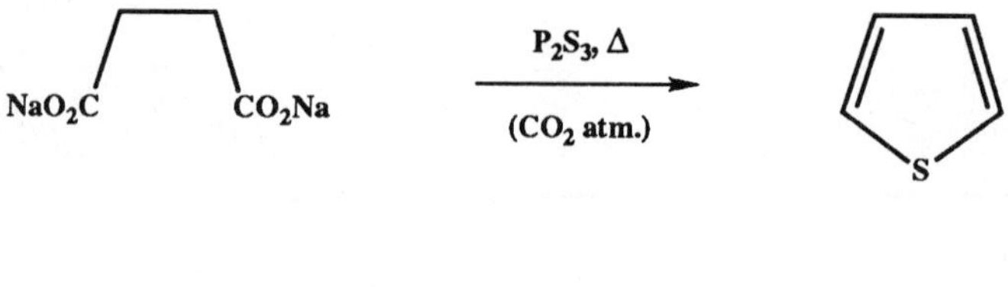

NaO_2C, CO_2Na → P_2S_3, Δ / (CO_2 atm.)

S

OS, CV 3, 76 (1955)

NH_2, Me — NaSCN, H_2SO_4; C_6H_5Cl, 100 °C — H, N, =NH, HS, Me — SO_2Cl_2; 50 °C — N, NH_2, S, Me

OS, CV 3, 95 (1955)

H, H_2N, N, NH_2, NH • H_2CO_3 — HCO_2H — H, H_2N, N, NH_2, NH • HCO_2H — 120 °C — NH_2, N, N, N, H

OS, CV 3, 96 (1955)

2 O, H, OEt — H_2NNH_2 • H_2O; EtOH, Δ — 2 O, H, N, H, NH_2 — 200 °C — NH_2, N, N, N

OS, CV 3, 106 (1955)

NH_2, NH_2 — HNO_2 — [N=N, OH, NH_2] — N, N, N, H

OS, *CV* 3, 151 (1955)

t-Bu-NH-C(=O)-NH$_2$

220 °C

N—t-Bu

OS, *CV* 3, 159 (1955)

n-Bu, NH, CH_3

1. Cl_2, aq. NaOH ligroin

2. H_2SO_4, H_2O

N—n-Bu

OS, *CV* 3, 323 (1955)

CH_3, OH, CH_3, CN + $(NH_4)_2CO_3$

Δ

CH_3, CH_3, H N, O, N H, O

OS, *CV* 3, 328 (1955)

CH_3, O, CO_2H

aq. CH_3NH_2
H_2, Ra (Ni)

(1000-2000 psi)
140 °C

CH_3, N, O, CH_3

OS, *CV* 3, 332 (1955)

SH, CH_3, NH + Cl, O, CH_3 → (benzene, Δ) S, CH_3, N, CH_3

OS, *CV* 3, 358 (1955)

Ph, CN, O, Ph → (H_2 (50 psi), Ra (Ni), MeOH, 85 °C) Ph, N, H, Ph

OS, *CV* 3, 394 (1955)

NH_2, NH_2 → (1. CS_2, Δ, EtOH, H_2O; 2. HCl, Δ) H, N, S, N, H

OS, *CV* 3, 460 (1955)

O, OH, H, HO, CH_2OH, HO, OH, H

d - fructose

→ (CH_2O, NH_4OH, air; $CuCO_3 \cdot Cu(OH)_2$, Δ) CH_2OH, HN, N → (HCl, H_2O; C_6H_6, 80 °C) CH_2OH, HN, N • HCl

***OS*, *CV* 3, 471 (1955)**

HO_2C, ONO_2, CO_2H, ONO_2 — $CH_2{=}O$ / NH_4OH → HO_2C, HO_2C, N, N H

***OS*, *CV* 3, 475 (1955)**

CO_2H, N—NH_2 • HCl, H — HCl, H_2O / Δ → OH, N, N H

***OS*, *CV* 3, 479 (1955)**

CH_3, O, N H, H — 1. *t*-BuOK, Δ / 2. H_2O → N H

***OS*, *CV* 3, 499 (1955)**

CH_3 — SO_2, CH_3OH / hydroquinone 85 °C → CH_3, SO_2

OS, CV 3, 513 (1955)

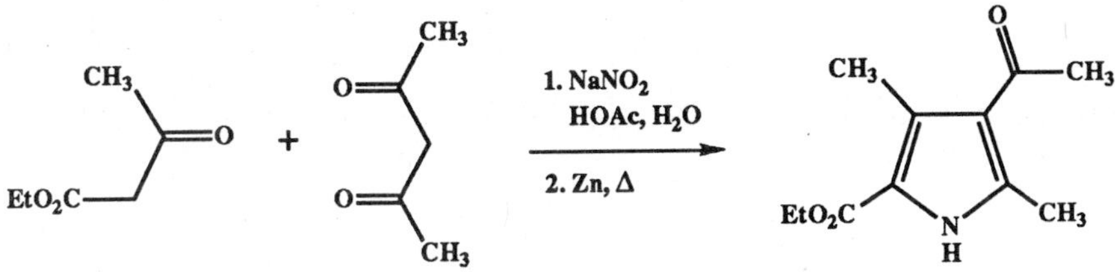

1. $NaNO_2$, HOAc, H_2O
2. Zn, Δ

OS, CV 3, 595 (1955)

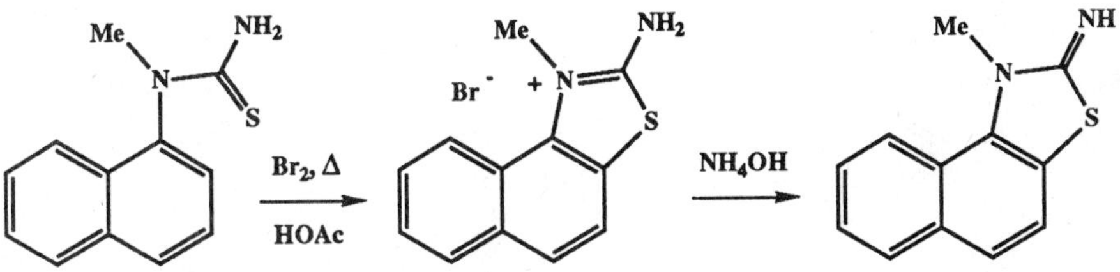

Br_2, Δ, HOAc

NH_4OH

OS, CV 3, 597 (1955)

1. $NaNH_2$, Et_2O, Δ
2. EtOH, H_2O

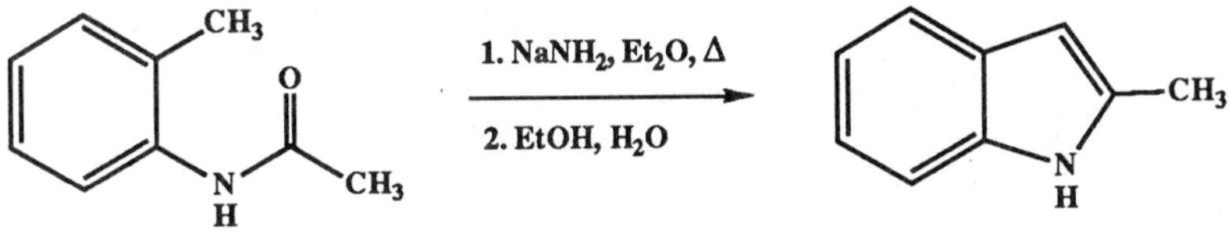

OS, CV 3, 660 (1955)

$NaNO_2$, aq. HOAc

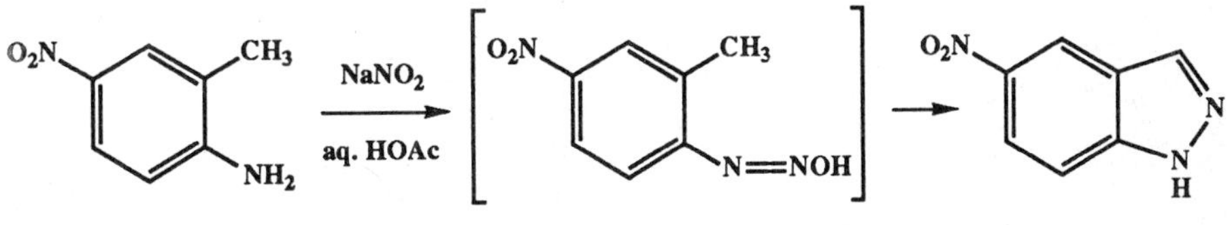

OS, CV 3, 708 (1955)

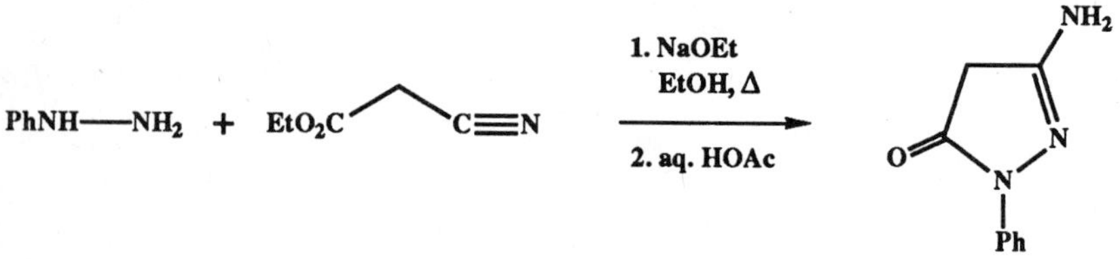

PhNH—NH$_2$ + EtO$_2$C–CH$_2$–C≡N → (1. NaOEt, EtOH, Δ; 2. aq. HOAc) → 3-amino-1-phenyl-2-pyrazolin-5-one (NH$_2$, N, N–Ph, O)

OS, CV 3, 725 (1955)

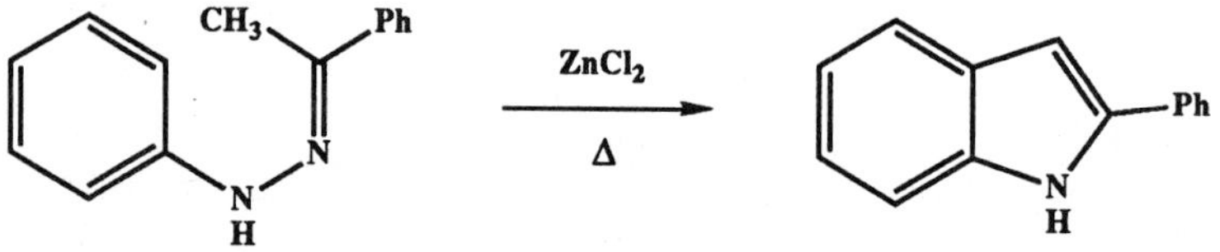

Acetophenone phenylhydrazone (CH$_3$, Ph, N, N–H) → (ZnCl$_2$, Δ) → 2-phenylindole (Ph, N–H)

OS, CV 3, 751 (1955)

ClCH$_2$CO$_2$Et + H$_2$N–C(=NH)–SH → (1. EtOH, Δ; 2. aq. NaOAc, Δ) → 2-imino-4-thiazolidinone (O, H–N, =NH, S)

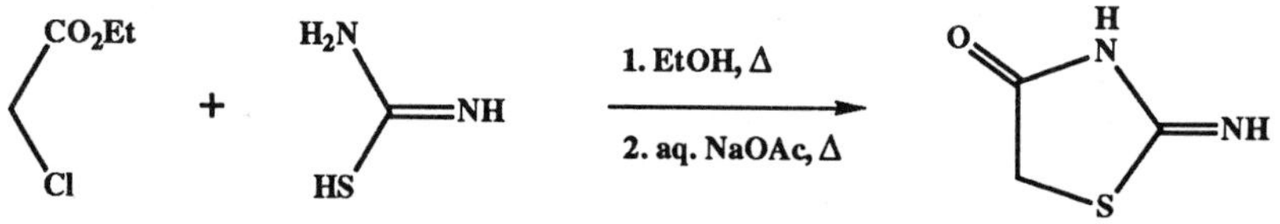

OS, CV 3, 763 (1955)

ClCH$_2$CO$_2$Na + NH$_4$S–C(=S)–H$_2$N → (1. H$_2$O; 2. HCl) → rhodanine (S, =S, O, N–H)

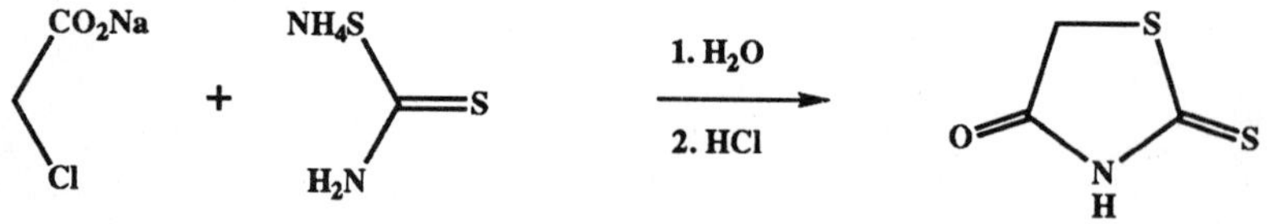

OS, CV 4, 6 (1963)

AcNH, NH_2 + S, S, S, CO_2H, CO_2H → (H_2O, Δ) AcNH, O, N, S, S

OS, CV 4, 10 (1963)

Cl, O + O, CH_3, CO_2Et → (1. Na, EtOH 45 - 50 °C; 2. HOAc) Cl, O, O, O, CH_3

OS, CV 4, 74 (1963)

NH_2, NO_2 → (NaOCl, KOH; EtOH, 0 °C) N, O, N+, O-

OS, CV 4, 75 (1963)

NO_2, N_3 → (Δ; toluene) N, O, N+, O-

OS, CV 4, **106 (1963)**

O, O, O

$H_2NCH_2CH_2OH$

Δ

O, NCH_2CH_2OH, O

OS, CV 4, **242 (1963)**

OH, HO, CO_2H, CO_2H

l - isomer

Ac_2O, Δ

H_2SO_4

OAc, AcO, O, O, O

OS, CV 4, **243 (1963)**

NC, CN, NC, CN

H_2S, pyridine

acetone, 0 °C

NC, CN, H_2N, S, NH_2

OS, CV 4, **342 (1963)**

CH_3, NOH, CH_3, NOH

succinic anhydride

Δ

CH_3, CH_3, N, N, O

OS, CV 4, 350 (1963)

CH_3 CH_3 OH HO $(CH_2)_4CH_3$ —— H_3PO_4, Δ ——> CH_3 CH_3 O $(CH_2)_4CH_3$

OS, CV 4, 351 (1963)

O O CH_3 CH_3 —— H_2NNH_2 / aq. NaOH ——> CH_3 CH_3 N N H

OS, CV 4, 357 (1963)

CH_3 CH_3 NO_2 O OCH_3 —— H_2 (1000 psi), Raney nickel / EtOH, 55 - 60 °C ——> CH_3 CH_3 N H O

OS, CV 4, 380 (1963)

Ph C≡N + N_3—Ph —— NaOMe / EtOH, 2 - 5 °C ——> Ph NH_2 N N N Ph

OS, CV 4, 444 (1963)

n-Bu, CO_2Et, Et, Me

H_2SO_4, Δ

ethylene glycol

n-Bu, Me, Et, O, O

OS, CV 4, 506 (1963)

CHO
H—C—OH
HO—C—H
H—C—OH
CH_2OH

d - gulonic acid

1. NaCN
aq. NH_4Cl
2. $Ba(OH)_2$, Δ
3. H_2SO_4

d - gulonic acid, γ - lactone

OS, CV 4, 534 (1963)

HO, OH, OH

p-TsOH

180 - 220 °C

OH, O

OS, CV 4, 536 (1963)

O, CHOH

H_2NNH_2

N, N, H

***OS*, *CV* 4, 569 (1963)**

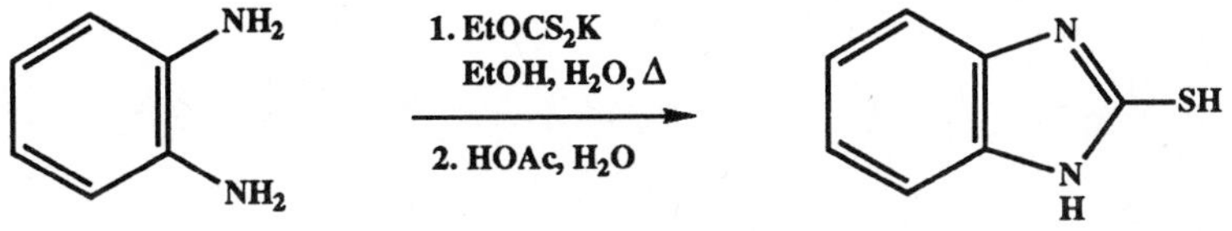

NH_2 / NH_2 — 1. $EtOCS_2K$, EtOH, H_2O, Δ; 2. HOAc, H_2O → N, SH, N H

***OS*, *CV* 4, 590 (1963)**

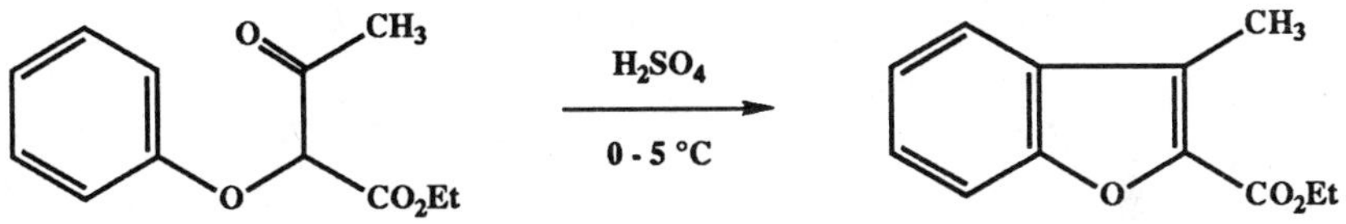

O, CH_3, O, CO_2Et — H_2SO_4, 0 - 5 °C → CH_3, O, CO_2Et

***OS*, *CV* 4, 620 (1963)**

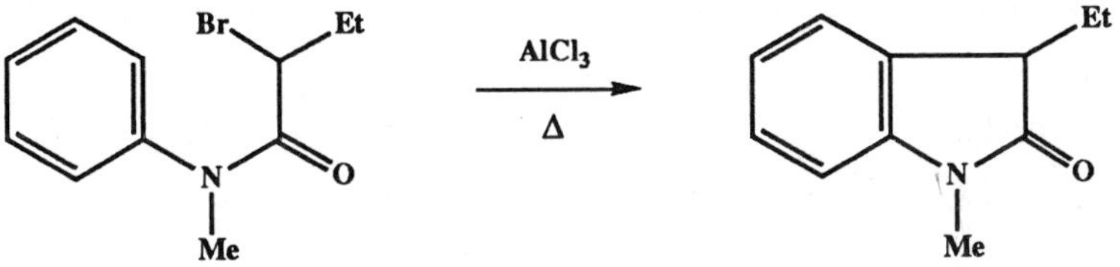

Br, Et, N, O, Me — $AlCl_3$, Δ → Et, N, O, Me

***OS*, *CV* 4, 649 (1963)**

MeO, MeO, CH_3, CO_2Me, O — 160 °C → CH_3, O, CO_2Me

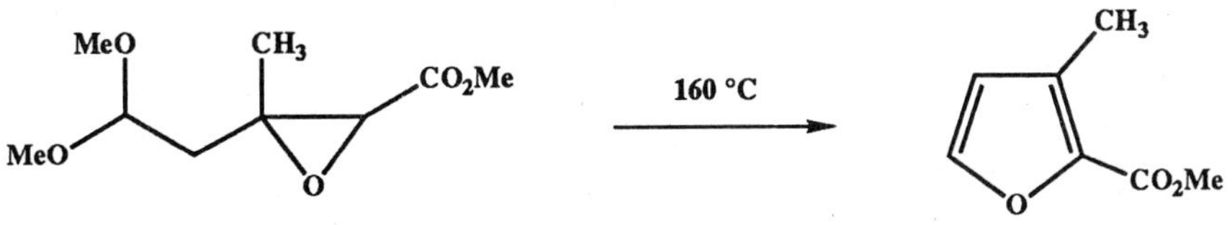

***OS*, *CV 4*, 657 (1963)**

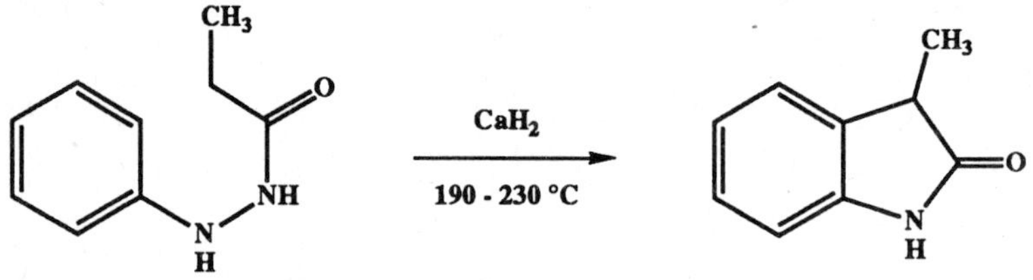

CH3 · O · NH · N H

CaH2 / 190 - 230 °C

CH3 · O · N H

***OS*, *CV 4*, 671 (1963)**

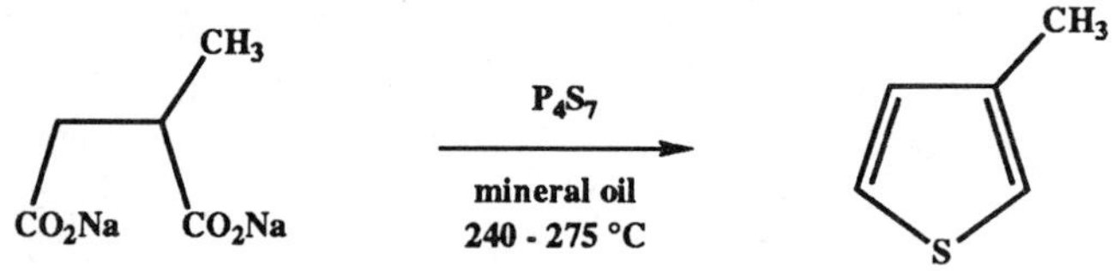

CH3 · CO2Na · CO2Na

P4S7 / mineral oil / 240 - 275 °C

CH3 · S

***OS*, *CV 4*, 744 (1963)**

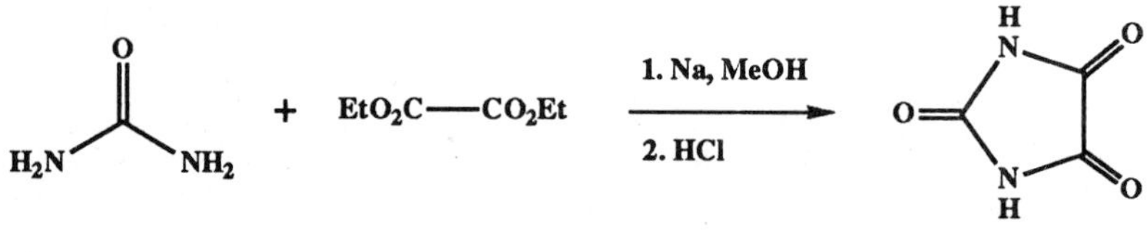

O · H2N · NH2 + EtO2C—CO2Et

1. Na, MeOH / 2. HCl

H N · O · O · N H · O

***OS*, *CV 4*, 788 (1963)**

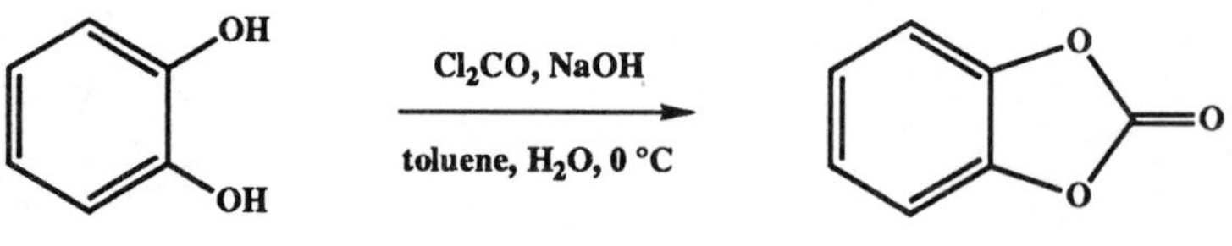

OH · OH

Cl2CO, NaOH / toluene, H2O, 0 °C

O · O · O

OS, CV 4, **884 (1963)**

NH$_2$ N H + O HOAc, Δ N H

OS, CV 4, **892 (1963)**

ClCH$_2$ CH$_2$Cl Na$_2$S DMF, H$_2$O, Δ S

OS, CV 5, **12 (1973)**

Cl O CH$_3$ N H O 1. KNH$_2$, NH$_3$ 2. HCl O CH$_3$ N H OH

OS, CV 5, **39 (1973)**

H N H$_2$N CN H$_2$SO$_4$ EtOH H$_2$N N H NH • H$_2$SO$_4$

OS, CV 5, 60 (1973)

NH_2 / SO_2Na —($NaNO_2$, H_2SO_4 / H_2O, glycerol, - 15 °C)→ N, N, SO_2

OS, CV 5, 80 (1973)

O, Ph, H + O, Ph, CO_2H —(NaOAc / Ac_2O, Δ)→ O, PhCH, O, Ph

OS, CV 5, 251 (1973)

CHO / OCH_2CO_2H —(NaOAc, Δ / Ac_2O, HOAc)→ O

OS, CV 5, 567 (1973)

OK, CH, CO_2Et, NO_2 —(H_2 (30 psi), PtO_2 / HOAc)→ CO_2Et, N H

***OS*, *CV 5*, 650 (1973)**

N_2O_3, Ac_2O; H_2O, 40 °C

CH$_3$ N H COCH$_3$ → [CH$_3$ N NO COCH$_3$] → N N H

***OS*, *CV 5*, 692 (1973)**

HO CH$_3$ AcO

1. $Pb(OAc)_4$, I_2 C_6H_{12}, light, Δ

2. CrO_3, H_2SO_4 acetone, 0 °C

O O CH$_3$ AcO

***OS*, *CV 5*, 716 (1973)**

MeO O OMe + NH$_2$ CO$_2$Me

HOAc, Δ

N CO$_2$Me

***OS*, *CV 5*, 787 (1973)**

CH$_3$

Ph—PCl$_2$

Cl Ph—P Cl CH$_3$

H_2O

O P Ph CH$_3$

***OS*, *CV 5*, 829 (1973)**

170 °C

o-$C_6H_4Cl_2$

NO_2 N_3 NO_2 NH

***OS*, *CV 5*, 941 (1973)**

CHO NO_2 $PhNH_2$ Δ CH═NPh NO_2 $(EtO)_3P$ 150 °C N—Ph N

***OS*, *CV 5*, 944 (1973)**

O O O $PhNH_2$ Et_2O O N H Ph CO_2H Ac_2O NaOAc, Δ O N—Ph O

***OS*, *CV 5*, 946 (1973)**

CO_2H O N H Ph Ac_2O, 80 °C O O N Ph

OS, *CV 5*, 957 (1973)

H, Ph, C=N⁺, n-Pr, O⁻ + maleimide N—Ph —EtOH→ Ph—N, O, N—Ph, n-Pr

OS, *CV 5*, 962 (1973)

Ph, N, NO, O, OH —Ac_2O, Δ→ Ph, N, N, O, H, ±, O

OS, *CV 5*, 973 (1973)

O, O, O —*l* - phenylalanine / Et_3N, toluene, Δ→ O, N, O, CO_2H, *, CH_2Ph

l - (-) isomer

OS, *CV 5*, 1022 (1973)

CH_3, O, CH_3, NOH + O, CH_3, CH_3, O, CH_3 —Zn / HOAc, Δ→ CH_3, CH_3, CH_3, CH_3, N, H

OS, CV 5, 1024 (1973)

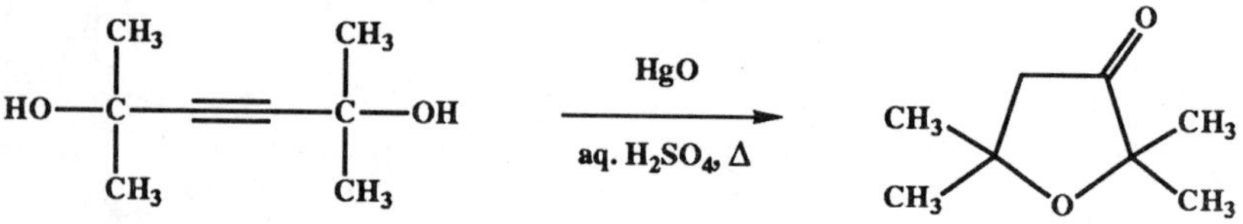

HgO; aq. H_2SO_4, Δ

OS, CV 5, 1064 (1973)

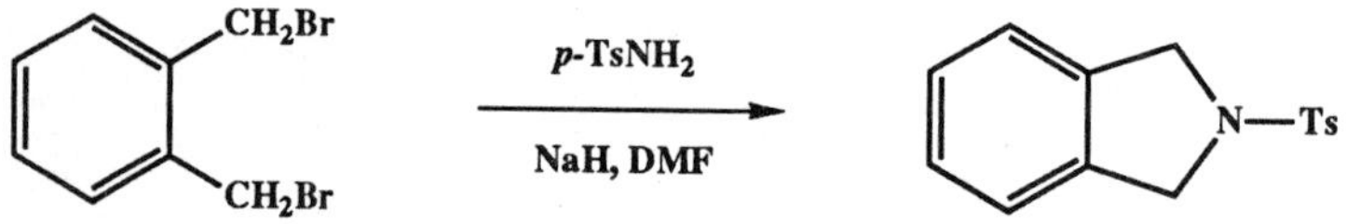

p-$TsNH_2$; NaH, DMF

OS, CV 5, 1070 (1973)

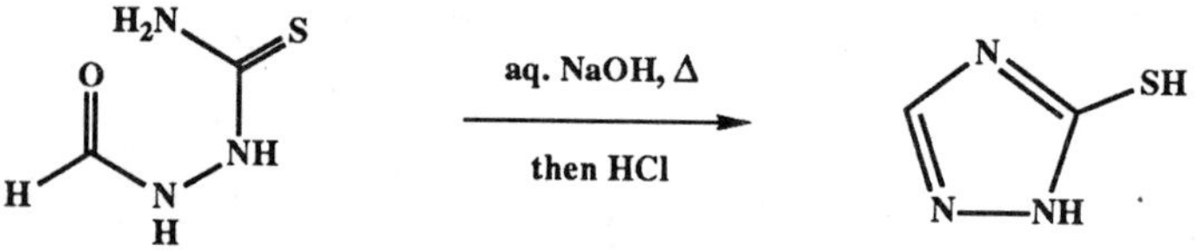

aq. NaOH, Δ; then HCl

OS, CV 5, 1124 (1973)

60 °C

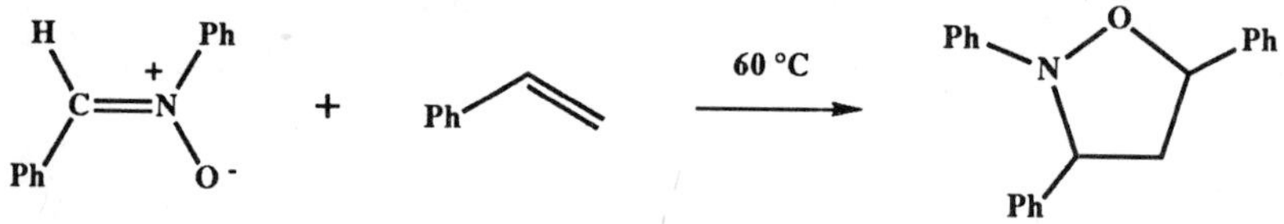

OS, CV 6, 31 (1988)

$Me_2\overset{+}{S}CH_2C{\equiv}CH\ Br^-$; NaOEt, EtOH, Δ

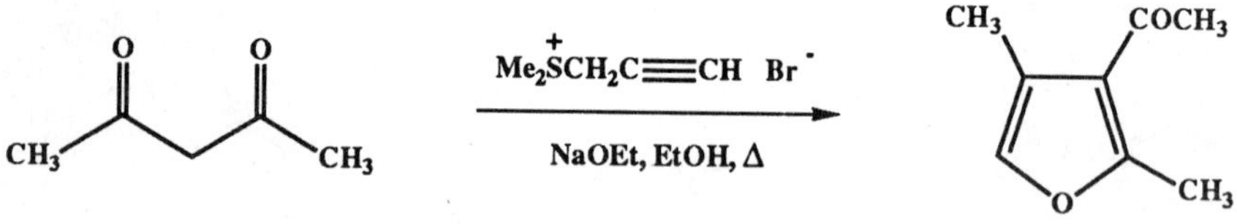

***OS*, *CV 6*, 278 (1988)**

$CH_3C(=NOH)C_6H_4$-*p*-Cl → (*n*-BuLi, 0 °C) → $CH_2LiC(=NOLi)C_6H_4$-*p*-Cl → (MeO_2C-C_6H_4-*p*-OMe, then aq. HCl, Δ) → 3-(C_6H_4-*p*-Cl)-5-(C_6H_4 - *p*-OMe)isoxazole

***OS*, *CV 6*, 312 (1988)**

erythro cyclobutanecarboxylate ester of $CH_3CH(O)CH(OSO_2CH_3)CH_3$ → ($NaBH_4$, pyridine, 115 °C) → 2-cyclobutyl-4,5-dimethyl-1,3-dioxolane (CH_3, CH_3)

***OS*, *CV 6*, 482 (1988)**

cis-1,2-bis(CH_2OSO_2Me)cyclohexane (H, H) → (Na_2S, DMSO, 120 °C) → *cis*-octahydrobenzo[c]thiophene (S; H, H)

***OS*, *CV 6*, 590 (1988)**

2-(CHOH)cyclohexanone (O) + $TsS(CH_2)_2S$-*p*-Ts (S—*p*-Ts, S—*p*-Ts) → (KOAc, MeOH, 65 °C) → spiro 1,3-dithiolane cyclohexanone (O, S, S)

***OS*, *CV* 6, 592 (1988)**

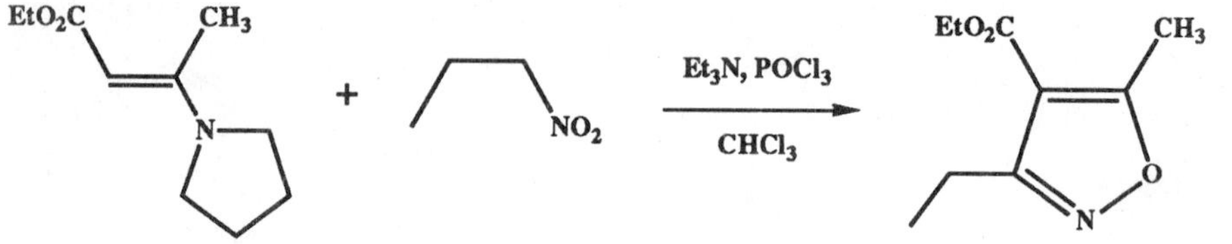

EtO_2C CH$_3$ + NO_2 $\xrightarrow[CHCl_3]{Et_3N,\ POCl_3}$ EtO_2C CH$_3$ N O

***OS*, *CV* 6, 601 (1988)**

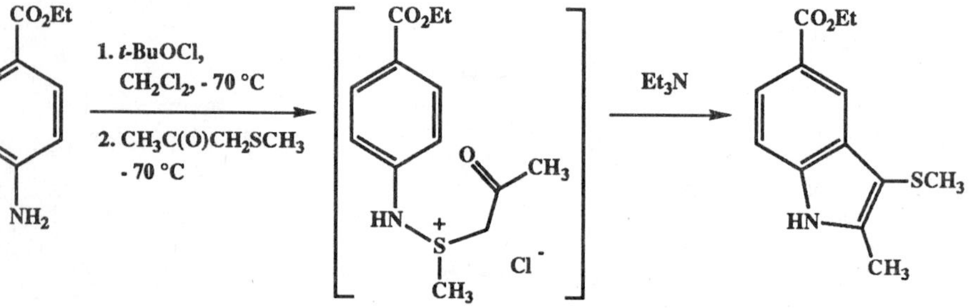

CO_2Et NH_2 1. *t*-BuOCl, CH_2Cl_2, - 70 °C 2. $CH_3C(O)CH_2SCH_3$ - 70 °C [CO_2Et O CH_3 HN S^+ CH_3 Cl^-] $\xrightarrow{Et_3N}$ CO_2Et SCH_3 HN CH_3

***OS*, *CV* 6, 620 (1988)**

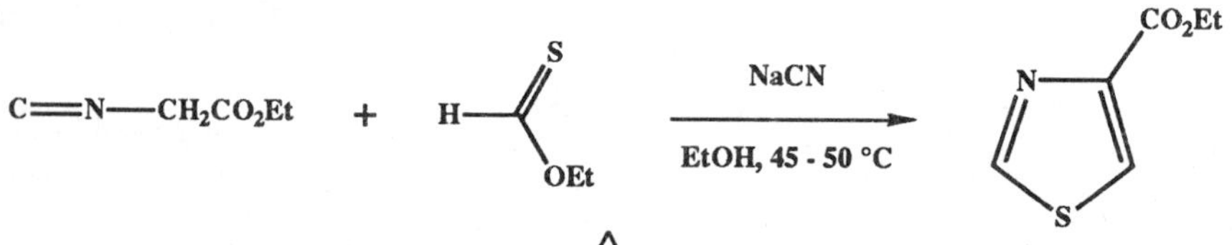

C═N—CH_2CO_2Et + H—C(=S)OEt $\xrightarrow[EtOH,\ 45 - 50\ °C]{NaCN}$ CO_2Et N S

***OS*, *CV* 6, 670 (1988)**

H O $\xrightarrow[toluene,\ \Delta]{MeNHOH \cdot HCl,\ NaOMe}$ [H ^-O N^+ Me] ⟶ O N Me H

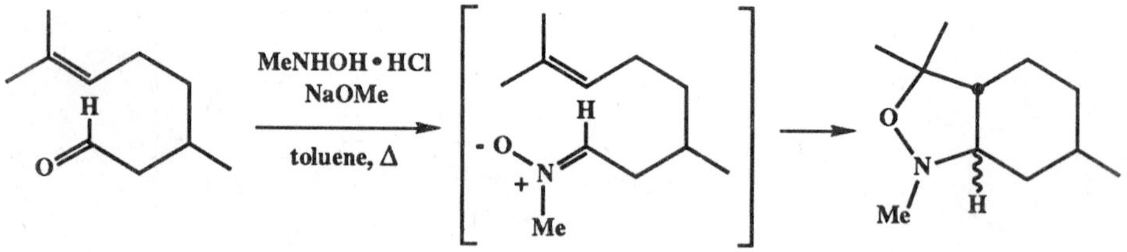

OS, CV 6, 791 (1988)

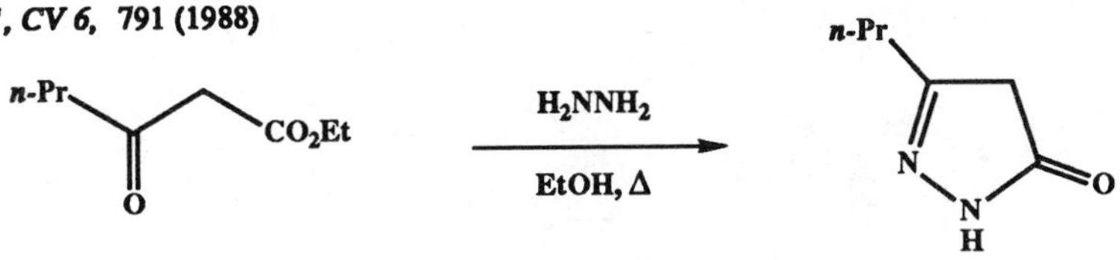
n-Pr
CO2Et
O
H2NNH2
EtOH, Δ
n-Pr
N
N
H
O

OS, CV 6, 916 (1988)

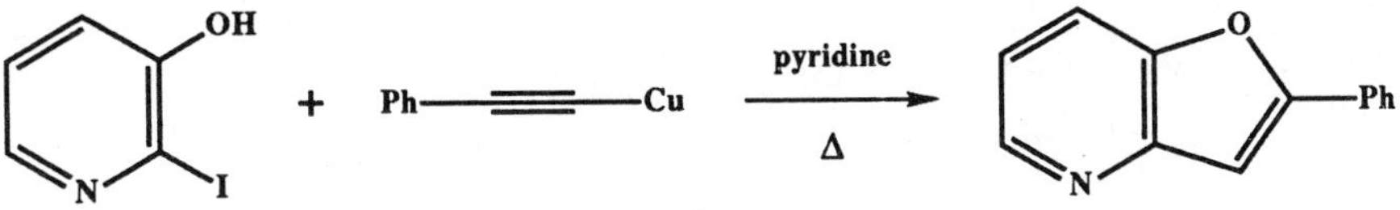
OH
N
I
+
Ph
Cu
pyridine
Δ
O
N
Ph

OS, CV 6, 936 (1988)

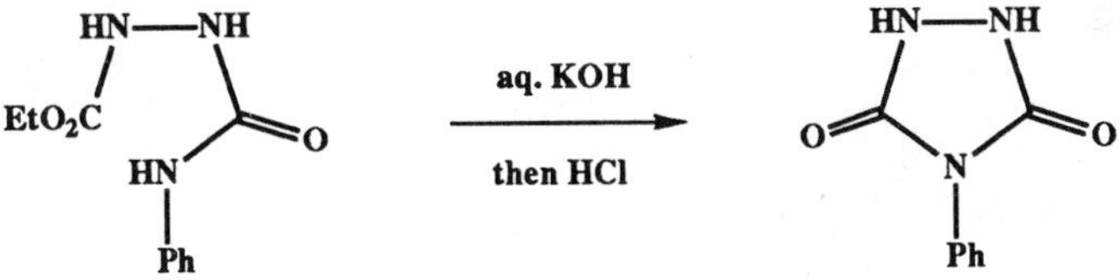
HN
NH
EtO2C
O
HN
Ph
aq. KOH
then HCl
HN
NH
O
O
N
Ph

OS, CV 7, 411 (1990); *60,* 66 (1981)

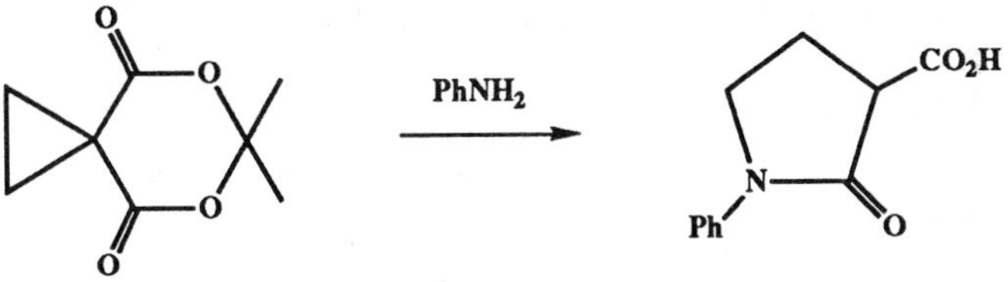
O
O
O
O
PhNH2
CO2H
N
Ph
O

OS, CV 7, 400 (1990); *61,* 22 (1983)

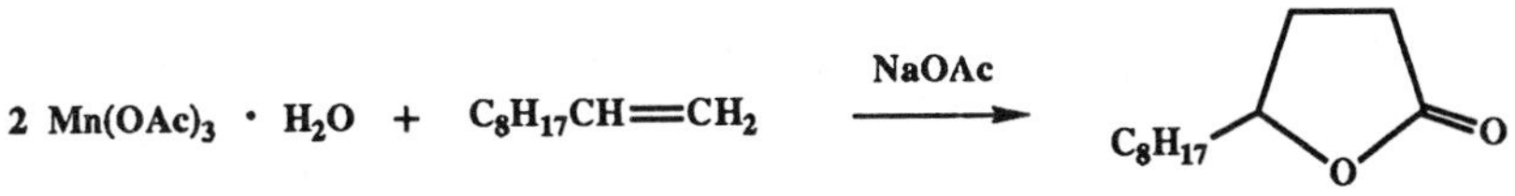
2 Mn(OAc)3 · H2O + C8H17CH=CH2
NaOAc
C8H17
O
O

OS, *CV 7*, 501 (1990); *62*, 48 (1984)

NHSO$_2$Tol

Pd^{+2}, Na_2CO_3, LiCl

O=⟨⟩=O

SO_2Tol

N

Me

OS, *CV 7*, 406 (1990); *63*, 10 (1985)

CH$_2$OH

CH$_2$OH

meso

horse-liver alcohol dehydrogenase
β-nicotinamide adenine dinucleotide

flavin mononucleotide
pH 9, 20 °C

H

O

H

O

(+) - (1*R*, 6*S*)

OS, *CV 7*, 99 (1990); *63*, 121 (1985)

HO$_2$C

H

H$_2$N

CO$_2$H

HNO$_2$

O

O

H

CO$_2$H

S - (+)

OS, *CV 7*, 34 (1990); *63*, 214 (1985)

OCH$_2$Ph

N

NO$_2$

Ra(Ni)
H$_2$N-NH$_2$

OCH$_2$Ph

N
H

OS, *CV* 7, 164 (1990); 64, 175 (1986)

Ph, CO_2H; I_2, $NaHCO_3$, $CHCl_3$; I_2, MeCN; Ph, O, I

OS, 65, 146 (1987)

NO_2, Me, NH_2; $HC(OEt)_3$, *p*-TsOH; NO_2, Me, N, OEt; $(CO_2Et)_2$, KOEt, DMF, DMSO; NO_2, N, H

OS, 66, 108 (1987)

MeO_2C, O, O, O; Bu_4NF; HO, O, O, O

OS, 67, 133 (1988)

Me_3SiCH_2—N—CH_2OMe, CH_2Ph + O, N, O, Ph; LiF, MeCN, ultrasound; CH_2Ph, N, O, N, O, Ph

OS, CV 1, 360 (1941)

AcO, OAc, AcO; O, O, CH_3, OEt; H_2SO_4, H_2O, 80 °C; HO, HO, O, O, CH_3

OS, CV 1, 478 (1941)

NH_2 + OH, OH, OH; nitrobenzene; $FeSO_4$, H_2SO_4, Δ; N

OS, CV 1, 552 (1941)

2 HO, O, O; Δ; O, O

OS, CV 2, 15 (1943)

Cl, CO_2H; NH_2; CuO, K_2CO_3, Δ; H, N, CO_2H

OS, CV 2, 60 (1943)

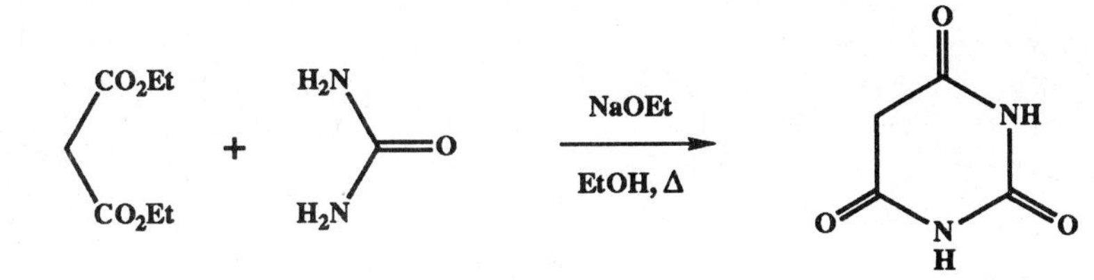

CO_2Et CO_2Et + H_2N H_2N O

NaOEt

EtOH, Δ

O NH O N H O

OS, CV 2, 79 (1943)

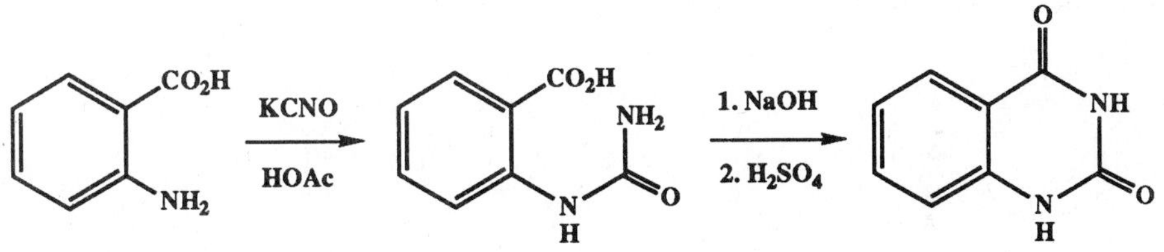

CO_2H NH_2

KCNO

HOAc

CO_2H NH_2 N H O

1. NaOH

2. H_2SO_4

O NH N H O

OS, CV 2, 126 (1943)

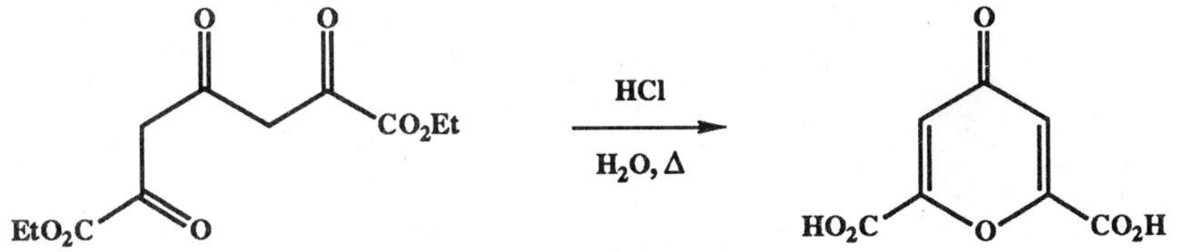

O O CO_2Et EtO_2C O

HCl

H_2O, Δ

O HO_2C O CO_2H

OS, CV 2, 214 (1943)

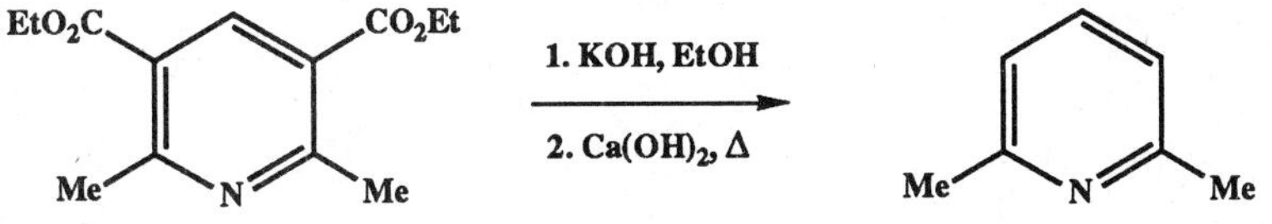

EtO_2C CO_2Et Me N Me

1. KOH, EtOH

2. $Ca(OH)_2$, Δ

Me N Me

OS, CV 2, **422 (1943)**

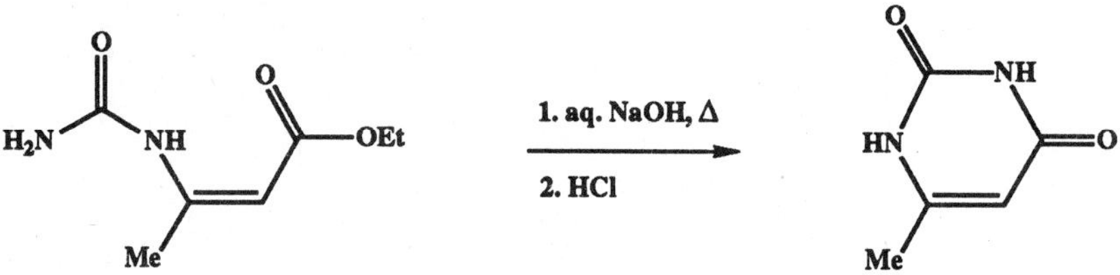

1. aq. NaOH, Δ
2. HCl

OS, CV 2, **485 (1943)**

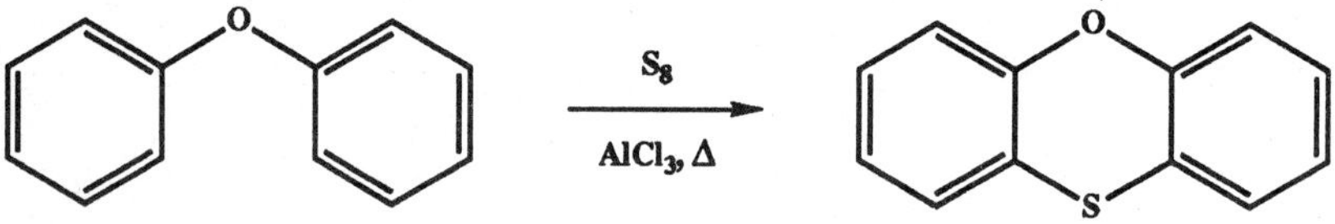

S_8
$AlCl_3$, Δ

OS, CV 2, **610 (1943)**

3 $CH_2{=}O$ + 3 H_2S → HCl

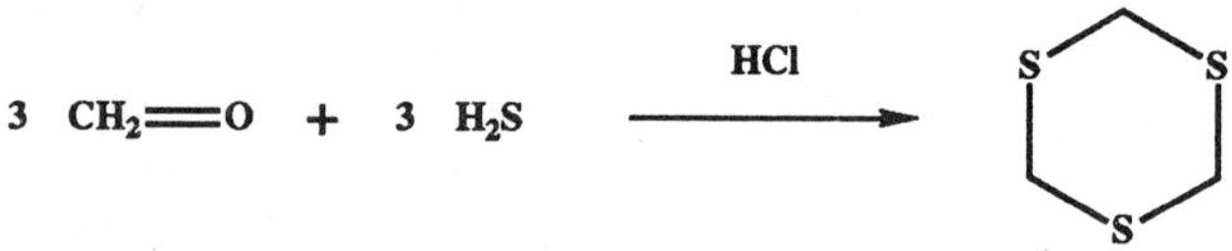

OS, CV 3, **53 (1955)**

1. $(NH_4)_2CO_3$ phenol, Δ
2. NaOH, H_2O

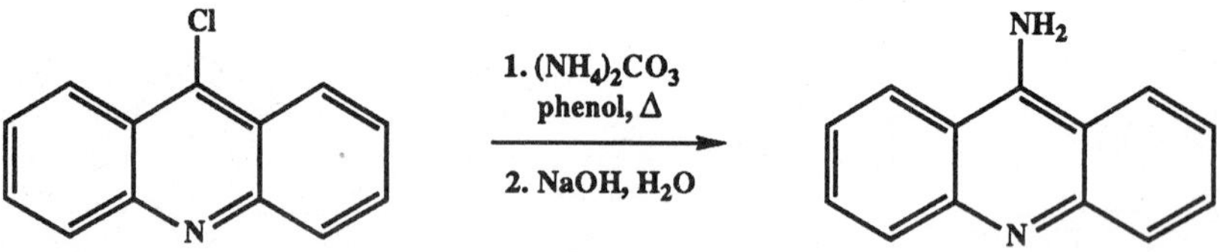

OS, *CV* 3, 71 (1955)

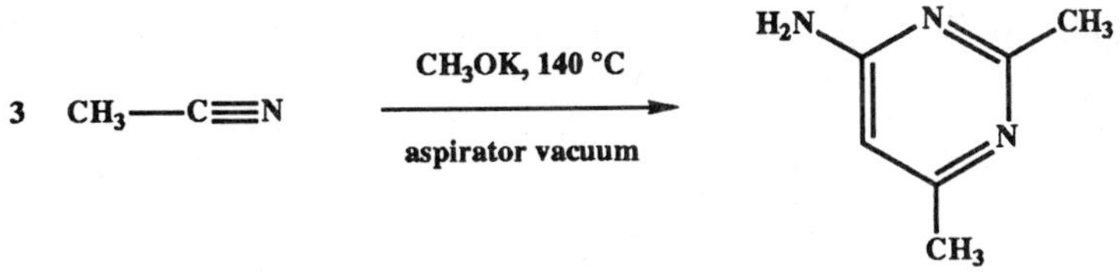

3 $CH_3—C≡N$ → (CH_3OK, 140 °C; aspirator vacuum)

H_2N N CH_3 N CH_3

OS, *CV* 3, 151 (1955)

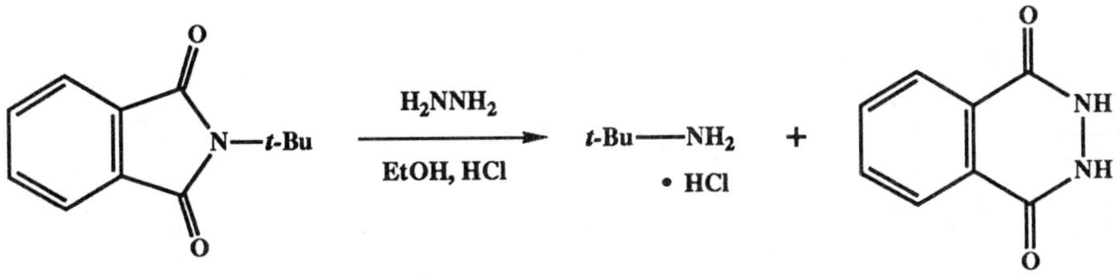

O N—*t*-Bu O → (H_2NNH_2; EtOH, HCl) *t*-Bu—NH_2 • HCl + O NH NH O

OS, *CV* 3, 165 (1955)

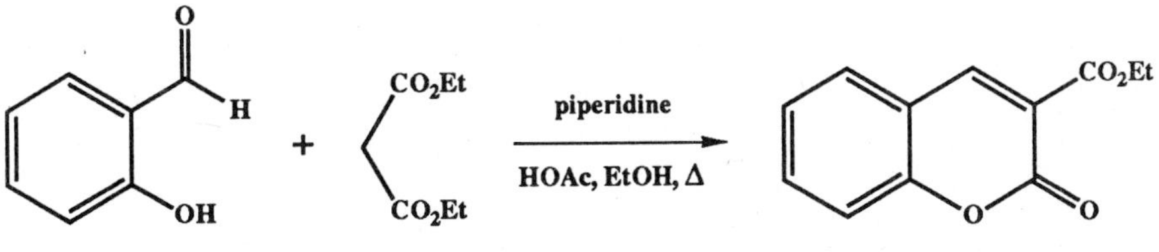

O H OH + CO_2Et CO_2Et → (piperidine; HOAc, EtOH, Δ) CO_2Et O O

***OS*, *CV 3*, 231 (1955)**

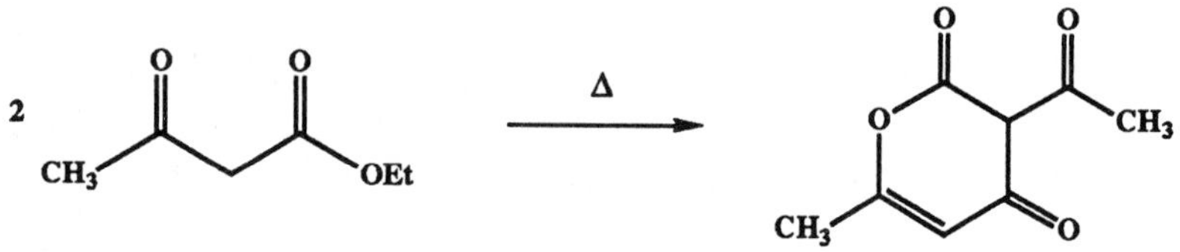

2 $CH_3C(O)CH_2C(O)OEt$ —Δ→ (pyranone product)

CH₃ OEt O O ; Δ ; O O O CH₃ CH₃ O

***OS*, *CV 3*, 272 (1955)**

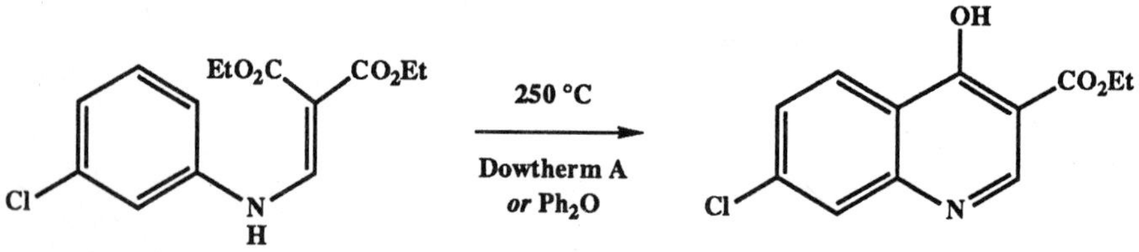

Cl, EtO_2C, CO_2Et, N H —250 °C, Dowtherm A *or* Ph_2O→ OH, CO_2Et, Cl, N

***OS*, *CV 3*, 281 (1955)**

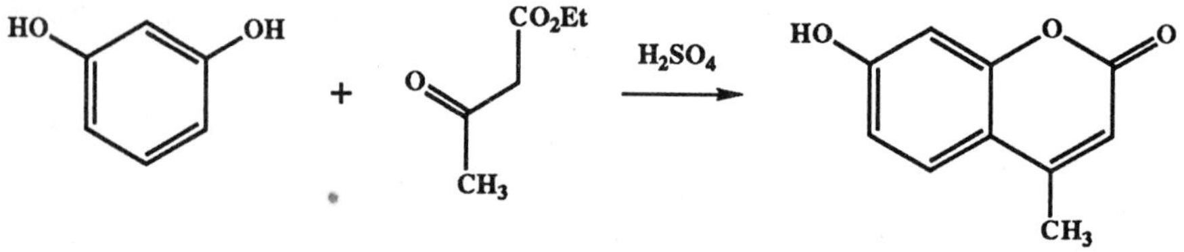

HO, OH + CO_2Et, O, CH_3 —H_2SO_4→ HO, O, O, CH_3

***OS*, *CV 3*, 329 (1955)**

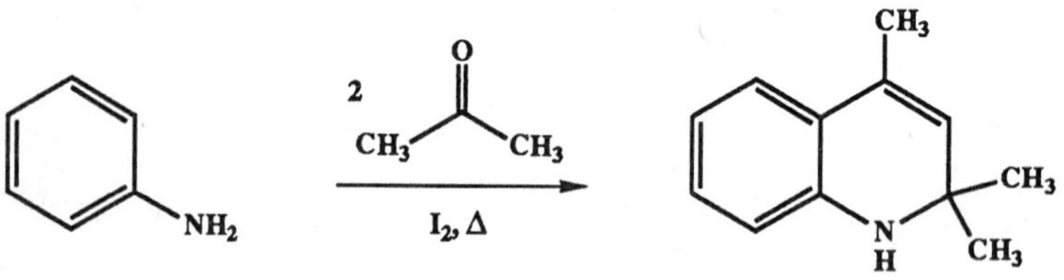

NH_2 —2 $CH_3C(O)CH_3$, I_2, Δ→ CH_3, CH_3, CH_3, N H

***OS, CV 3,* 387 (1955)**

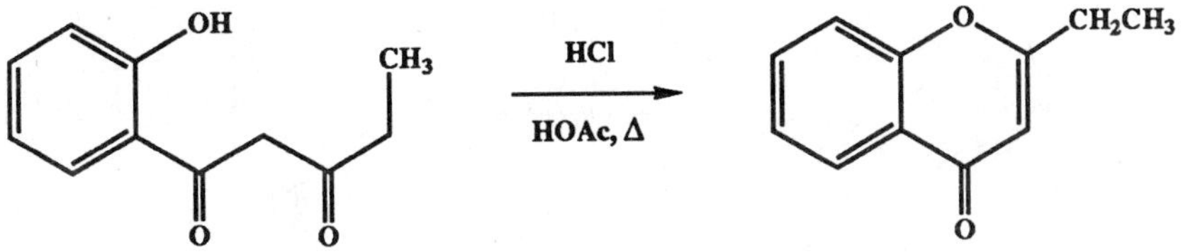

OH, CH_3, O, O → HCl / HOAc, Δ → O, CH_2CH_3, O

***OS, CV 3,* 449 (1955)**

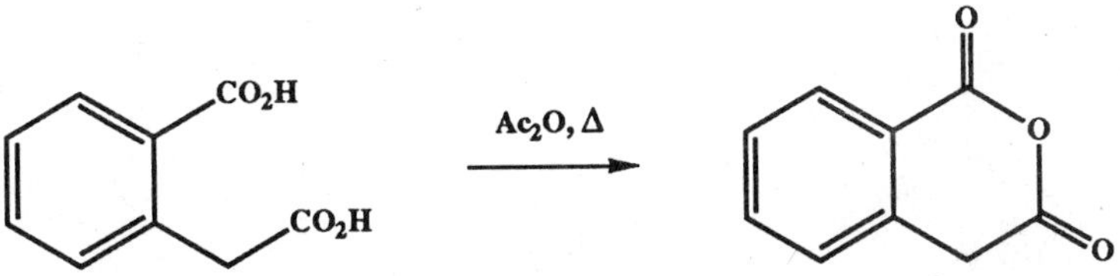

CO_2H, CO_2H → Ac_2O, Δ → O, O, O

***OS, CV 3,* 456 (1955)**

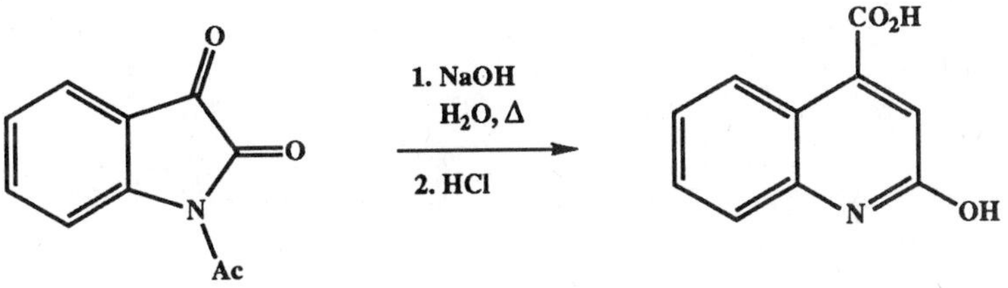

O, O, N, Ac → 1. NaOH H_2O, Δ 2. HCl → CO_2H, N, OH

***OS, CV 3,* 488 (1955)**

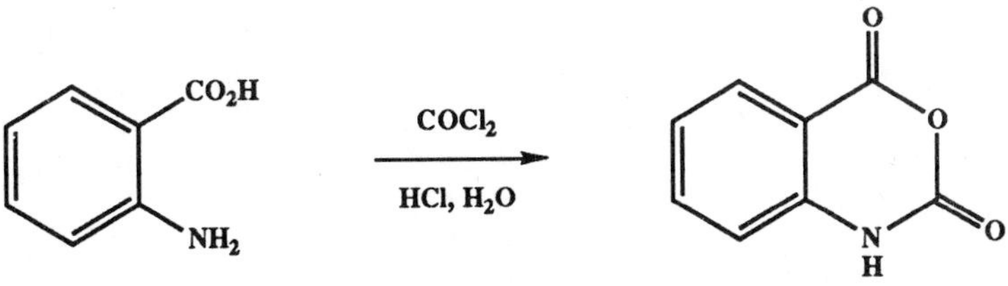

CO_2H, NH_2 → $COCl_2$ / HCl, H_2O → O, O, N H, O

OS, CV 3, 504 (1955)

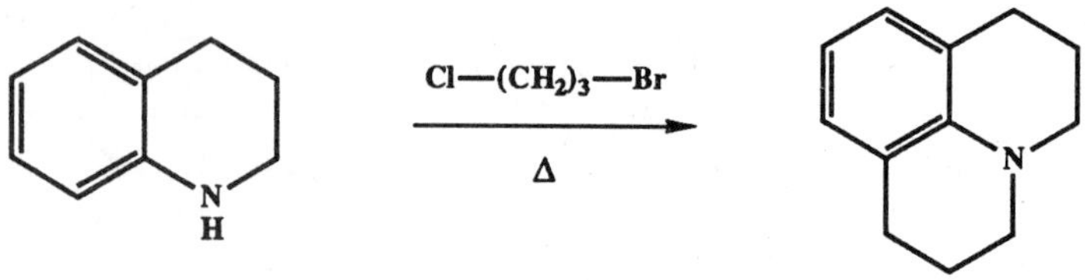

Cl—$(CH_2)_3$—Br

Δ

N
H

N

OS, CV 3, 568 (1955)

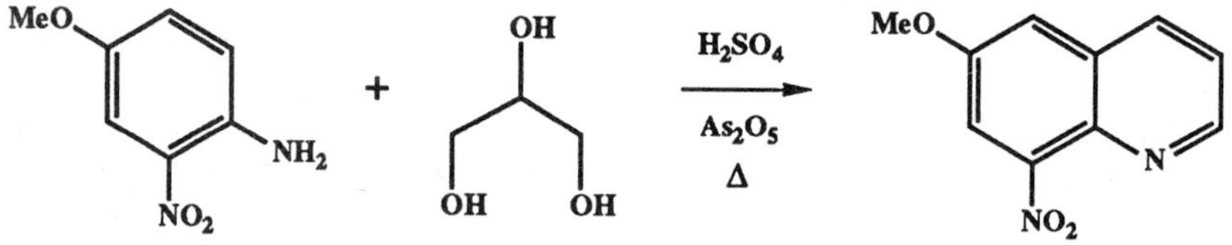

MeO

NH_2

NO_2

+

OH

OH OH

H_2SO_4

As_2O_5

Δ

MeO

N

NO_2

OS, CV 3, 580 (1955)

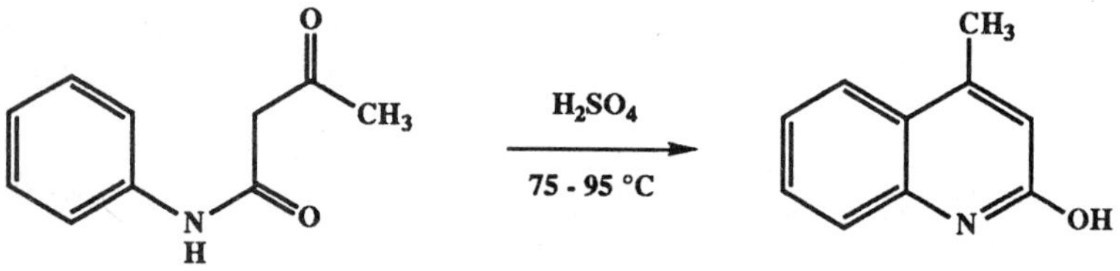

O

CH_3

N
H

O

H_2SO_4

75 - 95 °C

CH_3

N

OH

OS, CV 3, 581 (1955)

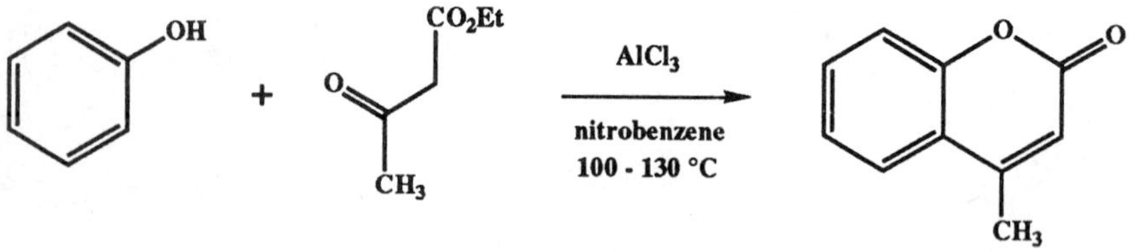

OH

+

CO_2Et

O

CH_3

$AlCl_3$

nitrobenzene

100 - 130 °C

O

O

CH_3

OS, CV 3, 593 (1955)

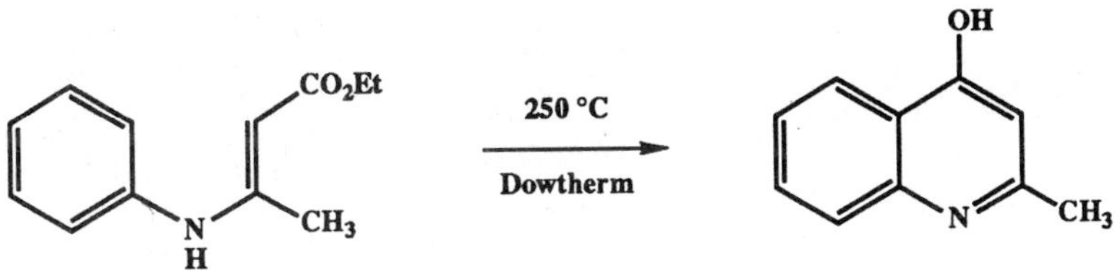

CO$_2$Et, N H, CH$_3$ → 250 °C, Dowtherm → OH, N, CH$_3$

OS, CV 3, 656 (1955)

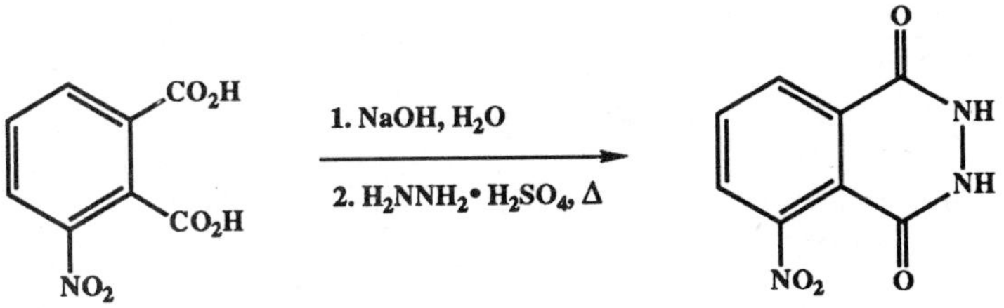

CO$_2$H, CO$_2$H, NO$_2$ → 1. NaOH, H$_2$O; 2. H$_2$NNH$_2$• H$_2$SO$_4$, Δ → O, NH, NH, NO$_2$, O

OS, CV 3, 753 (1955)

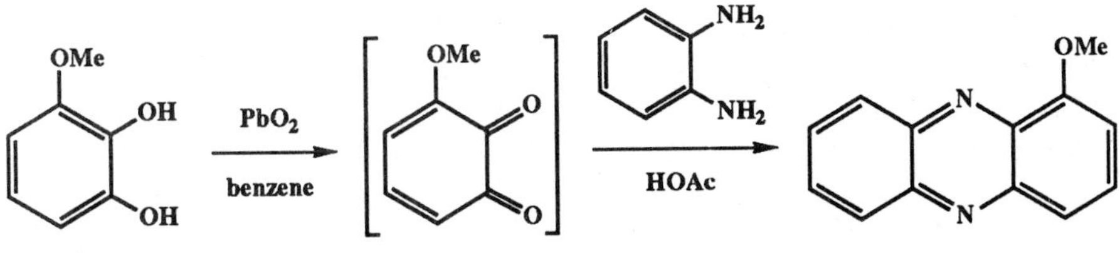

OMe, OH, OH → PbO$_2$, benzene → [OMe, O, O] → NH$_2$, NH$_2$; HOAc → OMe, N, N

OS, CV 4, 29 (1963)

Ph—NH–C(=NH)–NH–C(=NH)–NH_2 + $ClCH_2CO_2Et$ (Na, MeOH) → 2-anilino-4-amino-6-(chloromethyl)-1,3,5-triazine

OS, CV 4, 68 (1963)

3 PhMgBr (1. 3 $B(OMe)_3$, Et_2O, - 60 °C; 2. H_2O, H_2SO_4, 0 °C) → 3 $PhB(OH)_2$ (110 °C) → triphenylboroxine

OS, CV 4, 78 (1963)

Ph—CN + NC–NH–C(=NH)–NH_2 (KOH, Δ, $MeOCH_2CH_2OH$) → 2,4-diamino-6-phenyl-1,3,5-triazine

OS, CV 4, 201 (1963)

2 HO_2C–CH(OH)–CH_2–CO_2H (fuming H_2SO_4, Δ) → 2-oxo-2H-pyran-5-carboxylic acid

OS, CV 4, 210 (1963)

O, CH_3, CH_3 → HCO_2Et, NaOMe, Et_2O → [O, CH_3, CH=CH—ONa] → O, H_2N, CN; piperidine acetate, Δ, then HOAc → CN, CH_3, N, H, O

OS, CV 4, 245 (1963)

HN, NH_2, NH_2 + CO_2Et, N≡C → Na, EtOH, Δ; then aq. HOAc → H_2N, N, OH, N, NH_2

OS, CV 4, 247 (1963)

O, NH_2, NH_2 + C≡N, CO_2Et → Na, EtOH, Δ; then aq. HOAc → HO, N, NH_2, N, OH

OS, CV 4, 311 (1963)

CH_3, H, O + CH_2, OCH_3 → 200 °C; hydroquinone → CH_3, O, OCH_3

OS, CV 4, 396 (1963)

SH SH + Br Br — Na, EtOH, Δ → S S

OS, CV 4, 441 (1963)

2 CN OEt O + Et Me O — 1. NH_4OAc, 0 °C NH_3, EtOH; 2. HCl → NC O Et NH Me NC O

OS, CV 4, 451 (1963)

4 CH_3 O CH_3 O O CH_3 — 3 NH_4OH, H_2O; NH_4OAc, 230 °C, 800 - 3000 psi → 3 Et N CH_3

OS, CV 4, 478 (1963)

Ph O O CH_3 O — Δ, glycerol → O Ph O

OS, CV 4, **479 (1963)**

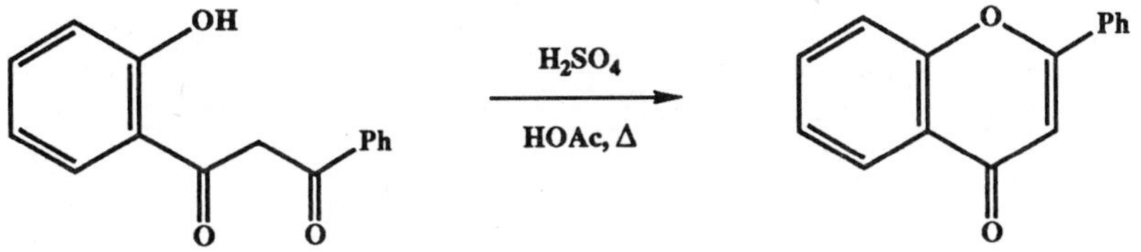

OH
Ph
O O

H_2SO_4

HOAc, Δ

O Ph
O

OS, CV 4, **496 (1963)**

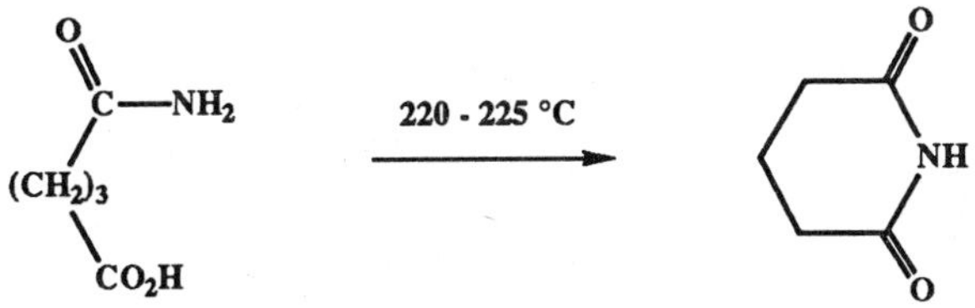

O
C—NH_2
$(CH_2)_3$
CO_2H

220 - 225 °C

O
NH
O

OS, CV 4, **515 (1963)**

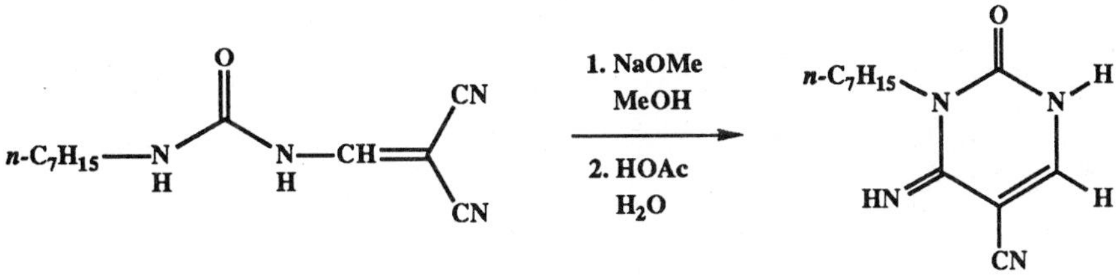

O
n-C_7H_{15}—N—H N—H—CH=C(CN)CN

1. NaOMe MeOH

2. HOAc H_2O

O
n-C_7H_{15} N N H
HN H
CN

OS, CV 4, **518 (1963)**

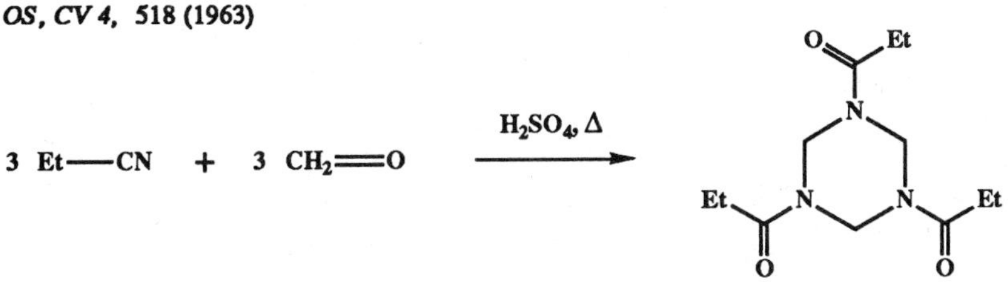

3 Et—CN + 3 CH_2=O

H_2SO_4, Δ

O Et
N
Et N N Et
O O

OS, CV 4, 529 (1963)

$O[(CH_2)_4SO_3Na]_2$ → 1. Amberlite resin (H^+); 2. 130 - 150 °C → 2 (cyclic O–SO_2 sultone)

OS, CV 4, 532 (1963)

CO_2Me → 1. NH_4OH, H_2O; 2. aq. NaOH, Δ then HCl → CO_2H, HO, N

OS, CV 4, 549 (1963)

2 CH_3, CO_2Et → H_2SO_4 → CH_3, CO_2Et, CH_3 + CH_3, CO_2H, CH_3

OS, CV 4, 566 (1963)

EtO—CH= C(CN)(CO_2Et) → H_2N–C(=S)–NH_2, Na, EtOH → EtO_2C, NH_2, N, N, SH + NC, OH, N, N, SH

ca. 9 : 1

OS, CV 4, 630 (1963)

CO_2H

CH_3

CO_2H

Ac_2O, Δ

O

CH_3

O

O

OS, CV 4, 638 (1963)

O

CO_2Et +

CH_3

S

H_2N NH_2

NaOMe

MeOH, Δ

CH_3 OH

N N

SH

OS, CV 4, 662 (1963)

CH_3 CN

\+

Ph CO_2Et

O

NH_2

CN

1. Na, EtOH

2. HCl, H_2O

Ph CH_3

NC CN

O N O

H

OS, CV 4, 677 (1963)

CH_3

OH OH

Cu chromite

200 °C

CH_3

O O

***OS*, *CV* 4, 786 (1963)**

Ph (styrene) → [2 CH_2=O; H_2SO_4, Δ] → 4-phenyl-1,3-dioxane (Ph)

***OS*, *CV* 4, 790 (1963)**

Ph—C(CN)(CO$_2$Et)—CH_2CH_2CN → [1. HCl, H_2O, HOAc, Δ; 2. Ac_2O, Δ] → 3-phenylglutaric anhydride (Ph)

***OS*, *CV* 4, 795 (1963)**

tetrahydropyran (O) → [$PhNH_2$; Al_2O_3, 340 °C] → N-phenylpiperidine (N—Ph)

***OS*, *CV* 4, 824 (1963)**

CHO—CHO • 2 $NaHSO_3$ + H_2N, H_2N (o-phenylenediamine) → [H_2O, Δ; then Na_2CO_3] → quinoxaline (N, N)

OS, *CV 5*, 450 (1973)

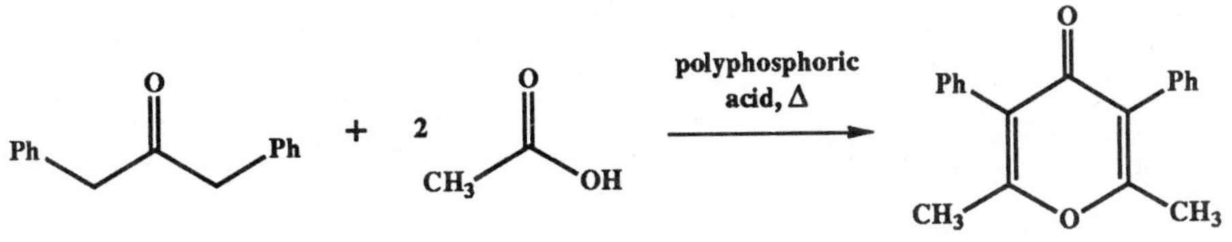

Ph–CH₂–CO–CH₂–Ph + 2 CH_3CO_2H → (polyphosphoric acid, Δ) → 2,6-dimethyl-3,5-diphenyl-4-pyrone

polyphosphoric acid, Δ

OS, *CV 5*, 623 (1973)

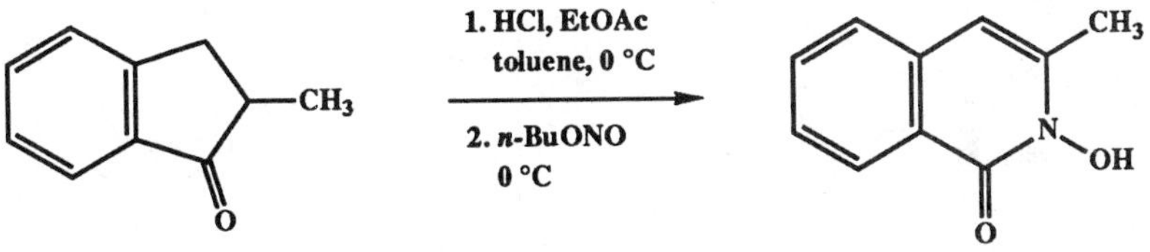

1. HCl, EtOAc toluene, 0 °C
2. *n*-BuONO 0 °C

CH_3 ... N–OH ... O

OS, *CV 5*, 635 (1973)

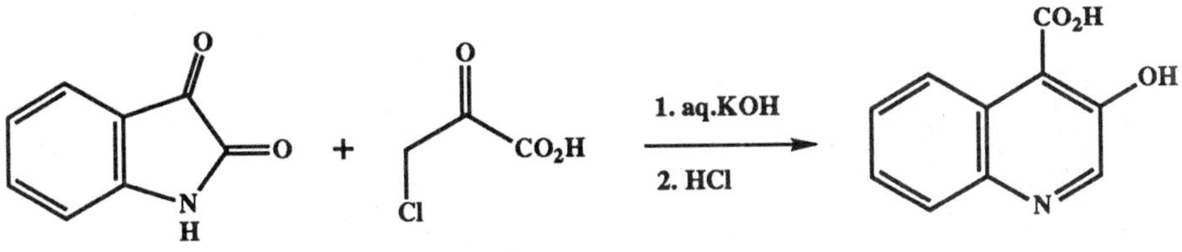

1. aq.KOH
2. HCl

CO_2H, OH, N, Cl, CO_2H

OS, *CV 5*, 703 (1973)

$CH(OEt)_2$ / $CH(OEt)_2$ + H_2N–C(=S)–NH_2

1. HCl, EtOH, Δ
2. NaOH, H_2O

N, N, SH

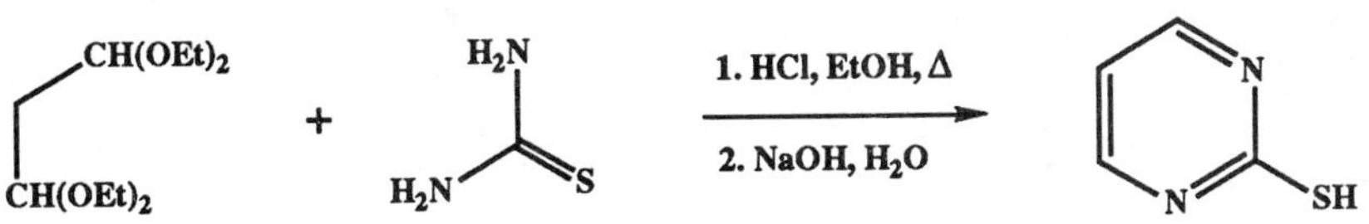

OS, *CV 5*, 721 (1973)

Ph–CO–CH$_2$–CO–CH$_2$–CO–C$_6$H$_4$-*p*-OMe → (H_2SO_4, 0 °C) → 2-Ph-6-(C$_6$H$_4$-*p*-OMe)-4H-pyran-4-one

OS, *CV 5*, 727 (1973)

2-aminobiphenyl (H_2N) → (1. BCl_3, $AlCl_3$, xylene, Δ; 2. H_2O) → B(OH)–N(H) dibenzo-fused azaborine (B$^-$=N$^+$)

OS, *CV 5*, 794 (1973)

CH_3–CO–CH$_2$–CH(OCH_3)$_2$ + 2 H–CO–NH_2 → (NH_4Cl, H_2O, Δ) → 4-methylpyrimidine (CH_3)

OS, *CV 5*, 1051 (1973)

Ph–CH$_2$–CO–N=C=S → ($AlCl_3$, CS_2, Δ; then aq. HCl) → 1-thioxo-1,4-dihydro-3(2H)-isoquinolinone (NH, O, S)

OS, CV 5, **1106 (1973)**

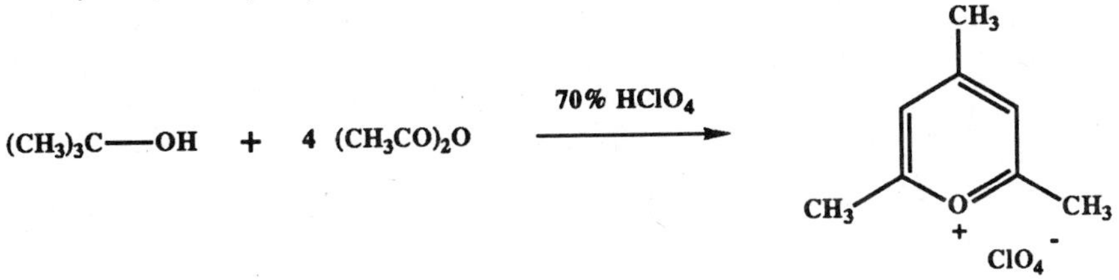
(CH3)3C—OH + 4 (CH3CO)2O
70% HClO4
CH3
CH3
CH3
O
+
ClO4 -

OS, CV 5, **1108 (1973)**

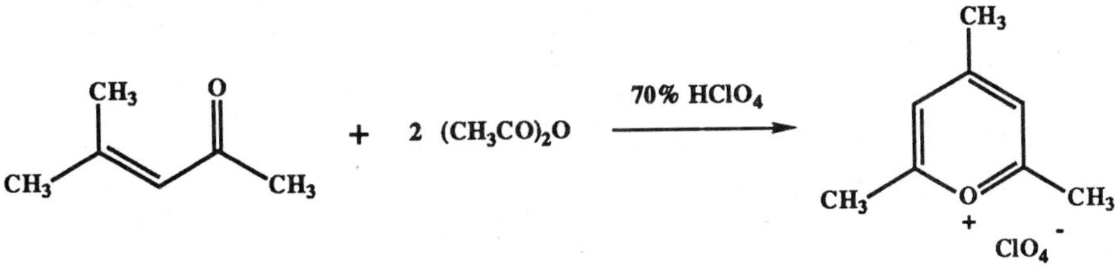
CH3
O
CH3
CH3
+ 2 (CH3CO)2O
70% HClO4
CH3
CH3
CH3
O
+
ClO4 -

OS, CV 5, **1112 (1973)**

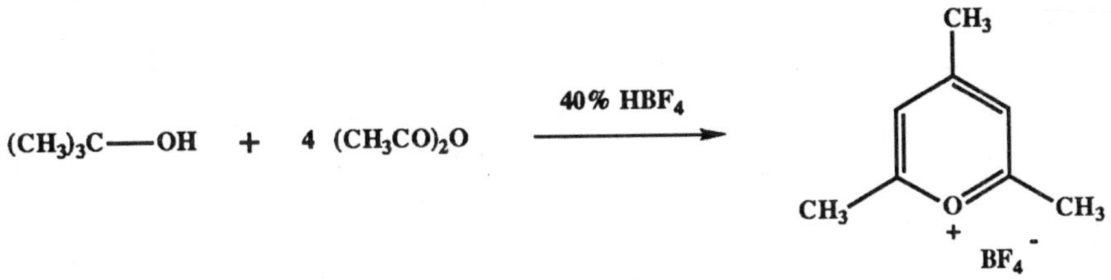
(CH3)3C—OH + 4 (CH3CO)2O
40% HBF4
CH3
CH3
CH3
O
+
BF4 -

OS, CV 5, **1114 (1973)**

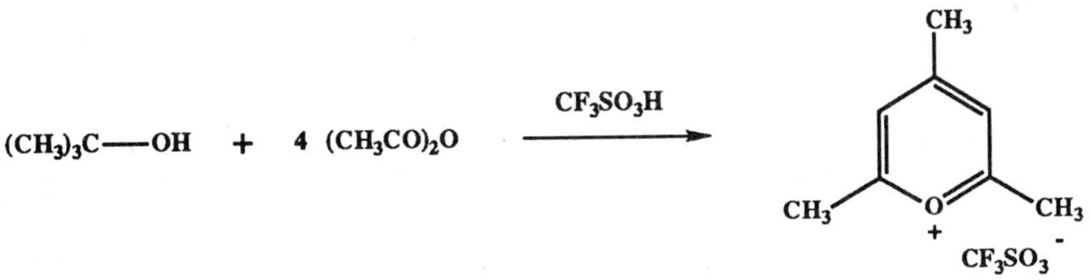
(CH3)3C—OH + 4 (CH3CO)2O
CF3SO3H
CH3
CH3
CH3
O
+
CF3SO3 -

OS, CV 5, 1135 (1973)

2 Ph / O / Ph + O / Ph / CH_3 → 52% HBF_4 / $C_2H_4Cl_2$, Et_2O, Δ → Ph / Ph / O / + / Ph / BF_4^-

OS, CV 6, 1 (1988)

MeO / MeO / $NHCOCH_3$ → $POCl_3$ / toluene, Δ → MeO / MeO / N / CH_3

OS, CV 6, 462 (1988)

HO / O + CH_2=O → H_2SO_4 / HOAc, Δ → O / O

OS, CV 6, 471 (1988)

MeO / MeO / O / OH → CH_2=O / HOAc, HCl / 80 - 90 °C → MeO / MeO / O / O

***OS*, *CV* 6, 496 (1988)**

CH3 H CH3 N CH3 O CH3 CH3 CH3 O N CH3

***OS*, *CV* 6, 512 (1988)**

O Br Br + O 2 NaI, 2 Cu CH_3CN, 50 °C O O

***OS*, *CV* 6, 556 (1988)**

HS SH + MeO OMe $BF_3 \cdot Et_2O$, Δ HOAc, $CHCl_3$ S S

***OS*, *CV* 6, 781 (1988)**

O CH3 O N Et H_2 (25 psi) Raney nickel EtOH, 25 °C HO H N Et CH3 O

***OS, CV 6,* 818 (1988)**

Cl
NCH$_3$
Ph
reflux
CH$_3$
N
Ph

***OS, CV 6,* 932 (1988)**

Ph—P
CN
CN
t-BuOK
toluene, Δ
Ph—P
NH$_2$
CN

***OS, CV 6,* 965 (1988)**

$CH_2CH_2\overset{+}{N}H_3$ Cl$^-$
N
H
H
OH • H_2O
O
O
aq. KOH, pH = 4
+ NH$_2$
CO$_2^-$
N
H

***OS, CV 6,* 976 (1988)**

SO$_2$
1. O$_3$, EtOH
CH$_2$Cl$_2$, - 78 °C
2. SO$_2$
EtO
O
OEt
SO$_2$
ca. 55 : 45 *cis* : *trans*
NH$_4$Cl
HOAc
Δ
H
N
SO$_2$

OS, *CV 6*, 1014 (1988)

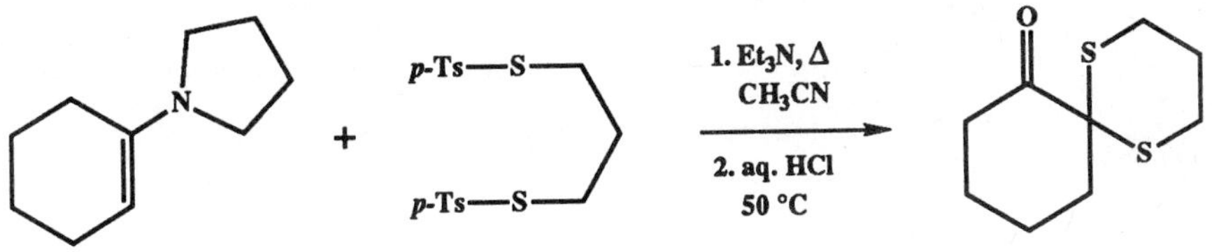

p-Ts—S ; p-Ts—S ; 1. Et_3N, Δ ; CH_3CN ; 2. aq. HCl ; 50 °C ; N ; O ; S ; S

OS, *CV 7*, 144 (1990); *60*, 34 (1981)

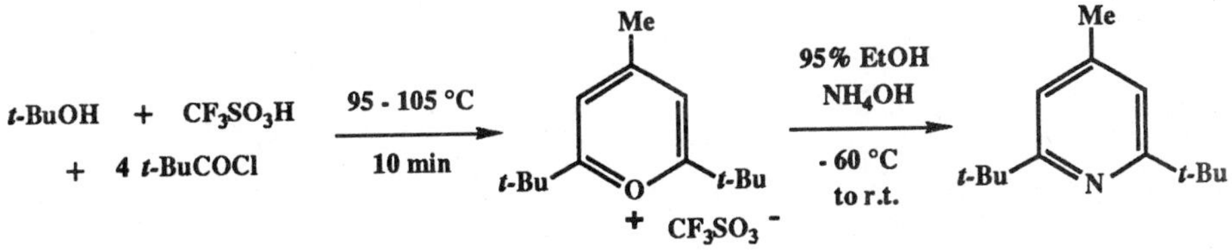

t-BuOH + CF_3SO_3H + 4 t-BuCOCl ; 95 - 105 °C ; 10 min ; Me ; t-Bu ; O ; t-Bu ; + ; $CF_3SO_3^-$; 95% EtOH ; NH_4OH ; - 60 °C to r.t. ; Me ; t-Bu ; N ; t-Bu

OS, *CV 7*, 386 (1990); *60*, 92 (1981)

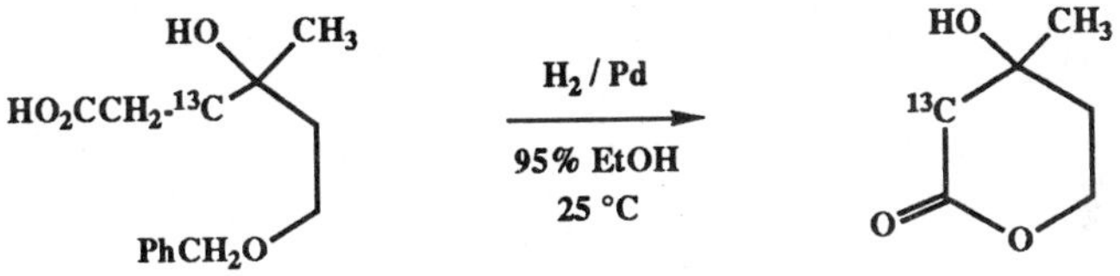

HO ; CH_3 ; HO_2CCH_2-^{13}C ; $PhCH_2O$; H_2 / Pd ; 95% EtOH ; 25 °C ; HO ; CH_3 ; ^{13}C ; O ; O

OS, *CV 7*, 326 (1990); *62*, 74 (1984)

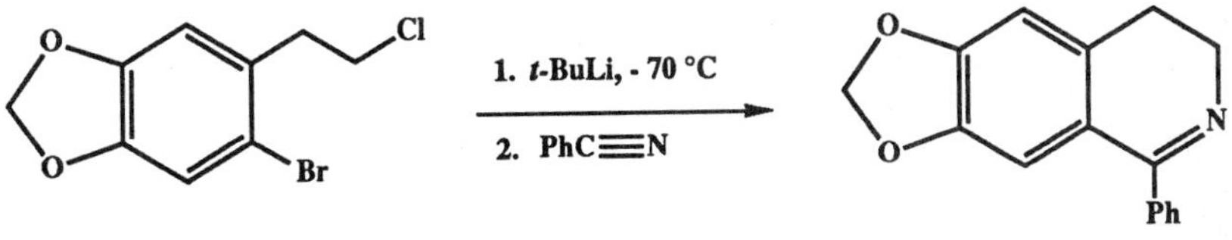

O ; O ; Cl ; Br ; 1. t-BuLi, - 70 °C ; 2. PhC≡N ; O ; O ; N ; Ph

OS, *65*, 98 (1987)

MeO, CO_2Me, CO_2Me, LDA, O, O, CO_2Me

OS, *65*, 215 (1987)

SH, OH, *R*, $(CH_2O)_n$, *p*-TsOH, C_6H_6, reflux, S, H, O, H, *R*

OS, *66*, 142 (1987)

EtO_2C, N, CO_2Et, EtO_2C, N, N, O, N, EtO_2C, N, CO_2Et, EtO_2C

OS, *68*, 188 (1989)

Me_3Si, H, N, OMe, $(CH_2O)_n$, *p*-TsOH, N, OMe

***OS*, *CV 2*, 371 (1943)**

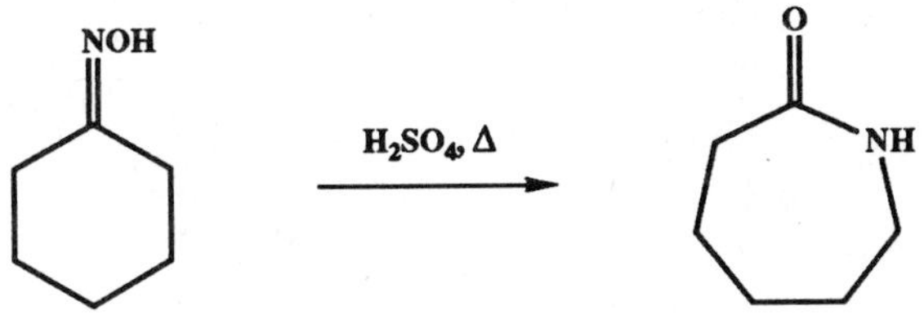

NOH

H_2SO_4, Δ

O

NH

***OS*, *CV 6*, 301 (1988)**

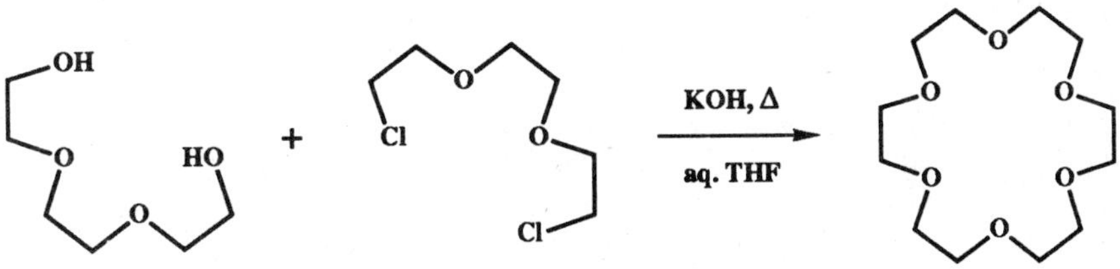

OH O HO O + Cl O O Cl

KOH, Δ

aq. THF

O O O O O O

***OS*, *CV 6*, 382 (1988)**

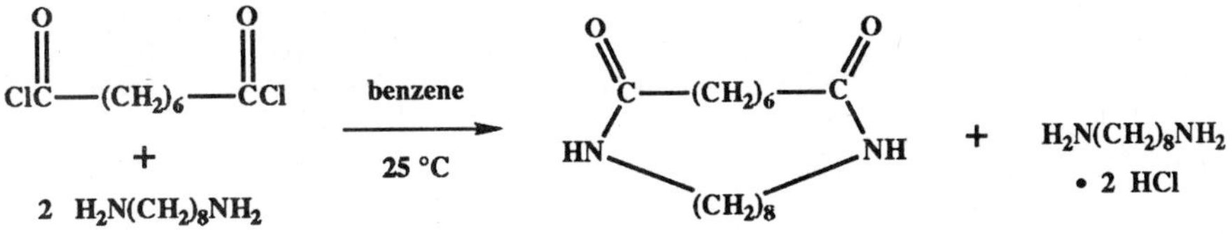

$ClC(=O)—(CH_2)_6—C(=O)Cl$ + 2 $H_2N(CH_2)_8NH_2$

benzene

25 °C

O=C—$(CH_2)_6$—C=O, HN—$(CH_2)_8$—NH + $H_2N(CH_2)_8NH_2$ • 2 HCl

***OS*, *CV 6*, 395 (1988)**

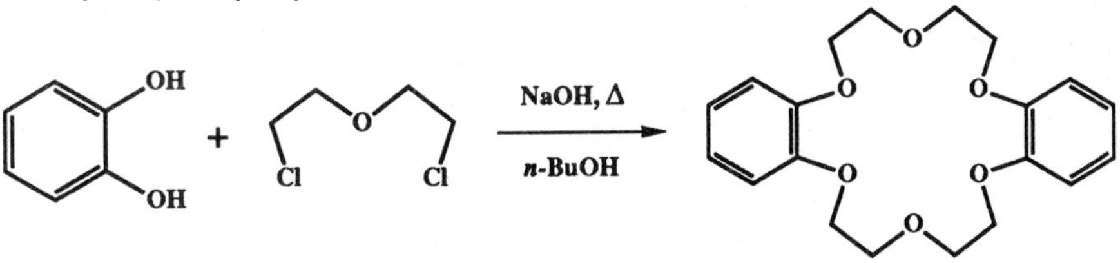

OH OH + Cl O Cl

NaOH, Δ

n-BuOH

O O O O O O

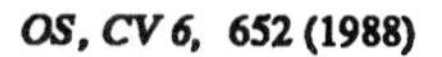
OS, CV 6, 652 (1988)

DMF

100 °C

OS, CV 6, 698 (1988)

Br—$(CH_2)_{10}$—COOH

K_2CO_3

DMSO, 100 °C

$(CH_2)_{10}$

OS, CV 6, 856 (1988)

HCl, $LiClO_4$

EtOH, 63 - 78 °C

OS, CV 7, 470 (1990); *63,* 192 (1985)

1. hν, PhSSPh, benzene
2. $(C_5H_4NS)_2$, Ph_3P benzene, 20 °C
3. $AgClO_4$, MeCN, reflux

R - (+)

OS, *65*, 150 (1987)

$HS(CH_2)_3SH$

1. 2 $ClCH_2CH_2OH$, NaOEt, EtOH
2. 2 H_2N-C(S)-NH_2 then conc. HCl, aq. KOH, aq. HCl
3. $Br(CH_2)_3Br$, 2 Cs_2CO_3, DMF, 55 - 60 °C

S S S S

OS, *CV 4*, 816 (1963)

CHO CHO + $MeNH_2$ + HO_2C HO_2C O

aq. NaOH
Na_2HPO_4, Δ

MeN O

OS, *CV 5*, 96 (1973)

+ N N CO_2Et CO_2Et

Et_2O, Δ

N—CO_2Et N CO_2Et

OS, *CV 5*, 115 (1973)

NHPh NHPh

$CH(OEt)_3$, Δ

Ph Ph N N N N Ph Ph

OS, *CV 5*, 670 (1973)

H_2N CO_2H

cis and *trans*

Dowtherm A

Δ

HN

O

OS, *CV 5*, 989 (1973)

EtO_2C—N—CH_2CO_2Et

1. KOEt, EtOH toluene, Δ

2. aq. HCl, 0 °C

N • HCl

O

OS, *CV 7*, 476 (1990); *64*, 189 (1986)

N O SMe SMe

1. N Me O

t-BuOK

2. NH_4OAc, HOAc

SMe N N N

OS, *68*, 206 (1989)

$PhCH_2NH_2$ • HCl

HCHO, H_2O

N CH_2Ph

OS, *CV* 2, 21 (1943)

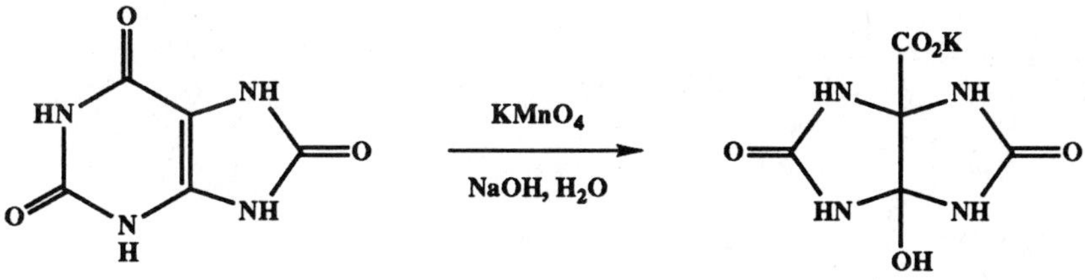
O
HN
NH
O
O
N
H
NH
KMnO4
NaOH, H2O
CO2K
HN
NH
O
O
HN
NH
OH

OS, *CV* 3, 209 (1955)

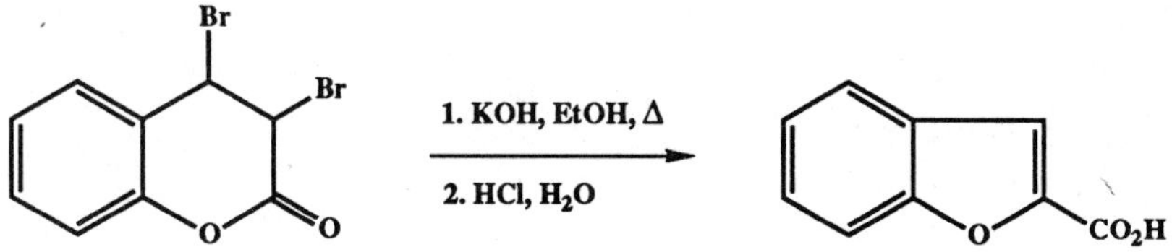
Br
Br
O
O
1. KOH, EtOH, Δ
2. HCl, H2O
O
CO2H

OS, *CV* 4, 594 (1963)

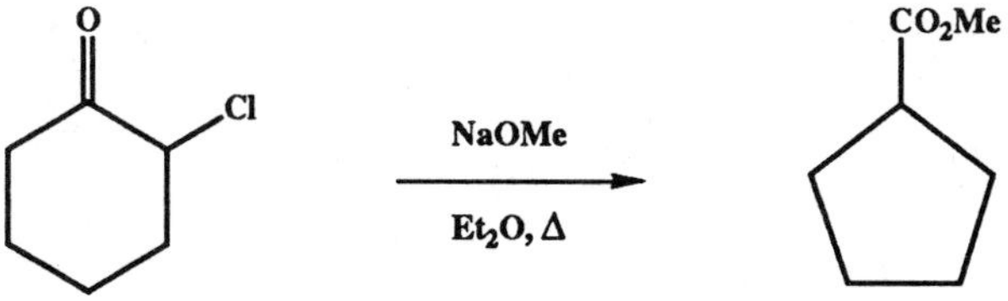
O
Cl
NaOMe
Et2O, Δ
CO2Me

OS, CV 4, 957 (1963)

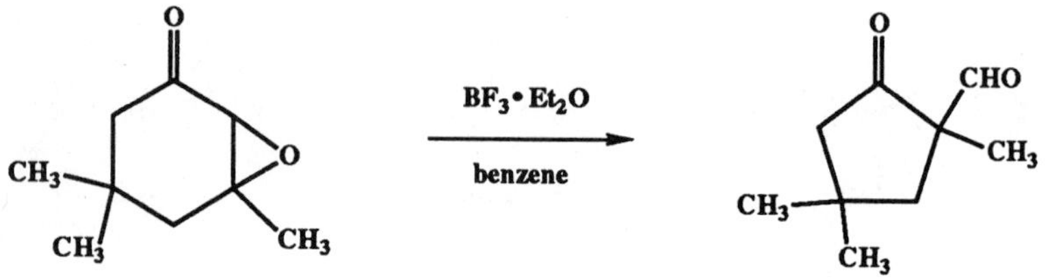
O
CH3
CH3
O
CH3
BF3 • Et2O
benzene
O
CHO
CH3
CH3
CH3

OS, CV 5, 320 (1973)

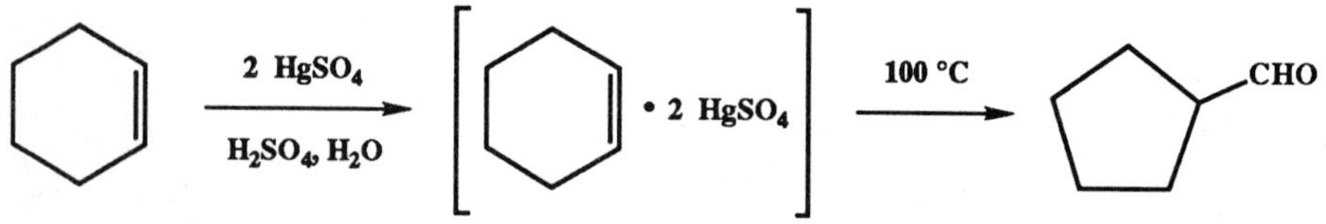
2 HgSO4
H2SO4, H2O
• 2 HgSO4
100 °C
CHO

OS, CV 6, 368 (1988)

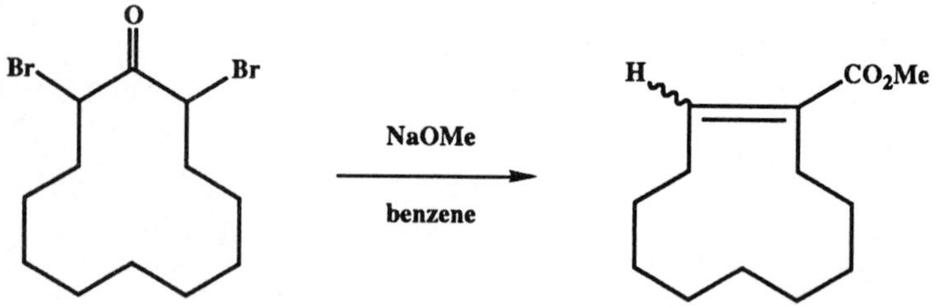
O
Br
Br
NaOMe
benzene
H
CO2Me

OS, CV 6, 840 (1988)

O

$=N_2$

hν, dioxane

Et_2O, H_2O

CO_2H

H

HO H H

HO H H

d isomer mixture

OS, CV 7, 129 (1990); *60*, 25 (1981)

OH

$BF_3 \cdot (n\text{-}Bu)_2O$

CHO

OH

OS, CV 7, 135 (1990); *62*, 191 (1984)

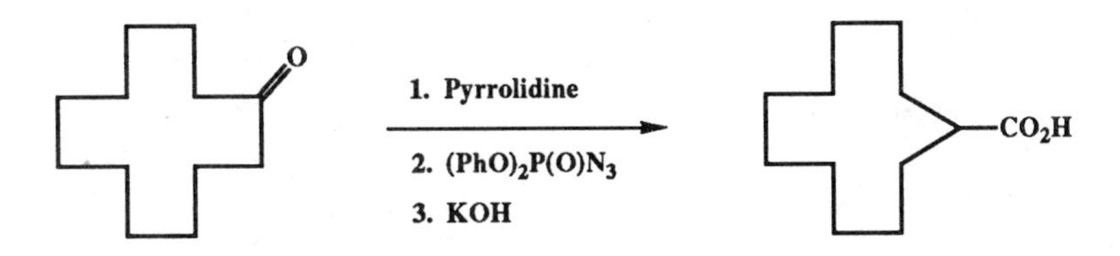

OS, CV 2, 76 (1943)

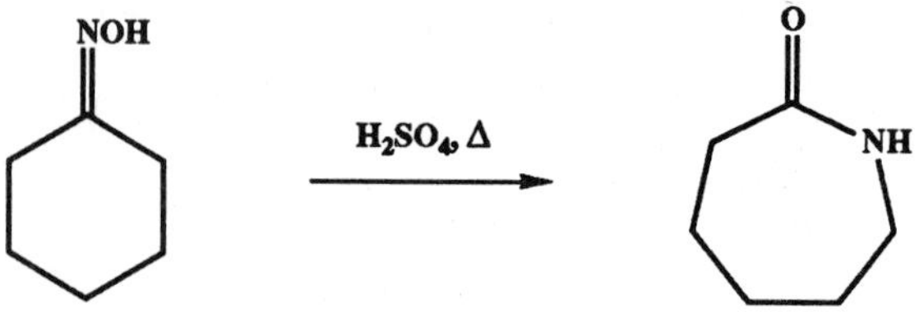

NOH

H_2SO_4, Δ

O

NH

OS, CV 2, 371 (1943)

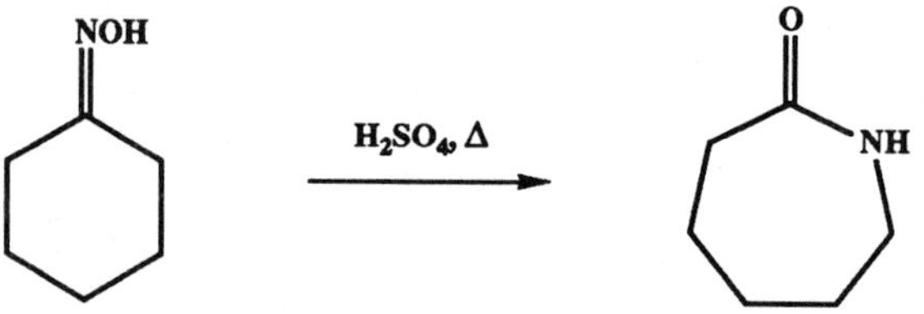

NOH

H_2SO_4, Δ

O

NH

OS, CV 3, 276 (1955)

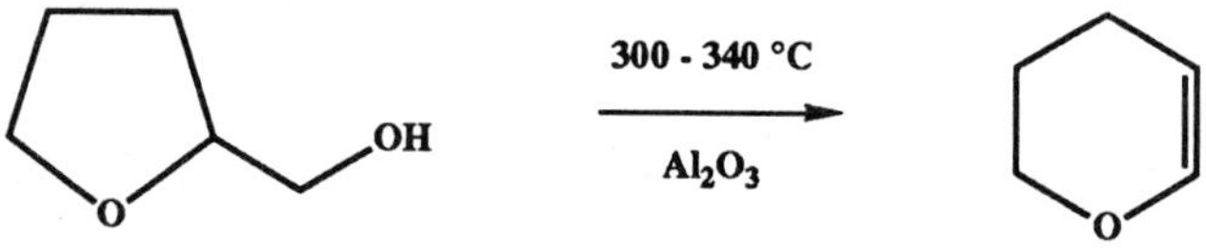

OH

O

300 - 340 °C

Al_2O_3

O

OS, CV 4, 221 (1963)

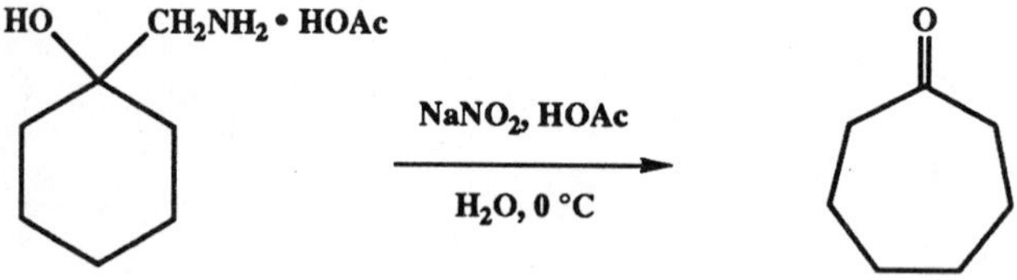

HO CH_2NH_2 • HOAc

$NaNO_2$, HOAc

H_2O, 0 °C

O

OS, CV 4, 225 (1963)

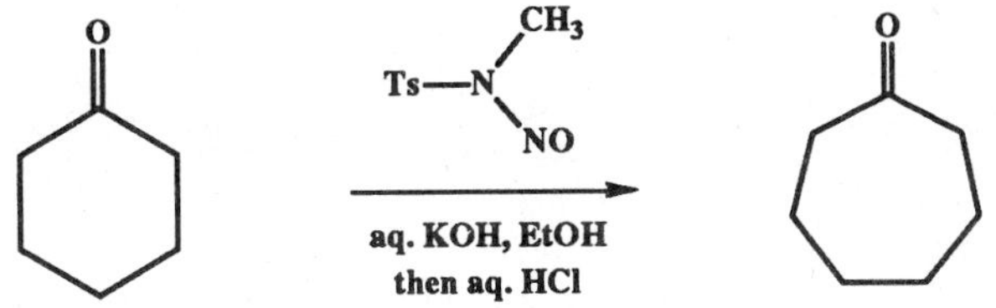

Ts—N(CH$_3$)(NO)

aq. KOH, EtOH
then aq. HCl

OS, CV 4, 780 (1963)

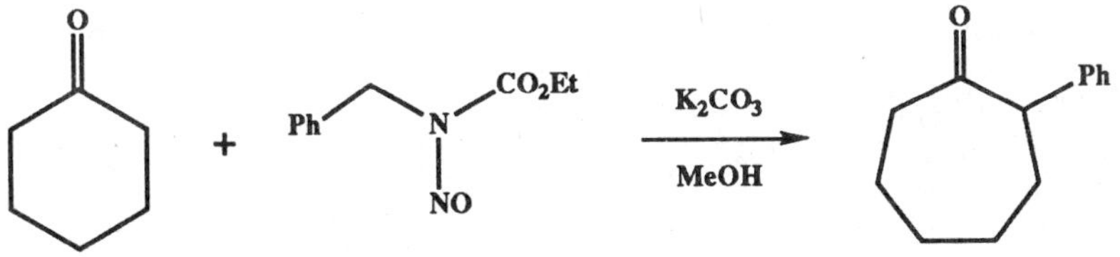

+ Ph CO$_2$Et N NO Ph

K_2CO_3

MeOH

OS, CV 5, 306 (1973)

Br Br

CH_3Li

Et_2O, - 35 °C

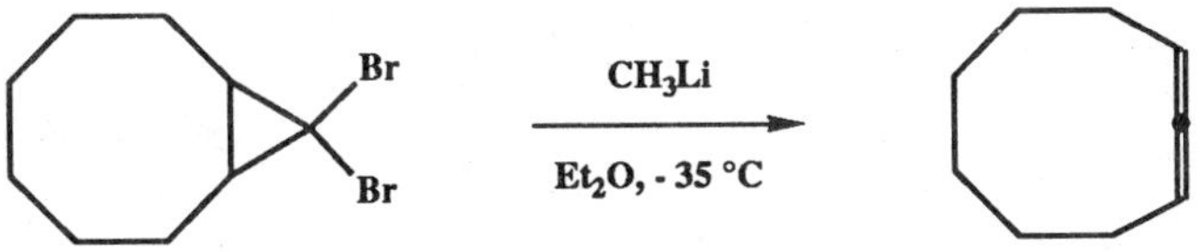

OS, CV 5, 408 (1973)

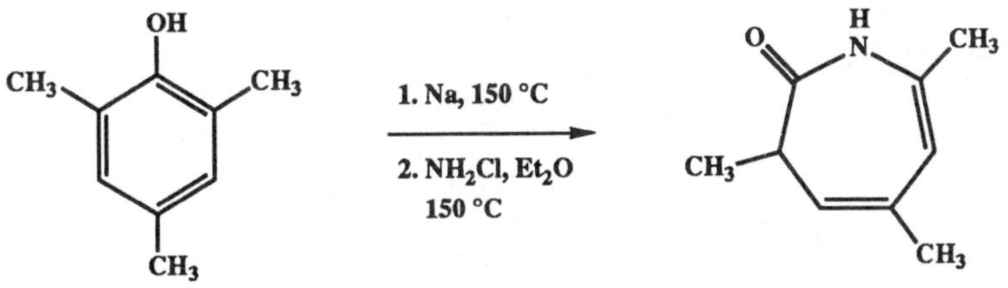

OH CH$_3$ CH$_3$ CH$_3$

1. Na, 150 °C

2. NH_2Cl, Et_2O
150 °C

O H N CH$_3$ CH$_3$ CH$_3$

***OS*, *CV 6*, 142 (1988)**

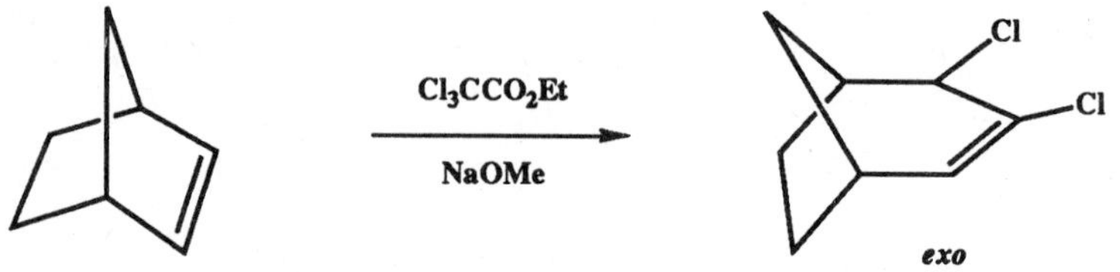

Cl_3CCO_2Et

NaOMe

exo

***OS*, *CV 6*, 327 (1988)**

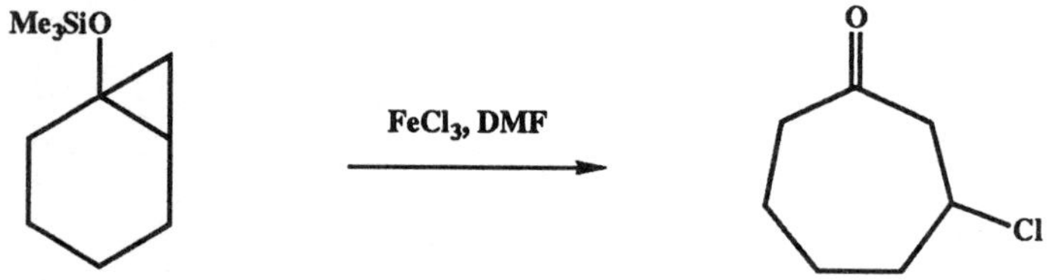

Me_3SiO

$FeCl_3$, DMF

***OS*, *CV 6*, 1037 (1988)**

NaOH

HOAc, Δ

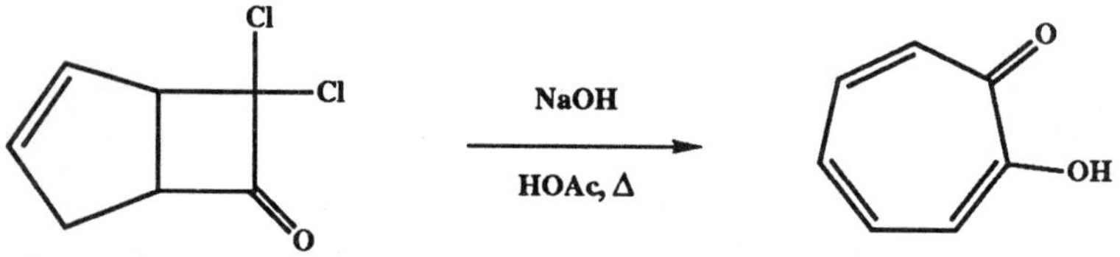

***OS*, *CV 7*, 114 (1990); *60*, 20 (1981)**

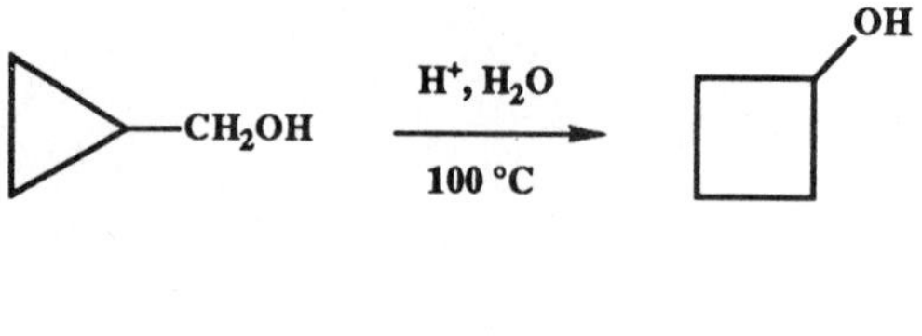

CH_2OH

H^+, H_2O

100 °C

OS, *CV* 7, 23 (1990); *61*, 98 (1983)

hv

benzene

N^+

O^-

N

O

OS, *CV* 7, 254 (1990); *63*, 188 (1985)

O

+ $H_2N—O—SO_3H$

HCO_2H

reflux

O

NH

OS, *CV* 7, 117 (1990); *64*, 50 (1986)

—CH_2OH

aq. HCl

100 °C

OH

OS, *65*, 17 (1987)

OEt $OSiMe_3$

O

CF_3CO_2H

O

O

OS, *66*, 185 (1987)

N–OH

$MeSO_2Cl$, Et_3N, CH_2Cl_2

N–OSO_2Me

Pr_3Al, CH_2Cl_2, hexane

N, Pr

$(i\text{-}Bu)_2AlH$, hexane

H, N, Pr

OS, *67*, 210 (1988)

OEt, OH

HBF_4

H_2O, Et_2O

O

REARRANGEMENT

REARRANGEMENT

***OS*, *CV 1*, 89 (1941)**

$NaBrO_3$, NaOH

Δ

***OS*, *CV 1*, 462 (1941)**

H_2SO_4

Δ

***OS*, *CV 2*, 19 (1943)**

KOH, Br_2

H_2O

***OS*, *CV 2*, 44 (1943)**

NaOCl, NaOH

H_2O

OS, *CV* 2, 73 (1943)

Ph–C(OH)(Ph)–C(OH)(Ph)–Ph → (I_2, HOAc, Δ) → PhC(=O)–C(Ph)(Ph)–Ph

OS, *CV* 2, 165 (1943)

CH_3–N(NO)–C(=O)–NH_2 → (KOH, H_2O, Et_2O) → CH_2N_2 + KCNO

OS, *CV* 2, 341 (1943)

2 (2-hydroxybenzoate, CO_2K) → (1. K_2CO_3, 230 °C (- CO_2); 2. HCl, H_2O) → (HO_2C, OH) + (OH)

OS, *CV* 2, 371 (1943)

(cyclohexanone oxime, NOH) → (H_2SO_4, Δ) → (caprolactam, O, NH)

OS, *CV 2*, 462 (1943)

2 $CH_3C(O)NH_2$ —[Br_2 / NaOH]→ $CH_3NHC(O)NHC(O)CH_3$ —[1. aq. HCl, Δ; 2. $NaNO_2$]→ $CH_3N(NO)C(O)NH_2$

OS, *CV 2*, 543 (1943)

—[$AlCl_3$ / CS_2, Δ]→ OH

$R_1 = H$
$R_2 = C(O)CH_2CH_3$

$R_1 = C(O)CH_2CH_3$
$R_2 = H$

OS, *CV 3*, 209 (1955)

Br Br O O —[1. KOH, EtOH, Δ; 2. HCl, H_2O]→ O CO_2H

OS, *CV 3*, 276 (1955)

O OH —[300 - 340 °C / Al_2O_3]→ O

REARRANGEMENT

***OS*, *CV* 3,** **281 (1955)**

AcO, O, O, CH_3 — $AlCl_3$, Δ → HO, O, CH_3, O, O, CH_3

***OS*, *CV* 3,** **356 (1955)**

Ph, O, Ph, N, NH_2 — $HgO, CaSO_4$, benzene → Ph, Ph, C=C=O

***OS*, *CV* 3,** **418 (1955)**

O, OCH_3 — Δ → OH, OCH_3

***OS*, *CV* 3,** **846 (1955)**

$n\text{-}C_{11}H_{23}$—C(O)Cl — NaN_3, H_2O, acetone → [$n\text{-}C_{11}H_{23}$—C(O)N_3] — Δ, benzene → $n\text{-}C_{11}H_{23}$—N=C=O

OS, CV 4, **45 (1963)**

O NH$_2$ N $\xrightarrow[\text{aq. NaOH}]{\text{Br}_2\text{, 70 °C}}$ NH$_2$ N

OS, CV 4, **148 (1963)**

Cl NO$_2$ $\xrightarrow[\text{H}_2\text{SO}_4\text{, aq. HOAc}]{\text{electricity (Cu cathode)}}$ [Cl NHOH] $\longrightarrow$ HO Cl NH$_2$

OS, CV 4, **221 (1963)**

HO CH$_2$NH$_2$ • HOAc $\xrightarrow[\text{H}_2\text{O, 0 °C}]{\text{NaNO}_2\text{, HOAc}}$ O

OS, CV 4, **375 (1963)**

Ph Ph O $\xrightarrow[\text{benzene}]{\text{BF}_3 \bullet \text{OEt}_2}$ Ph Ph O H

REARRANGEMENT

***OS*, *CV 4*, 479 (1963)**

Ph O O CH$_3$ O

KOH / pyridine

OH Ph O O

***OS*, *CV 4*, 585 (1963)**

NMe$_3$ + I$^-$

Na, NH$_3$

NMe$_2$ Me

***OS*, *CV 4*, 594 (1963)**

O Cl

NaOMe / Et$_2$O, Δ

CO$_2$Me

***OS*, *CV 4*, 819 (1963)**

O O (CH$_2$)$_4$ H$_2$NNH NHNH$_2$

NaNO$_2$ / HCl, H$_2$O

[O O (CH$_2$)$_4$ N$_3$ N$_3$]

Δ / benzene

O=C=N—(CH$_2$)$_4$—N=C=O

HCl, H$_2$O

HCl • H$_2$N—(CH$_2$)$_4$—NH$_2$ • HCl

OS, *CV 5*, 16 (1973)

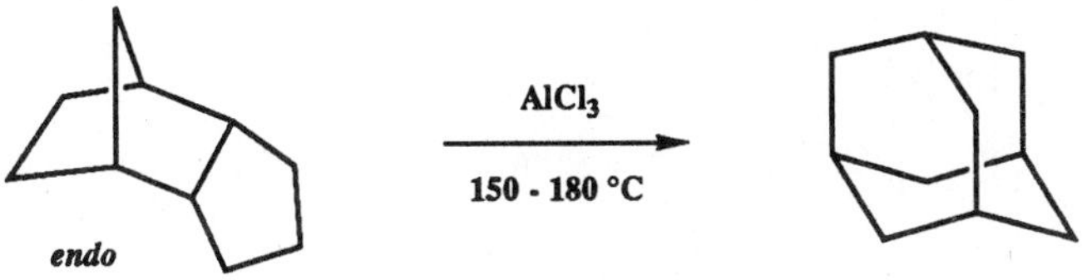

endo

$AlCl_3$

150 - 180 °C

OS, *CV 5*, 25 (1973)

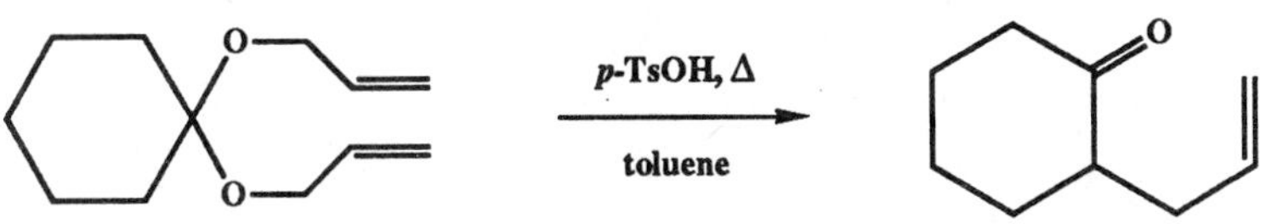

O

O

p-TsOH, Δ

toluene

O

OS, *CV 5*, 332 (1973)

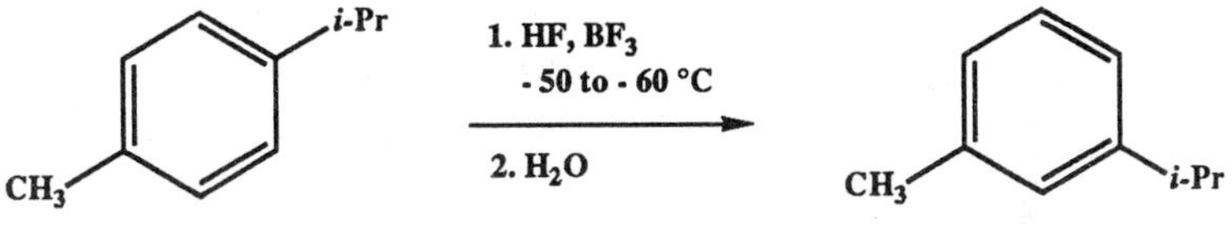

i-Pr

CH_3

1. HF, BF_3
- 50 to - 60 °C

2. H_2O

CH_3

i-Pr

OS, *CV 5*, 336 (1973)

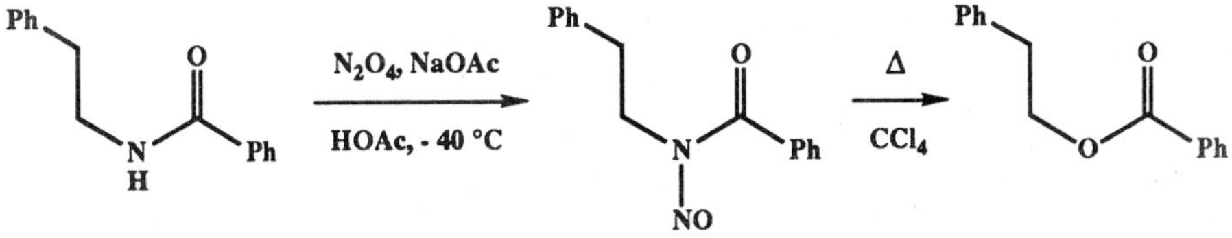

Ph

O

N
H

Ph

N_2O_4, NaOAc

HOAc, - 40 °C

Ph

O

N

NO

Ph

Δ

CCl_4

Ph

O

O

Ph

OS, CV 5, 456 (1973)

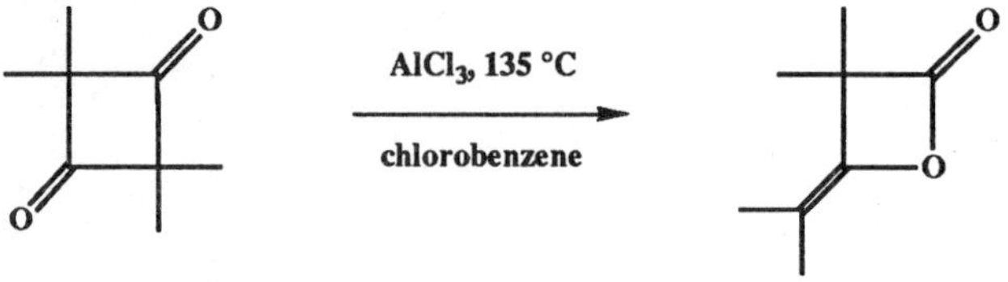

$AlCl_3$, 135 °C

chlorobenzene

OS, CV 5, 467 (1973)

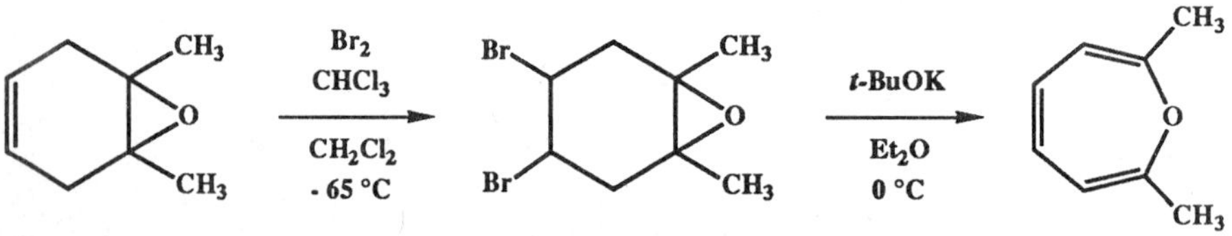

CH_3 CH_3 O

Br_2
$CHCl_3$
CH_2Cl_2
- 65 °C

Br Br CH_3 CH_3 O

t-BuOK
Et_2O
0 °C

CH_3 O CH_3

OS, CV 5, 598 (1973)

CF_3CO_3H
$BF_3 \bullet Et_2O$
CH_2Cl_2
0 - 5 °C

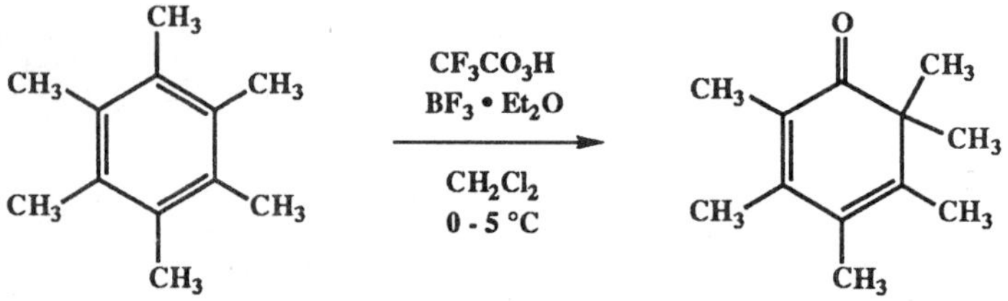

OS, CV 5, 813 (1973)

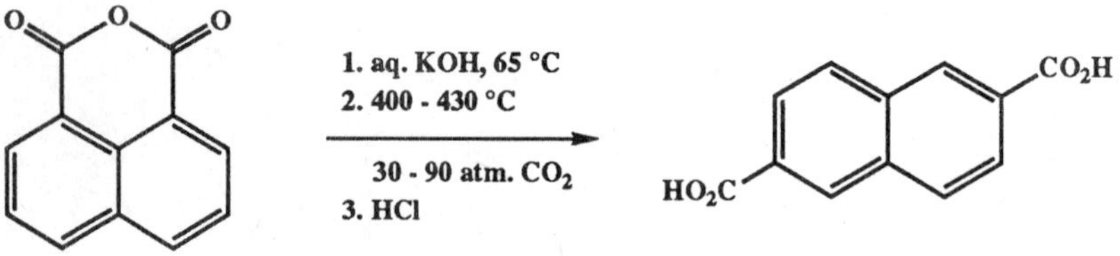

1. aq. KOH, 65 °C
2. 400 - 430 °C
30 - 90 atm. CO_2
3. HCl

HO_2C CO_2H

REARRANGEMENT

***OS*, *CV 5*, 937 (1973)**

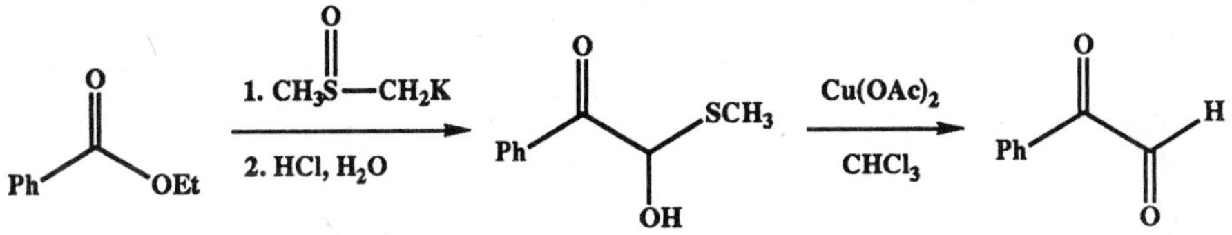

1. $CH_3S(O)—CH_2K$
2. HCl, H_2O

$Cu(OAc)_2$, $CHCl_3$

***OS*, *CV 6*, 39 (1988)**

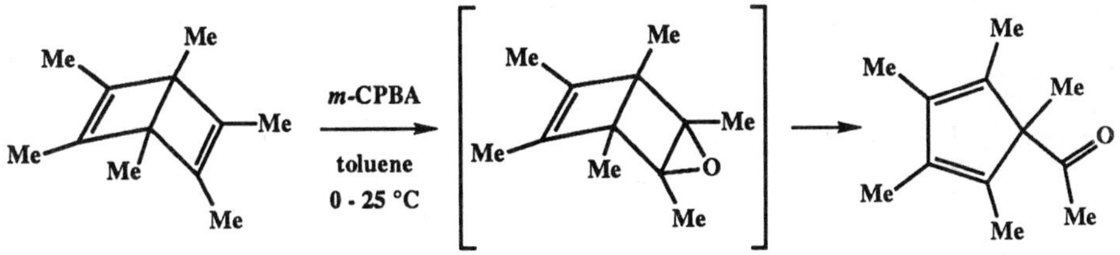

m-CPBA, toluene, 0 - 25 °C

***OS*, *CV 6*, 78 (1988)**

$(EtNH)_2SO_2$ → NaOCl, aq. NaOH → [HN(Et)—HN(Et)] → NaOCl, aq. NaOH → Et—N=N—Et

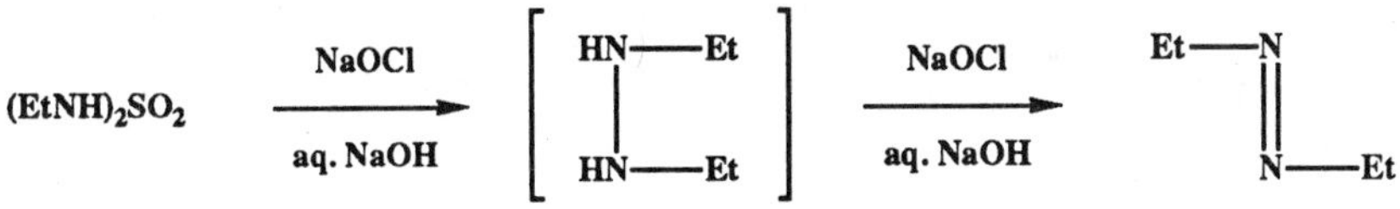

***OS*, *CV 6*, 95 (1988)**

$PhCH_2OH$, toluene, Δ

$NHCO_2CH_2Ph$

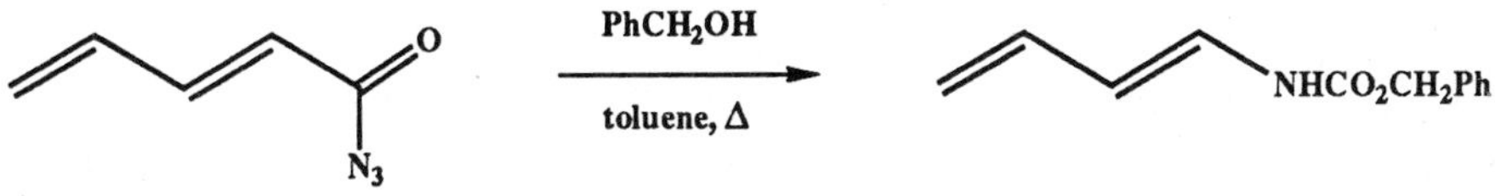

OS, CV 6, 226 (1988)

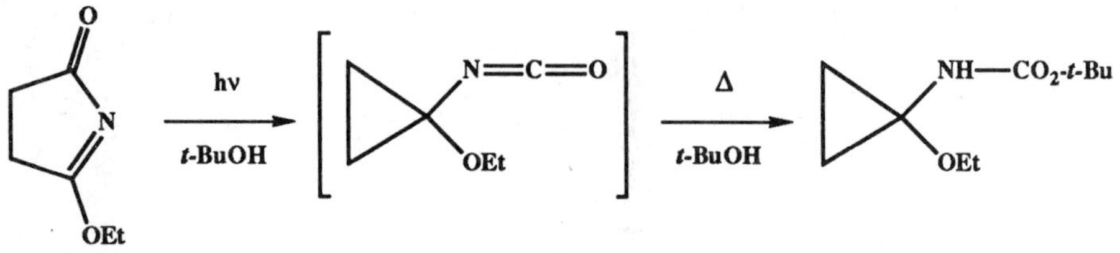

hν
t-BuOH
N═C═O
OEt
Δ
t-BuOH
NH—CO_2-t-Bu
OEt

OS, CV 6, 298 (1988)

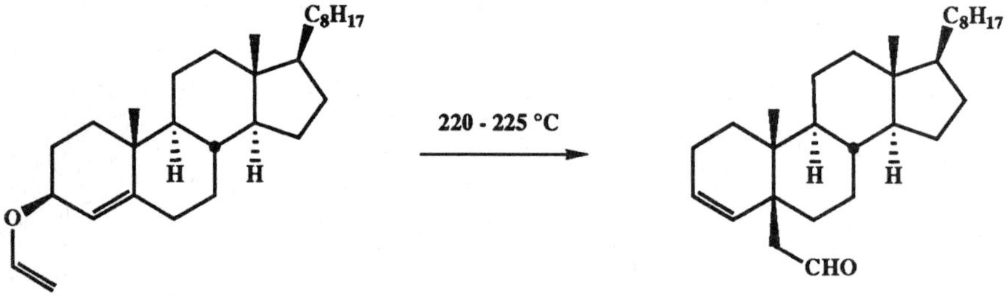

C_8H_{17}
H H
220 - 225 °C
C_8H_{17}
H H
CHO

OS, CV 6, 320 (1988)

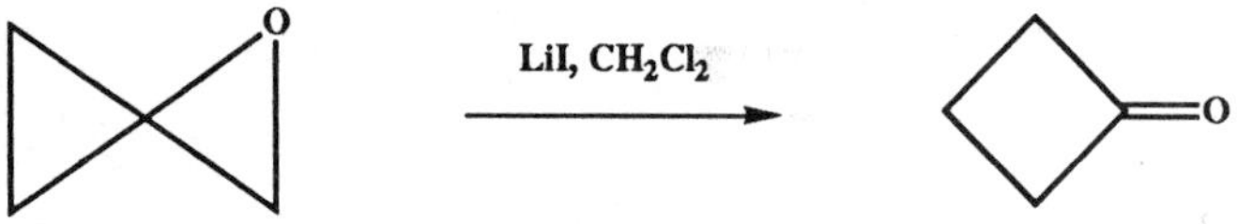

LiI, CH_2Cl_2

OS, CV 6, 378 (1988)

$C_{14}H_{20}$
"Tetrahydro - Binor - S"
$AlBr_3$
CS_2 or C_6H_{12}

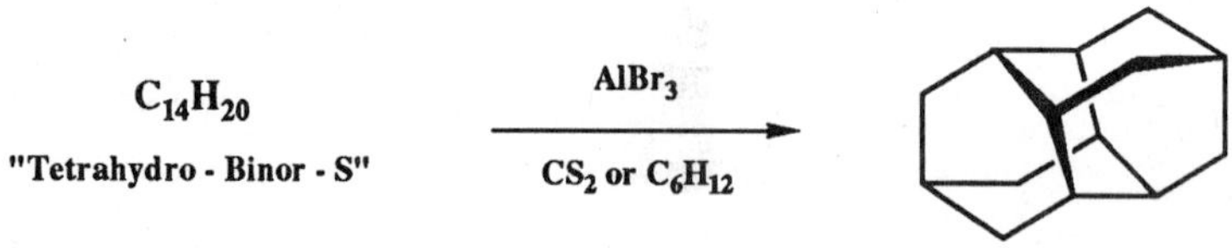

***OS*, *CV 6*, 491 (1988)**

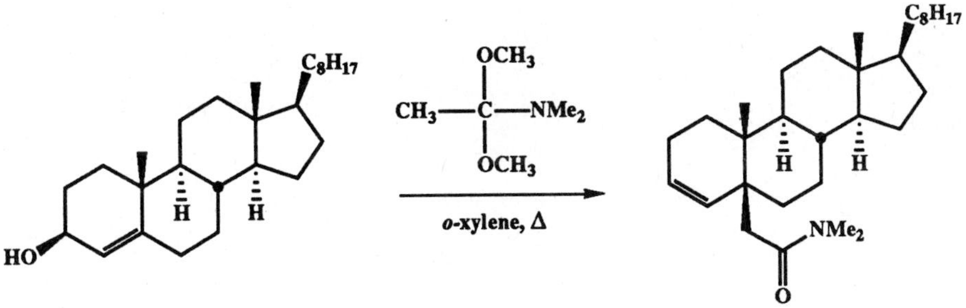

$$\xrightarrow[o\text{-xylene},\ \Delta]{CH_3C(OCH_3)_2NMe_2}$$

***OS*, *CV 6*, 507 (1988)**

$$\xrightarrow[140\ ^\circ C]{\text{xylene}}$$

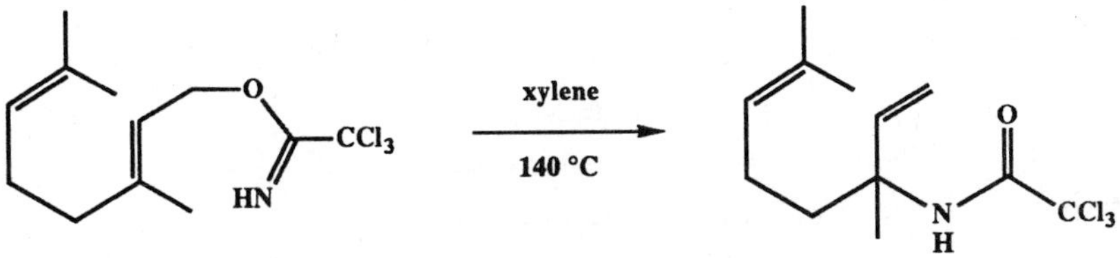

***OS*, *CV 6*, 584 (1988)**

$$\xrightarrow[\Delta]{CH_3C(OEt)_3}$$

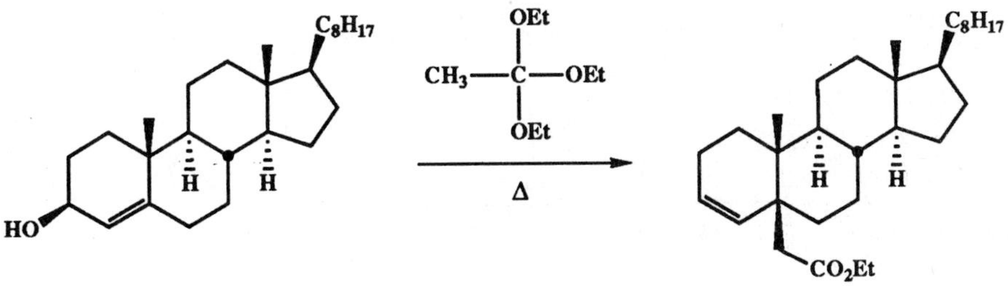

OS, CV 6, 606 (1988)

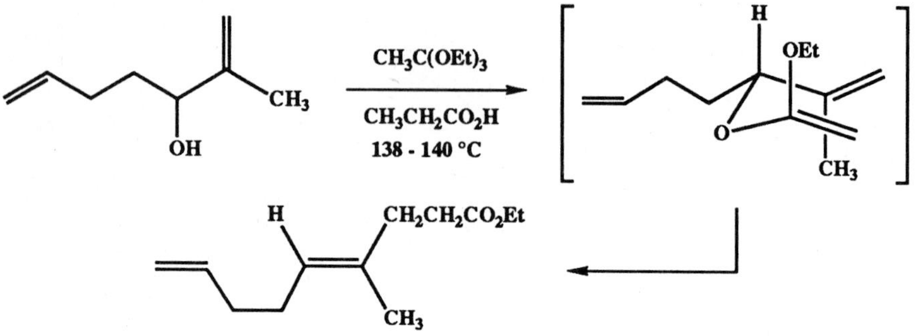

$CH_3C(OEt)_3$

$CH_3CH_2CO_2H$

138 - 140 °C

OH CH$_3$ H OEt O CH$_3$ H $CH_2CH_2CO_2Et$ CH_3

OS, CV 6, 613 (1988)

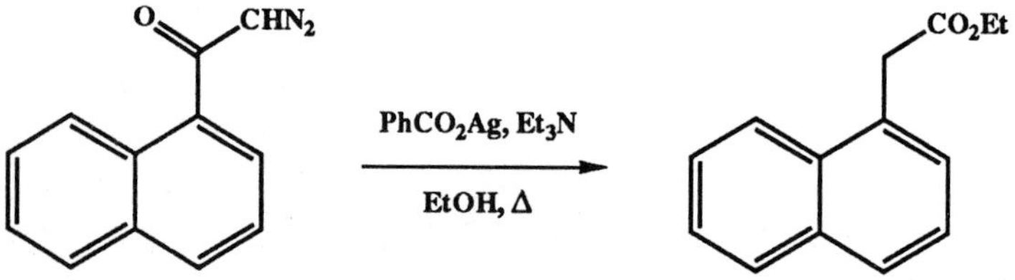

O CHN$_2$

$PhCO_2Ag$, Et_3N

EtOH, Δ

CO_2Et

OS, CV 6, 711 (1988)

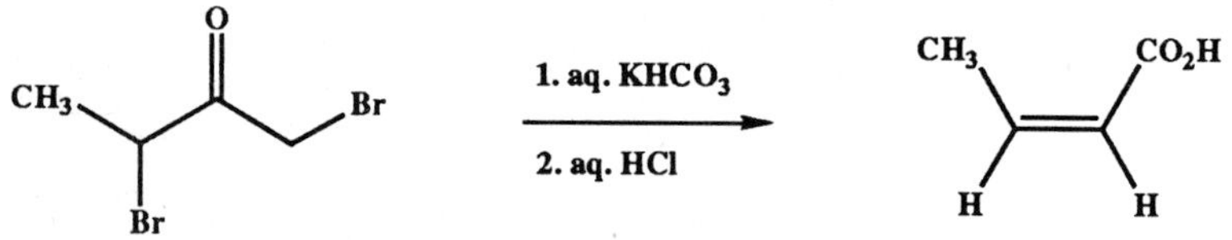

O CH$_3$ Br Br

1. aq. $KHCO_3$

2. aq. HCl

CH_3 CO_2H H H

OS, CV 6, 824 (1988)

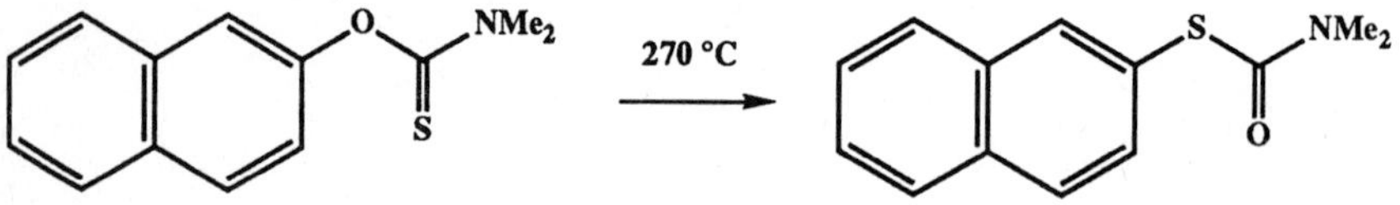

O NMe$_2$ S

270 °C

S NMe$_2$ O

OS, CV 6, 910 (1988)

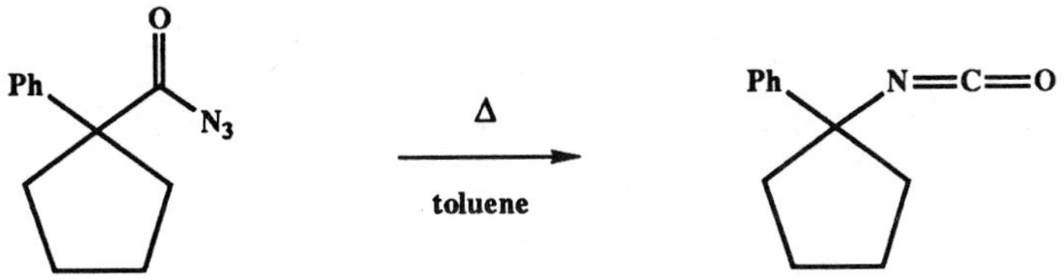

Ph, O, N_3 —Δ, toluene→ Ph, N=C=O

OS, CV 7, 12 (1990); *60*, 6 (1981)

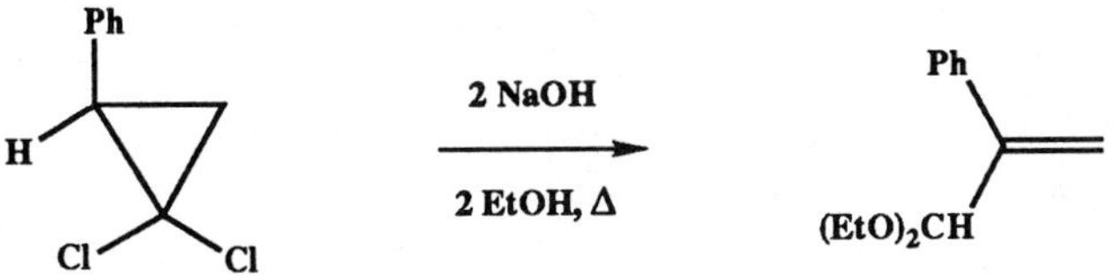

Ph, H, Cl, Cl —2 NaOH, 2 EtOH, Δ→ Ph, $(EtO)_2CH$

OS, CV 7, 177 (1990); *62*, 125 (1984)

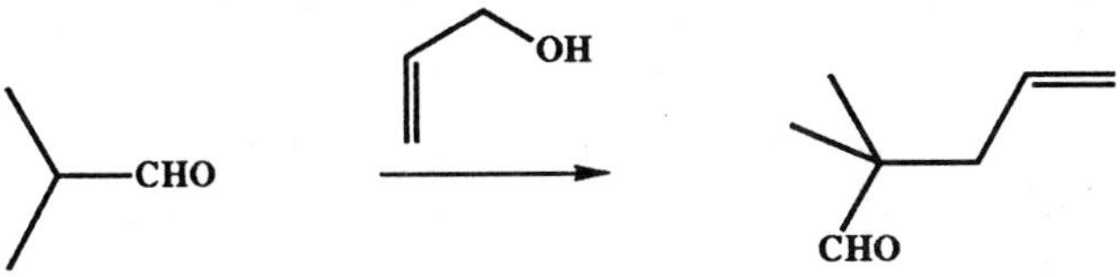

CHO —OH→ CHO

OS, CV 7, 164 (1990); *64*, 175 (1986)

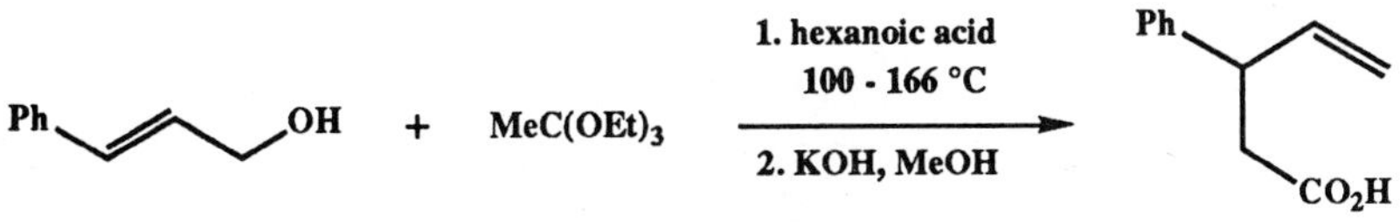

Ph, OH + $MeC(OEt)_3$ —1. hexanoic acid 100 - 166 °C; 2. KOH, MeOH→ Ph, CO_2H

OS, *65*, 90 (1987)

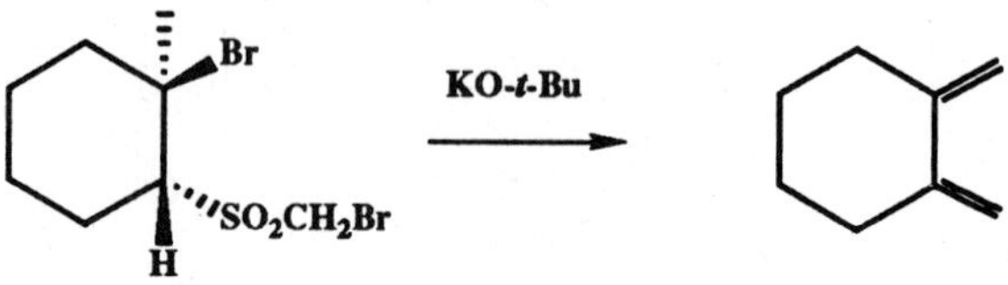

KO-*t*-Bu

Br

SO_2CH_2Br

H

OS, *65*, 173 (1987)

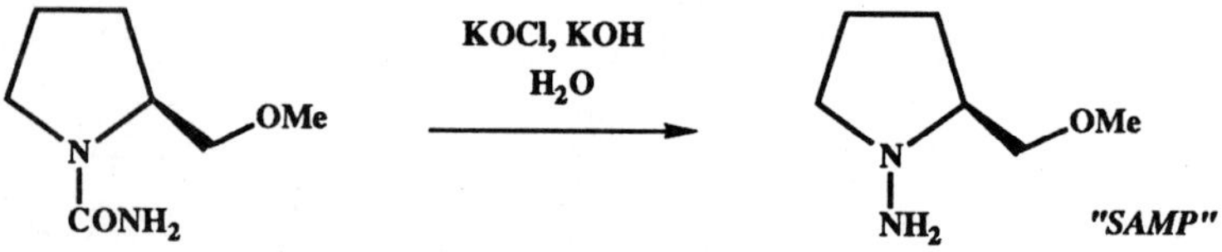

KOCl, KOH

H_2O

OMe

N

$CONH_2$

NH_2

"SAMP"

OS, *66*, 29 (1987)

CO_2H

O

Ph

160 - 180 °C

- CO_2

CHO

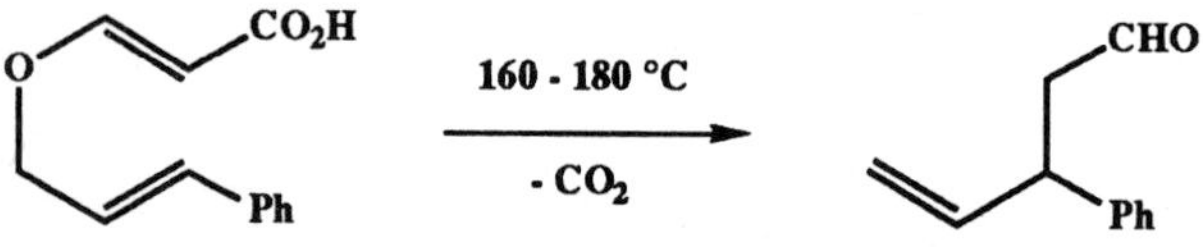

OS, *66*, 132 (1987)

O

C—NH_2

1. $(CF_3CO_2)_2$ PhI

2. H_2O

$\overset{+}{N}H_3$

Cl^-

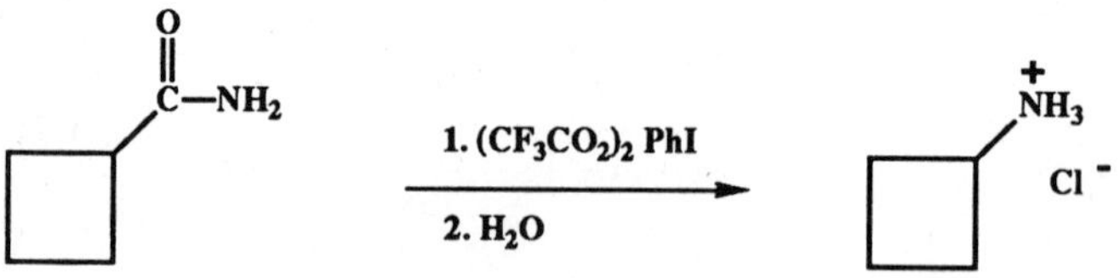

OS, CV 2, 140 (1943)

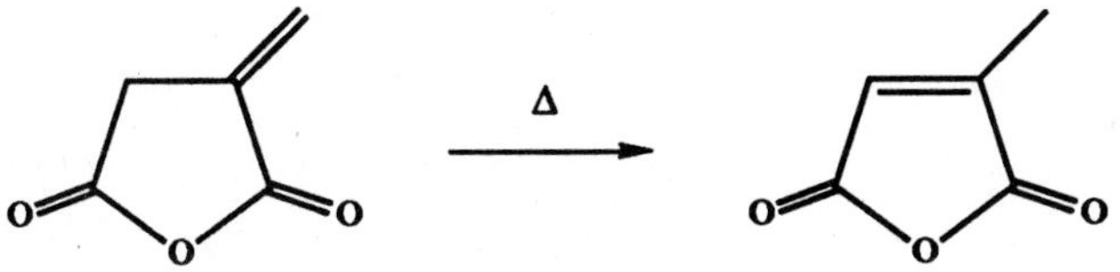

Δ

OS, CV 2, 382 (1943)

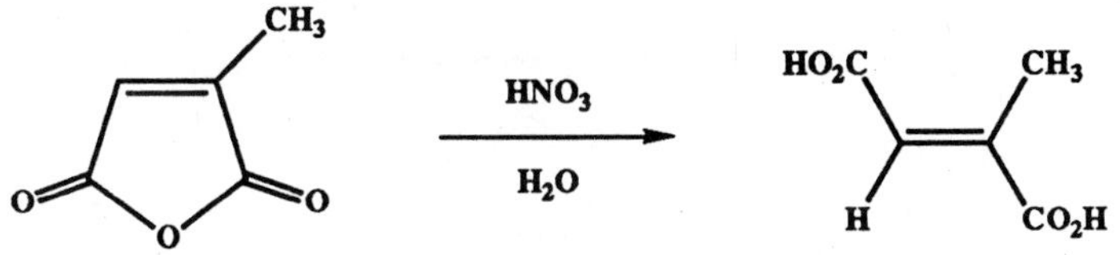

CH_3

HNO_3

H_2O

HO_2C CH_3

H CO_2H

OS, CV 3, 422 (1955)

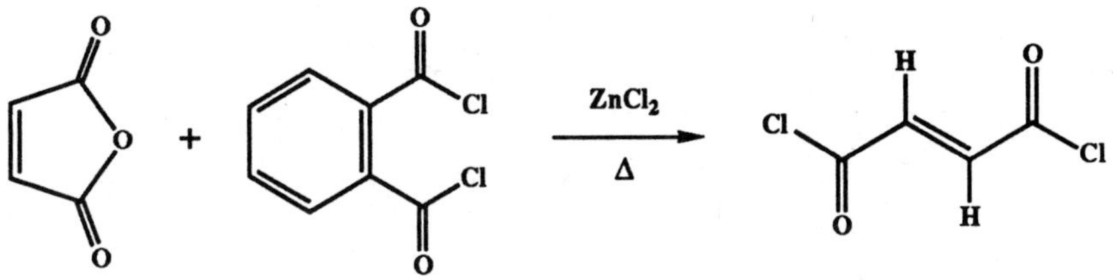

O O

O + Cl

Cl

O O

$ZnCl_2$

Δ

H O

Cl Cl

O H

OS, CV 4, 195 (1963)

C_8H_{17}

oxalic acid

95% EtOH, Δ

C_8H_{17}

O O

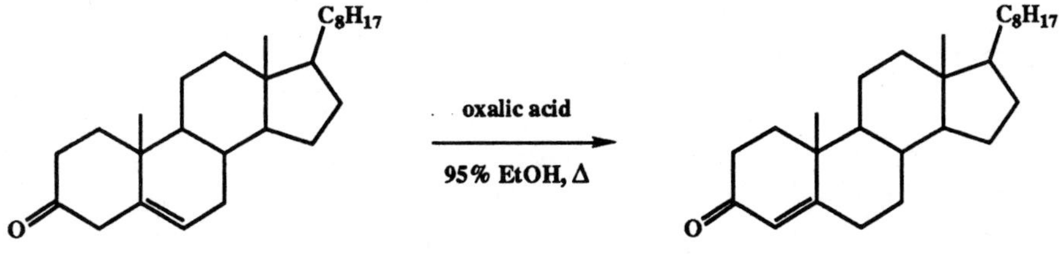

***OS*, *CV 6*, 68 (1988)**

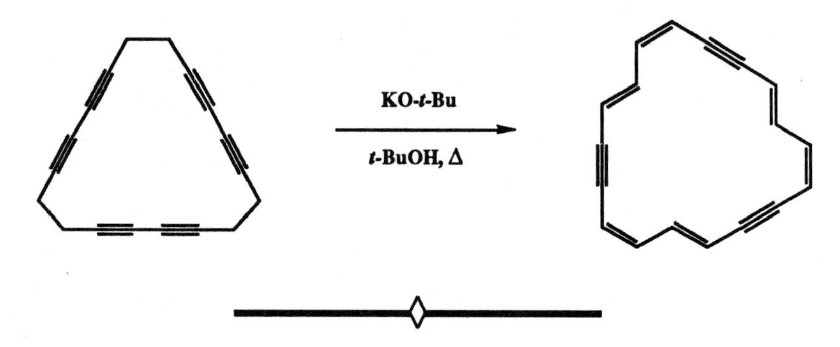

***OS*, *CV 6*, 87 (1988)**

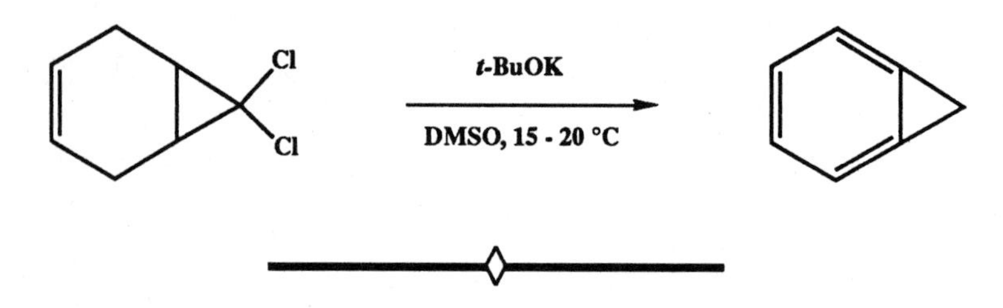

***OS*, *CV 6*, 145 (1988)**

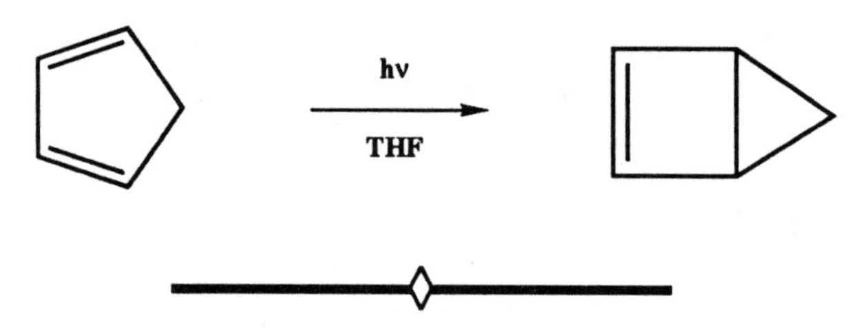

***OS*, *CV 6*, 925 (1988)**

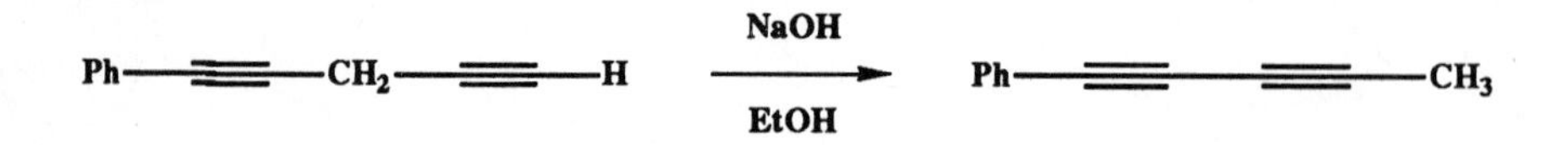

OS, CV 6, 962 (1988)

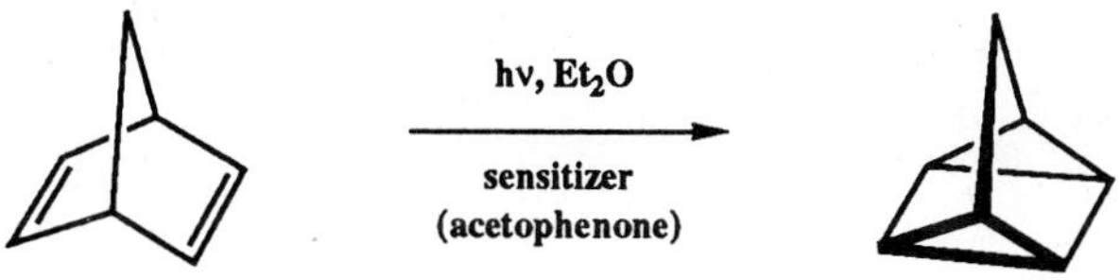

hν, Et_2O

sensitizer (acetophenone)

OS, CV 7, 470 (1990); *63,* 192 (1985)

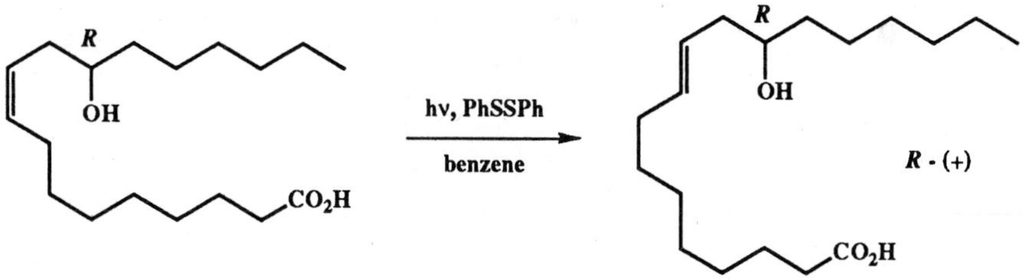

R

OH

CO_2H

hν, PhSSPh

benzene

R

OH

R - (+)

CO_2H

OS, 65, 224 (1987)

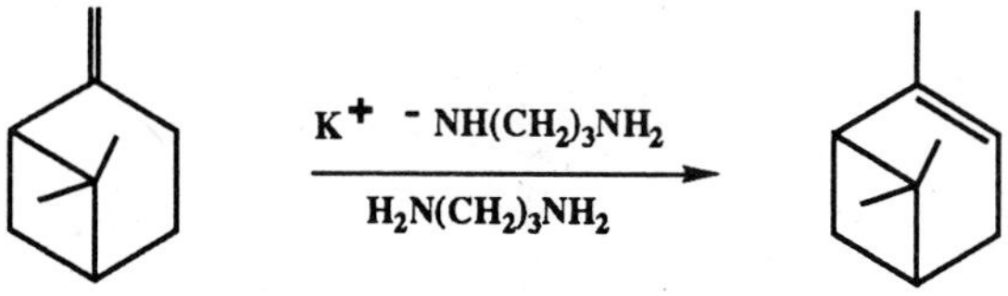

K^+ $^-NH(CH_2)_3NH_2$

$H_2N(CH_2)_3NH_2$

(-) - β - pinene

(-) - α - pinene

OS, *66*, 22 (1987)

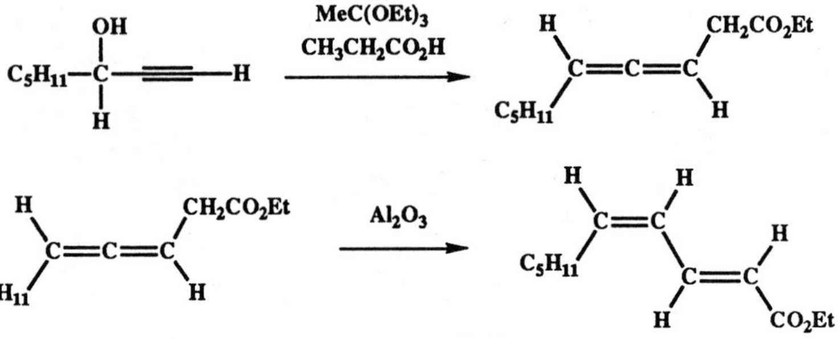
OH
C5H11—C—H
H
MeC(OEt)3
CH3CH2CO2H
H
CH2CO2Et
C=C=C
C5H11
H
H
CH2CO2Et
C=C=C
C5H11
H
Al2O3
H
H
C=C
C5H11
H
C=C
H
CO2Et

OS, *66*, 127 (1987)

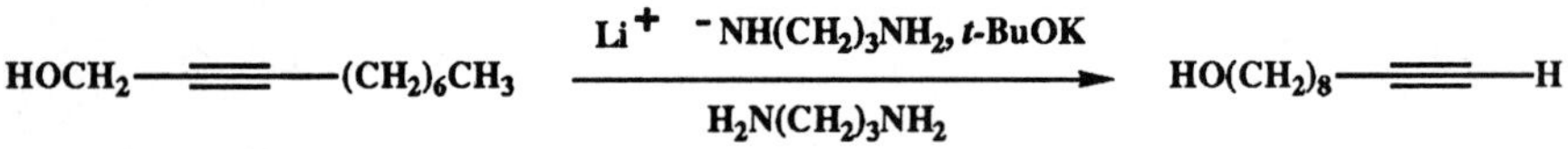
HOCH2—(CH2)6CH3
Li+ -NH(CH2)3NH2, t-BuOK
H2N(CH2)3NH2
HO(CH2)8—H

OS, *67*, 33 (1988)

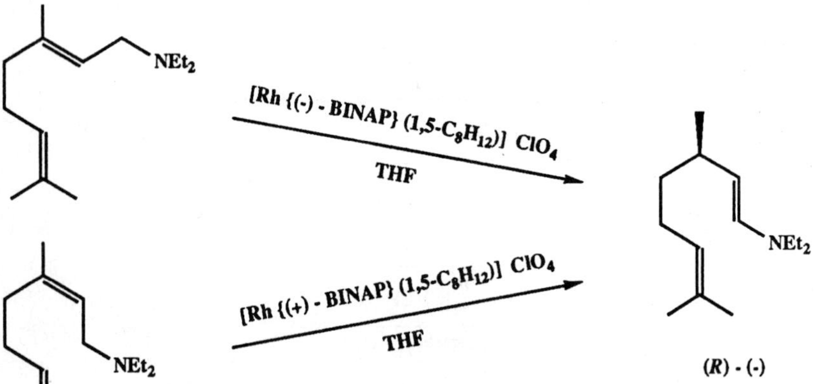
NEt2
[Rh {(-) - BINAP} (1,5-C8H12)] ClO4
THF
NEt2
[Rh {(+) - BINAP} (1,5-C8H12)] ClO4
THF
NEt2
(R) - (-)

REARRANGEMENT

OS, *67*, 205 (1988)

HO, S, S — $HgO, BF_3 \cdot Et_2O$ / H_2O, THF → CHO, OH

OXIDATION

OS, *CV 1*, 177 (1941)

$$Ph{-}N(H){-}OH \xrightarrow[\text{NH}_3\text{, Et}_2\text{O; 0 - 10 °C}]{n\text{-BuO—NO}} Ph{-}N(NO){-}O^- \; NH_4^+$$

OS, *CV 1*, 309 (1941)

$$2\ NH_4OH + NaOCl \xrightarrow[\text{(viscolizer)}]{\text{gelatine, }\Delta} H_2N{-}NH_2 \xrightarrow[0\ °C]{H_2SO_4} H_2N{-}NH_2 \cdot H_2SO_4$$

OS, *CV 1*, 506 (1941)

$$2\ CS_2 \xrightarrow{5\ Cl_2} 2\ CSCl_4 \xrightarrow[\Delta]{\text{Sn, HCl}} 2\ Cl_2C{=}S$$

OS, *CV 2*, 111 (1943)

$$(n\text{-Bu})_2SO_3 \xrightarrow[\Delta]{SO_2Cl_2} n\text{-BuOSO}_2\text{Cl} \xrightarrow{(n\text{-Bu})_2SO_3} (n\text{-Bu})_2SO_4$$

OS, *CV 2*, 240 (1943)

$$Ph{-}Se{-}Ph \xrightarrow{HNO_3\text{, HCl}} Ph{-}SeCl_2{-}Ph$$

OS, CV 2, **290 (1943)**

Me-C$_6$H$_4$-NHEt • HBr → ($NaNO_2$, aq. HCl) → Me-C$_6$H$_4$-N(NO)Et → ($SnCl_2$, HCl) → Me-C$_6$H$_4$-NHEt

OS, CV 2, **460 (1943)**

Ph-NH-CH_3 → ($NaNO_2$, HCl, H_2O) → Ph-N(NO)-CH_3

OS, CV 2, **461 (1943)**

CH_3—NH_2 → (H_2N-CO-NH_2, HCl, H_2O, Δ) → CH_3-NH-CO-NH_2 → ($NaNO_2$, H_2SO_4) → CH_3-N(NO)-CO-NH_2

OS, CV 2, **462 (1943)**

2 CH_3-CO-NH_2 → (Br_2, NaOH) → CH_3-NH-CO-NH-CO-CH_3 → (1. aq. HCl, Δ; 2. $NaNO_2$) → CH_3-N(NO)-CO-NH_2

OS, CV 2, **464 (1943)**

CH_3-NH-CO-OEt → ($NaNO_2$, HNO_3, H_2O, Et_2O) → CH_3-N(NO)-CO-OEt

OS, *CV 2*, 471 (1943)

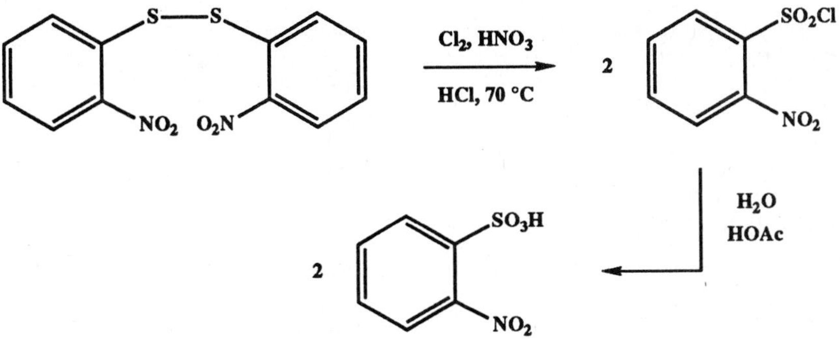

bis(2-nitrophenyl) disulfide → (Cl_2, HNO_3; HCl, 70 °C) → 2 2-nitrobenzenesulfonyl chloride (SO_2Cl, NO_2) → (H_2O, HOAc) → 2 2-nitrobenzenesulfonic acid (SO_3H, NO_2)

OS, *CV 2*, 496 (1943)

PhC(O)C(=N–NH_2)Ph → (HgO, KOH (cat.); Na_2SO_4, Et_2O) → PhC(O)C(=N_2)Ph

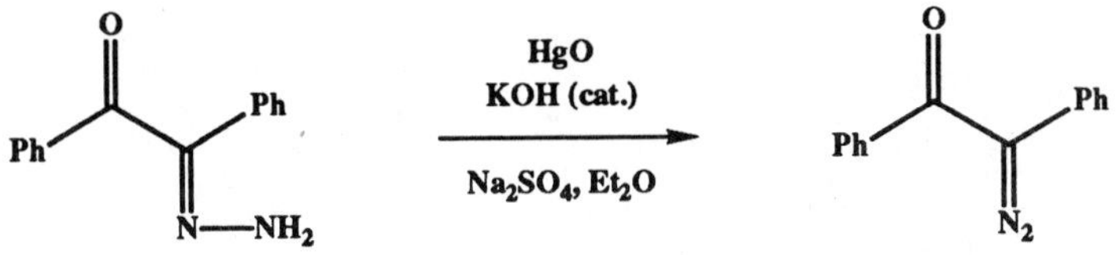

OS, *CV 3*, 86 (1955)

2 4-H_2N-C_6H_4-SNa → (H_2O_2; 70 °C) → H_2N-C_6H_4-S–S-C_6H_4-NH_2

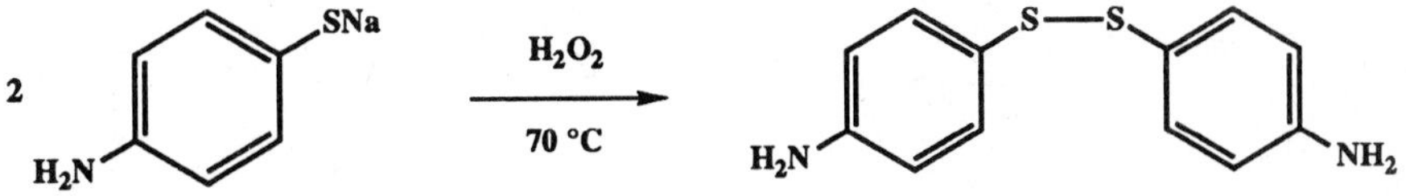

OS, CV 3, 116 (1955)

2 PhC(O)Cl —(KSH / EtOH)→ 2 PhC(O)SK —(I_2)→ PhC(O)S–SC(O)Ph

OS, CV 3, 226 (1955)

HO_2CCH(NH_2)CH_2S–SCH_2CH(NH_2)CO_2H —(HCl, Br_2 / H_2O)→ 2 $HO_3S$$CH_2$CH($NH_2$)$CO_2H \cdot H_2O$

OS, CV 3, 244 (1955)

$(CH_3)_2C(NHCH_3)CH_2C(O)CH_3$ —($NaNO_2$ / HOAc)→ $(CH_3)_2C(N(CH_3)NO)CH_2C(O)CH_3$ —(NaOR / ROH / Et_2O, Δ)→ CH_2N_2

R = *i*-Pr, C_6H_{11}

OS, CV 3, 334 (1955)

2-methyl-4-nitroaniline (CH_3, NH_2, O_2N) —($K_2S_2O_8$ / H_2SO_4)→ (CH_3, NO, O_2N) —($K_2Cr_2O_7$ / H_2SO_4)→ (CO_2H, NO_2, O_2N)

OS, CV 3, 351 (1955)

$Ph_2C{=}N{-}NH_2 \xrightarrow[\text{pet. ether}]{\text{HgO}} Ph_2C{=}N_2$

OS, CV 3, 356 (1955)

$PhC(=N{-}NH_2){-}C(=O)Ph \xrightarrow[\text{benzene}]{\text{HgO, CaSO}_4} Ph_2C{=}C{=}O$

OS, CV 3, 360 (1955)

$2\ (PhNHNH)_2C{=}S \xrightarrow[\text{MeOH}]{\text{KOH}} (PhN{=}N)(PhNHNH)C{=}S + (PhNHNH)(H_2N)C{=}S + PhNH_2$

OS, CV 3, 375 (1955)

$EtO_2C{-}NH{-}NH{-}CO_2Et \xrightarrow[\text{benzene}]{\text{H}_2\text{O, Cl}_2} EtO_2C{-}N{=}N{-}CO_2Et$

***OS*, *CV 3*, 482 (1955)**

$$C_6H_5I + Cl_2 \xrightarrow{CHCl_3} C_6H_5ICl_2$$

***OS*, *CV 3*, 485 (1955)**

$$2\ C_6H_5IO \xrightarrow[\text{steam-distill}]{H_2O} C_6H_5IO_2 + C_6H_5I$$

***OS*, *CV 3*, 486 (1955)**

$$C_6H_5ICl_2 \xrightarrow[\text{HOAc, 70 °C}]{\text{NaOCl, } H_2O} C_6H_5IO_2$$

***OS*, *CV 3*, 668 (1955)**

$$\text{Ph—NHOH} \xrightarrow[\text{aq. } H_2SO_4\text{, - 5 °C}]{Na_2Cr_2O_7} \text{Ph—NO}$$

***OS*, *CV 3*, 710 (1955)**

$$\text{Ph—NH—NH}_2 \xrightarrow[Et_2O\text{, } H_2O]{NaNO_2\text{, HCl}} \text{Ph—N}_3$$

OS, *CV* 4, 66 (1963)

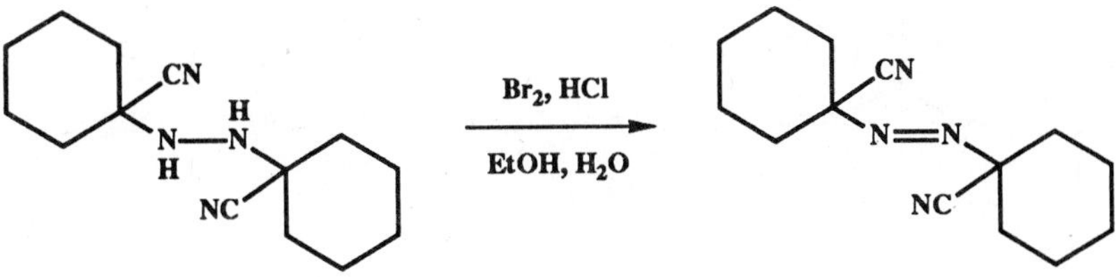

Br_2, HCl

EtOH, H_2O

OS, *CV* 4, 104 (1963)

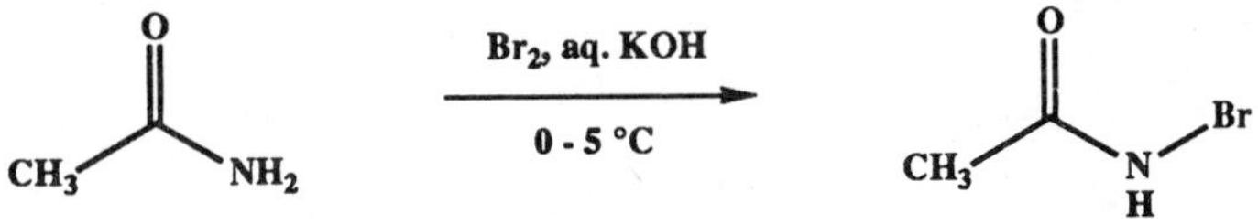

$CH_3C(O)NH_2$ → $CH_3C(O)NHBr$

Br_2, aq. KOH

0 - 5 °C

OS, *CV* 4, 125 (1963)

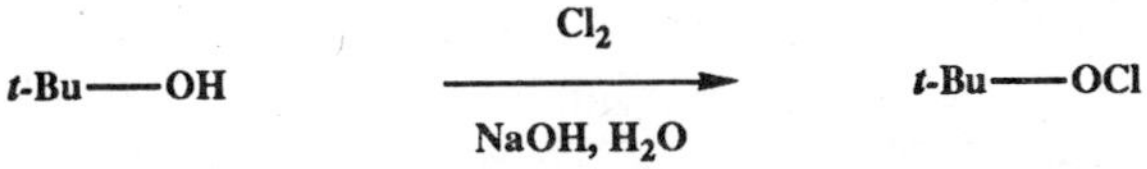

t-Bu—OH → *t*-Bu—OCl

Cl_2

NaOH, H_2O

OS, *CV* 4, 411 (1963)

EtO_2C—NH—NH—CO_2Et → EtO_2C—N=N—CO_2Et

fuming HNO_3

0 - 5 °C

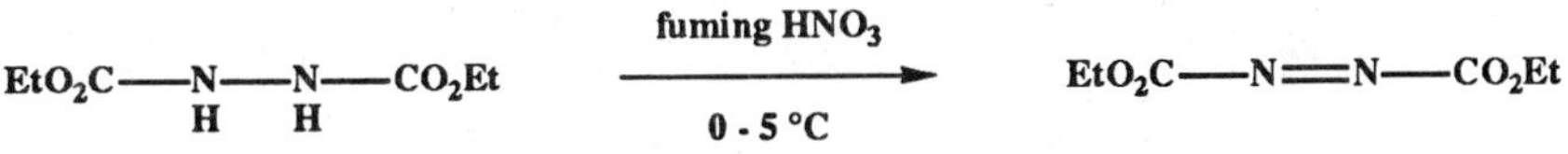

OS, *CV* 4, 612 (1963)

NMe_2 → $\overset{+}{N}Me_2$—O^-

H_2O_2

MeOH

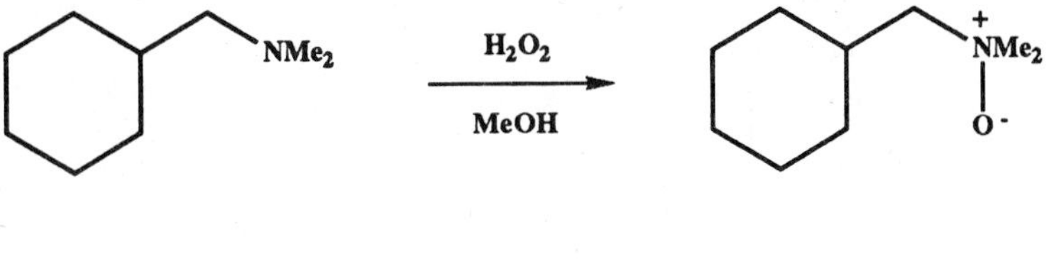

OS, *CV* 4, 704 (1963)

30% H_2O_2

HOAc, Δ

NH2

NH2

O

O

N

+N

O-

OS, *CV* 4, 780 (1963)

$NaNO_2$, HNO_3

H_2O, Et_2O

Ph

N

H

CO_2Et

Ph

N

NO

CO_2Et

OS, *CV* 4, 819 (1963)

$NaNO_2$

HCl, H_2O

O

O

H_2NNH

$(CH_2)_4$

$NHNH_2$

O

O

N_3

$(CH_2)_4$

N_3

Δ

benzene

HCl, H_2O

HCl • H_2N—$(CH_2)_4$—NH_2 • HCl

O=C=N—$(CH_2)_4$—N=C=O

OS, *CV* 4, 828 (1963)

peracetic acid

N

+N

O-

• HOAc

HCl, Δ

Δ, 1 mm Hg

pyridine N-oxide hydrochloride

pyridine N-oxide

OS, CV 4, 910 (1963)

3 C_6H_5Cl —(Na, $AsCl_3$; benzene, Δ)→ $(C_6H_5)_3As$ —(H_2O_2; acetone)→ $(C_6H_5)_3AsO$ —(1. C_6H_5MgBr, Δ; 2. HCl, Δ)→ $(C_6H_5)_4AsCl$ • HCl

OS, CV 4, 934 (1963)

CH_3–C_6H_4–SH —(Cl_2, CCl_4; dark)→ CH_3–C_6H_4–SCl

OS, CV 4, 943 (1963)

$CH_3C_6H_4SO_2Cl$ —(CH_3NH_2, Δ; aq. NaOH)→ $CH_3C_6H_4SO_2$–NH–CH_3 —($NaNO_2$; aq. HOAc)→ $CH_3C_6H_4SO_2$–N(NO)–CH_3

OS, CV 5, 43 (1973)

Pyridine —(H_2NOSO_3H, Δ; K_2CO_3, H_2O)→ N^+–NH^- (pyridinium ylide) —(HI, EtOH; -20 °C)→ N^+–NH_2 I^-

OS, *CV 5*, 157 (1973)

H_2N–NH–C(=O)–O-*t*-Bu → ($NaNO_2$ / HOAc, H_2O) → N_3–C(=O)–O-*t*-Bu

OS, *CV 5*, 160 (1973)

t-Bu-O_2C—NH—NH—CO_2-*t*-Bu → (*N*-bromo-succinimide / pyridine, CH_2Cl_2) → *t*-Bu-O_2C—N=N—CO_2-*t*-Bu

OS, *CV 5*, 184 (1973)

t-BuOH → (NaOCl / HOAc) → *t*-BuOCl

OS, *CV 5*, 208 (1973)

NH_2 (cyclohexylamine) → (2 *t*-BuOCl / benzene) → NCl_2 (cyclohexyl) → (KOAc / EtOH, Δ) → NCl (cyclohexanone N-chloroimine)

OS, CV 5, 336 (1973)

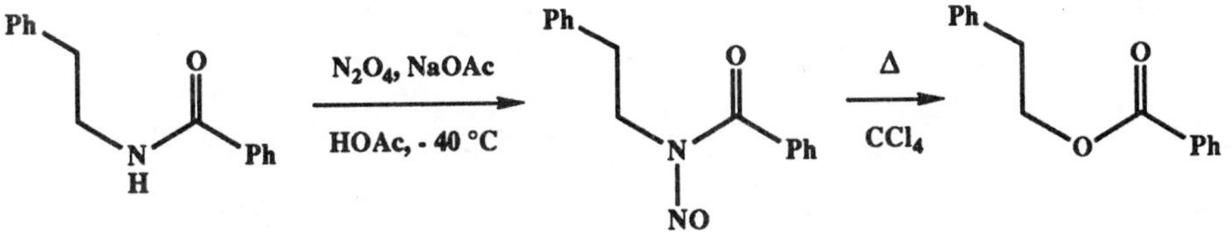

Ph
O
N
H
Ph
N_2O_4, NaOAc
HOAc, - 40 °C
Ph
O
N
Ph
NO
Δ
CCl_4
Ph
O
O
Ph

OS, CV 5, 341 (1973)

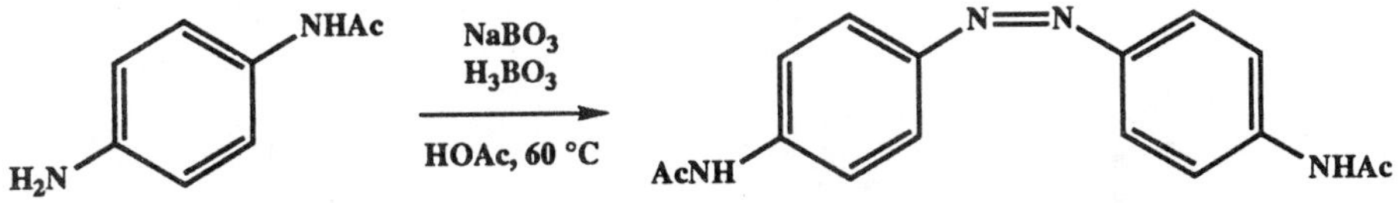

NHAc
H_2N
$NaBO_3$
H_3BO_3
HOAc, 60 °C
N=N
AcNH
NHAc

OS, CV 5, 367 (1973)

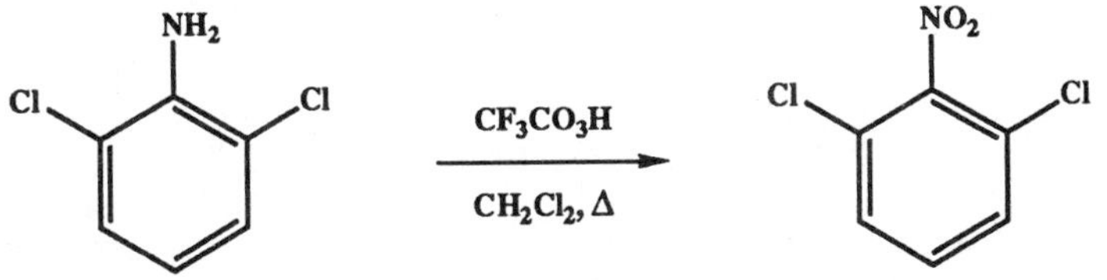

NH_2
Cl
Cl
CF_3CO_3H
CH_2Cl_2, Δ
NO_2
Cl
Cl

OS, CV 5, 660 (1973)

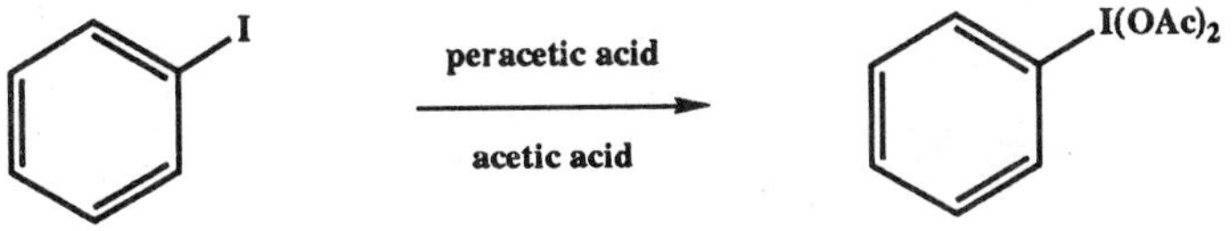

I
peracetic acid
acetic acid
$I(OAc)_2$

OS, CV 5, 663 (1973)

I_2, dioxane

NAg → NI

OS, CV 5, 665 (1973)

2 CH_3CO_3H

then H_2O, 35 - 100 °C

I → IO_2

OS, CV 5, 723 (1973)

PhS—SPh → 2 PhS(=O)—OMe

Pb(OAc)$_4$, MeOH

$CHCl_3$, Δ

OS, CV 5, 791 (1973)

Ph—S—CH_3 → Ph—S(=O)—CH_3

$NaIO_4$

H_2O, 0 °C

OS, CV 5, **839 (1973)**

OH, CN; HNO_3 / Ac_2O; ONO_2, CN; 2 O NH / Δ; NO_2, N, O; +; CN, N, O

OS, CV 5, **845 (1973)**

NH_2; $KMnO_4$, $MgSO_4$ / acetone, H_2O; NO_2

OS, CV 5, **897 (1973)**

O; NH_2OSO_3H / NH_4OH, 0 °C; NH, NH; Ag_2O / Et_2O; N, N

OS, CV 5, **909 (1973)**

NH_2, CH_3; 1. *t*-BuOCl benzene, 5 °C; 2. NaOMe, MeOH, Δ; 3. HCl, H_2O, Δ; O, NH_3Cl

OS, CV 6, 23 (1988)

2-acetyl-2-(phenylseleno)cyclohexanone → 2-acetyl-2-cyclohexen-1-one

H_2O_2, H_2O; CH_2Cl_2, 30 - 35 °C

CH₃ SePh

OS, CV 6, 78 (1988)

$(EtNH)_2SO_2$ —NaOCl, aq. NaOH→ [HN—Et / HN—Et] —NaOCl, aq. NaOH→ Et—N=N—Et

OS, CV 6, 161 (1988)

$(CF_3)_2C{=}NNH_2$ —$Pb(OAc)_4$→ $(CF_3)_2C{=}N_2$

OS, CV 6, 163 (1988)

2 $Ph(CF_3)_2C—OK$ + Ph_2S —Br_2, CCl_4, 25 °C→ $Ph_2S[OC(CF_3)_2Ph]_2$

OS, CV 6, 334 (1988)

NC, H H N–N–CO_2Me (cyclohexane) →[Br_2, CH_2Cl_2 / aq. $NaHCO_3$, 25 °C] NC, N=N–CO_2Me (cyclohexane)

OS, CV 6, 342 (1988)

CH_3–N (morpholine, O) →[30% aq. H_2O_2 / 75 °C] CH_3, O^-, N^+ (morpholine, O) • H_2O

OS, CV 6, 392 (1988)

$(CH_3)_2C{=}NNH_2$ →[HgO / KOH] $(CH_3)_2C{=}N_2$

OS, CV 6, 403 (1988)

Ph–$CHCO_2H$ / S / Ph–$CHCO_2H$ →[H_2O_2 / HOAc] Ph–$CHCO_2H$ / SO_2 / Ph–$CHCO_2H$ →[Br_2, KBr / aq. HOAc, light, 80 °C] Ph–CHBr / SO_2 / Ph–CHBr

OS, CV 6, 404 (1988)

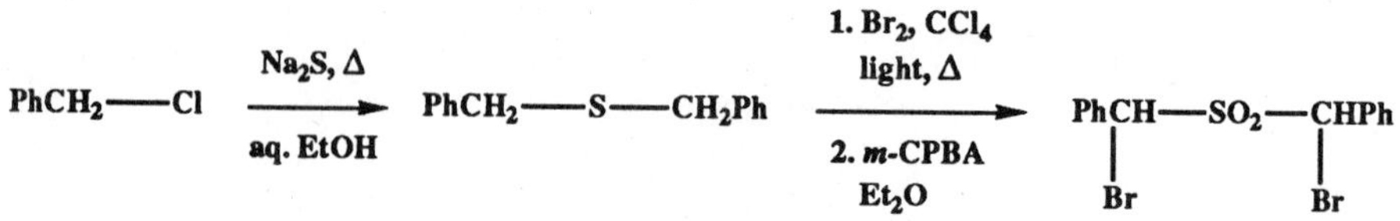

$PhCH_2$—Cl →(Na_2S, Δ; aq. EtOH) $PhCH_2$—S—CH_2Ph →(1. Br_2, CCl_4, light, Δ; 2. *m*-CPBA, Et_2O) PhCH(Br)—SO_2—CHPh(Br)

OS, CV 6, 482 (1988)

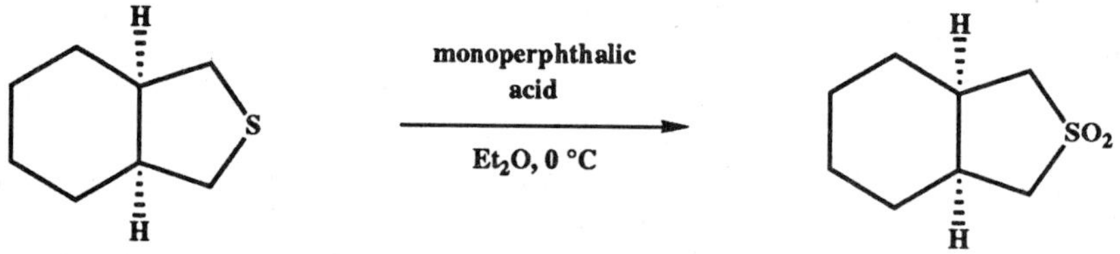

H, H, S →(monoperphthalic acid; Et_2O, 0 °C) H, H, SO_2

OS, CV 6, 501 (1988)

n-$C_{12}H_{25}$—NMe_2 →(*t*-Bu—O—OH; $VO(acac)_2$, Δ) *n*-$C_{12}H_{25}$—$\overset{+}{N}Me_2$—O^- + *t*-Bu—OH

$(acac)_2$ = $(CH_3COCH_2COCH_2)_2$

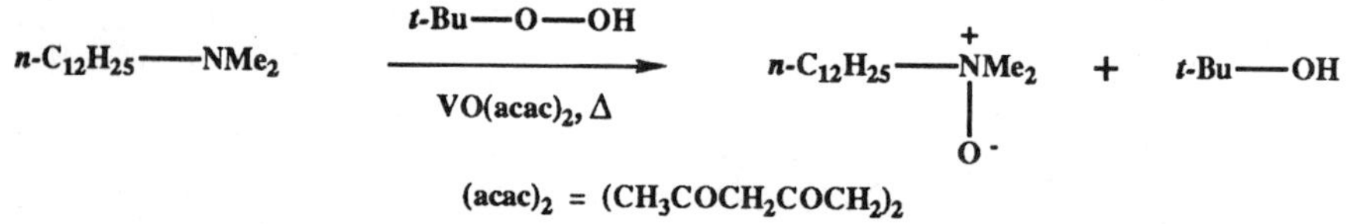

OS, CV 6, 791 (1988)

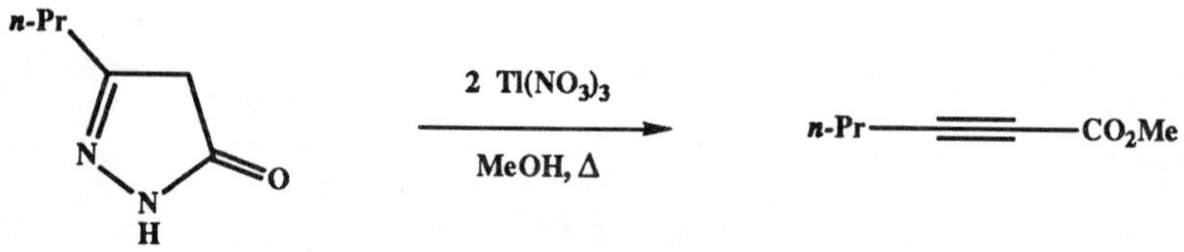

n-Pr, N, N, H, O →(2 $Tl(NO_3)_3$; MeOH, Δ) *n*-Pr—≡—CO_2Me

OS, CV 6, 803 (1988)

$$t\text{-Bu—NH}_2 \xrightarrow[\text{H}_2\text{O, 55 °C}]{\text{KMnO}_4} t\text{-Bu—NO}_2$$

$$t\text{-Bu—NHOH} \xrightarrow[\text{- 20 to 25 °C}]{\text{NaOBr, H}_2\text{O}} t\text{-Bu—N=O} \rightleftharpoons t\text{-Bu}(O^-)N^+{=}N^+(O^-)t\text{-Bu}$$

OS, CV 6, 936 (1988)

HN—NH, O, O, N, Ph $\xrightarrow[\text{EtOAc}]{t\text{-BuOCl}}$ N=N, O, O, N, Ph

OS, CV 6, 968 (1988)

N—H $\xrightarrow[\text{2. Ca(OCl)}_2\text{, H}_2\text{O, 0 °C}]{\text{1. HOAc, 10 °C}}$ N—Cl

OS, CV 6, 981 (1988)

$$p\text{-TsCH}_2\text{—N(H)—CO}_2\text{Et} \xrightarrow[\text{pyridine, 0 °C}]{\text{Cl—NO}} p\text{-TsCH}_2\text{—N(NO)—CO}_2\text{Et} \xrightarrow[\text{Et}_2\text{O, 10 - 15 °C}]{\text{alumina}} p\text{-TsCH=N}_2$$

OS, *CV* 7, 491 (1990); *62*, 210 (1984)

H_2O_2 / WO_3; Cl_2 / hν; Cl_2 / hν

S; SO_2; Cl, SO_2; Cl, Cl, SO_2

OS, *CV* 7, 8 (1990); *64*, 96 (1986)

1. $Ph_2P(O)ONH_2$
2. H^+, PhCHO
3. H^+, steam dist'n.
(- PhCHO)

O, Me, N, O, N, Me, Na^+, N, N; O, Me, N, O, N, Me, NH_2, N, N

OS, *CV* 7, 453 (1990); *64*, 157 (1986)

SPh → *m*-CPBA, CH_2Cl_2 → SOPh

SPh → H_2O_2, HOAc → SO_2Ph

OS, *68*, 49 (1989)

CH_3—S—CH_3

Ti(O-*i*-Pr)$_4$
Diethyl (2*S*, 3*S*)-tartrate

80% cumene hydroperoxide
H_2O, CH_2Cl_2

CH_3, O, S, CH_3

S - (-) isomer

OS, CV 1, 494 (1941)

$PhCO_3H$

$CHCl_3$, 0 °C

OS, CV 4, 552 (1963)

H_2O_2, NaOH

MeOH, H_2O

CH_3 CH_3 CH_3

CH_3 CH_3 CH_3

OS, CV 4, 860 (1963)

Ph H H Ph

trans

peracetic acid
NaOAc

HOAc, CH_2Cl_2

Ph H H Ph

trans

OS, CV 5, 414 (1973)

CH_3CO_3H
Na_2CO_3

CH_2Cl_2

H_2O

OH OH + HO OH

OS, CV 5, 467 (1973)

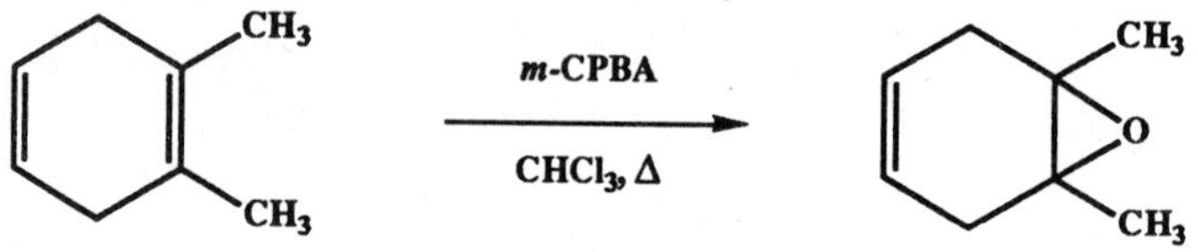

CH_3 CH_3 — *m*-CPBA / $CHCl_3$, Δ → CH_3 O CH_3

OS, CV 5, 1007 (1973)

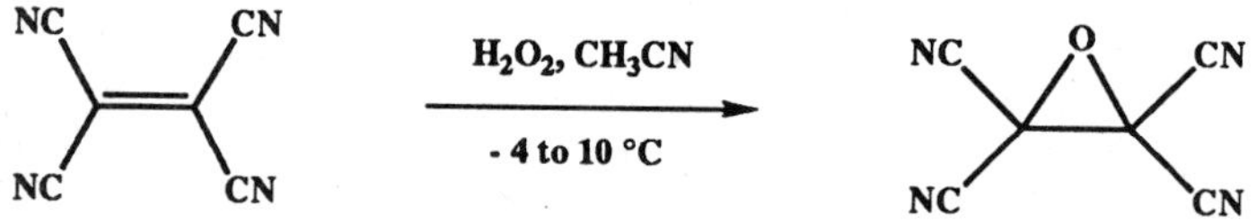

NC CN NC CN — H_2O_2, CH_3CN / - 4 to 10 °C → NC O CN NC CN

OS, CV 6, 39 (1988)

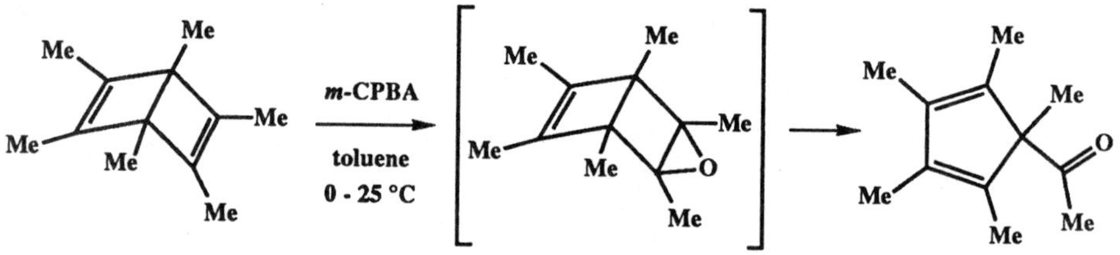

Me Me Me Me Me Me — *m*-CPBA / toluene / 0 - 25 °C → [Me Me Me Me Me O Me] → Me Me Me O Me Me Me

OS, CV 6, 320 (1988)

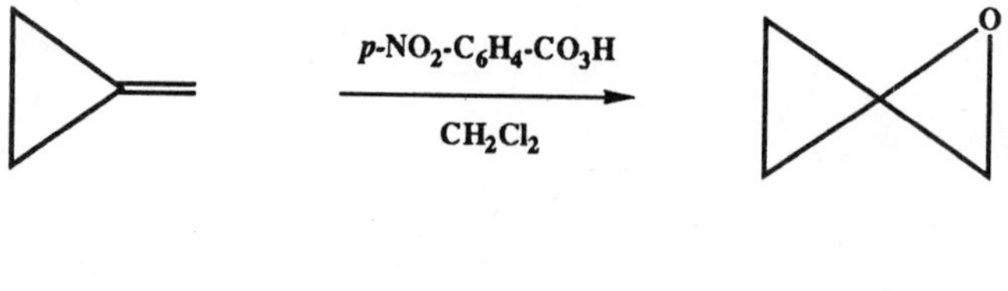

p-NO_2-C_6H_4-CO_3H / CH_2Cl_2 → O

OS, *CV 6*, 679 (1988)

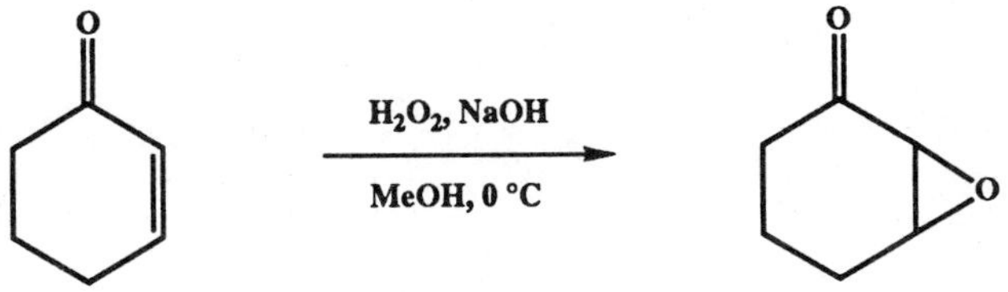

H_2O_2, NaOH

MeOH, 0 °C

OS, *CV 6*, 862 (1988)

peracetic acid

CH_2Cl_2, 15 °C

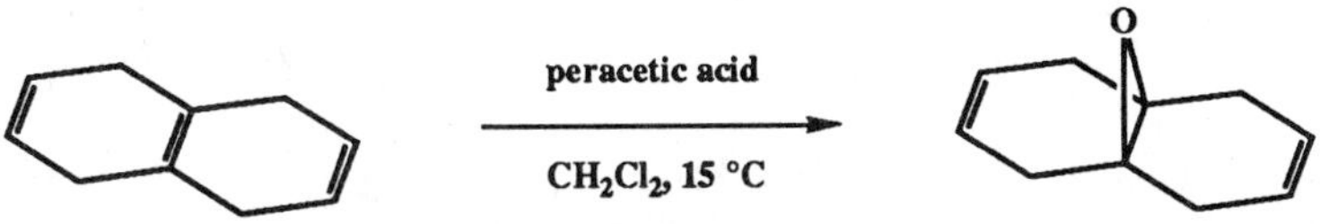

OS, *CV 7*, 126 (1990); *60*, 63 (1981)

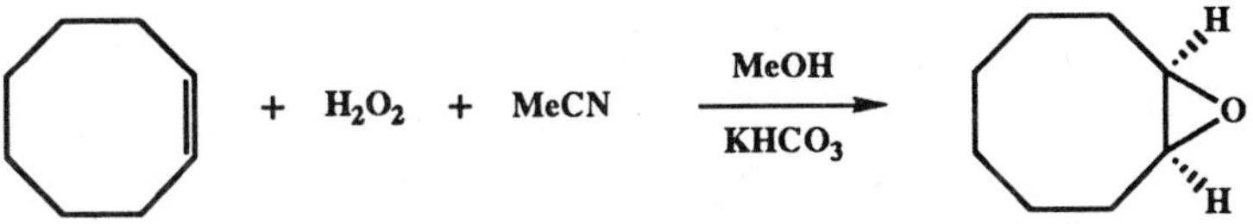

+ H_2O_2 + MeCN

MeOH

$KHCO_3$

OS, *CV 7*, 461 (1990); *63*, 66 (1985)

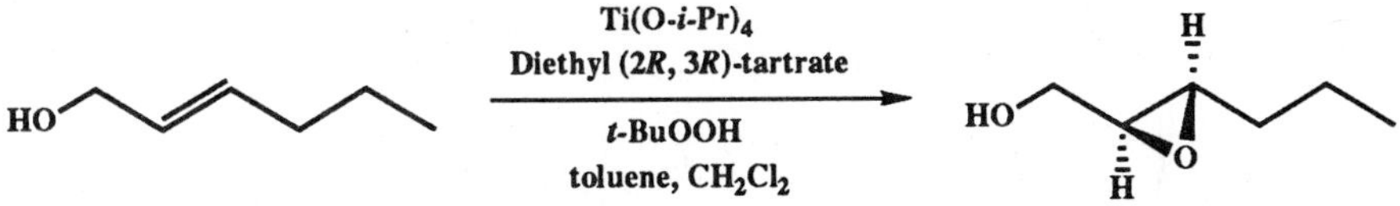

$Ti(O\text{-}i\text{-}Pr)_4$

Diethyl (2*R*, 3*R*)-tartrate

t-BuOOH

toluene, CH_2Cl_2

(2*S*, 3*S*)

OS, CV 1, 115 (1941)

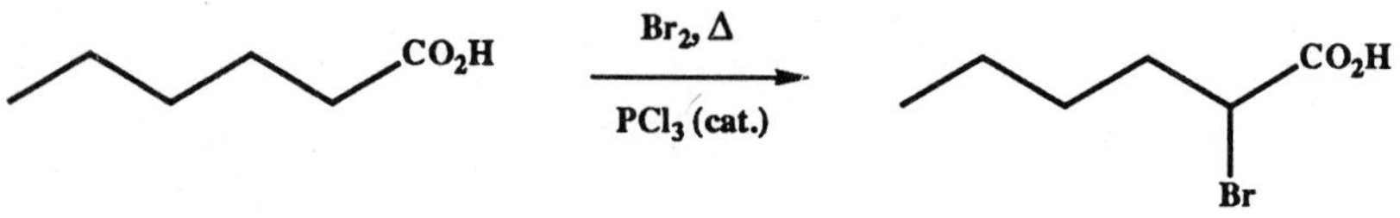

CO_2H —[Br_2, Δ / PCl_3 (cat.)]→ CO_2H, Br

OS, CV 1, 127 (1941)

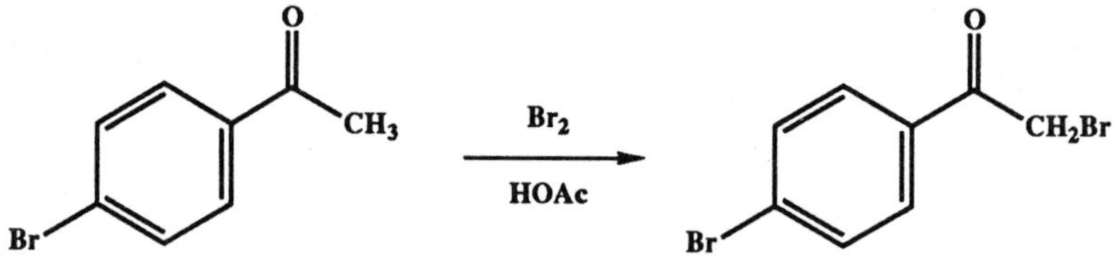

O, CH_3, Br —[Br_2 / HOAc]→ O, CH_2Br, Br

OS, CV 1, 155 (1941)

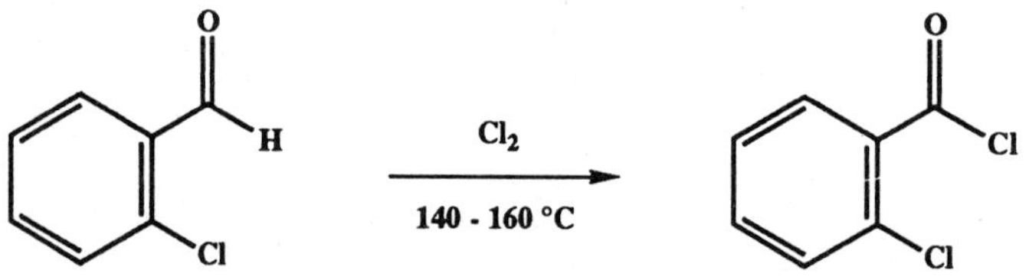

O, H, Cl —[Cl_2 / 140 - 160 °C]→ O, Cl, Cl

OS, CV 1, 245 (1941)

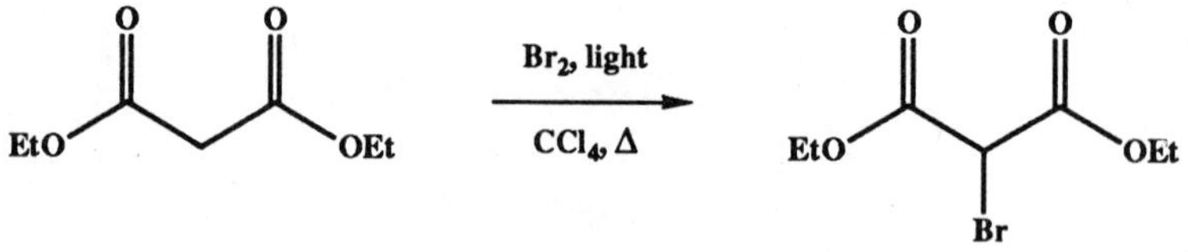

O, O, EtO, OEt —[Br_2, light / CCl_4, Δ]→ O, O, EtO, OEt, Br

OS, *CV* 2, 74 (1943)

Ph–C(=O)–N(H)–$(CH_2)_4$–CH_2CO_2H —(1. P, Br_2; 2. H_2O)→ Ph–C(=O)–N(H)–$(CH_2)_4$–CH(Br)CO_2H

OS, *CV* 2, 87 (1943)

$(HC(=O)CH_3)_3$ + 9 Br_2 —(S_8 (cat.))→ 3 HC(=O)CBr_3

OS, *CV* 2, 88 (1943)

$CH_3C(=O)CH_3$ —(Br_2; HOAc, H_2O)→ $CH_3C(=O)CH_2Br$

OS, *CV* 2, 89 (1943)

4-Br–C_6H_4–CH_3 —(Br_2, Δ; light)→ 4-Br–C_6H_4–$CHBr_2$ —(H_2O, Δ; $CaCO_3$)→ 4-Br–C_6H_4–C(=O)H

OS, CV 2, 93 (1943)

Me, Me, CO_2Et, CO_2Et — 1. KOH, H_2O; 2. HCl, Br_2 → Me, Me, CO_2H, Br, CO_2H — Δ → Me, Me, CO_2H, Br

OS, CV 2, 133 (1943)

Cl, CH_3 — Cl_2, PCl_5; light, Δ → Cl, $CHCl_2$ — H_2SO_4; H_2O → Cl, O, H

OS, CV 2, 244 (1943)

O, O, Ph, Ph — Br_2; $CHCl_3$ → O, O, Ph, Ph, Br, Br — NaOAc; AcOH; Δ → O, O, O, Ph, Ph

OS, CV 2, 357 (1943)

S — HgO, I_2; benzene → S, I

***OS*, *CV* 2,** **443 (1943)**

O_2N CH$_3$ — Br_2, 140 - 150 °C → O_2N CH$_2$Br

***OS*, *CV* 2,** **480 (1943)**

O CH$_3$ — Br_2, $AlCl_3$, Et_2O → O CH$_2$Br

***OS*, *CV* 2,** **549 (1943)**

O O H O — PCl_5 → [Cl Cl O O H Cl Cl] — H_2O → OH OH H O

***OS*, *CV* 3,** **188 (1955)**

O — Cl_2, H_2O → O Cl

OXIDATION - HALOGENATION

OS, CV 3, **343 (1955)**

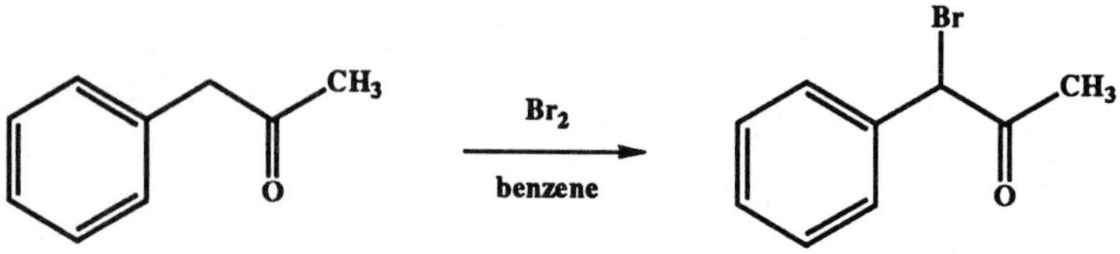

CH_3 O

Br_2

benzene

Br CH_3 O

OS, CV 3, **347 (1955)**

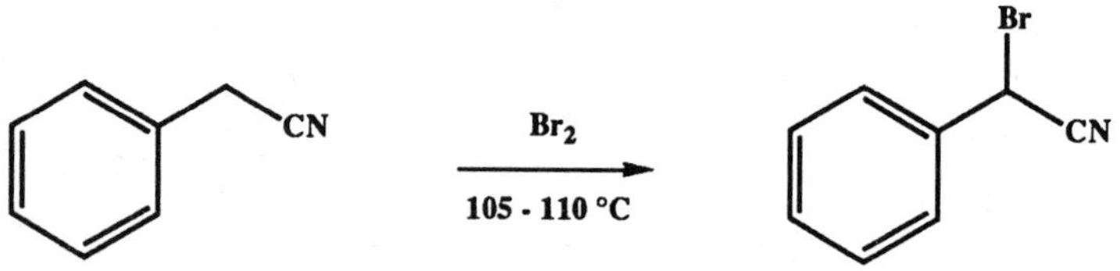

CN

Br_2

105 - 110 °C

Br CN

OS, CV 3, **381 (1955)**

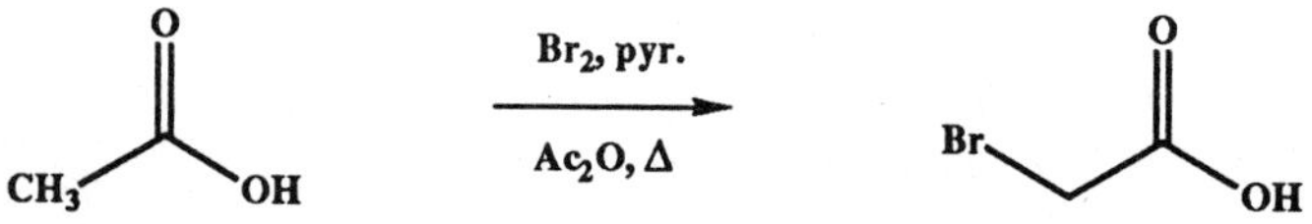

O CH_3 OH

Br_2, pyr.

Ac_2O, Δ

O Br OH

OS, CV 3, **495 (1955)**

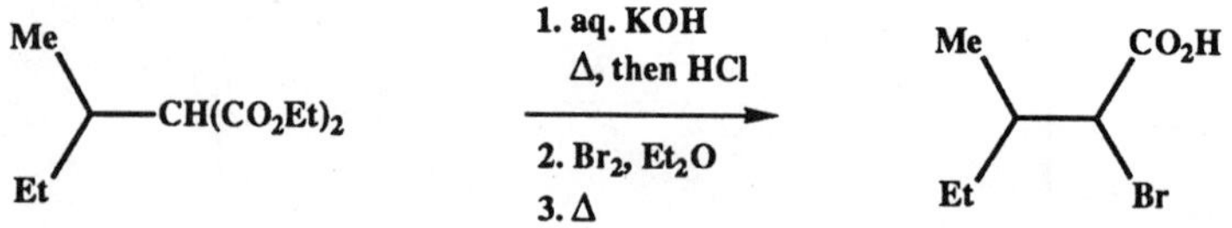

Me $-CH(CO_2Et)_2$ Et

1. aq. KOH
   Δ, then HCl
2. Br_2, Et_2O
3. Δ

Me CO_2H Et Br

OS, CV 3, 523 (1955)

i-Pr–CH₂CH₂CO₂H → (Br_2; PCl_3, Δ) → i-Pr–CH₂CH(Br)CO₂H → (NH_4OH) → i-Pr–CH₂CH(NH₂)CO₂H

$$i\text{-Pr}CH_2CH_2CO_2H \xrightarrow[PCl_3,\ \Delta]{Br_2} i\text{-Pr}CH_2CH(Br)CO_2H \xrightarrow{NH_4OH} i\text{-Pr}CH_2CH(NH_2)CO_2H$$

OS, CV 3, 538 (1955)

$$PhC(O)CH_3 \xrightarrow[HOAc]{Cl_2} PhC(O)CHCl_2 \xrightarrow[\text{2. aq. HCl}]{\text{1. aq. NaOH, 65 °C}} PhCH(OH)CO_2H$$

OS, CV 3, 623 (1955)

$$HO_2C(CH_2)_4CO_2H \xrightarrow[\Delta]{SOCl_2} ClOC(CH_2)_4COCl \xrightarrow[\text{2. EtOH}]{\text{1. } Br_2} EtO_2CCH(Br)CH_2CH_2CH(Br)CO_2Et$$

OS, CV 3, 705 (1955)

$$PhCH_2CH(CO_2H)_2 \xrightarrow[Et_2O]{Br_2} PhCH_2C(Br)(CO_2H)_2$$

OS, *CV 3*, 737 (1955)

$\xrightarrow[140\ °C]{Br_2}$ Br $\xrightarrow[\Delta]{H_2O}$ CO_2H CHO

OS, *CV 3*, 788 (1955)

CH_3 CH_3 $\xrightarrow[\Delta]{Br_2,\ light}$ $CHBr_2$ $CHBr_2$ $\xrightarrow[70-110\ °C]{95\%\ H_2SO_4}$ CHO CHO

OS, *CV 3*, 848 (1955)

CH_3 CH_3 CO_2H $\xrightarrow[PCl_3,\ \Delta]{Br_2}$ CH_3 CH_3 CO_2H Br $\xrightarrow{NH_4OH}$ CH_3 CH_3 CO_2H NH_2

dl - valine

OS, *CV 4*, 108 (1963)

CH_3 CH_3 $\xrightarrow[(PhCO)_2O,\ CCl_4,\ \Delta]{N\text{-bromo-succinimide}}$ CH_3 CH_3 Br

OS, *CV 4*, 110 (1963)

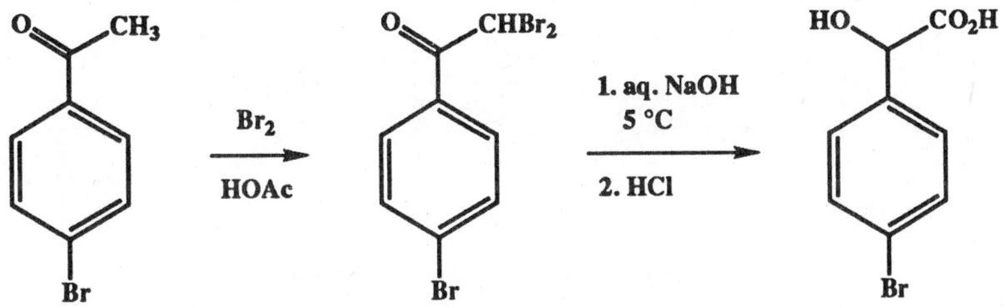
O
CH3
Br
Br2
HOAc
O
CHBr2
Br
1. aq. NaOH
5 °C
2. HCl
HO
CO2H
Br

OS, *CV 4*, 162 (1963)

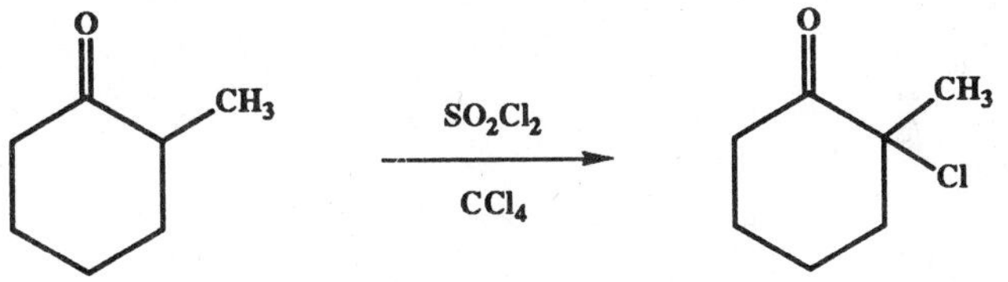
O
CH3
SO2Cl2
CCl4
O
CH3
Cl

OS, *CV 4*, 254 (1963)

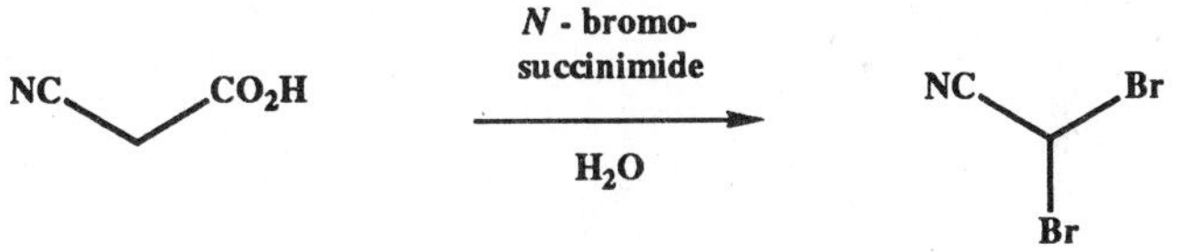
NC
CO2H
N - bromo-
succinimide
H2O
NC
Br
Br

OS, *CV 4*, 348 (1963)

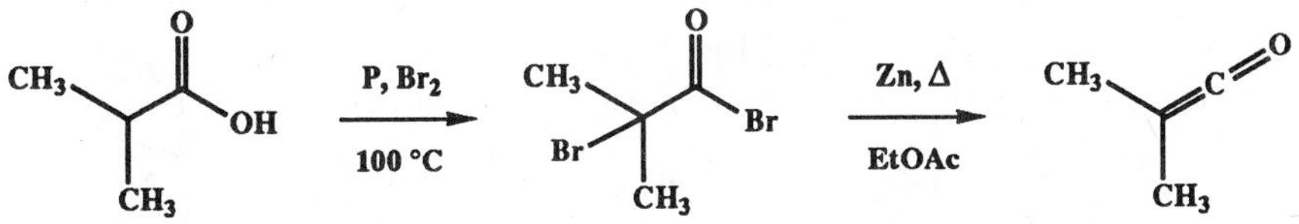
O
CH3
OH
CH3
P, Br2
100 °C
O
CH3
Br
Br
CH3
Zn, Δ
EtOAc
CH3
C
O
CH3

OS, CV 4, 398 (1963)

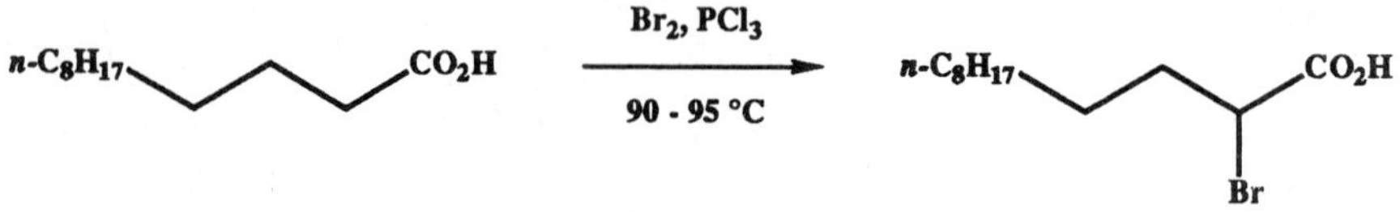

n-C_8H_{17} ... CO_2H → (Br_2, PCl_3; 90 - 95 °C) → n-C_8H_{17} ... CO_2H, Br

OS, CV 4, 545 (1963)

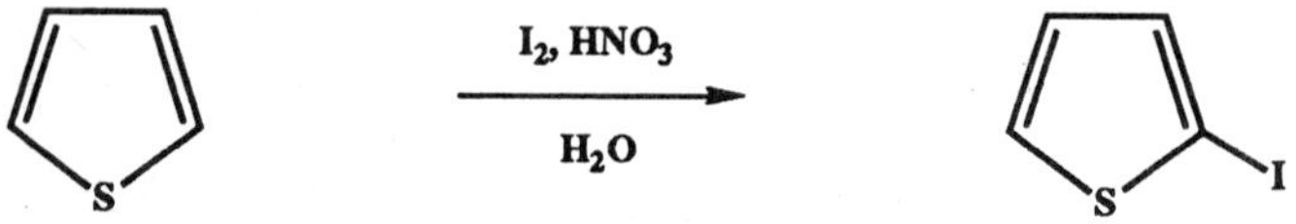

S → (I_2, HNO_3; H_2O) → S, I

OS, CV 4, 590 (1963)

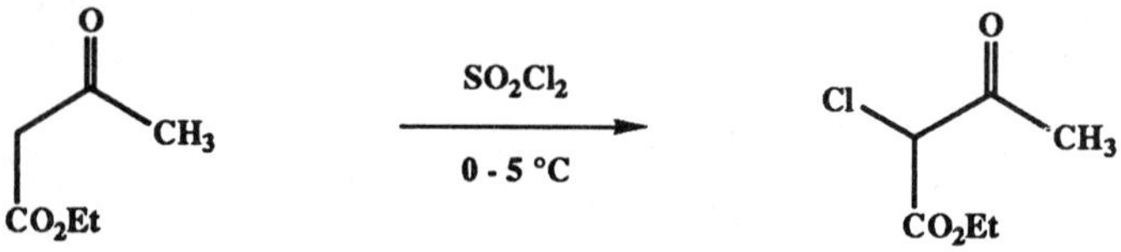

O, CH_3, CO_2Et → (SO_2Cl_2; 0 - 5 °C) → Cl, O, CH_3, CO_2Et

OS, CV 4, 608 (1963)

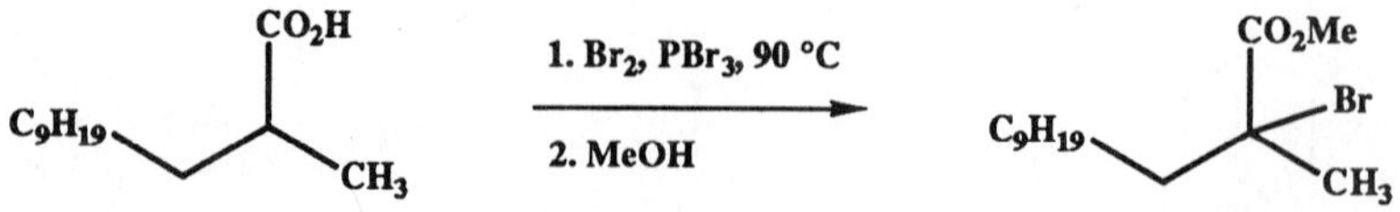

CO_2H, C_9H_{19}, CH_3 → (1. Br_2, PBr_3, 90 °C; 2. MeOH) → CO_2Me, Br, C_9H_{19}, CH_3

OS, CV 4, 616 (1963)

$C_9H_{19}CH_2CH(CH_3)CO_2H$ → (Br_2, PBr_3; 90 °C) → $C_9H_{19}CH_2CBr(CH_3)COBr$

OS, CV 4, 807 (1963)

o-xylene (ring with two CH_3) → (4 Br_2; light, Δ) → ring with two $CHBr_2$ → ($(CO_2K)_2$; aq. EtOH, Δ) → ring with two CHO

OS, CV 4, 921 (1963)

3-methylthiophene (CH_3, S) → (*N*-bromo-succinimide; benzoyl peroxide, benzene, Δ) → 3-thienyl CH_2Br (S)

OS, CV 4, 984 (1963)

o-xylene (ring with two CH_3) → (2 Br_2; light, 125 °C) → ring with two CH_2Br

OS, *CV 5*, 145 (1973)

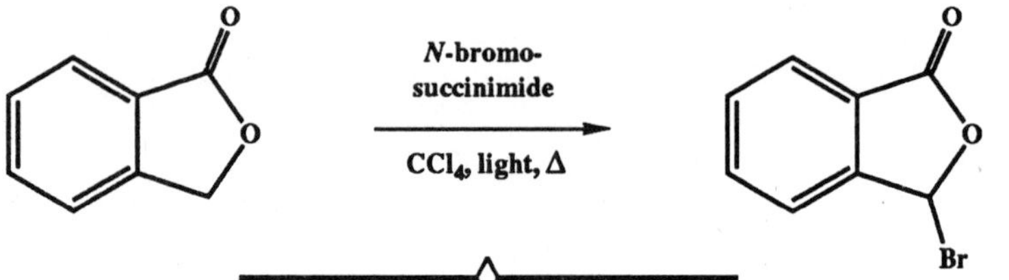

N-bromo-succinimide

CCl_4, light, Δ

Br

OS, *CV 5*, 221 (1973)

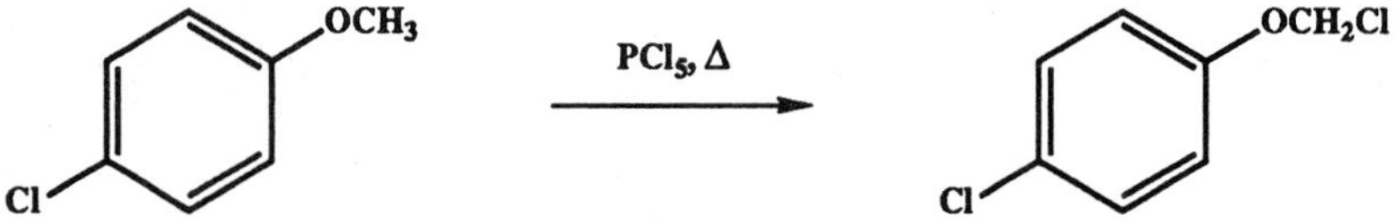

OCH_3 → PCl_5, Δ → OCH_2Cl

Cl

OS, *CV 5*, 255 (1973)

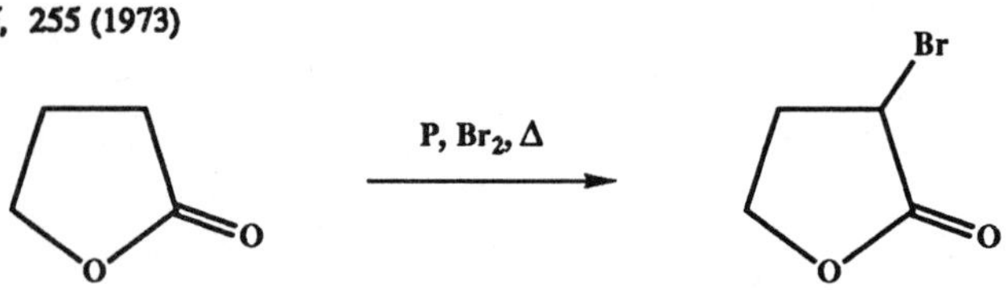

P, Br_2, Δ

Br

OS, *CV 5*, 328 (1973)

Ph Br

N-bromosuccinimide
benzoyl peroxide

CCl_4, Δ

Ph Br Br

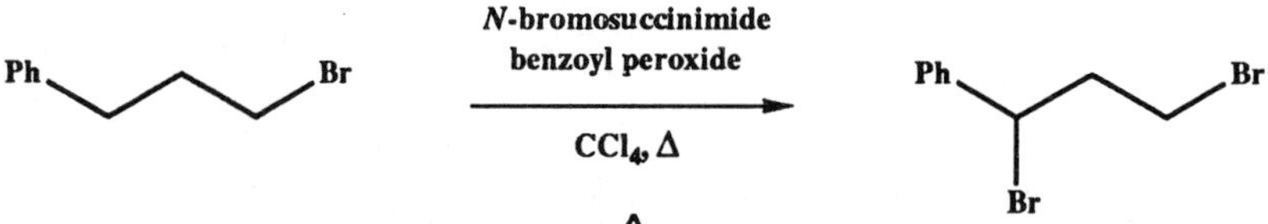

OS, *CV 5*, 514 (1973)

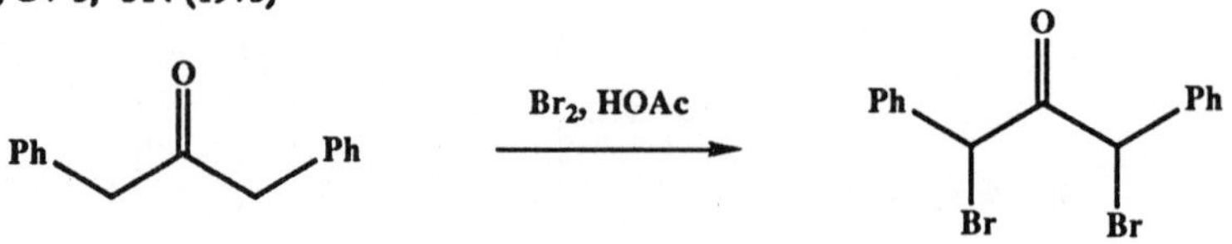

Ph Ph

Br_2, HOAc

Ph Ph Br Br

OS, *CV 5*, 825 (1973)

2-nitrotoluene (CH_3, NO_2) → N-bromosuccinimide / benzoyl peroxide, CCl_4, Δ → 2-nitrobenzyl bromide (CH_2Br, NO_2)

OS, *CV 5*, 921 (1973)

$$Ph-C\equiv C-H \xrightarrow{Br_2,\ aq.\ NaOH} Ph-C\equiv C-Br$$

OS, *CV 6*, 90 (1988)

$$PhC(=O)-NH(CH_2)_4-CH_2-C(=O)OH \xrightarrow[ClCH_2CH_2Cl,\ \Delta]{Cl_2,\ O_2,\ ClSO_3H} PhC(=O)-NH(CH_2)_4-CH(Cl)-C(=O)OH$$

OS, *CV 6*, 175 (1988)

6-methoxy-2-naphthyl (CH_3O) $COCH_3$ → "PTT", THF → CH_3O-naphthyl-$COCH_2Br$

6-methoxy-2-naphthyl (CH_3O) $COCH_3$ → 2 "PTT", THF → CH_3O-naphthyl-$COCHBr_2$

"PTT" = $PhN(Me)_3Br_3$

OS, CV 6, 184 (1988)

Ph(H)C=C(H)Ph → (N-bromosuccinimide; H_2O, DMSO) → Br, H, Ph, Ph, H, OH

OS, CV 6, 190 (1988)

$CH_3CH_2CH_2CH_2CH_2—CO_2H$ → (1. $SOCl_2$, CCl_4, 65 °C; 2. *N*-bromosuccinimide, HBr, 85 °C) → $CH_3CH_2CH_2CH_2CH(Br)—COCl$

OS, CV 6, 193 (1988)

$CH_3C(O)$-*i*-Pr → (Br_2, MeOH; 0 - 10 °C) → $BrCH_2C(O)$-*i*-Pr

OS, CV 6, 210 (1988)

2,5-di-*t*-Bu-quinone → (1. Cl_2, HOAc; 2. Et_2NH, Et_2O; 3. Cl_2, HOAc; 4. Et_2NH, Et_2O) → Cl, *t*-Bu, *t*-Bu, Cl quinone → (NaN_3; MeOH) → N_3, *t*-Bu, *t*-Bu, N_3 quinone

OS, CV 6, 271 (1988)

CO_2H / CO_2H → [SO_2Cl_2 / $(PhCO_2)_2$] → Cl— / CO_2H / CO_2H

OS, CV 6, 368 (1988)

O → [Br_2 / benzene, Et_2O] → Br / O / Br

OS, CV 6, 403 (1988)

$PhCH_2CO_2H$ → [Br_2, PCl_3 / benzene, Δ] → Ph—$CHCO_2H$ (Br)

OS, CV 6, 404 (1988)

$PhCH_2$—Cl → [Na_2S, Δ / aq. EtOH] → $PhCH_2$—S—CH_2Ph → [1. Br_2, CCl_4 light, Δ / 2. *m*-CPBA Et_2O] → PhCH—SO_2—CHPh (Br) (Br)

***OS*, *CV 6*, 427 (1988)**

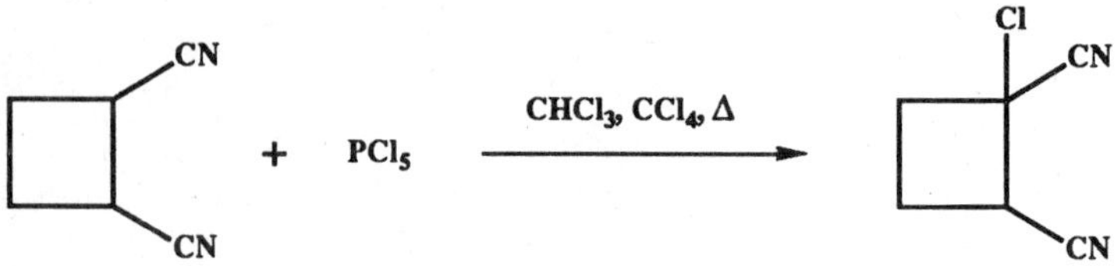

CN, CN + PCl_5 → ($CHCl_3$, CCl_4, Δ) Cl, CN, CN

***OS*, *CV 6*, 462 (1988)**

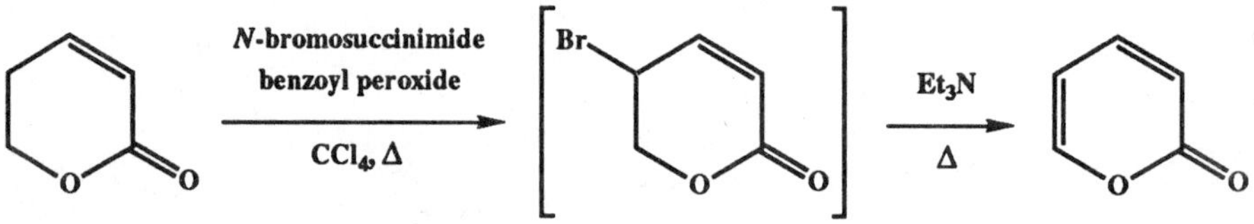

O, O → (*N*-bromosuccinimide, benzoyl peroxide, CCl_4, Δ) [Br, O, O] → (Et_3N, Δ) O, O

***OS*, *CV 6*, 512 (1988)**

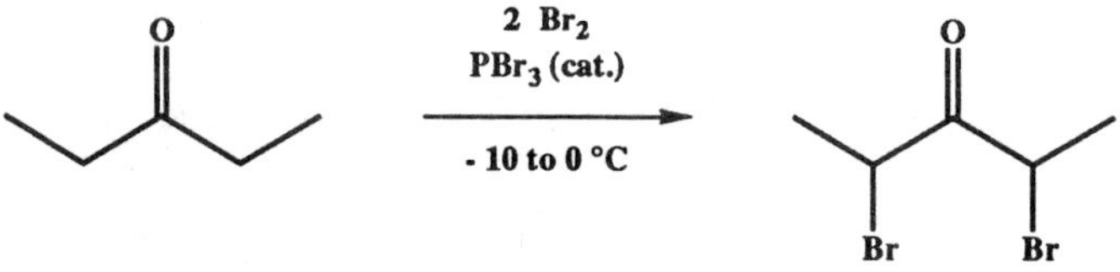

O → (2 Br_2, PBr_3 (cat.), - 10 to 0 °C) O, Br, Br

***OS*, *CV 6*, 520 (1988)**

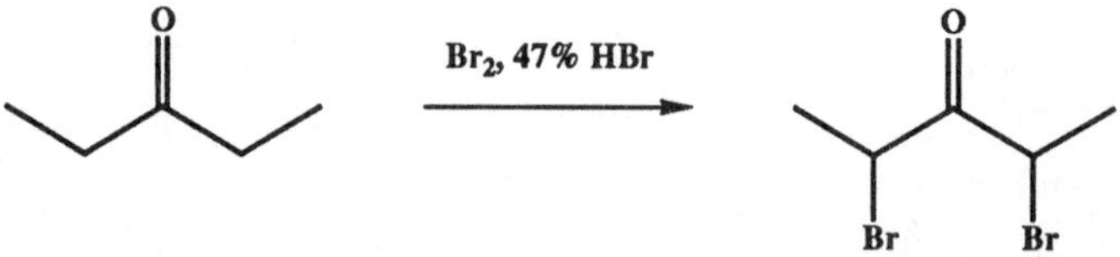

O → (Br_2, 47% HBr) O, Br, Br

OXIDATION - HALOGENATION

***OS, CV 6,* 560 (1988)**

OAc

N-bromosuccinimide

t-BuOH, H_2O, 12 - 25 °C

OAc

Br

HO

***OS, CV 6,* 711 (1988)**

O

CH_3 CH_3

2 Br_2, 5 °C

48% HBr

O

CH_3 Br

Br

***OS, CV 6,* 799 (1988)**

H H

H CO_2Me

I_2, N_2O_4

Et_2O, 0 °C

$O_2N—CH_2—CH—CO_2Me$

I

***OS, CV 6,* 893 (1988)**

Ph H

H $CH(OMe)_2$

ICl, NaN_3

CH_3CN

I

Ph $CH(OMe)_2$

N_3

OS, CV 6, 954 (1988)

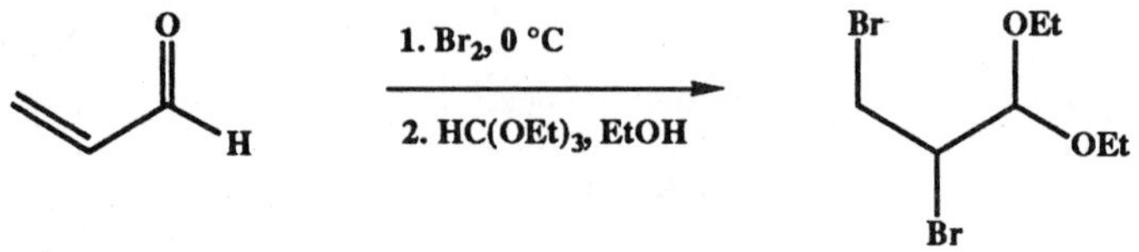

1. Br_2, 0 °C
2. $HC(OEt)_3$, EtOH

Br OEt OEt Br

OS, CV 6, 991 (1988)

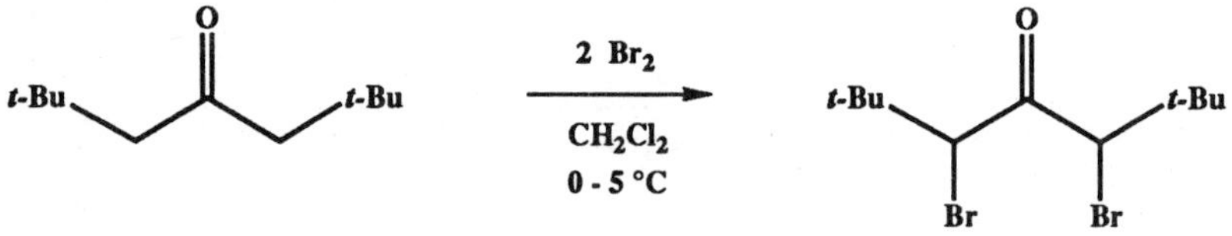

t-Bu *t*-Bu

2 Br_2

CH_2Cl_2

0 - 5 °C

t-Bu *t*-Bu Br Br

OS, CV 7, 271 (1990); *61,* 65 (1983)

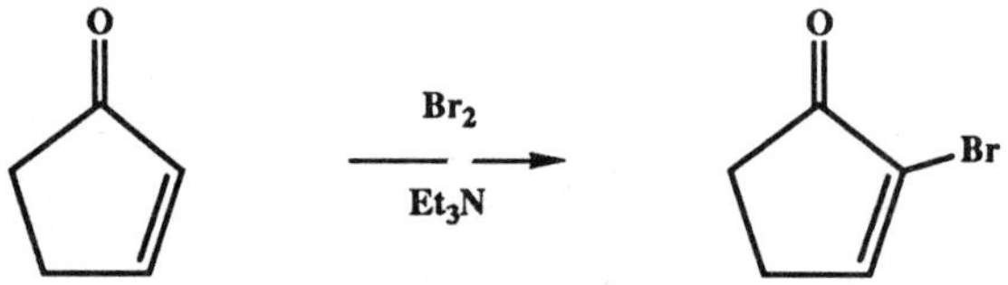

Br_2

Et_3N

Br

OS, CV 7, 491 (1990); *62,* 210 (1984)

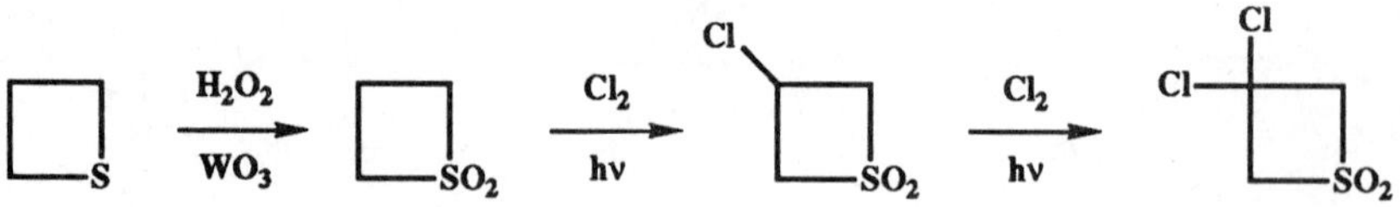

S H_2O_2 WO_3 SO_2 Cl_2 hν Cl SO_2 Cl_2 hν Cl Cl SO_2

OS, *65*, 243 (1987)

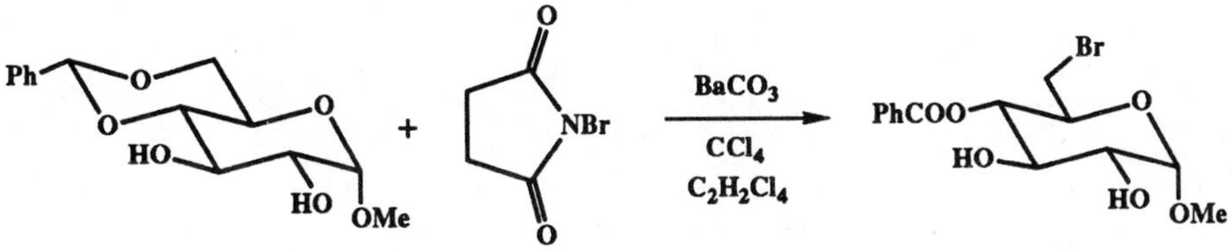

Ph O O O HO HO OMe + NBr O O → $BaCO_3$ / CCl_4 / $C_2H_2Cl_4$ → PhCOO Br O HO HO OMe

OS, *66*, 194 (1987)

O O O → 1. LDA THF, - 75 °C / 2. C_2Cl_6 THF, - 50 °C → Cl O O O

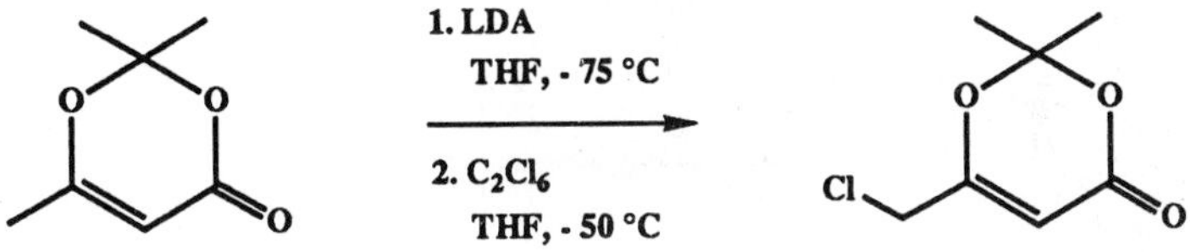

***OS*, *CV 1*, 138 (1941)**

6 *n*-BuOH → (2 $Na_2Cr_2O_7$; 8 H_2SO_4, H_2O) → 3 O—*n*-Bu

***OS*, *CV 1*, 211 (1941)**

OH, Cl, Cl → ($Na_2Cr_2O_7$; H_2SO_4, H_2O) → O, Cl, Cl

***OS*, *CV 1*, 340 (1941)**

OH → ($Na_2Cr_2O_7$ (*or* $K_2Cr_2O_7$); aq. H_2SO_4) → O

l - menthone

***OS*, *CV 1*, 383 (1941)**

NH_2 • HCl, OH → ($K_2Cr_2O_7$; H_2SO_4, H_2O, Δ) → O, O

OS, CV 1, 392 (1941)

$$\text{O}_2\text{N-C}_6\text{H}_4\text{-CH}_3 \xrightarrow[\text{aq. H}_2\text{SO}_4,\ \Delta]{\text{Na}_2\text{Cr}_2\text{O}_7} \text{O}_2\text{N-C}_6\text{H}_4\text{-COOH}$$

OS, CV 1, 482 (1941)

$$\text{HO-C}_6\text{H}_4\text{-OH} \xrightarrow[\text{H}_2\text{SO}_4,\ \text{H}_2\text{O}]{\text{Na}_2\text{Cr}_2\text{O}_7} \text{O=C}_6\text{H}_4\text{=O}$$

OS, CV 1, 543 (1941)

$$2,4,6\text{-(NO}_2)_3\text{C}_6\text{H}_2\text{CH}_3 \xrightarrow[\text{H}_2\text{SO}_4]{\text{Na}_2\text{Cr}_2\text{O}_7} 2,4,6\text{-(NO}_2)_3\text{C}_6\text{H}_2\text{CO}_2\text{H}$$

OS, CV 2, 139 (1943)

$$\text{HO-steroid-C}_8\text{H}_{17} \xrightarrow[\text{H}_2\text{O, benzene}]{\text{Na}_2\text{Cr}_2\text{O}_7,\ \text{H}_2\text{SO}_4,\ \text{HOAc}} \text{O=steroid-C}_8\text{H}_{17}$$

OS, CV 2, 336 (1943)

CrO_3

HOAc, H_2O

OS, CV 2, 441 (1943)

CH_3 O_2N

CrO_3, Ac_2O

H_2SO_4, HOAc

$CH(OAc)_2$ O_2N

OS, CV 2, 541 (1943)

CH_3 OH

$K_2Cr_2O_7$
H_2SO_4, H_2O

n-PrOH, Δ

CH_3 O H

OS, CV 3, 1 (1955)

$Na_2Cr_2O_7$

HOAc

OS, *CV 3*, 234 (1955)

AcO, Ph, Ph, AcO

1. CrO_3 HOAc
2. KOH
3. HCl

HO, O, OH, HO

OS, *CV 3*, 334 (1955)

CH_3, NH_2, O_2N

$K_2S_2O_8$ / H_2SO_4

CH_3, NO, O_2N

$K_2Cr_2O_7$ / H_2SO_4

CO_2H, NO_2, O_2N

OS, *CV 3*, 420 (1955)

CH_3, O

$Na_2Cr_2O_7$ / HOAc / Ac_2O, Δ

O, OH, O

OS, *CV 3*, 449 (1955)

$K_2Cr_2O_7$ / H_2SO_4, H_2O / 65 °C

CO_2H, CO_2H

OS, *CV* 3, 641 (1955)

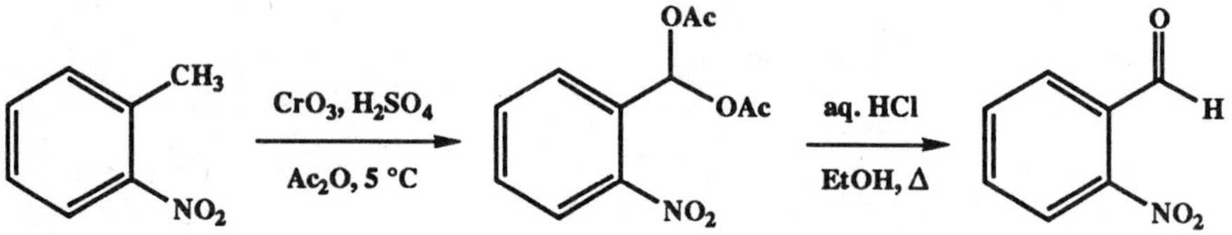

CH_3 NO_2

CrO_3, H_2SO_4
Ac_2O, 5 °C

OAc OAc NO_2

aq. HCl
EtOH, Δ

O H NO_2

OS, *CV* 4, 19 (1963)

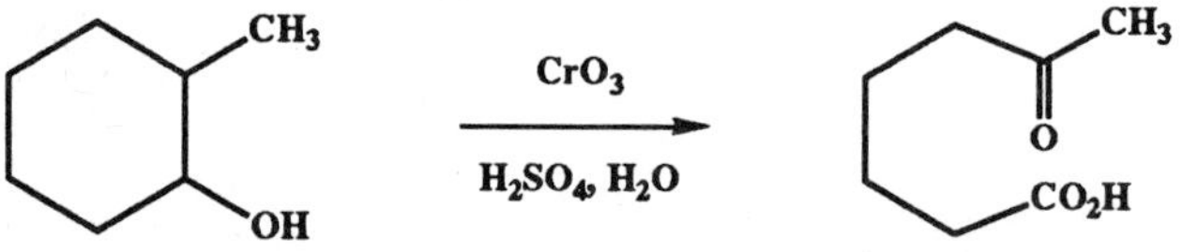

CH_3 OH

CrO_3
H_2SO_4, H_2O

CH_3 O CO_2H

OS, *CV* 4, 23 (1963)

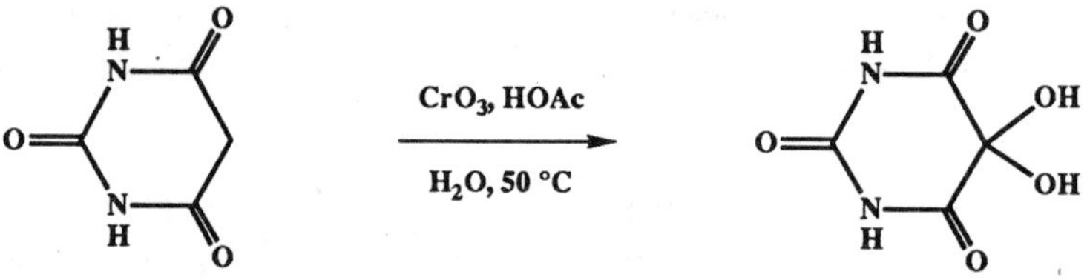

H N O O N H O

CrO_3, HOAc
H_2O, 50 °C

H N O OH O OH N H O

OS, *CV* 4, 148 (1963)

NH_2 Cl OH

$Na_2Cr_2O_7$
H_2SO_4, H_2O

O Cl O

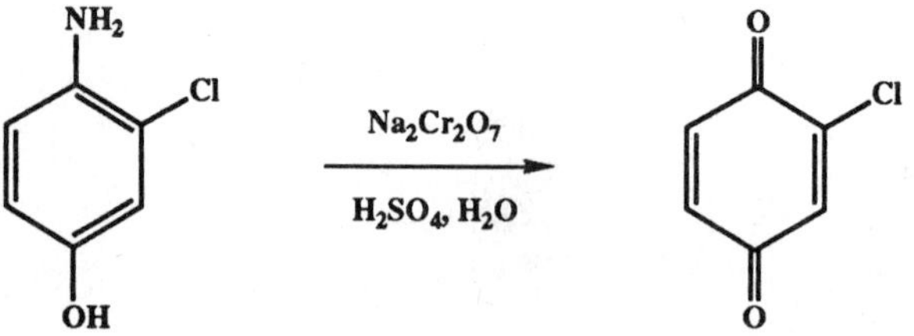

***OS*, *CV* 4,** **189 (1963)**

C_8H_{17}

HO

$Na_2Cr_2O_7$

HOAc
benzene

C_8H_{17}

O

O

***OS*, *CV* 4,** **195 (1963)**

C_8H_{17}

HO

Br

Br

$Na_2Cr_2O_7$

HOAc

C_8H_{17}

O

Br

Br

***OS*, *CV* 4,** **698 (1963)**

CrO_3

HOAc, H_2O

O

O

***OS*, *CV* 4,** **713 (1963)**

CH_3

O_2N

CrO_3, H_2SO_4

Ac_2O, 0 - 10 °C

$CH(OAc)_2$

O_2N

***OS*, *CV 4*, 757 (1963)**

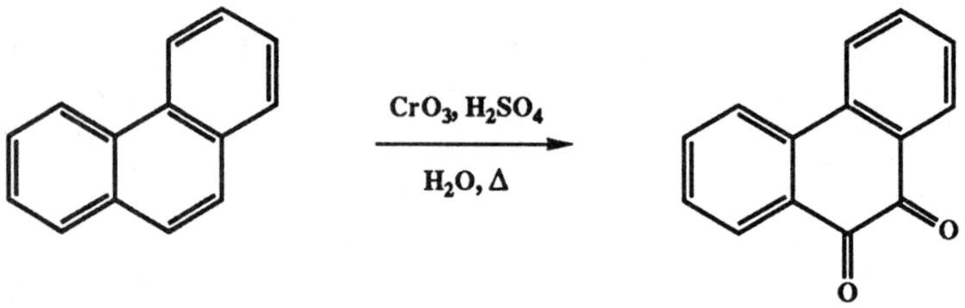

CrO_3, H_2SO_4

H_2O, Δ

O

O

***OS*, *CV 4*, 813 (1963)**

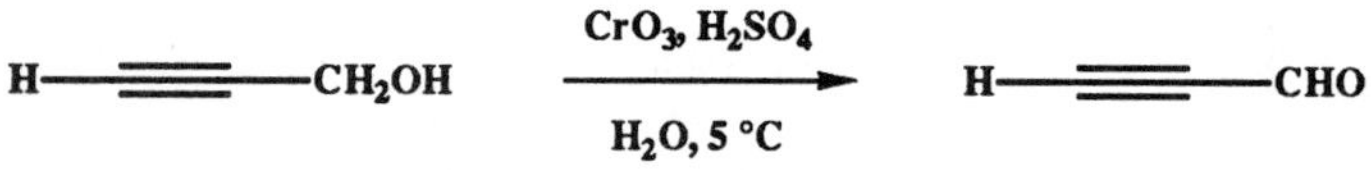

H—≡—CH_2OH → H—≡—CHO

CrO_3, H_2SO_4

H_2O, 5 °C

***OS*, *CV 5*, 310 (1973)**

OH

CrO_3, H_2SO_4

H_2O, acetone

O

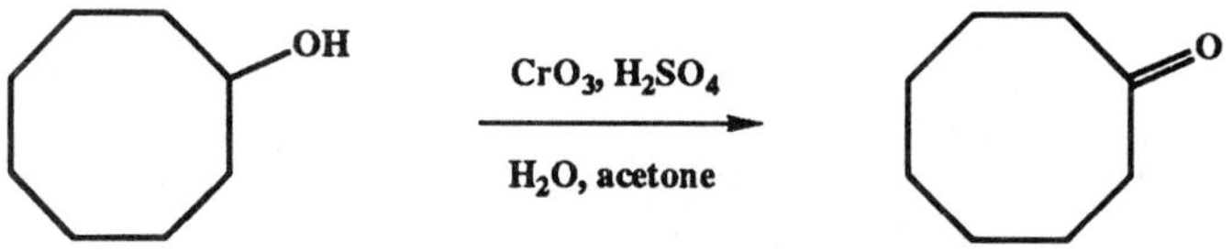

***OS*, *CV 5*, 324 (1973)**

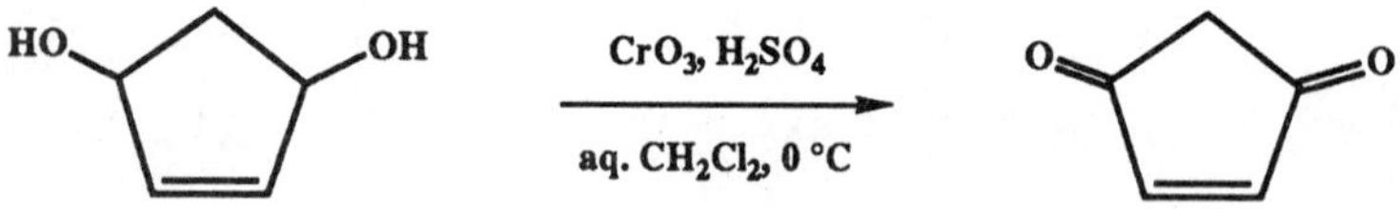

HO OH

CrO_3, H_2SO_4

aq. CH_2Cl_2, 0 °C

O O

OS, CV 5, **810 (1973)**

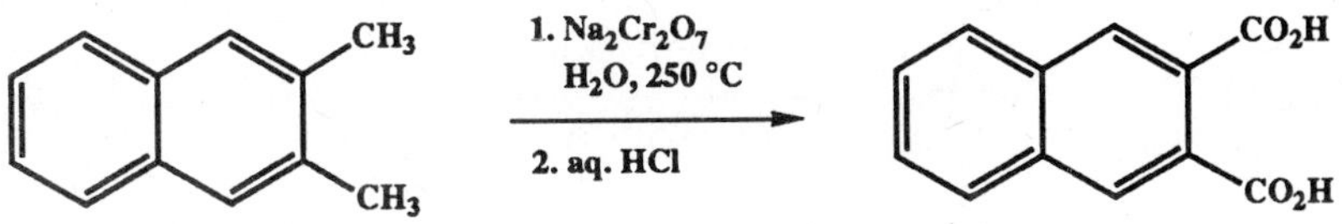
CH3
CH3
1. Na2Cr2O7
H2O, 250 °C
2. aq. HCl
CO2H
CO2H

OS, CV 5, **852 (1973)**

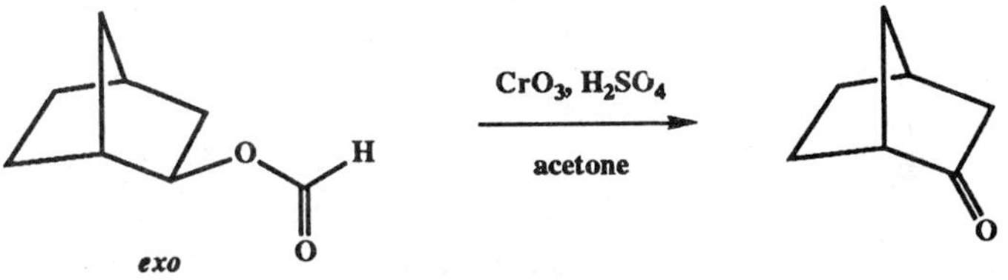
O
H
O
exo
CrO3, H2SO4
acetone
O

OS, CV 5, **866 (1973)**

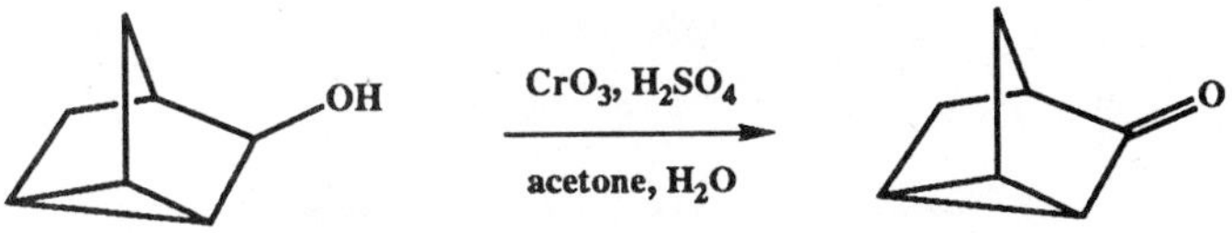
OH
CrO3, H2SO4
acetone, H2O
O

OS, CV 6, **373 (1988)**

$$CH_3(CH_2)_8{-}CH_2OH \xrightarrow[CH_2Cl_2]{CrO_3 \bullet (pyridine)_2} CH_3(CH_2)_8{-}CHO$$

***OS*, *CV 6*, 644 (1988)**

$CH_3(CH_2)_5—CH_2OH \xrightarrow[CH_2Cl_2,\ 25\ °C]{CrO_3\ (pyridine)_2} CH_3(CH_2)_5—CHO$

***OS*, *CV 6*, 1028 (1988)**

CH3 | t-Bu | 1. CrO_2Cl_2 | CH_2Cl_2, 0 - 5 °C | 2. Zn, H_2O | CH3 | t-Bu | O | H

***OS*, *CV 6*, 1033 (1988)**

CH3 | OH | SiMe3 | H_2CrO_4, H_2SO_4 | acetone, 0 °C | CH3 | O | SiMe3

***OS*, *CV 3*, 39 (1955)**

H | N | O | O | =CH—Ph | N | H | O | CrO_3, HOAc | H_2O, 50 - 60 °C | H | N | O | OH | O | OH | N | H | O

OS, *CV 7*, 114 (1990); *60*, 20 (1981)

OH

2 CrO_3
2 $(COOH)_2$
6 H^+

H_2O
10 - 15 °C

O

OS, *CV 7*, 177 (1990); *62*, 125 (1984)

OH

Cr^{+6}

O

OS, *65*, 81 (1987)

HO

R

R

$K_2Cr_2O_7$,
H_2SO_4

Et_2O - H_2O
25 °C

O

R

R

OS, *68*, 175 (1989)

HO

OH

CrO_3, H_2SO_4

acetone

O

O

OS, CV 1, 159 (1941)

ClHg–C_6H_4–CH_3 $\xrightarrow[\text{aq. NaOH, }\Delta\text{ then HCl}]{KMnO_4}$ ClHg–C_6H_4–CO_2H

OS, CV 1, 241 (1941)

Ph–CH(OH)–C(=O)OH $\xrightarrow[\text{aq. NaOH, - 3 °C}]{KMnO_4}$ Ph–C(=O)–C(=O)OH

OS, CV 2, 53 (1943)

$CH_3(CH_2)_5$–CH(OH)–CH_2–CH=CH–$(CH_2)_7CO_2H$ $\xrightarrow[\text{2. } H_2SO_4]{\text{1. } KMnO_4\text{, KOH, } H_2O}$ $HO_2C(CH_2)_7CO_2H$

OS, CV 2, 135 (1943)

2-Cl–C_6H_4–CH_3 $\xrightarrow[\text{2. HCl}]{\text{1. } KMnO_4\text{, } H_2O}$ 2-Cl–C_6H_4–CO_2H

OS, CV 2, 307 (1943)

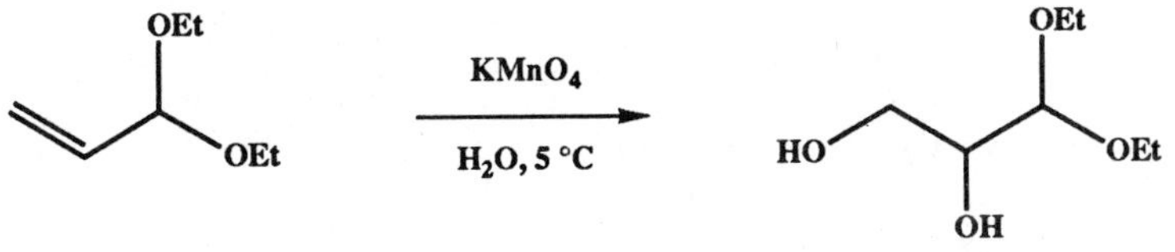

$KMnO_4$ / H_2O, 5 °C

OEt, OEt → HO, OH, OEt, OEt

dl - glyceraldehyde acetal

OS, CV 2, 315 (1943)

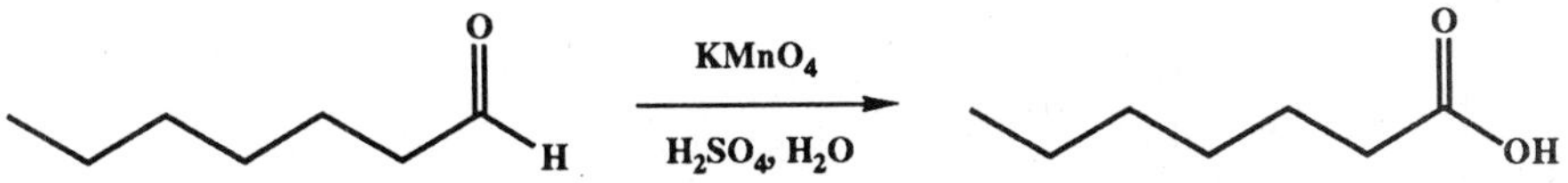

$KMnO_4$ / H_2SO_4, H_2O

O, H → O, OH

OS, CV 2, 523 (1943)

$KMnO_4$ / aq. NaOH, Δ → O, CO_2H, CO_2H

$NaHSO_3$ / HCl, H_2O → CHO, CO_2H

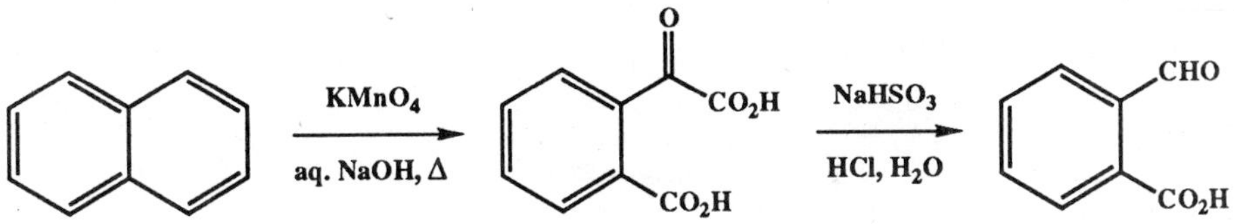

OS, CV 2, 538 (1943)

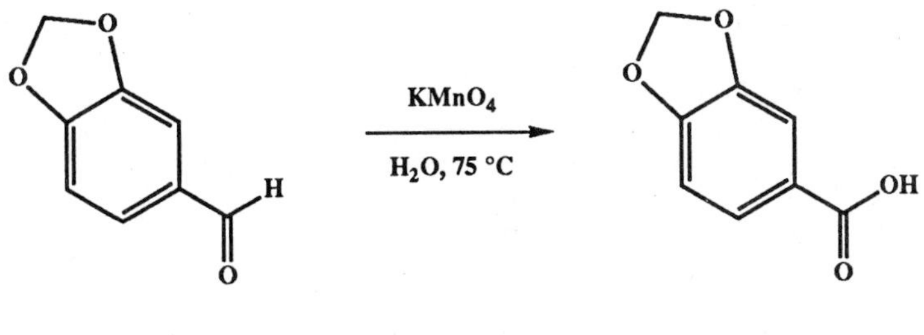

O, O, H, O → $KMnO_4$ / H_2O, 75 °C → O, O, OH, O

OS, CV 3, 740 (1955)

2-Methylpyridine (CH_3) $\xrightarrow[H_2O, \Delta]{KMnO_4}$ pyridine-2-carboxylic acid (CO_2H) $\xrightarrow{HCl}$ pyridinium-2-carboxylic acid chloride ($\overset{+}{N}$–H Cl^-, CO_2H)

OS, CV 3, 791 (1955)

4-Methylacetophenone ($COCH_3$, CH_3) $\xrightarrow[\text{2. } KMnO_4, NaOH, \Delta \text{ then } H_2SO_4]{\text{1. } HNO_3, H_2O, \Delta}$ terephthalic acid (CO_2H, CO_2H)

OS, CV 4, 467 (1963)

$CH_3CH(OH)CO_2Et$ $\xrightarrow[\text{pet. ether}]{KMnO_4 \text{, aq. } NaH_2PO_4}$ $CH_3C(O)CO_2Et$

OS, CV 4, 824 (1963)

Quinoxaline $\xrightarrow[\text{2. HCl}]{\text{1. aq. } KMnO_4, \Delta}$ pyrazine-2,3-dicarboxylic acid (CO_2H, CO_2H)

OS, *CV 5*, 393 (1973)

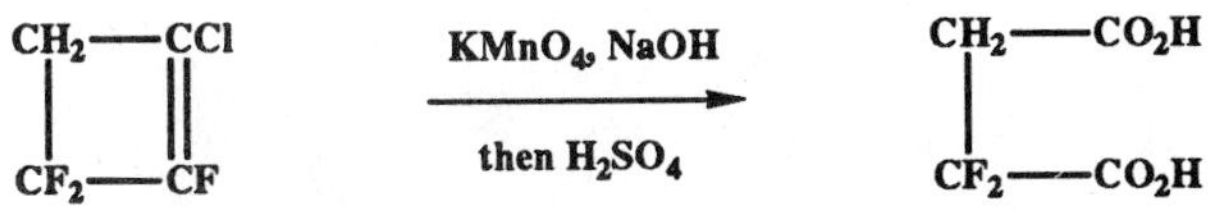

CH_2—CCl / CF_2—CF (cyclobutene) $\xrightarrow[\text{then } H_2SO_4]{KMnO_4,\ NaOH}$ CH_2—CO_2H / CF_2—CO_2H

OS, *CV 5*, 689 (1973)

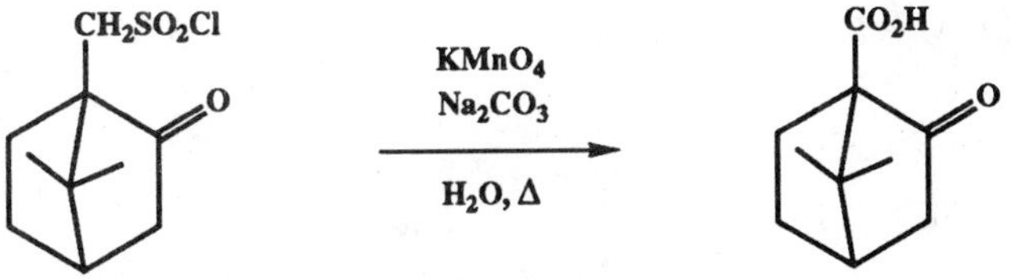

CH_2SO_2Cl, O $\xrightarrow[H_2O,\ \Delta]{KMnO_4,\ Na_2CO_3}$ CO_2H, O

OS, *CV 6*, 690 (1988)

OAc, O $\xrightarrow[KMnO_4]{NaIO_4}$ OH, O, HO, O

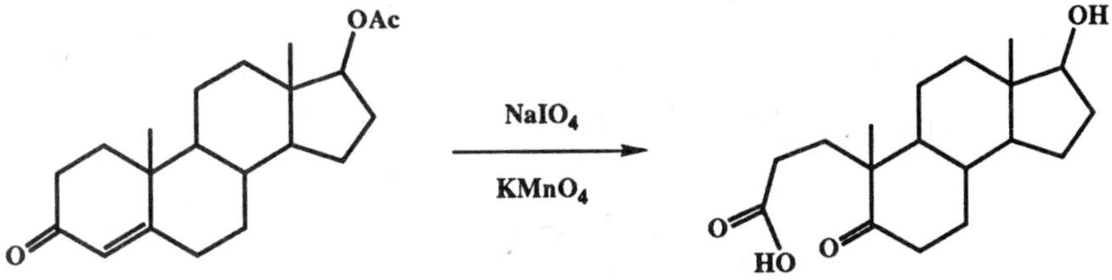

OS, *CV 7*, 397 (1990); *60*, 11 (1981)

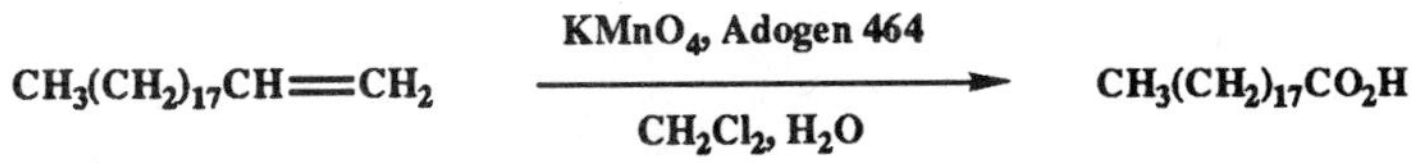

$$CH_3(CH_2)_{17}CH{=}CH_2 \xrightarrow[CH_2Cl_2,\ H_2O]{KMnO_4,\ \text{Adogen 464}} CH_3(CH_2)_{17}CO_2H$$

***OS*, *CV 7*, 102 (1990); *62*, 111 (1984)**

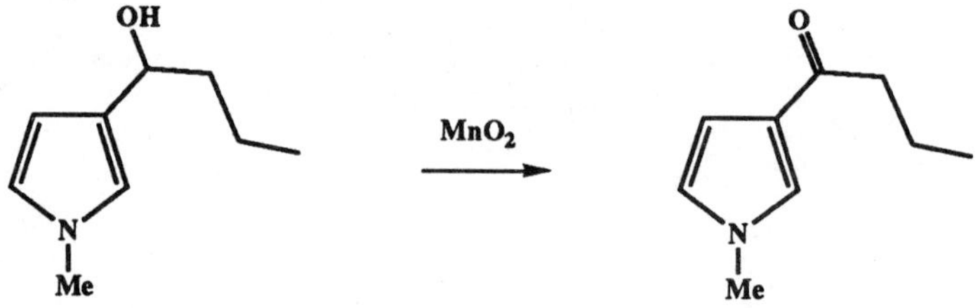

OH

MnO_2

O

N Me

N Me

***OS*, *68*, 109 (1989)**

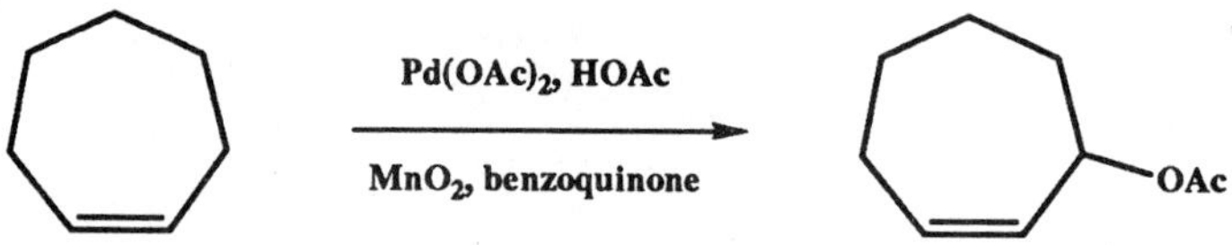

$Pd(OAc)_2$, HOAc

MnO_2, benzoquinone

OAc

OXIDATION - GENERAL

OS, CV 1, 10 (1941)

HO CO_2H

HO_2C CO_2H

fuming H_2SO_4

- 5 to 10 °C

O

HO_2C CO_2H

OS, CV 1, 18 (1941)

OH

50% HNO_3, Δ

NH_4VO_3 (cat.)

$HO_2C—(CH_2)_4—CO_2H$

OS, CV 1, 54 (1941)

3

OH

CO_2H

$CH_2{=}O$

4 $NaNO_2$

4 H_2SO_4

(dil. NH_4OH)

HO CO_2H

CO_2H

=O

HO CO_2H

(as ammonium salt)

OS, CV 1, 87 (1941)

OH

Ph Ph

O

$CuSO_4$, pyridine

H_2O, Δ

O

Ph Ph

O

OS, CV 1, 89 (1941)

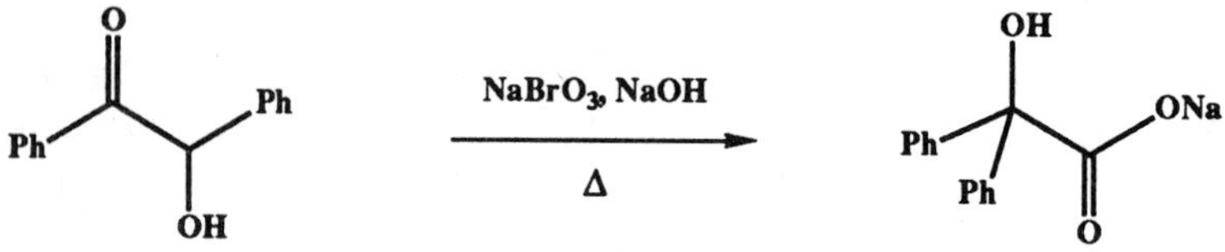
O
Ph
Ph
OH
NaBrO3, NaOH
Δ
OH
ONa
Ph
Ph
O

OS, CV 1, 104 (1941)

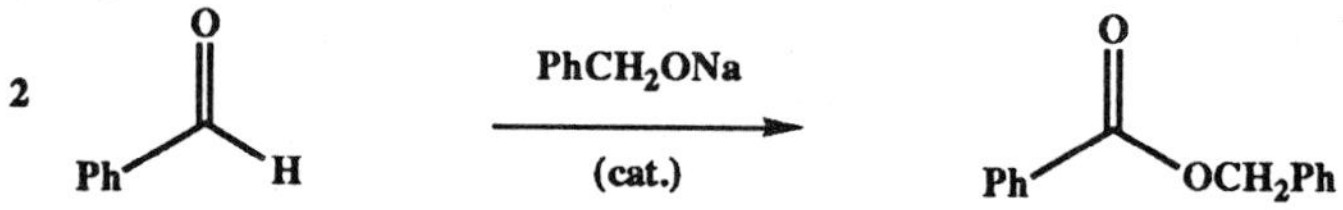
2
O
Ph
H
PhCH2ONa
(cat.)
O
Ph
OCH2Ph

OS, CV 1, 149 (1941)

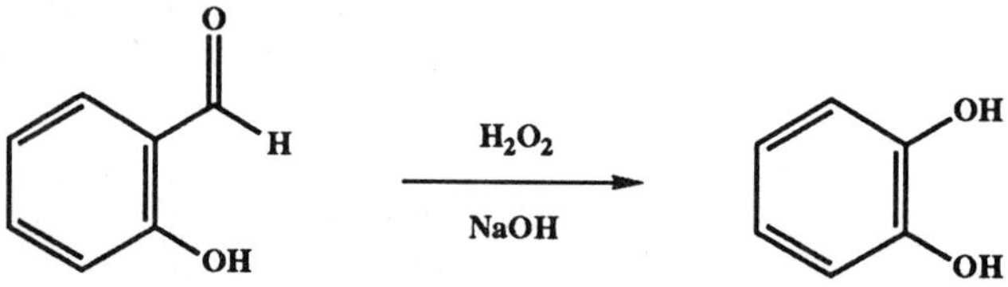
O
H
OH
H2O2
NaOH
OH
OH

OS, CV 1, 158 (1941)

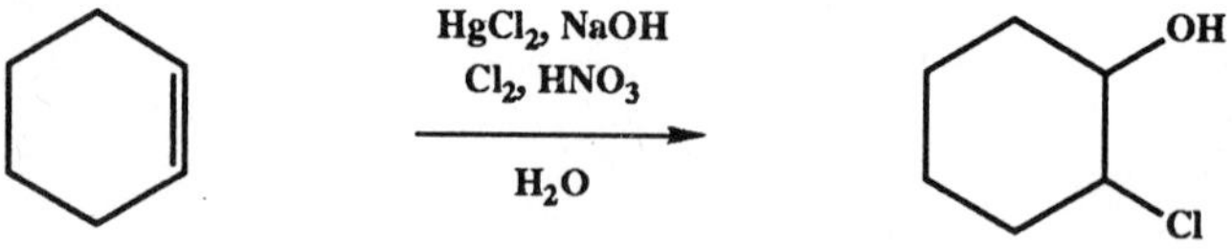
HgCl2, NaOH
Cl2, HNO3
H2O
OH
Cl

OS, CV 1, **166 (1941)**

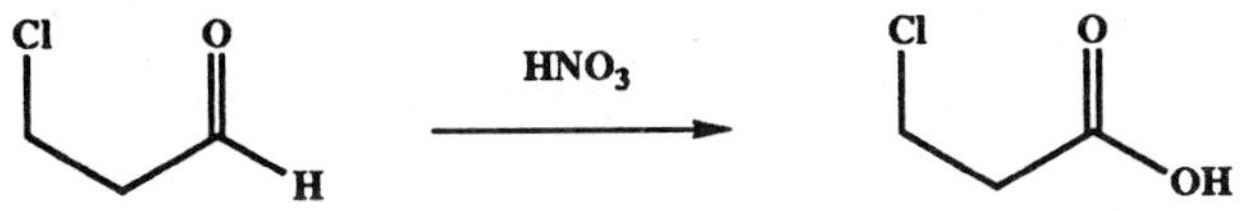

HNO_3

OS, CV 1, **168 (1941)**

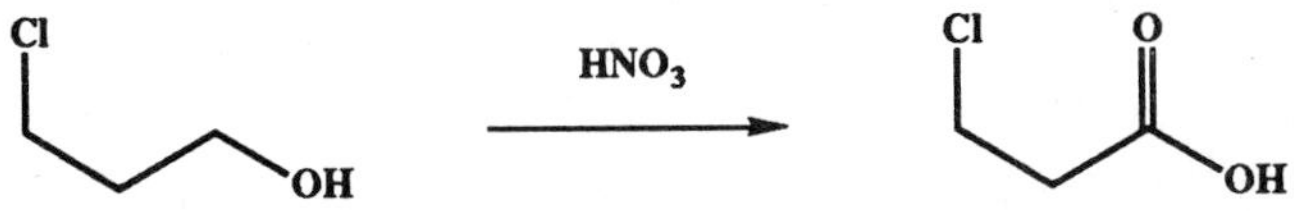

HNO_3

OS, CV 1, **266 (1941)**

2 N_2O_3

(As_2O_3, HNO_3)

0 °C

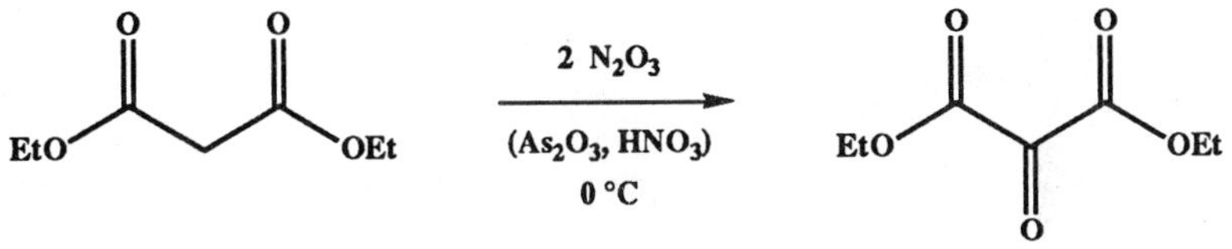

OS, CV 1, **276 (1941)**

1. 33% NaOH

2. H_2SO_4

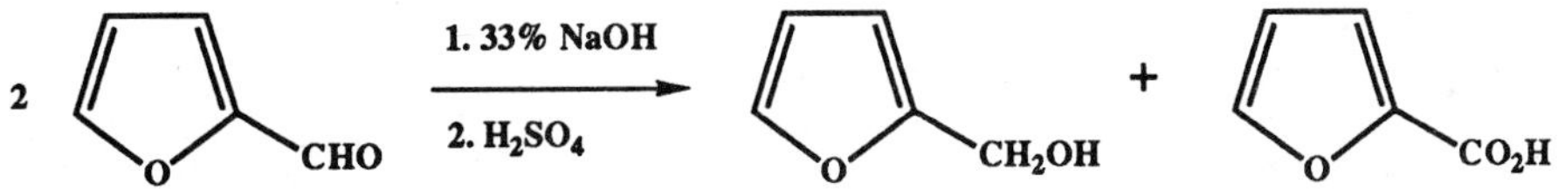

OS, *CV 1*, 385 (1941)

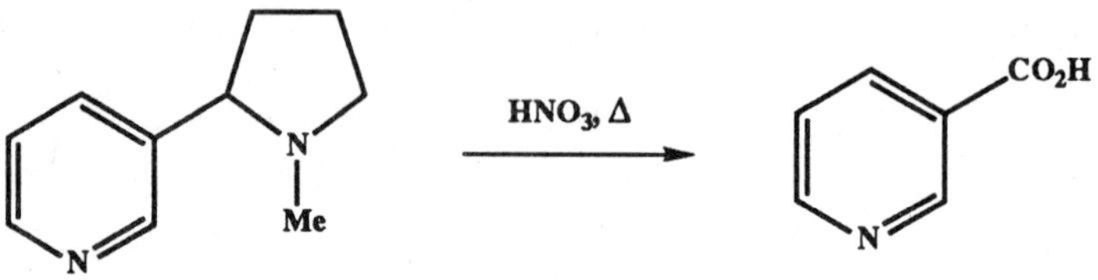

Me

HNO_3, Δ

CO_2H

OS, *CV 1*, 390 (1941)

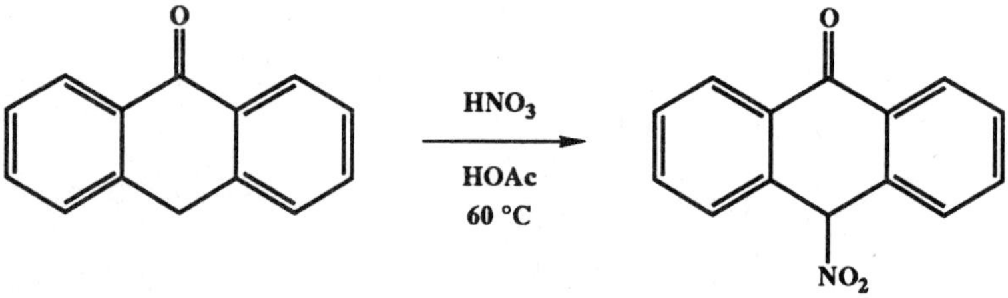

HNO_3

HOAc
60 °C

NO_2

OS, *CV 1*, 511 (1941)

Me
NH_2
HO

$NaNO_2$

aq. H_2SO_4, Δ

Me

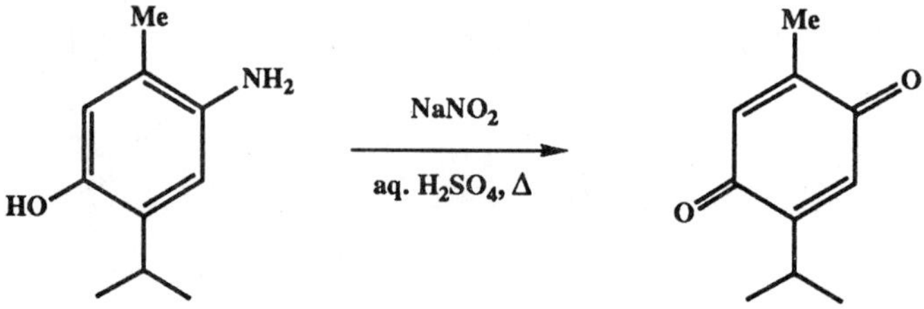

OS, *CV 1*, 526 (1941)

t-Bu
CH_3

1. 4 NaOH, 3 Br_2
2. H_2SO_4

t-Bu
OH

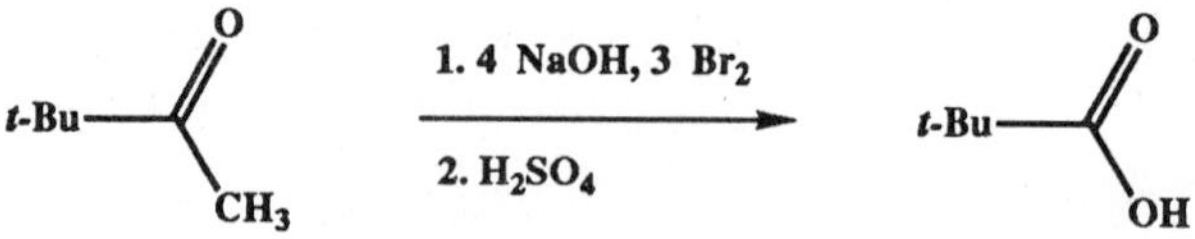

OS, *CV 2*, 175 (1943)

Br, OH, Br, NO_2 — Sn, HCl / H_2O, Δ → [Br, OH, Br, NH_3]$_2$ $SnCl_6$ — NaOH / Cl_2 → Br, O, Br, NCl

OS, *CV 2*, 202 (1943)

Me, O, O, OEt — $NaNO_2$ / HOAc, H_2O → Me, O, O, OEt, NO

OS, *CV 2*, 204 (1943)

CH_3, O, CH_3 — EtONO, HCl → CH_3, O, CH_3, NOH

OS, *CV 2*, 214 (1943)

EtO_2C, CO_2Et, Me, N H, Me — HNO_3 / H_2SO_4, Δ → EtO_2C, CO_2Et, Me, N, Me

***OS, CV 2,* 254 (1943)**

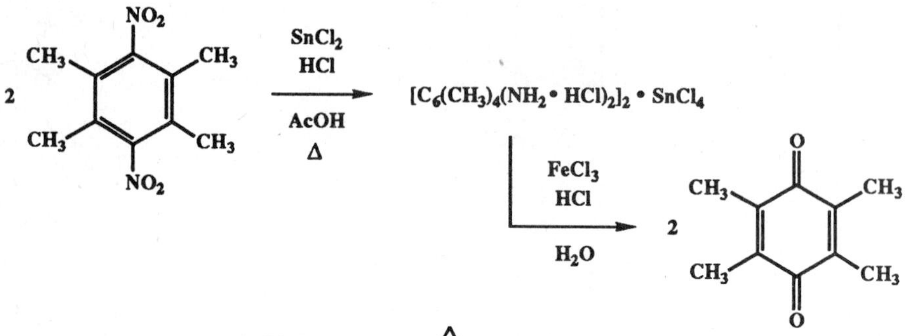

2 (1,4-dinitro-2,3,5,6-tetramethylbenzene) → $SnCl_2$, HCl, AcOH, Δ → $[C_6(CH_3)_4(NH_2 \cdot HCl)_2]_2 \cdot SnCl_4$ → $FeCl_3$, HCl, H_2O → 2 (2,3,5,6-tetramethyl-1,4-benzoquinone)

***OS, CV 2,* 302 (1943)**

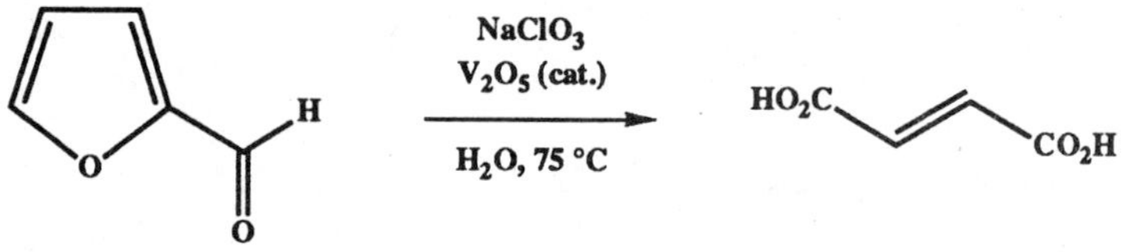

Furfural → $NaClO_3$, V_2O_5 (cat.), H_2O, 75 °C → HO_2C–CH=CH–CO_2H

***OS, CV 2,* 363 (1943)**

Ph–CO–CH_2–CH_3 → CH_3ONO, HCl, Et_2O → Ph–CO–C(=NOH)–CH_3

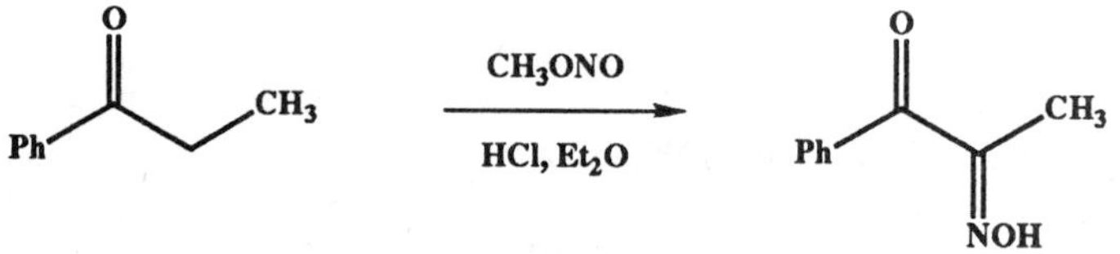

***OS, CV 2,* 419 (1943)**

N-Methylpyridinium $^-SO_4Me$ → $K_3Fe(CN)_6$, NaOH, H_2O → N-Me-2-pyridone

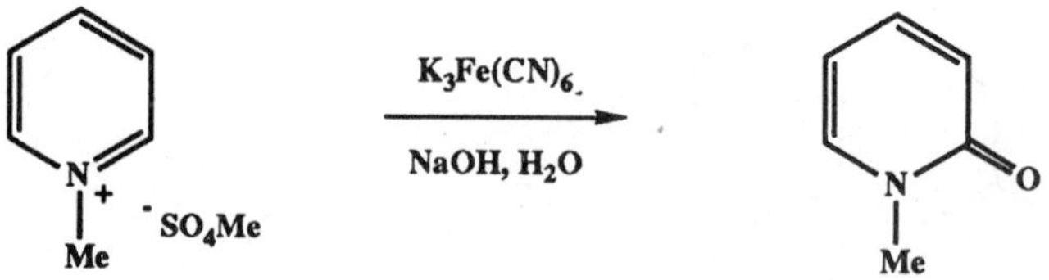

OS, CV 2, 423 (1943)

S_8

250 °C

OS, CV 2, 430 (1943)

NH_3Cl

OH

$FeCl_3$

HCl, H_2O

O

O

OS, CV 2, 440 (1943)

O

NH

O

N
H

O

fuming
HNO_3

O_2N

O

NH

O

N
H

O

OS, CV 2, 466 (1943)

S

HNO_3

Ac_2O, HOAc

S

NO_2

OS, CV 2, 509 (1943)

SeO_2, H_2O

dioxane, Δ

OS, CV 2, 512 (1943)

NO_2Na

CN

CH_3ONO_2

Na, EtOH

1. NaOH H_2O, Δ

2. HCl

NO_2

OS, CV 2, 553 (1943)

OH

OH

$NaClO_3$

V_2O_5

H_2SO_4, H_2O

O

O

OS, CV 3, 3 (1955)

Pb_3O_4

HOAc

60 - 70 °C

OAc

OS, CV 3, 37 (1955)

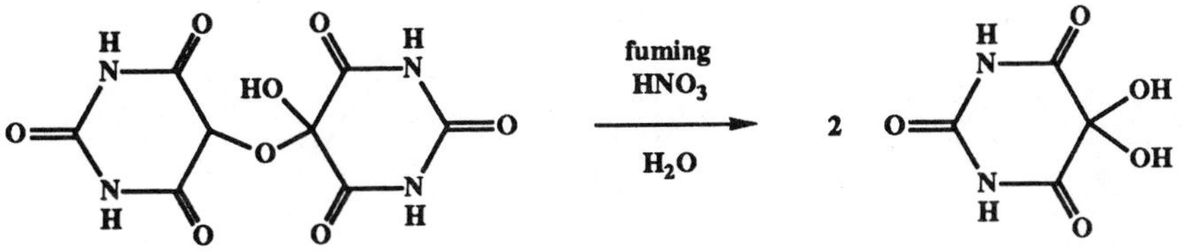

H N O O H N HO O N H O O N H O O

fuming HNO_3

H_2O

2 O H N O OH OH N H O

OS, CV 3, 191 (1955)

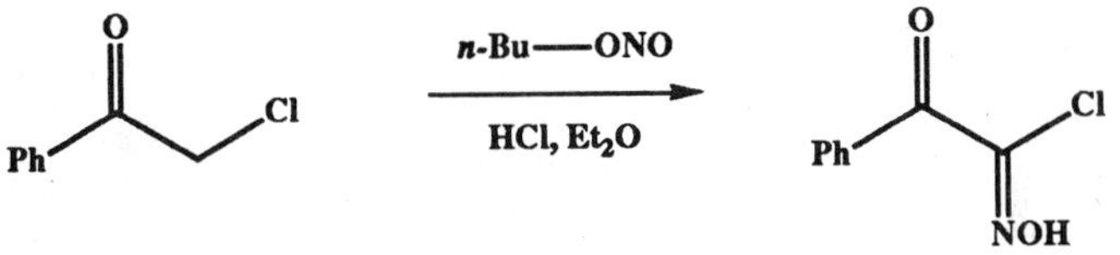

O Ph Cl

n-Bu—ONO

HCl, Et_2O

O Ph Cl NOH

OS, CV 3, 207 (1955)

C_8H_{17} HO

$(t\text{-BuO})_3Al$

acetone

benzene, Δ

C_8H_{17} O

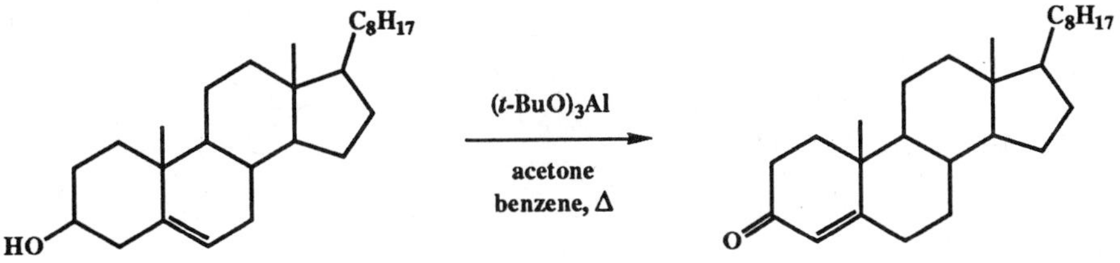

OS, CV 3, 217 (1955)

H_2O_2

HCO_2H

H_2O

OH OH

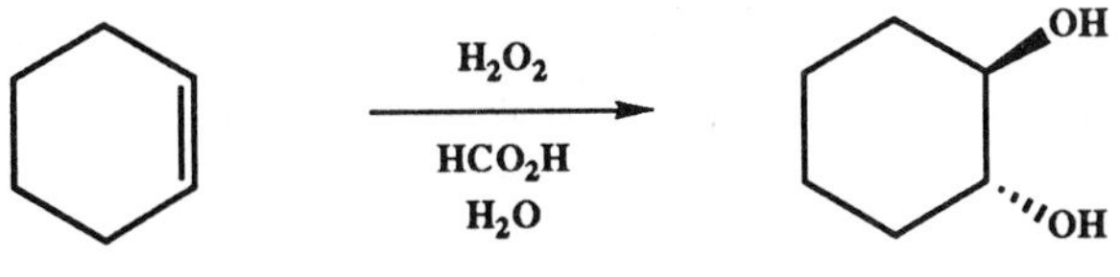

OS, CV 3, 302 (1955)

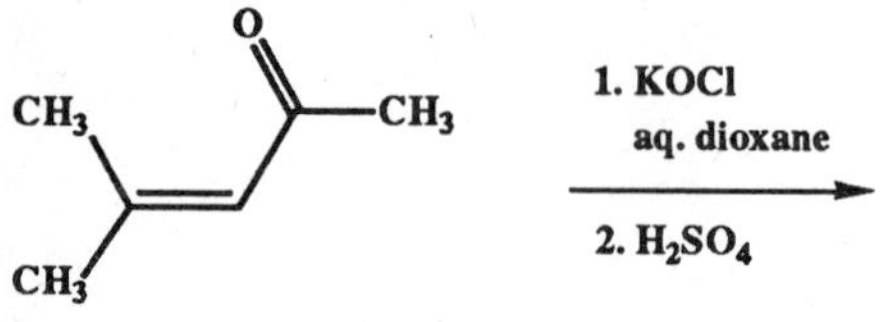

CH_3, CH_3, O, CH_3 — 1. KOCl, aq. dioxane; 2. H_2SO_4 → CH_3, CH_3, O, OH

OS, CV 3, 310 (1955)

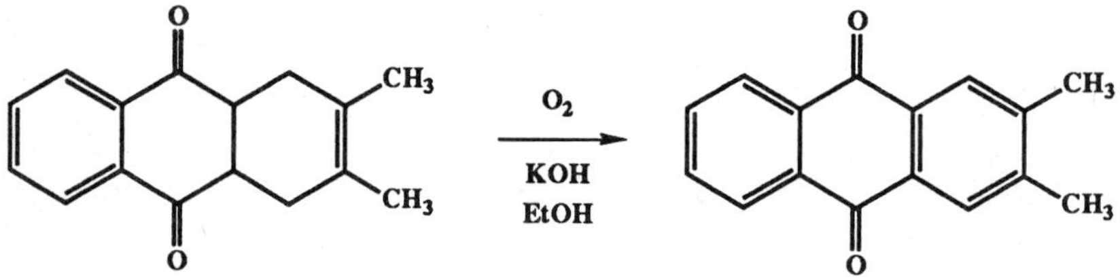

O, O, CH_3, CH_3 — O_2, KOH, EtOH → O, O, CH_3, CH_3

OS, CV 3, 358 (1955)

Ph, Ph, N H — Se, Δ → Ph, Ph, N H

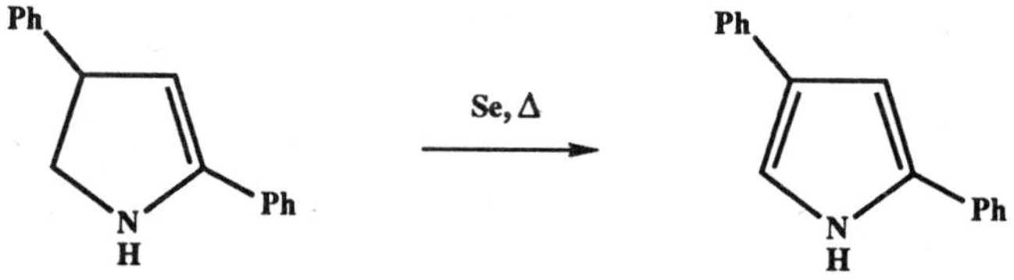

OS, CV 3, 392 (1955)

HCl • H_2N CO_2Et — $NaNO_2$, NaOAc; H_2O, Et_2O, 0 °C → N_2, H, CO_2Et

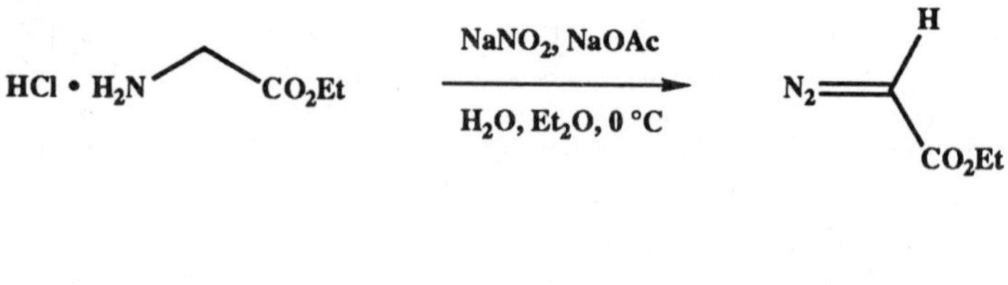

OS, *CV* 3, 438 (1955)

$(CH_3CHO)_3$ → (3 H_2SeO_3; aq. HOAc, dioxane, Δ) → 3 OHC–CHO → ($NaHSO_3$; H_2O, EtOH) → 3 OHC–CHO • 2 $NaHSO_3$ • H_2O

OS, *CV* 3, 619 (1955)

1. H_2O_2, NaOH, - 10 °C
2. H_2SO_4, H_2O, - 10 °C

O–OH
CO_2H

OS, *CV* 3, 633 (1955)

NH_2, OH, SO_3H → (HNO_3; H_2O) → SO_3H

(converted to K^+ and NH_4^+ salts)

OS, *CV* 3, 649 (1955)

2 O_2N–C_6H_4–COCl → (Na_2O_2; toluene, 0 °C) → O_2N–C_6H_4–CO–O–O–CO–C_6H_4–NO_2

OS, CV 3, 729 (1955)

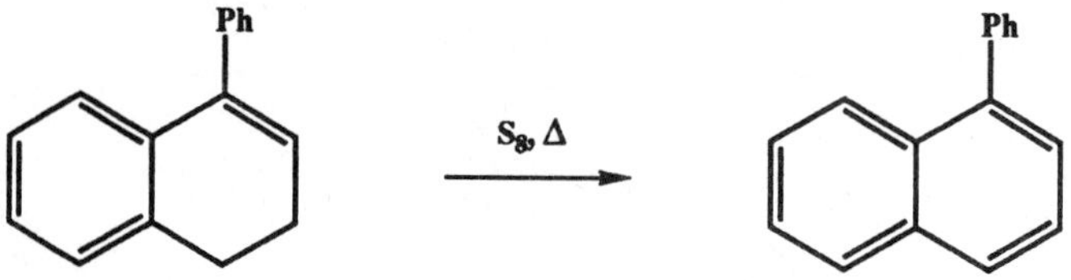

Ph

S_8, Δ

Ph

OS, CV 3, 745 (1955)

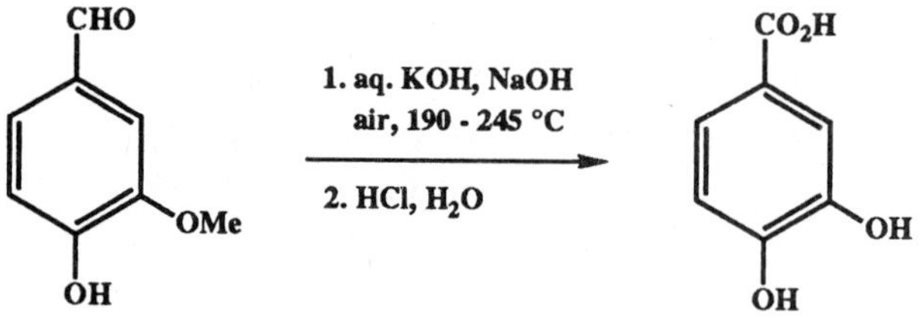

CHO

OMe

OH

1. aq. KOH, NaOH
air, 190 - 245 °C

2. HCl, H_2O

CO_2H

OH

OH

OS, CV 3, 759 (1955)

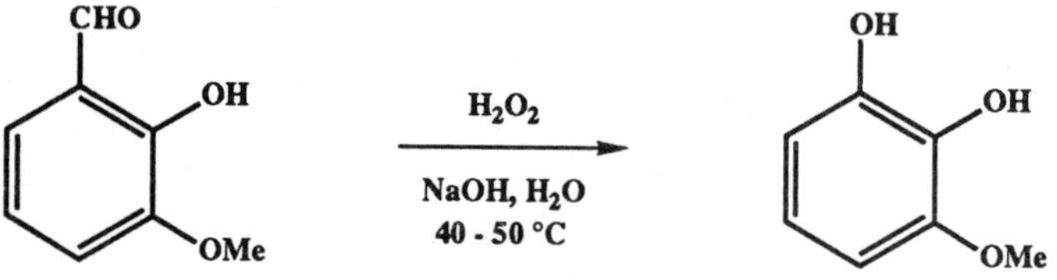

CHO

OH

OMe

H_2O_2

NaOH, H_2O
40 - 50 °C

OH

OH

OMe

OS, CV 3, 803 (1955)

$$4\ (CH_3CO)_2O\ +\ 4\ HNO_3 \longrightarrow C(NO_2)_4\ +\ 7\ CH_3CO_2H\ +\ CO_2$$

OS, CV 3, 807 (1955)

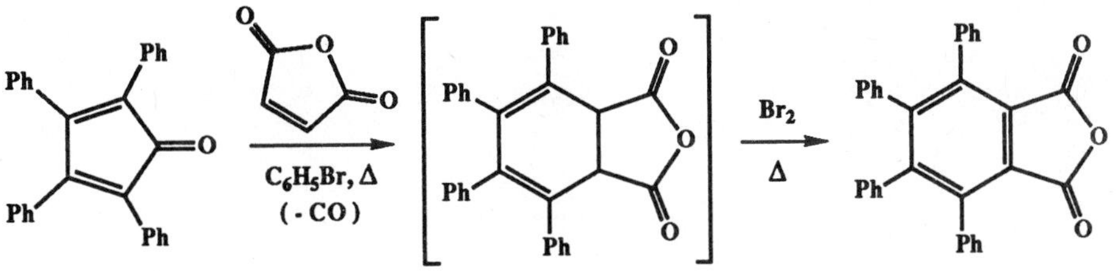

C_6H_5Br, Δ
(- CO)

Br_2
Δ

OS, CV 3, 811 (1955)

Δ
$CHCl_3$

H_2O, Δ

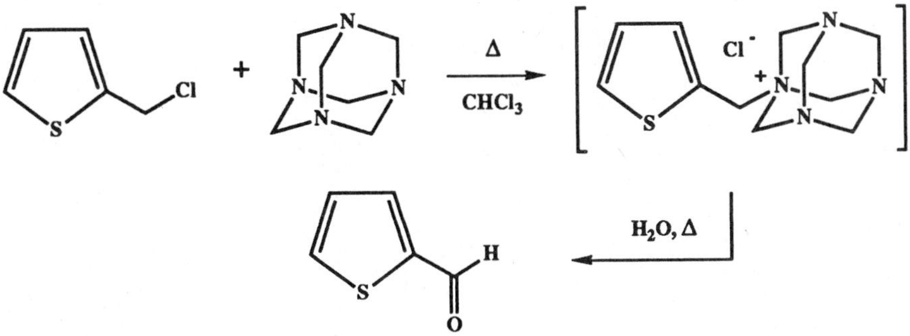

OS, CV 3, 820 (1955)

HNO_3
H_2O, Δ

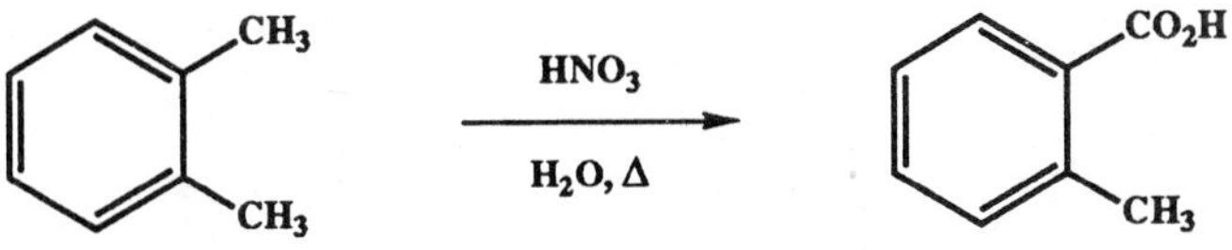

OS, CV 3, 822 (1955)

CH_3–C$_6$H$_4$–$CH(CH_3)_2$ → (HNO_3; H_2O, Δ) → CH_3–C$_6$H$_4$–CO_2H

OS, CV 4, 31 (1963)

O_2N–C$_6$H$_4$–CH_3 → (Na_2S, S_8; aq. NaOH, EtOH, Δ) → H_2N–C$_6$H$_4$–CHO

OS, CV 4, 124 (1963)

t-Bu-O_2C–CH(OH)–CH(OH)–CO_2-t-Bu (*d* isomer) → ($Pb(OAc)_4$; benzene) → 2 H–C(=O)–CO_2-t-Bu

OS, CV 4, 136 (1963)

2-naphthol (OH) → (40% CH_3CO_3H; HOAc) → 2-(CO_2H)–C$_6$H$_4$–CH=CH–CO_2H

OS, CV 4, 192 (1963)

C_8H_{17} HO cyclohexanone $Al(O\text{-}i\text{-}Pr)_3$ toluene C_8H_{17} O

OS, CV 4, 229 (1963)

O H_2SeO_3 *or* SeO_2 dioxane, H_2O O O

OS, CV 4, 345 (1963)

CH_3 CH_3 O O 1. NaOH, Cl_2 KOH, H_2O 2. HCl CH_3 CH_3 CO_2H CO_2H

OS, CV 4, 424 (1963)

• HCl H_2N O OEt $NaNO_2$, H_2SO_4 H_2O, CH_2Cl_2 N_2 O OEt

***OS*, *CV* 4, 484 (1963)**

O_3, DMF

aq. HOAc

CO_2H CHO

***OS*, *CV* 4, 493 (1963)**

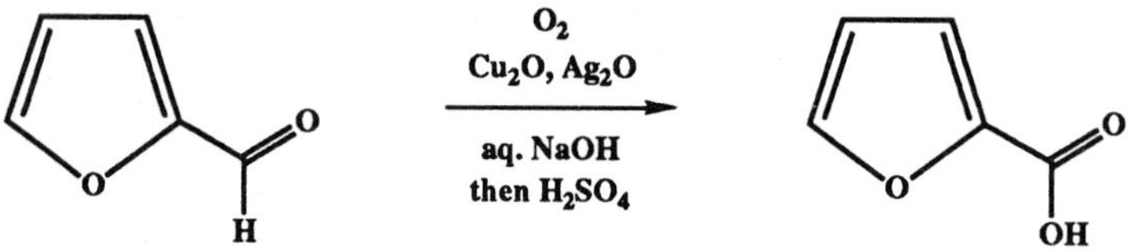

O_2
Cu_2O, Ag_2O

aq. NaOH
then H_2SO_4

***OS*, *CV* 4, 499 (1963)**

CHO
OH

$NaNO_2$, HNO_3

0 - 10 °C

HO_2C CO_2H

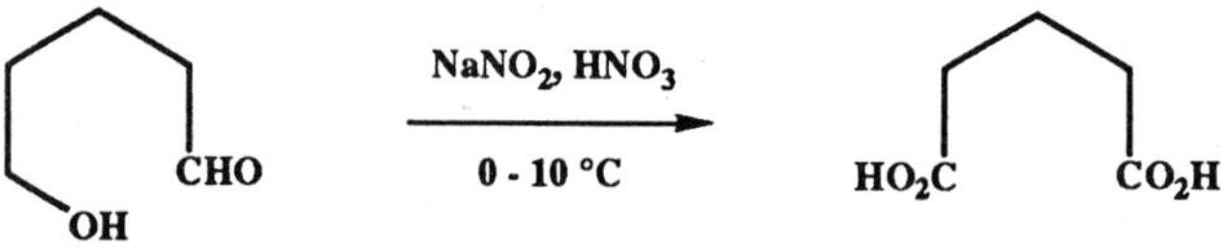

***OS*, *CV* 4, 536 (1963)**

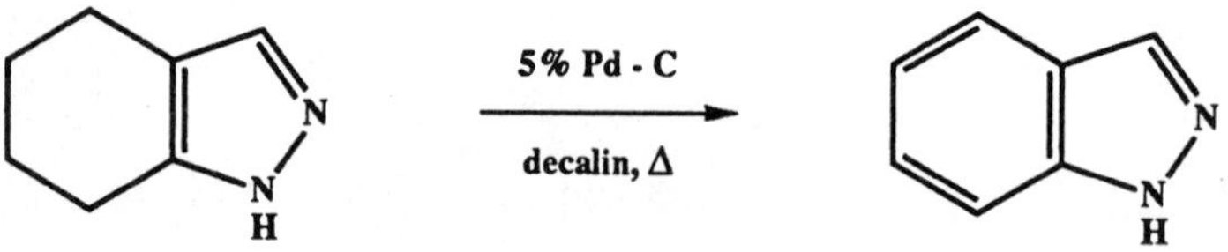

5% Pd - C

decalin, Δ

OS, *CV* 4, 579 (1963)

MeO_2C–C_6H_4–CH_2CH_3 → (air, Cr_2O_3, $CaCO_3$; 140 - 150 °C) → MeO_2C–C_6H_4–$C(O)CH_3$

OS, *CV* 4, 677 (1963)

3-methyl-1,5-pentanediol (CH_3, OH, OH) → (Cu chromite; 200 °C) → 4-methyltetrahydropyran-2-one (CH_3, O, O)

OS, *CV* 4, 690 (1963)

1-(CH_2Cl)naphthalene + hexamethylenetetramine → (HOAc, Δ; then HCl) → 1-naphthaldehyde (CHO)

OS, *CV* 4, 838 (1963)

$(CH_2)_8$ cyclic α-hydroxy ketone (O, OH) → ($Cu(OAc)_2$, MeOH; aq. HOAc, Δ) → $(CH_2)_8$ cyclic 1,2-dione (O, O)

OS, CV 4, 846 (1963)

$$Ph{-}CH{=}CH_2 \xrightarrow[\text{2. NaOH}]{\text{1. dioxane} \bullet SO_3} Ph{-}CH{=}CH{-}SO_3Na$$

$$\xrightarrow[\Delta]{PCl_5} Ph{-}CH{=}CH{-}SO_2Cl$$

OS, CV 4, 862 (1963)

$$CH_3(CH_2)_{13}{-}CH_2{-}CO_2H \xrightarrow[CCl_4]{SO_3} CH_3(CH_2)_{13}{-}CH(SO_3H){-}CO_2H$$

OS, CV 4, 895 (1963)

O₂ → 70 °C; product with OOH

$$\text{tetralin} \xrightarrow[70\ °C]{O_2} \text{1-tetralyl hydroperoxide (OOH)}$$

OS, CV 4, 918 (1963)

$$\text{3-(CH}_2\text{Br)thiophene} \xrightarrow[\text{2. distill salt}]{\text{1. hexamethylenetetramine}, \Delta} \text{3-(CHO)thiophene}$$

OS, CV 4, 919 (1963)

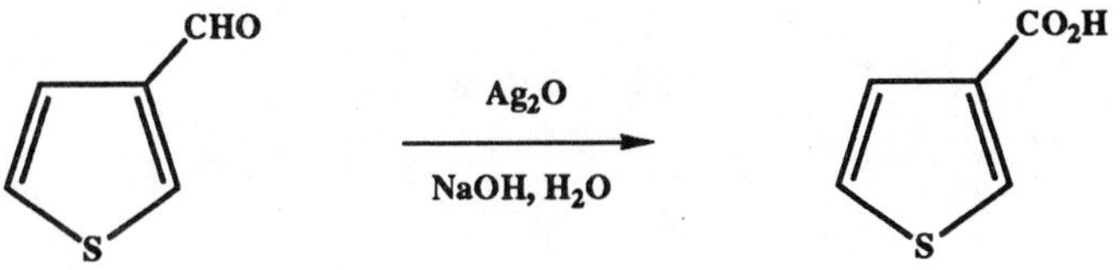

Thiophene-3-carboxaldehyde (CHO) → thiophene-3-carboxylic acid (CO_2H); reagents: Ag_2O; NaOH, H_2O

OS, CV 4, 932 (1963)

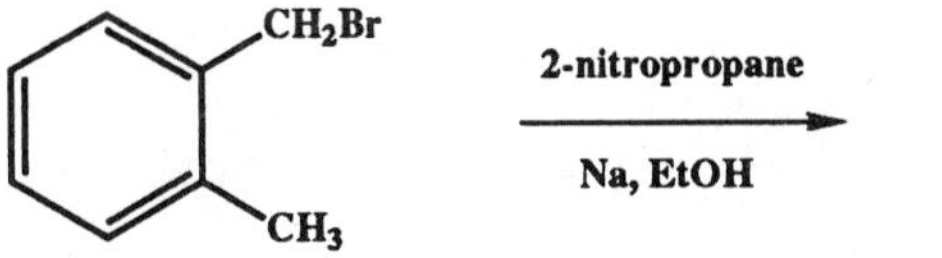

2-methylbenzyl bromide (CH_2Br, CH_3) → 2-methylbenzaldehyde (CHO, CH_3); reagents: 2-nitropropane; Na, EtOH

OS, CV 4, 972 (1963)

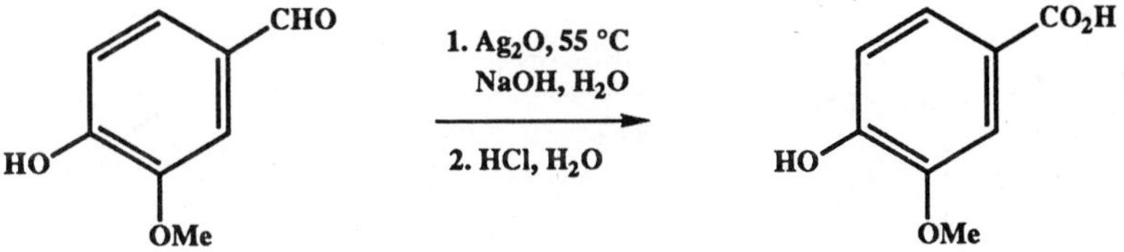

4-hydroxy-3-methoxybenzaldehyde (CHO, HO, OMe) → 4-hydroxy-3-methoxybenzoic acid (CO_2H, HO, OMe); reagents: 1. Ag_2O, 55 °C, NaOH, H_2O; 2. HCl, H_2O

OS, CV 4, 974 (1963)

4-hydroxy-3-methoxybenzaldehyde (CHO, HO, OMe) → 4-hydroxy-3-methoxybenzoic acid (CO_2H, HO, OMe); reagents: 1. KOH, NaOH, 180 - 195 °C; 2. HCl, H_2O

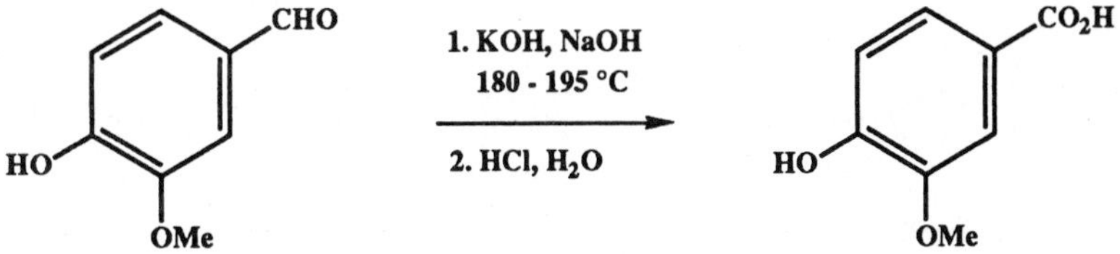

OS, CV 5, 8 (1973)

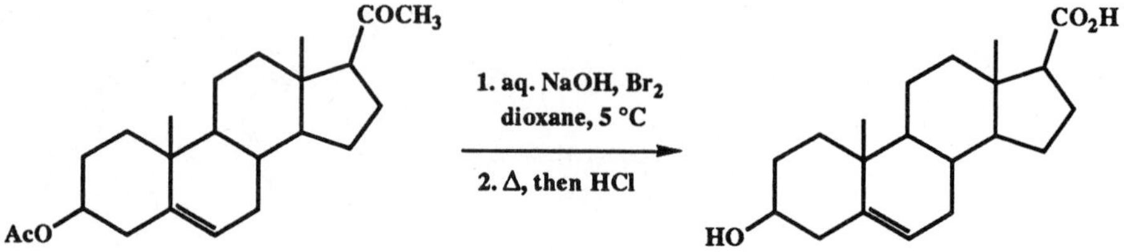

1. aq. NaOH, Br_2
dioxane, 5 °C

2. Δ, then HCl

AcO, $COCH_3$ → HO, CO_2H

OS, CV 5, 32 (1973)

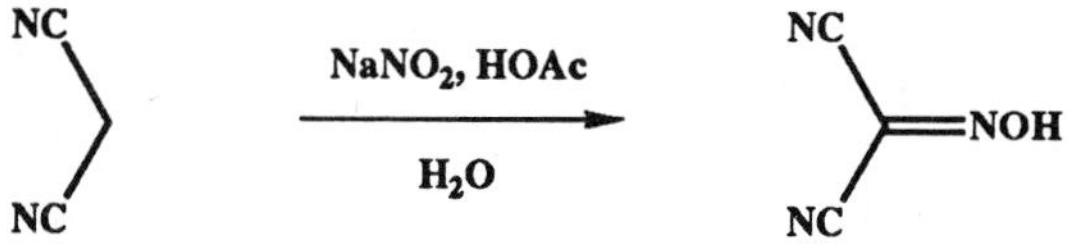

NC, NC → NC, NC, =NOH

$NaNO_2$, HOAc

H_2O

OS, CV 5, 35 (1973)

$-CH_3$ → CH_3, NH_2

NCl_3, $AlCl_3$

CH_2Cl_2, - 5 to 5 °C
then aq. HCl

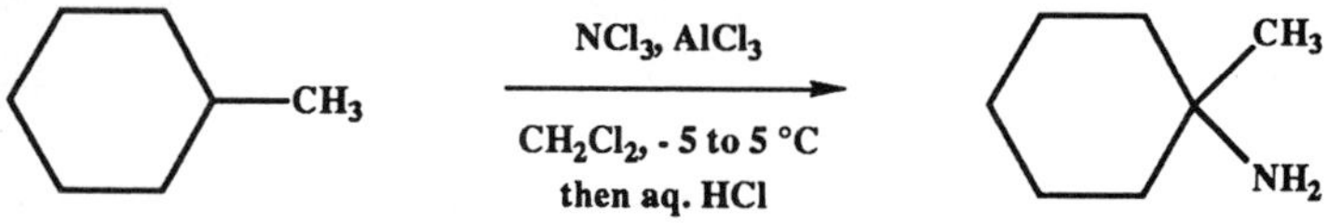

OS, CV 5, 46 (1973)

OH, C_6H_4-*p*-NMe_2, OMe → O, H, OMe

sulfanilic acid
$NaNO_2$, HCl

Na_2CO_3, H_2O

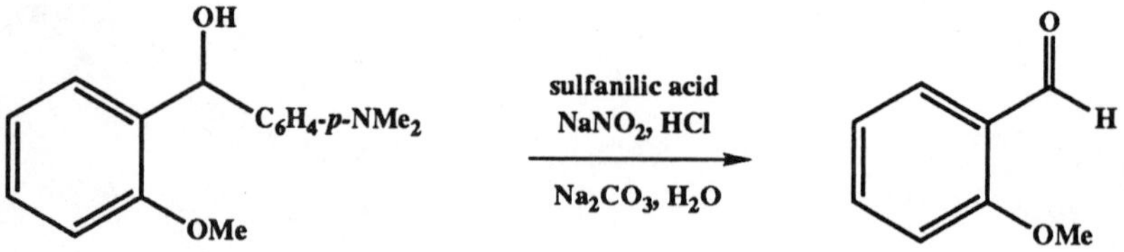

OS, CV 5, **70 (1973)**

Ph, O—O-*t*-Bu, O; Cu_2Br_2, Δ; O, Ph, O

OS, CV 5, **151 (1973)**

$PhCO_2O$-*t*-Bu; CuBr; benzene, Δ; O-*t*-Bu

OS, CV 5, **179 (1973)**

O, O, CH_3, O-*t*-Bu; *p*-TsN_3; Et_3N; CH_3CN; O, O, CH_3, O-*t*-Bu, N_2

OS, CV 5, **194 (1973)**

O; H_2SO_4; Ac_2O; SO_3H, O

dl - camphor

dl - camphor-sulfonic acid

***OS, CV 5,* 242 (1973)**

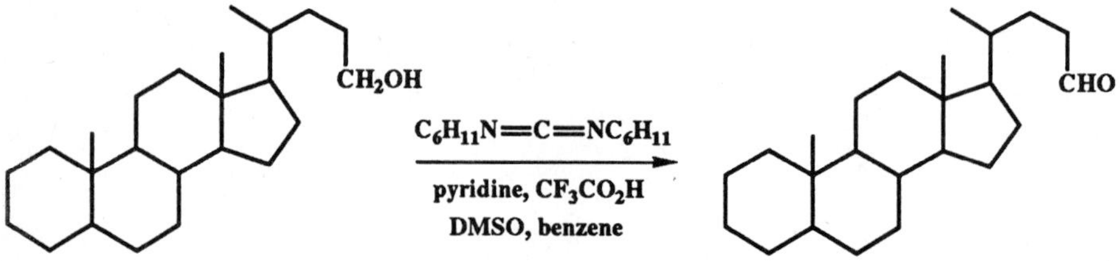

CH_2OH

$C_6H_{11}N{=}C{=}NC_6H_{11}$

pyridine, CF_3CO_2H
DMSO, benzene

CHO

***OS, CV 5,* 373 (1973)**

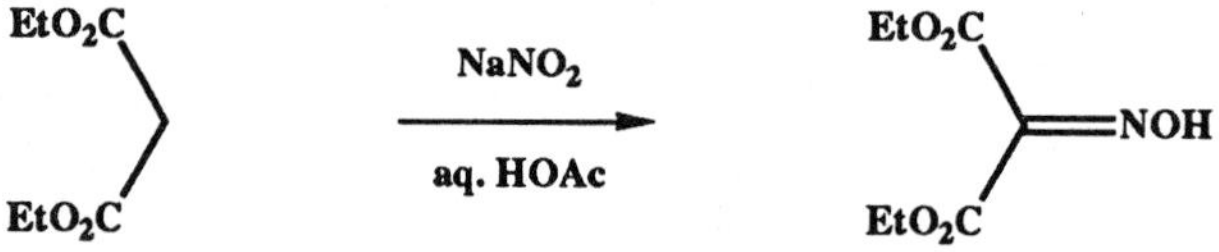

EtO_2C

EtO_2C

$NaNO_2$

aq. HOAc

EtO_2C

=NOH

EtO_2C

***OS, CV 5,* 379 (1973)**

EtO_2C

Et

EtO_2C

1. NaH, benzene

2. benzoyl peroxide
0 °C

EtO_2C OCOPh

EtO_2C Et

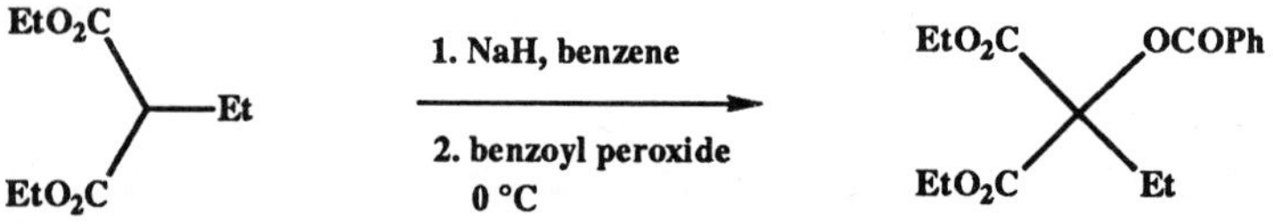

***OS, CV 5,* 403 (1973)**

O

Br_2, MeOH
Na_2CO_3

C_6H_6, - 15 to 0 °C

MeO O OMe

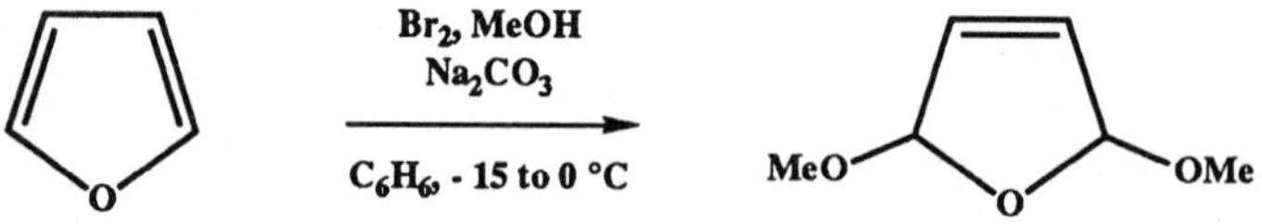

***OS*, *CV 5*, 428 (1973)**

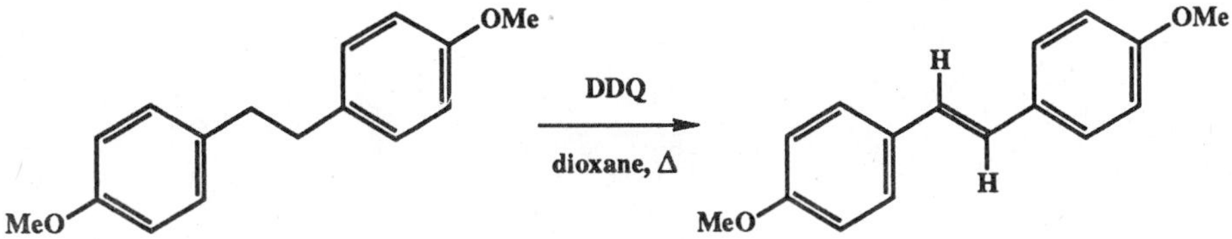

DDQ

dioxane, Δ

MeO, OMe, H, H

***OS*, *CV 5*, 489 (1973)**

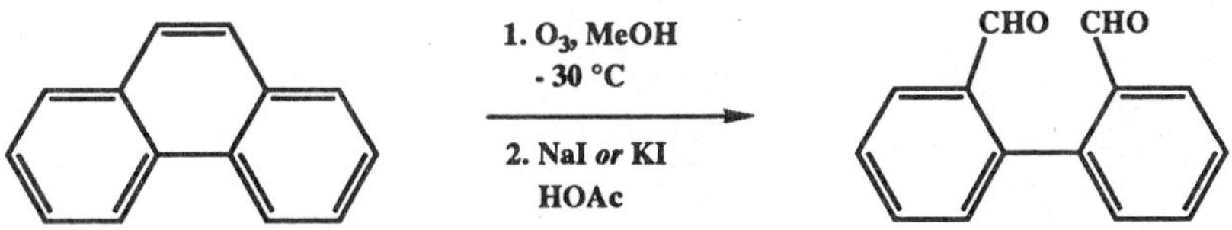

1. O_3, MeOH
- 30 °C

2. NaI *or* KI
HOAc

CHO CHO

***OS*, *CV 5*, 493 (1973)**

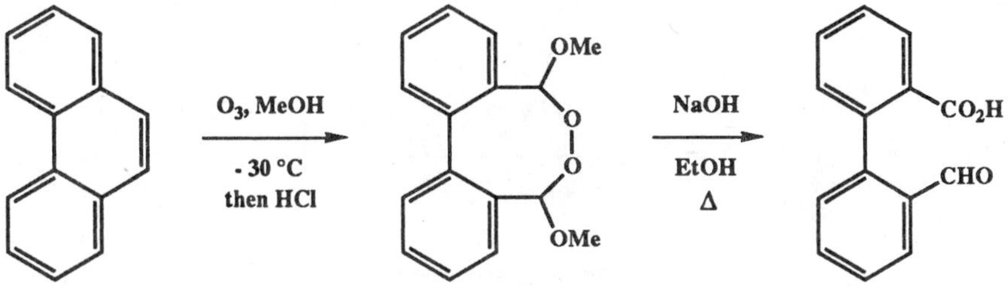

O_3, MeOH

- 30 °C
then HCl

OMe, O, O, OMe

NaOH

EtOH
Δ

CO_2H, CHO

***OS*, *CV 5*, 580 (1973)**

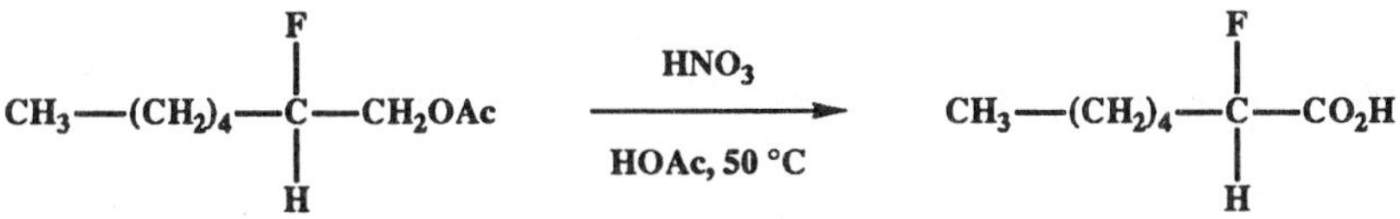

$$CH_3-(CH_2)_4-\underset{H}{\overset{F}{C}}-CH_2OAc \xrightarrow[\text{HOAc, 50 °C}]{HNO_3} CH_3-(CH_2)_4-\underset{H}{\overset{F}{C}}-CO_2H$$

OS, *CV 5*, 598 (1973)

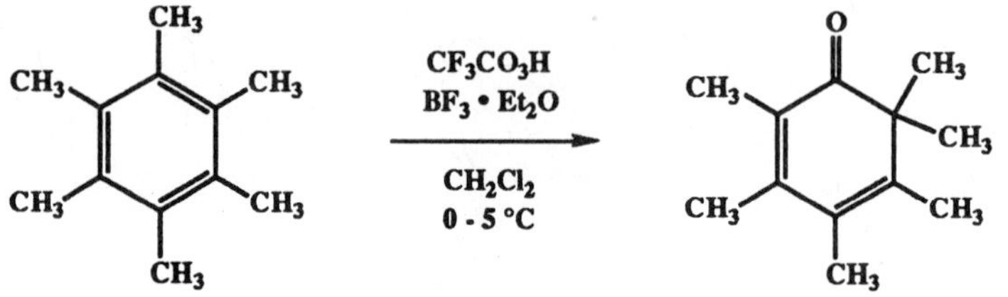
CH3
CH3
CH3
CH3
CH3
CH3
CF3CO3H
BF3 • Et2O
CH2Cl2
0 - 5 °C
O
CH3
CH3
CH3
CH3
CH3
CH3

OS, *CV 5*, 617 (1973)

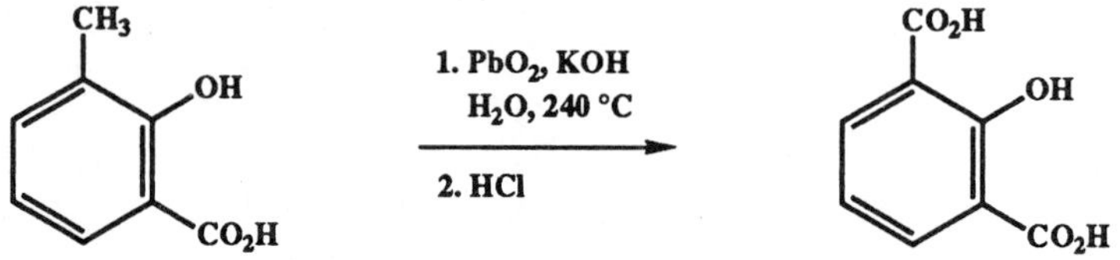
CH3
OH
CO2H
1. PbO2, KOH
H2O, 240 °C
2. HCl
CO2H
OH
CO2H

OS, *CV 5*, 668 (1973)

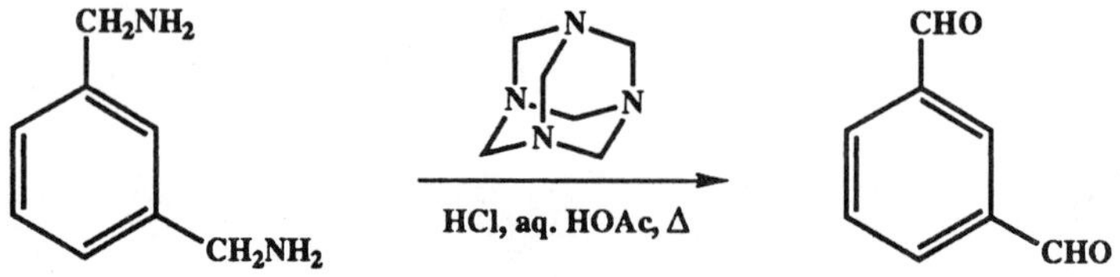
CH2NH2
CH2NH2
HCl, aq. HOAc, Δ
CHO
CHO

OS, *CV 5*, 805 (1973)

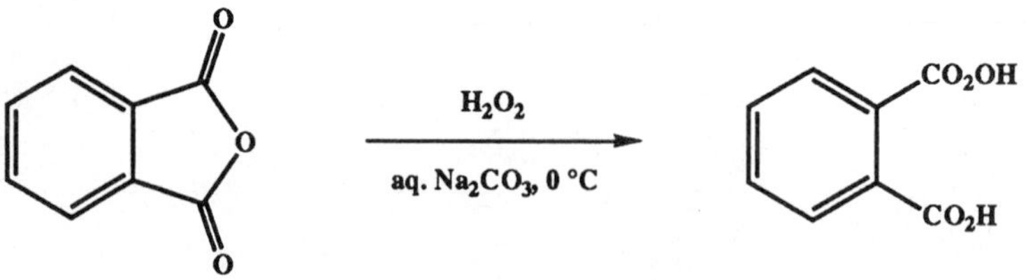
O
O
O
H2O2
aq. Na2CO3, 0 °C
CO2OH
CO2H

OS, CV 5, 825 (1973)

N-bromosuccinimide, benzoyl peroxide, CCl_4, Δ

1. pyridine, EtOH, Δ
2. $p\text{-}Me_2NC_6H_4NO$, NaOH, EtOH, 5 °C

H_2SO_4, H_2O

CH_3 / NO_2 → CH_2Br / NO_2 → $CH{=}N^{+}(O^{-})\text{—}C_6H_4NMe_2$ / NO_2 → CHO / NO_2

OS, CV 5, 872 (1973)

$$n\text{-}C_7H_{15}\text{—}CH_2I \xrightarrow[CHCl_3,\ \Delta]{Me_3\overset{+}{N}\text{—}O^{-}} n\text{-}C_7H_{15}\text{—}CHO$$

OS, CV 5, 904 (1973)

$$C_6H_5CO_2H \xrightarrow[CH_3SO_3H]{70\%\ H_2O_2} C_6H_5CO_3H$$

OS, CV 5, 1138 (1973)

$$2\ C_7H_8 \xrightarrow[CCl_4]{3\ PCl_5} (C_7H_7^{+}\ PCl_6^{-})(C_7H_7^{+}\ Cl^{-}) \xrightarrow[EtOH]{2\ HBF_4} 2\ C_7H_7^{+}\ BF_4^{-}$$

OXIDATION - GENERAL

***OS*, *CV* 6, 23 (1988)**

CH$_3$

1. NaH, THF, 0 °C
2. PhSeCl

CH$_3$
SePh

***OS*, *CV* 6, 43 (1988)**

O_3
silica gel, - 65 to - 45 °C

OH

***OS*, *CV* 6, 48 (1988)**

98% H_2SO_4
80 °C

O

***OS*, *CV* 6, 137 (1988)**

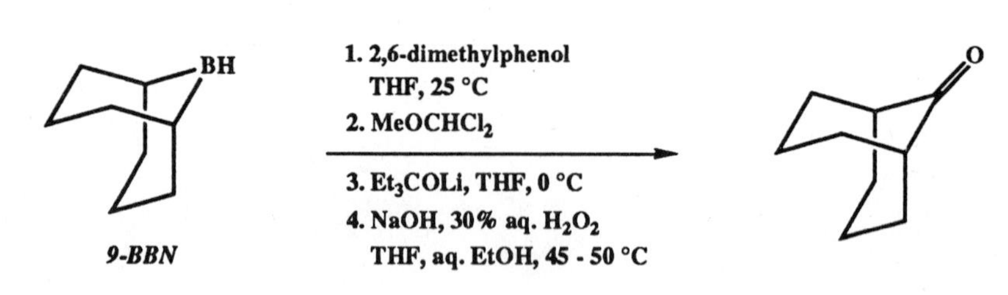

OS, CV 6, 218 (1988)

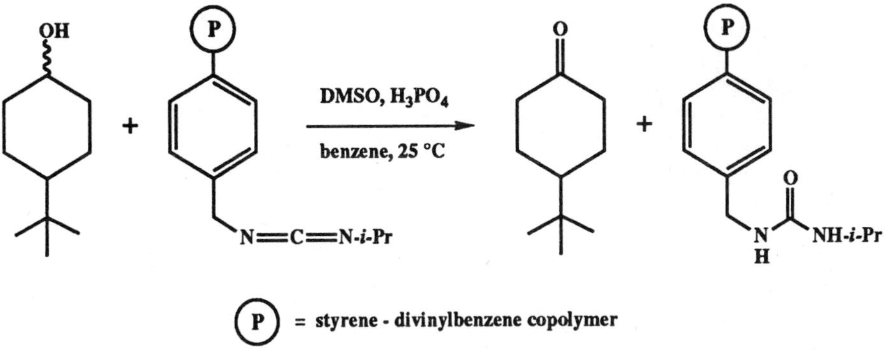

DMSO, H_3PO_4

benzene, 25 °C

OH

P

N=C=N-*i*-Pr

O

NH-*i*-Pr

N
H

P = styrene - divinylbenzene copolymer

OS, CV 6, 220 (1988)

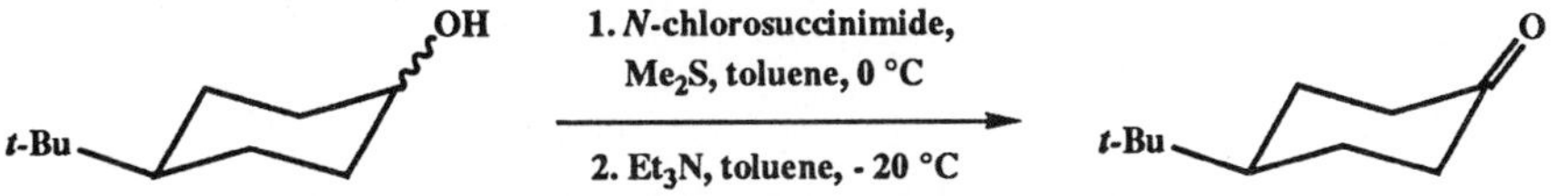

t-Bu

OH

1. *N*-chlorosuccinimide, Me_2S, toluene, 0 °C

2. Et_3N, toluene, - 20 °C

t-Bu

O

OS, CV 6, 276 (1988)

m-ClC_6H_4 C(=O) Cl

1. H_2O_2, aq. NaOH $MgSO_4$, dioxane

2. aq. H_2SO_4

m-ClC_6H_4 C(=O) O—OH

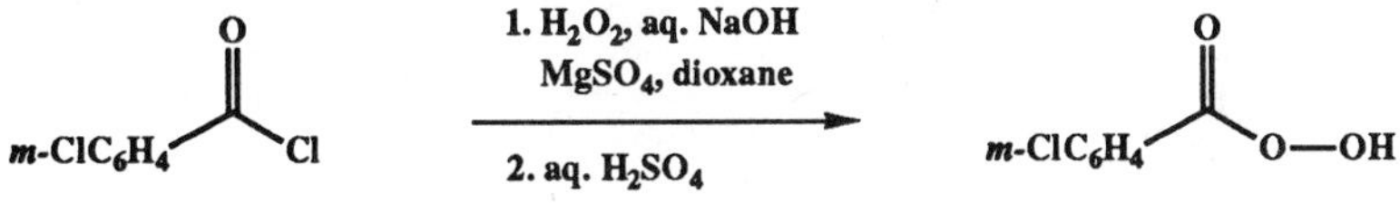

OS, CV 6, 342 (1988)

CH_3 N^+ O^- O • H_2O

OsO_4

H_2O, acetone

OH

OH

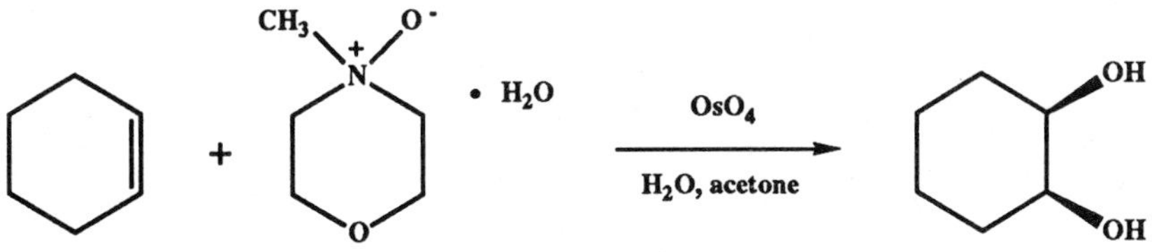

***OS*, *CV* 6,** **412 (1988)**

OH, t-Bu, t-Bu → O, t-Bu, t-Bu, O

O_2, salcomine

DMF, 30 - 50 °C

***OS*, *CV* 6,** **414 (1988)**

O, O, t-Bu-O, O-t-Bu → O, O, t-Bu-O, O-t-Bu, N_2

p-TsN_3

$CH_3N(n\text{-}C_8H_{17})_3^+$ Cl^-

aq. NaOH, CH_2Cl_2

***OS*, *CV* 6,** **480 (1988)**

OH, CH_3, CH_3 → O, O, CH_3, CH_3

2 :Ȯ—N̈($SO_3K)_2$

H_2O, Et_2O

***OS*, *CV* 6,** **648 (1988)**

O, NO_2 → O, O

1. NaOMe, MeOH, 0 °C

2. O_3, MeOH, - 78 °C

OS, CV 6, 662 (1988)

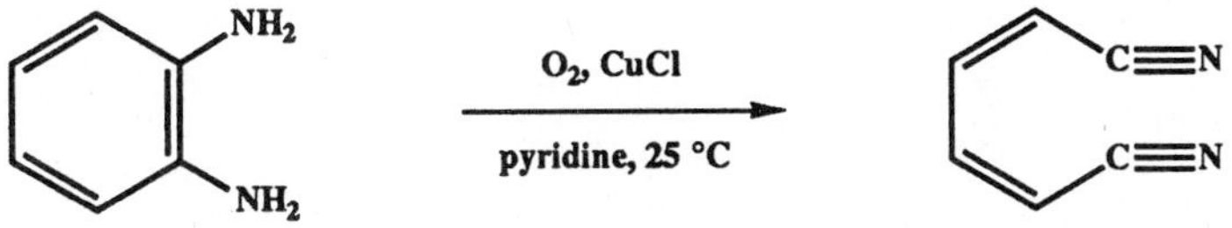

NH$_2$ NH$_2$ — O_2, CuCl / pyridine, 25 °C → C≡N C≡N

OS, CV 6, 719 (1988)

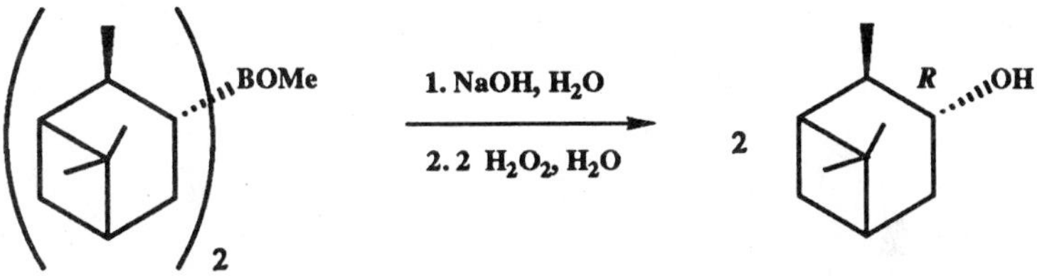

BOMe, 2, (-) isomer — 1. NaOH, H_2O / 2. 2 H_2O_2, H_2O → 2 *R* OH, *R* - (-) isomer

OS, CV 6, 731 (1988)

DDQ, HOAc / dioxane, Δ

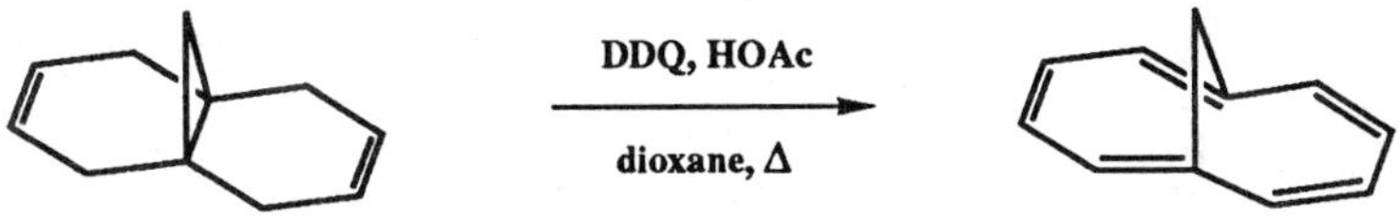

OS, CV 6, 766 (1988)

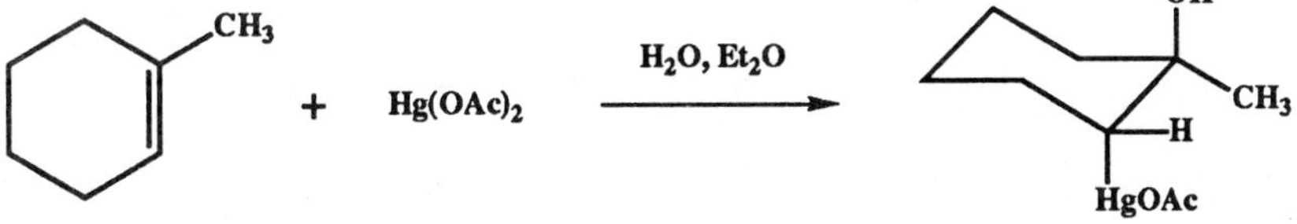

CH$_3$ + Hg(OAc)$_2$ — H_2O, Et_2O → OH, CH$_3$, H, HgOAc

OS, CV 6, **795 (1988)**

I—N=C=O

Et_2O

NCO

I

OS, CV 6, **815 (1988)**

Ph—HgOAc +

CH_3

OH

$Pd(OAc)_2$

CH_3CN, Et_2O

CH_3

Ph

OH

PdOAc

CH_3

Ph

O

H

OS, CV 6, **837 (1988)**

N_2O_4, O_2

Et_2O, 10 °C

NO_2

X

Et_3N

NO_2

X = NO_2 and ONO and ONO_2

OS, CV 6, **840 (1988)**

O

H H

HO

Isoamyl nitrite

t-BuOK, *t*-BuOH

25 °C

O

=NOH

H H

HO

OS, CV 6, 919 (1988)

$(C_6H_{13})_3B$ —NaOH / H_2O_2→ n-$C_6H_{13}OH$ + OH / n-C_4H_9 / CH_3

CH_3 CH_3 CH_3 H —1. BH_3, THF; 2. 1-octene; 3. NaOH, H_2O_2→ CH_3 CH_3 CH_3 OH

OS, CV 6, 943 (1988)

(±) - α - pinene —$NaBH_4$, $BF_3 \cdot Et_2O$ / diglyme, 0 °C→ BH 2 —$\overset{+}{N}H_3OSO_3^-$ / diglyme, 100 °C→ NH_2

OS, CV 6, 946 (1988)

(-) - β - pinene —H_2O_2, SeO_2 / *t*-BuOH, H_2O, 40 - 50 °C→ OH (+) - *trans* - pinocarveol

OS, CV 6, 971 (1988)

MeO OMe —2 MeOH, KOH / electricity (Pt anode), 8 - 14 °C→ MeO OMe MeO OMe

OS, CV 6, 979 (1988)

1. *n*-BuLi, - 30 °C THF-pentane
2. S, - 70 to - 10 °C
3. aq. H_2SO_4

SH

OS, CV 6, 1010 (1988)

OH, CH_3, CH_3, CH_3

2 :Ö—N̈(SO_3Na)$_2$

H_2O, heptane, 12 °C

O, CH_3, CH_3, CH_3, O

OS, CV 6, 1014 (1988)

N

+

p-Ts—S

p-Ts—S

1. Et_3N, Δ CH_3CN
2. aq. HCl 50 °C

O, S, S

OS, CV 6, 1033 (1988)

H, $SiMe_3$

1. Br_2, - 78 °C
2. Et_2NH, Δ

Br, $SiMe_3$

CH_3, OH, $SiMe_3$

H_2CrO_4, H_2SO_4

acetone, 0 °C

CH_3, O, $SiMe_3$

OS, *CV* 7, 112 (1990); *60*, 18 (1981)

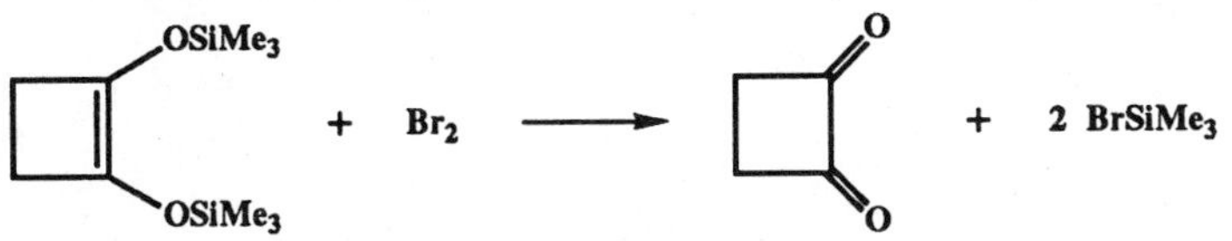

OSiMe3 OSiMe3 + Br_2 → O O + 2 $BrSiMe_3$

OS, *CV* 7, 27 (1990); *61*, 14 (1983)

$PhCH_2NH$– (tetrazole, N, N, N, N–H) → NaOBr, H_2O / CH_2Cl_2, 0 °C → $PhCH_2N{\equiv}C$

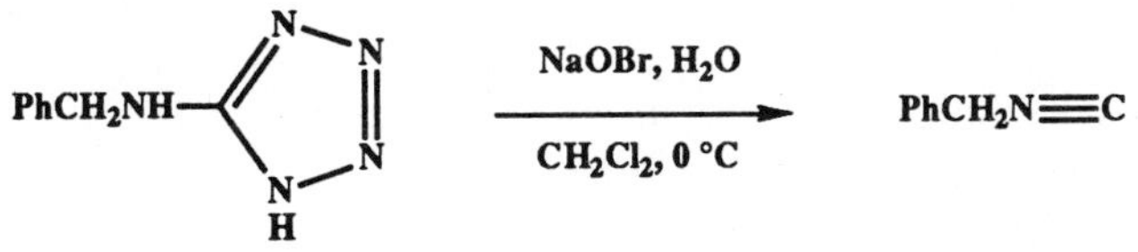

OS, *CV* 7, 56 (1990); *61*, 17 (1983)

$H_2NNH_2 \cdot H_2O$ + 2 $ClCO_2CH_2CCl_3$ → 1. Na_2CO_3, EtOH 2. fuming HNO_3 → $Cl_3CCH_2O_2C{-}N{=}N{-}CO_2CH_2CCl_3$

OS, *CV* 7, 375 (1990); *61*, 85 (1983)

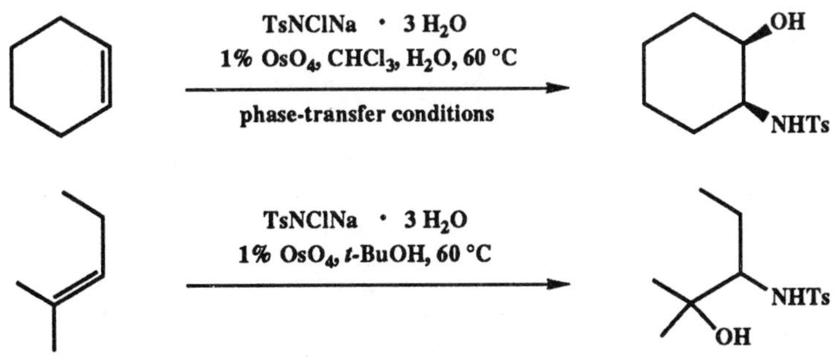

(cyclohexene) → TsNClNa · 3 H_2O, 1% OsO_4, $CHCl_3$, H_2O, 60 °C / phase-transfer conditions → OH NHTs

(alkene) → TsNClNa · 3 H_2O, 1% OsO_4, *t*-BuOH, 60 °C → NHTs OH

OS, CV 7, **223 (1990);** ***61,*** **93 (1983)**

Ph–CH=CH–Ph + $EtO_2CNClNa$ + $AgNO_3$ + H_2O —(1% OsO_4, MeCN)→ Ph–CH($NHCO_2Et$)–CH(OH)–Ph

OS, CV 7, **473 (1990);** ***61,*** **129 (1983)**

DDQ, dioxane, reflux

OS, CV 7, **137 (1990);** ***62,*** **9 (1984)**

$$n\text{-}C_8H_{17}CH{=}CH_2 \xrightarrow[\text{DMF, } H_2O]{PdCl_2,\ CuCl,\ O_2} n\text{-}C_8H_{17}COCH_3$$

OS, CV 7, **266 (1990);** ***62,*** **58 (1984)**

OH —(1. *n*-BuLi, TMEDA; 2. Me_3SiCl)→ Me_3Si … $OSiMe_3$ —(H_2SO_4)→ Me_3Si … OH

OS, CV 7, **4 (1990);** ***62,*** **149 (1984)**

Ac, N, O, O —(1. Cl_2, hν, CCl_4; 2. 120 - 150 °C)→ Ac, N, O, O

OS, *CV 7*, 406 (1990); *63*, 10 (1985)

CH_2OH / CH_2OH — *meso* — horse-liver alcohol dehydrogenase, β-nicotinamide adenine dinucleotide, flavin mononucleotide, pH 9, 20 °C → (+) - (1*R*, 6*S*)

OS, *CV 7*, 185 (1990); *63*, 89 (1985)

OH, O, $OSiMe_3$ — H_5IO_6, THF, 25 °C → OH, COOH

OS, *CV 7*, 251 (1990); *63*, 154 (1985)

S_8 + $CF_3CF{=}CF_2$ —KF, DMF→ $(CF_3)_2C$(S)(S)$C(CF_3)_2$ —KF, KIO_3→ $CF_3{-}C({=}O){-}CF_3$

OS, *CV 7*, 307 (1990); *63*, 206 (1985)

N, CO_2Me — -2 e⁻, MeOH → N, CO_2Me, OMe

OS, *CV* 7, 149 (1990); *64*, 19 (1986)

N—OH

1. *p*-TsCl pyridine

2. EtOK

N

H^+

EtOH

OEt

C—CH_2NH_2

OEt

OS, *CV* 7, 195 (1990); *64*, 92 (1986)

Me

+N I⁻

N

Me

S_8

K_2CO_3, MeOH

Me

N

S

N

Me

OS, *CV* 7, 282 (1990); *64*, 118 (1986)

$OSiMe_3$

1. *m*-CPBA, hexane, - 15 °C

2. $Et_3N \cdot HF$, CH_2Cl_2

O

HO

OS, *CV* 7, 277 (1990); *64*, 127 (1986)

O

1. LDA

2. $MoO_5 \cdot py \cdot (Me_2N)_3PO$

O

OH

5 : 1
endo : *exo*

OS, *CV* 7, 263 (1990); *64*, 138 (1986)

Me

KOH, COOH, IO

MeOH, 5 - 10 °C

MeO OMe

OH

5% H_2SO_4

CH_2Cl_2

O

OH

OS, *CV* 7, 168 (1990); *64*, 150 (1986)

1. O_3, MeOH, CH_2Cl_2
2. *p*-TsOH
3. $NaHCO_3$
4. Me_2S

$(MeO)_2CH-(CH_2)_4-CHO$

1. O_3, MeOH, CH_2Cl_2
2. Ac_2O, Et_3N

$MeO_2C-(CH_2)_4-CHO$

1. O_3, MeOH, CH_2Cl_2
2. *p*-TsOH
3. Ac_2O, Et_3N

$MeO_2C-(CH_2)_4-CH(OMe)_2$

OS, *CV* 7, 258 (1990); *64*, 164 (1986)

OH

$(COCl)_2$, DMSO

CH_2Cl_2

- 50 to - 60° C

H

O

OS, *65*, 159 (1987)

$MeO_2CN{=}S{=}NCO_2Me$ → 1. (2-methyl-2-butene); 2. KOH, MeOH → $NHCO_2Me$

OS, *66*, 14 (1987)

OH → 1. *n*-BuLi; 2. Me_3SiCl; 3. *t*-BuLi; 4. NH_4Cl → OH, $SiMe_3$ → $(COCl)_2$, DMSO, Et_3N → O, $SiMe_3$

OS, *66*, 180 (1987)

OH, OH → O_2, CuCl, MeOH, pyridine → CO_2H, CO_2Me

OS, *66*, 203 (1987)

$PhSO_2N{=}CHPh$ → *m*-CPBA, $(Et_3NCH_2Ph)^+ Cl^-$, $NaHCO_3$ → O, N, C, Ph, H, $PhSO_2$

OS, *67*, 121 (1988)

CHO

CuCl, $PdCl_2$

H_2O, DMF, O_2

O

CHO

OS, *67*, 125 (1988)

n-$C_8H_{17}CH_2CO_2Et$

1. LDA, THF, - 78 °C

2. $Ph_2MeSiCl$

n-C_8H_{17}—CH—CO_2Et

$SiMePh_2$

OS, *67*, 157 (1988)

+ $PhSeSO_2Ph$

1. hν, CCl_4

2. H_2O_2, CH_2Cl_2

SO_2Ph

OS, *68*, 1 (1989)

Me_3SiCH_2MgCl

1. $(PhO)_2P(O)N_3$

2. H_2O

Me_3SiCHN_2 + $(PhO)_2P(O)NH_2$

OS, *68*, 25 (1989)

H OMe

H H

1. Me_3CLi

2. Me_3SiCl

H OMe

H $SiMe_3$

OS, *68*, 41 (1989)

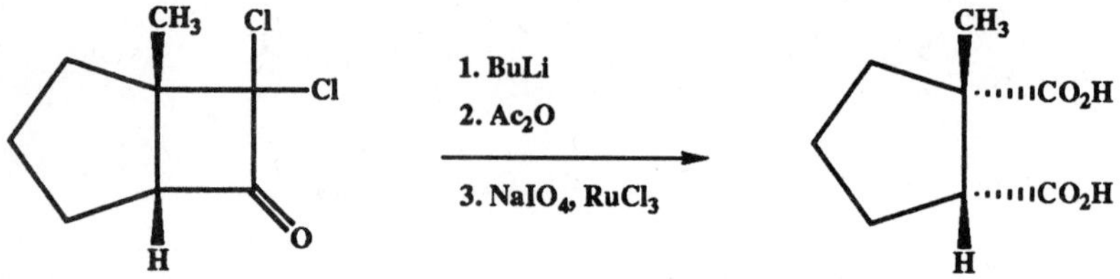

OS, *68*, 162 (1989)

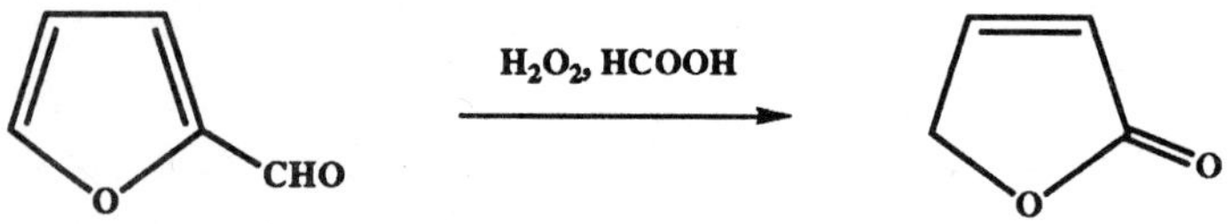

REDUCTION

OS, CV 1, 49 (1941)

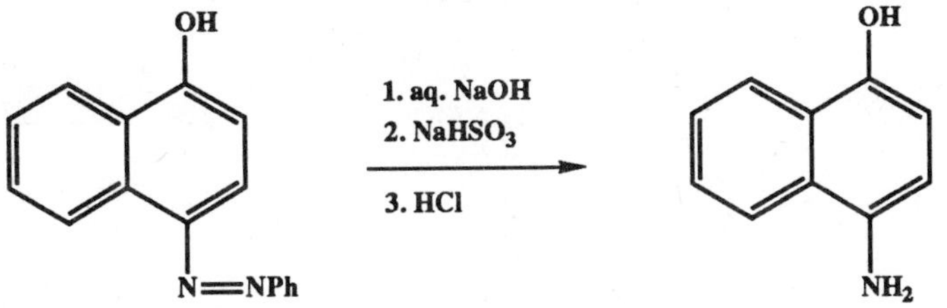

1. aq. NaOH
2. NaHSO$_3$
3. HCl

OH

N=NPh

OH

NH$_2$

OS, CV 1, 52 (1941)

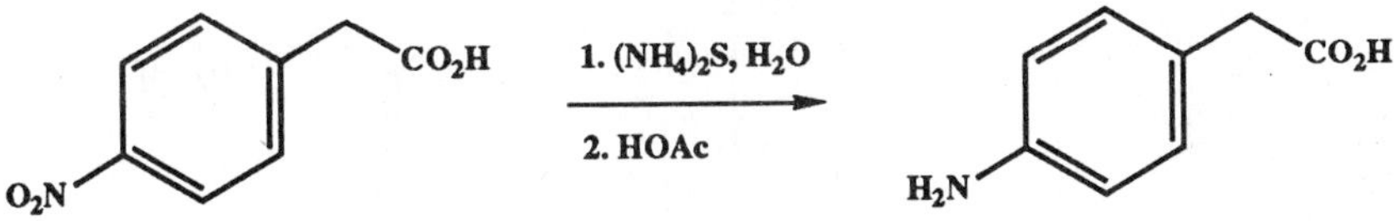

CO_2H

O_2N

1. $(NH_4)_2S$, H_2O
2. HOAc

CO_2H

H_2N

OS, CV 1, 73 (1941)

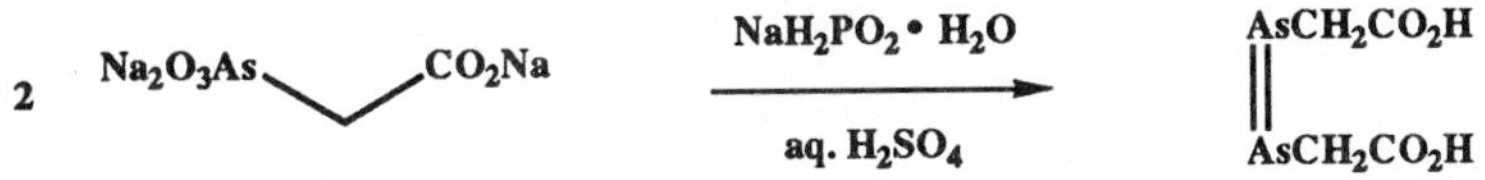

2 Na_2O_3As CO_2Na

$NaH_2PO_2 \bullet H_2O$

aq. H_2SO_4

$AsCH_2CO_2H$

$AsCH_2CO_2H$

OS, CV 1, 445 (1941)

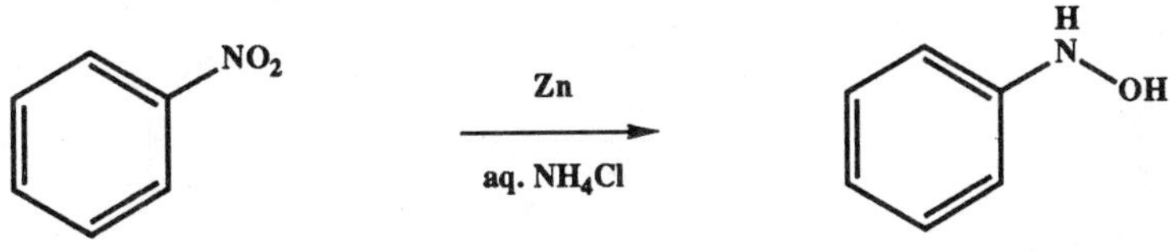

NO_2

Zn

aq. NH_4Cl

H
N
OH

***OS, CV 1*, 455 (1941)**

CO_2H

O_2N NO_2

NO_2

Sn, HCl

Δ

CO_2H

H_2N NH_2

NH_2

***OS, CV 1*, 485 (1941)**

H

H_2N N NO_2

O

electricity

(Hg cathode)

H_2SO_4

H

H_2N N $NH_2 \cdot H_2SO_4$

O

***OS, CV 1*, 492 (1941)**

2 SO_2Cl

Me

3 Zn

H_2O, Δ

(*p*-Me-$C_6H_4SO_2)_2Zn$

Na_2CO_3

2 SO_2Na

Me

***OS, CV 1*, 504 (1941)**

SO_2Cl

Zn, H_2SO_4

H_2O

SH

OS, *CV 1*, 511 (1941)

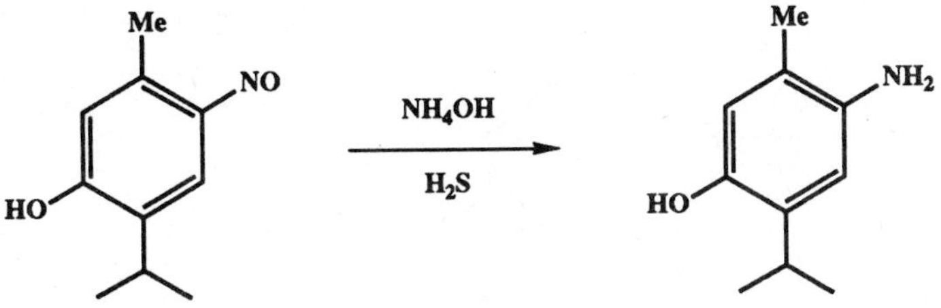

Me, NO, HO — NH_4OH / H_2S → Me, NH_2, HO

OS, *CV 2*, 33 (1943)

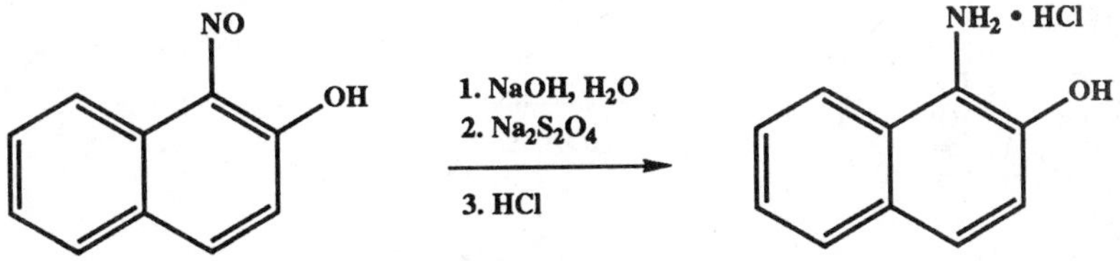

NO, OH — 1. NaOH, H_2O 2. $Na_2S_2O_4$ 3. HCl → NH_2 • HCl, OH

OS, *CV 2*, 42 (1943)

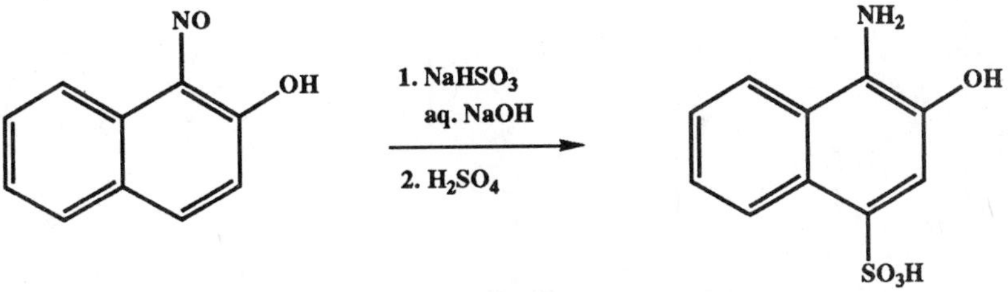

NO, OH — 1. $NaHSO_3$ aq. NaOH 2. H_2SO_4 → NH_2, OH, SO_3H

OS, *CV 2*, 57 (1943)

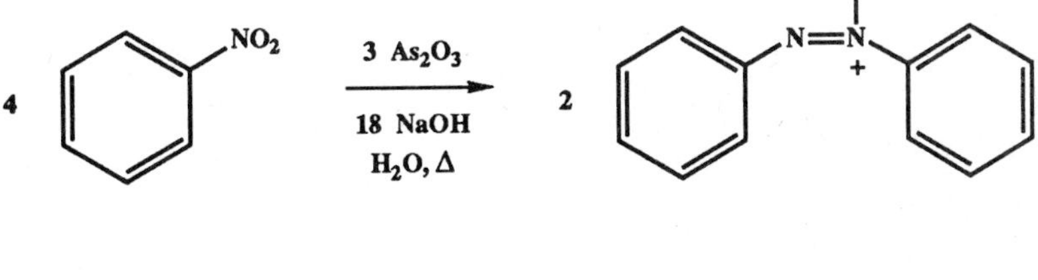

4 NO_2 — 3 As_2O_3 / 18 NaOH H_2O, Δ → 2 N=N⁺–O⁻

OS, CV 2, 130 (1943)

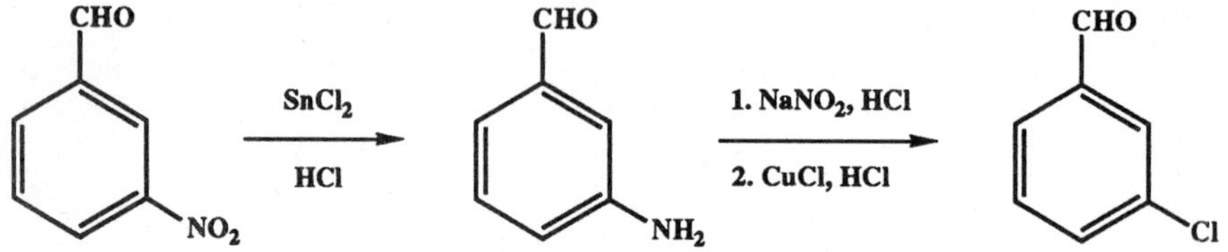

CHO
NO_2
$SnCl_2$
HCl
CHO
NH_2
1. $NaNO_2$, HCl
2. CuCl, HCl
CHO
Cl

OS, CV 2, 160 (1943)

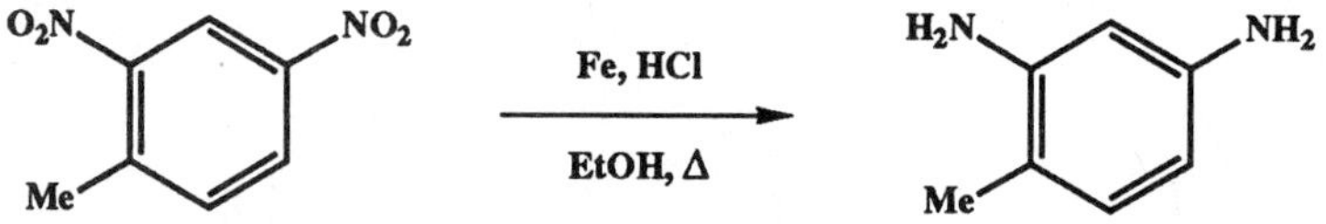

O_2N
NO_2
Me
Fe, HCl
EtOH, Δ
H_2N
NH_2
Me

OS, CV 2, 175 (1943)

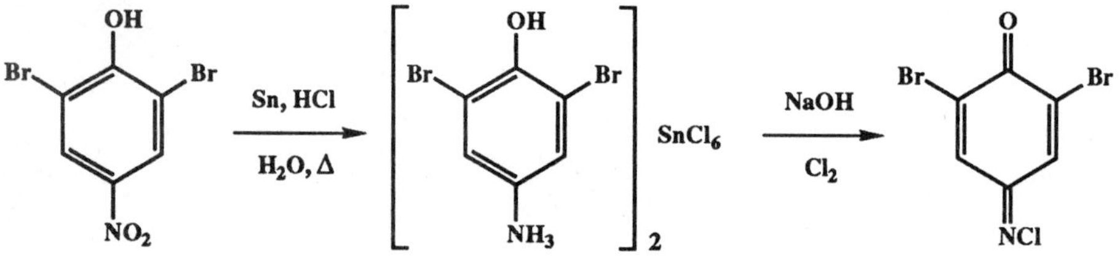

OH
Br
Br
NO_2
Sn, HCl
H_2O, Δ
OH
Br
Br
NH_3
2
$SnCl_6$
NaOH
Cl_2
O
Br
Br
NCl

OS, CV 2, 202 (1943)

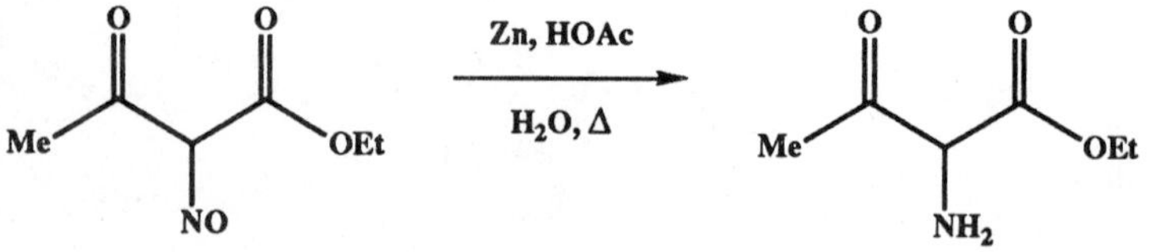

O
O
Me
OEt
NO
Zn, HOAc
H_2O, Δ
O
O
Me
OEt
NH_2

OS, CV 2, 211 (1943)

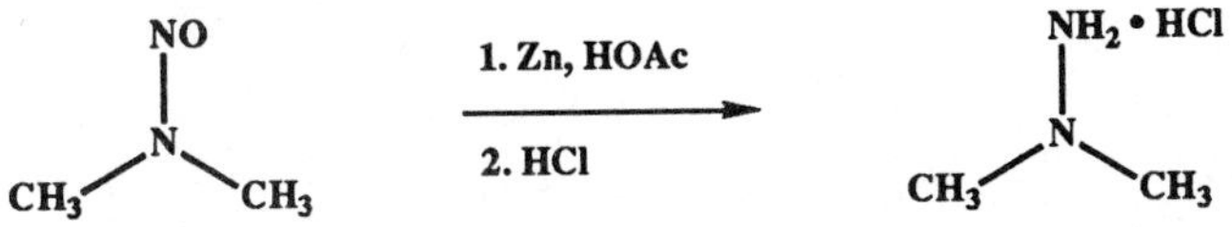
NO
N
CH3
CH3
1. Zn, HOAc
2. HCl
NH2 • HCl
N
CH3
CH3

OS, CV 2, 234 (1943)

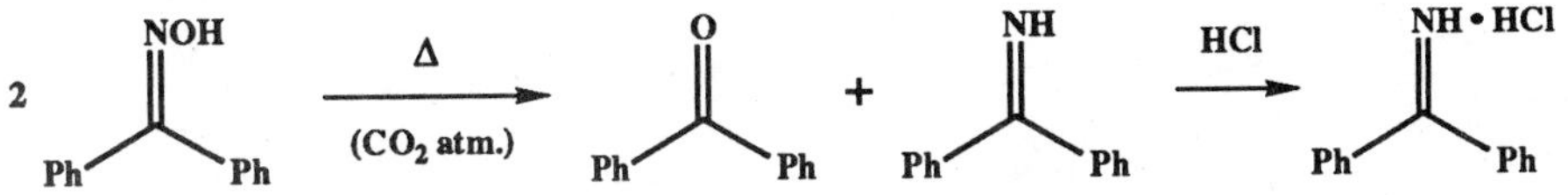
2
NOH
Ph
Ph
Δ
(CO2 atm.)
O
Ph
Ph
+
NH
Ph
Ph
HCl
NH • HCl
Ph
Ph

OS, CV 2, 254 (1943)

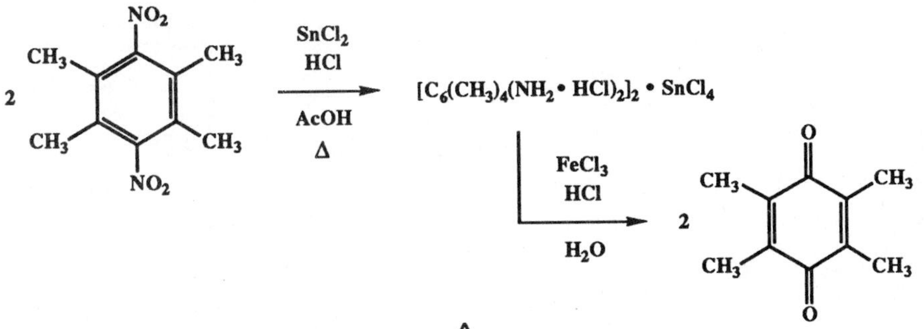
2
NO2
CH3
CH3
CH3
CH3
NO2
SnCl2
HCl
AcOH
Δ
[C6(CH3)4(NH2 • HCl)2]2 • SnCl4
FeCl3
HCl
H2O
2
O
CH3
CH3
CH3
CH3
O

OS, CV 2, 290 (1943)

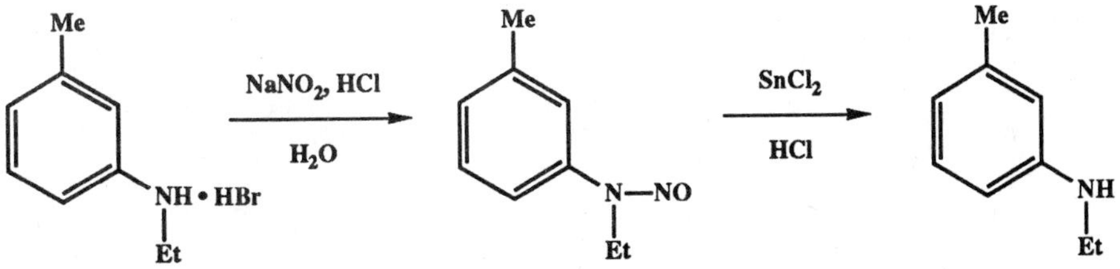
Me
NH • HBr
Et
NaNO2, HCl
H2O
Me
N—NO
Et
SnCl2
HCl
Me
NH
Et

***OS*, *CV* 2,** **418 (1943)**

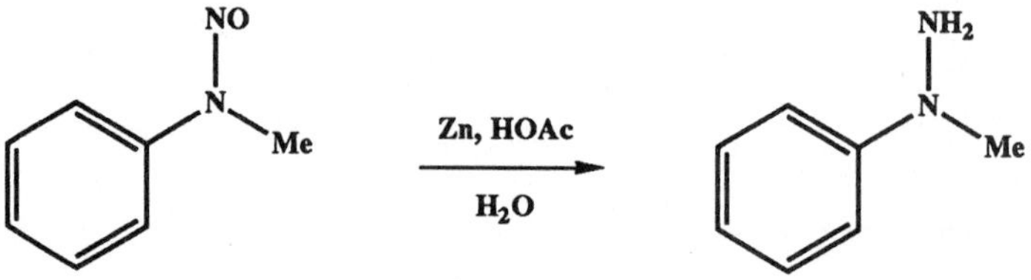

Zn, HOAc

H_2O

NO N Me → NH_2 N Me

***OS*, *CV* 2,** **447 (1943)**

NO_2 → NH_2

Zn, $CaCl_2$

aq. EtOH, Δ

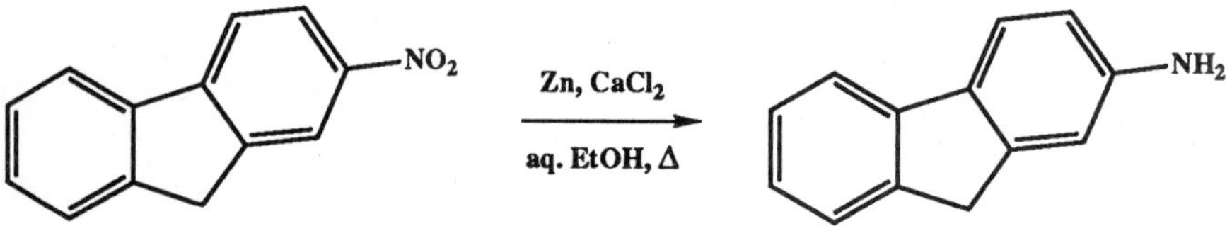

***OS*, *CV* 2,** **471 (1943)**

SO_3H NO_2 → SO_3H NH_2

Fe, HOAc

H_2O, Δ

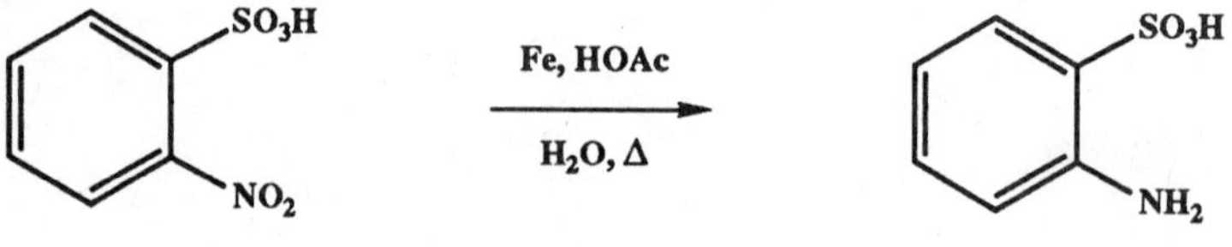

OS, *CV* 2, 501 (1943)

NH_2 NO_2 — Zn, EtOH / aq. NaOH, Δ → NH_2 NH_2

OS, *CV* 2, 580 (1943)

CO_2H HO_2C S—S — Zn, HOAc → 2 CO_2H SH

OS, *CV* 2, 617 (1943)

O HN NO_2 O N H O — Sn, HCl / Δ → O HN NH_2 O N H O

OS, *CV* 3, 56 (1955)

NO_2 H O — $FeSO_4$, HCl / NH_4OH, H_2O / 90 °C → NH_2 H O

***OS*, *CV* 3, 59 (1955)**

H_2 (1000 psi)
Ra (Ni)
MeOH

OMe OMe NO_2 → OMe OMe NH_2

***OS*, *CV* 3, 63 (1955)**

H_2 (1000 psi)
Ra (Ni)
EtOH
100 - 120 °C

CH_3 *i*-Pr NO_2 → CH_3 *i*-Pr NH_2

***OS*, *CV* 3, 69 (1955)**

$Na_2S_2O_4$
H_2O, Δ

O NH NH NO_2 O → O NH NH NH_2 O

***OS*, *CV* 3, 73 (1955)**

H_2N H N NO_2 NH —Zn / HOAc→ H_2N H N NH_2 NH • HOAc —$NaHCO_3$→ H_2N H N NH_2 NH • H_2CO_3

OS, *CV* 3, 82 (1955)

OH

O_2N NO_2

Na_2S, NH_4Cl, H_2O

NH_4OH, 80 - 85 °C

OH

O_2N NH_2

OS, *CV* 3, 86 (1955)

Cl

O_2N

Na_2S, H_2O, Δ

SNa

H_2N

OS, *CV* 3, 91 (1955)

Ph—CH=CH—CO_2H

NH_2OH

H_2O, Δ

[Ph CO_2H NHOH]

NH_2OH

Ph CO_2H

NH_2

OS, *CV* 3, 103 (1955)

NO_2

Zn, NaOH

aq. MeOH, Δ

N=N

OS, CV 3, 239 (1955)

AcHN, SO_2, NO_2; $SnCl_2$, HCl; H_2N, SO_2, NH_2

OS, CV 3, 242 (1955)

NH_2, NO_2, NO_2; H_2S, NH_4OH, EtOH; NH_2, NH_2, NO_2

OS, CV 3, 360 (1955)

2 PhNHNH, PhNHNH, S; KOH, MeOH; PhN=N, PhNHNH, S + PhNHNH, H_2N, S + $PhNH_2$

OS, CV 3, 453 (1955)

H, O, NO_2; $SnCl_2$, HCl; H, O, NH_2 • $(SnCl_4)_x$; 1. $NaNO_2$, HCl, H_2O; 2. H_2O, Δ; H, O, OH

OS, CV 3, 485 (1955)

$$2\ C_6H_5IO \xrightarrow[\text{steam-distill}]{H_2O} C_6H_5IO_2 + C_6H_5I$$

OS, CV 3, 668 (1955)

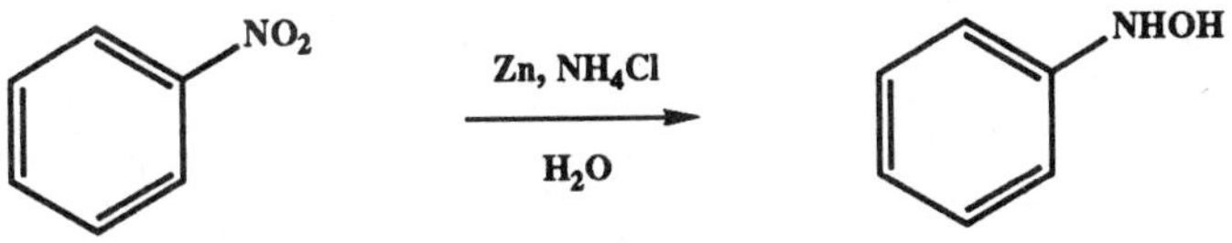

NO_2 → (Zn, NH_4Cl / H_2O) → NHOH

OS, CV 4, 31 (1963)

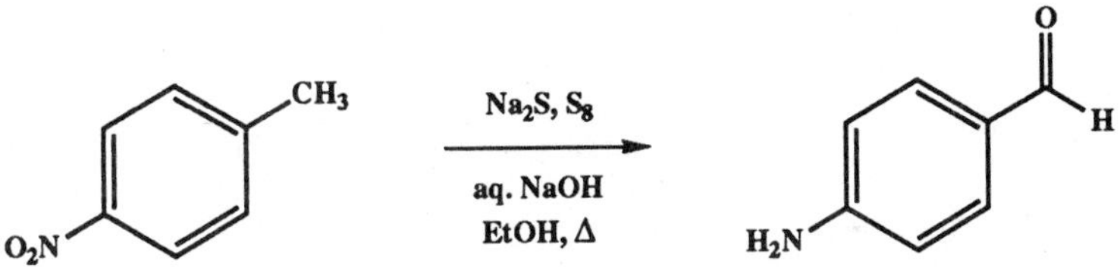

CH_3, O_2N → (Na_2S, S_8 / aq. NaOH / EtOH, Δ) → O, H, H_2N

OS, CV 4, 148 (1963)

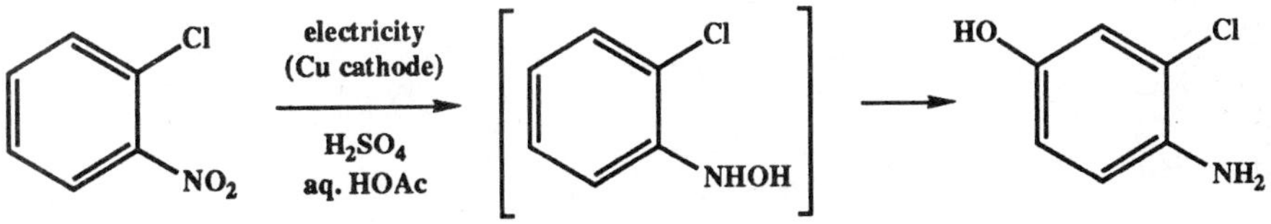

Cl, NO_2 → (electricity (Cu cathode) / H_2SO_4 / aq. HOAc) → [Cl, NHOH] → HO, Cl, NH_2

***OS*, *CV 4*, 166 (1963)**

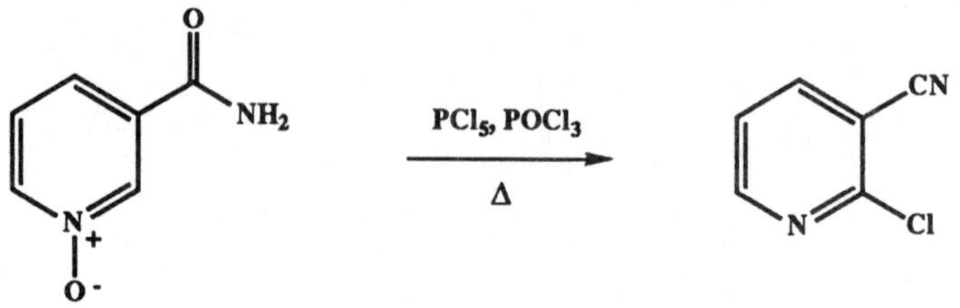

O
NH$_2$
N$^+$
O$^-$

PCl$_5$, POCl$_3$
Δ

CN
N
Cl

***OS*, *CV 4*, 295 (1963)**

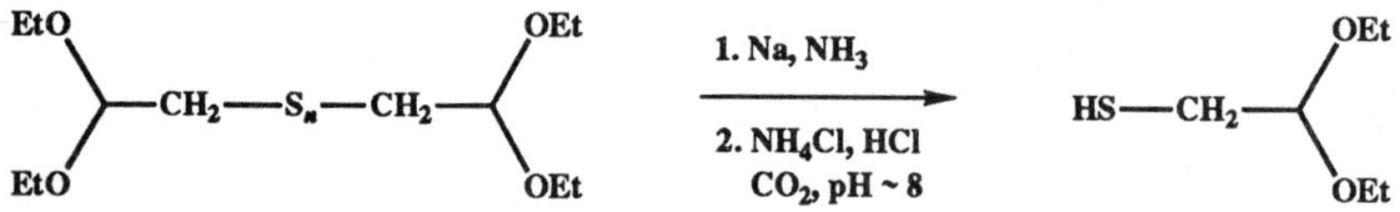

EtO
OEt
—CH$_2$—S$_n$—CH$_2$—
EtO
OEt

1. Na, NH$_3$
2. NH$_4$Cl, HCl
CO$_2$, pH ~ 8

OEt
HS—CH$_2$—
OEt

***OS*, *CV 4*, 695 (1963)**

SO$_2$Cl
ClO$_2$S

Zn amalgam
33% H$_2$SO$_4$, Δ

SH
HS

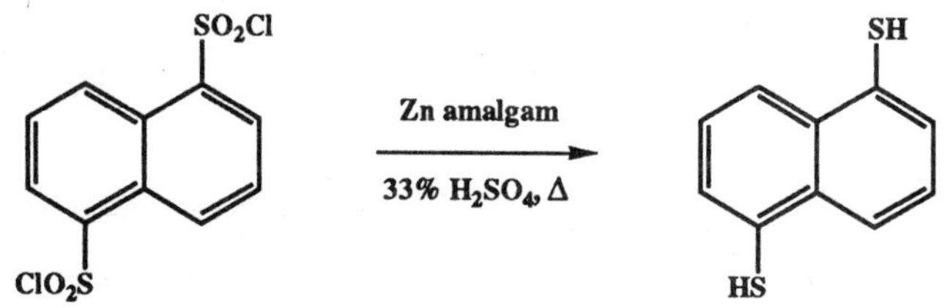

***OS*, *CV 5*, 30 (1973)**

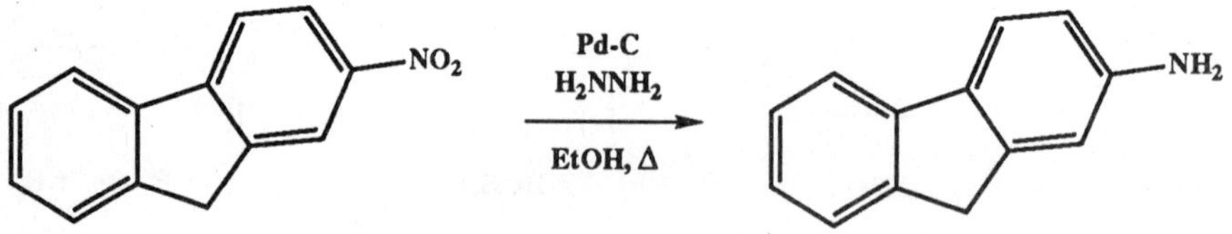

NO$_2$

Pd-C
H$_2$NNH$_2$
EtOH, Δ

NH$_2$

OS, *CV 5*, 32 (1973)

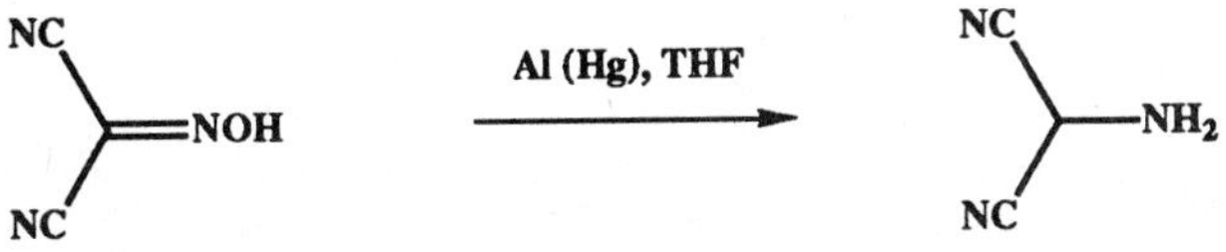

Al (Hg), THF

NC, NC, =NOH → NC, NC, NH_2

OS, *CV 5*, 269 (1973)

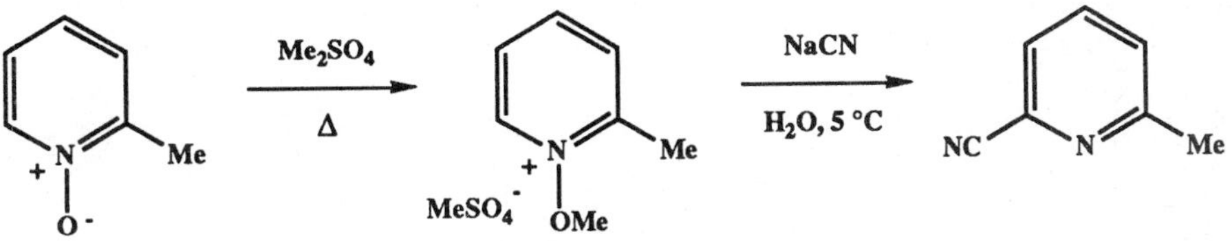

Me_2SO_4, Δ; NaCN, H_2O, 5 °C

Me, +N, O-; $MeSO_4^-$, +N, OMe, Me; NC, N, Me

OS, *CV 5*, 355 (1973)

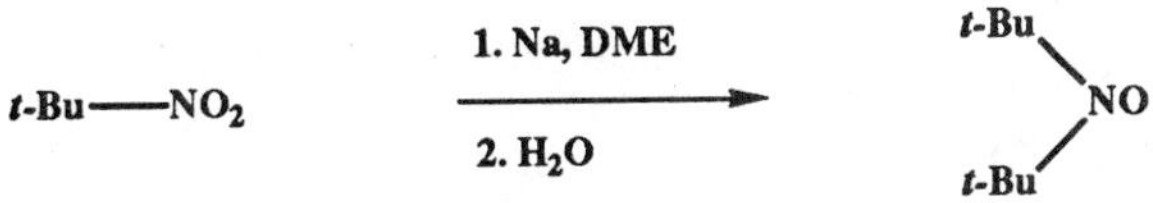

t-Bu—NO_2 → 1. Na, DME; 2. H_2O → *t*-Bu, *t*-Bu, NO

OS, *CV 5*, 419 (1973)

S-Bu, S-Bu → 1. Na, NH_3; 2. NH_4Cl → SH, SH

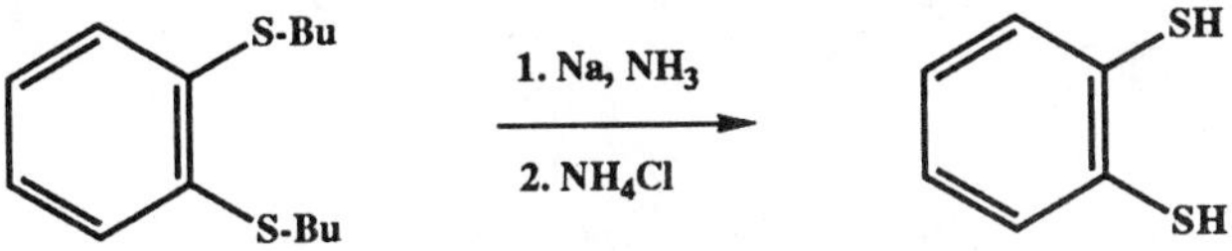

OS, *CV 5*, 552 (1973)

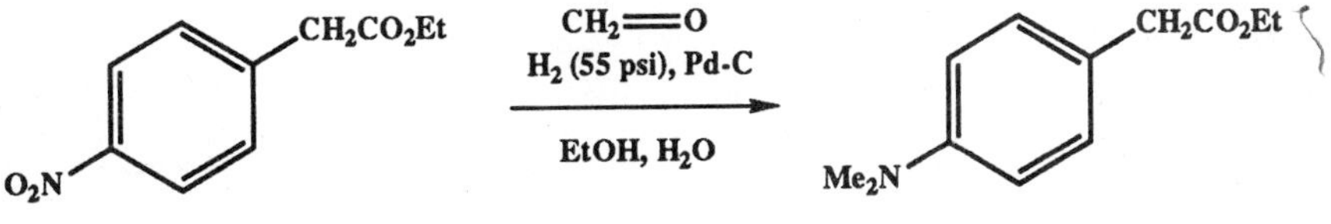

CH_2CO_2Et, O_2N → CH_2=O; H_2 (55 psi), Pd-C; EtOH, H_2O → CH_2CO_2Et, Me_2N

OS, *CV 5*, 586 (1973)

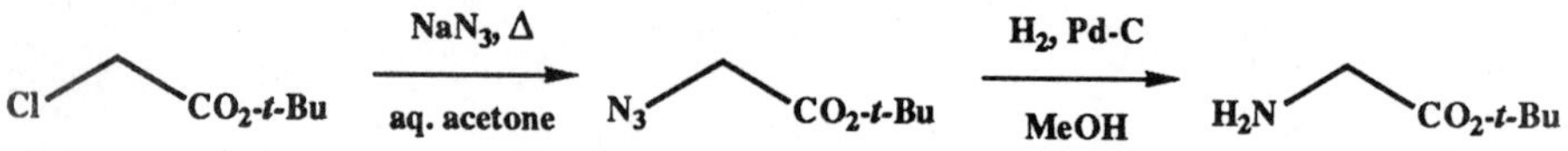

Cl–CH₂–CO_2-*t*-Bu → (NaN_3, Δ / aq. acetone) → N_3–CH₂–CO_2-*t*-Bu → (H_2, Pd-C / MeOH) → H_2N–CH₂–CO_2-*t*-Bu

OS, *CV 5*, 673 (1973)

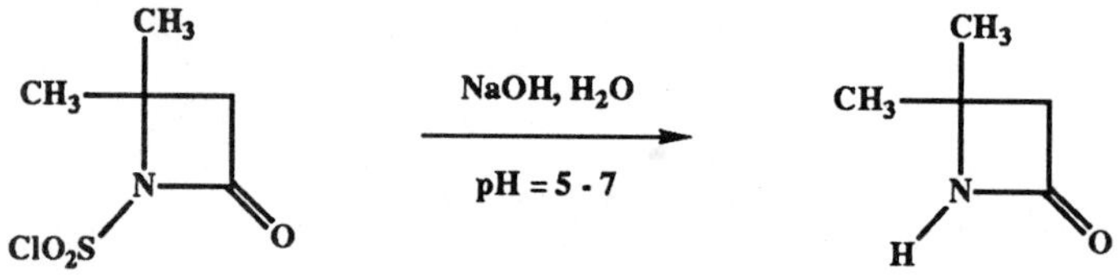

CH_3, CH_3, ClO_2S–N, O → (NaOH, H_2O / pH = 5 - 7) → CH_3, CH_3, H–N, O

OS, *CV 5*, 829 (1973)

NO_2 → (H_2 (25 - 50 psi), 5% Pd - C / EtOH) → NH_2

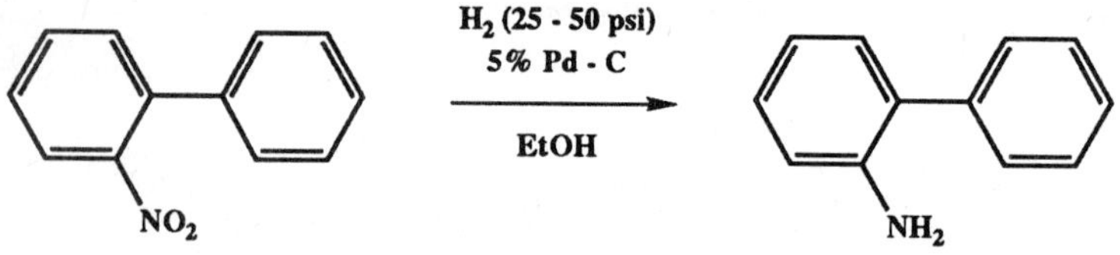

OS, *CV 5*, 843 (1973)

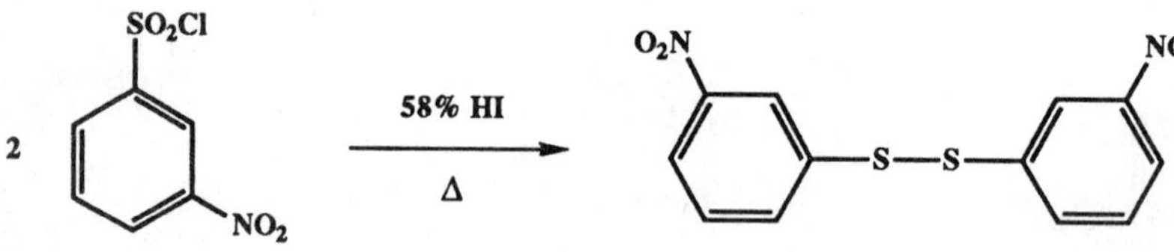

2 SO_2Cl, NO_2 → (58% HI / Δ) → O_2N, S—S, NO_2

OS, CV 5, 1067 (1973)

H_2N, NH_2, O_2N, NO_2 (benzene ring) → Na_2S_2 / H_2O, Δ → H_2N, NH_2, O_2N, NH_2 (benzene ring)

OS, CV 5, 1130 (1973)

Ph, Ph, Ph, NO_2 (benzene ring) → H_2 (1000 psi), Ra (Ni) / dioxane → Ph, Ph, Ph, NH_2 (benzene ring)

OS, CV 6, 130 (1988)

$PhCH_2—S—S—CH_2Ph$ → $(Me_2N)_3P$ / benzene, Δ → $PhCH_2—S—CH_2Ph$ + $(Me_2N)_3PS$

OS, CV 6, 803 (1988)

t-Bu—NO_2 → Al - Hg / H_2O, Et_2O → *t*-Bu—NHOH

OS, *CV 7*, 124 (1990); *61*, 74 (1983)

1. 3 equiv. *n*-BuLi, Et_2O, 25 °C
2. H_2O

SH, H

OS, *CV 7*, 34 (1990); *63*, 214 (1985)

OCH_2Ph, N, NO_2

Ra(Ni), $H_2N\text{-}NH_2$

OCH_2Ph, N, H

OS, *65*, 166 (1987)

t-Bu—CH_2—C(Me)(Me)—N(O-*t*-Bu)-*t*-Bu

Sodium naphthalenide, THF

t-Bu—CH_2—C(Me)(Me)—N(H)-*t*-Bu

OS, *67*, 187 (1988)

NO_2

5% Rh-C, $H_2NNH_2 \cdot H_2O$, THF, 30 °C

NHOH

OS, *CV* 4, 271 (1963)

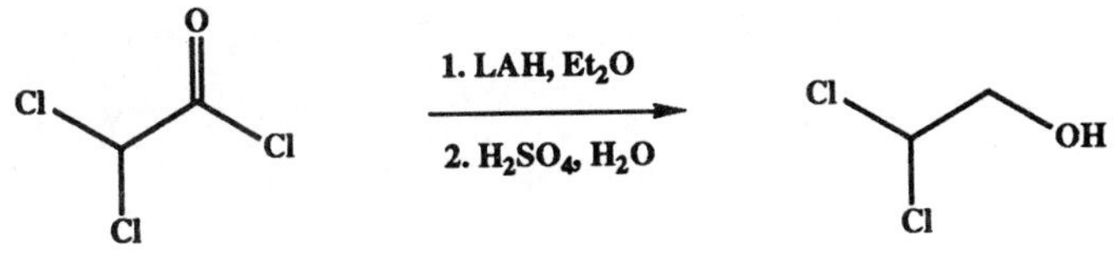

O
Cl
Cl
Cl

1. LAH, Et_2O
2. H_2SO_4, H_2O

Cl
OH
Cl

OS, *CV* 4, 339 (1963)

O
NMe_2

LAH, Et_2O
Δ

NMe_2

OS, *CV* 4, 354 (1963)

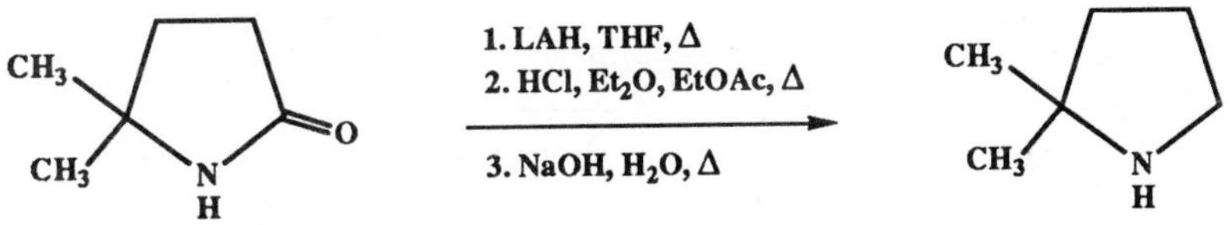

CH_3
CH_3
O
N
H

1. LAH, THF, Δ
2. HCl, Et_2O, EtOAc, Δ
3. NaOH, H_2O, Δ

CH_3
CH_3
N
H

OS, *CV* 4, 564 (1963)

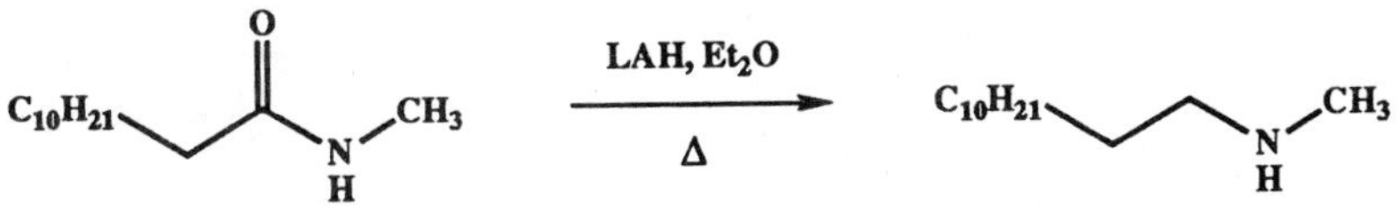

O
$C_{10}H_{21}$
CH_3
N
H

LAH, Et_2O
Δ

$C_{10}H_{21}$
CH_3
N
H

OS, CV 4, 834 (1963)

CO$_2$Et, N, CH$_3$ — 1. LAH, Et$_2$O, Δ; 2. HCl then NaOH → CH$_2$OH, N, CH$_3$

OS, CV 5, 175 (1973)

O, *t*-Bu — 1. LAH, AlCl$_3$, Et$_2$O; 2. *t*-BuOH, Δ; 3. excess ketone, Δ; 4. H$_2$SO$_4$, H$_2$O → *t*-Bu, OH

trans

OS, CV 5, 294 (1973)

O, OEt — LAH, Et$_2$O, Δ → OH, OEt

OS, CV 5, 303 (1973)

O, O — 1. LAH, AlCl$_3$ Et$_2$O, Δ; 2. H$_2$SO$_4$, H$_2$O → OCH$_2$CH$_2$OH

OS, *CV 5*, 692 (1973)

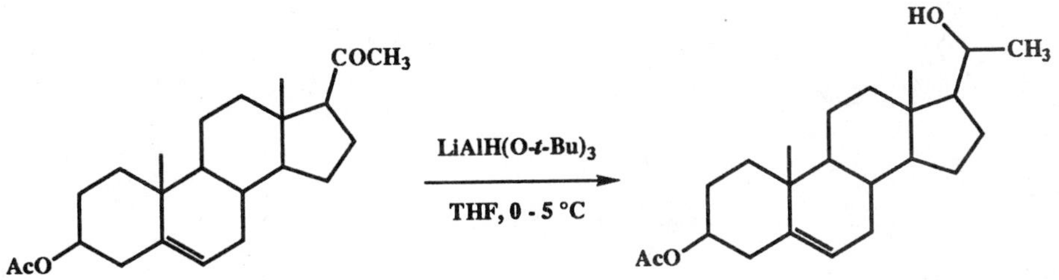

COCH$_3$

AcO

LiAlH(O-*t*-Bu)$_3$

THF, 0 - 5 °C

HO

CH$_3$

AcO

OS, *CV 6*, 62 (1988)

OH

p-Ts-NHN

H H H

NaBH$_4$

MeOH, Δ

OH

H H H

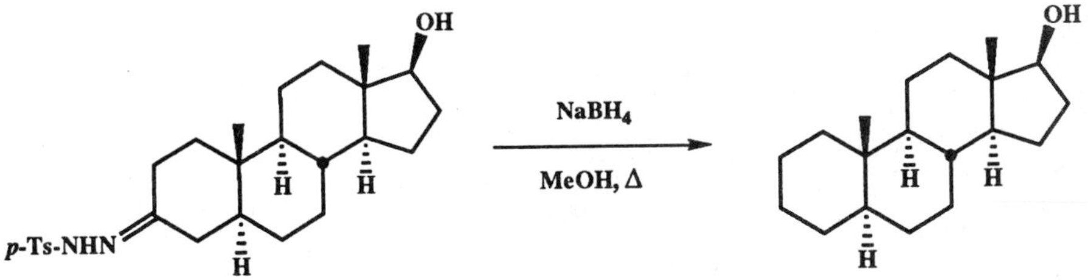

OS, *CV 6*, 109 (1988)

S S Ph

N H

LAH

ZnCl$_2$, CuCl$_2$

THF, Δ

Ph

N H

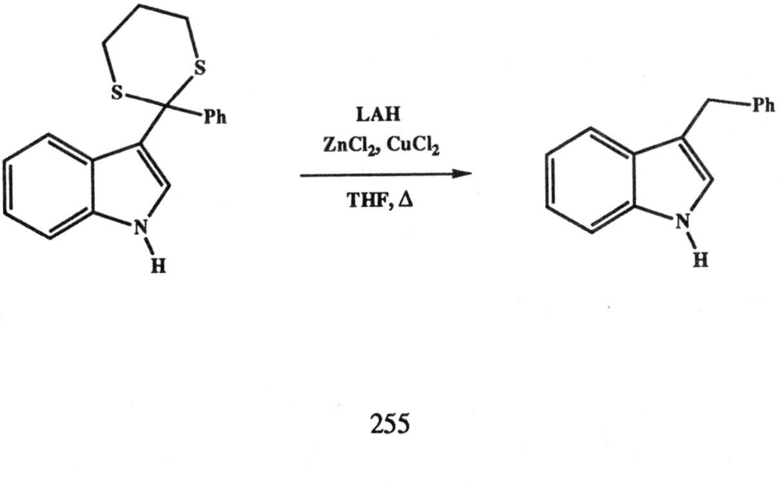

OS, *CV* 6, 142 (1988)

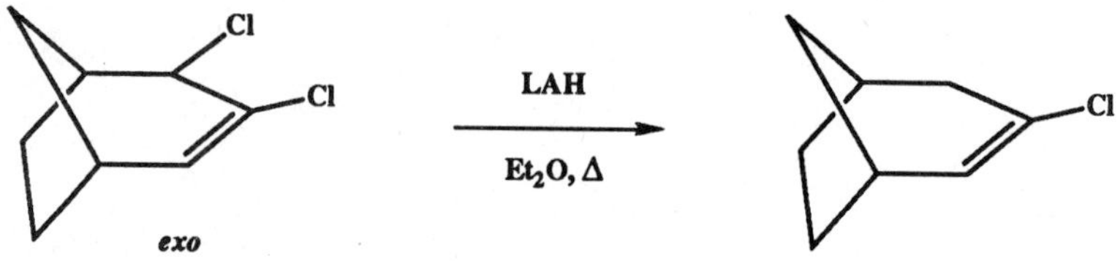

Cl
Cl
exo
LAH
Et_2O, Δ
Cl

OS, *CV* 6, 312 (1988)

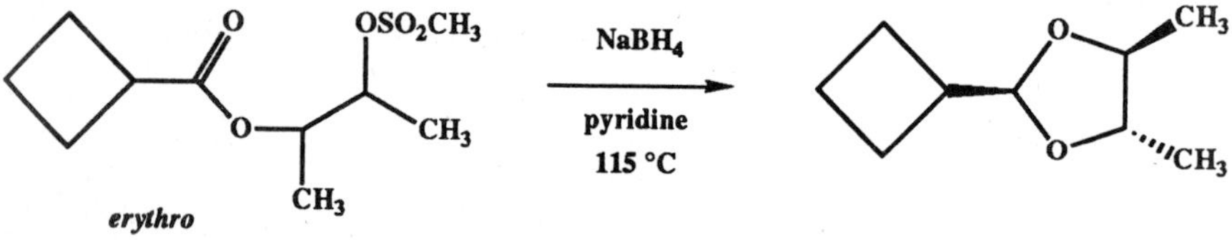

O
OSO_2CH_3
O
CH_3
CH_3
erythro
$NaBH_4$
pyridine
115 °C
O
O
CH_3
CH_3

OS, *CV* 6, 376 (1988)

$$CH_3(CH_2)_8—CH_2I \xrightarrow[\text{HMPA, 70 °C}]{NaBH_3CN} CH_3(CH_2)_8—CH_3$$

$$CH_3(CH_2)_{10}—CH_2OTs \xrightarrow[\text{HMPA, 80 °C}]{NaBH_3CN} CH_3(CH_2)_{10}—CH_3$$

OS, *CV* 6, 382 (1988)

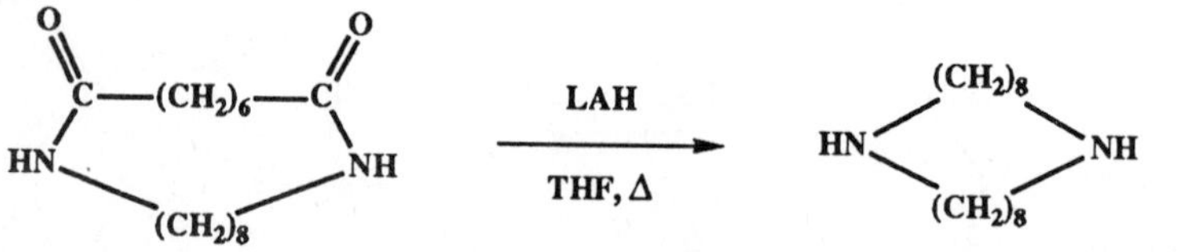

O
O
C—$(CH_2)_6$—C
HN
NH
$(CH_2)_8$
LAH
THF, Δ
$(CH_2)_8$
HN
NH
$(CH_2)_8$

OS, CV 6, 482 (1988)

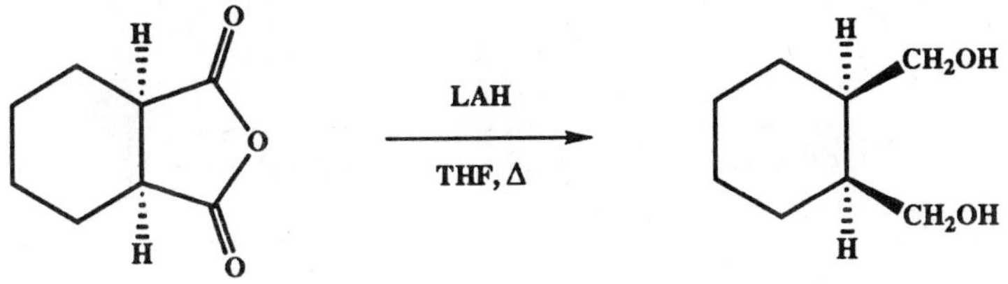

LAH

THF, Δ

CH_2OH

CH_2OH

OS, CV 6, 499 (1988)

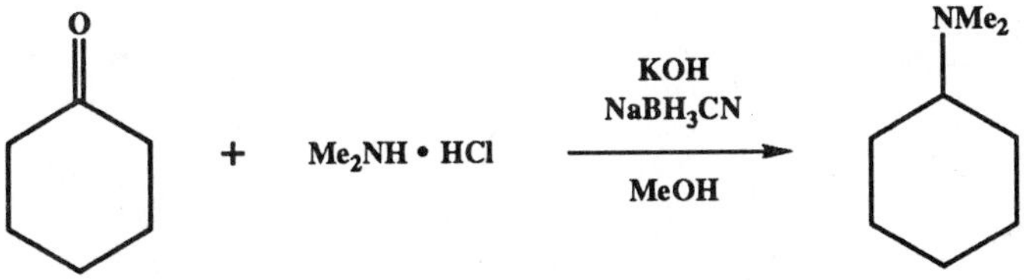

+ $Me_2NH \cdot HCl$

KOH
$NaBH_3CN$

MeOH

NMe_2

OS, CV 6, 529 (1988)

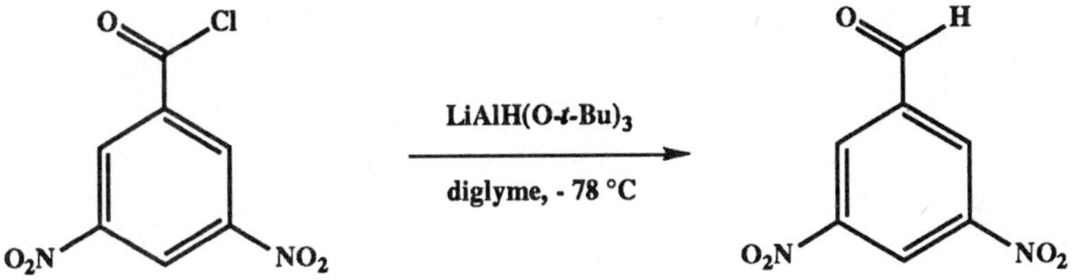

$LiAlH(O\text{-}t\text{-}Bu)_3$

diglyme, - 78 °C

O_2N NO_2

OS, CV 6, 766 (1988)

OH CH_3 H HgOAc

$NaBH_4$, NaOH

H_2O, Et_2O

OH CH_3

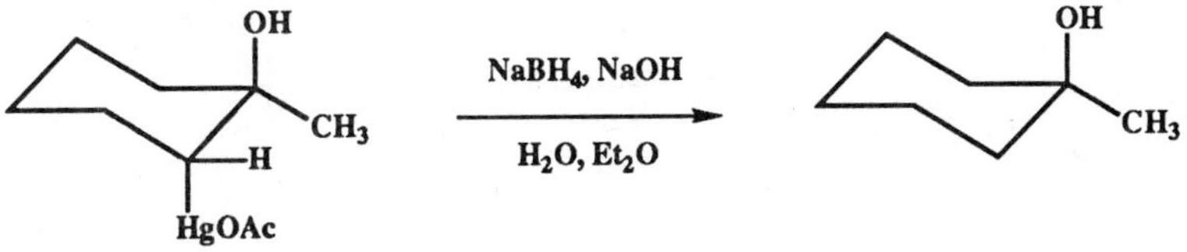

***OS*, *CV 6*, 769 (1988)**

LAH; Et_2O, 0 °C

Zn amalgam, HCl; Et_2O, - 15 °C

CH_3 CH_3 CH_3 OH O

***OS*, *CV 6*, 781 (1988)**

EtO_2C CH_3 Et N O

LAH; Et_2O, Δ

HO CH_3 Et N O

***OS*, *CV 6*, 887 (1988)**

LAH; Et_2O, Δ

O O H OH H OH

***OS*, *CV 6*, 905 (1988)**

CH_3 CH_3 CH_3 N O Ph

$NaBH_4$; 9 *N* HCl, - 35 to - 45 °C

CH_3 CH_3 CH_3 N H O Ph

OS, CV 7, 129 (1990); *60*, 25 (1981)

LAH

OH OH

OH

OS, CV 7, 393 (1990); *60*, 108 (1981)

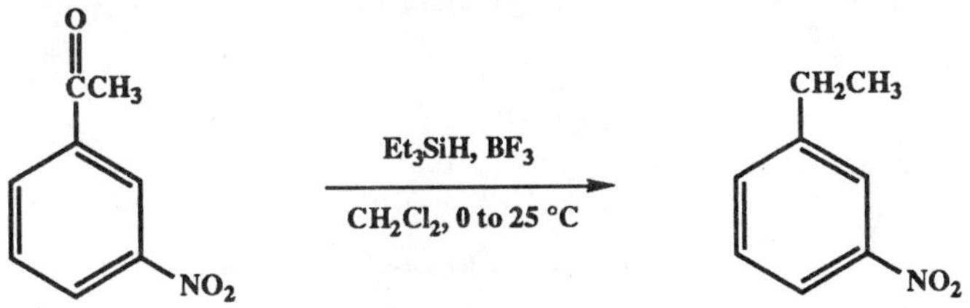

O
CCH_3 NO_2

Et_3SiH, BF_3
CH_2Cl_2, 0 to 25 °C

CH_2CH_3 NO_2

OS, CV 7, 41 (1990); *61*, 24 (1983)

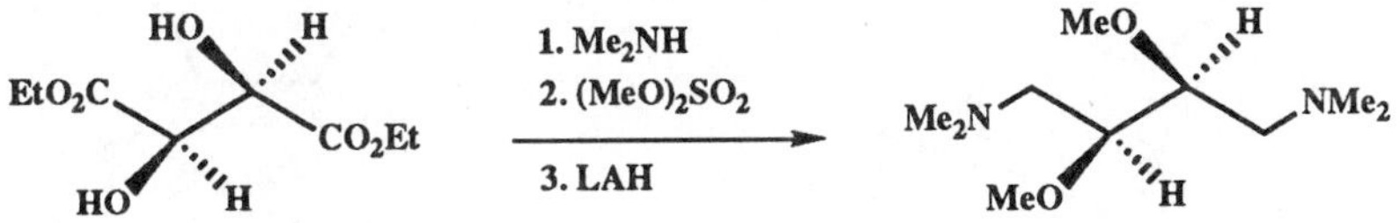

HO H EtO_2C CO_2Et HO H

(+) - (*R*, *R*)

1. Me_2NH
2. $(MeO)_2SO_2$
3. LAH

MeO H Me_2N NMe_2 MeO H

(+) - (*S*, *S*) "DDB"

OS, CV 7, 530 (1990); *63*, 136 (1985)

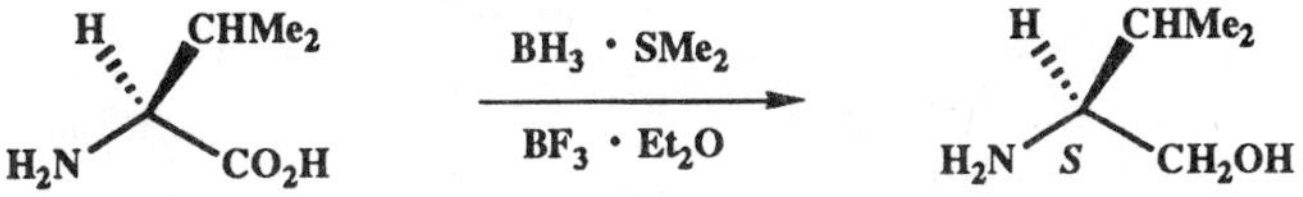

H $CHMe_2$ H_2N CO_2H

$BH_3 \cdot SMe_2$
$BF_3 \cdot Et_2O$

H $CHMe_2$ H_2N *S* CH_2OH

OS, *CV 7*, 356 (1990); *63*, 140 (1985)

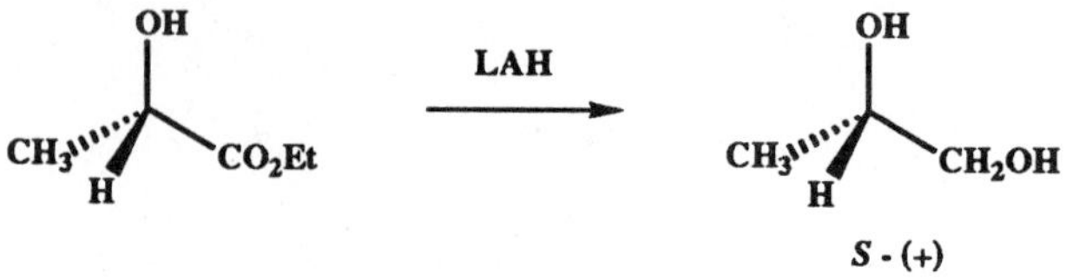

OH
LAH
CH_3 H CO_2Et
OH
CH_3 H CH_2OH
S - (+)

OS, *CV 7*, 456 (1990); *64*, 10 (1986)

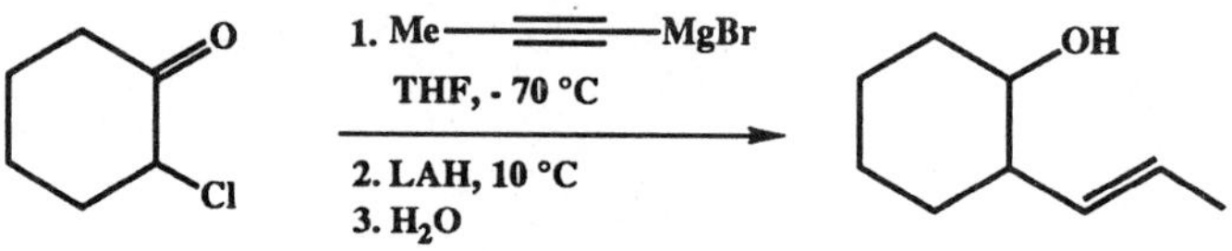

O
Cl
1. Me—≡—MgBr
THF, - 70 °C
2. LAH, 10 °C
3. H_2O
OH

OS, *CV 7*, 139 (1990); *64*, 57 (1986)

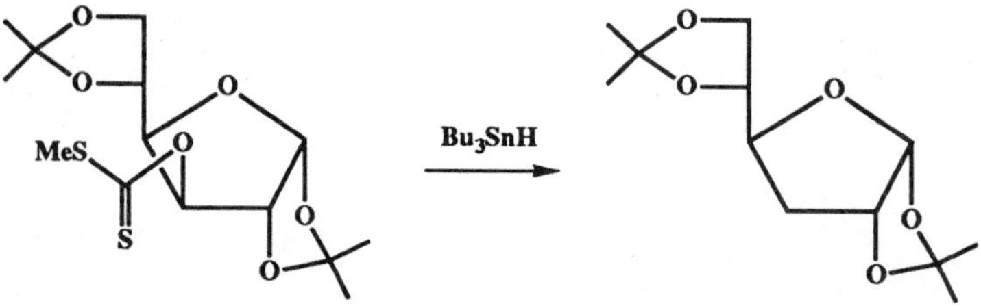

MeS
S
Bu_3SnH

OS, *CV 7*, 241 (1990); *64*, 73 (1986)

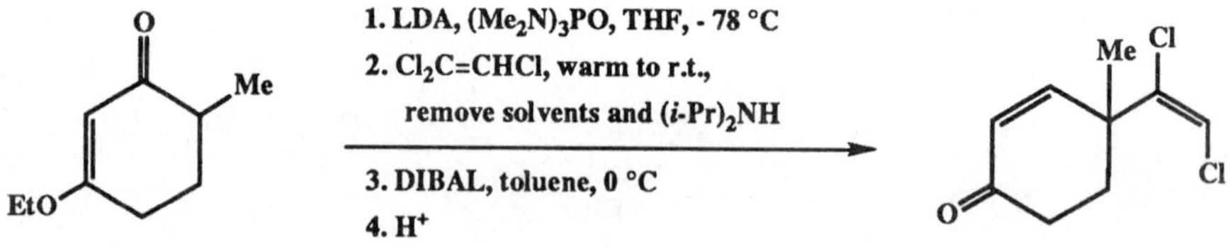

O
Me
EtO
1. LDA, $(Me_2N)_3PO$, THF, - 78 °C
2. $Cl_2C{=}CHCl$, warm to r.t., remove solvents and $(i\text{-}Pr)_2NH$
3. DIBAL, toluene, 0 °C
4. H^+
Me Cl
Cl
O

OS, *CV 7*, 221 (1990); *64*, 104 (1986)

HO_2C–CH=CH–CO_2Et → ($BH_3 \cdot THF$; - 10 °C to r.t.) → $HOCH_2$–CH=CH–CO_2Et

OS, *CV 7*, 524 (1990); *64*, 182 (1986)

Me_3Si–C≡C–CH_2OH → ($NaAlH_2(OCH_2CH_2OMe)_2$ ("SMEAH"); Et_2O, toluene, 20 °C) → Me_3Si(H)C=C(H)CH_2OH

OS, *CV 7*, 476 (1990); *64*, 189 (1986)

SMe

N N N → ($NaBH_4$; $NiCl_2 \cdot 6\ H_2O$) → N N N

OS, *65*, 173 (1987)

S [or *R*] N H CO_2H → (LAH; THF) → *S* [or *R*] N H OH

OS, *66*, 121 (1987)

$C_4H_9C(O)(CH_2)_4CO_2H$ → [Cl–CH=N⁺(Me)Me Cl⁻] → $C_4H_9C(O)(CH_2)_4COCl$ → [$LiAlH(t\text{-}BuO)_3$] → $C_4H_9C(O)(CH_2)_4CHO$

OS, *66*, 160 (1987)

CO_2H, Cl, H, R → [LAH, Et_2O] → CH_2OH, Cl, H, R → [KOH, H_2O] → H_2C, O, H, R

S isomers

R = Me, *i*-Pr, *i*-Bu, (*S*) - *sec*-Bu

R isomers

OS, *66*, 185 (1987)

N–OSO_2Me → [Pr_3Al, CH_2Cl_2, hexane] → [N, Pr] → [$(i\text{-}Bu)_2AlH$, hexane] → H, N, Pr

OS, *67*, 69 (1988)

BocNH—CH(*S*)—CON(Me)—OMe, CH_2CHMe_2 → [LAH, ether] → BocNH—CH(*S*)—CHO, CH_2CHMe_2

***OS*, *68*, 77 (1989)**

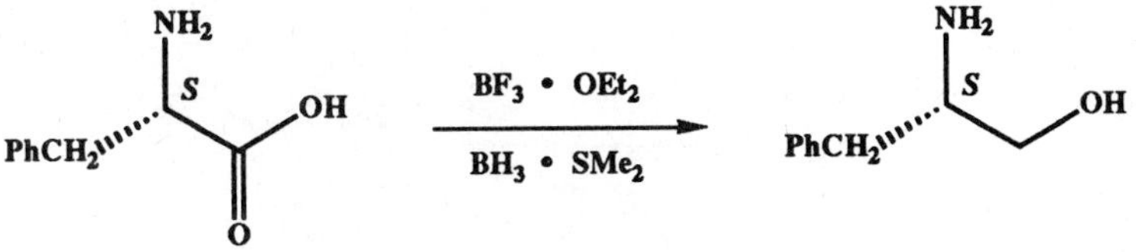

NH$_2$ S OH PhCH$_2$ O

BF$_3$ • OEt$_2$

BH$_3$ • SMe$_2$

NH$_2$ S OH PhCH$_2$

***OS*, *68*, 92 (1989)**

MeO$_2$C CO$_2$Me O O Me Me

l isomer

LAH, Et$_2$O

HOCH$_2$ CH$_2$OH O O Me Me

***OS, CV 1*, 101 (1941)**

H_2 (3 atm)
PtO_2

EtOAc

***OS, CV 1*, 240 (1941)**

H_2 (3 atm)
PtO_2

EtOH

***OS, CV 2*, 191 (1943)**

H_2, PtO_2, HOAc
(R = H) *or*

1. H_2, PtO_2, HOAc
2. NaOH, aq. EtOH
(R = Ac)

***OS, CV 2*, 325 (1943)**

1. H_2 (2000-3000 psi)
$CuCr_2O_4$ (cat.)
255 °C

2. KOH, EtOH, Δ

OS, CV 2, 491 (1943)

H_2 (40 psi), PtO_2, HOAc

OS, CV 2, 566 (1943)

H_2 (100 psi), PdO

OS, CV 3, 148 (1955)

H_2SO_4, H_2O, Δ; NaOH, H_2O, Δ; H_2, Ra (Ni), dioxane, 60 °C

OS, CV 3, 229 (1955)

$NC—(CH_2)_8—CN \xrightarrow[\text{NH}_3\text{, EtOH, 125 °C}]{\text{H}_2\text{ (1500 psi), Ra (Ni)}} H_2NCH_2—(CH_2)_8—CH_2NH_2$

OS, CV 3, 278 (1955)

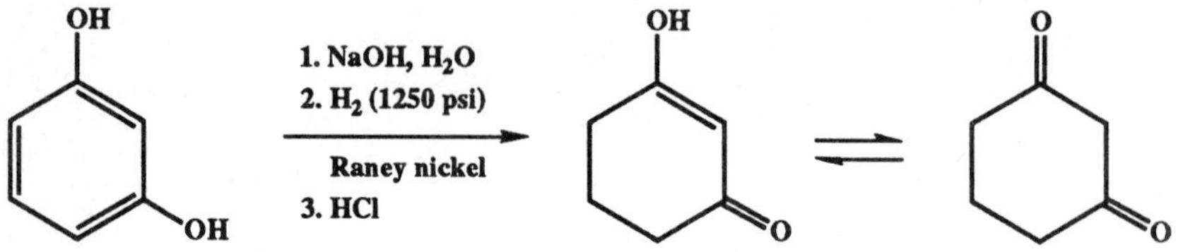

1. NaOH, H_2O
2. H_2 (1250 psi)

Raney nickel

3. HCl

OH OH OH O O O

OS, CV 3, 328 (1955)

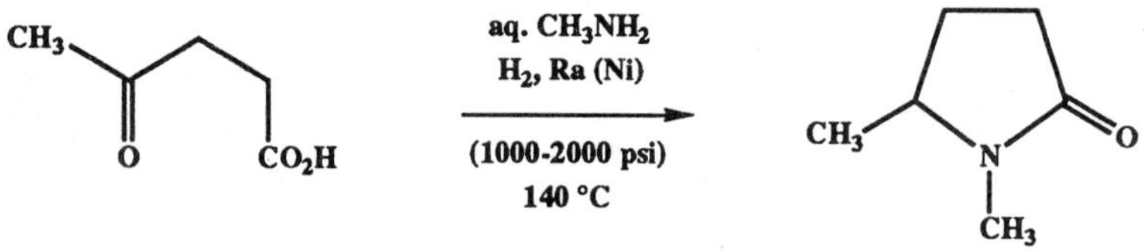

aq. CH_3NH_2

H_2, Ra (Ni)

(1000-2000 psi)

140 °C

CH_3 O CO_2H CH_3 N O CH_3

OS, CV 3, 358 (1955)

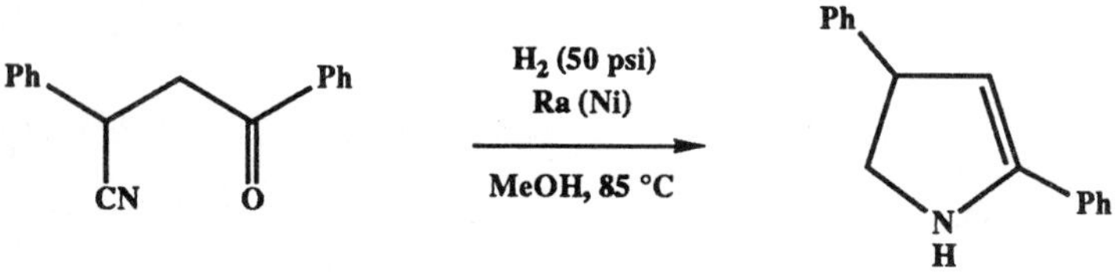

H_2 (50 psi)

Ra (Ni)

MeOH, 85 °C

Ph Ph CN O Ph Ph N H

OS, CV 3, 385 (1955)

H_2 (15 - 30 psi)

Pd-C

piperidine

HOAc

O n-C_3H_7 H + CO_2Et CN n-C_3H_7 CO_2Et CN

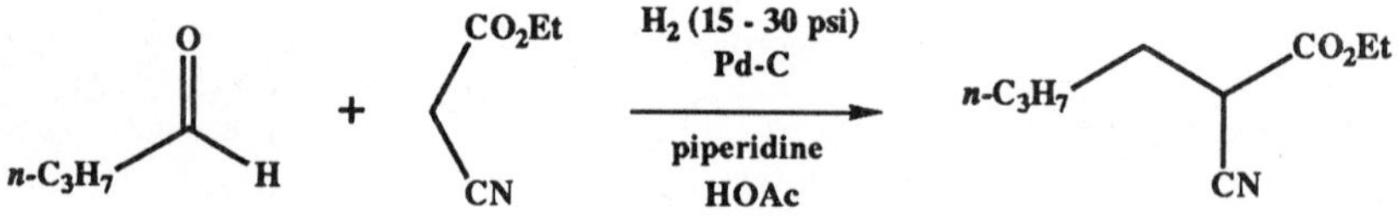

OS, *CV* 3, 501 (1955)

CH_3COCH_3 + $H_2NCH_2CH_2OH$ → [H_2 (25 psi), PtO_2, EtOH] → $(CH_3)_2CHNHCH_2CH_2OH$

OS, *CV* 3, 519 (1955)

2-chloro-4-methylquinoline → [H_2 (30 psi), Pd-C, NaOAc, HOAc, 55 - 70 °C] → 4-methylquinoline

OS, *CV* 3, 551 (1955)

2,4,6-trimethylbenzoyl chloride → [H_2, Pd - $BaSO_4$, xylene, Δ] → 2,4,6-trimethylbenzaldehyde

OS, *CV* 3, 627 (1955)

2-naphthoyl chloride → [H_2, Pd - $BaSO_4$, NaOH, xylene, 140 - 150 °C] → 2-naphthaldehyde

OS, CV 3, 693 (1955)

OH → H_2 (3300-6200 psi), $CuCr_2O_4$, 250 - 300 °C → OH, OH

OS, CV 3, 717 (1955)

O, Ph, CH_3 → H_2, Ra (Ni) (3500-5000 psi), NH_3, 150 °C → NH_2, Ph, CH_3

OS, CV 3, 720 (1955)

Ph, CN → H_2, Ra (Ni) (2000 psi), NH_3, 130 °C → Ph, NH_2

OS, CV 3, 742 (1955)

2 O, $CH{=}CHCO_2H$ → Ra (Ni), Δ, aq. NaOH → O, $CH_2CH_2CO_2H$ + O, O, $CH_2CH_2CH_3$

OS, CV 3, 794 (1955)

O → H_2 (40 psi), Raney nickel, Et_2O → O

OS, *CV 3*, 827 (1955)

CH_3 NH_2 —PhCHO→ CH_3 N=CHPh —H_2 (1000 psi), Ra (Ni), Et_2O→ CH_3 $NHCH_2Ph$

OS, *CV 4*, 136 (1963)

CO_2H CO_2H —Raney nickel, aq. NaOH, Δ→ CO_2H CO_2H

OS, *CV 4*, 216 (1963)

$(CH_2)_8$ =O OH —H_2 (135 atm), Cu chromite, 150 °C→ $(CH_2)_8$ OH OH

~ 1.7 : 1 *cis* : *trans*

OS, *CV 4*, 221 (1963)

HO CH_2NO_2 —H_2 (40 psi), Raney nickel, HOAc→ HO CH_2NH_2 • HOAc

***OS, CV* 4, 298 (1963)**

$$EtO{-}CH{=}C(CO_2Et)_2 \xrightarrow[\text{EtOH, 45 °C}]{H_2,\ \text{Ra (Ni)}\ (1000\text{-}1500\ \text{psi})} EtOCH_2CH(CO_2Et)_2 \xrightarrow{\Delta} CH_2{=}C(CO_2Et)_2$$

***OS, CV* 4, 304 (1963)**

H_2 (2 atm), Pt *or* Pd-C, EtOH

CO2Et, CO2Et

***OS, CV* 4, 313 (1963)**

H_2 (2000-2900 psi), Cu - Cr oxide cat., cyclohexane, 150 °C

***OS, CV* 4, 357 (1963)**

H_2 (1000 psi), Raney nickel, EtOH, 55 - 60 °C

CH_3, CH_3, NO_2, O, OCH_3 → CH_3, CH_3, N, H, O

OS, CV 4, 408 (1963)

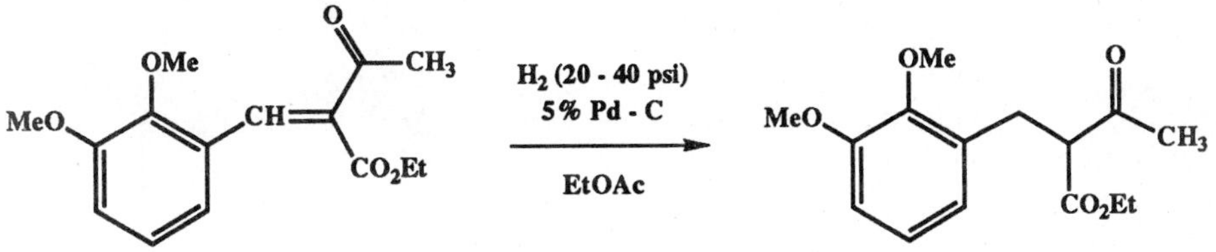

MeO, OMe substituted benzylidene acetoacetate (CH=C(COCH$_3$)CO$_2$Et) → saturated product

H_2 (20 - 40 psi)
5% Pd - C
EtOAc

OS, CV 4, 603 (1963)

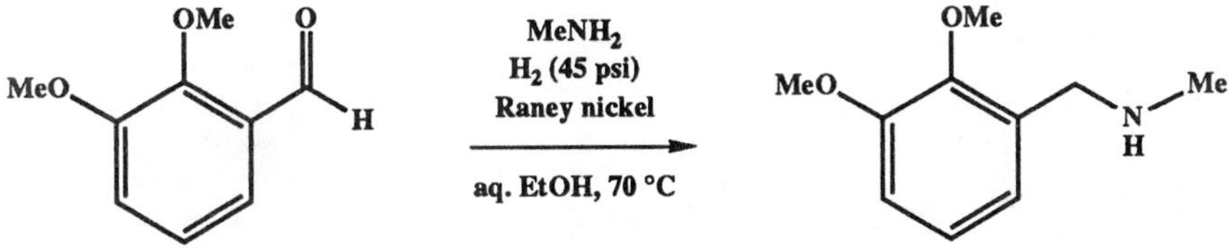

MeNH$_2$
H_2 (45 psi)
Raney nickel
aq. EtOH, 70 °C

OS, CV 4, 638 (1963)

Ra (Ni)
NH_4OH, Δ

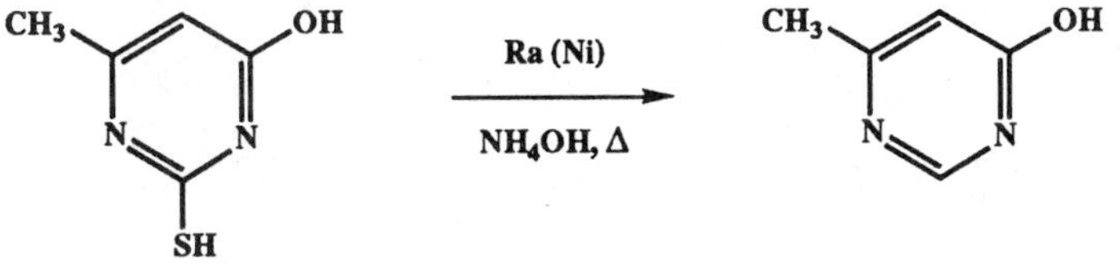

OS, CV 4, 660 (1963)

H_2 (1625 psi)
Raney nickel
125 °C

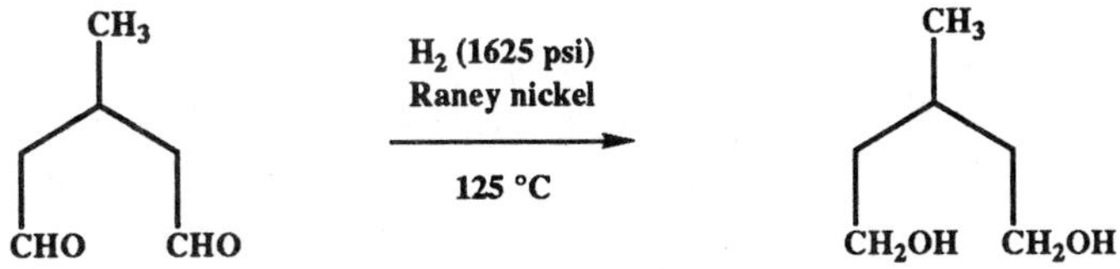

OS, CV 4, 887 (1963)

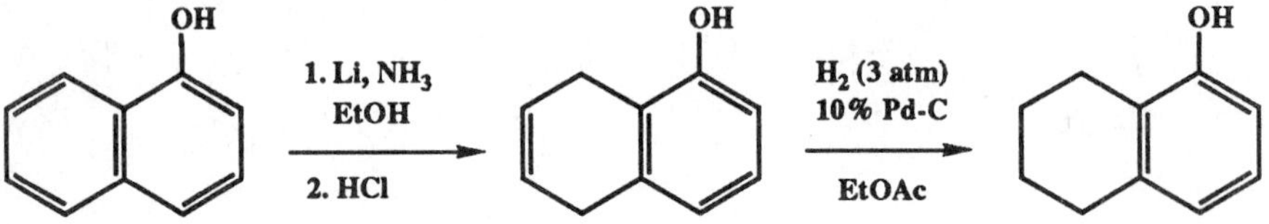

OH → 1. Li, NH_3, EtOH; 2. HCl → OH → H_2 (3 atm), 10% Pd-C, EtOAc → OH

OS, CV 5, 16 (1973)

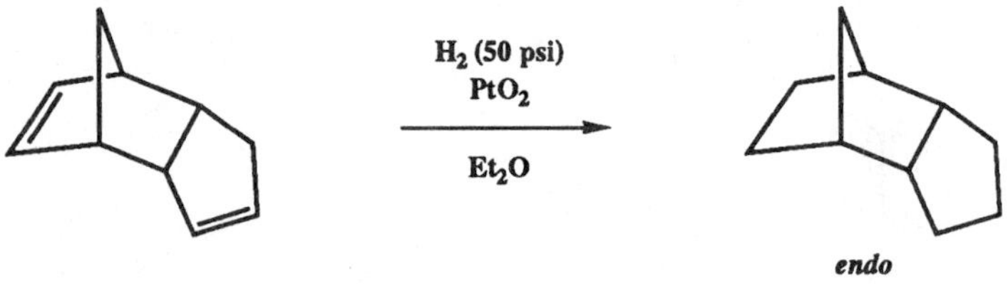

H_2 (50 psi), PtO_2, Et_2O

endo

OS, CV 5, 96 (1973)

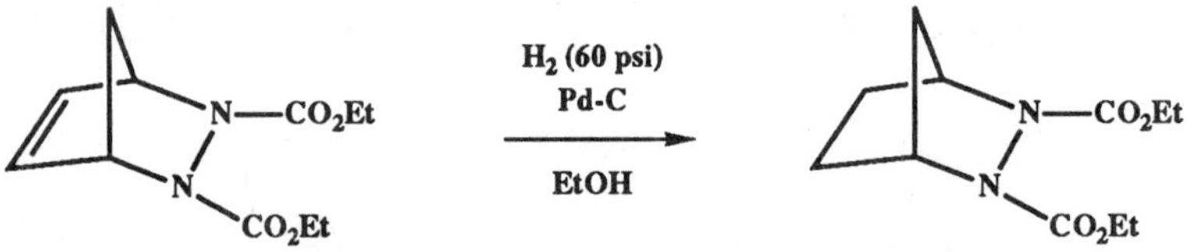

N—CO_2Et, N, CO_2Et → H_2 (60 psi), Pd-C, EtOH → N—CO_2Et, N, CO_2Et

OS, CV 5, 277 (1973)

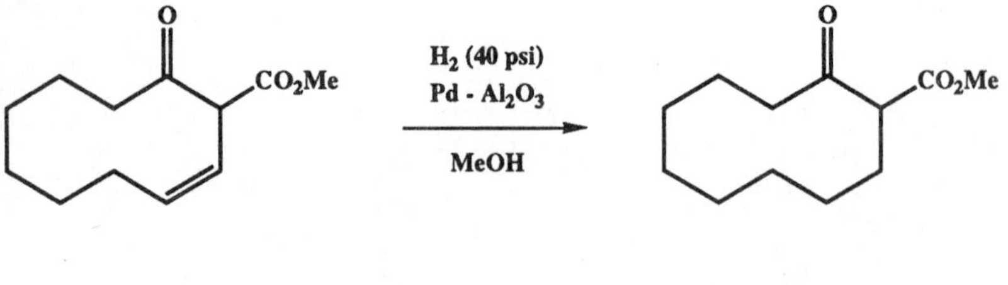

O, CO_2Me → H_2 (40 psi), Pd - Al_2O_3, MeOH → O, CO_2Me

OS, *CV 5*, 346 (1973)

Br, NO_2, N, NH_2 — Fe, HCl, Δ / aq. EtOH → Br, NH_2, N, NH_2 — H_2 / Pd - $SrCO_3$ / aq. NaOH → NH_2, N, NH_2

OS, *CV 5*, 376 (1973)

EtO_2C, NOH, EtO_2C — 1. H_2 (50 - 60 psi) Pd-C, EtOH / 2. HCl, Et_2O → EtO_2C, NH_2 • HCl, EtO_2C

OS, *CV 5*, 591 (1973)

HO, HO, CO_2H, HO — H_2 (2200 psi) / Rh - Al_2O_3 / EtOH, 100 °C → HO, HO, CO_2H, HO

OS, *CV 5*, 670 (1973)

CO_2H, H_2N — H_2 (50 psi) / 10% Rh-C / H_2O → CO_2H, H_2N

cis and *trans*

OS, CV 5, **743 (1973)**

OH, OH

1. H_2 (1900 psi), Ra (Ni)
aq. NaOH, 50 °C

2. CH_3I, dioxane, Δ

O, CH_3, O

OS, CV 5, **880 (1973)**

H

H_2 (1 atm)
"Conditioned
Pd / $CaCO_3$"

quinoline, hexane

OS, CV 5, **989 (1973)**

CO_2Et, N

$BrCH_2CO_2Et$

EtOH, Δ

CO_2Et, N^+, Br^-, CH_2CO_2Et

1. H_2, Pd-C
(100 atm, 90 °C)

2. aq. K_2CO_3
$CHCl_3$, 0 °C

CO_2Et, N, CH_2CO_2Et

OS, CV 6, **68 (1988)**

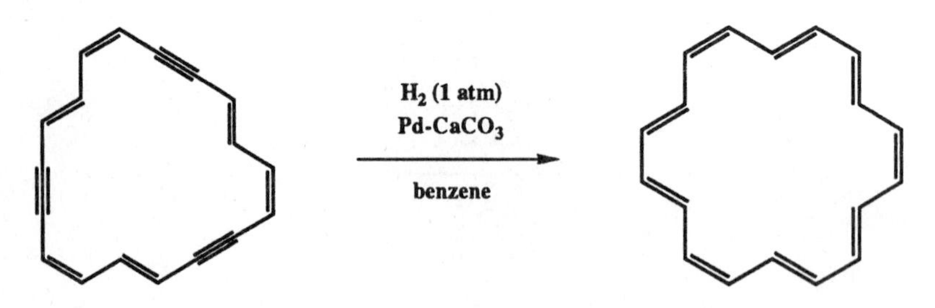

OS, CV 6, 150 (1988)

H_2 (40 psi)
5% Pd-C
benzene
40 °C

N—N N N O Ph

OS, CV 6, 252 (1988)

PhCH$_2$O—C(=O)—NH—CH(—(CH$_2$)$_2$SMe)—CO$_2$H → H$_2$N—CH(—(CH$_2$)$_2$SMe)—CO$_2$H

H_2 (1 atm)
Pd, Et_3N
Me_2NCOMe
liq. NH_3, - 33 °C

l isomer

l - methionine

PhCH$_2$O—C(=O)—NH—CH(CH$_2$O-*t*-Bu)—C(=O)—NH—CH(CH$_2$S-*t*-Bu)—CO$_2$-*t*-Bu —(same)→ H$_2$N—CH(CH$_2$O-*t*-Bu)—C(=O)—NH—CH(CH$_2$S-*t*-Bu)—CO$_2$-*t*-Bu

l, l isomer

OS, CV 6, 371 (1988)

OH

H_2 (60 psi)
Rh - Al_2O_3
EtOH, HOAc

OH

***OS*, *CV 6*, 378 (1988)**

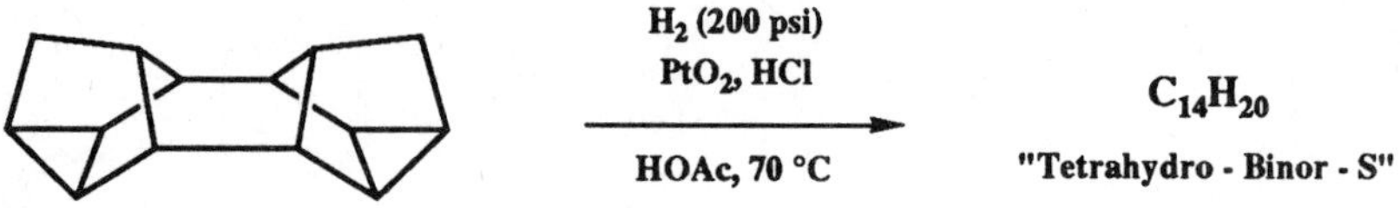

H_2 (200 psi)
PtO_2, HCl
HOAc, 70 °C

$C_{14}H_{20}$
"Tetrahydro - Binor - S"

***OS*, *CV 6*, 395 (1988)**

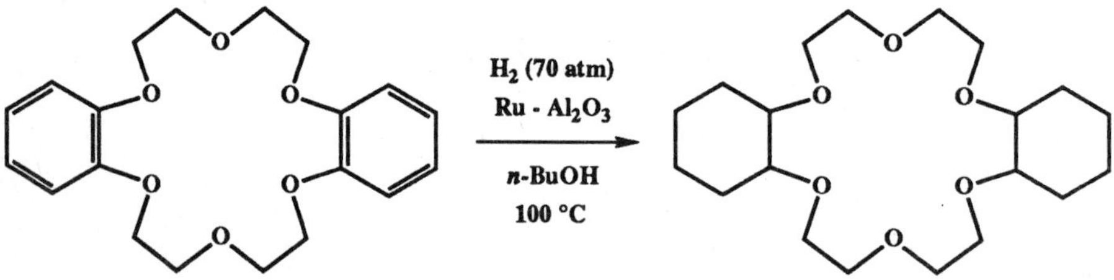

O O O O O O

H_2 (70 atm)
Ru - Al_2O_3
n-BuOH
100 °C

O O O O O O

***OS*, *CV 6*, 459 (1988)**

O

H_2, $(Ph_3P)_3RhCl$
benzene

O

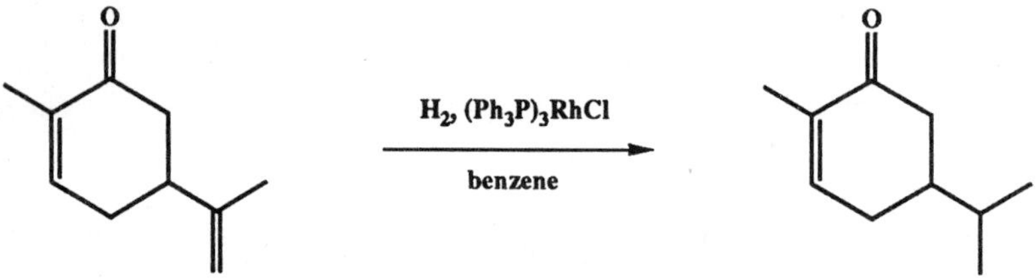

***OS*, *CV 6*, 581 (1988)**

CO_2Et
NH_2 SCH_3

Raney nickel
EtOH, 25 °C

CO_2Et
CH_3
NH_2

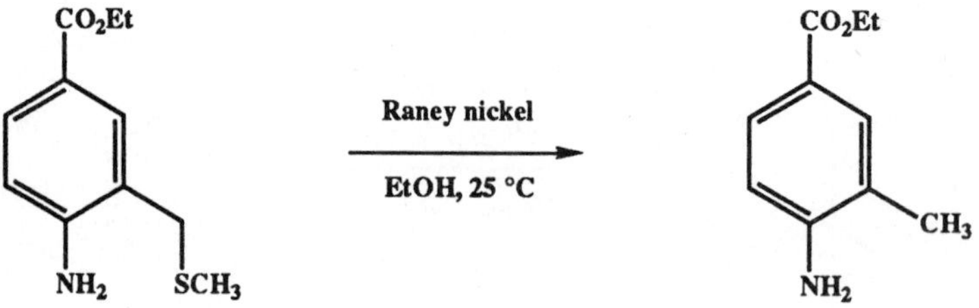

***OS*, *CV 6*, 601 (1988)**

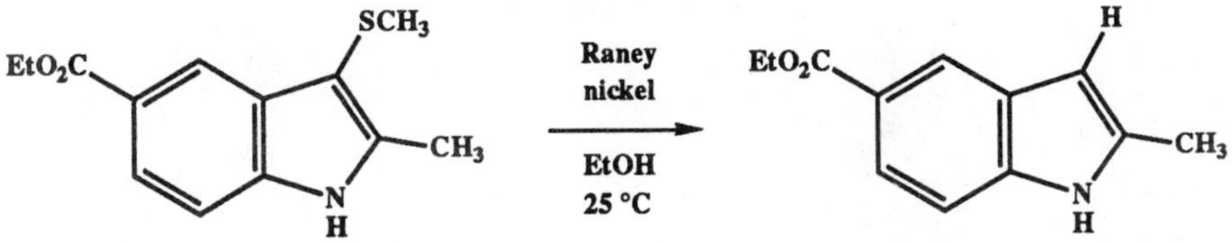

Raney nickel

EtOH

25 °C

***OS*, *CV 6*, 631 (1988)**

Raney nickel

HCO_2H, Δ

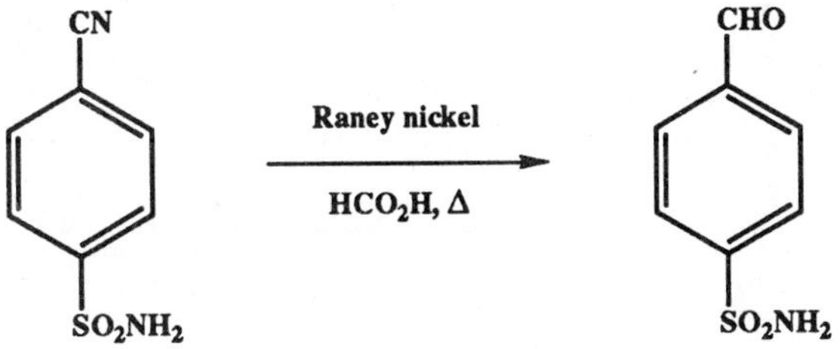

***OS*, *CV 6*, 856 (1988)**

H_2 (170 atm.)

5% Pd-C

EtOH, 105 °C

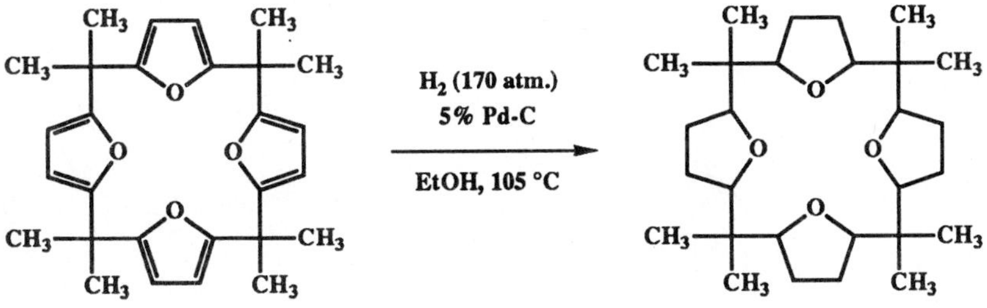

OS, *CV 6*, 1007 (1988)

MeO, MeO, MeO, O, Cl → H_2 (50 psi), Pd-C, "Quinoline S", NaOAc, toluene → MeO, MeO, MeO, O, H

OS, *CV 7*, 287 (1990); *60*, 72 (1981)

+ Cl^-, NH, Bz, N, H, N, H, N, Bz → H_2, Pd/C, EtOH, 50 psi → + Cl^-, NH, Bz, N, H, N, H, N, Bz

OS, *CV 7*, 433 (1990); *60*, 104 (1981)

Ph, N_3 → 1. Ra (Ni), 2. HCl → Ph, NH_3^+, Cl^-

OS, *CV 7*, 27 (1990); *61*, 14 (1983)

PhCHO + H_2N–(N, N, N, N, H) → 1. Et_3N, MeOH, 50 °C; 2. H_2 (500 psi), Pd / C → $PhCH_2NH$–(N, N, N, N, H)

OS, *CV 7*, 417 (1990); *63*, 18 (1985)

$+ H_2$ → "[(*S*,*S*)-BPPM]Rh(COD)Cl", 50 atm, 30 - 50 °C → *D* - (-)

OS, *CV 7*, 226 (1990); *64*, 108 (1986)

Me—≡—CO_2Et → H_2 (1 atm), Pd / $BaSO_4$, quinoline, Et_2O → Me, CO_2Et, H, H

OS, *68*, 64 (1989)

CH_2, MeO_2C, OH + Ph_2P, Rh, Ph_2P (0.4 mol %), $CF_3SO_3^-$ → H_2, MeOH → CH_3, MeO_2C, OH

OS, *68*, 182 (1989)

Me_3Si—≡—OH → H_2 (1 atm), Pd / $BaSO_4$, quinoline → Me_3Si, OH

OS, *68*, 227 (1989)

$PhCH_2$—N, N—CH_2Ph → H_2 (60 psi), Pd-C, EtOH → NH, HN

OS, CV 1, 60 (1941)

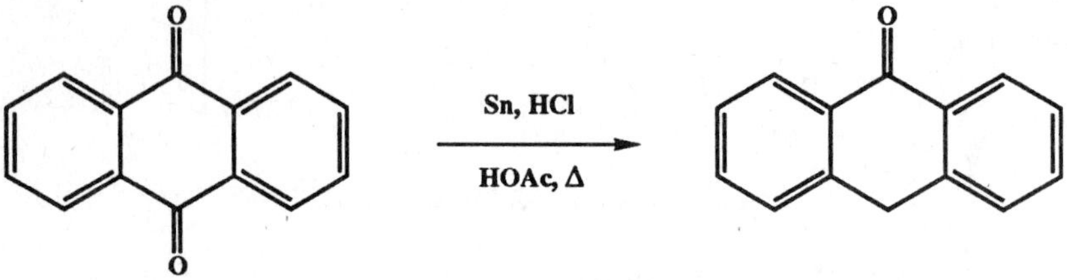

Sn, HCl

HOAc, Δ

OS, CV 1, 90 (1941)

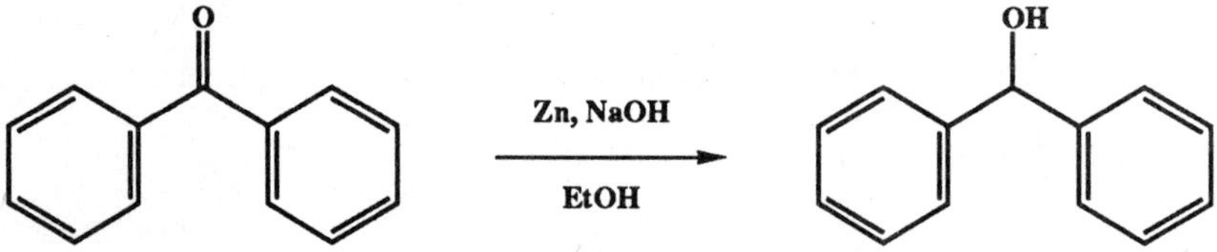

Zn, NaOH

EtOH

OS, CV 1, 99 (1941)

Na, EtOH

Δ

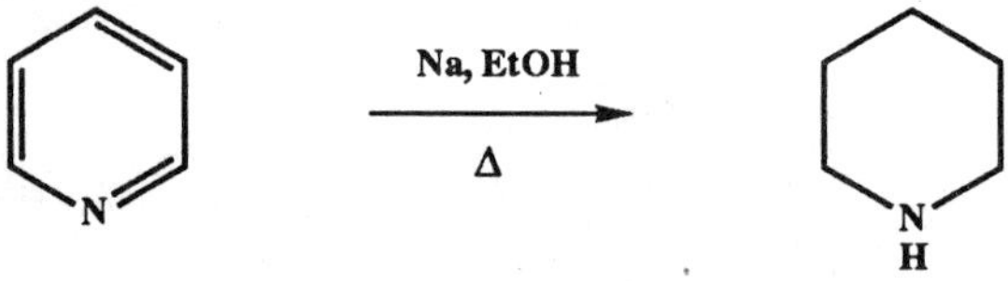

OS, CV 1, 133 (1941)

1. 2 H_2SO_4
$NaNO_2$, EtOH

2. Cu bronze

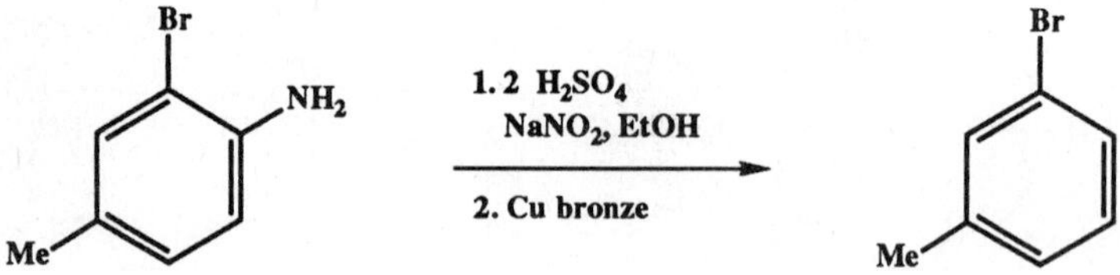

OS, CV 1, 304 (1941)

$$CH_3(CH_2)_5{-}CHO \xrightarrow[H_2O,\ \Delta]{Fe,\ HOAc} CH_3(CH_2)_5{-}CH_2OH$$

OS, CV 1, 499 (1941)

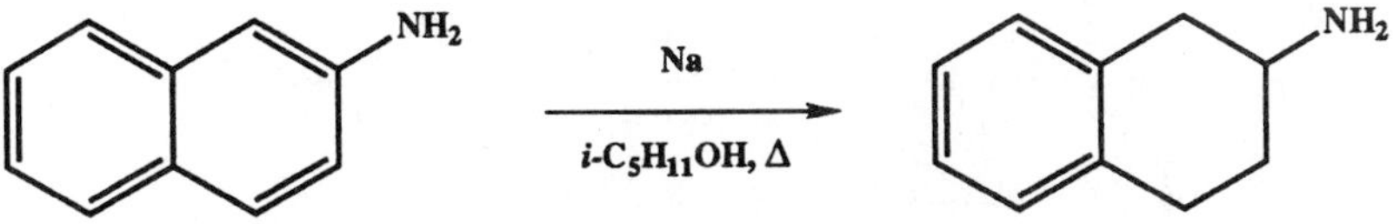

Na

i-$C_5H_{11}OH$, Δ

NH₂

OS, CV 1, 506 (1941)

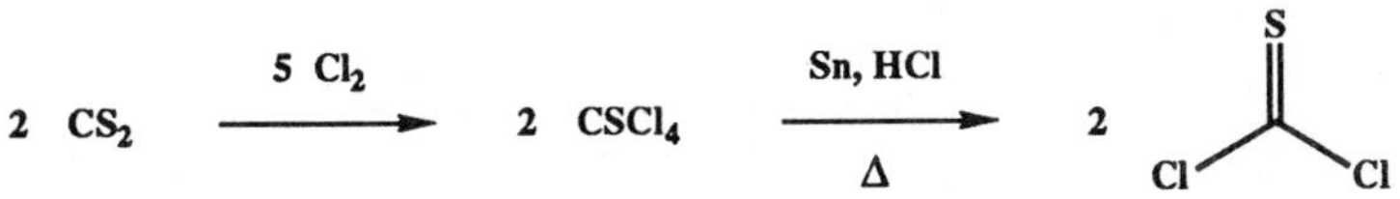

$$2\ CS_2 \xrightarrow{5\ Cl_2} 2\ CSCl_4 \xrightarrow[\Delta]{Sn,\ HCl} 2\ Cl_2C{=}S$$

OS, CV 1, 554 (1941)

Na (Hg)

EtOH

O

OH

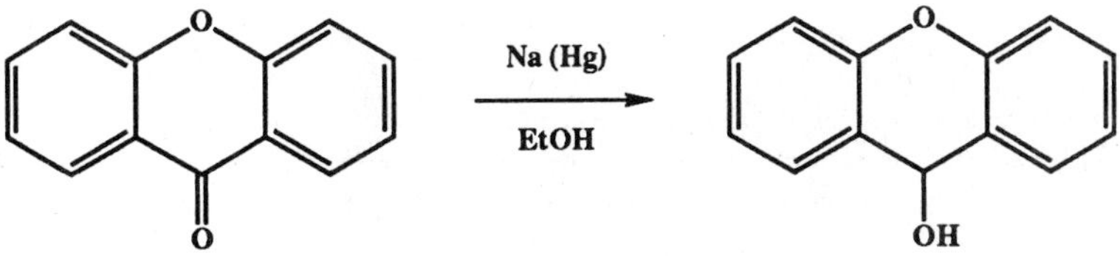

***OS*, *CV* 2,** **62 (1943)**

Cu

H_2SO_4

O

O

O

***OS*, *CV* 2,** **154 (1943)**

$EtO_2C—(CH_2)_8—CO_2Et$ —Na / EtOH→ $HO—(CH_2)_{10}—OH$

***OS*, *CV* 2,** **317 (1943)**

O

CH_3

Na, EtOH

H_2O

OH

CH_3

***OS*, *CV* 2,** **318 (1943)**

NOH

H

Na, EtOH

Δ

NH_2

H

***OS*, *CV* 2,** **320 (1943)**

$C_{16}H_{33}—I$ —Zn, HCl / HOAc, Δ→ $C_{16}H_{33}—H$

OS, CV 2, 353 (1943)

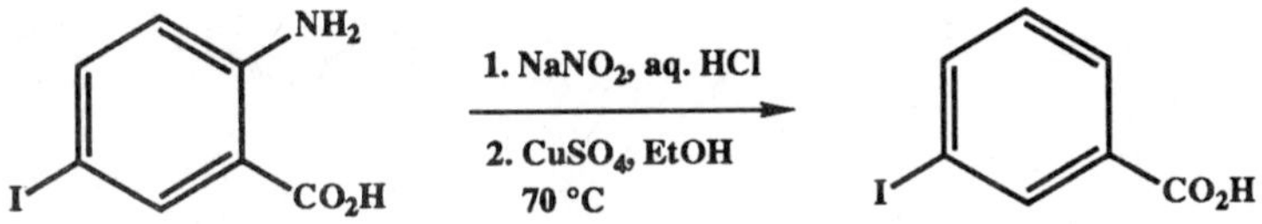

1. $NaNO_2$, aq. HCl
2. $CuSO_4$, EtOH, 70 °C

I, NH_2, CO_2H → I, CO_2H

OS, CV 2, 372 (1943)

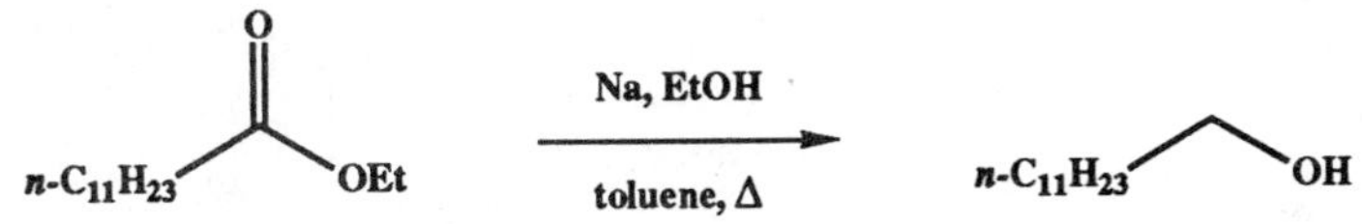

Na, EtOH; toluene, Δ

O, n-$C_{11}H_{23}$, OEt → n-$C_{11}H_{23}$, OH

OS, CV 2, 393 (1943)

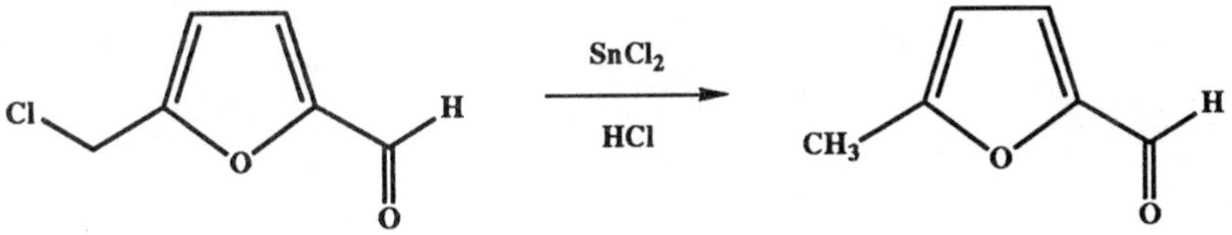

$SnCl_2$; HCl

Cl, O, H, O → CH_3, O, H, O

OS, CV 2, 468 (1943)

Na, n-BuOH; Δ

O, $C_{17}H_{33}$, O-n-Bu → $C_{17}H_{33}$, OH

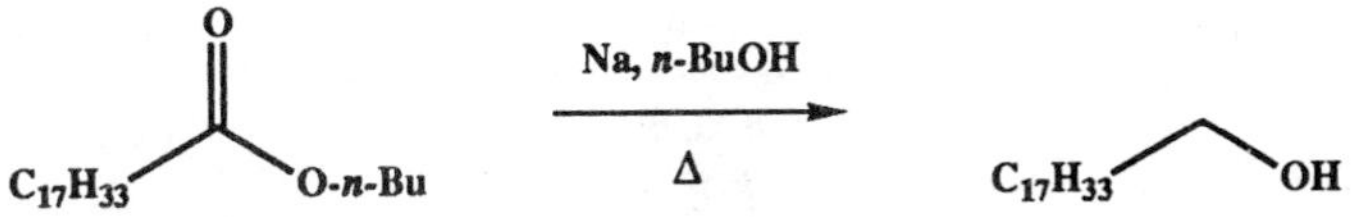

OS, CV 2, 478 (1943)

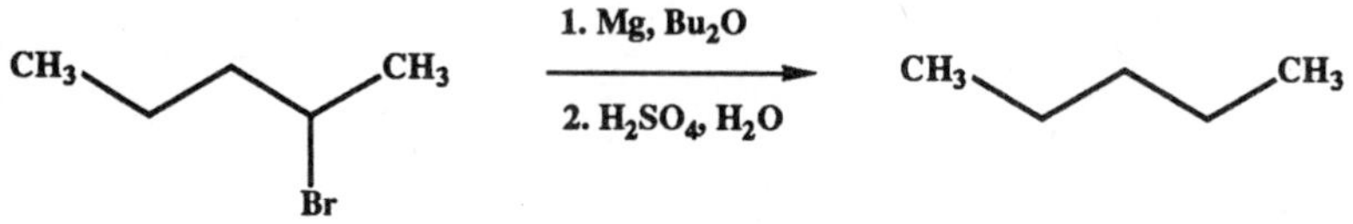

1. Mg, Bu_2O
2. H_2SO_4, H_2O

CH_3 CH_3 Br

CH_3 CH_3

OS, CV 2, 499 (1943)

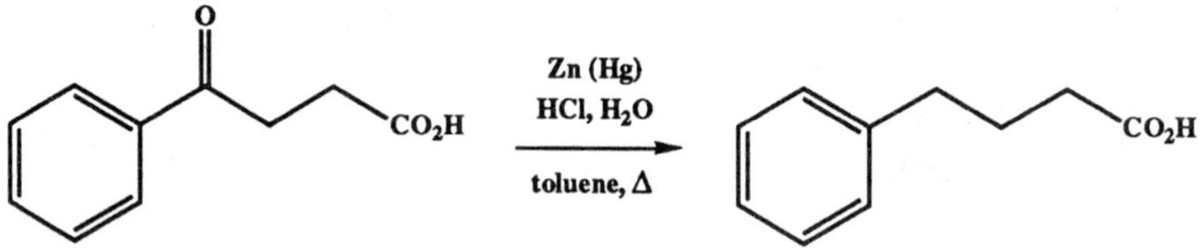

O CO_2H

Zn (Hg)
HCl, H_2O
toluene, Δ

CO_2H

OS, CV 2, 526 (1943)

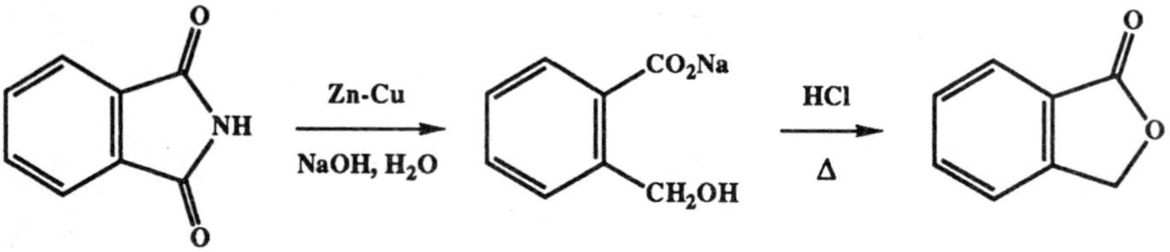

O NH O

Zn-Cu
NaOH, H_2O

CO_2Na CH_2OH

HCl
Δ

O O

OS, CV 2, 535 (1943)

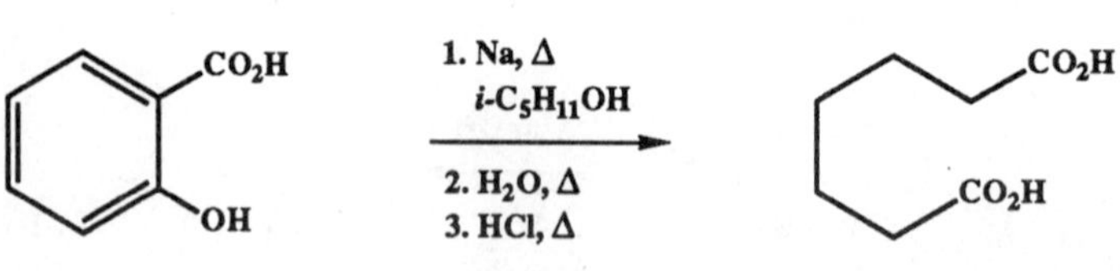

CO_2H OH

1. Na, Δ
i-$C_5H_{11}OH$
2. H_2O, Δ
3. HCl, Δ

CO_2H CO_2H

OS, CV 3, 132 (1955)

OH
Br_2
HOAc
Br
OH
Br
Sn
Δ
Br
OH

OS, CV 3, 444 (1955)

OH O
$(CH_2)_5CH_3$
mossy
Zn (Hg)
aq. HCl
EtOH, Δ
OH
$(CH_2)_5CH_3$

OS, CV 3, 586 (1955)

HO
MeO
O
H
N
=NH
N
Me
Na (Hg)
H_2O
HO
MeO
O
H
N
=NH
N
Me

OS, CV 3, 626 (1955)

CN
HCl
$SnCl_2$
Et_2O
[CH=NH • HCl • $SnCl_4$]
H_2O, Δ
CHO

OS, *CV 3*, 671 (1955)

$$CH_3(CH_2)_7—CH=CH—(CH_2)_7CO_2Et \xrightarrow[EtOH, \Delta]{Na} CH_3(CH_2)_7—CH=CH—(CH_2)_7CH_2OH$$

OS, *CV 3*, 786 (1955)

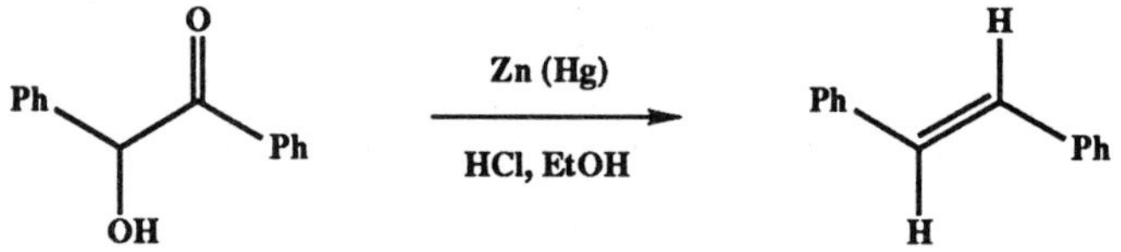

Ph, O, Ph, OH → Zn (Hg) / HCl, EtOH → Ph, H, H, Ph

OS, *CV 3*, 818 (1955)

O, NHPh, CH_3 → 1. PCl_5, benzene, Δ; 2. $SnCl_2$, HCl, Et_2O; 3. HCl, H_2O → O, H, CH_3

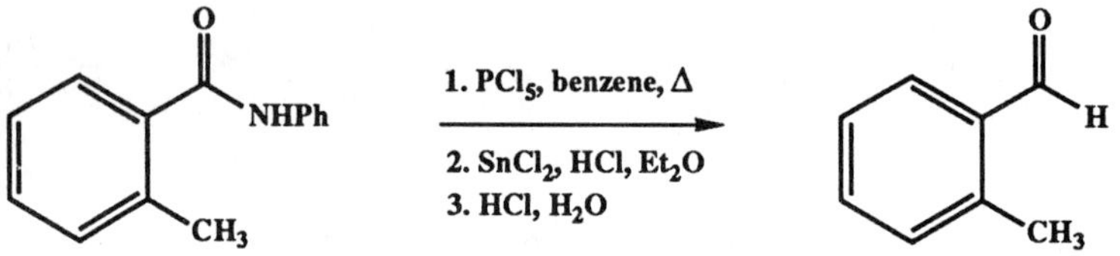

OS, *CV 4*, 195 (1963)

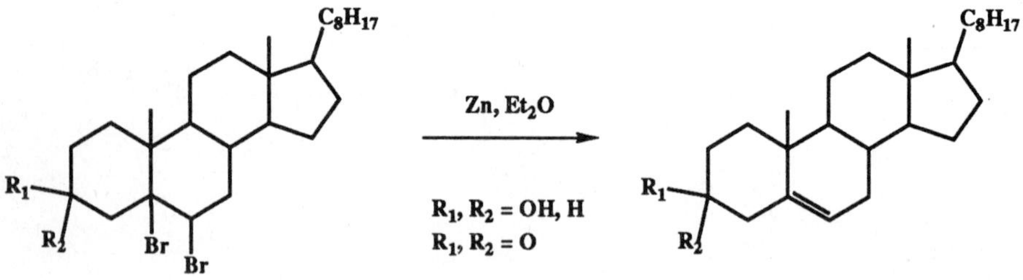

C_8H_{17}, R_1, R_2, Br, Br → Zn, Et_2O → C_8H_{17}, R_1, R_2

R_1, R_2 = OH, H
R_1, R_2 = O

OS, CV 4, 203 (1963)

HO, OCH_3, O, H → Zn (Hg) / HCl, Δ → HO, OCH_3, CH_3

OS, CV 4, 218 (1963)

$(CH_2)_8$, =O, OH → Zn, HCl / HOAc → $(CH_2)_8$, =O, H

OS, CV 4, 348 (1963)

CH_3, O, OH, CH_3 → P, Br_2 / 100 °C → CH_3, O, Br, Br, CH_3 → Zn, Δ / EtOAc → CH_3, C=O, CH_3

OS, CV 4, 508 (1963)

$CH_2\overset{+}{N}Me_3$ I^-, CH_3, CH_3 → Na (Hg) / H_2O, Δ → CH_3, CH_3, CH_3

OS, CV 4, 798 (1963)

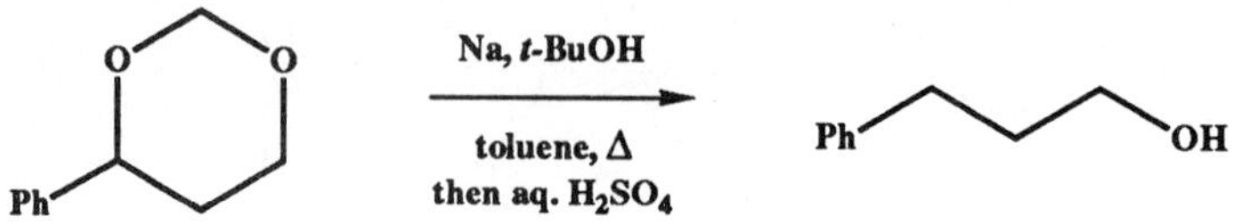

Na, *t*-BuOH

toluene, Δ
then aq. H_2SO_4

Ph OH

OS, CV 4, 887 (1963)

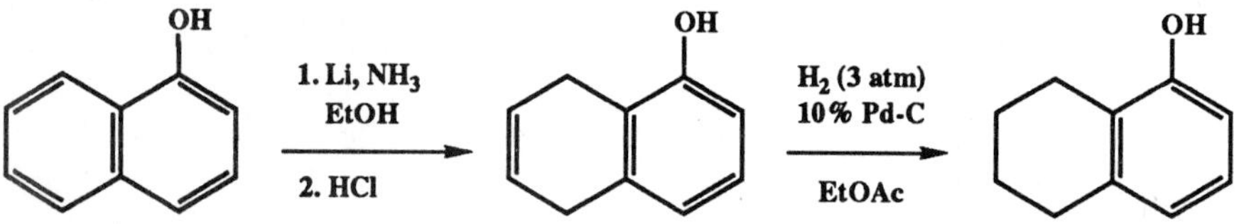

OH

1. Li, NH_3
EtOH

2. HCl

OH

H_2 (3 atm)
10% Pd-C

EtOAc

OH

OS, CV 4, 903 (1963)

OEt

1. Na, EtOH, Δ

2. H_2O, HCl, Δ

O

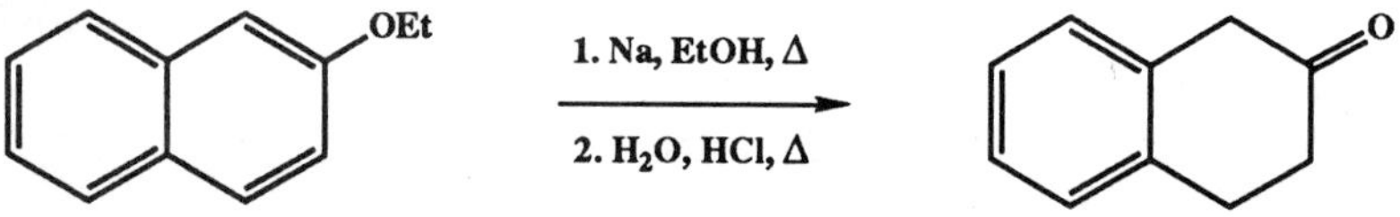

OS, CV 5, 149 (1973)

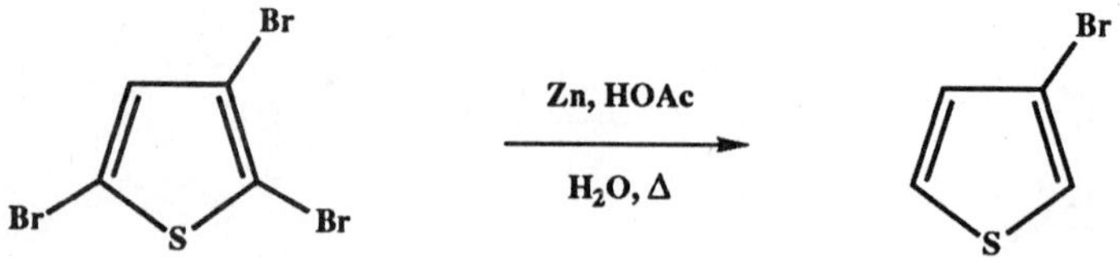

Br

Br S Br

Zn, HOAc

H_2O, Δ

Br

S

OS, *CV 5*, 339 (1973)

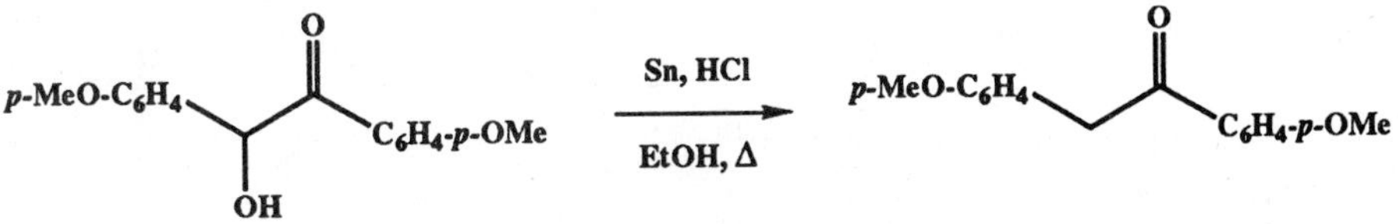

p-MeO-C6H4–CH(OH)–C(=O)–C6H4-p-OMe → (Sn, HCl / EtOH, Δ) → p-MeO-C6H4–CH2–C(=O)–C6H4-p-OMe

OS, *CV 5*, 346 (1973)

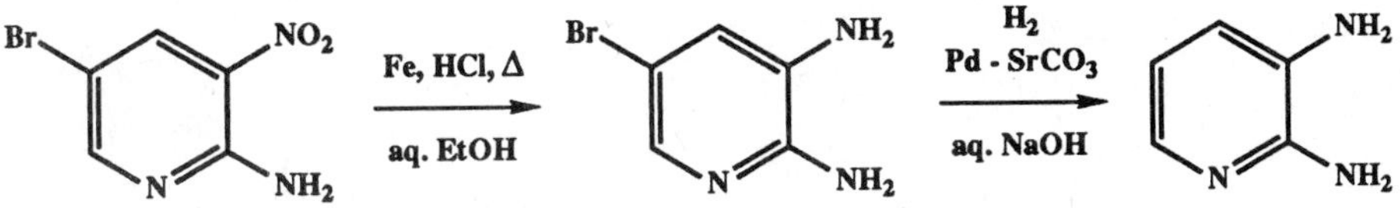

Br, NO2, NH2, N → (Fe, HCl, Δ / aq. EtOH) → Br, NH2, NH2, N → (H_2, Pd - $SrCO_3$ / aq. NaOH) → NH2, NH2, N

OS, *CV 5*, 373 (1973)

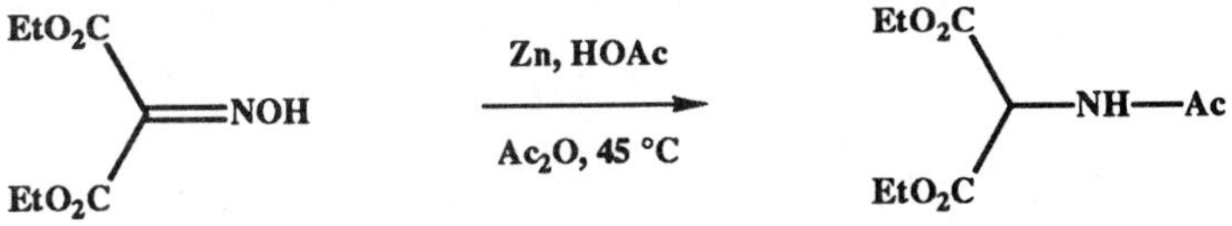

EtO_2C, EtO_2C, =NOH → (Zn, HOAc / Ac_2O, 45 °C) → EtO_2C, EtO_2C, –NH—Ac

OS, *CV 5*, 393 (1973)

CH_2—CCl_2, CF_2—CFCl → (Zn / EtOH, Δ) → CH_2—CCl, CF_2—CF

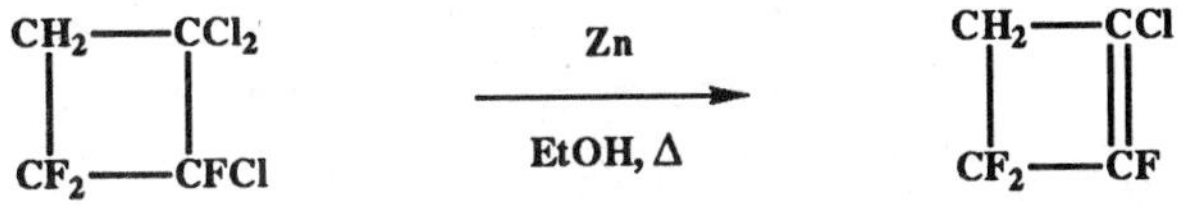

OS, *CV 5*, 398 (1973)

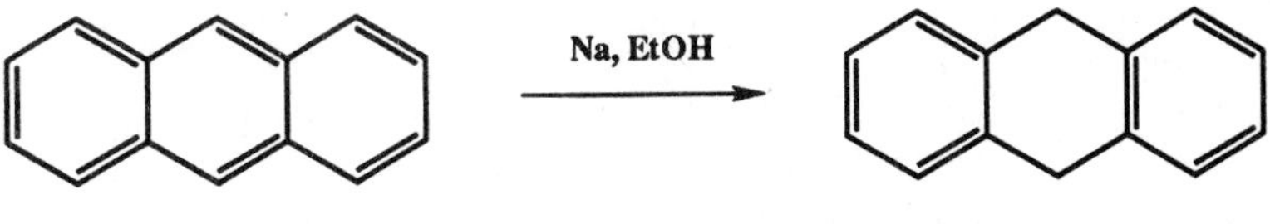

→ (Na, EtOH) →

OS, CV 5, **400 (1973)**

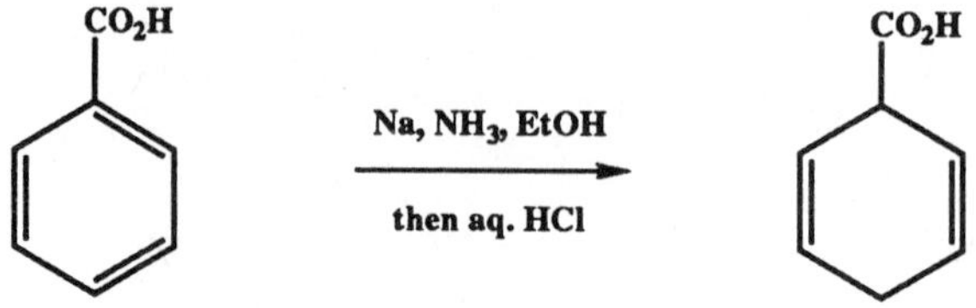
CO2H
Na, NH3, EtOH
then aq. HCl
CO2H

OS, CV 5, **424 (1973)**

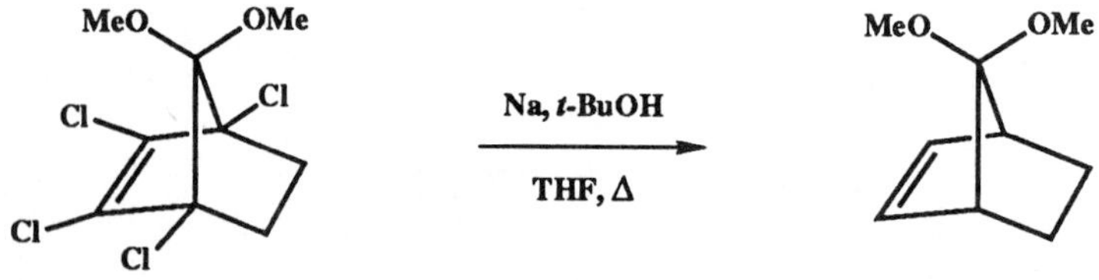
MeO
OMe
Cl
Cl
Cl
Cl
Na, t-BuOH
THF, Δ
MeO
OMe

OS, CV 5, **467 (1973)**

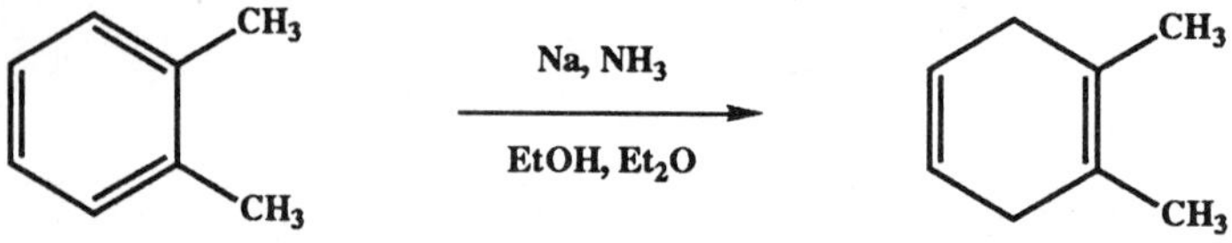
CH3
CH3
Na, NH3
EtOH, Et2O
CH3
CH3

OS, CV 5, **595 (1973)**

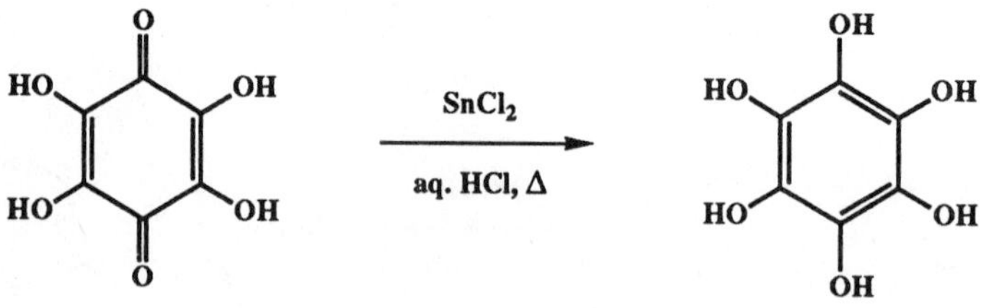
O
HO
OH
HO
OH
O
SnCl2
aq. HCl, Δ
OH
HO
OH
HO
OH
OH

***OS*, *CV 5*, 993 (1973)**

EtO$_2$C–CH=CH–CO$_2$Et → (aq. Cr^{+2}, DMF) → EtO$_2$C–CH$_2$CH$_2$–CO$_2$Et

***OS*, *CV 6*, 51 (1988)**

3-methylcyclohex-2-enone → (2 Li, H$_2$O; NH$_3$, Et$_2$O, - 33 °C) → [lithium enolate, O$^-$ Li$^+$, CH$_3$] → (allyl bromide (Br); NH$_3$, Et$_2$O, - 33 °C) → 2-allyl-3-methylcyclohexanone (CH$_3$)

***OS*, *CV 6*, 82 (1988)**

Tetrachloro (Cl, Cl, Cl, Cl) benzobarrelene → (Na, *t*-BuOH; THF, Δ) → benzobarrelene

***OS*, *CV 6*, 133 (1988)**

1-bromo-3-chlorocyclobutane (Cl, Br) → (2 Na; dioxane, Δ) → bicyclo[1.1.0]butane

OS, CV 6, 289 (1988)

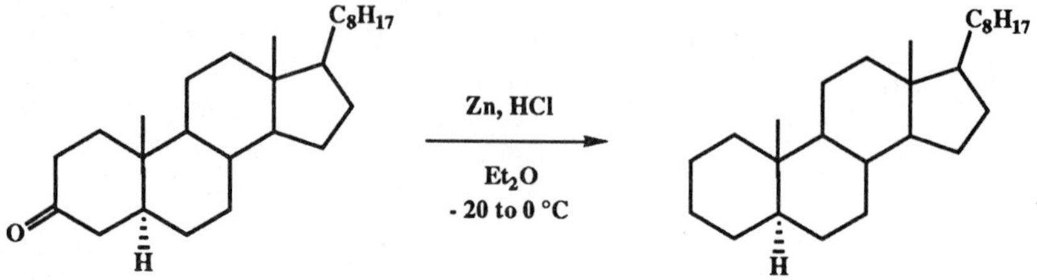

C_8H_{17}

Zn, HCl

Et_2O
- 20 to 0 °C

C_8H_{17}

OS, CV 6, 461 (1988)

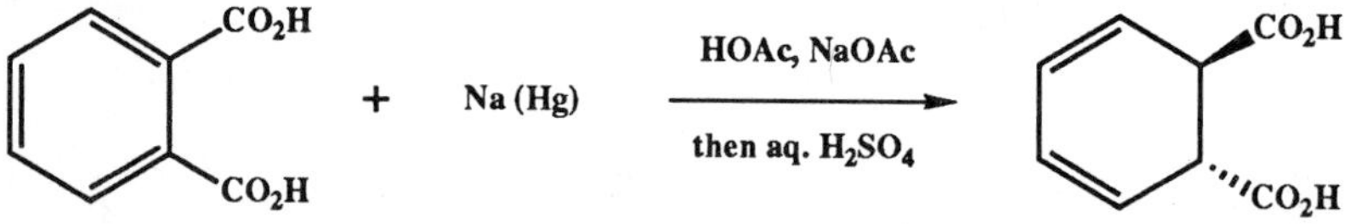

CO_2H CO_2H + Na (Hg) HOAc, NaOAc then aq. H_2SO_4 CO_2H CO_2H

OS, CV 6, 537 (1988)

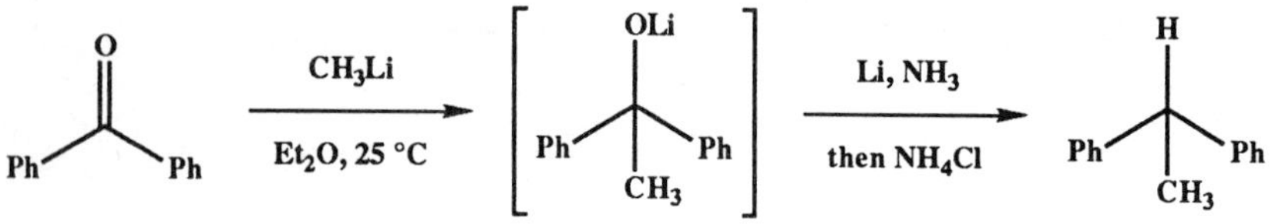

Ph(C=O)Ph — CH_3Li, Et_2O, 25 °C → [Ph, Ph, OLi, CH_3] — Li, NH_3 then NH_4Cl → Ph, Ph, H, CH_3

OS, CV 6, 731 (1988)

Na, NH_3

EtOH, Et_2O

Cl Cl

Na, NH_3

Et_2O

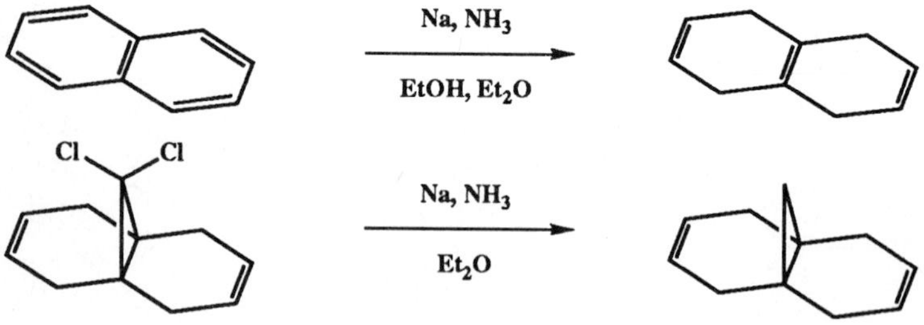

OS, CV 6, 762 (1988)

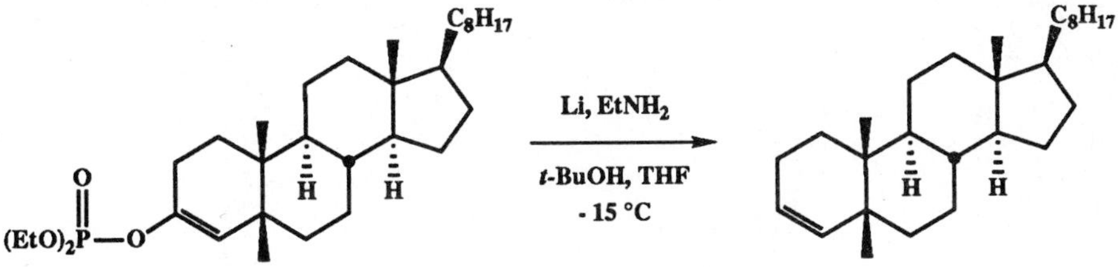

$(EtO)_2P(O)-O$ ... C_8H_{17}

Li, $EtNH_2$

t-BuOH, THF

- 15 °C

C_8H_{17}

OS, CV 6, 769 (1988)

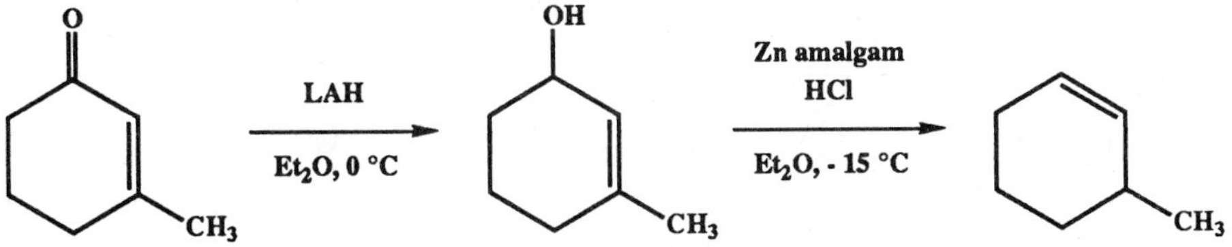

O, CH_3

LAH

Et_2O, 0 °C

OH, CH_3

Zn amalgam

HCl

Et_2O, - 15 °C

CH_3

OS, CV 6, 821 (1988)

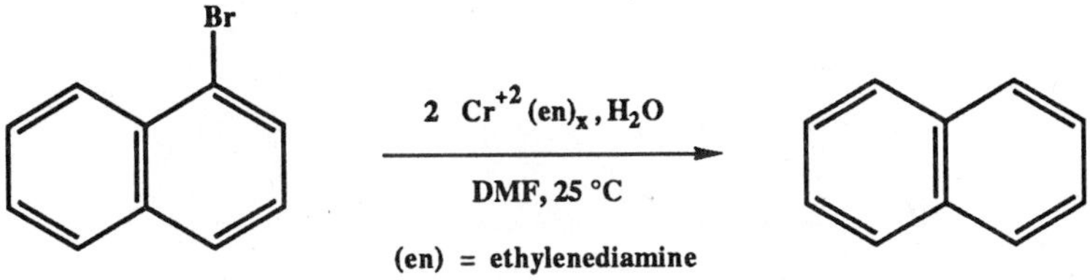

Br

2 $Cr^{+2}(en)_x$, H_2O

DMF, 25 °C

(en) = ethylenediamine

OS, CV 6, 852 (1988)

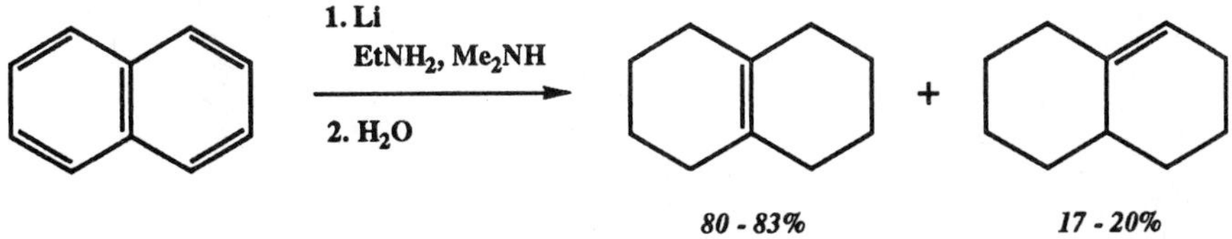

1. Li

$EtNH_2$, Me_2NH

2. H_2O

+

80 - 83%

17 - 20%

OS, *CV 6*, 996 (1988)

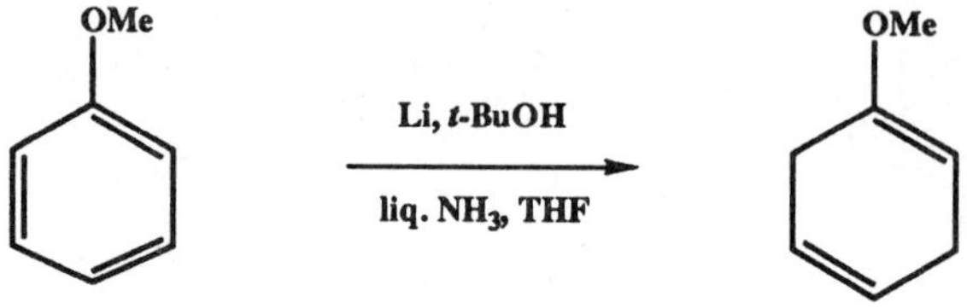

OMe

Li, *t*-BuOH

liq. NH_3, THF

OMe

OS, *CV 7*, 249 (1990); *61*, 59 (1983)

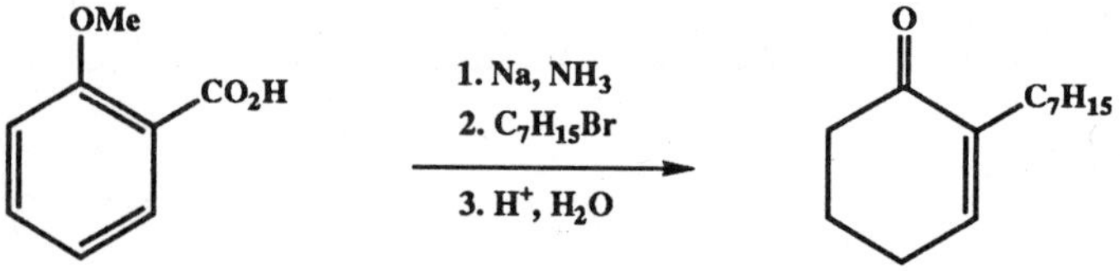

OMe

CO_2H

1. Na, NH_3

2. $C_7H_{15}Br$

3. H^+, H_2O

O

C_7H_{15}

OS, *CV 7*, 66 (1990); *61*, 116 (1983)

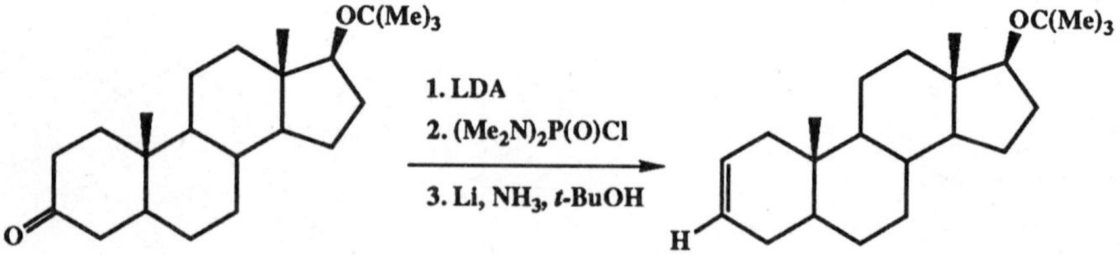

OC(Me)$_3$

O

1. LDA

2. $(Me_2N)_2P(O)Cl$

3. Li, NH_3, *t*-BuOH

OC(Me)$_3$

H

OS, *65*, 203 (1987)

(*R*, *S*) ~ 87 : 13 (*R*, *R*)

Na, *i*-PrOH; toluene, reflux

~ 87 : 13

OS, *65*, 215 (1987)

SCH_2Ph

Na, NH_3; MeOH, - 78 °C

SH, OH, *R*

OS, *68*, 32 (1989)

Bu, Cl, Cl

Zn; TMEDA-HOAc; EtOH

Bu

***OS*, *CV 1*, 104 (1941)**

2 PhCHO → ($PhCH_2ONa$ (cat.)) → $PhCO_2CH_2Ph$

***OS*, *CV 1*, 224 (1941)**

$Ph_2C(OH)CO_2H$ → (P, I_2, H_2O *or* aq. HI, P; HOAc, Δ) → Ph_2CHCO_2H

***OS*, *CV 1*, 276 (1941)**

2 furan-2-CHO → (1. 33% NaOH; 2. H_2SO_4) → furan-2-CH_2OH + furan-2-CO_2H

***OS*, *CV 1*, 311 (1941)**

$C_6H_5CH{=}CH{-}C(O){-}ONa$ → (1. electricity (Hg cathode) Na_2SO_4, NaOH; 2. H_2SO_4) → $C_6H_5CH_2CH_2CO_2H$ (OH)

OS, CV 1, **347 (1941)**

$$2\ CH_2{=}O + NH_4Cl \xrightarrow{\Delta} CH_3NH_2 \cdot HCl + HCO_2H$$

OS, CV 1, **357 (1941)**

$$CHBr_3 + Na_3AsO_3 + NaOH \xrightarrow{\Delta} CH_2Br_2 + Na_3AsO_4 + NaBr$$

OS, CV 1, **358 (1941)**

$$CHI_3 + Na_3AsO_3 + NaOH \xrightarrow{65\ °C} CH_2I_2 + Na_3AsO_4 + NaI$$

OS, CV 1, **388 (1941)**

NH_2, NO_2, SO_3H (4-amino-3-nitrobenzenesulfonic acid) $\xrightarrow[\Delta]{aq.\ H_2SO_4}$ NH_2, NO_2 (2-nitroaniline)

OS, CV 1, **415 (1941)**

NH_2, NO_2, Me $\xrightarrow[EtOH,\ H_2O]{NaNO_2,\ H_2SO_4}$ Me, NO_2

OS, CV 1, 528 (1941)

$$3\ (CH_2{=}O)_3 \xrightarrow[\Delta]{2\ NH_4Cl} 2\ (CH_3)_3N \cdot HCl \xrightarrow{NaOH} 2\ (CH_3)_3N$$

OS, CV 1, 531 (1941)

$$3\ (CH_2{=}O)_3 \xrightarrow[\Delta]{2\ NH_4Cl} 2\ (CH_3)_3N \cdot HCl$$

OS, CV 1, 548 (1941)

$$(C_6H_5)_3CCl \cdot AlCl_3 \xrightarrow[Et_2O]{HCl} (C_6H_5)_3CH$$

OS, CV 2, 489 (1943)

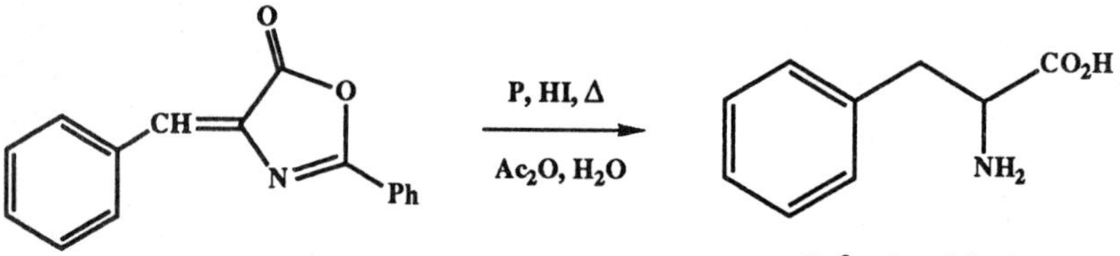

CH= (oxazolone, O, N, Ph) → P, HI, Δ / Ac_2O, H_2O → CO_2H, NH_2

dl - β - phenylalanine

OS, CV 2, 503 (1943)

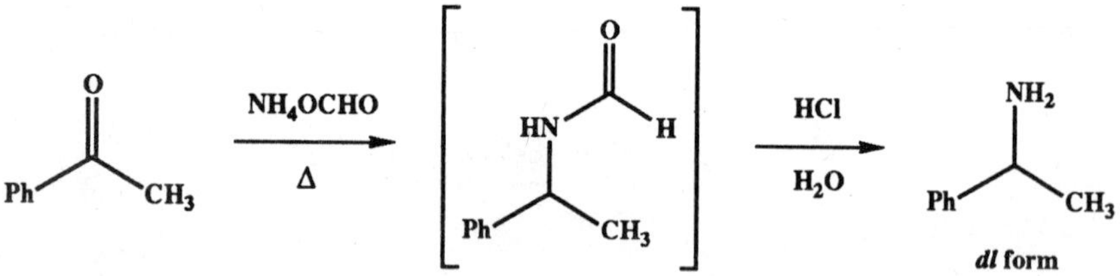

Ph, O, CH_3 → NH_4OCHO / Δ → [HN, O, H, Ph, CH_3] → HCl / H_2O → NH_2, Ph, CH_3

dl form

OS, CV 2, 523 (1943)

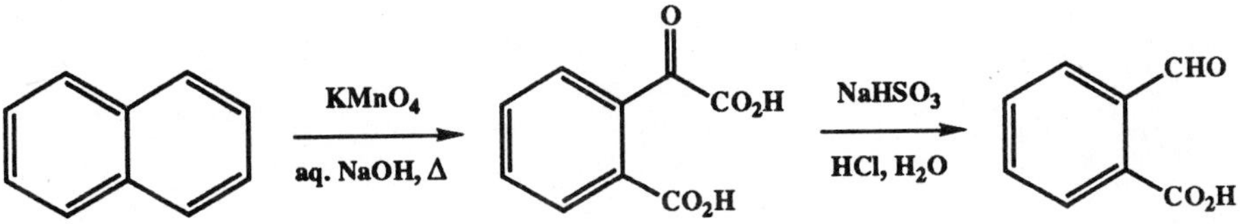
KMnO4
aq. NaOH, Δ
O
CO2H
CO2H
NaHSO3
HCl, H2O
CHO
CO2H

OS, CV 2, 545 (1943)

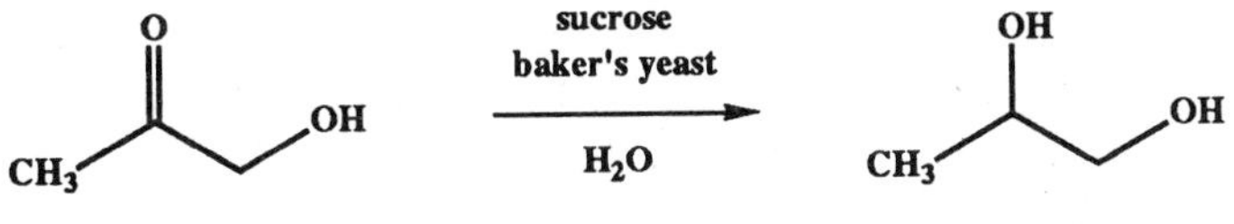
O
OH
CH3
sucrose
baker's yeast
H2O
OH
OH
CH3

l-(-) isomer

OS, CV 2, 590 (1943)

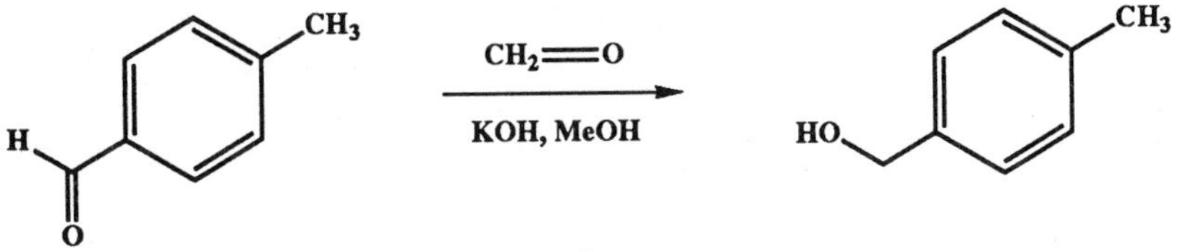
CH3
H
O
CH2=O
KOH, MeOH
CH3
HO

OS, CV 2, 592 (1943)

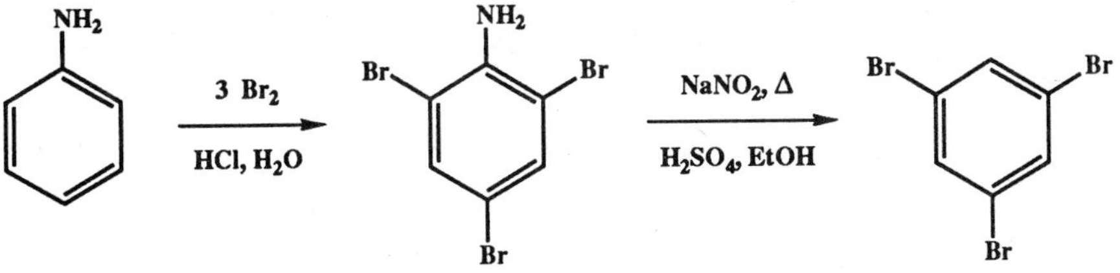
NH2
3 Br2
HCl, H2O
NH2
Br
Br
Br
NaNO2, Δ
H2SO4, EtOH
Br
Br
Br

OS, *CV 2*, 598 (1943)

1. $Al(OEt)_3$, EtOH, Δ
2. H_2SO_4, H_2O

OS, *CV 3*, 42 (1955)

H_2S

OS, *CV 3*, 60 (1955)

electricity
(Pb electrodes)

OS, *CV 3*, 295 (1955)

1. $NaNO_2$, HCl
2. H_3PO_2, 0 °C

R = CH_3, OCH_3

OS, CV 3, 475 (1955)

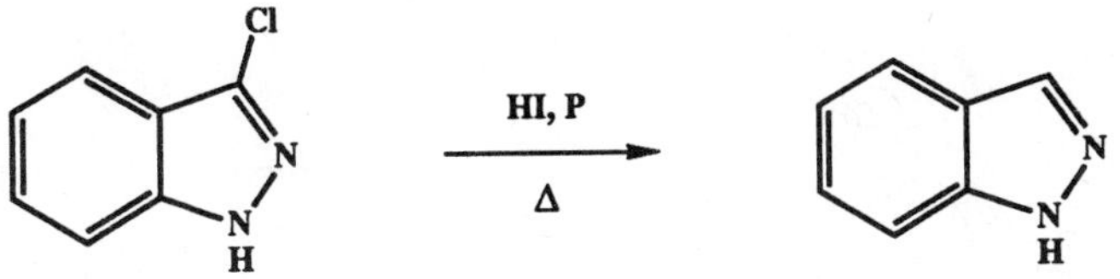

Cl

N

N
H

HI, P

Δ

N

N
H

OS, CV 3, 513 (1955)

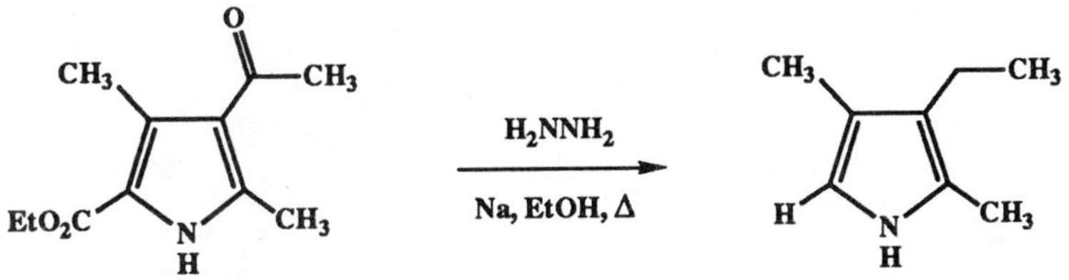

O

CH_3 CH_3

EtO_2C N CH_3
H

H_2NNH_2

Na, EtOH, Δ

CH_3 CH_3

H N CH_3
H

OS, CV 4, 15 (1963)

O

CH_3 O

O

$Na_2S_2O_4$

aq. $CHCl_3$

O

CH_3 OH

OH

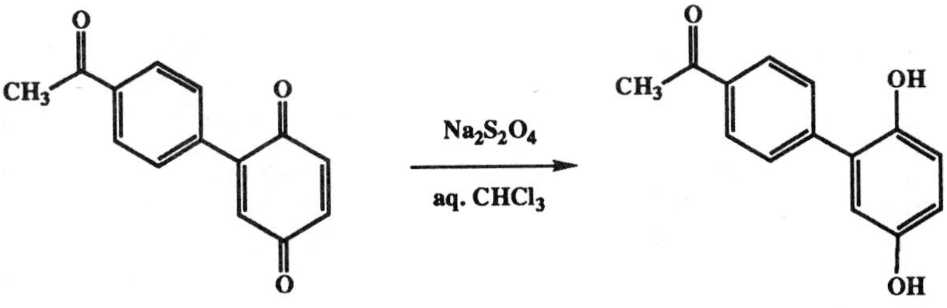

OS, *CV* 4, 25 (1963)

H N O OH OH O N H O

H_2S

CS_2

H N O O OH N H O

OS, *CV* 4, 114 (1963)

Br

CH_3

NO_2

KCN, H_2O, Δ

then HCl

Br

CH_3

CO_2H

OS, *CV* 4, 377 (1963)

O

Ph

Ph

O

H_2NNH_2

n-PrOH, Δ

NNH_2

Ph

Ph

NNH_2

HgO

benzene

Ph—≡—Ph

OS, CV 4, 510 (1963)

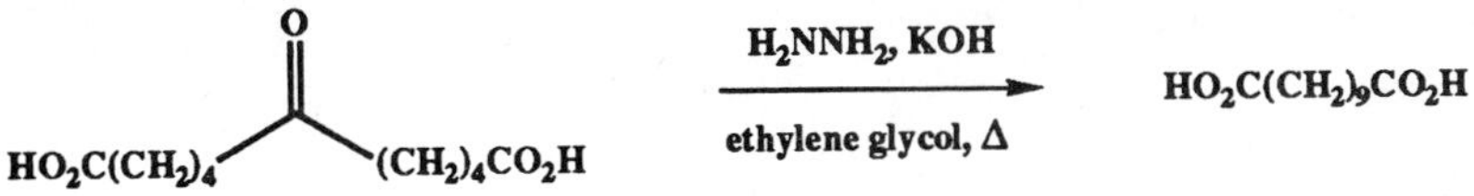

$HO_2C(CH_2)_4C(=O)(CH_2)_4CO_2H$ —[H_2NNH_2, KOH; ethylene glycol, Δ]→ $HO_2C(CH_2)_9CO_2H$

OS, CV 4, 947 (1963)

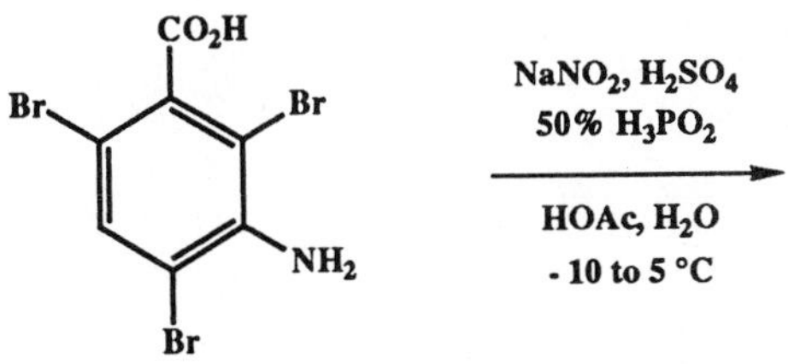

CO₂H / Br / Br / NH₂ / Br —[$NaNO_2$, H_2SO_4; 50% H_3PO_2; HOAc, H_2O; - 10 to 5 °C]→ CO₂H / Br / Br / Br

OS, CV 5, 281 (1973)

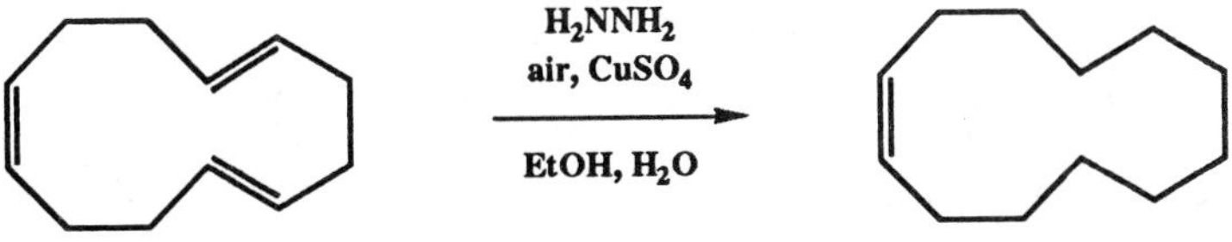

—[H_2NNH_2; air, $CuSO_4$; EtOH, H_2O]→

OS, CV 5, 533 (1973)

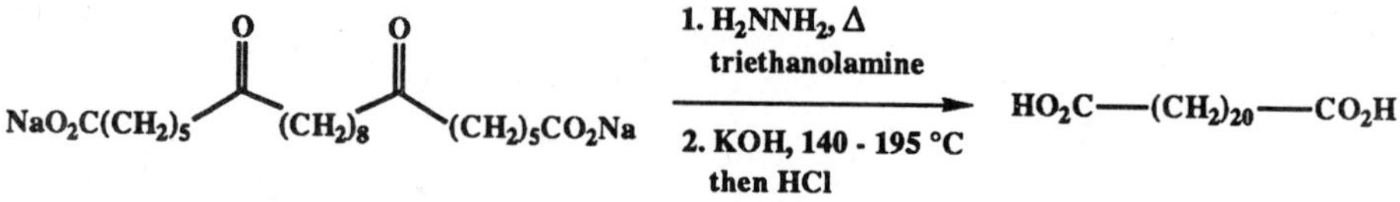

$NaO_2C(CH_2)_5C(=O)(CH_2)_8C(=O)(CH_2)_5CO_2Na$ → $HO_2C—(CH_2)_{20}—CO_2H$

1. H_2NNH_2, Δ, triethanolamine
2. KOH, 140 - 195 °C, then HCl

OS, CV 5, 747 (1973)

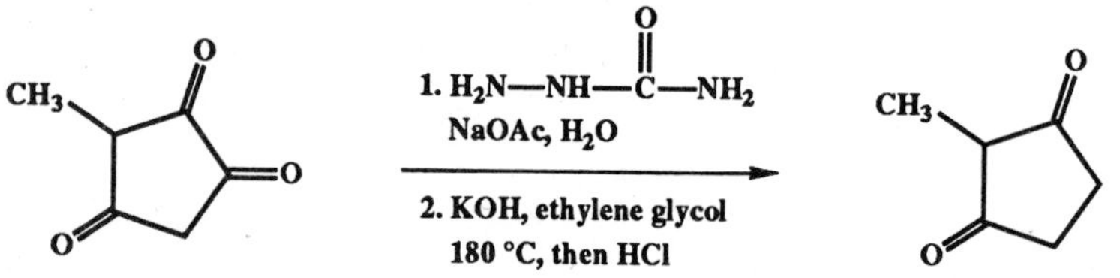

1. $H_2N—NH—C(=O)—NH_2$, NaOAc, H_2O
2. KOH, ethylene glycol, 180 °C, then HCl

OS, CV 5, 901 (1973)

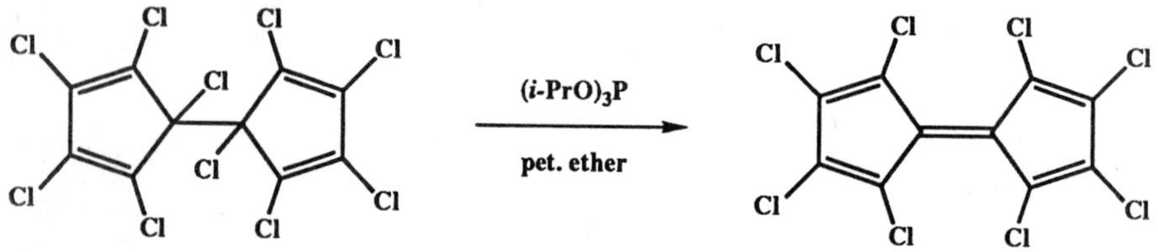

$(i\text{-PrO})_3P$

pet. ether

OS, CV 5, 998 (1973)

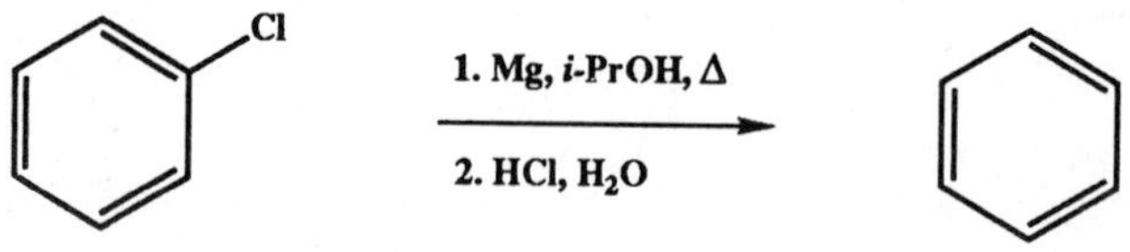

1. Mg, *i*-PrOH, Δ
2. HCl, H_2O

OS, *CV 5*, 1070 (1973)

SH

N

N—NH

HNO_3, H_2O

45 °C

N

N—NH

OS, *CV 6*, 41 (1988)

O

+ *p*-Ts-CH_2N≡C

t-BuOK, EtOH

DME, 5 - 35 °C

C≡N

OS, *CV 6*, 153 (1988)

$C(CH_2Br)_4$ + 2e$^-$

Hg cathode

(-1.8 V *vs*. SCE)

$(n\text{-}Bu)_4NBr$, DMF

CH_2Br

CH_2Br

+ 2 Br$^-$

OS, *CV 6*, 215 (1988)

O

t-Bu

$IrCl_4$, HCl, $P(OMe)_3$

aq. *i*-PrOH, Δ

OH

t-Bu

OS, *CV 6*, 223 (1988)

C≡N

t-Bu

C≡N

1. B_2H_6, THF, then HCl

2. aq. NaOH

NH_2

t-Bu

NH_2

***OS*, *CV* 6, 293 (1988)**

p-TsNHN — C_8H_{17}

1. (catecholborane: O, BH, O) $CHCl_3$, 0 °C

2. NaOAc $CHCl_3$, Δ

C_8H_{17} — H H H H

***OS*, *CV* 6, 334 (1988)**

NC N=N—CO_2Me

NaOMe, MeOH

0 - 10 °C

NC H

***OS*, *CV* 6, 747 (1988)**

CO_2H

1. $HSiCl_3$ CH_3CN, Δ

2. $(n\text{-}Pr)_3N$, Δ

CH_2SiCl_3

1. MeOH, Δ

2. aq. KOH MeOH, Δ

CH_3

***OS*, *CV* 6, 776 (1988)**

n-Pr — O — CH_3 — O — S — CH_3

$C_6H_5P[(CH_2)_3NMe_2]_2$ LiBr

CH_3CN, 70 °C

n-Pr — O — CH_3 — CH_3 — O

***OS, CV 7*, 121 (1990); *60*, 29 (1981)**

WCl$_6$, 2 *n*-BuLi

THF, 25 °C

***OS, CV 7*, 18 (1990); *62*, 165 (1984)**

HI, P

HOAc

***OS, CV 7*, 215 (1990); *63*, 1 (1985)**

CO_2Et

Baker's yeast

H_2O, sucrose, 25 - 30 °C

OH

CO_2Et

+

OH

CO_2Et

ca. 93% *S* - (+)

ca. 7% *R* - (-)

***OS, CV 7*, 402 (1990); *63*, 57 (1985)**

O

C_5H_{11}

H

1. 0 - 25 °C

2. H_2O_2, NaOH

HO H

C_5H_{11}

H

ca. 93% *R* - (+)

***OS*, *68*, 56 (1989)**

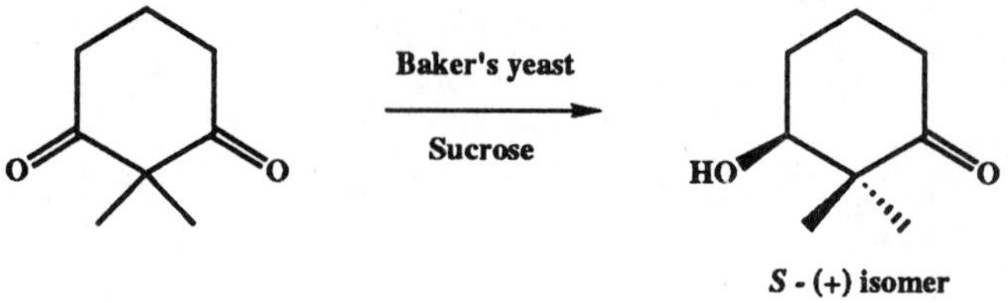

Baker's yeast

Sucrose

S - (+) isomer

***OS*, *68*, 138 (1989)**

$Me_2CH(CH_2)_3$

1. $(CF_3SO_2)_2O$, Me, *t*-Bu, N, *t*-Bu

2. $Pd(OAc)_2$, Ph_3P, $(n\text{-}Bu)_3N$, HCOOH

$Me_2CH(CH_2)_3$

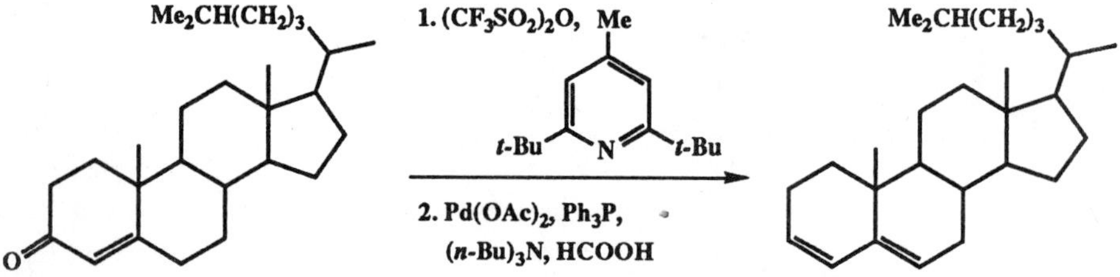

ADDITION

OS, CV 1, 14 (1941)

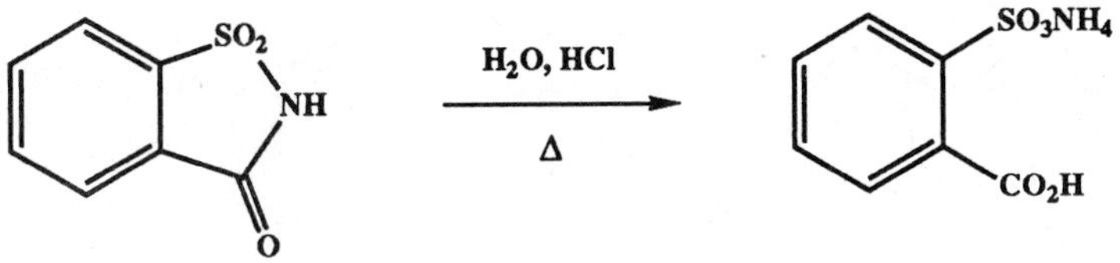

H_2O, HCl, Δ

SO₂ NH O → SO₃NH₄ CO₂H

OS, CV 1, 21 (1941)

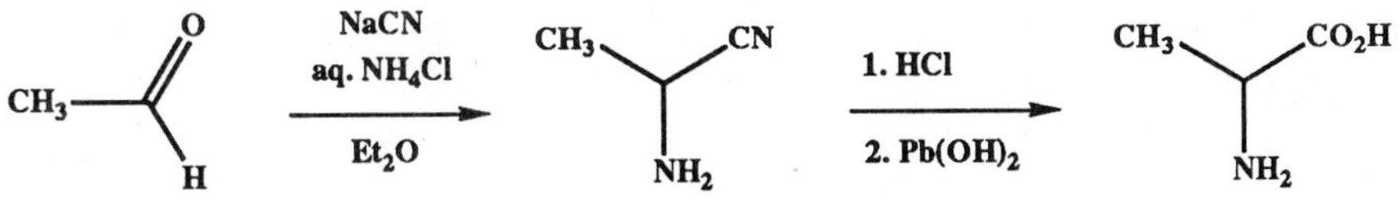

CH_3CHO → NaCN, aq. NH_4Cl, Et_2O → $CH_3CH(NH_2)CN$ → 1. HCl, 2. $Pb(OH)_2$ → $CH_3CH(NH_2)CO_2H$

OS, CV 1, 131 (1941)

HO–CH_2CH_2–CN → 2 HBr, H_2O, Δ → Br–CH_2CH_2–CO_2H

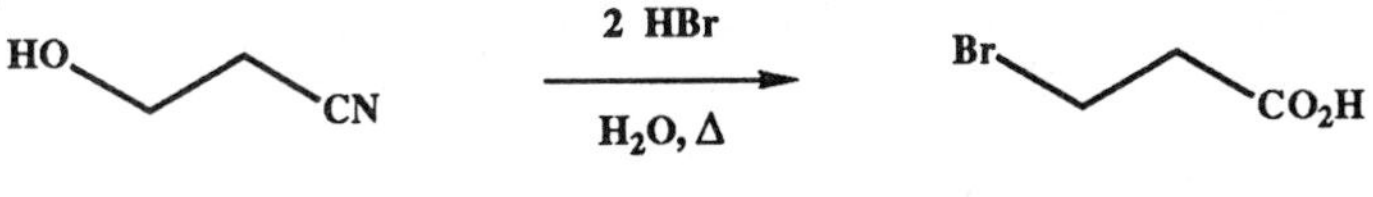

OS, CV 1, 270 (1941)

$PhCH_2CN$ → H_2SO_4, H_2O, EtOH, Δ → $PhCH_2CO_2Et$

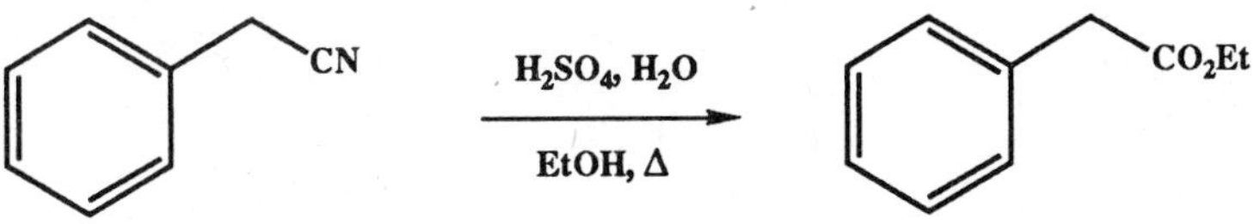

OS, CV 1, 289 (1941)

NC–$(CH_2)_3$–CN $\xrightarrow[\Delta]{HCl,\ H_2O}$ HO_2C–$(CH_2)_3$–CO_2H

OS, CV 1, 298 (1941)

$(H_2NCH_2CN) \cdot H_2SO_4 \xrightarrow[\Delta]{Ba(OH)_2} (H_2NCH_2CO_2)_2Ba \xrightarrow{H_2SO_4} 2\ H_2N—CH_2CO_2H$

OS, CV 1, 321 (1941)

HO–CH_2CH_2–CN $\xrightarrow[2.\ aq.\ H_2SO_4]{1.\ aq.\ NaOH,\ \Delta}$ HO–CH_2CH_2–CO_2H

OS, CV 1, 336 (1941)

PhCHO $\xrightarrow[H_2O]{NaHSO_3,\ NaCN}$ PhCH(OH)CN $\xrightarrow[H_2O]{HCl}$ PhCH(OH)CO_2H

OS, CV 1, 391 (1941)

O, OMe, 1. aq. NaOH, Δ, 2. HCl, O, OH, NO_2, NO_2

OS, CV 1, 406 (1941)

CN, H_2SO_4, H_2O, Δ, CO_2H, O_2N, O_2N

OS, CV 1, 436 (1941)

CN, H_2SO_4, H_2O, Δ, CO_2H

OS, CV 1, 451 (1941)

CO_2Et, CH=, CN, NaCN, EtOH, HCl, CN, CO_2Et, CN, HCl, Δ, CO_2H, CO_2H

OS, CV 1, **523 (1941)**

EtO_2C, CO_2Et, CO_2Et, EtO_2C → (HCl, H_2O; Δ) → HO_2C, CO_2H, CO_2H

OS, CV 2, **1 (1943)**

PhCH, O, N, O, Me → (H_2O; acetone, Δ) → CO_2H, PhCH=, Me, N H, O

OS, CV 2, **28 (1943)**

O, NH → (HCl, H_2O; Δ) → HO_2C—NH_2 • HCl

OS, CV 2, **76 (1943)**

O, NH → (NaOH; H_2O) → $H_2N—(CH_2)_5—CO_2H$ → (PhCOCl) → PhC(=O)—N(H)—$(CH_2)_5—CO_2H$

OS, *CV* 2, 284 (1943)

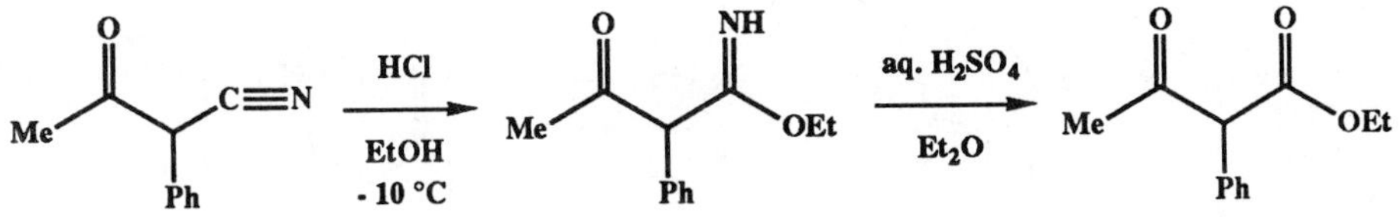
O
Me
C≡N
Ph
HCl
EtOH
- 10 °C
O
NH
Me
OEt
Ph
aq. H_2SO_4
Et_2O
O
O
Me
OEt
Ph

OS, *CV* 2, 292 (1943)

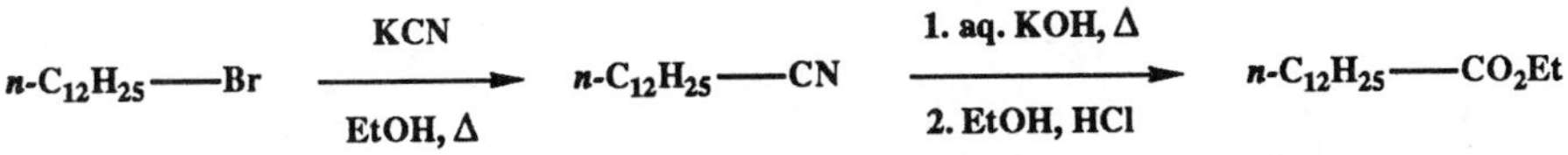
n-$C_{12}H_{25}$—Br
KCN
EtOH, Δ
n-$C_{12}H_{25}$—CN
1. aq. KOH, Δ
2. EtOH, HCl
n-$C_{12}H_{25}$—CO_2Et

OS, *CV* 2, 299 (1943)

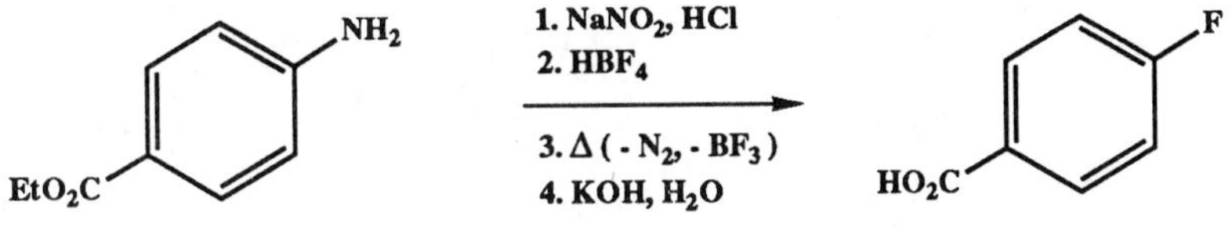
NH_2
EtO_2C
1. $NaNO_2$, HCl
2. HBF_4
3. Δ (- N_2, - BF_3)
4. KOH, H_2O
F
HO_2C

OS, *CV* 2, 310 (1943)

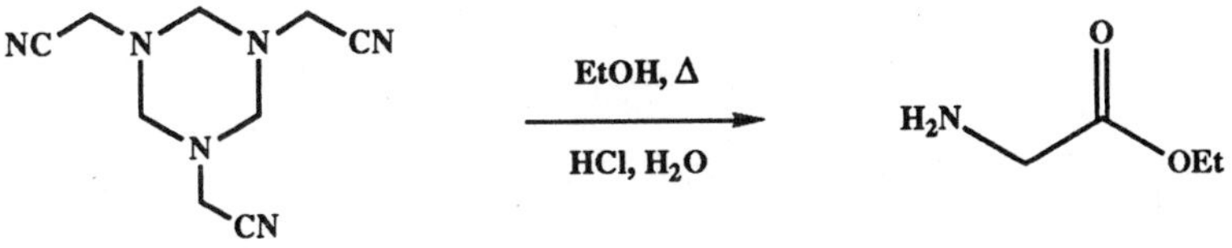
NC
N
N
CN
N
CN
EtOH, Δ
HCl, H_2O
O
H_2N
OEt

***OS*, *CV 2*, 333 (1943)**

NaOH
H_2O, Δ

MeO, OMe, CH=, O, N, Ph, CO_2H

***OS*, *CV 2*, 368 (1943)**

H_2O, Δ

HO_2C, CO_2H

***OS*, *CV 2*, 376 (1943)**

Cl, CO_2H

1. Na_2CO_3
2. NaCN

CN, CO_2Na

aq. NaOH

CO_2Na, CO_2Na

aq. $CaCl_2$

CO_2^-, Ca^{+2}, CO_2^-

HCl

CO_2H, CO_2H

***OS*, *CV 2*, 382 (1943)**

CH_3

HNO_3
H_2O

HO_2C, CH_3, H, CO_2H

OS, CV 2, 489 (1943)

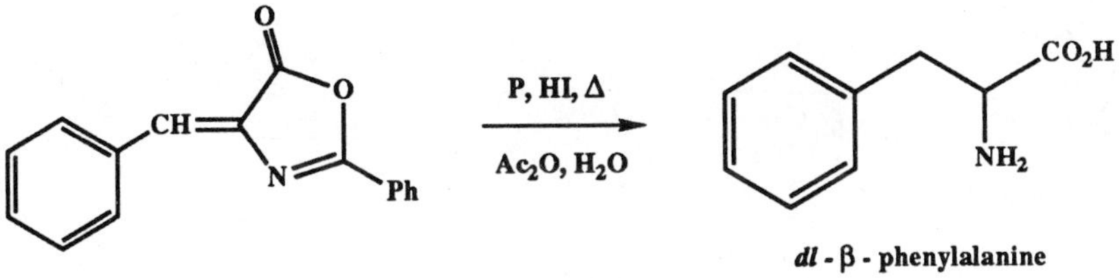

P, HI, Δ

Ac_2O, H_2O

CH= O N Ph O

CO_2H NH_2

dl - β - phenylalanine

OS, CV 2, 588 (1943)

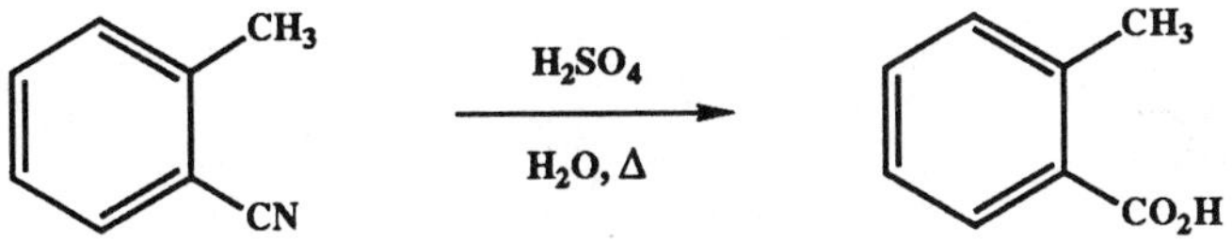

CH_3 CN

H_2SO_4

H_2O, Δ

CH_3 CO_2H

OS, CV 3, 20 (1955)

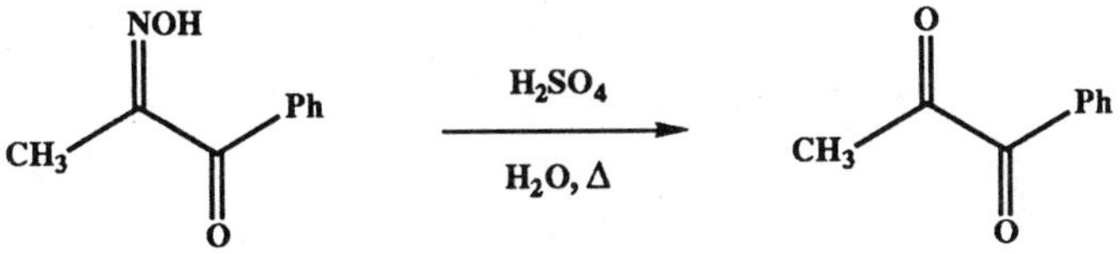

NOH Ph CH_3 O

H_2SO_4

H_2O, Δ

O Ph CH_3 O

OS, CV 3, 34 (1955)

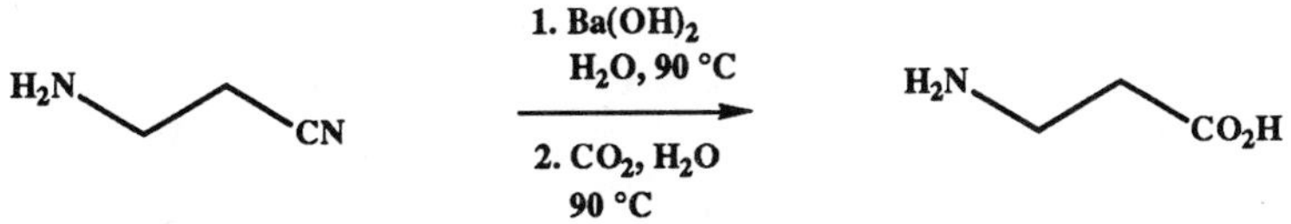

H_2N CN

1. $Ba(OH)_2$
H_2O, 90 °C

2. CO_2, H_2O
90 °C

H_2N CO_2H

OS, CV 3, 84 (1955)

Ph–CHO → (NaCN, NH_4Cl / H_2O, MeOH) → H_2N, CN, Ph, H → (1. aq. HCl, Δ / 2. NH_4OH) → H_2N, CO_2H, Ph, H

OS, CV 3, 88 (1955)

Ph–CO–Me → (NaCN, NH_4Cl / H_2O, EtOH) → H_2N, CN, Ph, Me → (1. aq. HCl, Δ / 2. pyridine) → H_2N, CO_2H, Ph, Me

OS, CV 3, 114 (1955)

PhCO–CN → (HCl, H_2O) → PhCO–CO_2H

OS, CV 3, 154 (1955)

t-Bu–NH–CO–NH_2 → (aq. NaOH, Δ / ethylene glycol) → *t*-Bu—NH_2

OS, CV 3, 221 (1955)

Cl–(CH₂)₃–CN → (NaOH, Δ) → cyclopropyl–CN → (H_2SO_4, H_2O) → cyclopropyl–CO_2H

OS, CV 3, 557 (1955)

2,4,6-trimethylphenyl–CH_2CN → (H_2SO_4, H_2O, Δ) → 2,4,6-trimethylphenyl–CH_2CO_2H

OS, CV 3, 615 (1955)

CH_3—CH=CH—CO_2Et → (1. NaCN, Δ, H_2O, EtOH; 2. $Ba(OH)_2$, Δ then HNO_3) → $CH_3CH(CO_2H)CH_2CO_2H$

OS, CV 3, 851 (1955)

CH_2=CH–CH_2–CN → (HCl, H_2O, Δ) → CH_2=CH–CH_2–CO_2H

OS, *CV* 4, 39 (1963)

NH, aq. HCl, CO_2H, $NH_2 \cdot HCl$, Amberlite resin, *(free amine)*

OS, *CV* 4, 58 (1963)

HO, CN, Ph, CH_3; H_2O / HCl; HO, O, C—NH_2, Ph, CH_3; 1. NaOH 2. HCl; HO, O, C—OH, Ph, CH_3

OS, *CV* 4, 93 (1963)

Me, Et, $PhCH_2$, CN; 1. KOH, Δ, 6 hr ethylene glycol 2. HCl; Me, Et, $PhCH_2$, CO_2H

OS, *CV* 4, 499 (1963)

O; 0.2 *N* HNO_3, Δ; OH, CHO

OS, *CV* 4, 660 (1963)

CH_3 ... OMe — HCl, H_2O → CH_3, CHO, CHO

OS, *CV* 4, 688 (1963)

CHO — Br_2 / H_2O, Δ → Br, Br, HO_2C, CHO

OS, *CV* 4, 760 (1963)

Ph, CN — HCl, H_2O → Ph, O, NH_2

OS, *CV* 4, 790 (1963)

CN, Ph—C—CH_2CH_2CN, CO_2Et — 1. HCl, H_2O HOAc, Δ / 2. Ac_2O, Δ → Ph, O, O, O

OS, *CV 4*, 804 (1963)

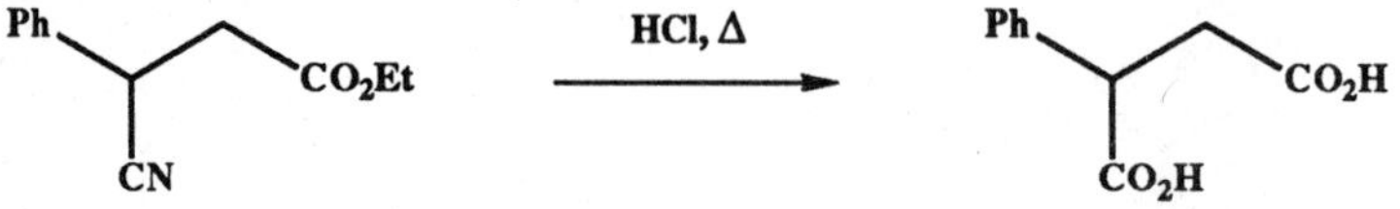

Ph, CN, CO_2Et → HCl, Δ → Ph, CO_2H, CO_2H

OS, *CV 4*, 816 (1963)

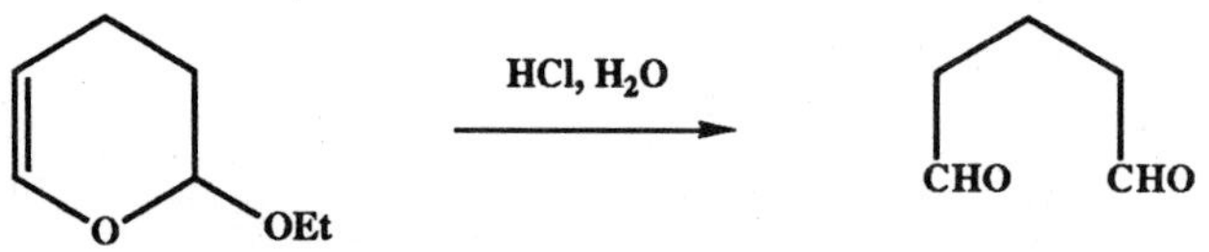

OEt, O → HCl, H_2O → CHO, CHO

OS, *66*, 37 (1987)

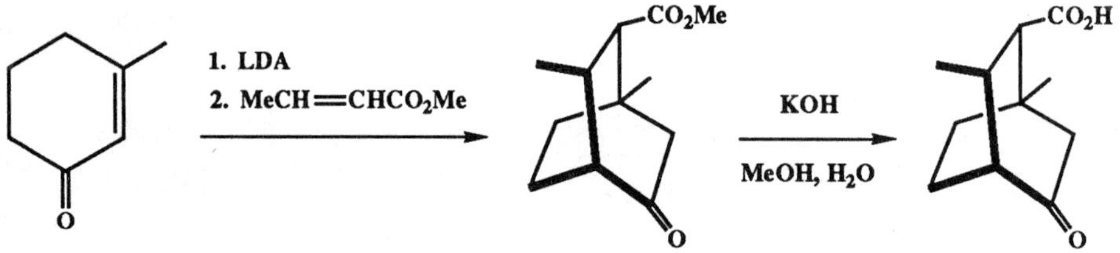

O → 1. LDA 2. $MeCH{=}CHCO_2Me$ → CO_2Me, O → KOH, $MeOH, H_2O$ → CO_2H, O

OS, *CV 1*, 5 (1941)

$CH_3—C≡N$ —(EtOH, HCl)→ $CH_3—C(OEt)=NH \cdot HCl$

OS, *CV 1*, 80 (1941)

PhCHO —($PhNH_2$)→ PhCH=NPh

OS, *CV 1*, 140 (1941)

n-BuOH + H_2NC(O)NH_2 —(Δ)→ H_2NC(O)O—*n*-Bu

OS, *CV 1*, 158 (1941)

cyclohexene —($HgCl_2$, NaOH, Cl_2, HNO_3 / H_2O)→ 2-chlorocyclohexanol (OH, Cl)

OS, CV 1, 166 (1941)

HCl

- 10 to - 15 °C

Cl O H

OS, CV 1, 196 (1941)

O CH_3 CH_3 CH_3

aq. NH_3

aq. oxalic acid

EtOH

O CH_3 CH_3 CH_3 NH_2 • $H_2C_2O_4$ • H_2O

OS, CV 1, 302 (1941)

H_2N H N C N NH

NH_4NO_3

160 °C

H_2N H N NH_2 • HNO_3 NH NH

NH_4NO_3

2 H_2N NH_2 • HNO_3 NH

OS, CV 1, 317 (1941)

O O

2 Ac_2O

H_2SO_4 (cat.)

OAc OAc OAc

OS, CV 1, 318 (1941)

$$HO{-}N(SO_3Na)_2 \xrightarrow[H_2O,\ 70\ ^\circ C]{CH_3COCH_3} HO{-}N{=}C(CH_3)_2 \xrightarrow[H_2O]{HCl} HO{-}NH_2 \cdot HCl$$

OS, CV 1, 330 (1941)

$$C_6H_5NH_2 \xrightarrow[Et_2O]{[CH_2{=}C{=}O]} C_6H_5NHC(O)CH_3$$

OS, CV 1, 377 (1941)

$$CH_2{=}O + CH_3OH + HCl \longrightarrow ClCH_2OCH_3$$

OS, CV 1, 447 (1941)

$$C_6H_5NH_2 \xrightarrow[NH_4OH]{CS_2} C_6H_5NHC(=S)S^- \ NH_4^+$$

***OS*, *CV* 2,** **70 (1943)**

O, NOH

$NH_2OH \cdot HCl$

NaOH, EtOH, Δ

***OS*, *CV* 2,** **137 (1943)**

O, H, EtOH, HCl, OEt, Cl, OEt

***OS*, *CV* 2,** **204 (1943)**

O, CH_3, CH_3, NOH

NaO_3SNHOH

H_2O, Δ

NOH, CH_3, CH_3, NOH

***OS*, *CV* 2,** **284 (1943)**

O, C≡N, Me, Ph

HCl

EtOH

- 10 °C

O, NH, Me, OEt, Ph

aq. H_2SO_4

Et_2O

O, O, Me, OEt, Ph

OS, *CV* 2, 313 (1943)

$NH_2OH \cdot HCl$, Na_2CO_3, H_2O

NOH

H

OS, *CV* 2, 336 (1943)

HCl

Cl

OS, *CV* 2, 395 (1943)

2 PhCHO

$H_2NNH_2 \cdot H_2SO_4$, NH_4OH → PhCH=N—N=CHPh

$(CH_3)_2SO_4$ → $CH_3NH—NH_2 \cdot H_2SO_4$

OS, *CV* 2, 419 (1943)

N+ Me $^-SO_4Me$

$K_3Fe(CN)_6$, NaOH, H_2O

N Me, O

OS, *CV 2*, 453 (1943)

NH$_2$ (4-nitroaniline, NO$_2$) —[$Cl_2C{=}O$ / EtOAc]→ [NHC(O)Cl, NO$_2$] —[Δ]→ N═C═O, NO$_2$

OS, *CV 2*, 503 (1943)

Ph(C=O)CH$_3$ —[NH_4OCHO / Δ]→ [HN(CHO)CH(Ph)CH$_3$] —[HCl / H_2O]→ NH$_2$, Ph, CH$_3$

dl form

OS, *CV 2*, 622 (1943)

MeO, MeO, CHO —[1. $NH_2OH \cdot HCl$, EtOH, H_2O; 2. Ac_2O, Δ]→ MeO, MeO, C≡N

OS, *CV 3*, 10 (1955)

PhNH$_2$ + diketene —[Δ / benzene]→ PhNHC(O)CH$_2$C(O)CH$_3$

OS, *CV* 3, 22 (1955)

OH

P$_2$O$_5$

benzene, Δ

O

CH$_3$

OS, *CV* 3, 66 (1955)

O

Et Et

NaCN, NH$_4$Cl
NH$_4$OH, H$_2$O

MeOH, 60 °C

H$_2$N CN

Et Et

1. H$_2$O, HCl
then Δ

2. Pb(OH)$_2$

H$_2$N CO$_2$H

Et Et

OS, *CV* 3, 91 (1955)

Ph—CH═CH—CO$_2$H

NH$_2$OH

H$_2$O, Δ

[Ph CO$_2$H NHOH]

NH$_2$OH

Ph CO$_2$H

NH$_2$

OS, *CV* 3, 93 (1955)

CN

NH$_4$OH

H$_2$N CN

CN

CN

HN

CN

OS, *CV 3*, 258 (1955)

CH_3NH_2 + 2 $CH_2{=}CHCO_2Et$ —(EtOH, H_2O)→ $CH_3N(CH_2CH_2CO_2Et)_2$

OS, *CV 3*, 275 (1955)

Et_2NH + $CH_2{=}O$ —($NaHSO_3$; NaCN, H_2O)→ $Et_2N{-}CH_2{-}CN$

OS, *CV 3*, 360 (1955)

2 $PhNHNH_2$ —(CS_2; Et_2O)→ $PhNHNH{-}C({=}S){-}HS \cdot PhNHNH_2$ —(Δ)→ $PhNHNH{-}C({=}S){-}NHNHPh$

OS, *CV 3*, 371 (1955)

$CH_2{=}CHCHO$ —(NH_4Cl; EtOH, Δ)→ $EtOCH_2CH_2CHO$ + $EtOCH_2CH_2CH(OEt)_2$

OS, *CV 3*, 374 (1955)

$CH_3COCH_2CO_2Et$ —($PhNH_2$, HOAc (cat.); benzene, Δ)→ $CH_3C(NHPh){=}CHCO_2Et$

OS, *CV* 3, 436 (1955)

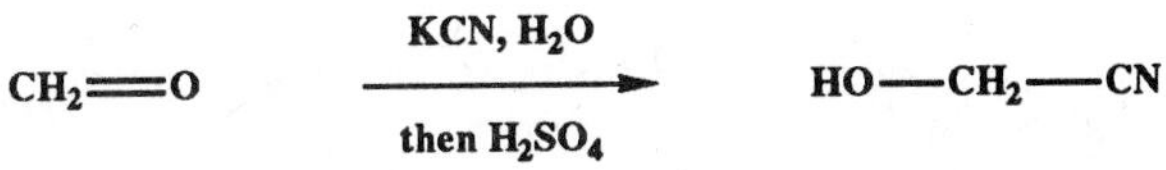

$CH_2{=}O$ → (KCN, H_2O; then H_2SO_4) → $HO{-}CH_2{-}CN$

OS, *CV* 3, 458 (1955)

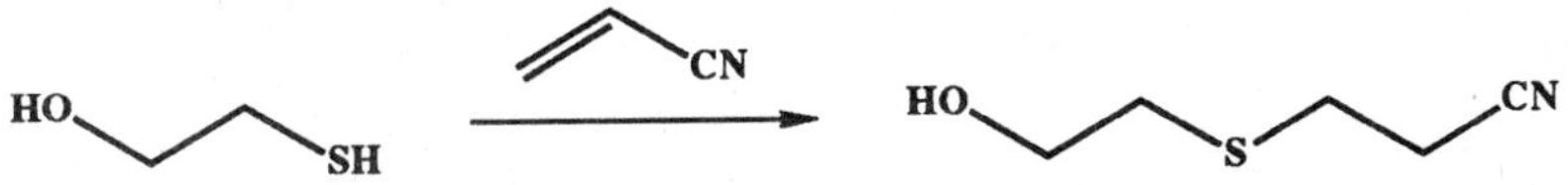

HO–CH₂CH₂–SH → (CH₂=CH–CN) → HO–CH₂CH₂–S–CH₂CH₂–CN

OS, *CV* 3, 576 (1955)

CH₂=CH–C(=O)OMe → (HBr, Et_2O) → Br–CH₂CH₂–C(=O)OMe

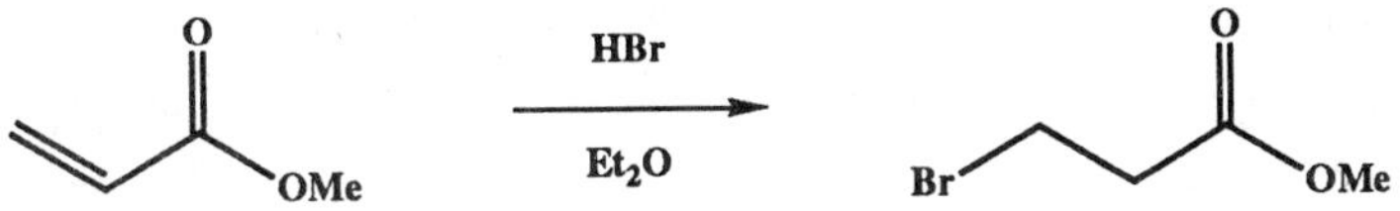

OS, *CV* 3, 609 (1955)

1-Naphthyl–N(CH₃)–CN → (H_2S, NH_3; EtOH) → 1-Naphthyl–N(CH₃)–C(=S)–NH₂

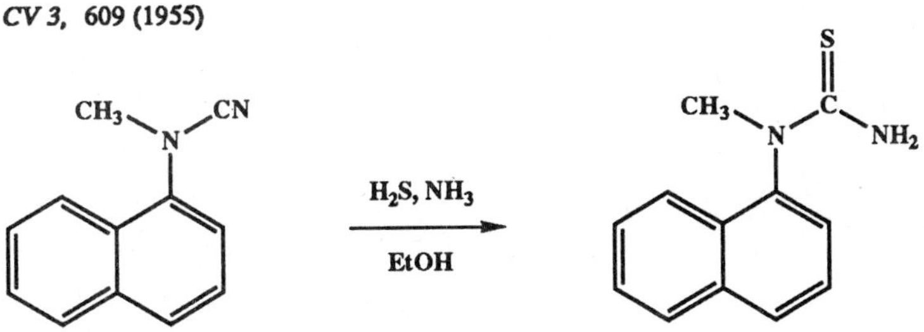

OS, CV 3, 617 (1955)

CH_3—N=C=S $\xrightarrow{NH_4OH}$ $CH_3NHC(=S)NH_2$

OS, CV 3, 735 (1955)

PhC(=O)N=C=S $\xrightarrow[\text{acetone}]{PhNH_2}$ [PhC(=O)NHC(=S)NHPh] $\xrightarrow[H_2O, \Delta]{NaOH}$ PhNHC(=S)NH_2

OS, CV 3, 774 (1955)

CH_2=CHCO_2Me

1. $Hg(OAc)_2$, MeOH
2. KBr, H_2O
3. Br_2, light, $CHCl_3$, 50 °C

→ MeOCH_2CH(Br)CO_2Me

OS, CV 3, 813 (1955)

CH_3CH=$CHCO_2H$

1. $Hg(OAc)_2$, CH_3OH
2. KBr, Br_2, light, 0 °C
3. 40% HBr

→ CH_3CH(OCH_3)CH(Br)CO_2H

OS, *CV 3*, 853 (1955)

$H-C\equiv C-H$

HgO, 40 - 50 °C

OS, *CV 4*, 13 (1963)

HgO, H_2SO_4

H_2O, 60 °C

OS, *CV 4*, 49 (1963)

NaNCO

HOAc, H_2O

OS, *CV 4*, 146 (1963)

Cu(OAc)$_2$• H_2O

95 - 130 °C

OS, *CV 4*, 157 (1963)

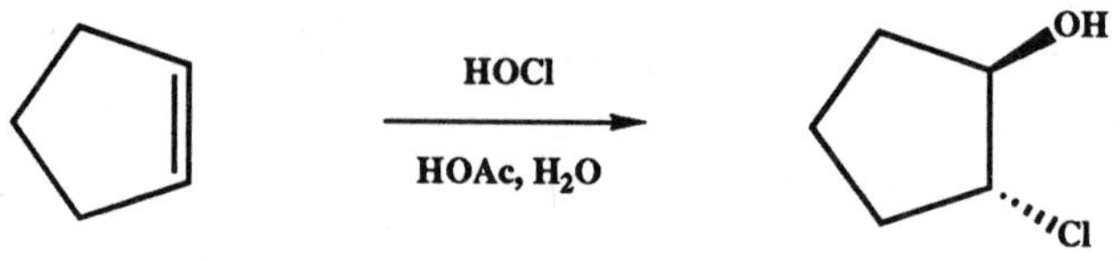
HOCl
HOAc, H_2O
OH
Cl

OS, *CV 4*, 180 (1963)

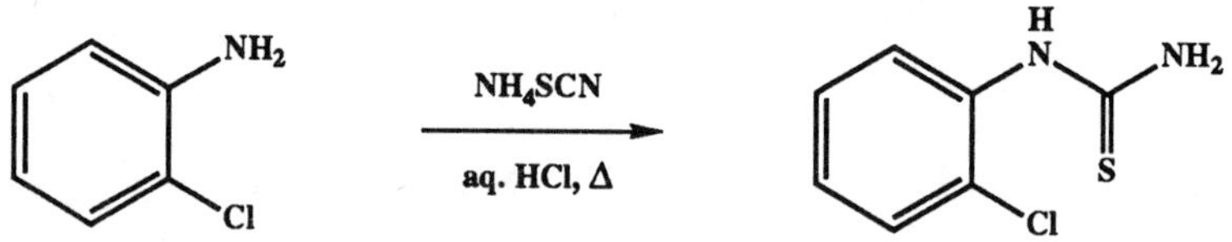
NH_2
Cl
NH_4SCN
aq. HCl, Δ
H
N
NH_2
S
Cl

OS, *CV 4*, 184 (1963)

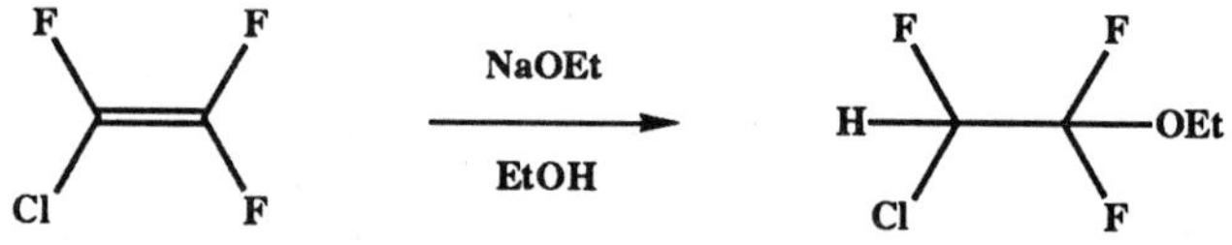
F
F
Cl
F
NaOEt
EtOH
F
F
H
OEt
Cl
F

OS, *CV 4*, 186 (1963)

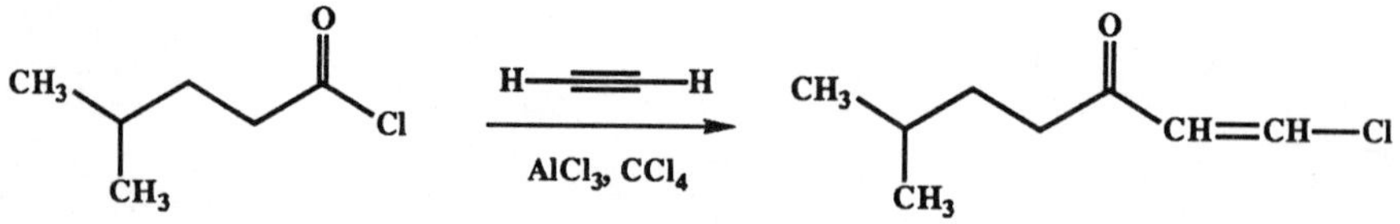
O
CH_3
Cl
CH_3
H—≡—H
$AlCl_3$, CCl_4
O
CH_3
CH=CH—Cl
CH_3

OS, *CV* 4, 205 (1963)

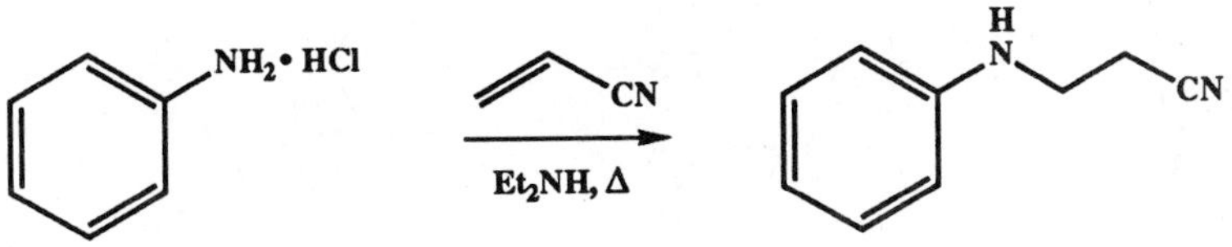

$NH_2 \cdot HCl$

$CH_2{=}CH{-}CN$

Et_2NH, Δ

H
N
CN

OS, *CV* 4, 213 (1963)

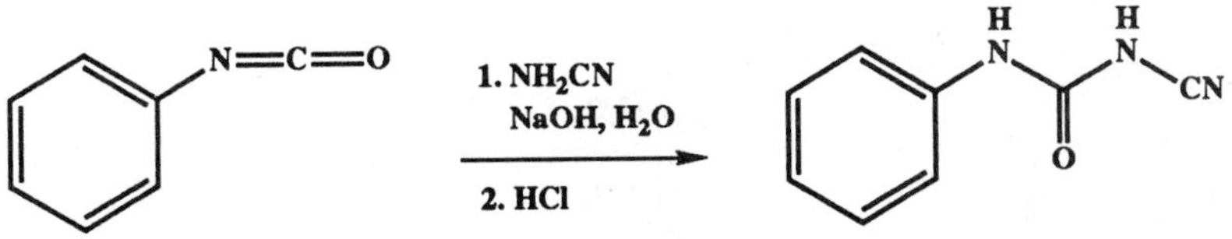

N=C=O

1. NH_2CN
   NaOH, H_2O

2. HCl

H H
N N
CN
O

OS, *CV* 4, 229 (1963)

O
O

2 $HONH_3^+ \ Cl^-$

aq. KOH, 0 °C

NOH
NOH

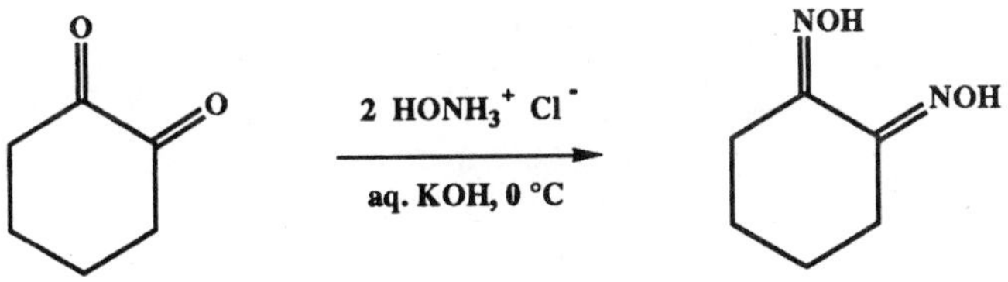

OS, *CV* 4, 238 (1963)

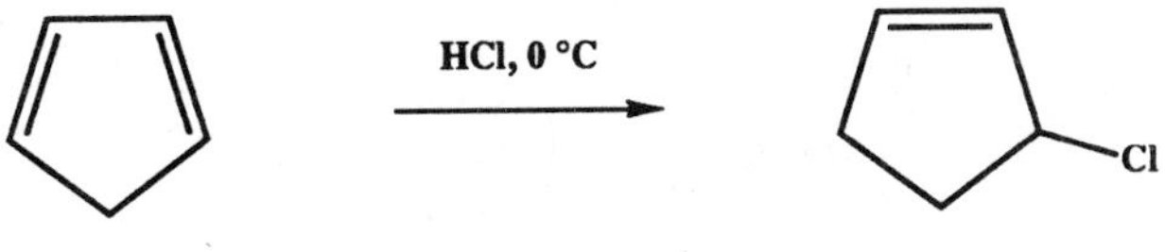

HCl, 0 °C

Cl

OS, *CV 4*, 261 (1963)

$$CH_2(CO_2H)_2 + 2\ CH_2{=}C(CH_3)_2 \xrightleftharpoons{H_2SO_4} CH_2(CO_2\text{-}t\text{-}Bu)_2$$

OS, *CV 4*, 317 (1963)

$$CH_3(CH_2)_7{-}CH{=}CH{-}(CH_2)_7CO_2H \xrightarrow[HCO_2H]{H_2O_2} CH_3(CH_2)_7{-}CH(OH){-}CH(OCHO){-}(CH_2)_7CO_2H$$

OS, *CV 4*, 359 (1963)

$$Me_2N{-}C{\equiv}N \xrightarrow[NH_4OH,\ H_2O]{H_2Se} Me_2N{-}C(=Se){-}NH_2$$

OS, *CV 4*, 430 (1963)

$$CH_3(CH_2)_5CHO + EtO_2C{-}CH{=}CH{-}CO_2Et \xrightarrow[\text{trichloroethylene},\ \Delta]{\text{benzoyl peroxide}} CH_3(CH_2)_5C(=O){-}CH(CO_2Et){-}CH_2CO_2Et$$

OS, *CV* 4, 489 (1963)

O, H, O

$\xrightarrow[H_2SO_4\ (cat.)]{Ac_2O}$

O, $CH(OAc)_2$

OS, *CV* 4, 502 (1963)

H_2N, H, N, CN, NH

$\xrightarrow[65 - 75\ °C]{H_2S,\ H_2O}$

H_2N, H, N, NH_2, NH, S

OS, *CV* 4, 515 (1963)

$n\text{-}C_7H_{15}\text{—}NH_2$

$\xrightarrow[aq.\ HCl,\ 75\ °C]{NaCNO}$

$n\text{-}C_7H_{15}\text{—}N(H)$, O, NH_2

OS, *CV* 4, 543 (1963)

$\xrightarrow[80\ °C]{KI,\ H_3PO_4}$

I

OS, CV 4, 630 (1963)

$CH_3-CH=CH-CO_2Me$ + $EtO_2C-CH_2-CO_2Et$

1. Na, EtOH, Δ
2. HCl, H_2O, Δ
3. 180 - 190 °C
(- CO_2)

→ $CH_3-CH(CH_2CO_2H)_2$

OS, CV 4, 633 (1963)

$CH_3C(=O)CH_2CO_2H$

$PhN_2^+ Cl^-$
aq. NaOAc, 50 °C
(- CO_2)

→ $CH_3C(=O)CH=N-NHPh$

OS, CV 4, 645 (1963)

$H_2N-C\equiv N$

HCl
CH_3OH

→ $H_2N-C(OCH_3)=NH \cdot HCl$

OS, CV 4, 652 (1963)

$(CH_3)_2CHNO_2$ + $CH_2=CH-CO_2Me$

$PhCH_2NMe_3^+ OH^-$
dioxane, H_2O, Δ

→ $(CH_3)_2C(NO_2)CH_2CH_2CO_2Me$

OS, *CV* 4, 669 (1963)

2 $CH_2{=}CHCO_2Me$ —(H_2S, NaOAc / aq. EtOH, Δ)→ $MeO_2C(CH_2)_2S(CH_2)_2CO_2Me$

OS, *CV* 4, 748 (1963)

CH_3CHO —(EtOH / HCl, - 5 °C)→ $CH_3CH(OEt)Cl$ —(Br_2)→ $BrCH_2CH(OEt)Br$

OS, *CV* 4, 769 (1963)

Ph—C≡N + Ph—NH_2 —($AlCl_3$ / 200 °C)→ PhC(=NH)NHPh

OS, *CV* 5, 73 (1973)

$PhCH_2OH$ —($CH_2{=}CHCN$ / H_2SO_4)→ $PhCH_2NHC(O)CH{=}CH_2$

OS, CV 5, **130 (1973)**

Ph–CHO → ($CHCl_3$, KOH) → **Ph–CH(OH)–CCl_3**

OS, CV 5, **136 (1973)**

$CH_3(CH_2)_4$–CH=CH_2 → (CH_3C(O)NHBr; HF, Et_2O) → **$CH_3(CH_2)_4$–CH(F)–CH_2Br**

OS, CV 5, **258 (1973)**

p-Ts—NH—NH_2 → (HO–CH(OH)–CO_2H; HCl, H_2O) → **p-Ts—NH—N=CH—CO_2H**

OS, CV 5, **266 (1973)**

cyclooctene → (NOCl, HCl; trichloroethylene) → **2-chlorocyclooctanone oxime (=NOH • HCl, Cl)**

OS, *CV* 5, 403 (1973)

Furan → 2,5-dimethoxy-2,5-dihydrofuran (MeO, OMe)

Br_2, MeOH
Na_2CO_3
C_6H_6, - 15 to 0 °C

OS, *CV* 5, 437 (1973)

Benzaldehyde → PhCH(NMe_2)CN

1. $NaHSO_3$, H_2O
2. Me_2NH
NaCN, H_2O

OS, *CV* 5, 555 (1973)

EtN=C=O $\xrightarrow[\text{CH}_2\text{Cl}_2,\ 5\ ^\circ\text{C}]{\text{H}_2\text{N(CH}_2\text{)}_3\text{NMe}_2}$ EtNH–C(=O)–NH(CH_2)_3NMe_2 $\xrightarrow[\text{Et}_3\text{N}]{p\text{-TsCl}}$ EtN=C=N(CH_2)_3NMe_2

OS, *CV* 5, 575 (1973)

Piperidine (N–H) → N-ethylpiperidine (N–CH_2CH_3)

CH_2=CH_2
400 psi, 100 °C

***OS*, *CV 5*, 647 (1973)**

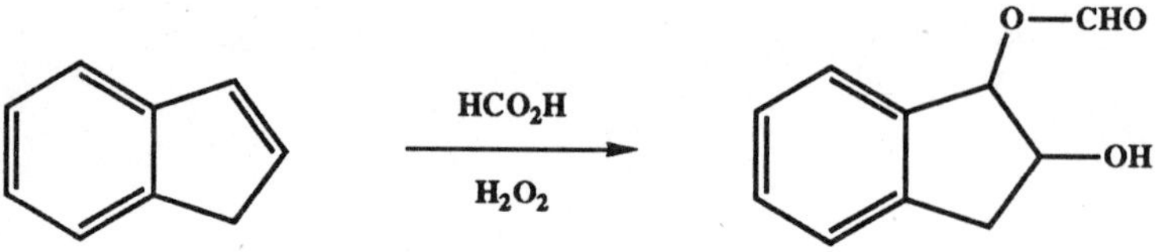

HCO_2H, H_2O_2

O—CHO
OH

***OS*, *CV 5*, 818 (1973)**

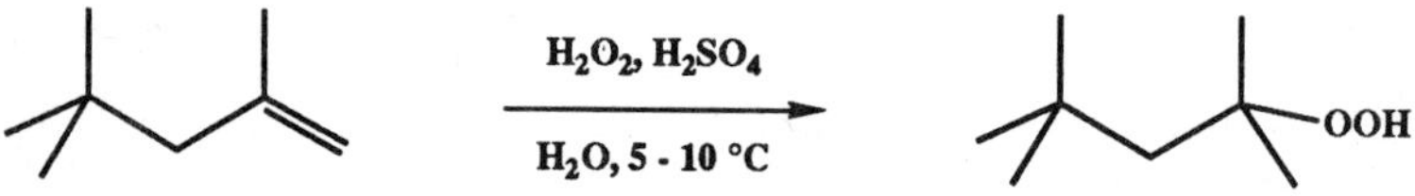

H_2O_2, H_2SO_4
H_2O, 5 - 10 °C

OOH

***OS*, *CV 5*, 852 (1973)**

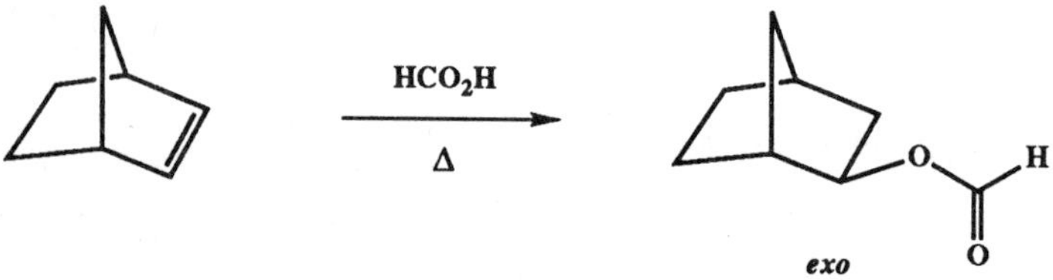

HCO_2H
Δ

exo

***OS*, *CV 5*, 966 (1973)**

Ph—N=C=S + H_2N–C(OMe)=NH • HCl → (aq. KOH, Δ; acetone) Ph–NH–C(=S)–NH–C(OMe)=NH

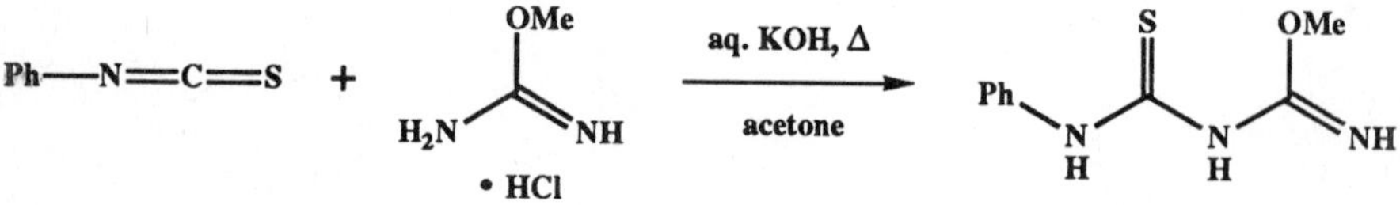

OS, *CV 5*, 1103 (1973)

$\xrightarrow[ZnCl_2]{HCl}$

OS, *CV 6*, 1 (1988)

MeO, MeO, NCOCH$_3$ $\xrightarrow[60 - 65\ °C]{5\%\ aq.\ HCl}$ MeO, MeO, NHCOCH$_3$, O, CH$_3$

OS, *CV 6*, 5 (1988)

$CH_3C(=O)-NH_2$ + $CH_2=O$ $\xrightarrow{K_2CO_3,\ H_2O}$ $CH_3C(=O)-NHCH_2OH$

OS, *CV 6*, 10 (1988)

$CH_3C(=O)CH_3$ $\xrightarrow{H_2NNH_2 \cdot H_2O}$ $(CH_3)_2C=NN=C(CH_3)_2$

$(CH_3)_2C=NN=C(CH_3)_2$ $\xrightarrow[100\ °C]{H_2NNH_2}$ $(CH_3)_2C=NNH_2$

***OS*, *CV 6*, 12 (1988)**

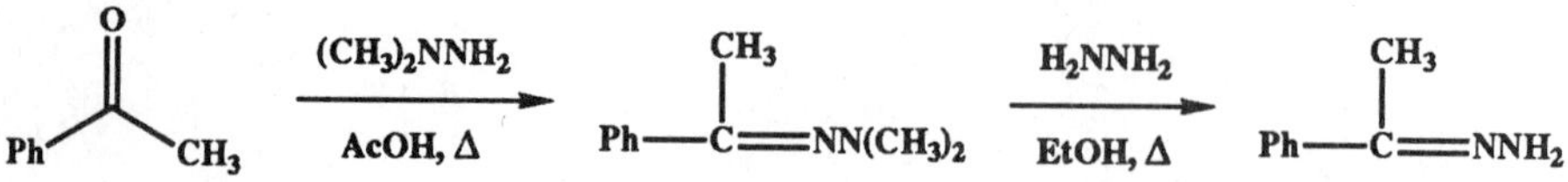

***OS*, *CV 6*, 62 (1988)**

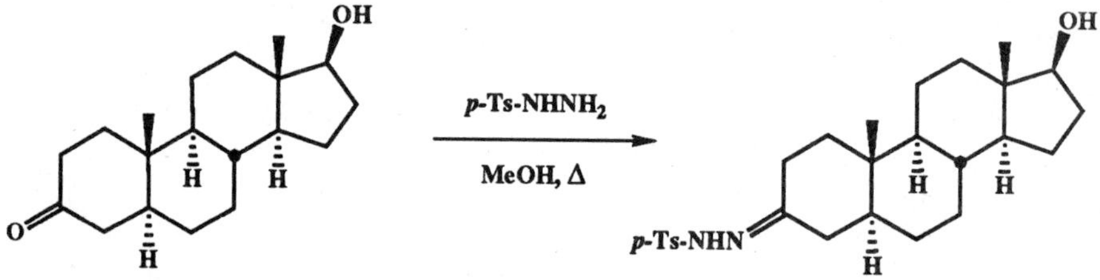

***OS*, *CV 6*, 75 (1988)**

$$HO(CH_2)_3NH_2 + 2\ CH_2{=}CHCO_2Et \longrightarrow HO(CH_2)_3N(CH_2CH_2CO_2Et)_2$$

***OS*, *CV 6*, 101 (1988)**

$$PhCH_2OH \xrightarrow{\text{HCHO, HCl, 25 °C}} PhCH_2OCH_2Cl$$

***OS*, *CV 6*, 163 (1988)**

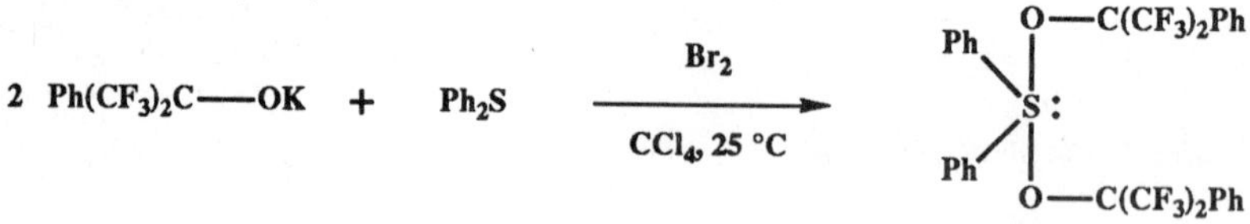

OS, *CV 6*, 184 (1988)

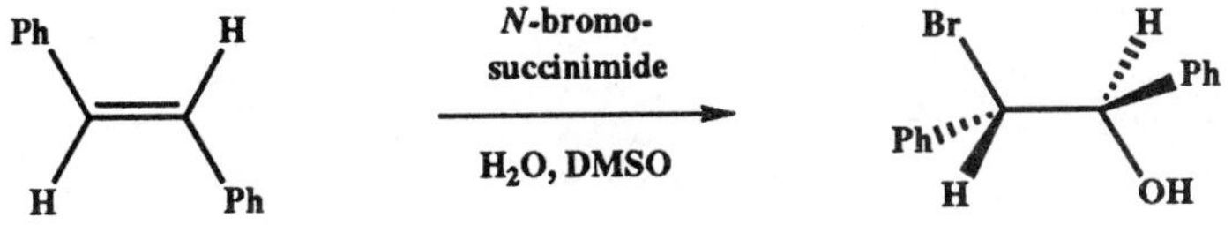

Ph H H Ph (alkene) → *N*-bromo-succinimide / H_2O, DMSO → Br, H, Ph, Ph, H, OH

OS, *CV 6*, 226 (1988)

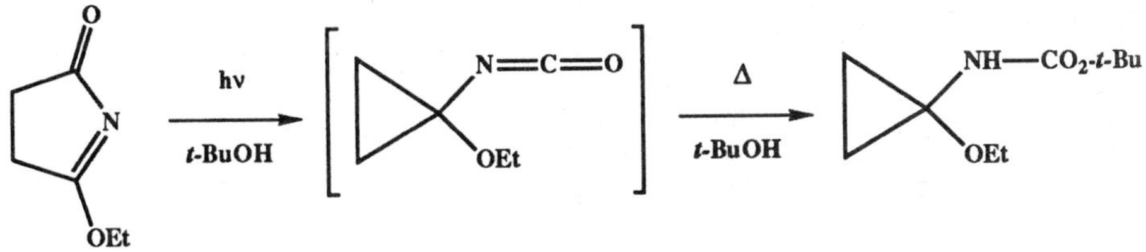

O, N, OEt → hν / *t*-BuOH → [N=C=O, OEt] → Δ / *t*-BuOH → NH—CO_2-*t*-Bu, OEt

OS, *CV 6*, 273 (1988)

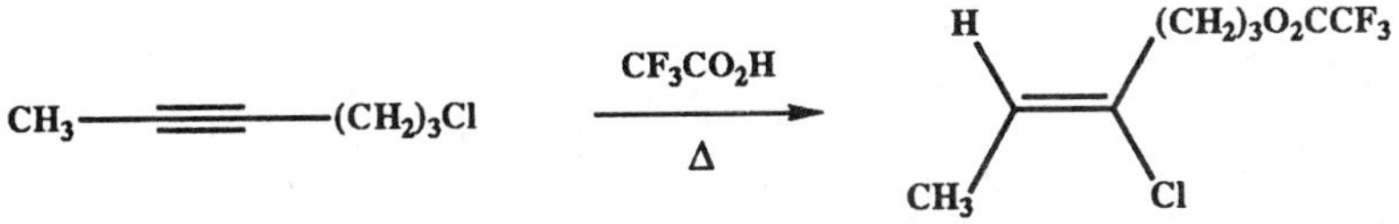

CH_3—≡—$(CH_2)_3Cl$ → CF_3CO_2H / Δ → H, $(CH_2)_3O_2CCF_3$, CH_3, Cl

OS, *CV 6*, 293 (1988)

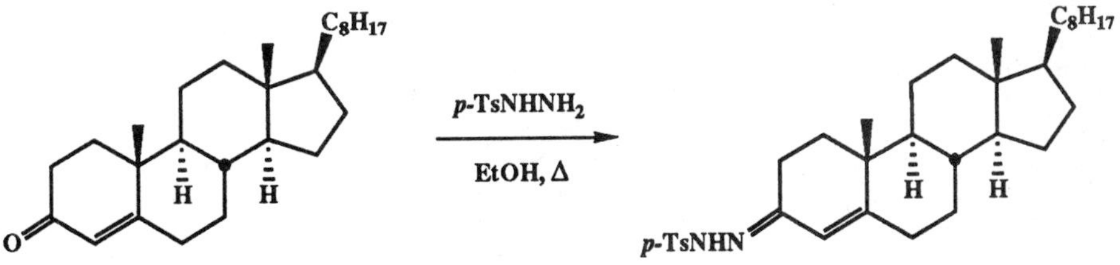

C_8H_{17}, H, H, O → *p*-$TsNHNH_2$ / EtOH, Δ → C_8H_{17}, H, H, *p*-TsNHN

OS, CV 6, 327 (1988)

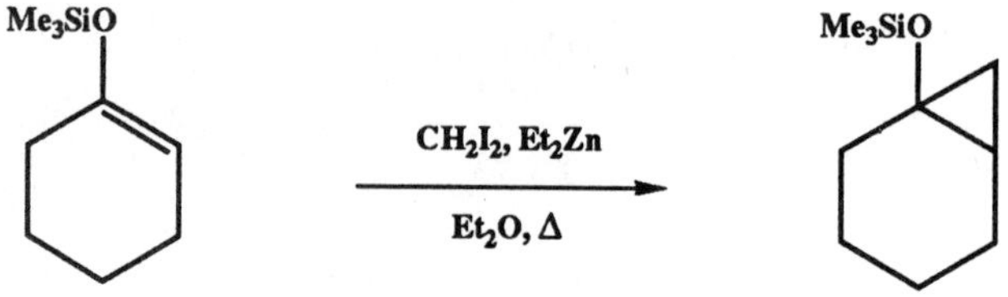

Me_3SiO (cyclohexene) → Me_3SiO (bicyclo[4.1.0]heptane)

CH_2I_2, Et_2Zn

Et_2O, Δ

OS, CV 6, 334 (1988)

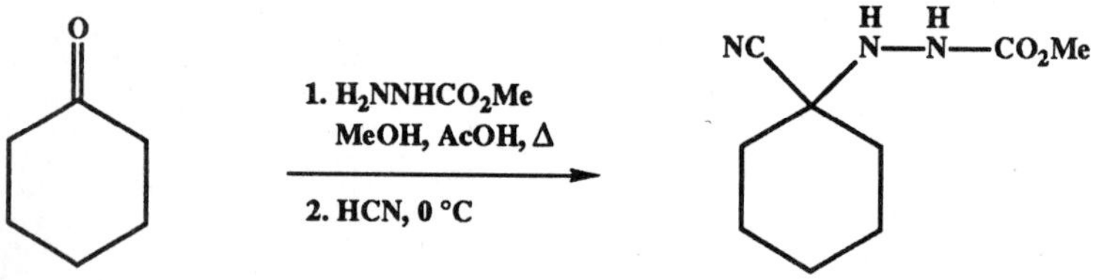

1. H_2NNHCO_2Me
MeOH, AcOH, Δ

2. HCN, 0 °C

NC, H H N—N—CO_2Me

OS, CV 6, 338 (1988)

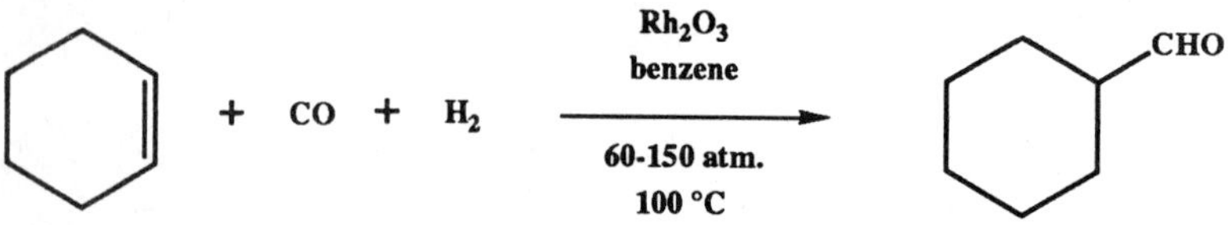

+ CO + H_2

Rh_2O_3
benzene

60-150 atm.
100 °C

CHO

OS, CV 6, 361 (1988)

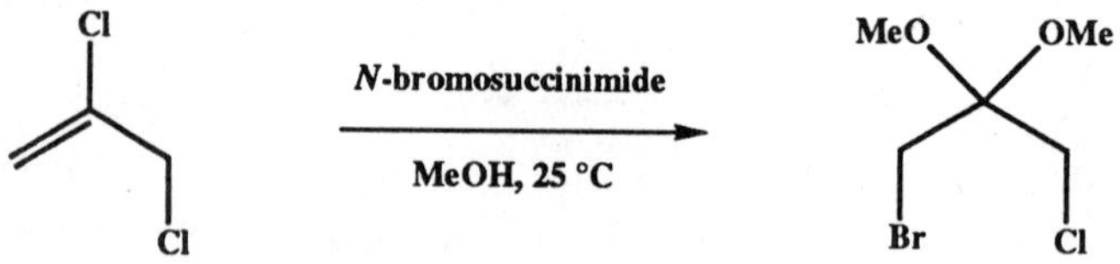

Cl Cl

N-bromosuccinimide

MeOH, 25 °C

MeO OMe Br Cl

***OS*, *CV 6*, 448 (1988)**

$$(EtO)_2P(=O)-CH_2CHO \xrightarrow[\text{MeOH}]{H_2N-C_6H_{11}} (EtO)_2P(=O)-CH{=}CH-NH-C_6H_{11}$$

***OS*, *CV 6*, 474 (1988)**

$$Me_2NH + CH_2{=}O \xrightarrow{H_2O,\ 25\ °C} Me_2NCH_2NMe_2$$

***OS*, *CV 6*, 496 (1988)**

$$(CH_3)_2CH-CHO + HN(CH_2)_4 \xrightarrow[(-H_2O)]{\Delta} (CH_3)_2C{=}CH-N(CH_2)_4$$

***OS*, *CV 6*, 507 (1988)**

$$R-CH_2OH \xrightarrow[2.\ CCl_3CN,\ 0\ °C]{1.\ NaH,\ Et_2O} R-CH_2-O-C(=NH)-CCl_3$$

OS, *CV 6*, 520 (1988)

O Ph CH$_3$ + O N H —(*p*-TsOH; benzene, Δ)→ O N CH$_2$ Ph

OS, *CV 6*, 526 (1988)

O H —(*t*-BuNH$_2$; K$_2$CO$_3$)→ N-*t*-Bu H —(EtMgBr; THF, Δ)→ [BrMg N-*t*-Bu H] —(1. PhCH$_2$Cl, Δ; 2. 10% HCl, Δ)→ PhCH$_2$ O H

OS, *CV 6*, 560 (1988)

OAc —(*N*-bromosuccinimide; *t*-BuOH, H$_2$O, 12 - 25 °C)→ OAc Br HO

OS, *CV 6*, 592 (1988)

EtO$_2$C CH$_3$ O + HN —(benzene; Δ)→ EtO$_2$C CH$_3$ N

OS, CV 6, 664 (1988)

$(CF_3)_2C{=}O$ → (1. NH_3, pyridine; 2. $POCl_3$, Δ) → $(CF_3)_2C{=}NH$

OS, CV 6, 679 (1988)

2-Cyclohexen-1-one → (H_2O_2, NaOH; MeOH, 0 °C) → 2,3-epoxycyclohexanone → (trans-2,3-diphenyl-1-aziridinamine, Ph / Ph / N—NH_2; benzene, 25 °C) → 2,3-epoxycyclohexanone N-(trans-2,3-diphenylaziridin-1-yl)imine (Ph, Ph, N, N, O)

OS, CV 6, 715 (1988)

$H_3\overset{+}{N}CH_2CH_2COOH\ Cl^-$ + $ClC(O)OCCl_3$ → (dioxane; 55 - 60 °C) → $O{=}C{=}N{-}CH_2CH_2{-}C(O)Cl$

OS, CV 6, 719 (1988)

2 (+) - α - pinene → ($Me_2S \bullet BH_3$; THF) → $(\text{isopinocampheyl})_2BH$ → (MeOH; THF) → $(\text{isopinocampheyl})_2BOMe$

(+) - α - pinene

(-) isomer

OS, *CV 6*, 737 (1988)

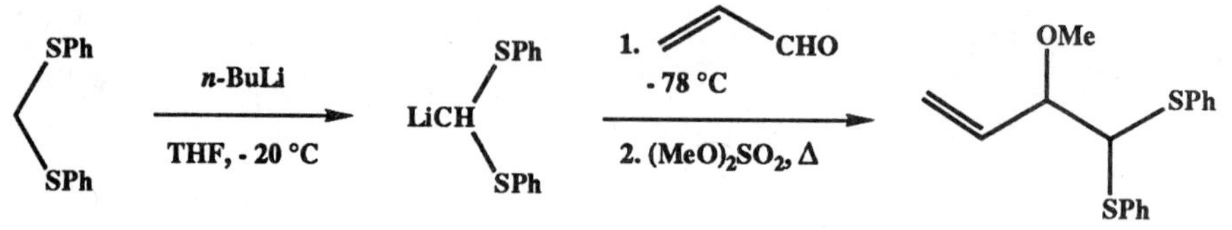

SPh, SPh; n-BuLi; THF, - 20 °C; LiCH(SPh)SPh; 1. CHO, - 78 °C; 2. $(MeO)_2SO_2$, Δ; OMe, SPh, SPh

OS, *CV 6*, 762 (1988)

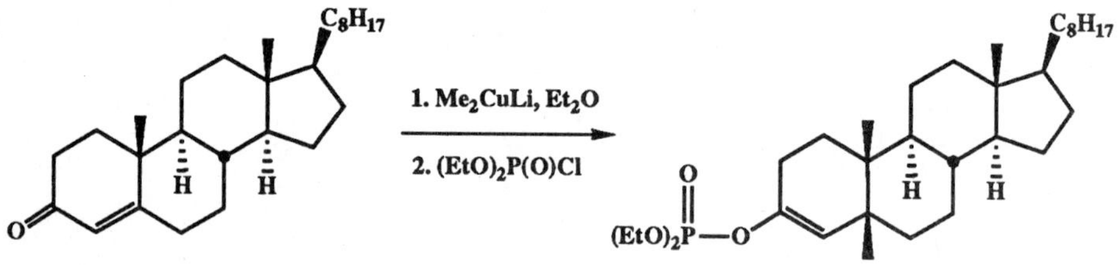

C_8H_{17}; O; H; H; 1. Me_2CuLi, Et_2O; 2. $(EtO)_2P(O)Cl$; C_8H_{17}; O; $(EtO)_2P$—O; H; H

OS, *CV 6*, 766 (1988)

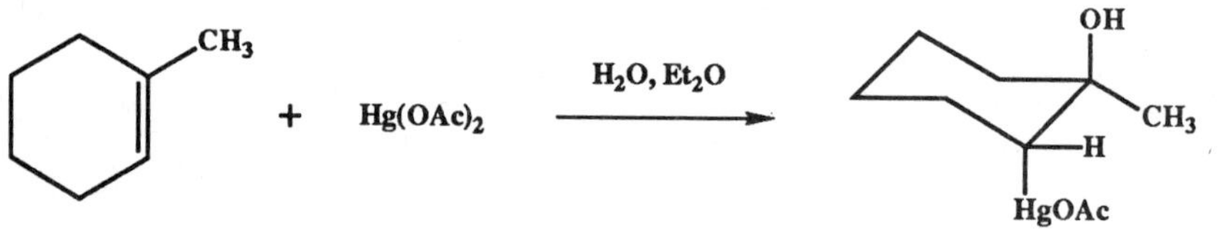

CH_3 + $Hg(OAc)_2$; H_2O, Et_2O; OH; CH_3; H; HgOAc

OS, *CV 6*, 795 (1988)

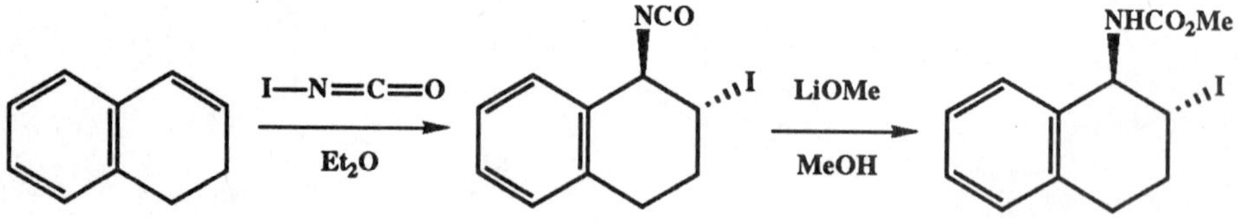

I—N=C=O; Et_2O; NCO; I; LiOMe; MeOH; $NHCO_2Me$; I

OS, CV 6, 799 (1988)

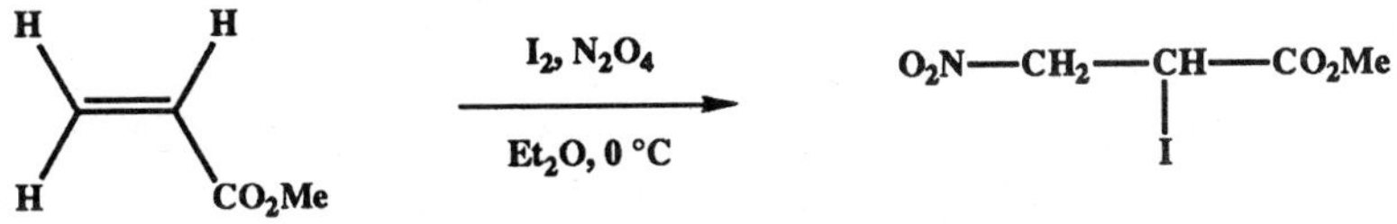

I_2, N_2O_4

Et_2O, 0 °C

$O_2N—CH_2—CH(I)—CO_2Me$

OS, CV 6, 818 (1988)

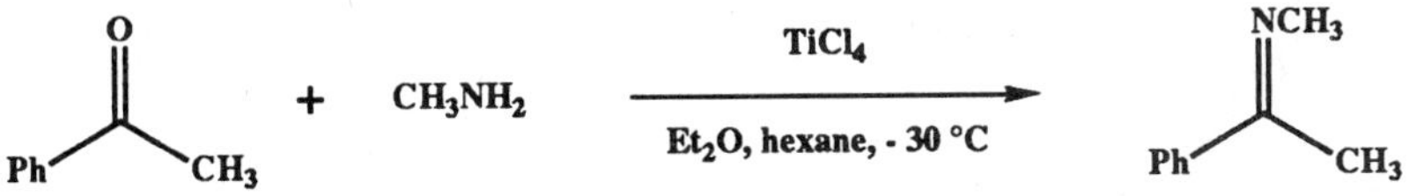

+ CH_3NH_2

$TiCl_4$

Et_2O, hexane, - 30 °C

OS, CV 6, 837 (1988)

N_2O_4, O_2

Et_2O, 10 °C

Et_3N

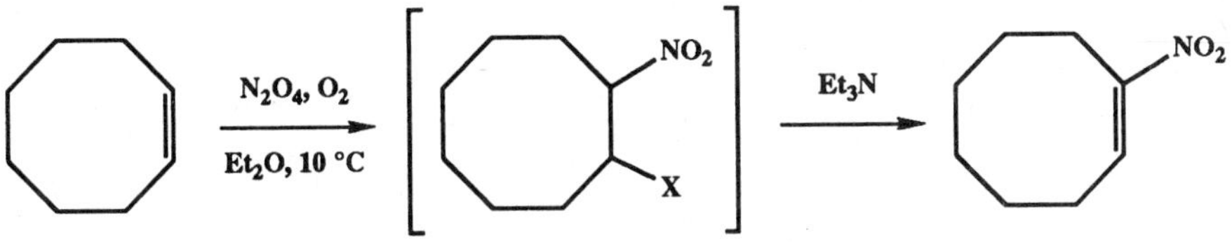

X = NO_2 and ONO and ONO_2

OS, CV 6, 852 (1988)

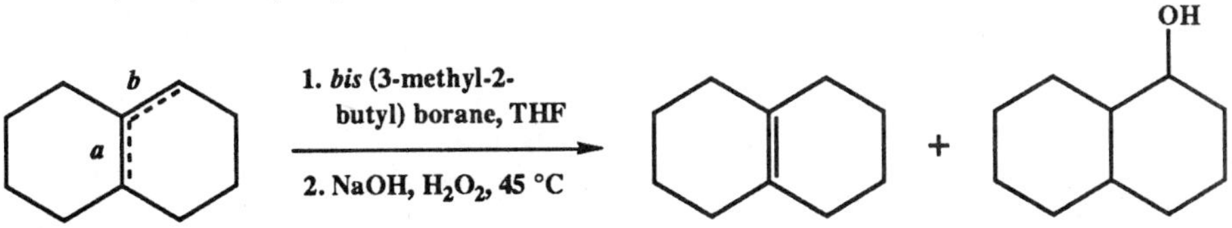

1. *bis* (3-methyl-2-butyl) borane, THF

2. NaOH, H_2O_2, 45 °C

ca. 82 : 18 *a* : *b*

OS, *CV 6*, 893 (1988)

Ph(H)C=C(H)CH(OMe)$_2$ —[ICl, NaN_3 / CH_3CN]→ PhCH(N_3)CH(I)CH(OMe)$_2$

OS, *CV 6*, 919 (1988)

$(CH_3)_2C=CH(CH_3)$ —[1. BH_3, THF; 2. 1-octene; 3. NaOH, H_2O_2]→ $(CH_3)_2CHCH(OH)CH_3$

OS, *CV 6*, 932 (1988)

$C_6H_5PH_2$ + 2 $CH_2=CHCN$ —[KOH / CH_3CN]→ $C_6H_5P(CH_2CH_2CN)_2$

OS, *CV 6*, 936 (1988)

EtO_2C–HN–NH_2 + Ph–N=C=O —[benzene]→ EtO_2C–HN–NH–C(=O)–HN–Ph

OS, CV 6, 943 (1988)

(±) - α - pinene

$NaBH_4$, $BF_3 \cdot Et_2O$, diglyme, 0 °C

BH

2

$\overset{+}{N}H_3OSO_3^-$, diglyme, 100 °C

NH_2

OS, CV 6, 951 (1988)

P

NH_2

i-Pr—N=C=O

THF, 25 °C to reflux

H N O NH-*i*-Pr

P = styrene - divinylbenzene copolymer

OS, CV 6, 981 (1988)

p-Ts⁻ Na⁺ $\xrightarrow[\text{HCO}_2\text{H, H}_2\text{O, 70 - 75 °C}]{\text{CH}_2\text{=O, H}_2\text{NCO}_2\text{Et}}$ *p*-Ts—CH_2—N(H)—CO_2Et

OS, CV 6, 987 (1988)

p-Ts⁻ Na⁺ $\xrightarrow[\text{HCO}_2\text{H, H}_2\text{O, 90 - 95 °C}]{\text{CH}_2\text{=O, H}_2\text{NCHO}}$ *p*-Ts—CH_2—N(H)—CHO

OS, *CV* 7, 20 (1990); *60*, 14 (1981)

$Ph_2C{=}O$ → (Me_3SiCN; ZnI_2, CH_2Cl_2) → $Ph_2C(OSiMe_3)(CN)$ → (3 *N* HCl; THF) → $Ph_2C(OH)(CN)$

OS, *CV* 7, 517 (1990); *60*, 126 (1981)

p-benzoquinone + Me_3SiCN → (18-crown-6, KCN; CCl_4) → 4-(trimethylsilyloxy)-4-cyano-2,5-cyclohexadienone (OSiMe3, CN)

OS, *CV* 7, 27 (1990); *61*, 14 (1983)

PhCHO + H_2N-(5-aminotetrazole, N–H) → (1. Et_3N, MeOH, 50 °C; 2. H_2 (500 psi), Pd / C) → $PhCH_2NH$-(tetrazole, N–H)

OS, *CV* 7, 375 (1990); *61*, 85 (1983)

cyclohexene → ($TsNClNa \cdot 3\,H_2O$; 1% OsO_4, $CHCl_3$, H_2O, 60 °C; phase-transfer conditions) → cis-2-(NHTs)cyclohexanol (OH, NHTs)

2-methyl-2-pentene → ($TsNClNa \cdot 3\,H_2O$; 1% OsO_4, *t*-BuOH, 60 °C) → product (NHTs, OH)

OS, *CV 7*, 223 (1990); *61*, 93 (1983)

Ph, Ph + $EtO_2CNClNa$ + $AgNO_3$ + H_2O —(1% OsO_4, MeCN)→ OH, Ph, Ph, $NHCO_2Et$

OS, *CV 7*, 427 (1990); *61*, 103 (1983)

$BH_3 \cdot Et_3N$, diglyme, 130-140 and 200 °C; H, H, B, H; 1. CO, $(CH_2OH)_2$; 2. H_2O_2, NaOH, 95% EtOH; H, OH, H, H

OS, *CV 7*, 304 (1990); *61*, 112 (1983)

(+) isomer; hν; C_6H_5OH; H^+, MeOH; OMe

OS, *CV 7*, 66 (1990); *61*, 116 (1983)

O, OH; Me, Me; O, $OC(Me)_3$

OS, *CV* 7, 59 (1990); *62*, 140 (1984)

CHO

1. HBr

2. $HO(CH_2)_3OH$, H^+

Br O O

1. HBr

2. HO HO , $(EtO)_3CH$

O O Br

OS, *CV* 7, 521 (1990); *62*, 196 (1984)

MeO CHO

Me_3SiCl

KCN

$Zn(CN)_2$

MeO CH $OSiMe_3$ CN

OS, *CV* 7, 339 (1990); *63*, 44 (1985)

(-) - α - pinene

$BH_3 \cdot THF$

BH 2

CH_2CO_2Me

H_2O_2, NaOH

HO R R CH_2CO_2Me

OS, *CV* 7, 381 (1990); *63*, 79 (1985)

EtCHO + NaCN

$NaHSO_3$

H_2O

0 °C

OH CN

OEt

HCl

O O CN

OS, *CV 7*, 294 (1990); *64*, 39 (1986)

Me_3SiCN

ZnI_2, CH_2Cl_2

$OSiMe_3$

$N{\equiv}C$

KF

MeOH

OH

$N{\equiv}C$

OS, *CV 7*, 160 (1990); *64*, 80 (1986)

$(EtO)_2P(O)H + (CH_2O)_n + Et_3N \longrightarrow (EtO)_2P(O)CH_2OH$

OS, *CV 7*, 258 (1990); *64*, 164 (1986)

$Ph_3P{=}CH_2$

THF

1. $(Me_2CHCH(Me))_2BH$

2. H_2O_2, NaOH

OH

OS, *65*, 90 (1987)

$BrCH_2SO_2Br$

hν

Br

SO_2CH_2Br

H

OS, *65*, 215 (1987)

$PhCH_2SH$; NaOH, THF, reflux → SCH_2Ph ketone; Na, NH_3, MeOH, - 78 °C → SH, OH; $(CH_2O)_n$, *p*-TsOH, C_6H_6, reflux → S, O, H, H, *R*

OS, *66*, 29 (1987)

CO_2Et; Me_3N, H_2O, CH_2Cl_2 → Me_3N^+, CO_2^-; HO, Ph; 1. NaH, THF 2. H_3O^+ → O, CO_2H, Ph

OS, *67*, 44 (1988)

Et_2NH; Li or BuLi → NEt_2

OS, *67*, 48 (1988)

Et_2NH; Li or BuLi → NEt_2

OS, *67*, 105 (1988)

LiCl, LiOAc • 2 H_2O
$Pd(OAc)_2$ (7.5 mol %)
p-benzoquinone
HOAc, 25 °C

AcO Cl

OS, *67*, 114 (1988)

O

AcOH, $Pd(PPh_3)_4$

HO OAc

OS, *67*, 157 (1988)

+ $PhSeSO_2Ph$

1. hν, CCl_4
2. H_2O_2, CH_2Cl_2

SO_2Ph

OS, *68*, 148 (1989)

$PhSO_2Na$, $HgCl_2$
DMSO-H_2O (1 : 5), r.t.

HgCl
SO_2Ph

OS, CV 1, **521 (1941)**

Br, Br_2, CCl_4, 0 °C → Br, Br, Br

OS, CV 2, **171 (1943)**

Br_2, EtOH, CCl_4, - 5 to - 1 °C → Br, Br

OS, CV 2, **177 (1943)**

HO_2C, CO_2H, Br_2, H_2O, Δ → HO_2C, Br, Br, CO_2H

OS, CV 2, **270 (1943)**

Ph—CH=CH—C(=O)—OEt, Br_2, CCl_4, 0 °C → Ph, Br, O, Br, OEt

OS, CV 2, 307 (1943)

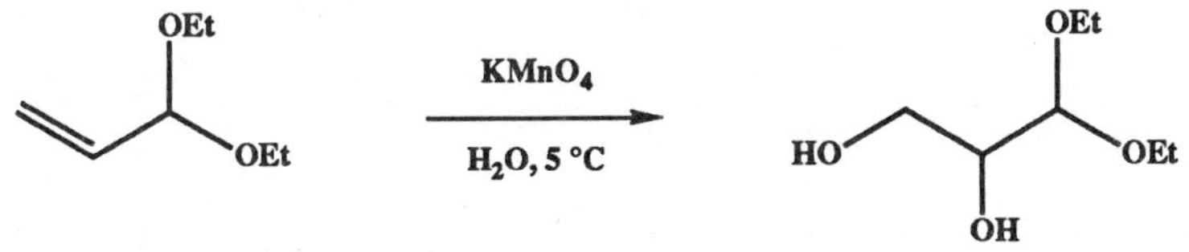

$KMnO_4$

H_2O, 5 °C

OEt OEt HO OEt OEt OH

dl - glyceraldehyde acetal

OS, CV 2, 408 (1943)

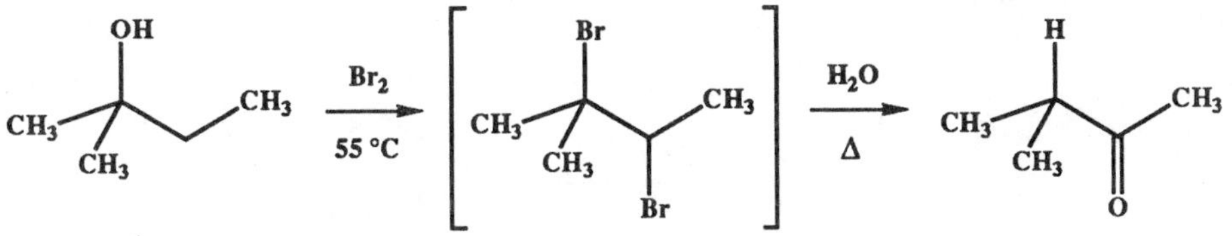

OH CH_3 CH_3 CH_3

Br_2

55 °C

Br CH_3 CH_3 CH_3 Br

H_2O

Δ

H CH_3 CH_3 CH_3 O

OS, CV 3, 105 (1955)

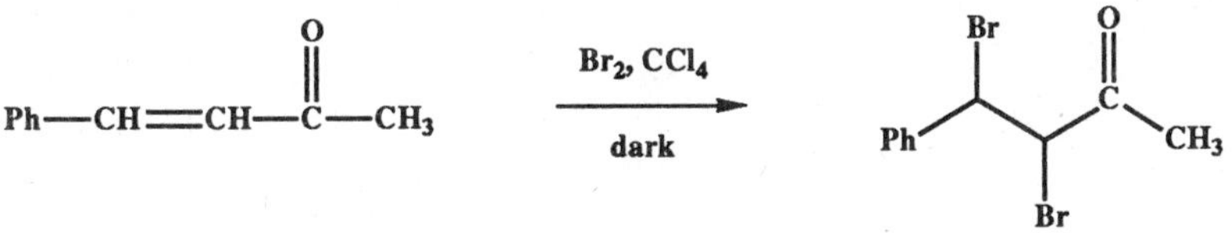

Ph—CH=CH—C(=O)—CH_3

Br_2, CCl_4

dark

Br O Ph C CH_3 Br

OS, CV 3, 123 (1955)

OAc

Br_2

EtOH

Br Br OAc

Br H O

OEt Br OEt

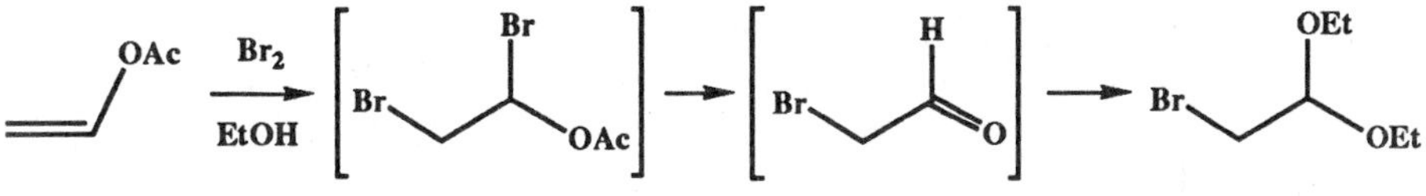

OS, *CV* 3, 127 (1955)

$CH_3(CH_2)_4CH_2CHO$ → (KOAc, Ac_2O, Δ) → $CH_3(CH_2)_4CH=C(OAc)H$ → (Br_2, CCl_4) → $CH_3(CH_2)_4CH(Br)CH(Br)OAc$ → (MeOH) → $CH_3(CH_2)_4CH(Br)CH(OMe)_2$

OS, *CV* 3, 209 (1955)

Coumarin → (Br_2, $CHCl_3$) → 3,4-dibromo-3,4-dihydrocoumarin

OS, *CV* 3, 217 (1955)

Cyclohexene → (H_2O_2, HCO_2H, H_2O) → trans-1,2-cyclohexanediol (OH, OH)

OS, *CV* 3, 350 (1955)

Ph–CH=CH–Ph → (Br_2, Et_2O) → Ph–CH(Br)–CH(Br)–Ph

OS, CV 3, 482 (1955)

$$C_6H_5I + Cl_2 \xrightarrow{CHCl_3} C_6H_5ICl_2$$

OS, CV 3, 526 (1955)

Fatty acids from sunflower-seed oil $\xrightarrow[\text{pet. ether}]{Br_2}$ $CH_3(CH_2)_4$–CHBr–CHBr–CH_2–CHBr–CHBr–$(CH_2)_7CO_2H$

OS, CV 3, 531 (1955)

Fatty acids from linseed oil $\xrightarrow[Et_2O]{Br_2}$ CH_3CH_2–CHBr–CHBr–CH_2–CHBr–CHBr–CH_2–CHBr–CHBr–$(CH_2)_7CO_2H$

OS, CV 3, 621 (1955)

2-Furoic acid (CO_2H, O) $\xrightarrow[H_2O, \Delta]{Br_2}$ HO_2C–CBr=CBr–CHO

OS, CV 3, 731 (1955)

$$PhCH{=}CH{-}CHO \xrightarrow[2.\ K_2CO_3,\ \Delta]{1.\ Br_2,\ HOAc} PhCH{=}CBr{-}CHO$$

***OS*, *CV* 3, 785 (1955)**

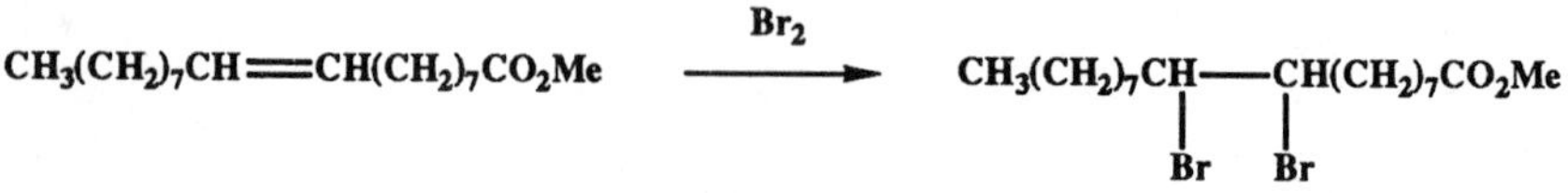

$CH_3(CH_2)_7CH{=}CH(CH_2)_7CO_2Me \xrightarrow{Br_2} CH_3(CH_2)_7CH(Br){-}CH(Br)(CH_2)_7CO_2Me$

***OS*, *CV* 4, 130 (1963)**

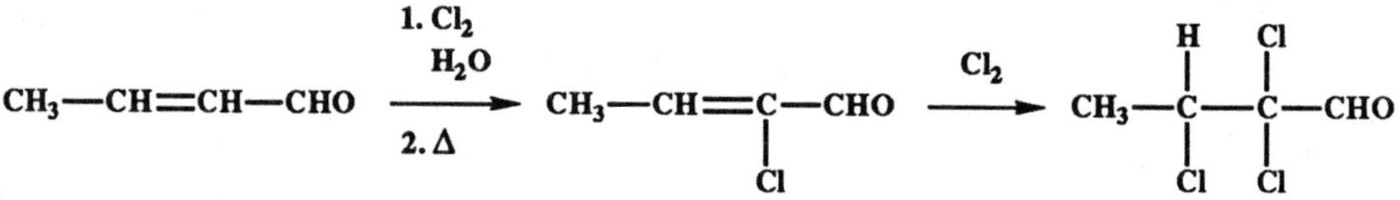

$CH_3{-}CH{=}CH{-}CHO \xrightarrow[2.\ \Delta]{1.\ Cl_2,\ H_2O} CH_3{-}CH{=}C(Cl){-}CHO \xrightarrow{Cl_2} CH_3{-}CHCl{-}CCl_2{-}CHO$

***OS*, *CV* 4, 195 (1963)**

Br₂, HOAc, NaOAc, Et₂O (forward); Zn, Et₂O (reverse)

HO, C_8H_{17}, Br, Br

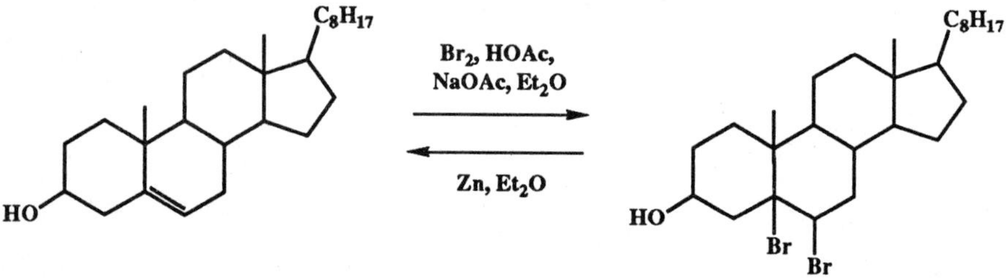

***OS*, *CV* 4, 688 (1963)**

CHO, O $\xrightarrow[H_2O,\ \Delta]{Br_2}$ HO_2C, Br, Br, CHO

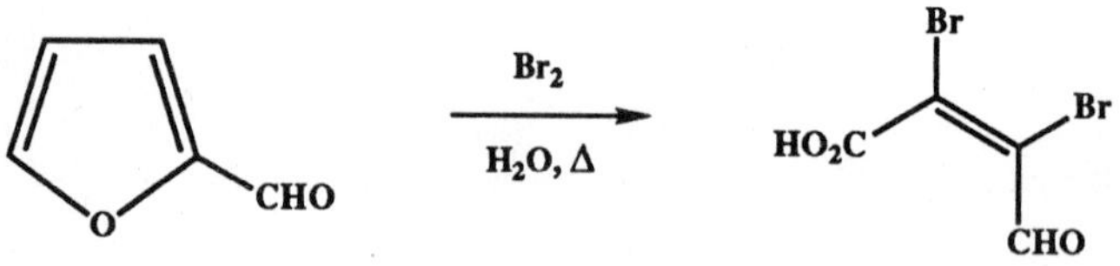

OS, CV 4, 851 (1963)

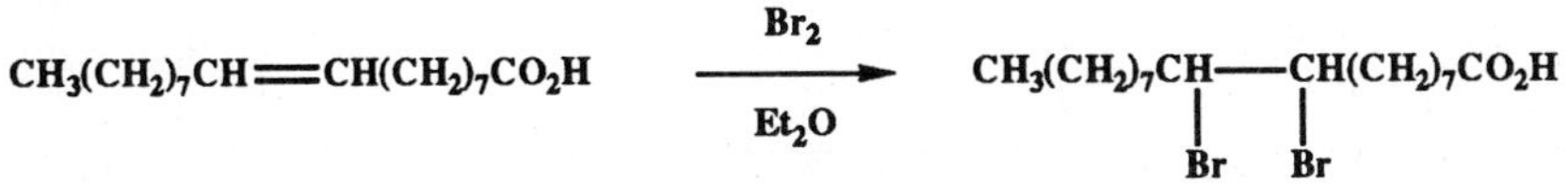

$$CH_3(CH_2)_7CH{=}CH(CH_2)_7CO_2H \xrightarrow[Et_2O]{Br_2} CH_3(CH_2)_7CH(Br){-}CH(Br)(CH_2)_7CO_2H$$

OS, CV 4, 969 (1963)

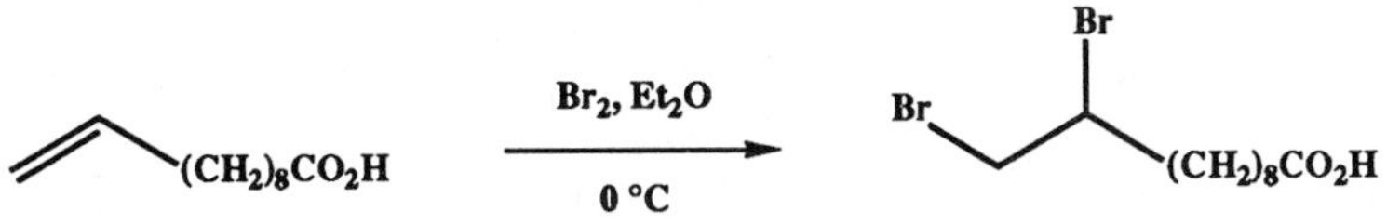

$$CH_2{=}CH(CH_2)_8CO_2H \xrightarrow[0\ °C]{Br_2,\ Et_2O} BrCH_2CH(Br)(CH_2)_8CO_2H$$

OS, CV 5, 467 (1973)

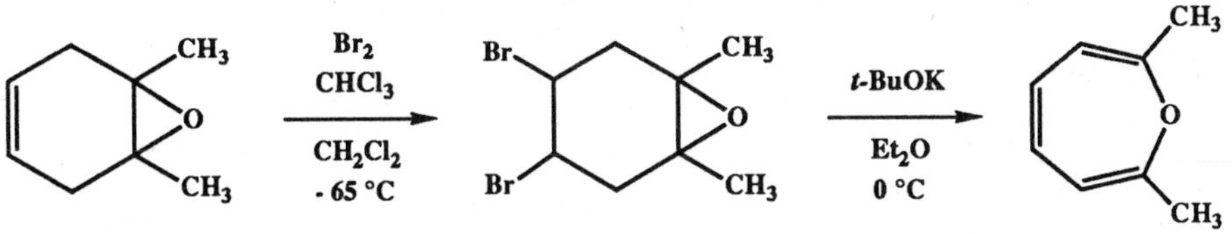

Br₂, CHCl₃, CH₂Cl₂, -65 °C; *t*-BuOK, Et₂O, 0 °C

OS, CV 5, 709 (1973)

$$CH_3S{-}SCH_3 \xrightarrow[-10\ to\ 0\ °C]{Cl_2,\ Ac_2O} 2\ CH_3S{-}Cl \xrightarrow{Cl_2} 2\ CH_3SCl_3 \xrightarrow{Ac_2O} 2\ CH_3S(O)Cl$$

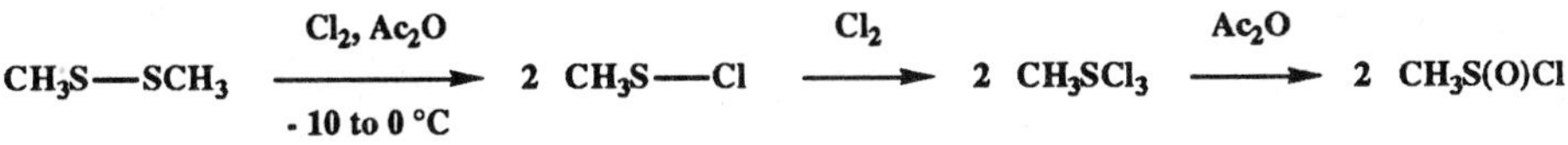

OS, CV 5, 710 (1973)

$$CH_3S{-}SCH_3 \xrightarrow[-10\ to\ 0\ °C]{Cl_2,\ HOAc} 2\ CH_3S{-}Cl \xrightarrow{Cl_2} 2\ CH_3SCl_3 \xrightarrow[r.t.]{HOAc} 2\ CH_3S(O)Cl$$

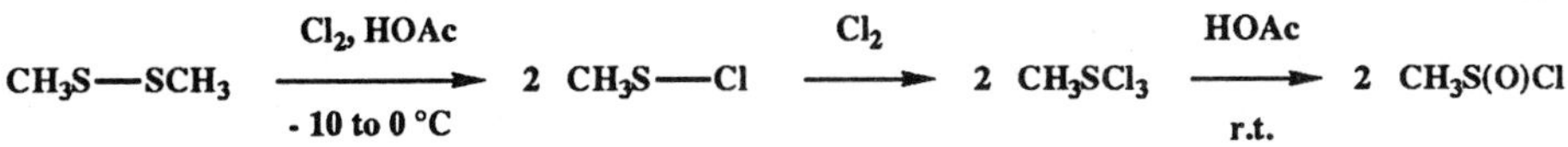

OS, CV 6, 196 (1988)

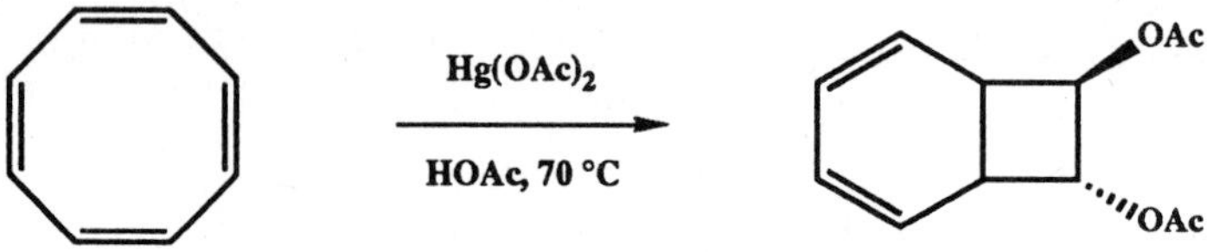

$Hg(OAc)_2$

HOAc, 70 °C

OAc

OAc

OS, CV 6, 342 (1988)

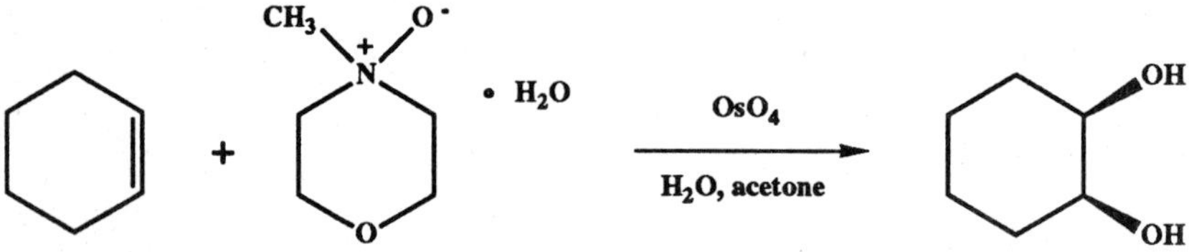

CH_3 N^+ O^- • H_2O

OsO_4

H_2O, acetone

OH

OH

OS, CV 6, 348 (1988)

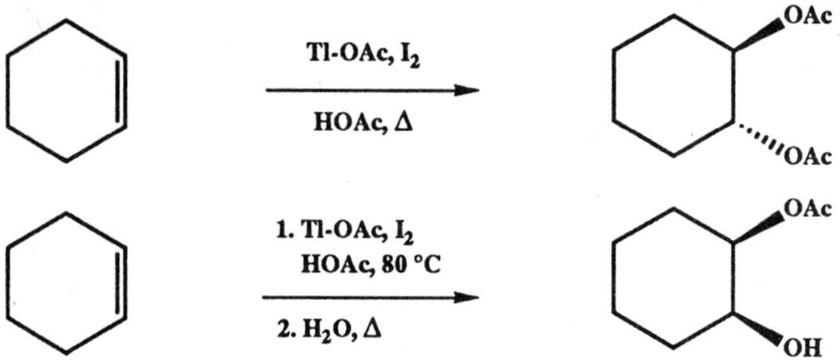

Tl-OAc, I_2

HOAc, Δ

OAc

OAc

1. Tl-OAc, I_2
HOAc, 80 °C

2. H_2O, Δ

OAc

OH

OS, CV 6, 422 (1988)

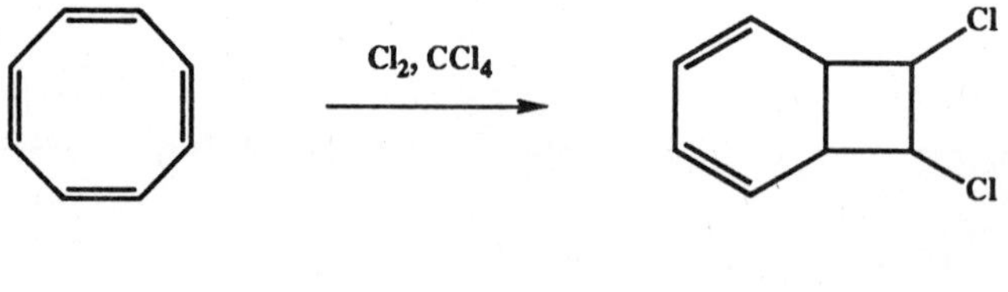

Cl_2, CCl_4

Cl

Cl

OS, *CV 6*, 424 (1988)

Ph_3PCl_2

benzene

$OPPh_3$ +

Cl^-

Δ

OS, *CV 6*, 862 (1988)

Br_2

Et_2O, - 78 °C

OS, *CV 6*, 954 (1988)

1. Br_2, 0 °C

2. $HC(OEt)_3$, EtOH

OS, *CV 7*, 12 (1990); *60*, 6 (1981)

$HCCl_3$, NaOH

40 - 60 °C

phase-transfer conditions

OS, *CV 7*, 203 (1990); *60*, 53 (1981)

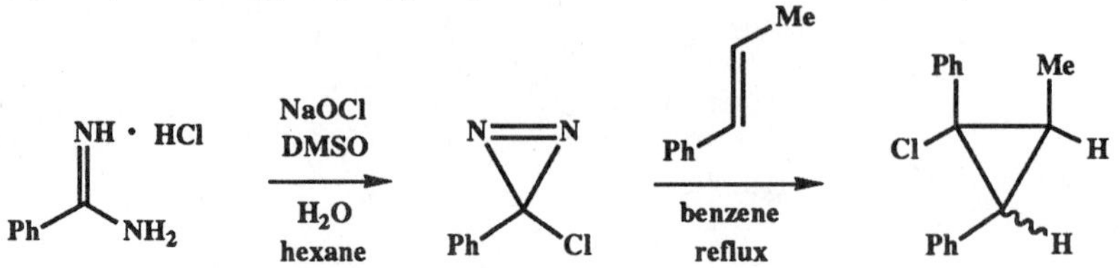

NaOCl, DMSO → H_2O, hexane

benzene, reflux

OS, *CV 7*, 200 (1990); *61*, 39 (1983)

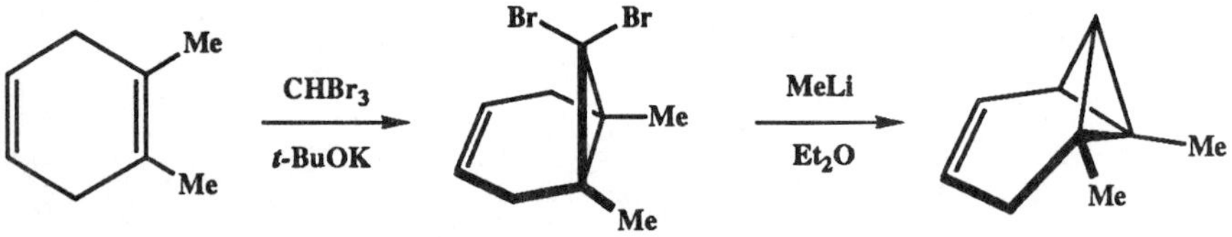

$CHBr_3$ → *t*-BuOK

MeLi → Et_2O

OS, *67*, 176 (1988)

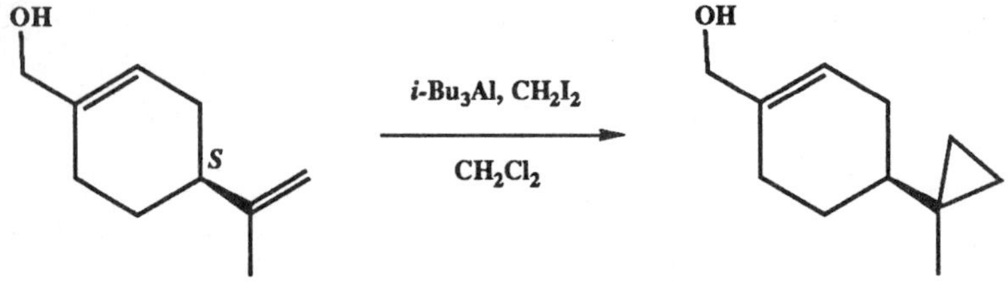

i-Bu_3Al, CH_2I_2 → CH_2Cl_2

OS, *68*, 220 (1989)

$CHBr_3$, 50% NaOH → TEBA, CH_2Cl_2

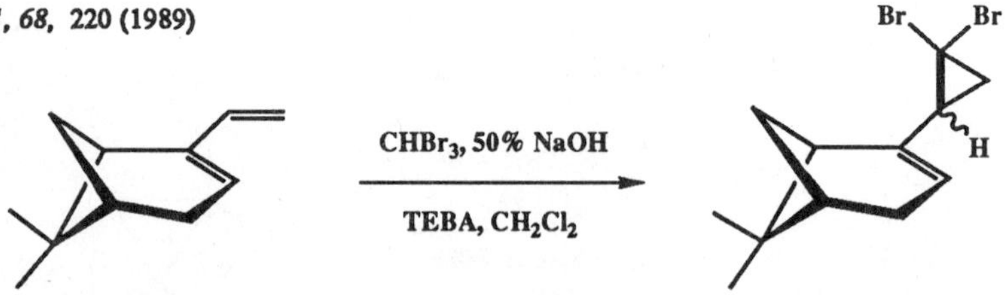

(*1R*) isomer

4 : 1 mixture

ELIMINATION

OS, CV 1, 15 (1941)

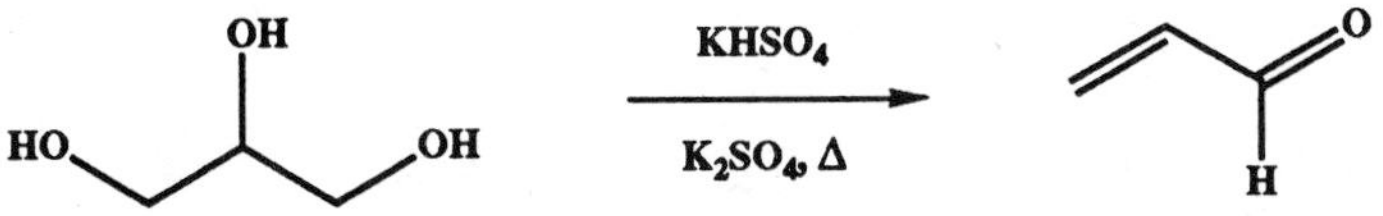

OH HO OH

$KHSO_4$

K_2SO_4, Δ

O H

OS, CV 1, 42 (1941)

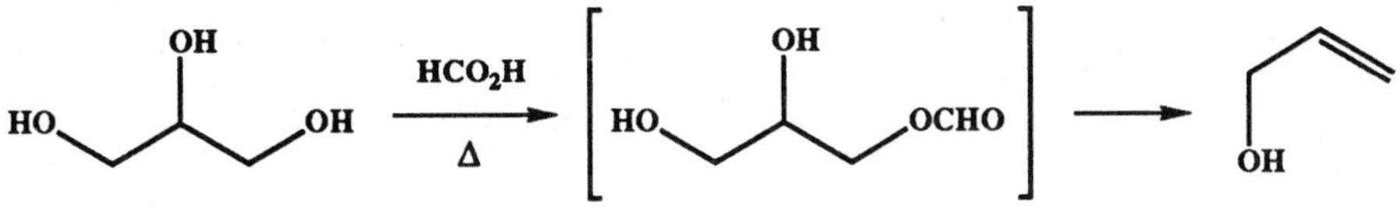

OH HO OH

HCO_2H

Δ

[OH HO OCHO]

OH

OS, CV 1, 183 (1941)

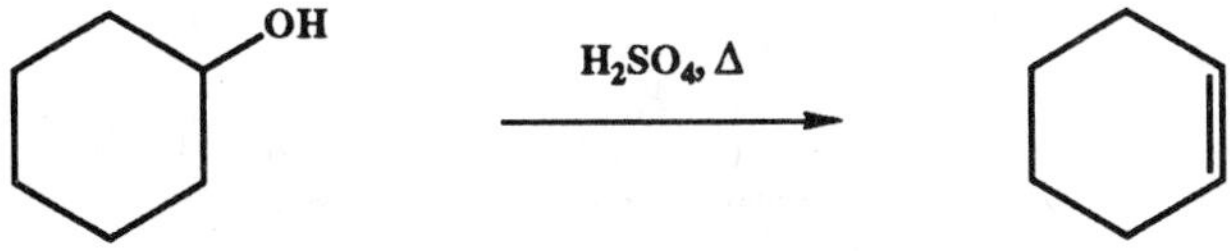

OH

H_2SO_4, Δ

OS, CV 1, 191 (1941)

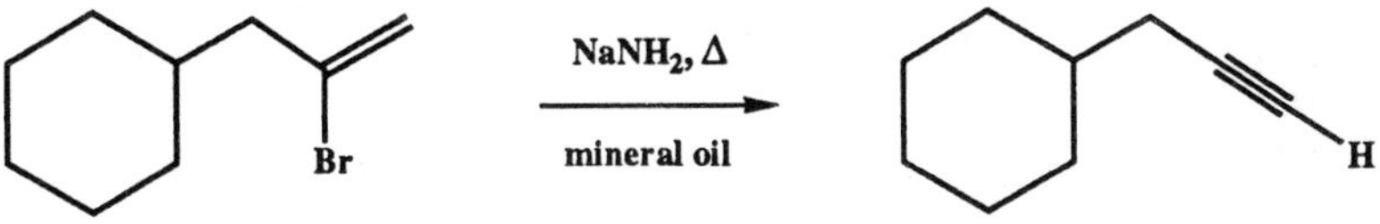

Br

$NaNH_2$, Δ

mineral oil

H

OS, CV 1, 205 (1941)

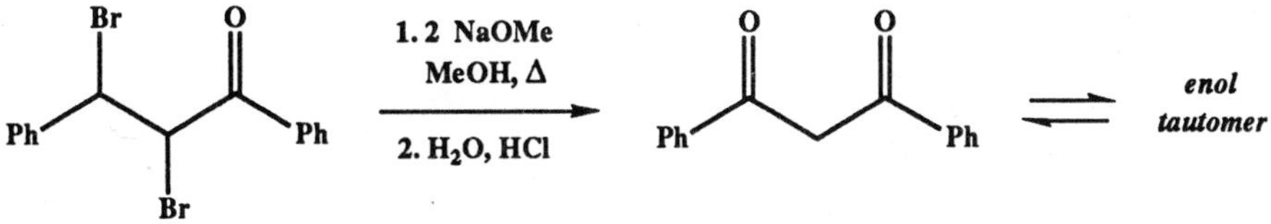

Br O Ph Ph Br

1. 2 NaOMe MeOH, Δ

2. H_2O, HCl

O O Ph Ph

enol tautomer

OS, CV 1, **209 (1941)**

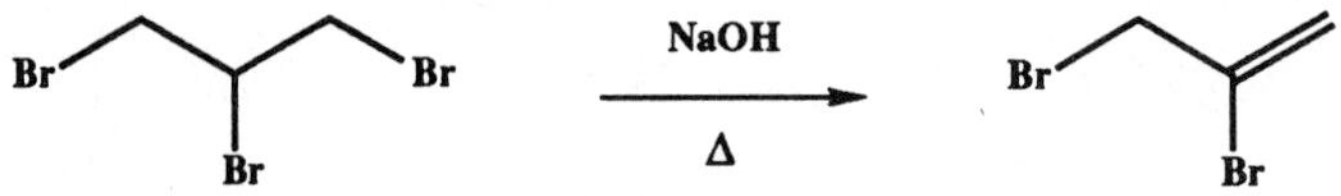

Br NaOH Δ Br Br Br Br

OS, CV 1, **226 (1941)**

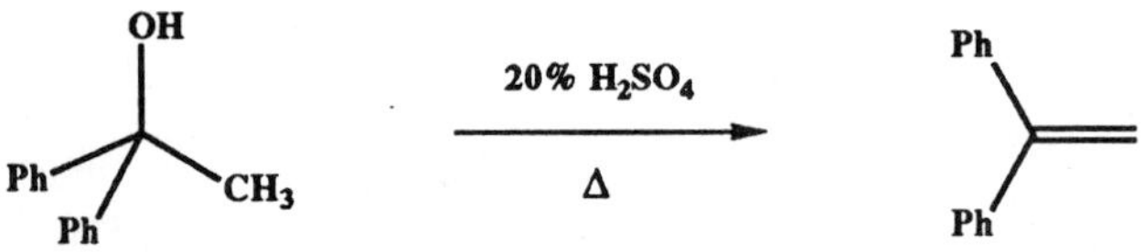

OH Ph Ph CH_3 20% H_2SO_4 Δ Ph Ph

OS, CV 1, **345 (1941)**

OH O CH_3 CH_3 CH_3 cat. I_2, Δ CH_3 O CH_3 CH_3

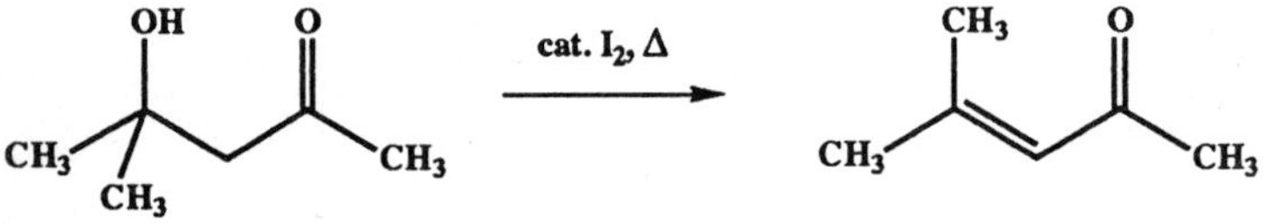

OS, CV 1, **430 (1941)**

$$CH_3CH_2—CH_2—CH(OH)—CH_3 \xrightarrow[\Delta]{H_2SO_4} CH_3CH_2—CH{=}CH—CH_3$$

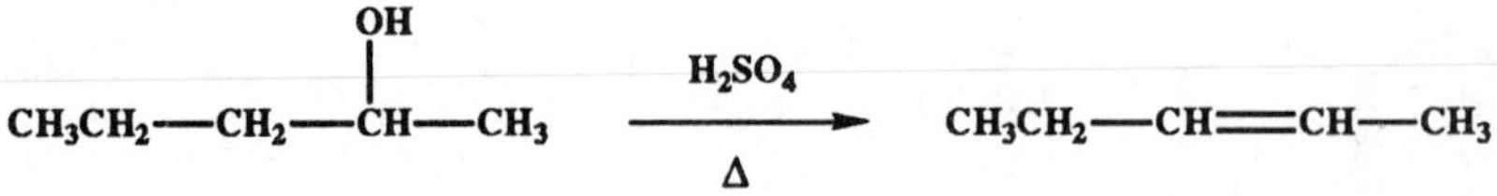

ELIMINATION

OS, CV 1, 438 (1941)

CH=CH—Br → (KOH, 200 °C) → phenylacetylene (C≡C—H)

OS, CV 1, 447 (1941)

N(H)—C(=S)—S⁻ NH_4^+ → ($Pb(NO_3)_2$, H_2O) → N=C=S

OS, CV 2, 10 (1943)

HO_2C—CH(Br)—CH(Br)—CO_2H → (1. KOH, MeOH, Δ; 2. H_2SO_4, H_2O) → HO_2C—C≡C—CO_2H

OS, CV 2, 12 (1943)

HO_2C—CH_2—C(OH)(CO_2H)—CH_2—CO_2H → (H_2SO_4, H_2O, Δ) → HO_2C—CH=C(CO_2H)—CH_2—CO_2H

OS, *CV* 2, 17 (1943)

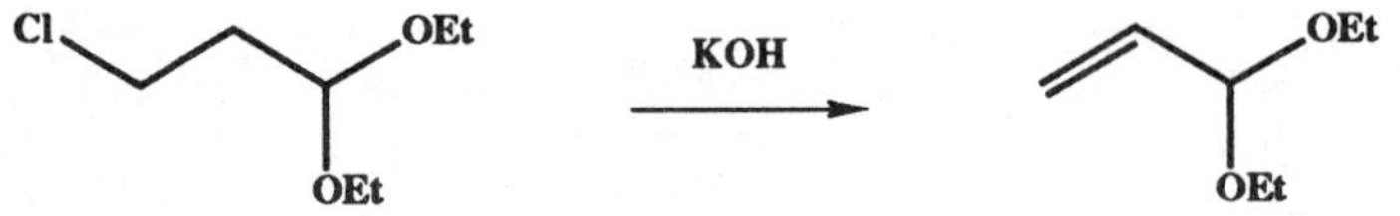

Cl / OEt / OEt — KOH → OEt / OEt

OS, *CV* 2, 165 (1943)

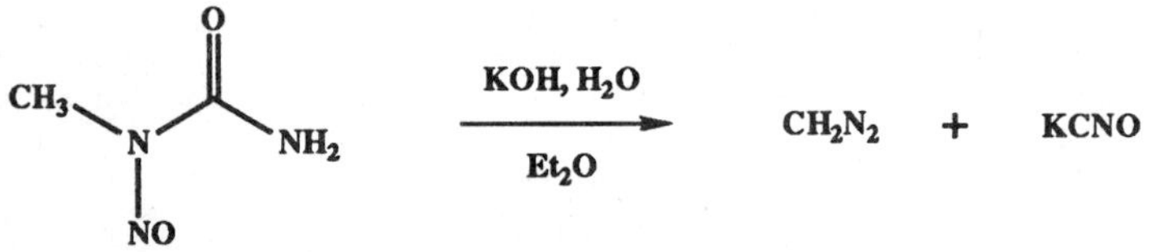

O / CH_3 / N / NH_2 / NO — KOH, H_2O / Et_2O → CH_2N_2 + KCNO

OS, *CV* 2, 181 (1943)

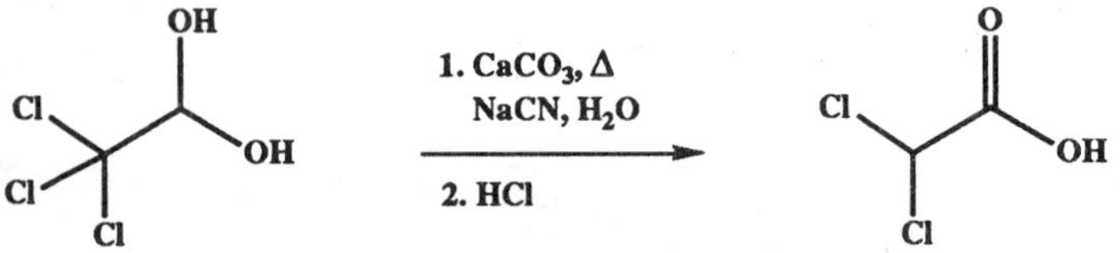

OH / Cl / Cl / Cl / OH — 1. $CaCO_3$, Δ NaCN, H_2O / 2. HCl → O / Cl / Cl / OH

OS, *CV* 2, 379 (1943)

O / NH_2 / CN — PCl_5, Δ → CN / CN

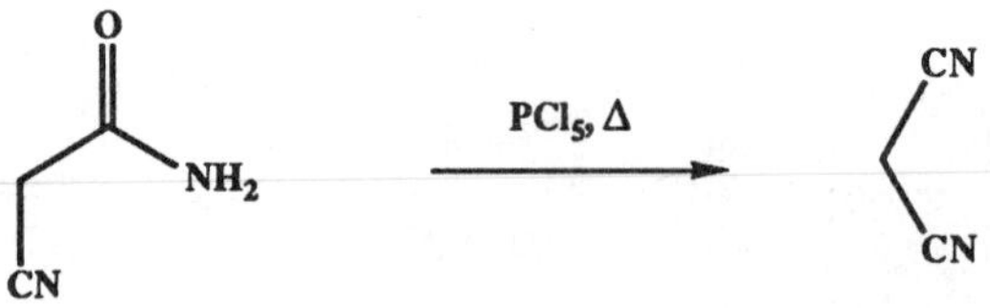

OS, CV 2, 408 (1943)

OH, CH_3, CH_3, CH_3 — Br_2, 55 °C → [Br, CH_3, CH_3, CH_3, Br] — H_2O, Δ → H, CH_3, CH_3, CH_3, O

OS, CV 2, 453 (1943)

NH_2, NO_2 — $Cl_2C{=}O$, EtOAc → [NHC(O)Cl, NO_2] — Δ → N=C=O, NO_2

OS, CV 2, 515 (1943)

Br, O, Ph, OEt, Br — 1. KOH, EtOH, Δ; 2. H_2SO_4, H_2O → Ph—C≡C—CO_2H

OS, CV 2, 606 (1943)

Ph, OH, Ph, Ph — H_2SO_4, H_2O, Δ → Ph, Ph, H, Ph

OS, CV 2, 622 (1943)

$$\text{3,4-(MeO)}_2\text{C}_6\text{H}_3\text{CHO} \xrightarrow[\text{2. Ac}_2\text{O, }\Delta]{\text{1. NH}_2\text{OH}\cdot\text{HCl, EtOH, H}_2\text{O}} \text{3,4-(MeO)}_2\text{C}_6\text{H}_3\text{C}\equiv\text{N}$$

OS, CV 3, 30 (1955)

$$\text{CH}_2{=}\text{CHCO}_2\text{Et} \xrightarrow{590\ ^\circ\text{C}} \text{CH}_2{=}\text{CHCO}_2\text{H}$$

OS, CV 3, 125 (1955)

$$\text{PhCHBrCHBrCOCH}_3 \xrightarrow[\text{EtOH, }\Delta]{\text{NaOAc}} \text{PhCH}{=}\text{CBrCOCH}_3$$

OS, CV 3, 204 (1955)

$$\text{3-ClC}_6\text{H}_4\text{CH(OH)CH}_3 \xrightarrow[\Delta]{\text{KHSO}_4} \text{3-ClC}_6\text{H}_4\text{CH}{=}\text{CH}_2$$

OS, CV 3, 237 (1955)

HO, HO, CO_2Me

1. PhMgBr, Δ Et_2O, C_6H_6
2. Ac_2O HOAc, Δ

AcO, AcO, Ph, Ph

OS, CV 3, 244 (1955)

CH_3, CH_3, CH_3, NH, CH_3, O

$NaNO_2$ / HOAc

CH_3, CH_3, CH_3, N, NO, CH_3, O

NaOR / ROH / Et_2O, Δ

CH_2N_2

R = *i*-Pr, C_6H_{11}

OS, CV 3, 260 (1955)

Cl_3C, OH, OH

NH_4OH / KCN, Et_2O

HCl_2C, O, NH_2

OS, CV 3, 270 (1955)

p-ClPh, *p*-ClPh, Cl, Cl, Cl

aq. KOH / diethylene glycol, Δ

[*p*-ClPh, *p*-ClPh, Cl, Cl]

1. 3 KOH
2. H_2SO_4

p-ClPh, *p*-ClPh, O, OH

OS, CV 3, 276 (1955)

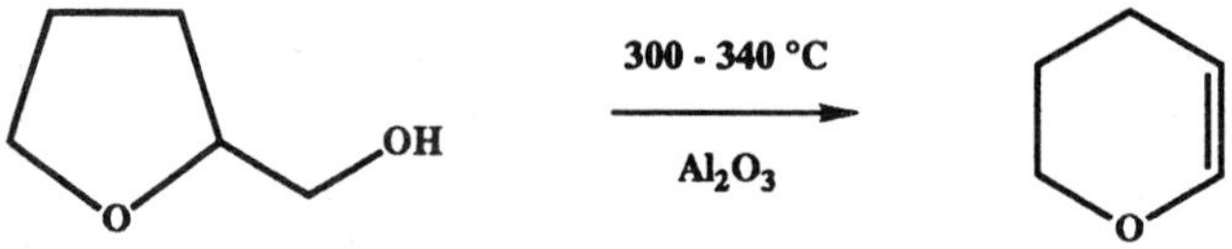

300 - 340 °C

Al_2O_3

OH

O

OS, CV 3, 312-313 (1955)

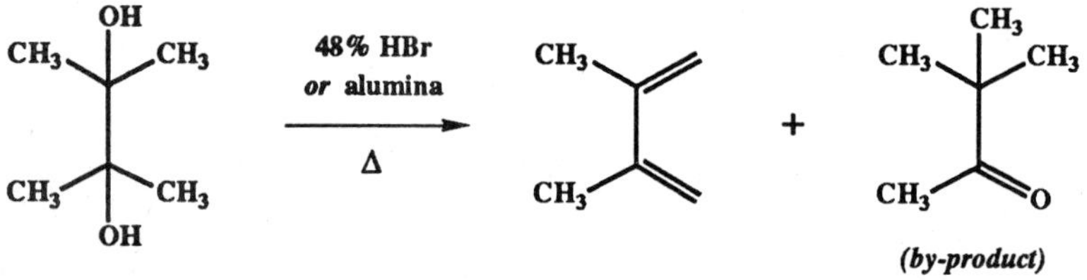

OH

CH_3 CH_3

CH_3 CH_3

OH

48% HBr
or alumina

Δ

CH_3

CH_3

+

CH_3

CH_3 CH_3

CH_3 O

(by-product)

OS, CV 3, 350 (1955)

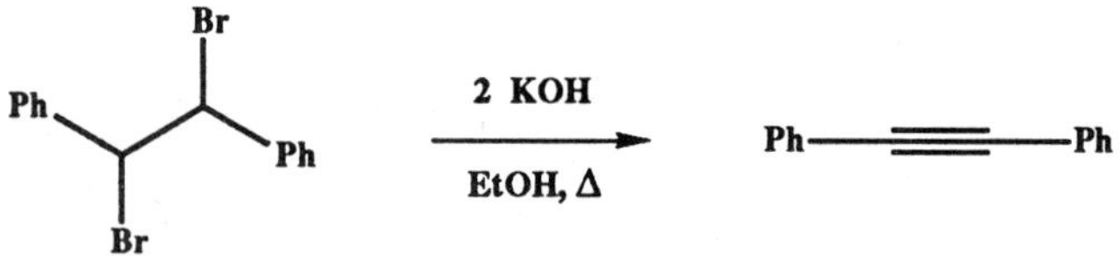

Br

Ph

Ph

Br

2 KOH

EtOH, Δ

Ph—≡—Ph

OS, CV 3, 493 (1955)

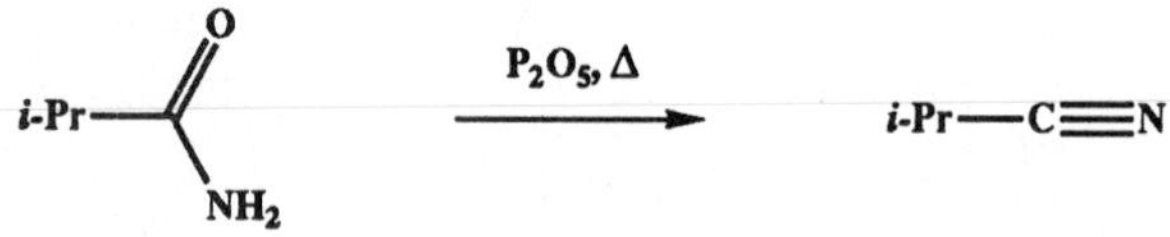

O

i-Pr

NH_2

P_2O_5, Δ

i-Pr—C≡N

OS, CV 3, 506 (1955)

Br, OEt, OEt → *t*-BuOK, Δ → OEt, OEt

OS, CV 3, 526 (1955)

Br, Br, $CH_3(CH_2)_4$, Br, Br, $(CH_2)_7CO_2H$ → Zn, HCl, EtOH, Δ → H, H, $CH_3(CH_2)_4$, H, H, $(CH_2)_7CO_2Et$

OS, CV 3, 531 (1955)

Br, Br, CH_3CH_2, Br, Br, Br, Br, $(CH_2)_7CO_2H$ → Zn, HCl, EtOH, Δ → CH_3CH_2, $(CH_2)_7CO_2Et$

OS, CV 3, 535 (1955)

O, NH_2, CN → $POCl_3$, Δ, ethylene dichloride → CN, CN

OS, *CV* 3, 560 (1955)

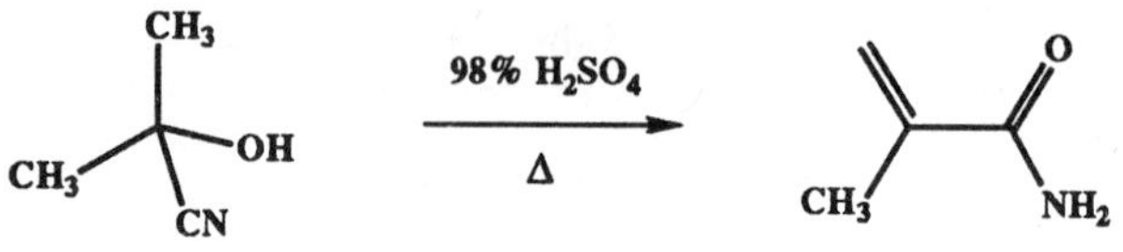

CH_3, CH_3, OH, CN; 98% H_2SO_4, Δ; CH_3, O, NH_2

OS, *CV* 3, 584 (1955)

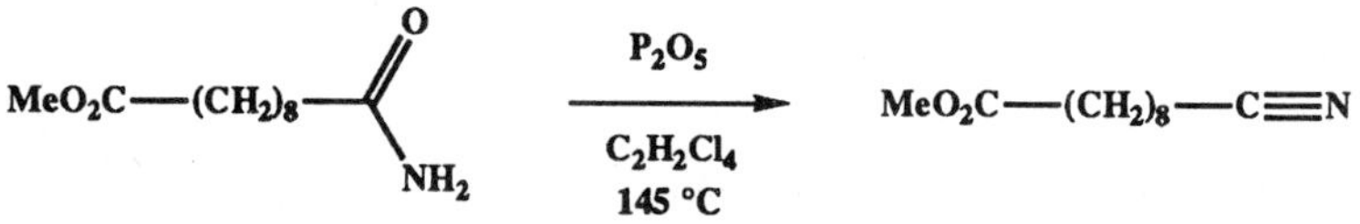

$MeO_2C—(CH_2)_8—C(=O)NH_2$ → (P_2O_5; $C_2H_2Cl_4$, 145 °C) → $MeO_2C—(CH_2)_8—C≡N$

OS, *CV* 3, 599 (1955)

$CH_3—NH_2$ → (CS_2; aq. NaOH) → [$CH_3NHC(=S)SNa$] → ($ClCO_2Et$) → [$CH_3NHC(=S)SCO_2Et$] → $CH_3—N=C=S$

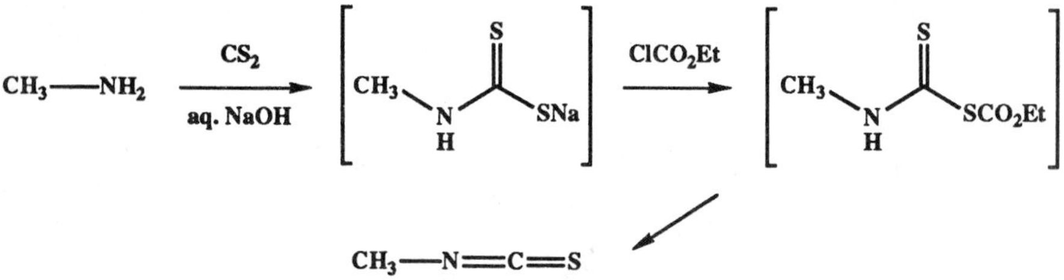

OS, *CV* 3, 623 (1955)

EtO_2C, Br, Br, CO_2Et; KOH, MeOH; Δ, then HCl; HO_2C, CO_2H

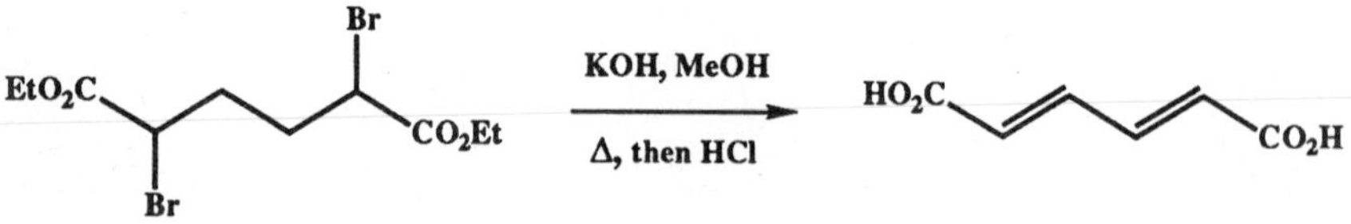

OS, *CV* 3, 646 (1955)

O_2N ... CO_2H → (p-$TsNH_2$; PCl_5, Δ) → O_2N ... C≡N

OS, *CV* 3, 690 (1955)

CH_2OH, OH, OH, H, CHO, HO, H, OH

d - glucose

NH_2OH
Na, MeOH
65 °C

CH_2OH, OH, NOH, OH, H, CH, HO, H, OH

Ac_2O
NaOAc
HOAc, Δ

CH_2OAc, OAc, N, OAc, H, C, AcO, H, OAc

d - isomer

OS, *CV* 3, 698 (1955)

Cl, O

1. Na, Et_2O
2. H_2O

HO

OS, *CV* 3, 729 (1955)

O

1. PhMgBr
Et_2O, Δ
2. aq. HCl

HO, Ph

Ac_2O
Δ

Ph

OS, *CV 3*, 731 (1955)

$$PhCH{=}CH{-}CHO \xrightarrow[2.\ K_2CO_3,\ \Delta]{1.\ Br_2,\ HOAc} PhCH{=}CBr{-}CHO \xrightarrow{HC(OEt)_3,\ \Delta} PhCH{=}CBr{-}CH(OEt)_2$$

$$PhCH{=}CBr{-}CH(OEt)_2 \xrightarrow[EtOH]{KOH,\ \Delta} PhC{\equiv}C{-}CH(OEt)_2 \xrightarrow[H_2SO_4]{H_2O,\ \Delta} PhC{\equiv}C{-}CHO$$

OS, *CV 3*, 768 (1955)

$$HO_2C(CH_2)_8CO_2H \xrightarrow[160\ °C]{H_2N{-}C(=O){-}NH_2} \left[H_2NOC(CH_2)_8CONH_2\right] \xrightarrow{220 - 340\ °C} NC(CH_2)_8CN + NC(CH_2)_8CO_2H$$

OS, *CV 3*, 785 (1955)

$$CH_3(CH_2)_7CH(Br){-}CH(Br)(CH_2)_7CO_2Me \xrightarrow[2.\ HCl,\ H_2O]{1.\ KOH,\ \Delta,\ n\text{-}C_5H_{11}OH} CH_3(CH_2)_7C{\equiv}C(CH_2)_7CO_2H$$

OS, *CV 3*, 846 (1955)

$$n\text{-}C_{11}H_{23}C(=O)Cl \xrightarrow[H_2O,\ acetone]{NaN_3} \left[n\text{-}C_{11}H_{23}C(=O)N_3\right] \xrightarrow[benzene]{\Delta} n\text{-}C_{11}H_{23}{-}N{=}C{=}O$$

OS, *CV* 4, 128 (1963)

$$CH_3-C(Cl)=CH-CH_2OH \xrightarrow[2.\ NH_4Cl]{1.\ NaNH_2,\ NH_3} CH_3-C\equiv C-CH_2OH$$

OS, *CV* 4, 130 (1963)

$$CH_3-CH=CH-CHO \xrightarrow[2.\ \Delta]{1.\ Cl_2,\ H_2O} CH_3-CH=C(Cl)-CHO$$

OS, *CV* 4, 144 (1963)

$$ClCH_2C(=O)NH_2 \xrightarrow[\text{trimethylbenzene}]{P_2O_5,\ \Delta} ClCH_2C\equiv N$$

OS, *CV* 4, 162 (1963)

$$\text{2-chloro-2-methylcyclohexanone (O, CH}_3\text{, Cl)} \xrightarrow[or\ \text{LiCl, DMF, }\Delta]{\text{collidine, }\Delta} \text{2-methyl-2-cyclohexenone (O, CH}_3\text{)}$$

OS, *CV 4*, 166 (1963)

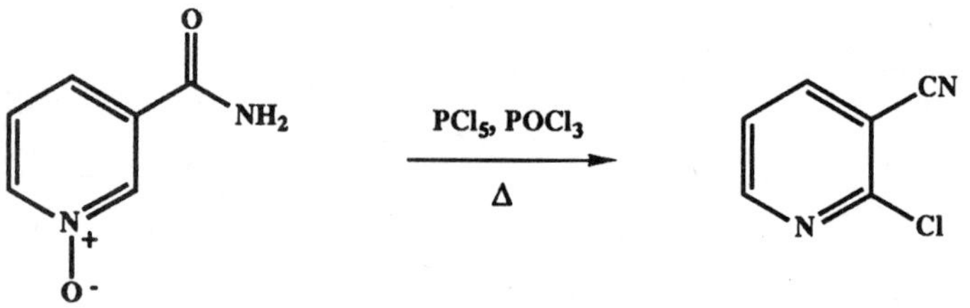

O, NH$_2$, N+, O- — PCl_5, $POCl_3$, Δ → CN, N, Cl

OS, *CV 4*, 172 (1963)

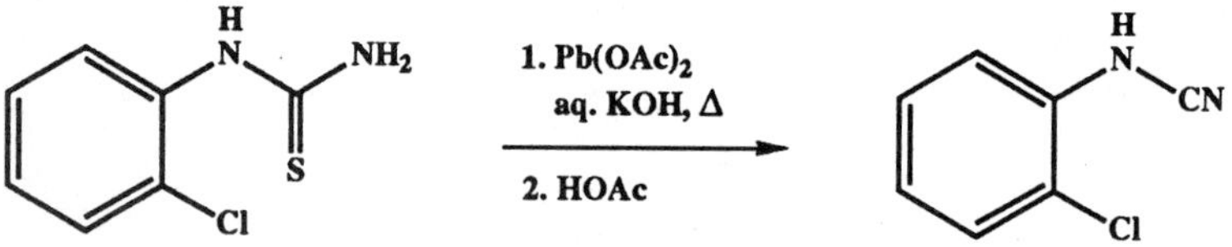

H N NH$_2$ S Cl — 1. $Pb(OAc)_2$ aq. KOH, Δ; 2. HOAc → H N CN Cl

OS, *CV 4*, 250 (1963)

NO, *p*-Ts—N, CH_3 — aq. KOH, Et_2O, 75 °C; diethylene glycol monomethyl ether → CH_2N_2

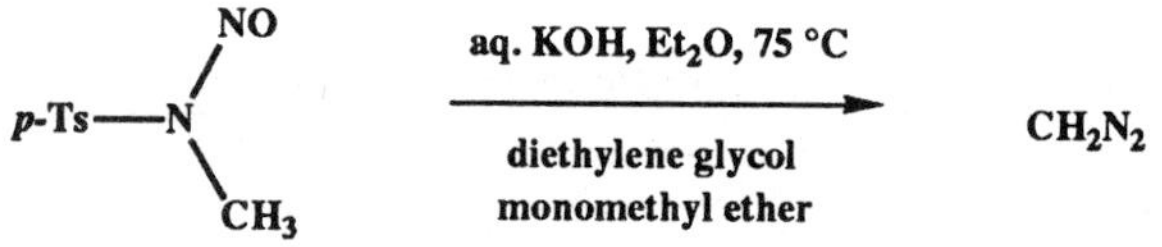

OS, *CV 4*, 268 (1963)

Cl, Cl, Cl, F, Cl, F — Zn, MeOH → Cl, Cl, F, F

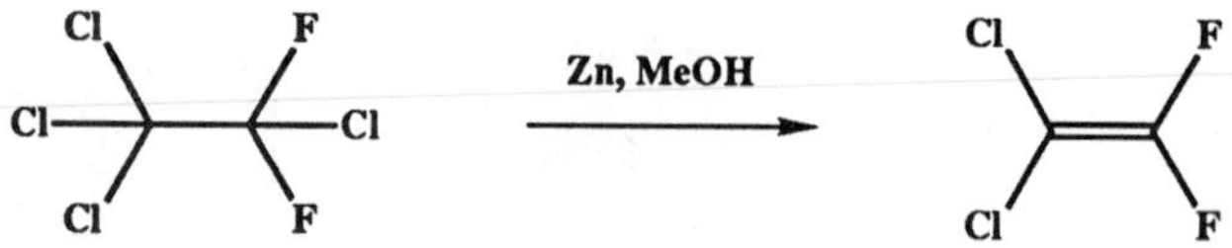

OS, CV 4, 298 (1963)

$EtO{-}CH{=}C(CO_2Et)_2 \xrightarrow[\text{EtOH, 45 °C}]{H_2,\ \text{Ra (Ni)}\ (1000\text{-}1500\ \text{psi})} EtOCH_2CH(CO_2Et)_2 \xrightarrow{\Delta} CH_2{=}C(CO_2Et)_2$

OS, CV 4, 398 (1963)

$n\text{-}C_8H_{17}CH_2CH_2CHBrCO_2H \xrightarrow[\text{2. } H_2SO_4,\ H_2O \text{; 3. } H_2SO_4,\ EtOH]{\text{1. } t\text{-BuOK, } \Delta} n\text{-}C_8H_{17}CH_2CH{=}CHCO_2H + n\text{-}C_8H_{17}CH{=}CHCH_2CO_2Et$

OS, CV 4, 404 (1963)

$ClCH_2CH(OEt)_2 \xrightarrow[\text{2. aq. NaCl}]{\text{1. Na, } NH_3} H{-}C{\equiv}C{-}OEt$

OS, CV 4, 436 (1963)

$CH_3CH_2CH_2CH_2CH(Et)C(=O)NH_2 \xrightarrow[\text{benzene, 80 °C}]{SOCl_2} CH_3CH_2CH_2CH_2CH(Et)C{\equiv}N$

OS, *CV* 4, 444 (1963)

n-Bu–CH(Et)–CH(OH)–CH(Me)–CO_2Et → ($POCl_3$, pyridine) → n-Bu–C(Et)=CH–C(Me)–CO_2Et (double bond isomers)

OS, *CV* 4, 486 (1963)

H_2N–CO–CH=CH–CO–NH_2 → (P_2O_5, Δ) → NC–CH=CH–CN

OS, *CV* 4, 608 (1963)

C_9H_{19}–CH_2–C(Br)(CO_2Me)(CH_3) → (quinoline, 160 - 170 °C) → C_9H_{19}–CH=C(CO_2Me)(CH_3)

OS, *CV* 4, 612 (1963)

cyclohexyl–CH_2–N^+Me_2–O^- → (160 °C, (then HCl)) → methylenecyclohexane (=CH_2) + Me_2NOH • HCl

OS, *CV* 4, 616 (1963)

$C_9H_{19}CH_2C(Br)(CH_3)COBr$ → (KO-*t*-Bu; *t*-BuOH, Δ) → $C_9H_{19}CH_2C(=CH_2)CO_2$-*t*-Bu

OS, *CV* 4, 683 (1963)

$ClCH_2$—CH=C(Cl)CH_3 → (KOH, 170 °C; ethylene glycol, *n*-butyl Cellosolve) → CH_2=CH—C≡C—H

OS, *CV* 4, 700 (1963)

1-Naphthyl-NH—C(=S)—NH_2 → (C_6H_5Cl; reflux) → 1-Naphthyl—N=C=S

OS, *CV* 4, 706 (1963)

3-Pyridyl—C(=O)—NH_2 → (P_2O_5, Δ) → 3-Pyridyl—CN

OS, *CV* 4, 727 (1963)

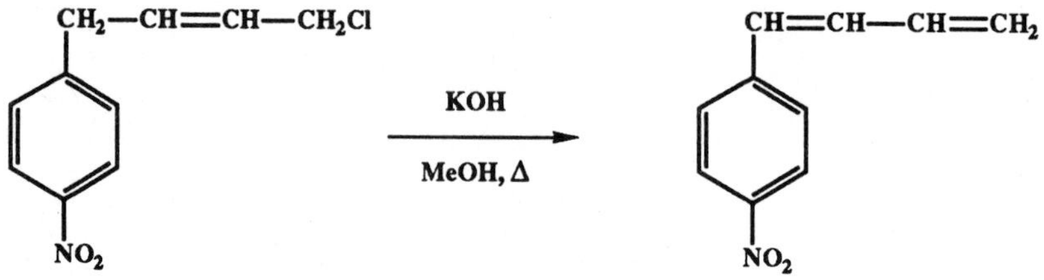

$CH_2—CH=CH—CH_2Cl$ (p-nitrophenyl) → KOH / MeOH, Δ → $CH=CH—CH=CH_2$ (p-nitrophenyl)

NO_2 NO_2

OS, *CV* 4, 746 (1963)

AcO ... OAc → 575 °C →

OS, *CV* 4, 748 (1963)

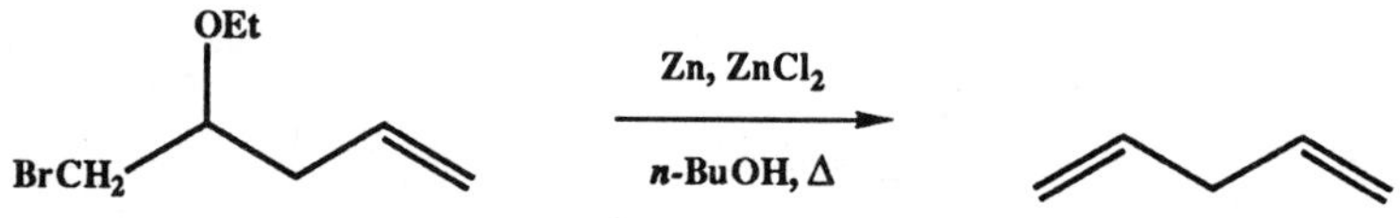

OEt

$BrCH_2$ → Zn, $ZnCl_2$ / *n*-BuOH, Δ →

OS, *CV* 4, 755 (1963)

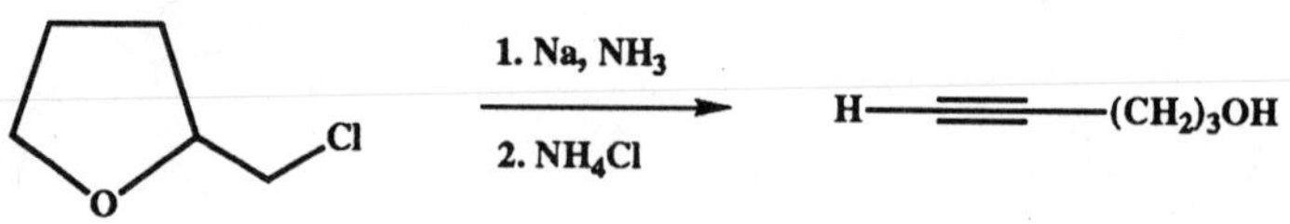

Cl, O → 1. Na, NH_3 / 2. NH_4Cl → $H—C≡C—(CH_2)_3OH$

ELIMINATION

***OS, CV* 4, 763 (1963)**

$$\mathrm{PhCH(Br)CH_2Br} \xrightarrow{\mathrm{Na,\ NH_3}} \mathrm{Ph{-}C{\equiv}C{-}H}$$

***OS, CV* 4, 851 (1963)**

$$\mathrm{CH_3(CH_2)_7CH(Br){-}CH(Br)(CH_2)_7CO_2H} \xrightarrow[\text{2. HCl}]{\text{1. 3 Na, NH}_3} \mathrm{CH_3(CH_2)_7C{\equiv}C(CH_2)_7CO_2H}$$

***OS, CV* 4, 969 (1963)**

$$\mathrm{BrCH_2CH(Br)(CH_2)_8CO_2H} \xrightarrow[\text{2. HCl, H}_2\text{O}]{\text{1. 3 NaNH}_2} \mathrm{H{-}C{\equiv}C{-}(CH_2)_8CO_2H}$$

***OS, CV* 4, 980 (1963)**

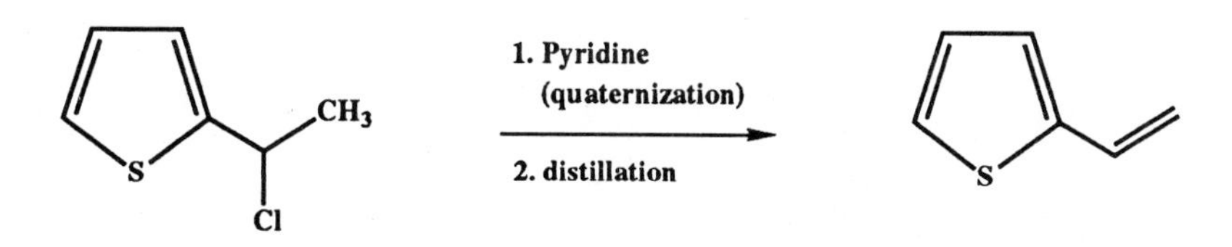

OS, *CV 5*, 22 (1973)

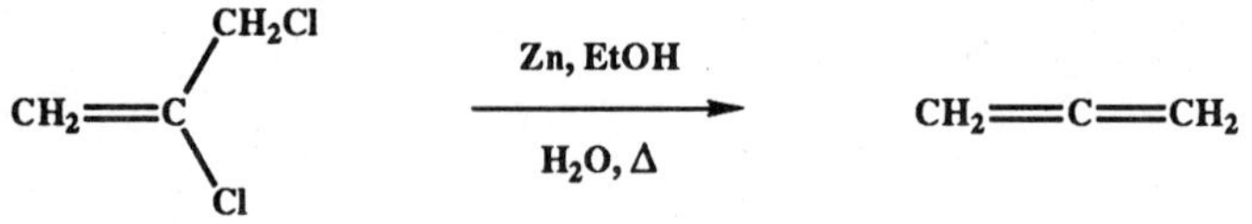
CH2Cl
CH2=C
Cl
Zn, EtOH
H2O, Δ
CH2=C=CH2

OS, *CV 5*, 39 (1973)

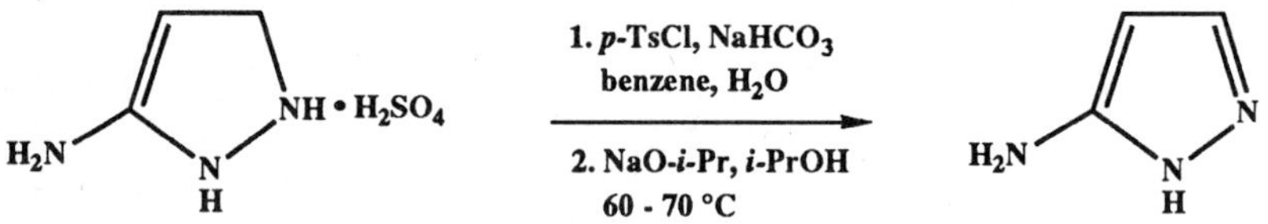
H2N
N
H
NH • H2SO4
1. p-TsCl, NaHCO3
benzene, H2O
2. NaO-i-Pr, i-PrOH
60 - 70 °C
H2N
N
N
H

OS, *CV 5*, 255 (1973)

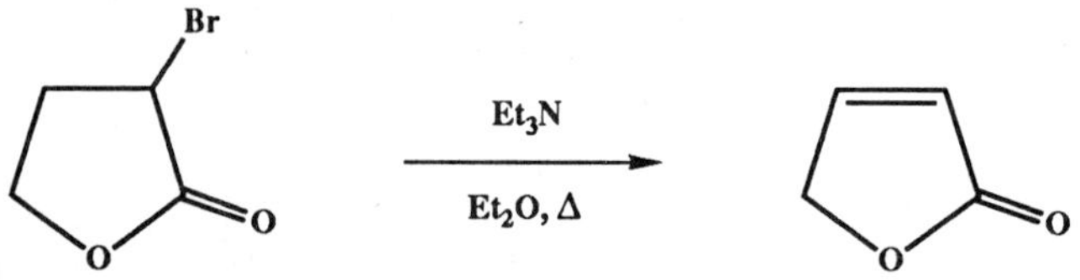
Br
O
O
Et3N
Et2O, Δ
O
O

OS, *CV 5*, 285 (1973)

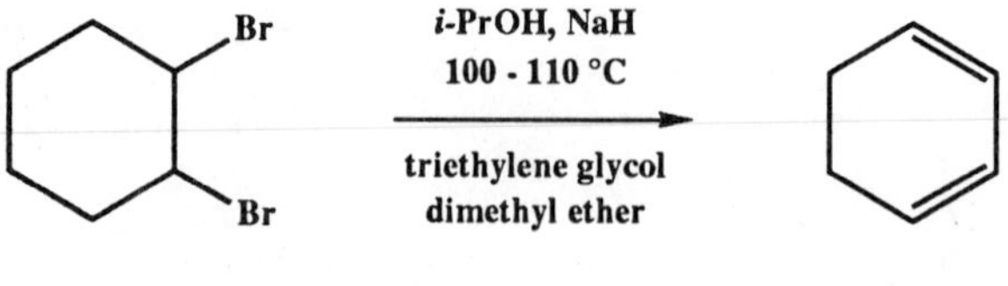
Br
Br
i-PrOH, NaH
100 - 110 °C
triethylene glycol
dimethyl ether

OS, *CV 5*, 300 (1973)

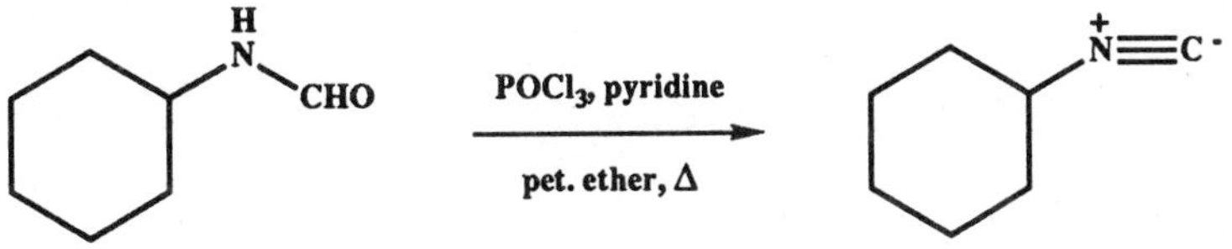
H
N
CHO
POCl3, pyridine
pet. ether, Δ
N≡C

OS, *CV 5*, 315 (1973)

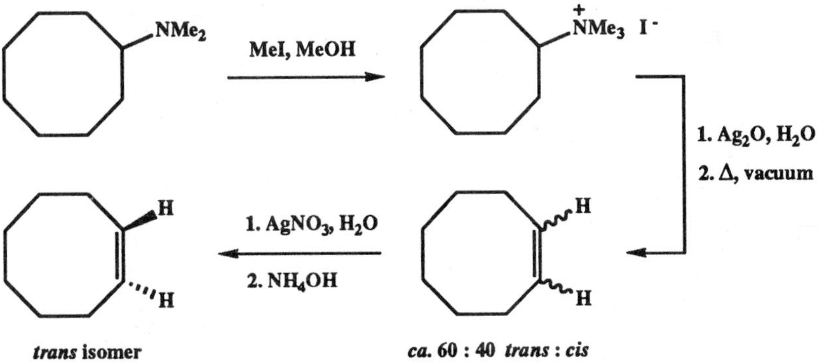
NMe2
MeI, MeOH
NMe3 I
1. Ag2O, H2O
2. Δ, vacuum
1. AgNO3, H2O
2. NH4OH
H
H
trans isomer
ca. 60 : 40 trans : cis

OS, *CV 5*, 326 (1973)

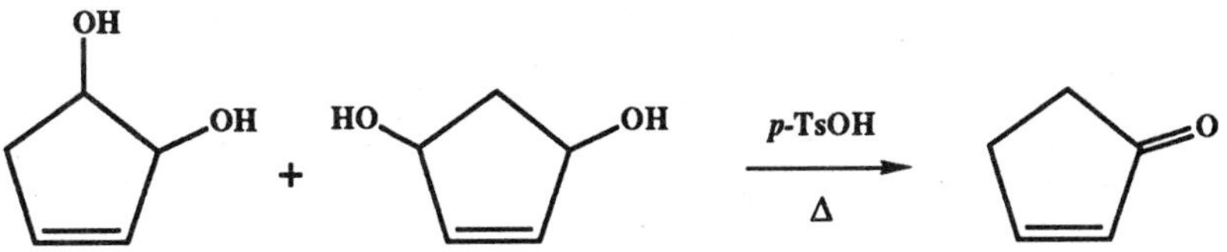
OH
OH
+
HO
OH
p-TsOH
Δ
O

***OS*, *CV 5*, 351 (1973)**

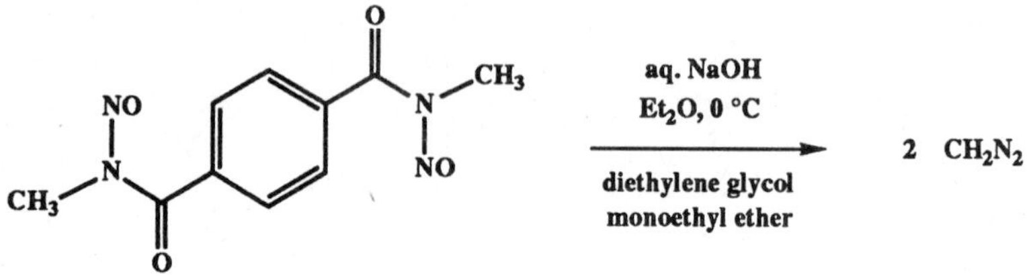

aq. NaOH, Et_2O, 0 °C; diethylene glycol monoethyl ether → 2 CH_2N_2

***OS*, *CV 5*, 393 (1973)**

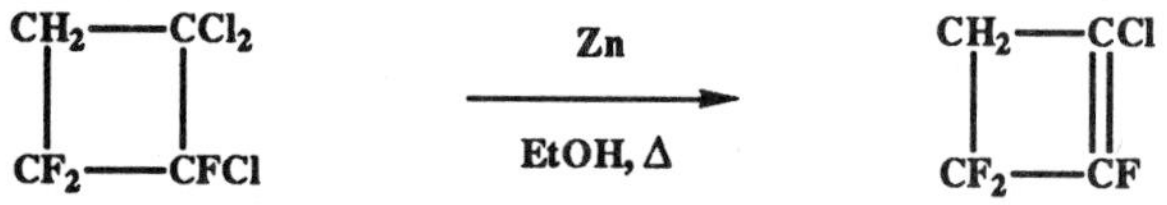

Zn; EtOH, Δ

***OS*, *CV 5*, 424 (1973)**

Na, *t*-BuOH; THF, Δ

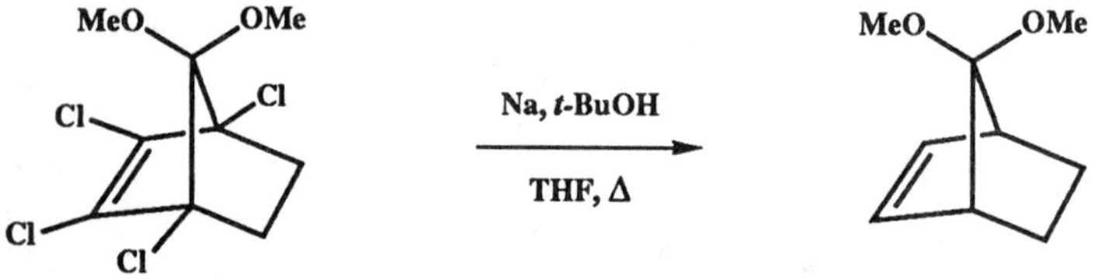

***OS*, *CV 5*, 467 (1973)**

Br_2, $CHCl_3$; CH_2Cl_2, - 65 °C

t-BuOK; Et_2O, 0 °C

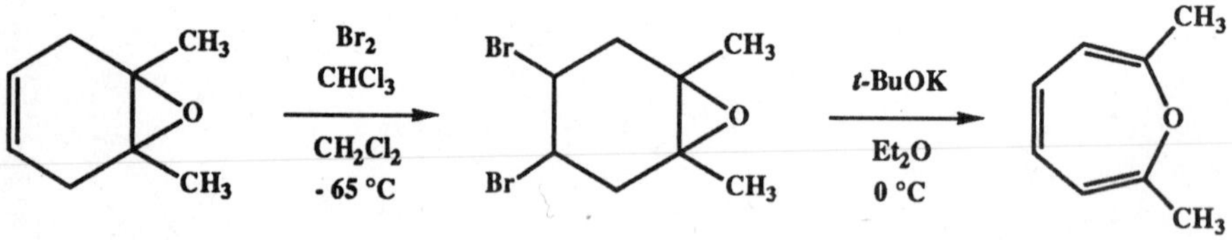

OS, *CV 5*, 555 (1973)

EtN=C=O → ($H_2N(CH_2)_3NMe_2$; CH_2Cl_2, 5 °C) → EtNH–C(=O)–$NH(CH_2)_3NMe_2$ → (*p*-TsCl; Et_3N) → EtN=C=$N(CH_2)_3NMe_2$

OS, *CV 5*, 608 (1973)

C_6H_9Br (mixture of bromohexadienes) → (Me_2NCH_2Ph; H_2O, 50 °C) → $C_6H_9\overset{+}{N}Me_2CH_2Ph$ Br^- → (NaOH; H_2O, Δ)

ca. 3 : 7 *cis* : *trans*

OS, *CV 5*, 647 (1973)

O—CHO

—OH

7% H_2SO_4

Δ

=O

OS, *CV 5*, 772 (1973)

CH_3–NH–CHO → (*p*-TsCl, quinoline; 75 °C, vacuum) → $CH_3—\overset{+}{N}\equiv\overset{-}{C}$

OS, CV 5, 877 (1973)

95 °C

d isomer

d isomer

OS, CV 5, 901 (1973)

$(i\text{-PrO})_3P$

pet. ether

OS, CV 5, 937 (1973)

1. $CH_3S(O)$—CH_2K

2. HCl, H_2O

$Cu(OAc)_2$

$CHCl_3$

OS, CV 5, 1060 (1973)

$POCl_3$, KO-t-Bu

t-BuOH

OS, CV 5, **1145 (1973)**

PhO–CH₂CH₂–Br → (Ph_3P; phenol, 90 °C) → PhO–CH₂CH₂–PPh_3^+ Br^- → (EtOAc, Δ) → CH₂=CH–PPh_3^+ Br^-

OS, CV 6, **23 (1988)**

2-acetyl-2-(phenylseleno)cyclohexanone → (H_2O_2, H_2O; CH_2Cl_2, 30 - 35 °C) → 2-acetyl-2-cyclohexen-1-one

OS, CV 6, **75 (1988)**

azetidine–$N(CH_2)_2CO_2Et$ → (KOH) → azetidine NH + CH_2=$CHCO_2K$ + EtOH

OS, CV 6, **82 (1988)**

hexachlorobenzene → (*n*-BuLi; Et_2O, - 78 to - 60 °C) → pentachlorophenyllithium → (benzene; Et_2O, - 60 to 25 °C) → tetrachlorobenzobarrelene

tetrachlorobenzobarrelene → (Na, *t*-BuOH; THF, Δ) → benzobarrelene

***OS*, *CV 6*, 87 (1988)**

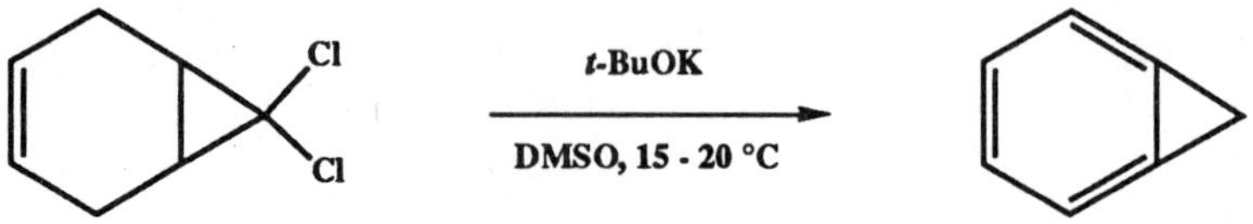

t-BuOK

DMSO, 15 - 20 °C

***OS*, *CV 6*, 172 (1988)**

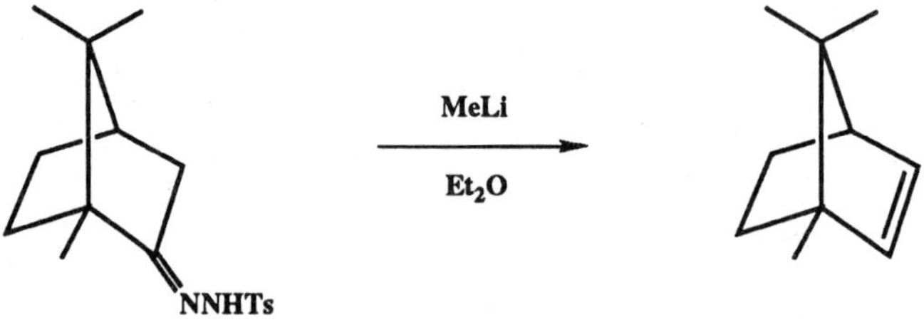

NNHTs

MeLi

Et_2O

***OS*, *CV 6*, 232 (1988)**

$t\text{-Bu—NH}_2$ $\xrightarrow[\text{H}_2\text{O, CH}_2\text{Cl}_2, \Delta]{\text{CHCl}_3\text{, 3 NaOH, [Et}_3\text{NCH}_2\text{Ph]}^+ \text{ Cl}^-}$ $t\text{-Bu—N}{\equiv}\text{C}$

***OS*, *CV 6*, 282 (1988)**

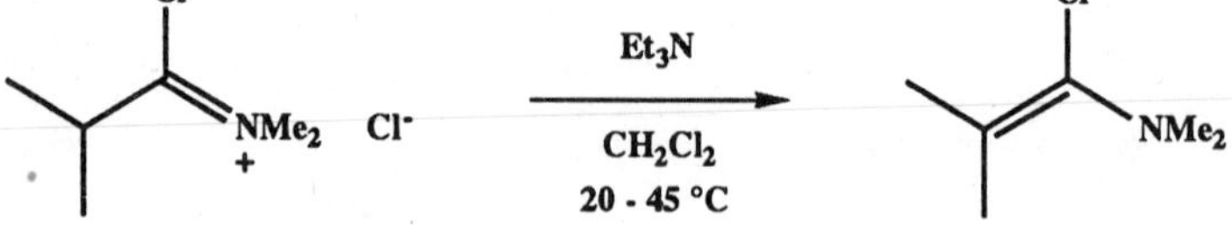

Cl

NMe_2^+ Cl^-

Et_3N

CH_2Cl_2

20 - 45 °C

Cl

NMe_2

OS, CV 6, 304 (1988)

C_6H_5–CH=CH–C(=O)–$NHSO_2Cl$ —(DMF)→ C_6H_5–CH=CH–C≡N

OS, CV 6, 307 (1988)

HO, CN; MeO —($KHSO_4$, 130 - 150 °C)→ CN; MeO

OS, CV 6, 310 (1988)

Cl, Cl —($Fe_2(CO)_9$ / benzene, 50 °C)→ $Fe(CO)_3$

OS, CV 6, 327 (1988)

O, Cl —(NaOAc / MeOH, Δ)→ O

***OS, CV 6,* 427 (1988)**

Et_3N

benzene, Δ

***OS, CV 6,* 465 (1988)**

DMF, 10 °C

***OS, CV 6,* 474 (1988)**

$Me_2NCH_2NMe_2$ $\xrightarrow[-10 \text{ to } -15\ °C]{CF_3CO_2H}$ $Me_2\overset{+}{N}{=}CH_2$ $CF_3CO_2^-$

***OS, CV 6,* 505 (1988)**

1. PCl_5, 25 °C
2. H_2SO_4, MeOH, Δ

Et_3N

THF, 0 - 5 °C

OS, CV 6, 549 (1988)

$$Ph_2CH{-}C(=O)Cl \xrightarrow[Et_2O,\ 0\ °C]{Et_3N} Ph_2C{=}C{=}O$$

OS, CV 6, 552 (1988)

$$Ph_2CH{-}O{-}CH_2CH_2{-}NMe_2 \xrightarrow[acetone]{MeI} Ph_2CH{-}O{-}CH_2CH_2{-}\overset{+}{N}Me_3\ I^- \xrightarrow[MeOH,\ \Delta]{Anion\text{-}exchange\ resin\ (OH^-)} Ph_2CH{-}O{-}CH{=}CH_2$$

OS, CV 6, 564 (1988)

$$Cl{-}CH_2{-}CH(OEt)_2 \xrightarrow[NH_3]{NaNH_2} Na^+\ {}^-C{\equiv}C{-}OEt \xrightarrow{EtBr} Et{-}C{\equiv}C{-}OEt$$

OS, CV 6, 620 (1988)

$$H{-}C(=O){-}NH{-}CH_2CO_2Et \xrightarrow[CH_2Cl_2,\ 0\ °C]{POCl_3,\ Et_3N} C{\equiv}N{-}CH_2CO_2Et$$

OS, CV 6, 664 (1988)

$$(CF_3)_2C{=}O \xrightarrow[2.\ POCl_3,\ \Delta]{1.\ NH_3,\ pyridine} (CF_3)_2C{=}NH$$

OS, CV 6, 675 (1988)

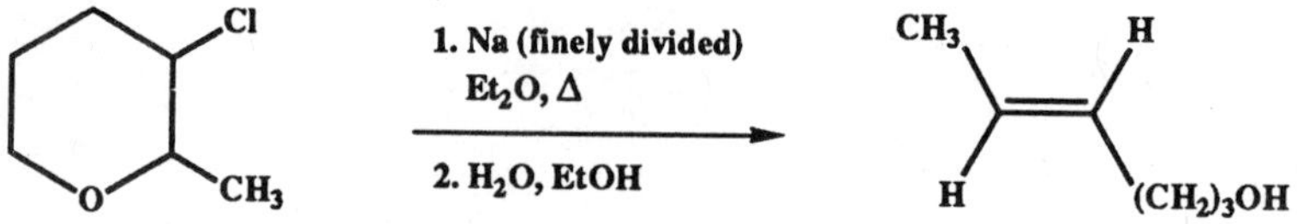

1. Na (finely divided)
Et_2O, Δ
2. H_2O, EtOH

Cl
CH_3
O
CH_3
H
H
$(CH_2)_3OH$

OS, CV 6, 683 (1988)

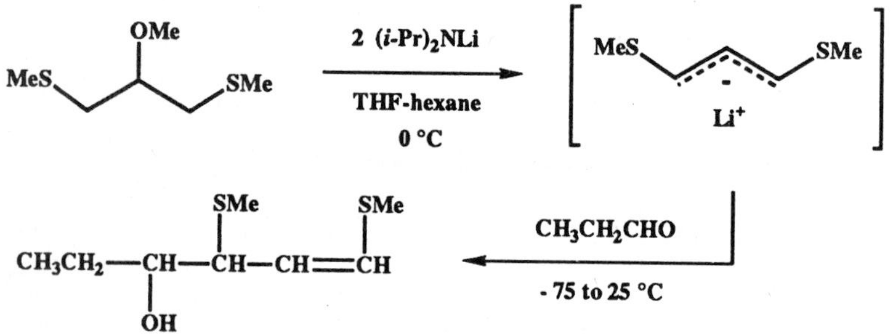

OMe
MeS
SMe
2 $(i\text{-}Pr)_2NLi$
THF-hexane
0 °C
MeS
SMe
Li^+

SMe
SMe
CH_3CH_2—CH—CH—CH═CH
OH
CH_3CH_2CHO
- 75 to 25 °C

OS, CV 6, 737 (1988)

OMe
SPh
SPh
$(CuOSO_2CF_3)_2 \cdot C_6H_6$
$(i\text{-}Pr)_2NEt$
radical inhibitor
benzene - THF, Δ
OMe
SPh
H

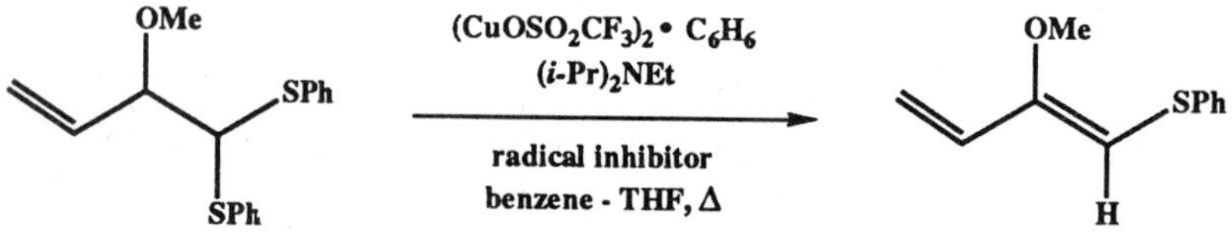

OS, CV 6, 751 (1988)

CH_3
t-Bu—CH_2—C—NH—
CH_3
O
H
$SOCl_2$, DMF
CH_3
t-Bu—CH_2—C—N≡C
CH_3

OS, *CV 6*, 791 (1988)

n-Pr

N

N
H

O

2 $Tl(NO_3)_3$

MeOH, Δ

n-Pr—≡—CO_2Me

OS, *CV 6*, 799 (1988)

$O_2N—CH_2—CH—CO_2Me$
I

NaOAc

Et_2O, 0 °C

O_2N H

H CO_2Me

OS, *CV 6*, 837 (1988)

N_2O_4, O_2

Et_2O, 10 °C

NO_2

X

Et_3N

NO_2

X = NO_2 and ONO and ONO_2

OS, *CV 6*, 862 (1988)

Br

Br

O

Br

Br

NaOMe

Et_2O, Δ

O

OS, CV 6, 883 (1988)

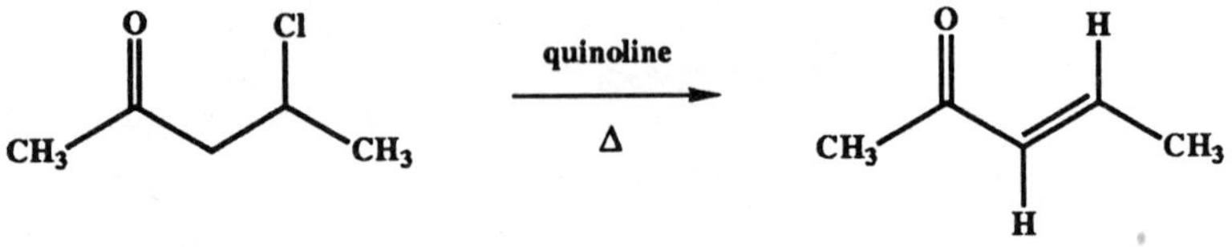

quinoline

Δ

OS, CV 6, 893 (1988)

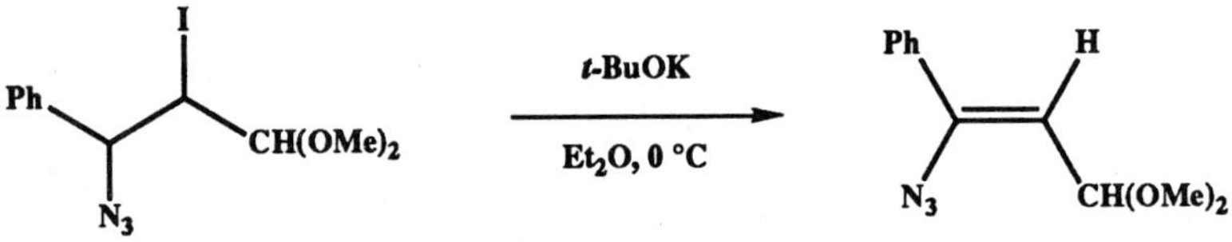

t-BuOK

Et_2O, 0 °C

OS, CV 6, 948 (1988)

Et_2NLi

Et_2O, Δ

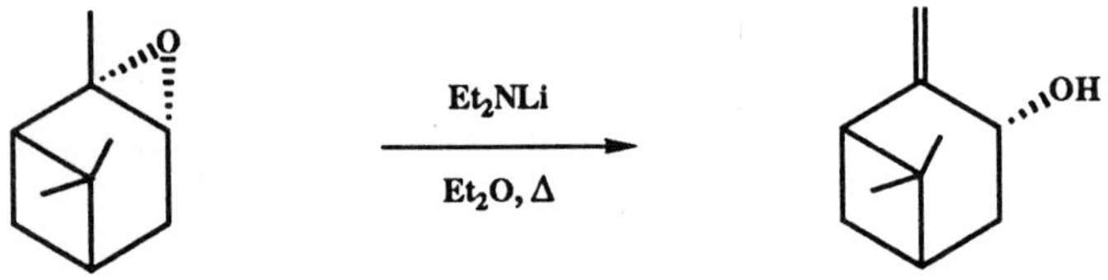

α - pinene oxide

trans - pinocarveol

OS, CV 6, 951 (1988)

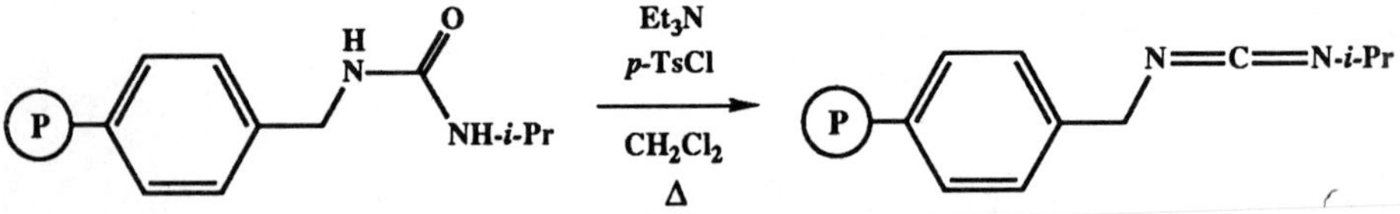

Et_3N

p-TsCl

CH_2Cl_2

Δ

(P) = styrene - divinylbenzene copolymer

OS, CV 6, 954 (1988)

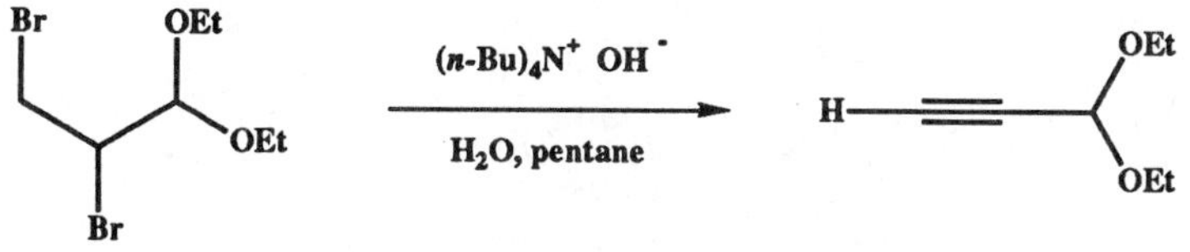

Br, OEt, OEt, Br → $(n\text{-Bu})_4N^+\ OH^-$ / H_2O, pentane → H—C≡C—CH(OEt)(OEt)

OS, CV 6, 968 (1988)

N—Cl (piperidine) → KOH, EtOH / Δ → N (tetrahydropyridine)

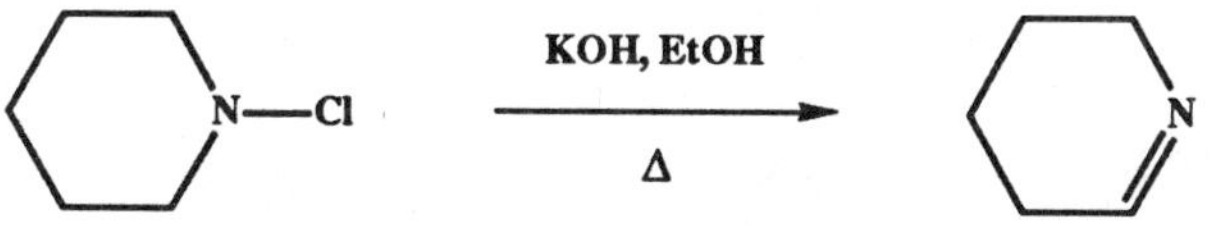

tripiperidein trimers
(monomer in solution)

OS, CV 6, 981 (1988)

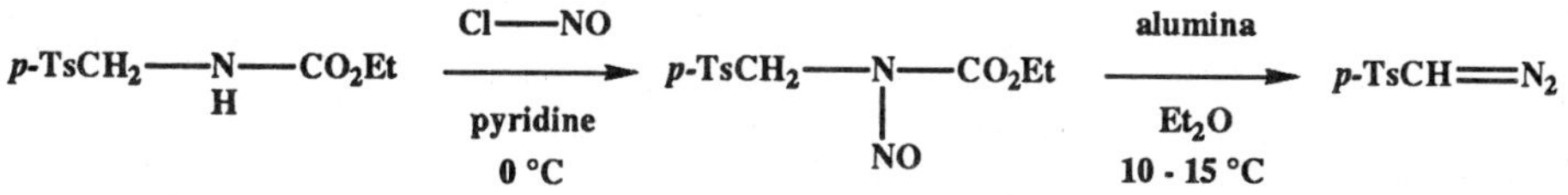

$p\text{-TsCH}_2\text{—N(H)—CO}_2\text{Et}$ → Cl—NO / pyridine / 0 °C → $p\text{-TsCH}_2\text{—N(NO)—CO}_2\text{Et}$ → alumina / Et_2O / 10 - 15 °C → $p\text{-TsCH=N}_2$

OS, CV 6, 987 (1988)

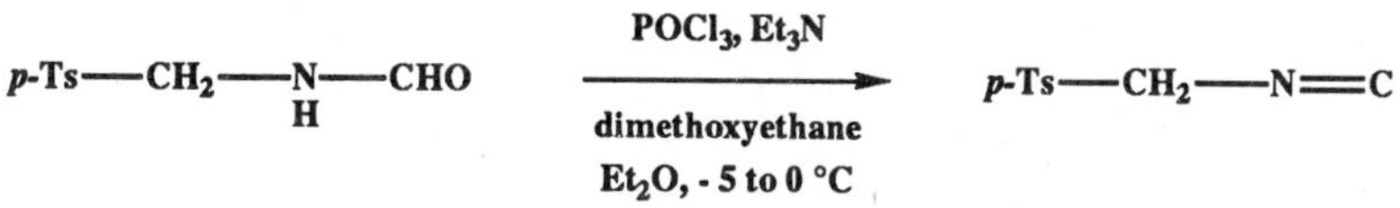

$p\text{-Ts—CH}_2\text{—N(H)—CHO}$ → $POCl_3$, Et_3N / dimethoxyethane / Et_2O, - 5 to 0 °C → $p\text{-Ts—CH}_2\text{—N≡C}$

OS, *CV* 7, 203 (1990); *60*, 53 (1981)

Ph, Me, Cl, H, Ph, H — t-BuOK / THF → Ph, Ph, Me, H

OS, *CV* 7, 396 (1990); *60*, 101 (1981)

$HOCH_2-CH(NO_2)-CH_3$ —phthalic anhydride→ $CH_2=C(CH_3)NO_2$

OS, *CV* 7, 210 (1990); *61*, 56 (1983)

$(HOCH_2)_2C(CO_2Et)_2$ —48% HBr, 85 - 90 °C→ $CH_2=C(CO_2H)CH_2Br$ —EtOH, H^+, reflux→ $CH_2=C(CO_2Et)CH_2Br$

OS, *CV* 7, 319 (1990); *61*, 77 (1983)

$(BrCH_2)_2CHCO_2Me$ + Et_3N → $CH_2=C(CO_2Me)CH_2Br$ + $Et_3NH^+ Br^-$

OS, *CV 7*, 491 (1990); *62*, 210 (1984)

Cl
Cl_2
hν
SO_2
Cl
Cl
SO_2
Et_3N
Et_3N
Cl
SO_2
SO_2

OS, *CV 7*, 63 (1990); *64*, 44 (1986)

O
1. LDA, THF
2. $(EtO)_2POCl$
3. LDA (2.25 equiv)
4. H_2O

OS, *CV 7*, 117 (1990); *64*, 50 (1986)

OH
p-TsCl
pyridine
OTs
t-BuOK
DMSO
70 °C

OS, *CV 7*, 241 (1990); *64*, 73 (1986)

Me Cl
Cl
O
1. $HOCH_2CH_2OH$, *p*-TsOH, C_6H_6, reflux; remove solvents, *p*-TsOH
2. 2 equiv. *n*-BuLi, THF, - 78 °C
Me
O
O

OS, *CV 7*, 453 (1990); *64*, 157 (1986)

$$PhSCH_2CH_2Br \xrightarrow[EtOH]{NaOEt} CH_2{=}CH{-}SPh$$

OS, *CV 7*, 438 (1990); *64*, 207 (1986)

$$PhCHO \xrightarrow{NH_2NHTs} PhCHNNHTs \xrightarrow[2.\ \Delta,\ vacuum]{1.\ NaOMe} PhCHN_2$$

OS, *65*, 12 (1987)

$$PhC(=O){-}CH_2{-}C(OH)Me_2 \xrightarrow[Et_3N]{(CF_3CO)_2O} PhC(=O){-}CH{=}CMe_2$$

OS, *65*, 32 (1987)

$$\text{2-(bromomethyl)-2-(chloromethyl)-1,3-dioxane} \xrightarrow{KNH_2,\ NH_3} \text{4,8-dioxaspiro[2.5]oct-1-ene}$$

OS, *65*, 68 (1987)

t-BuO(Cl)CH–CH(Cl)Ot-Bu → t-BuOK, hexane, 0 - 20 °C → t-BuO(Cl)C=CH(Ot-Bu) → $NaNH_2$, liq. NH_3, Et_2O → t-Bu-O—≡—O-t-Bu

OS, *65*, 90 (1987)

Br, SO_2CH_2Br, H → KO-t-Bu →

OS, *66*, 1 (1987)

$MeSO_3CH_2$—≡—$SiMe_3$ → MeMgCl, CuBr, LiBr, THF → $H_2C{=}C{=}C(SiMe_3)(Me)$

OS, *66*, 95 (1987)

O, OEt, $Fe(CO)_2C_5H_5$, CH_3 → $HBF_4 \cdot Et_2O$, Et_2O, - 78 °C → O, $^+Fe(CO)_2C_5H_5$, BF_4^-, CH_3 → CH_3CN, reflux → O, CH_3

***OS*, *66*, 173 (1987)**

Cl, C=C, CO_2H, CO_2Et, H

$\xrightarrow[90\ °C]{Et_3N}$

CO_2Et, H

***OS*, *67*, 125 (1988)**

$n\text{-}C_8H_{17}$—CH—CO_2Et (CH bearing $SiMePh_2$)

1. MeMgBr, THF
2. MeLi
3. KO-*t*-Bu

$n\text{-}C_8H_{17}CH{=}CMe_2$

***OS*, *67*, 157 (1988)**

\+ $PhSeSO_2Ph$

1. hν, CCl_4
2. H_2O_2, CH_2Cl_2

SO_2Ph

***OS*, *68*, 148 (1989)**

HgCl, SO_2Ph

2 *M* NaOH

Et_2O or CH_2Cl_2

r.t.

SO_2Ph

SUBSTITUTION

OS, CV 1, 7 (1941)

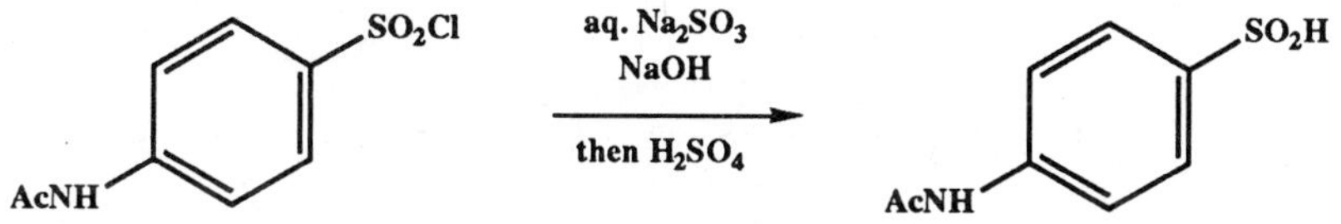

OS, CV 1, 73 (1941)

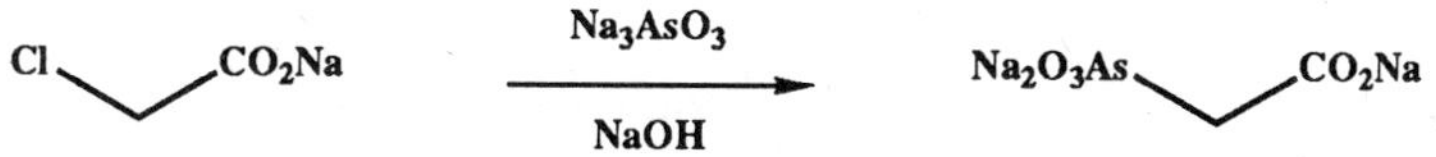

OS, CV 1, 84 (1941)

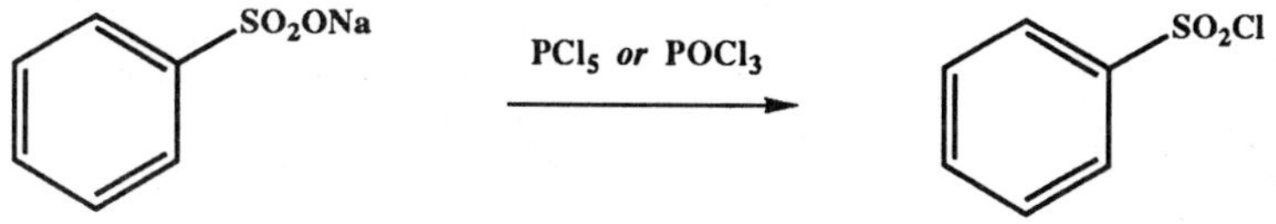

OS, CV 1, 99 (1941)

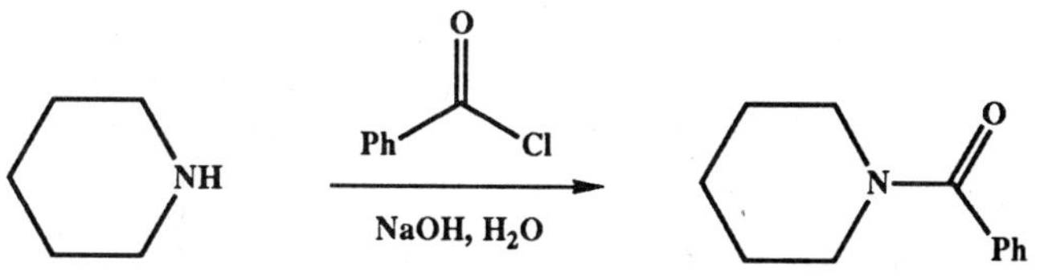

OS, *CV 1*, 201 (1941)

N—CN → (1. aq. H_2SO_4, Δ; 2. aq. NaOH) → N—H

OS, *CV 1*, 298 (1941)

NC, N, N, CN, N, CN → (EtOH, H_2SO_4) → $(H_2NCH_2CN) \bullet H_2SO_4$ → ($Ba(OH)_2$, Δ) → $(H_2NCH_2CO_2)_2Ba$

OS, *CV 1*, 318 (1941)

$HO—N(SO_3Na)_2$ → (O, CH_3, CH_3; H_2O, 70 °C) → HO—N=C(CH_3)(CH_3) → (HCl, H_2O) → $HO—NH_2 \bullet HCl$

OS, *CV 1*, 399 (1941)

H_2N, $NH_2 \bullet HNO_3$, NH → (H_2SO_4) → H_2N, H, N, NO_2, NH

OS, CV 1, 417 (1941)

O
H_2N $NH_2 \cdot HNO_3$
H_2SO_4
0 °C
O
H_2N N NO_2
H

OS, CV 1, 431 (1941)

O O
O—O
1. NaOMe, MeOH
$CHCl_3$, 0 °C
2. H_2SO_4, H_2O
O
O—OH

OS, CV 2, 11 (1943)

O
H_2N OH
Ac_2O, H_2O
H
N
Ac
O
OH

OS, CV 2, 25 (1943)

O
NK
O
Cl—$(CH_2)_3CN$
O
N—$(CH_2)_3CN$
O
H_2SO_4
H_2O
H_2N CO_2H

OS, CV 2, 76 (1943)

Caprolactam (NH) $\xrightarrow[H_2O]{NaOH}$ $H_2N—(CH_2)_5—CO_2H$ $\xrightarrow{PhCOCl}$ $PhC(=O)—NH—(CH_2)_5—CO_2H$

OS, CV 2, 106 (1943)

$$3\ n\text{-BuOH} + B(OH)_3 \xrightarrow{\Delta} B(O\text{-}n\text{-Bu})_3$$

OS, CV 2, 109 (1943)

$$3\ n\text{-BuOH} + POCl_3 \xrightarrow[\text{benzene}]{\text{pyridine}} PO(O\text{-}n\text{-Bu})_3$$

OS, CV 2, 310 (1943)

NC–CH₂ substituted triazinane (three N–CH₂CN) $\xrightarrow[HCl,\ H_2O]{EtOH,\ \Delta}$ $H_2NCH_2C(=O)OEt$

OS, CV 2, 328 (1943)

$ClCH_2CO_2H$ $\xrightarrow[H_2O]{NH_4OH}$ $H_2NCH_2CO_2H$ $\xrightarrow[\text{2. HCl}]{\text{1. PhCOCl; NaOH, } H_2O}$ $PhCOHNCH_2CO_2H$

OS, CV 2, 395 (1943)

2 PhCHO → ($H_2NNH_2 \cdot H_2SO_4$, NH_4OH) → PhCH=N–N=CHPh → ($(CH_3)_2SO_4$) → $CH_3NH–NH_2 \cdot H_2SO_4$

OS, CV 2, 455 (1943)

bis(2-nitrophenyl) disulfide (S–S, NO_2 O_2N) → (Cl_2, CCl_4, 55 °C) → 2 2-nitrophenyl–S–Cl (NO_2)

OS, CV 3, 172 (1955)

$(CH_3)_2C=N–ONa$ → (1. $BrCH_2CO_2Na$, 2. HCl) → $(CH_3)_2C=N–OCH_2CO_2H$ → (HCl, H_2O, Δ) → $(H_2NOCH_2CO_2H)_2$ HCl

OS, CV 3, 256 (1955)

N-$(CH_2)_3Br$ phthalimide → ($(n\text{-}Bu)_2NH$, xylene, Δ) → N-$(CH_2)_3N(n\text{-}Bu)_2$ phthalimide → (HCl, H_2O, Δ) → $H_2N(CH_2)_3N(n\text{-}Bu)_2$

OS, CV 3, 355 (1955)

$$C_6H_5IO + C_6H_5IO_2 \xrightarrow[H_2O]{NaOH} [(C_6H_5)_2I]\,IO_3 \xrightarrow[H_2O]{KI} [(C_6H_5)_2I]\,I$$

OS, CV 3, 375 (1955)

$$H_2N\text{—}NH_2 \xrightarrow[aq.\ Na_2CO_3,\ EtOH]{2\ Cl\text{—}CO_2Et} EtO_2C\text{—}NH\text{—}NH\text{—}CO_2Et$$

OS, CV 3, 404 (1955)

$$2\ N(CO_2Et)_3 + 5\ H_2NNH_2 \xrightarrow[\Delta]{H_2O} 3\ H_2NNHCO_2Et + HN(C(=O)NHNH_2)_2$$

OS, CV 3, 415 (1955)

$$H_2NCO_2Et + 2\ ClCO_2Et \xrightarrow[\Delta]{Na,\ Et_2O} N(CO_2Et)_3$$

OS, CV 3, 483 (1955)

$$C_6H_5ICl_2 \xrightarrow[Na_2CO_3]{NaOH,\ H_2O} C_6H_5IO$$

OS, CV 3, 608 (1955)

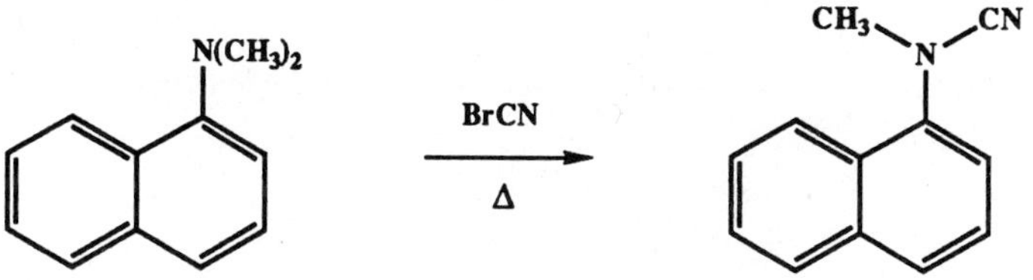

N(CH$_3$)$_2$

BrCN

Δ

CH$_3$ N CN

OS, CV 3, 711 (1955)

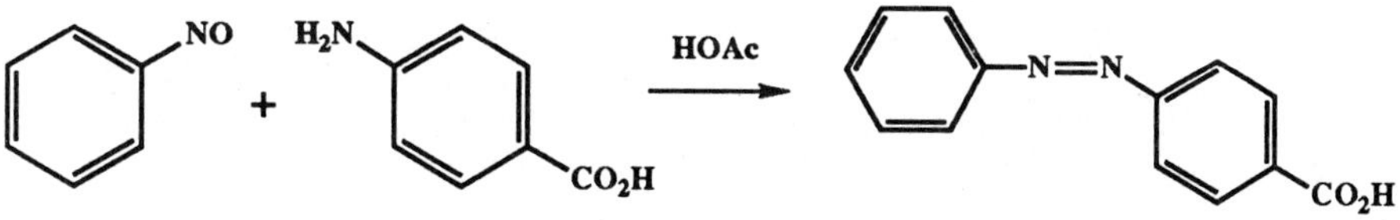

NO + H$_2$N CO$_2$H

HOAc

N=N CO$_2$H

OS, CV 3, 723 (1955)

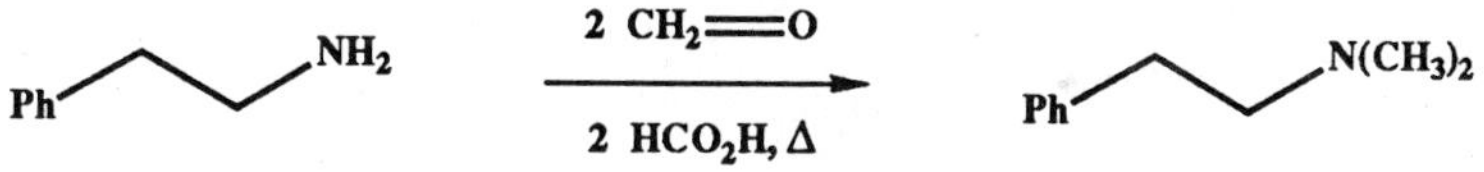

Ph NH$_2$

2 CH$_2$=O

2 HCO$_2$H, Δ

Ph N(CH$_3$)$_2$

OS, CV 3, 765 (1955)

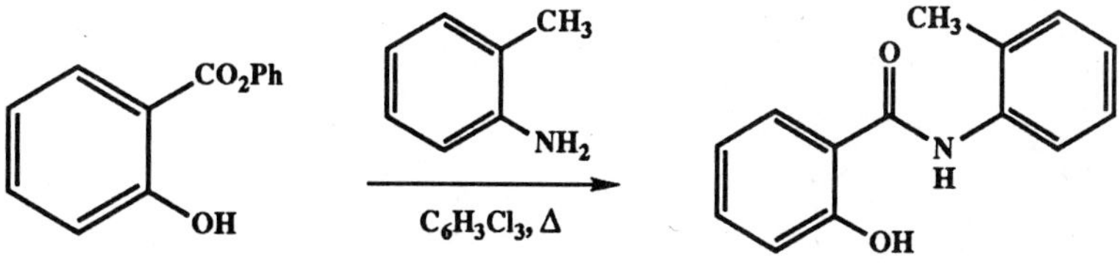

CO$_2$Ph OH

CH$_3$ NH$_2$

C$_6$H$_3$Cl$_3$, Δ

CH$_3$ O N H OH

OS, *CV* 4, 258 (1963)

$$(n\text{-Bu})_2SnCl_2 \xrightarrow[\text{THF}, \Delta]{2\ CH_2{=}CHMgBr} (n\text{-Bu})_2Sn(CH{=}CH_2)_2$$

OS, *CV* 4, 325 (1963)

$$(i\text{-PrO})_3P + CH_3I \xrightarrow{\Delta} (i\text{-PrO})_2P(=O){-}CH_3$$

OS, *CV* 4, 361 (1963)

$$H_2N{-}C(=O){-}NH{-}NO_2 \xrightarrow[H_2O, \Delta]{Me_2NH} Me_2N{-}C(=O){-}NH_2$$

OS, *CV* 4, 411 (1963)

$$H_2N{-}NH_2 \xrightarrow[H_2O,\ EtOH]{2\ ClCO_2Et,\ Na_2CO_3} EtO_2C{-}NH{-}NH{-}CO_2Et$$

OS, *CV* 4, 466 (1963)

$$\text{pyrrolidine (NH)} + Br{-}CH(CH_3){-}CO_2Et \xrightarrow[\text{benzene}]{\Delta} \text{pyrrolidin-1-yl}{-}CH(CH_3){-}CO_2Et$$

OS, CV 4, 473 (1963)

2 cyclopentadiene → ferrocene (Fe)

1. 2 Na, xylene
2. $FeCl_2$, THF

OS, CV 4, 476 (1963)

2 cyclopentadiene → ferrocene (Fe)

2 Et_2NH
$FeCl_2$, THF

OS, CV 4, 521 (1963)

$$H_2N{-}(CH_2)_6{-}NH_2 \cdot 2\ HCl \xrightarrow[\text{tetralin, 185 °C}]{2\ Cl_2CO} OCN{-}(CH_2)_6{-}NCO$$

OS, CV 4, 571 (1963)

$$CH_3SO_3H \xrightarrow{SOCl_2,\ \Delta} CH_3SO_2Cl$$

OS, CV 4, 582 (1963)

$$\text{2-}CH_3C_6H_4CH_2NMe_2 \xrightarrow[\text{EtOH, }\Delta]{\text{EtBr}} \text{2-}CH_3C_6H_4CH_2\overset{+}{N}Me_2Et\ Br^- \xrightarrow[\text{2. NaOH, }\Delta]{\text{1. NaOAc, AcOH, }\Delta} \text{2-}CH_3C_6H_4CH_2OH$$

OS, CV 4, 674 (1963)

p-$MeC_6H_4SO_2Cl$ → (1. Na_2SO_3, 75 °C, aq. $NaHCO_3$; 2. Me_2SO_4, Δ) → p-$MeC_6H_4SO_2Me$

OS, CV 4, 693 (1963)

SO_3Na / NaO_3S (naphthalene-1,5-disulfonate) → (2 PCl_5, 110 °C) → SO_2Cl / ClO_2S

OS, CV 4, 846 (1963)

Ph—CH=CH_2 → (1. dioxane • SO_3; 2. NaOH) → Ph—CH=CH—SO_3Na → (PCl_5, Δ) → Ph—CH=CH—SO_2Cl

OS, CV 4, 881 (1963)

$Sn(Cl)_4$ → (4 EtMgBr, Et_2O, Δ) → $Sn(Et)_4$

***OS, CV* 4, 937 (1963)**

$CH_3-C_6H_4-SO_2Na \xrightarrow{SOCl_2} CH_3-C_6H_4-SOCl$

***OS, CV* 4, 940 (1963)**

$2\ CH_3C_6H_4SO_2OH \xrightarrow[\text{kieselguhr, asbestos}]{P_2O_5,\ \Delta} CH_3C_6H_4-SO_2-O-SO_2-C_6H_4CH_3$

***OS, CV* 4, 943 (1963)**

$CH_3C_6H_4SO_2Cl \xrightarrow[\text{aq. NaOH}]{CH_3NH_2,\ \Delta} CH_3C_6H_4SO_2-NH-CH_3 \xrightarrow[\text{aq. HOAc}]{NaNO_2} CH_3C_6H_4SO_2-N(NO)-CH_3$

***OS, CV* 4, 955 (1963)**

$PCl_3 + 3\ EtOH \xrightarrow[\text{pet. ether, }\Delta]{PhNEt_2} P(OEt)_3$

OS, *CV 5*, 88 (1973)

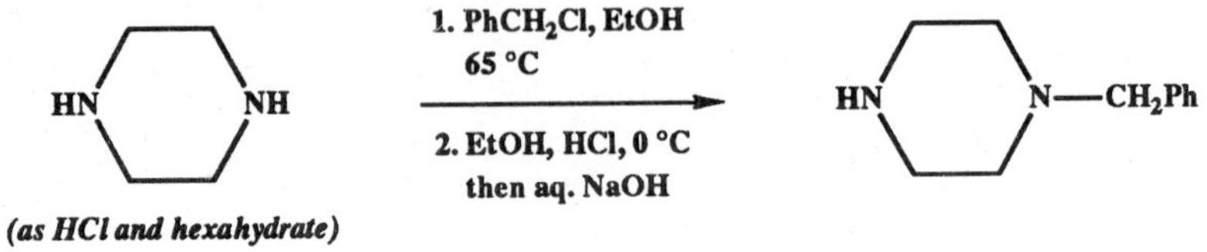

HN NH (as HCl and hexahydrate) → 1. $PhCH_2Cl$, EtOH 65 °C; 2. EtOH, HCl, 0 °C then aq. NaOH → HN N—CH_2Ph

OS, *CV 5*, 196 (1973)

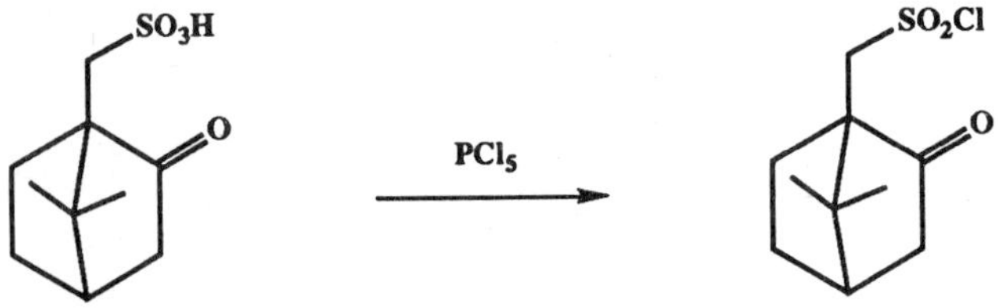

SO_3H, O → PCl_5 → SO_2Cl, O

OS, *CV 5*, 201 (1973)

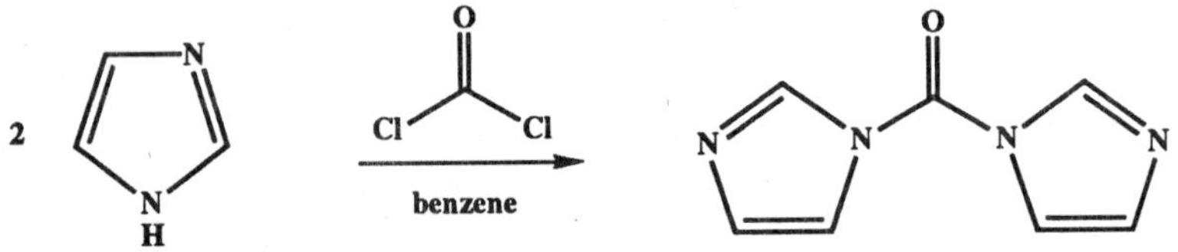

2 N, N H → Cl(C=O)Cl, benzene → N N (C=O) N N

OS, *CV 5*, 204 (1973)

$ClCH_2C(O)NH_2$ → ClC(O)C(O)Cl, ethylene dichloride, Δ → $ClCH_2C(O)N{=}C{=}O$

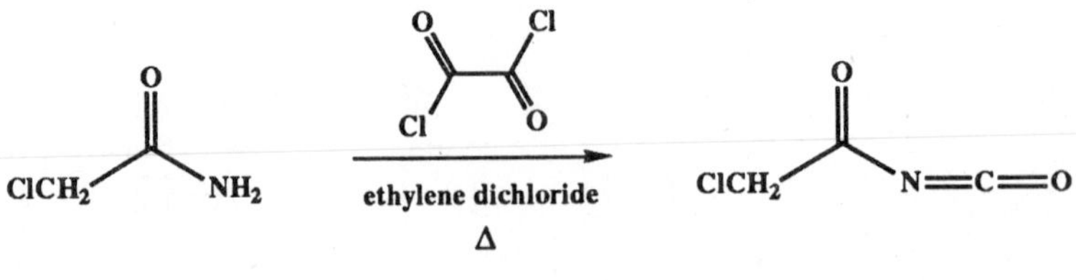

OS, CV 5, 208 (1973)

NH$_2$ (cyclohexylamine) → [2 *t*-BuOCl / benzene] → NCl$_2$ (N,N-dichlorocyclohexylamine) → [KOAc / EtOH, Δ] → NCl (N-chlorocyclohexanimine)

OS, CV 5, 211 (1973)

2 *i*-PrMgCl → [PCl_3, Et_2O] → $(i\text{-}Pr)_2PCl$

OS, CV 5, 218 (1973)

$ClCH_2—P(=O)Cl_2$ → [P_4S_{10} / 180 - 190 °C] → $ClCH_2—P(=S)Cl_2$

OS, CV 5, 226 (1973)

Cl—C≡N → [SO_3, Δ] → $Cl—SO_2—N{=}C{=}O$

OS, CV 5, 341 (1973)

H_2N–C$_6$H$_4$–NHAc → [$NaBO_3$, H_3BO_3 / HOAc, 60 °C] → AcNH–C$_6$H$_4$–N=N–C$_6$H$_4$–NHAc

OS, CV 5, **555 (1973)**

$$EtN{=}C{=}N(CH_2)_3NMe_2 \xrightarrow[CH_2Cl_2,\ Et_2O,\ 0\ °C]{pyridine\ HCl} EtN{=}C{=}N(CH_2)_3\overset{+}{N}HMe_2\ Cl^-$$

$$EtN{=}C{=}N(CH_2)_3NMe_2 \xrightarrow{MeI,\ Et_2O} EtN{=}C{=}N(CH_2)_3\overset{+}{N}Me_3\ I^-$$

OS, CV 5, **575 (1973)**

Piperidine (N–H) $\xrightarrow[400\ psi,\ 100\ °C]{CH_2{=}CH_2}$ N-ethylpiperidine (N–CH_2CH_3)

OS, CV 5, **602 (1973)**

$$6\ Me_2NH \xrightarrow{PCl_3,\ Et_2O} (Me_2N)_3P + 3\ Me_2NH_2^+\ Cl^-$$

OS, CV 5, **658 (1973)**

$PhI(OAc)_2 \xrightarrow{aq.\ NaOH} PhIO$

OS, CV 5, 663 (1973)

N-silver succinimide → (I_2, dioxane) → N-iodosuccinimide (NI)

OS, CV 5, 709 (1973)

$CH_3S—SCH_3$ → (Cl_2, Ac_2O; - 10 to 0 °C) → 2 $CH_3S—Cl$ → (Cl_2) → 2 CH_3SCl_3 → (Ac_2O) → 2 $CH_3S(O)Cl$

OS, CV 5, 710 (1973)

$CH_3S—SCH_3$ → (Cl_2, HOAc; - 10 to 0 °C) → 2 $CH_3S—Cl$ → (Cl_2) → 2 CH_3SCl_3 → (HOAc; r.t.) → 2 $CH_3S(O)Cl$

OS, CV 5, 727 (1973)

2 [dibenzo-azaborine, $\overset{+}{N}$—H, $\overset{-}{B}$—OH] → (80 °C; 200 mm Hg) → *bis* - ether → (2 CH_3MgBr; Et_2O, Δ) → 2 [dibenzo-azaborine, $\overset{+}{N}$—H, $\overset{-}{B}$—CH_3]

OS, CV 5, 736 (1973)

PhCHO —(n-BuNH$_2$ / C_6H_6, Δ)→ PhCH=NBu —(Me_2SO_4 / benzene)→ PhCH=$\overset{+}{N}$Bu(Me) —(1. H_2O, Δ / 2. NaOH)→ HN(Me)—Bu

OS, CV 5, 758 (1973)

PhCHO —($EtNH_2$)→ PhCH=NEt —(1. MeI, 100 °C / 2. H_2O, Δ / 3. NaOH, Δ)→ Me—NH—Et

OS, CV 5, 769 (1973)

indole —(CH_3I / Na, NH_3, Et_2O)→ N-CH_3 indole

OS, CV 5, 801 (1973)

cyclohexyl-NH_2 —(Si(NCO)$_4$ / benzene, Δ)→ Si (NH—C(=O)—NH—C_6H_{11})$_4$ —(H_2O, Δ / i-PrOH)→ cyclohexyl-NH—C(=O)—NH_2

OS, CV 5, 802 (1973)

2,6-dimethylaniline (CH_3, NH_2, CH_3) →[$Si(NCS)_4$ / benzene, Δ] Si (NH—C(=S)—NH—C_8H_9)$_4$ →[H_2O, Δ / *i*-PrOH] 2,6-dimethylphenylthiourea (CH_3, N(H)–C(=S)–NH_2, CH_3)

OS, CV 5, 839 (1973)

(CH₃)₂C(OH)CN →[HNO_3 / Ac_2O] (CH₃)₂C(ONO_2)CN →[2 morpholine (O, NH) / Δ] N-nitromorpholine (NO_2–N, O) + 2-methyl-2-morpholinopropionitrile (CN, N, O)

OS, CV 5, 843 (1973)

2 m-nitrobenzenesulfonyl chloride (SO_2Cl, NO_2) →[58% HI / Δ] bis(m-nitrophenyl) disulfide (O_2N–C₆H₄–S—S–C₆H₄–NO_2)

OS, CV 5, 959 (1973)

PhS—SPh →[AgF_2 / Freon-113] 2 $PhSF_3$

OS, CV 5, 969 (1973)

C_6H_5HgCl + Cl_3C—CO_2Na →[DME, Δ] $C_6H_5HgCCl_3$

OS, CV 5, 1001 (1973)

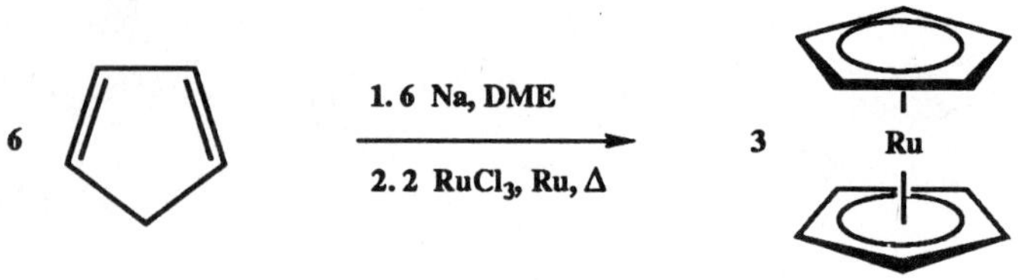

6 [cyclopentadiene] —(1. 6 Na, DME; 2. 2 $RuCl_3$, Ru, Δ)→ 3 [ruthenocene, Ru]

OS, CV 5, 1016 (1973)

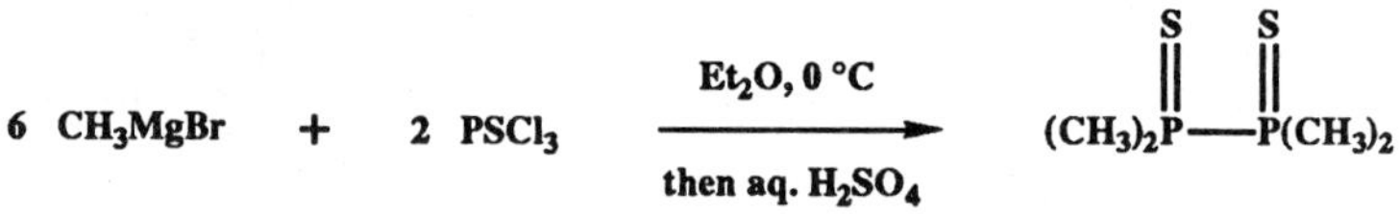

6 CH_3MgBr + 2 $PSCl_3$ —(Et_2O, 0 °C; then aq. H_2SO_4)→ $(CH_3)_2P(S)—P(S)(CH_3)_2$

OS, CV 5, 1055 (1973)

p-Ts—Cl —(H_2NNH_2; THF, H_2O)→ *p*-Ts—NH—NH_2

OS, CV 5, 1080 (1973)

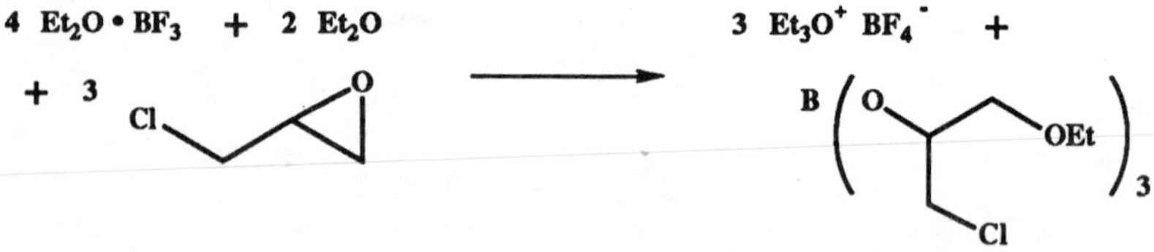

4 $Et_2O \cdot BF_3$ + 2 Et_2O + 3 [epichlorohydrin, Cl] → 3 Et_3O^+ BF_4^- + B(O–CH(CH_2Cl)–CH_2–OEt$)_3$

OS, CV 5, 1096 (1973)

$$2\ Et_3O^+\ BF_4^- + 3\ Me_2O \xrightarrow{CH_2Cl_2} 2\ Me_3O^+\ BF_4^- + 3\ Et_2O$$

OS, CV 5, 1099 (1973)

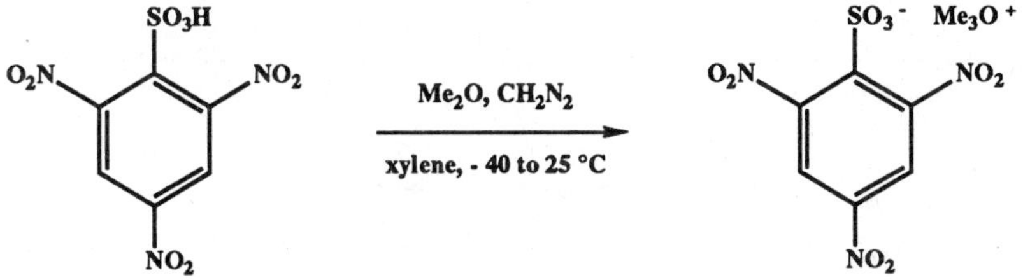

2,4,6-trinitrobenzenesulfonic acid (SO_3H; O_2N, NO_2, NO_2) $\xrightarrow[\text{xylene, - 40 to 25 °C}]{Me_2O,\ CH_2N_2}$ $SO_3^-\ Me_3O^+$ (O_2N, NO_2, NO_2)

OS, CV 6, 78 (1988)

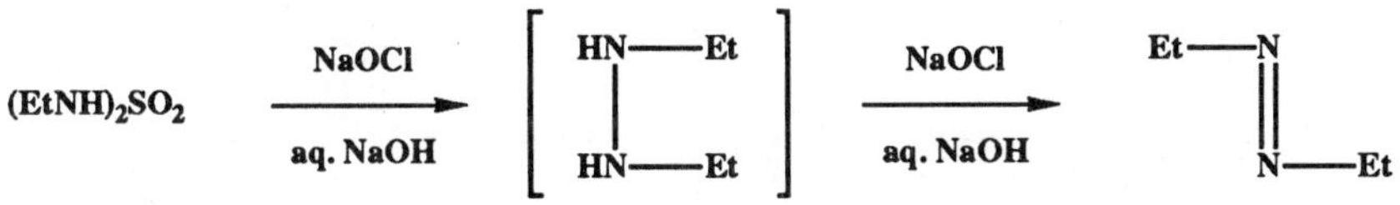

$(EtNH)_2SO_2$ $\xrightarrow[\text{aq. NaOH}]{NaOCl}$ [HN—Et / HN—Et] $\xrightarrow[\text{aq. NaOH}]{NaOCl}$ Et—N=N—Et

OS, CV 6, 115 (1988)

isoquinoline (N) $\xrightarrow[CH_2Cl_2,\ H_2O]{KCN,\ PhCOCl}$ 2-benzoyl-1,2-dihydroisoquinoline-1-carbonitrile (N, Ph, O, NC, H)

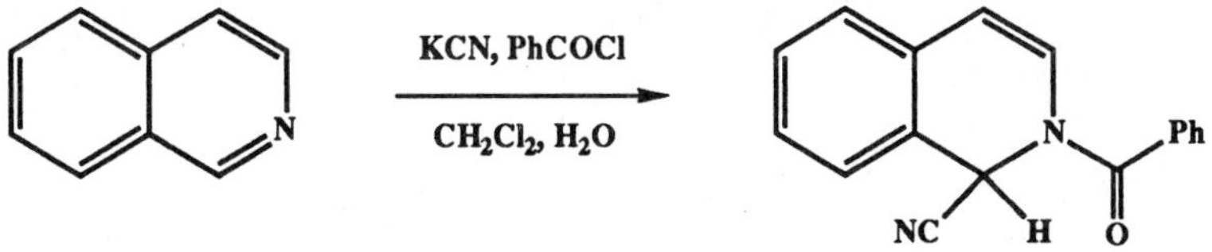

***OS*, *CV* 6, 232 (1988)**

$$t\text{-Bu—NH}_2 \xrightarrow[\text{H}_2\text{O, CH}_2\text{Cl}_2,\ \Delta]{\text{CHCl}_3,\ 3\ \text{NaOH},\ [\text{Et}_3\text{NCH}_2\text{Ph}]^+\ \text{Cl}^-} t\text{-Bu—N}{\equiv}\text{C}$$

◇

***OS*, *CV* 6, 235 (1988)**

$$i\text{-Pr-Br} \xrightarrow[\text{MeOH, H}_2\text{O},\ \Delta]{\text{Na}_2\text{S}_2\text{O}_3\bullet\ 5\ \text{H}_2\text{O}} i\text{-Pr—S—SO}_3\text{Na} \xrightarrow[\text{H}_2\text{O},\ 0-5\ °\text{C}]{sec\text{-BuSNa}} i\text{-Pr—S—S—}sec\text{-Bu}$$

◇

***OS*, *CV* 6, 310 (1988)**

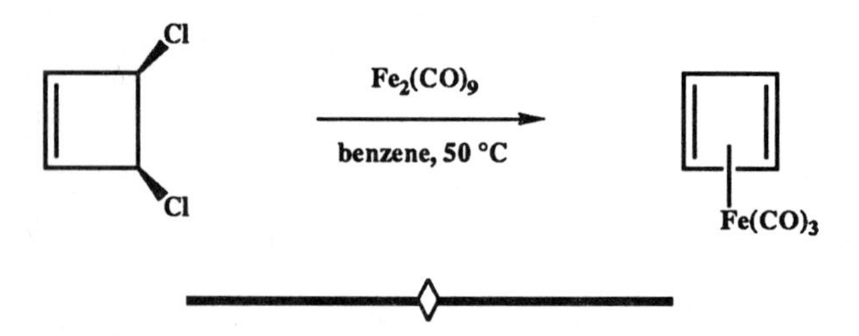

◇

***OS*, *CV* 6, 353 (1988)**

$$\text{Me}_3\text{Si—O—SiMe}_3 \xrightarrow[60-140\ °\text{C}]{\text{Al, I}_2} \text{Me}_3\text{Si-I}$$

◇

***OS*, *CV* 6, 436 (1988)**

$$\text{Et}_3\text{Al} + \text{HCN} \xrightarrow{\text{benzene}} \text{Et}_2\text{AlCN} + \text{Et-H}$$

OS, CV 6, **440 (1988)**

Et_2NSiMe_3 + SF_4 —(Freon 11; - 60 °C)→ Et_2NSF_3 + $FSiMe_3$

OS, CV 6, **727 (1988)**

CH_3SO_2Cl —(Na_2SO_3, $NaHCO_3$; H_2O, 25 °C)→ CH_3SO_2Na —(Cl—CN; H_2O, 10 - 15 °C)→ CH_3SO_2CN

OS, CV 6, **840 (1988)**

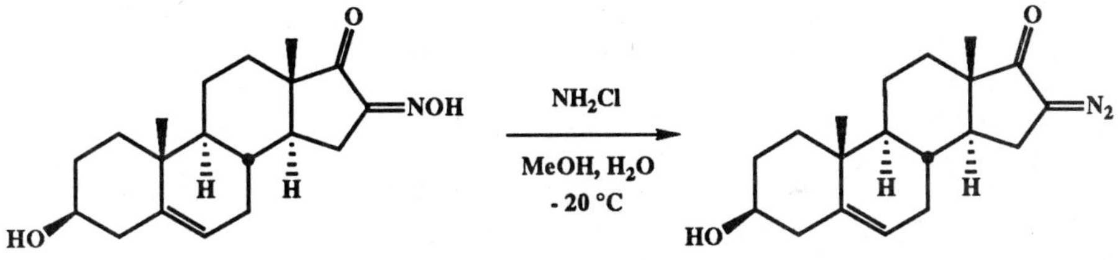

=NOH —(NH_2Cl; MeOH, H_2O; - 20 °C)→ $=N_2$

HO H H O HO H H O

OS, CV 6, **996 (1988)**

OMe —($Fe(CO)_5$; Bu_2O, Δ)→ OMe $Fe(CO)_3$ + OMe $Fe(CO)_3$

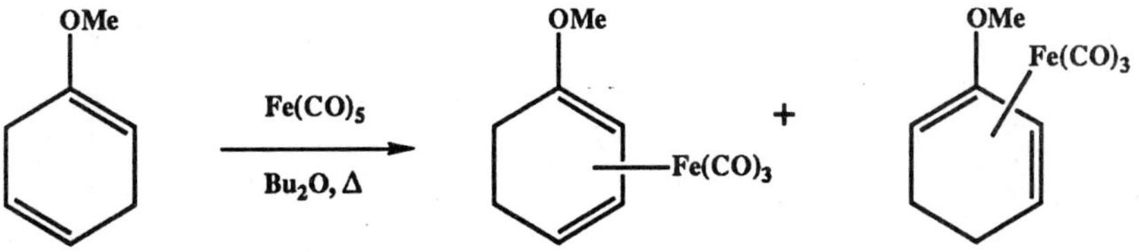

OS, *CV 6*, 1019 (1988)

4 $Et_2O \cdot BF_3$ + 6 Me_2O + 3 (epichlorohydrin, Cl) → 3 $Me_3O^+ BF_4^-$ + 4 Et_2O + B(O–CH(CH$_2$Cl)–CH$_2$OMe)$_3$

OS, *CV 6*, 1030 (1988)

Me_3SiCl —(NaN_3, diglyme)→ Me_3SiN_3

OS, *CV 7*, 27 (1990); *61*, 14 (1983)

PhCHO + H_2N–(5-aminotetrazole, N–H) —(1. Et_3N, MeOH, 50 °C; 2. H_2 (500 psi), Pd / C)→ $PhCH_2NH$–(tetrazole, N–H)

OS, *CV 7*, 206 (1990); *62*, 187 (1984)

$(PhO)_2P(O)$–Cl + NaN_3 → $(PhO)_2P(O)$–N_3

OS, *CV 7*, 30 (1990); *63*, 175 (1985)

CbzNH–CH(R_1)–C(O)–O–C(4-ClC$_6$H$_4$)=CH–C(O)–N(piperazine)NMe —(H_2N–CH(R_2)–CO_2Me)→ CbzNH–CH(R_1)–C(O)–NH–CH(R_2)–CO_2Me

OS, *CV* 7, 149 (1990); *64*, 19 (1986)

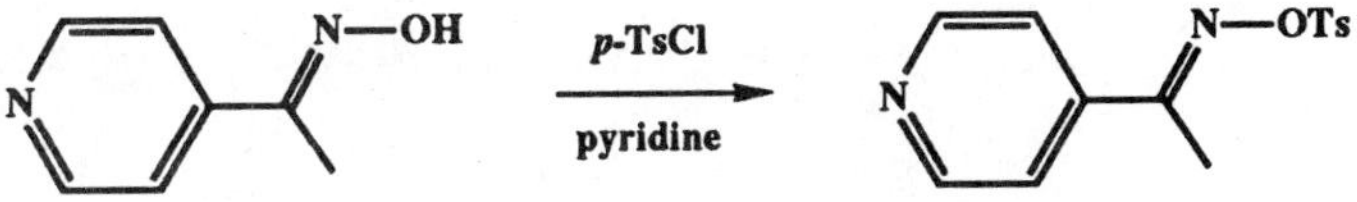

p-TsCl, pyridine

OS, *CV* 7, 197 (1990); *64*, 85 (1986)

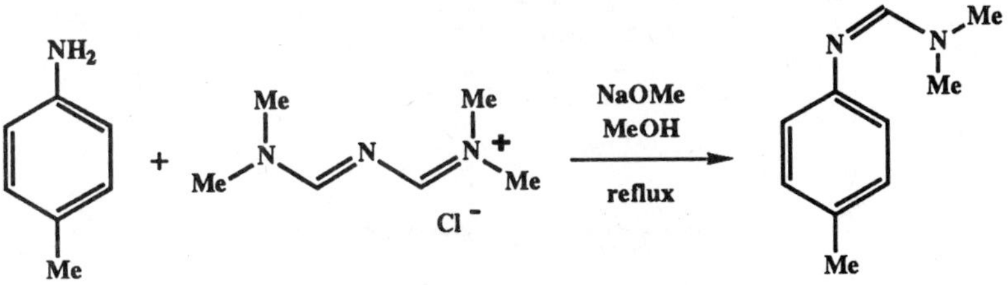

NaOMe, MeOH, reflux

OS, *CV* 7, 495 (1990); *64*, 196 (1986)

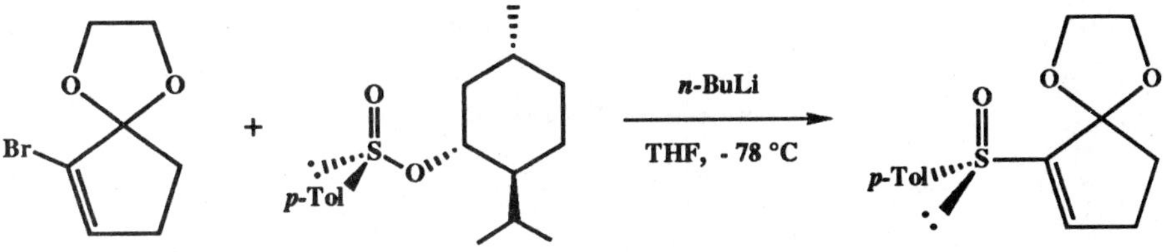

n-BuLi, THF, - 78 °C

S - (-) *S* - (+)

OS, *CV* 7, 506 (1990); *64*, 217 (1986)

$$CF_3CO_2H + CF_3SO_3H \xrightarrow{P_2O_5} CF_3CO_2SO_2CF_3$$

SUBSTITUTION - HETEROATOM

OS, CV 7, 528 (1990); *64,* 221 (1986)

$$3\ Me_3SiNMe_2 + SF_4 \xrightarrow[25\ ^\circ C]{Et_2O} (Me_2N)_3S^+\ F_2SiMe_3^- + 2\ FSiMe_3$$

OS, 65, 166 (1987)

$$t\text{-Bu}-CH_2-C(Me)_2-NH_2 \xrightarrow[\text{3. Sodium naphthalenide, THF}]{\text{1. } Na_2WO_4 \cdot 2\ H_2O,\ H_2O_2;\ \text{2. } t\text{-Bu-NHNH}_2,\ PbO_2} t\text{-Bu}-CH_2-C(Me)_2-NH\text{-}t\text{-Bu}$$

OS, 67, 149 (1988)

$$p\text{-TsCl} + Me_3Si-C\equiv C-SiMe_3 \xrightarrow[\text{2. } K_2CO_3,\ KHCO_3,\ MeOH,\ H_2O]{\text{1. } AlCl_3,\ CH_2Cl_2} p\text{-Ts}-C\equiv C-H$$

OS, 68, 104 (1989)

$$2\ CH_2{=}CHCH_2MgBr \xrightarrow{(n\text{-Bu})_3Sn-O-Sn(n\text{-Bu})_3} 2\ (n\text{-Bu})_3SnCH_2CH{=}CH_2$$

OS, CV 1, 135 (1941)

CH_3 / NH_2 → CH_3 / Br

1. 2 HBr
 $NaNO_2$
2. Cu powder

OS, CV 1, 136 (1941)

NH_2 / CH_3 → Br / CH_3

1. H_2SO_4
 $NaNO_2$, H_2O
2. HBr, CuBr

OS, CV 1, 162 (1941)

NO_2 / NH_2 → NO_2 / Cl

1. aq. HCl, $NaNO_2$
2. CuCl

OS, CV 1, 170 (1941)

NH_2 / R_1 / R_2 → Cl / R_1 / R_2

1. aq. HCl, $NaNO_2$
2. CuCl

R_1 = Me, R_2 = H
R_1 = H, R_2 = Me

OS, CV 1, 374 (1941)

NH_2, CO_2H — 1. HCl, $NaNO_2$; 2. $PhNMe_2$ → N=N, CO_2H, NMe_2

OS, CV 1, 404 (1941)

NH_2, NO_2 — 1. $NaNO_2$, aq. H_2SO_4; 2. H_2O, H_2SO_4, 160 °C → OH, NO_2

OS, CV 1, 442 (1941)

NH_2 — 1. $NaNO_2$, HCl; 2. NaOH, SO_2; 3. HCl, then NaOH → H, N, NH_2

OS, CV 2, 130 (1943)

CHO, NO_2 — $SnCl_2$, HCl → CHO, NH_2 — 1. $NaNO_2$, HCl; 2. CuCl, HCl → CHO, Cl

OS, CV 2, 145 (1943)

ClN_2 N_2Cl CH_3 CH_3 + 2 OH NH_2 SO_3Na SO_3Na

Na_2CO_3, H_2O

NH_2 OH NaO_3S N=N N=N OH NH_2 SO_3Na CH_3 CH_3 NaO_3S SO_3Na

OS, CV 2, 163 (1943)

NH_2 $NaNO_2$, HCl / H_2O N_2Cl NH_2 NaOAc N=N—N(H)—

OS, CV 2, 188 (1943)

H_2N • 2 HCl NH_2 1. $NaNO_2$ HCl, H_2O 2. HBF_4, Δ F F

SUBSTITUTION - DIAZONIUM

OS, CV 2, 225 (1943)

O_2N–C_6H_4–NH_2

1. $NaNO_2$, HBF_4, H_2O
2. $NaNO_2$, Cu, H_2O

O_2N–C_6H_4–NO_2

OS, CV 2, 238 (1943)

2 C_6H_5–NH_2

1. $NaNO_2$, HCl, H_2O
2. Se, KOH (K_2Se_x)

C_6H_5–Se–C_6H_5

OS, CV 2, 295 (1943)

C_6H_5–NH_2 • HCl

1. $NaNO_2$, aq. HCl
2. HBF_4
3. Δ (- N_2, - BF_3)

C_6H_5–F

OS, CV 2, 299 (1943)

EtO_2C–C_6H_4–NH_2

1. $NaNO_2$, HCl
2. HBF_4
3. Δ (- N_2, - BF_3)
4. KOH, H_2O

HO_2C–C_6H_4–F

OS, CV 2, 351 (1943)

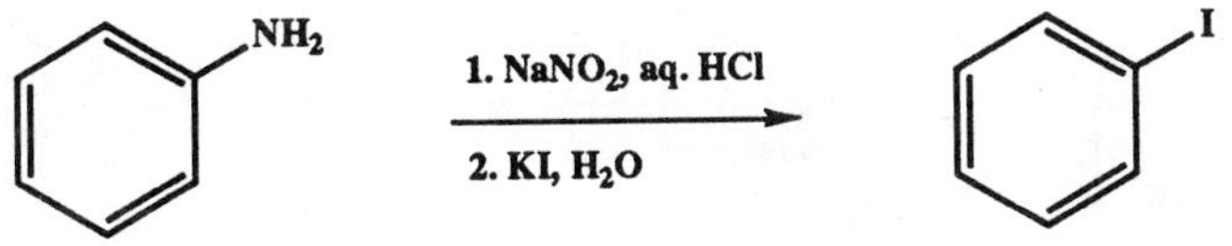

NH$_2$

1. $NaNO_2$, aq. HCl

2. KI, H_2O

I

OS, CV 2, 355 (1943)

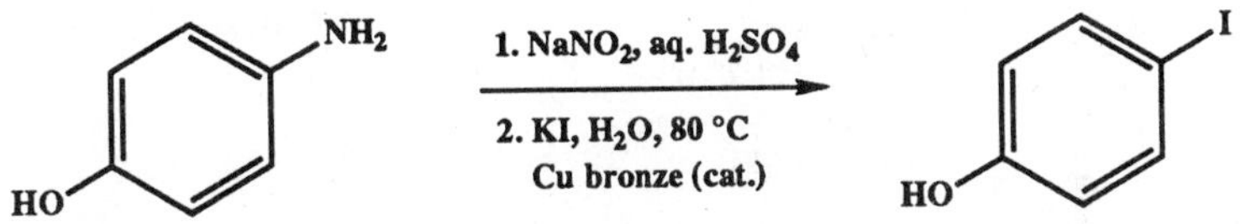

NH$_2$

HO

1. $NaNO_2$, aq. H_2SO_4

2. KI, H_2O, 80 °C
Cu bronze (cat.)

I

HO

OS, CV 2, 381 (1943)

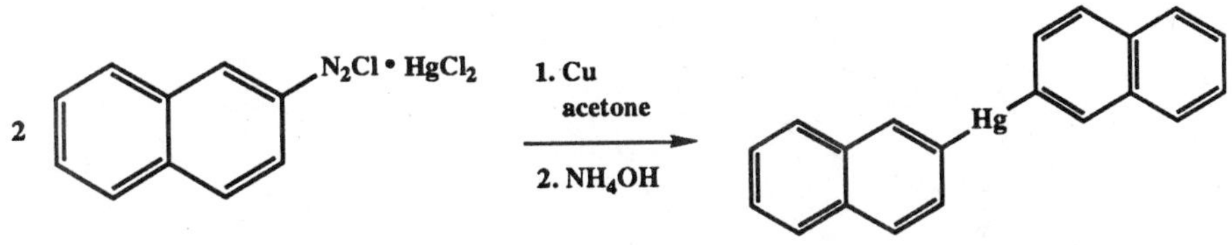

2

$N_2Cl \bullet HgCl_2$

1. Cu
acetone

2. NH_4OH

Hg

OS, CV 2, 432 (1943)

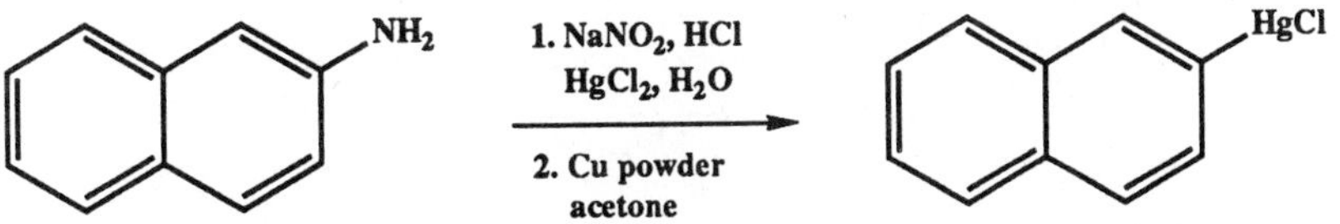

NH$_2$

1. $NaNO_2$, HCl
$HgCl_2$, H_2O

2. Cu powder
acetone

HgCl

OS, CV 2, 494 (1943)

1. $NaNO_2$, aq. HCl
2. Na_3AsO_3, $CuSO_4$
3. HCl

NH_2 → AsO_3H_2

OS, CV 2, 580 (1943)

1. $NaNO_2$, HCl, H_2O
2. Na_2S_2

2 CO_2H, NH_2 → CO_2H HO_2C, S—S

OS, CV 2, 604 (1943)

1. $NaNO_2$, H_2SO_4 then H_3PO_4
2. KI, H_2O

NH_2, I, I, NO_2 → I, I, I, NO_2

OS, CV 3, 130 (1955)

1. $NaNO_2$, H_2SO_4
2. H_2O

NH_2, CH_3, Br → OH, CH_3, Br

OS, CV 3, 185 (1955)

NH_2, Cl

1. $NaNO_2$, HBr, H_2O

2. CuBr, HBr, Δ

Br, Cl

OS, CV 3, 341 (1955)

NH_2, NO_2

1. $NaNO_2$
   H_2SO_4, HOAc

2. $NaNO_2$, H_2O
   Copper sulfites

NO_2, NO_2

OS, CV 3, 453 (1955)

H, O, NO_2

$SnCl_2$
HCl

H, O, NH_2 • $(SnCl_4)_x$

1. $NaNO_2$
   HCl, H_2O

2. H_2O, Δ

H, O, OH

OS, CV 3, 475 (1955)

CO_2H, NH_2

1. $NaNO_2$
   HCl, H_2O

2. aq. H_2SO_3
   then HCl

CO_2H, N(H)—NH_2 • HCl

OS, CV 3, 665 (1955)

O_2N–C_6H_4–N_2BF_4 → (NaAsO$_2$, Cu$_2$Cl$_2$; aq. NaOH then HCl) → O_2N–C_6H_4–AsO_3H_2

OS, CV 3, 809 (1955)

m-CH_3–C_6H_4–NH_2 → (1. NaNO$_2$, aq. HCl; 2. EtOCS$_2$K, 40 - 45 °C) → m-CH_3–C_6H_4–S–C(=S)–OEt → (1. KOH, EtOH, Δ; 2. H_2SO_4, H_2O) → m-CH_3–C_6H_4–SH

OS, CV 4, 75 (1963)

o-NO_2–C_6H_4–NH_2 • HCl → (1. NaNO$_2$, HCl, H_2O; 2. NaN$_3$, HCl, H_2O) → o-NO_2–C_6H_4–N_3

OS, CV 4, 160 (1963)

2,6-(O_2N)$_2$–C_6H_3–NH_2 → (1. NaNO$_2$, H_2SO_4, HOAc; 2. HCl, Cu$_2$Cl$_2$) → 2,6-(O_2N)$_2$–C_6H_3–Cl

OS, CV 4, 182 (1963)

1. $NaNO_2$, HCl, H_2O
2. NaOH, pH = 7, 0 °C

OS, CV 4, 718 (1963)

NH_2 / NO_2 → $N{=}N{-}NMe_2$ / NO_2

1. $NaNO_2$, HCl, H_2O
2. Me_2NH, Na_2CO_3, H_2O

OS, CV 5, 133 (1973)

NH_2 / Br → [$\overset{+}{N}_2$ PF_6^- / Br] → F / Br

1. $NaNO_2$, HCl, H_2O
2. HPF_6

165 °C

OS, CV 5, 829 (1973)

NH_2 → NO_2 / NH_2 → NO_2 / N_3

HNO_3, H_2SO_4

1. $NaNO_2$, H_2SO_4
2. NaN_3, H_2O

OS, *CV 5*, 1130 (1973)

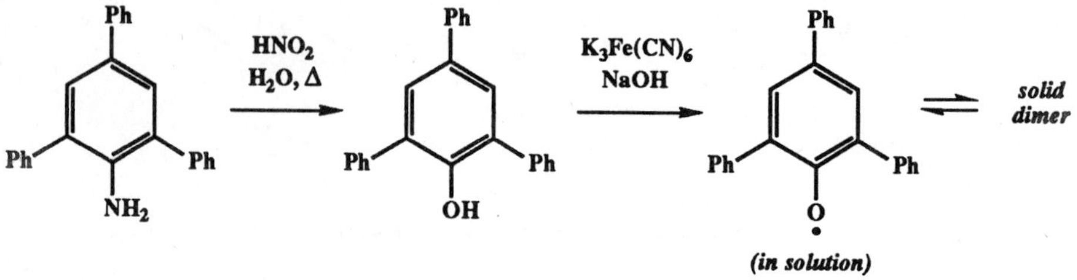

Ph, Ph, Ph, NH_2 — HNO_2, H_2O, Δ → Ph, Ph, Ph, OH — $K_3Fe(CN)_6$, NaOH → Ph, Ph, Ph, O• *(in solution)* ⇌ *solid dimer*

OS, *CV 7*, 508 (1990); *60*, 121 (1981)

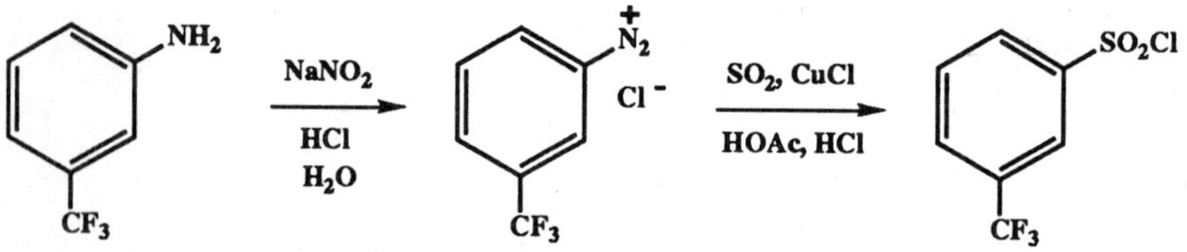

NH_2, CF_3 — $NaNO_2$, HCl, H_2O → $\overset{+}{N}_2$ Cl^-, CF_3 — SO_2, CuCl, HOAc, HCl → SO_2Cl, CF_3

***OS*, *CV 1*, 8 (1941)**

AcNH

$ClSO_3H$

SO_2Cl

AcNH

***OS*, *CV 1*, 49 (1941)**

OH

$PhN_2^+\ Cl^-$

7 - 10 °C

OH

N=NPh

***OS*, *CV 1*, 58 (1941)**

OH

NaOH

$(CH_3)_2SO_4$, Δ

OCH_3

***OS*, *CV 1*, 70 (1941)**

H_2N

H_3AsO_4

155 - 160 °C

AsO_3H_2

H_2N

***OS, CV 1,* 75 (1941)**

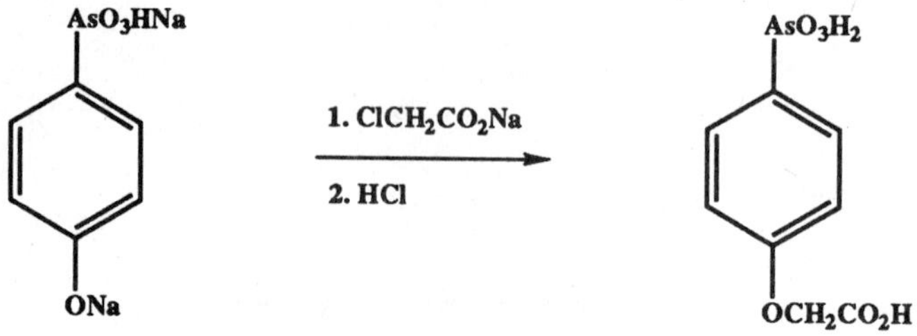

AsO_3HNa

ONa

1. $ClCH_2CO_2Na$

2. HCl

AsO_3H_2

OCH_2CO_2H

***OS, CV 1,* 82 (1941)**

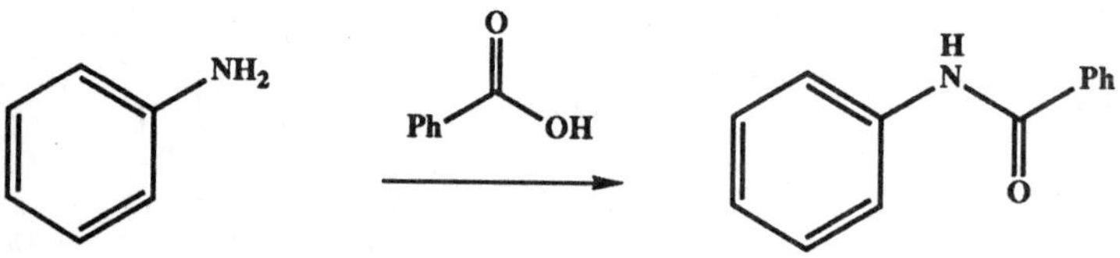

NH_2

O

Ph OH

H

N

Ph

O

***OS, CV 1,* 85 (1941)**

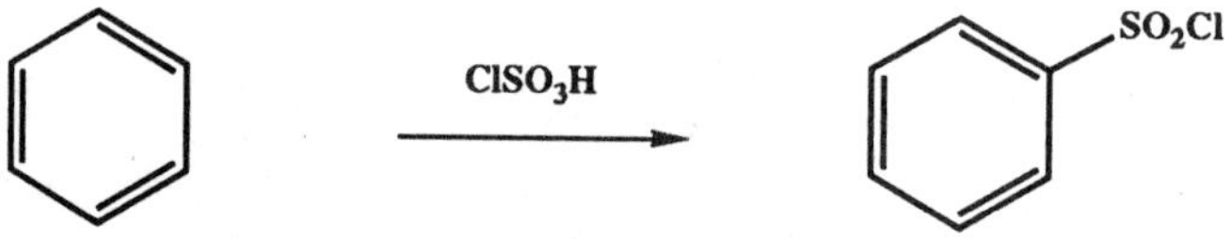

$ClSO_3H$

SO_2Cl

***OS, CV 1,* 102 (1941)**

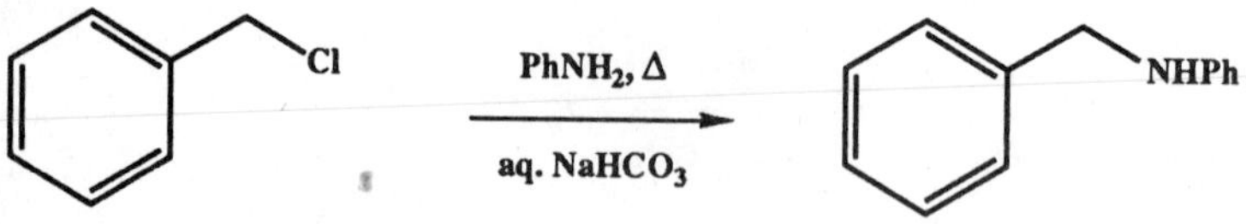

Cl

$PhNH_2$, Δ

aq. $NaHCO_3$

NHPh

OS, CV 1, 111 (1941)

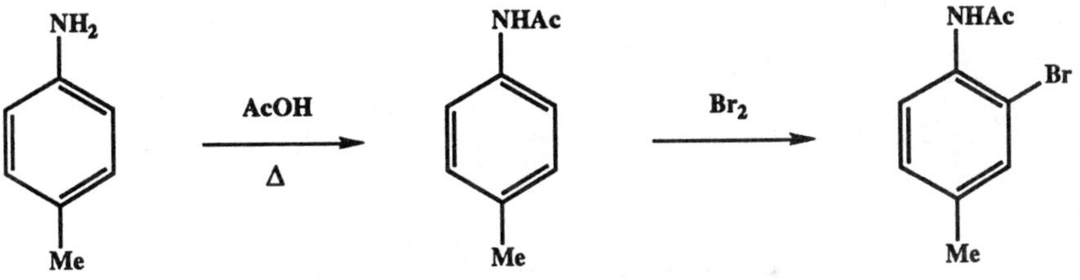

NHAc, Me, AcOH, Δ, Br_2, NHAc, Br, Me

OS, CV 1, 121 (1941)

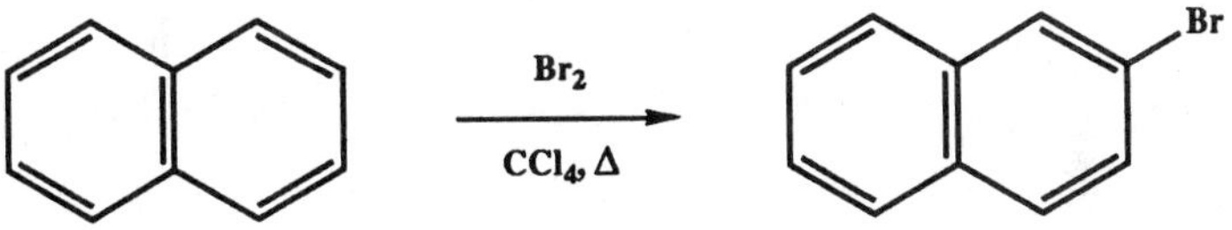

Br_2, CCl_4, Δ, Br

OS, CV 1, 123 (1941)

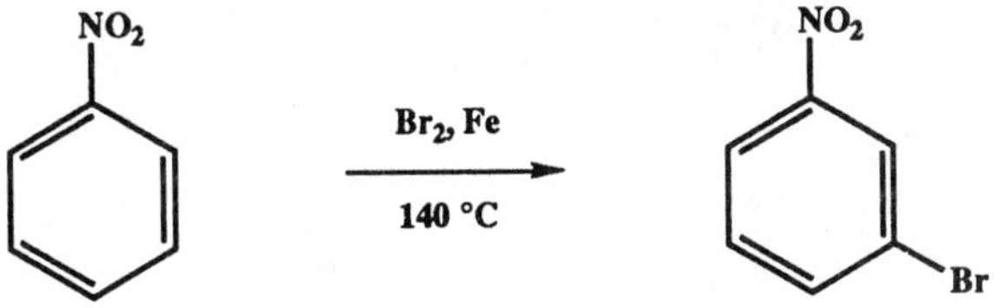

NO_2, Br_2, Fe, 140 °C, NO_2, Br

OS, CV 1, 125 (1941)

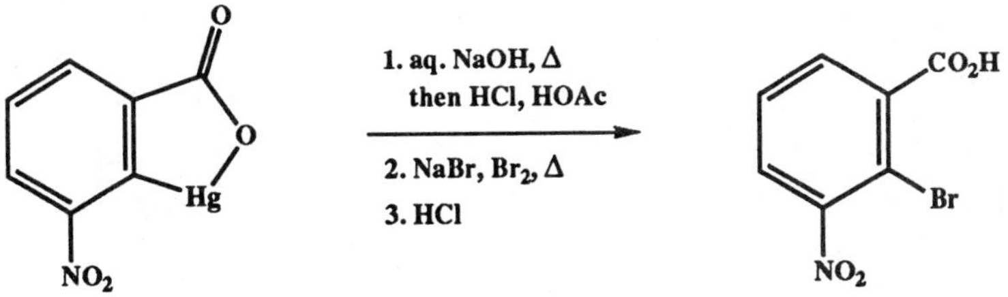

O, O, Hg, NO_2

1. aq. NaOH, Δ then HCl, HOAc
2. NaBr, Br_2, Δ
3. HCl

CO_2H, Br, NO_2

OS, CV 1, 128 (1941)

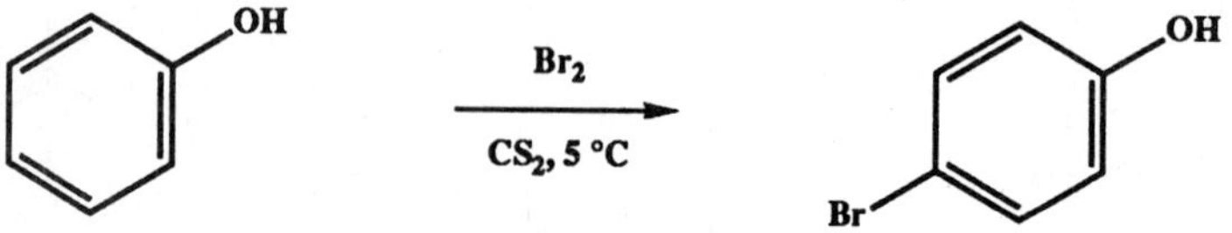

Br_2

CS_2, 5 °C

OH

Br

OS, CV 1, 161 (1941)

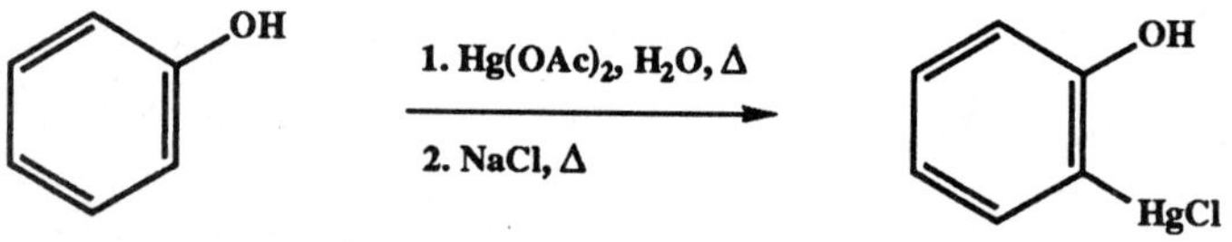

1. $Hg(OAc)_2$, H_2O, Δ

2. NaCl, Δ

OH

HgCl

OS, CV 1, 175 (1941)

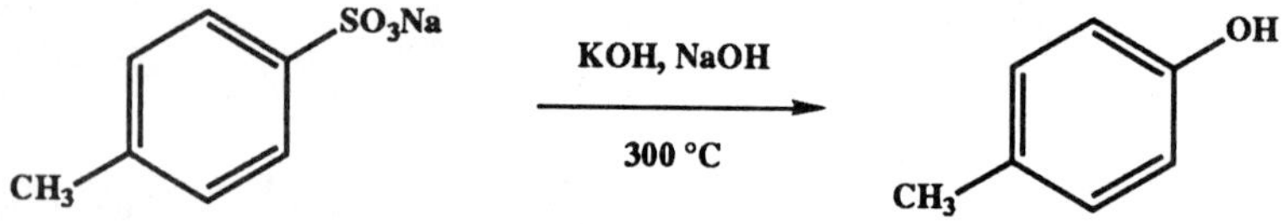

SO_3Na

CH_3

KOH, NaOH

300 °C

OH

CH_3

OS, CV 1, 207 (1941)

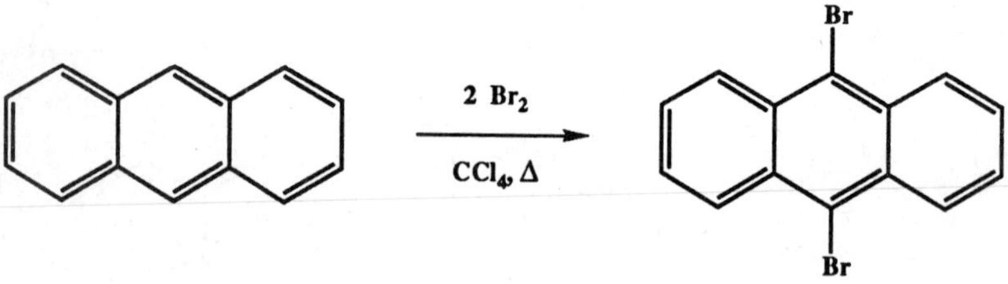

2 Br_2

CCl_4, Δ

Br

Br

OS, CV 1, **214 (1941)**

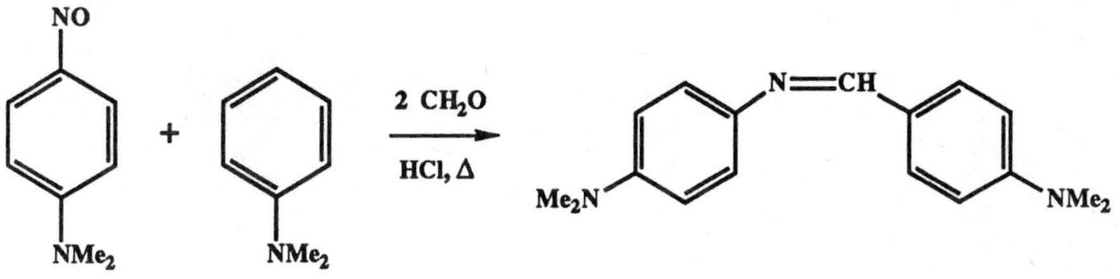

NO, NMe$_2$ + NMe$_2$ $\xrightarrow[\text{HCl, }\Delta]{2\ CH_2O}$ N=CH, Me$_2$N, NMe$_2$

OS, CV 1, **217 (1941)**

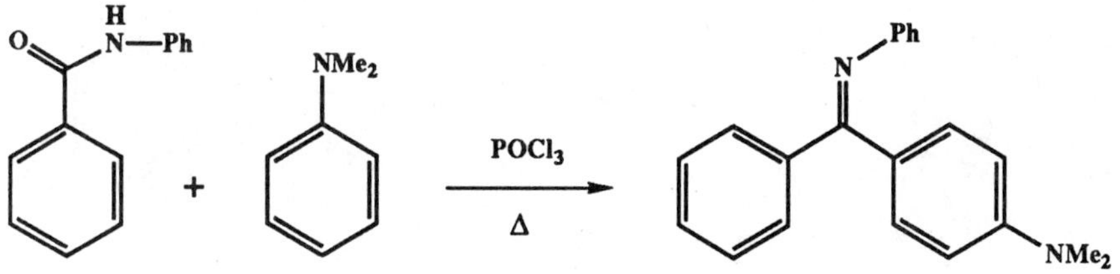

O, H, N—Ph + NMe$_2$ $\xrightarrow[\Delta]{POCl_3}$ Ph, N, NMe$_2$

OS, CV 1, **219 (1941)**

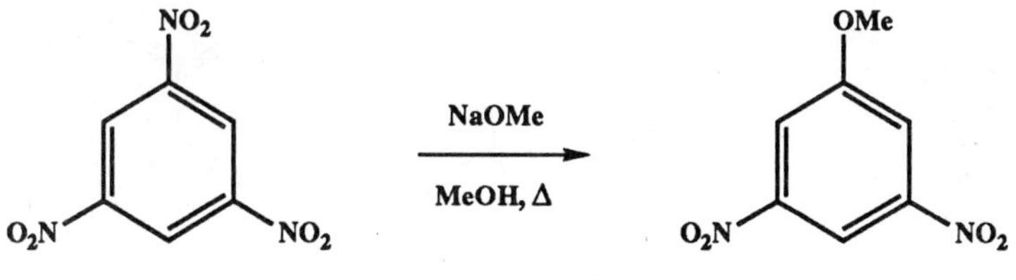

NO$_2$, O$_2$N, NO$_2$ $\xrightarrow[\text{MeOH, }\Delta]{NaOMe}$ OMe, O$_2$N, NO$_2$

OS, CV 1, **220 (1941)**

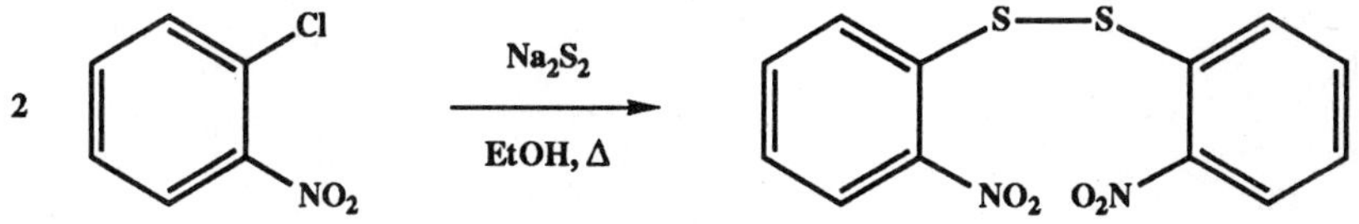

2 Cl, NO$_2$ $\xrightarrow[\text{EtOH, }\Delta]{Na_2S_2}$ S—S, NO$_2$ O$_2$N

***OS*, *CV 1*, 228 (1941)**

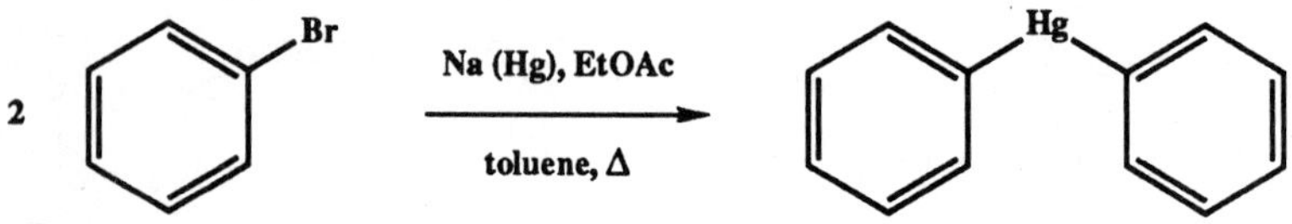

2 PhBr → Na (Hg), EtOAc / toluene, Δ → Ph_2Hg

***OS*, *CV 1*, 231 (1941)**

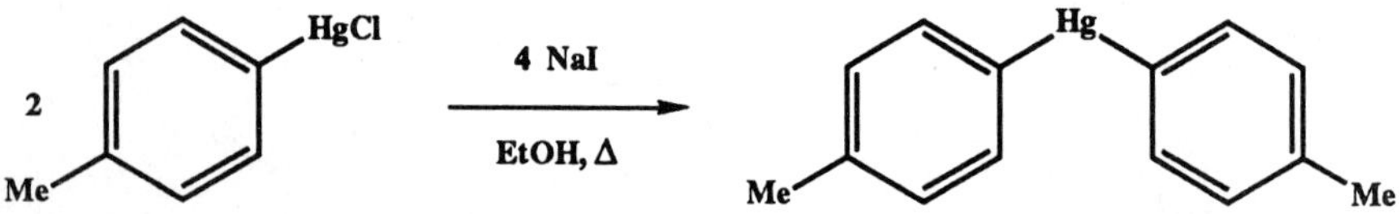

2 4-Me-C_6H_4-HgCl → 4 NaI / EtOH, Δ → $(4\text{-Me-}C_6H_4)_2Hg$

***OS*, *CV 1*, 323 (1941)**

C_6H_6 → I_2, HNO_3 → PhI

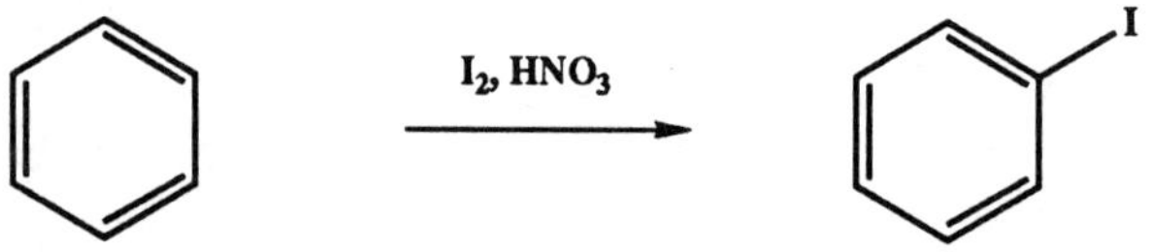

***OS*, *CV 1*, 325 (1941)**

4-ClHg-C_6H_4-CO_2H → 2 I_2 / EtOH, Δ → 4-I-C_6H_4-CO_2H

OS, CV 1, 326 (1941)

2 I_2, $CHCl_3$

OH, HgCl → OH, I

OS, CV 1, 372 (1941)

HNO_3, H_2SO_4, 5 - 15 °C

O, OMe → O, OMe, NO_2

OS, CV 1, 396 (1941)

HNO_3, H_2SO_4

CN → CN, O_2N

OS, CV 1, 408 (1941)

HNO_3, H_2SO_4, Δ

O, O, O → CO_2H, CO_2H, NO_2 + CO_2H, CO_2H, O_2N

***OS, CV 1*, 411 (1941)**

OH

$NaNO_2$, H_2SO_4

aq. NaOH, 0 °C

NO

OH

***OS, CV 1*, 455 (1941)**

CO_2H

H_2N

NH_2

NH_2

H_2O, HCl

Δ

HO

OH

OH

***OS, CV 1*, 488 (1941)**

NH_2

H_2O_3As

1. $ClCH_2CONH_2$, Δ
NaOH, $NaHCO_3$

2. HCl, H_2O
then 2 NaOH

H
N

O

NH_2

Na_2O_3As

***OS, CV 1*, 490 (1941)**

OH

H_3AsO_4

140 °C

OH

H_2O_3As

(as Na salt)

OS, CV 1, 511 (1941)

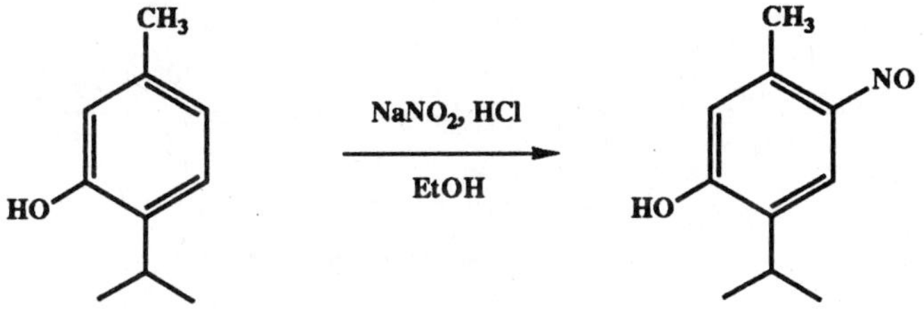

CH_3 HO NaNO$_2$, HCl EtOH CH_3 NO HO

OS, CV 1, 519 (1941)

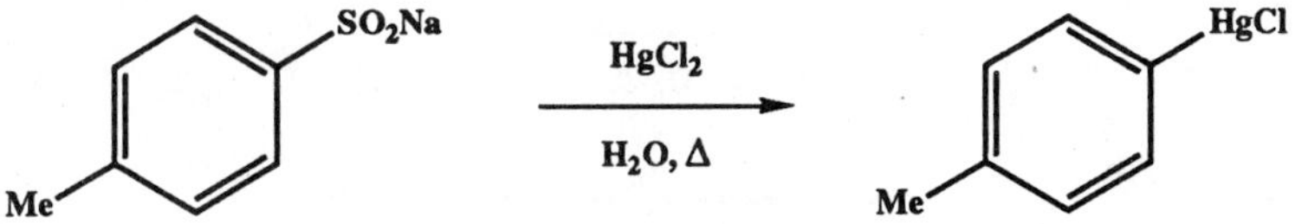

SO_2Na Me $HgCl_2$ H_2O, Δ HgCl Me

OS, CV 1, 537 (1941)

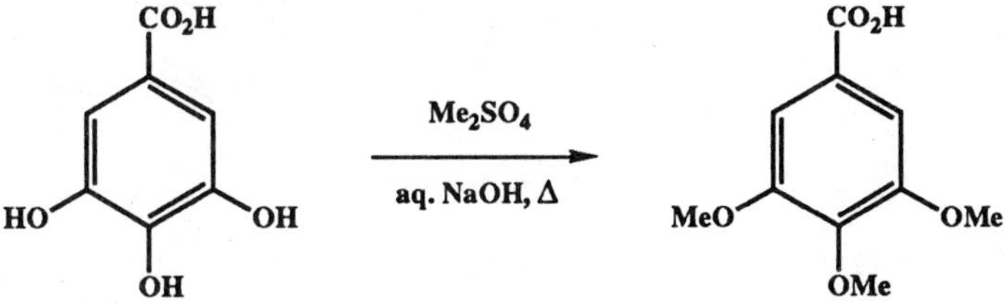

CO_2H HO OH OH Me_2SO_4 aq. NaOH, Δ CO_2H MeO OMe OMe

OS, CV 1, 544 (1941)

$$Ph_2NH + PhI \xrightarrow[\text{nitrobenzene, }\Delta]{K_2CO_3\text{, Cu powder}} Ph_3N$$

OS, CV 1, 550 (1941)

$$3\ C_6H_5\text{—}Br \xrightarrow{Mg,\ Et_2O} 3\ C_6H_5\text{—}MgBr \xrightarrow[Et_2O]{SbCl_3} (C_6H_5)_3Sb$$

OS, CV 2, 15 (1943)

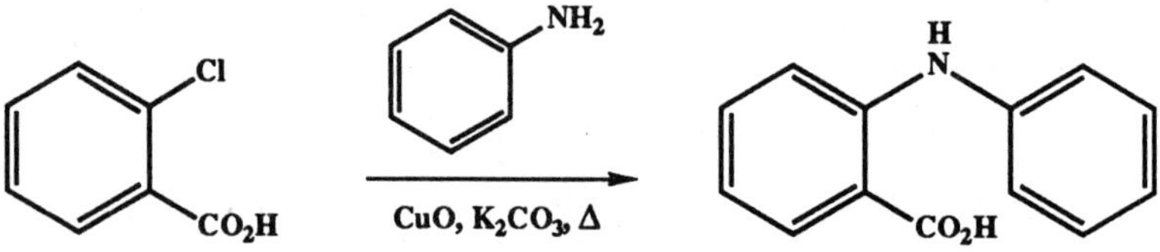

NH$_2$

Cl

CO$_2$H

CuO, K_2CO_3, Δ

H
N

CO$_2$H

OS, CV 2, 35 (1943)

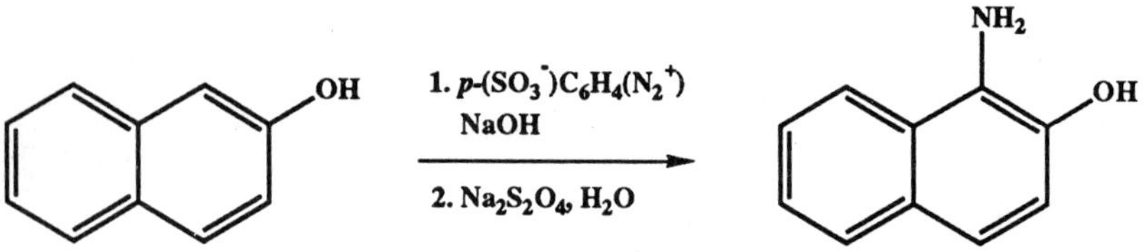

OH

1. p-$(SO_3^-)C_6H_4(N_2^+)$
NaOH

2. $Na_2S_2O_4$, H_2O

NH$_2$

OH

OS, CV 2, 39 (1943)

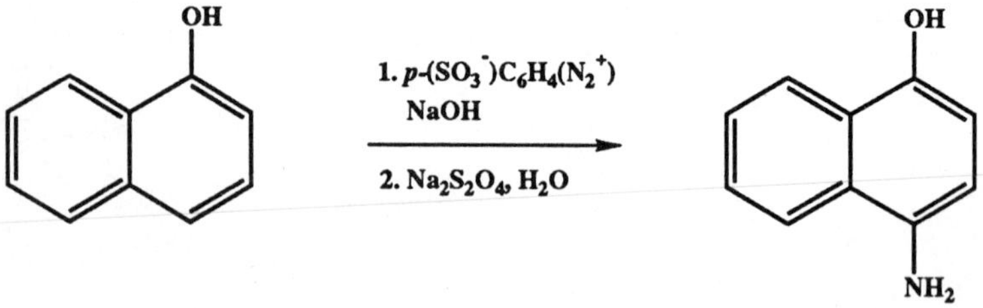

OH

1. p-$(SO_3^-)C_6H_4(N_2^+)$
NaOH

2. $Na_2S_2O_4$, H_2O

OH

NH$_2$

OS, CV 2, 42 (1943)

NO, OH

1. $NaHSO_3$
   aq. NaOH
2. H_2SO_4

NH_2, OH, SO_3H

OS, CV 2, 95 (1943)

CH_3, CH_3, CH_3

Br_2
CCl_4

CH_3, Br, CH_3, CH_3

OS, CV 2, 97 (1943)

OH

2 H_2SO_4
Δ

OH, SO_3H, SO_3H

1. aq. NaOH
2. Br_2
3. H_2SO_4

OH, Br, SO_3H, SO_3H

OS, CV 2, 100 (1943)

HO, OH, CO_2H

Br_2
HOAc

HO, OH, Br, CO_2H

OS, *CV* 2, 128 (1943)

O SO$_3$K

O

NaClO$_3$

aq. HCl, Δ

O Cl

O

OS, *CV* 2, 173 (1943)

OH

NO$_2$

Br_2, HOAc

OH

Br Br

NO$_2$

OS, *CV* 2, 196 (1943)

NH$_2$

NO$_2$

ICl, HOAc

Δ

NH$_2$

I I

NO$_2$

OS, *CV* 2, 221 (1943)

Cl

NO$_2$

NO$_2$

NH_3, NH_4OAc

Δ

NH$_2$

NO$_2$

NO$_2$

OS, *CV* 2, 223 (1943)

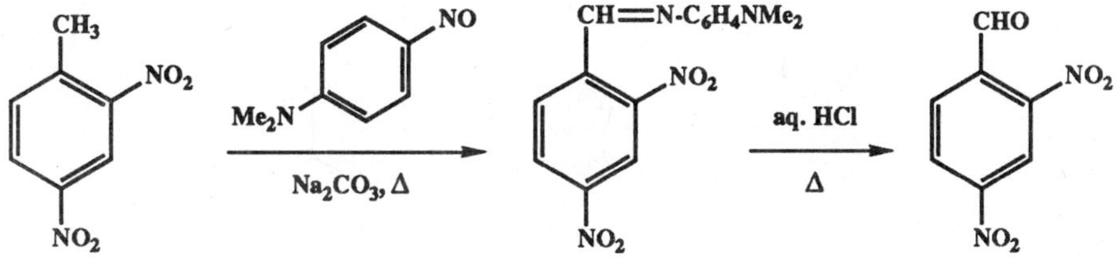

2,4-dinitrotoluene (CH_3, NO_2, NO_2) + Me_2N-C_6H_4-NO → (Na_2CO_3, Δ) → $CH{=}N$-$C_6H_4NMe_2$ (2,4-dinitro) → (aq. HCl, Δ) → CHO, NO_2, NO_2

OS, *CV* 2, 228 (1943)

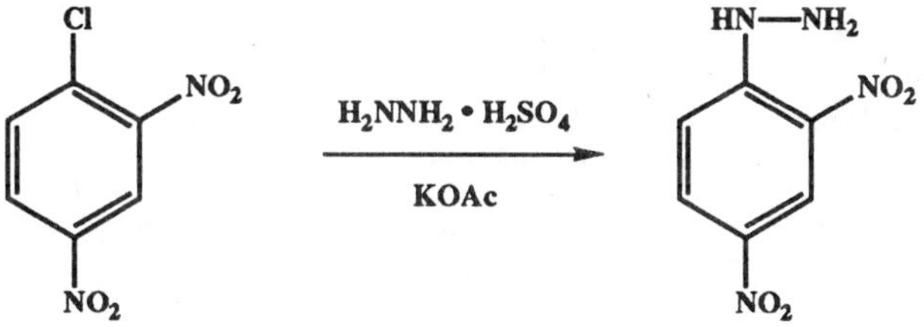

Cl, NO_2, NO_2 → ($H_2NNH_2 \cdot H_2SO_4$, KOAc) → HN—NH_2, NO_2, NO_2

OS, *CV* 2, 240 (1943)

Ph_2SeCl_2 + benzene → ($AlCl_3$) → Ph_3SeCl

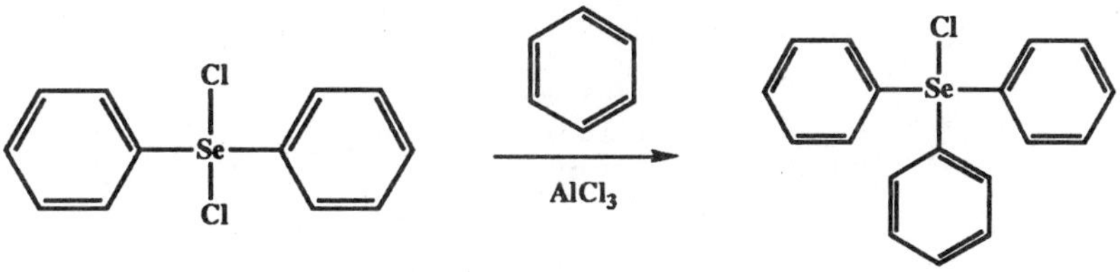

OS, *CV* 2, 242 (1943)

2 benzene → (S_2Cl_2, $AlCl_3$) → Ph–S–Ph

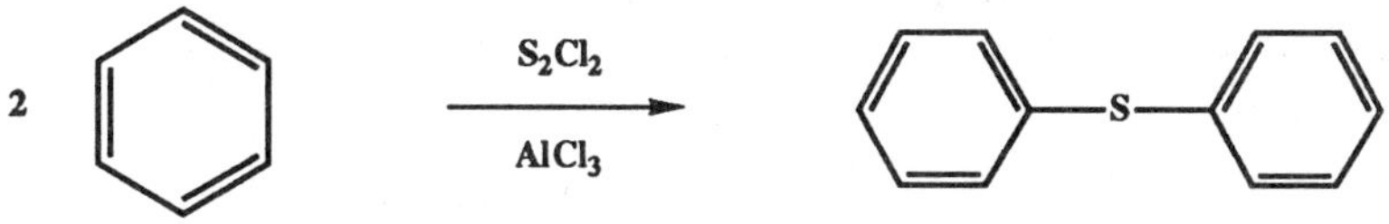

***OS*, *CV* 2, 254 (1943)**

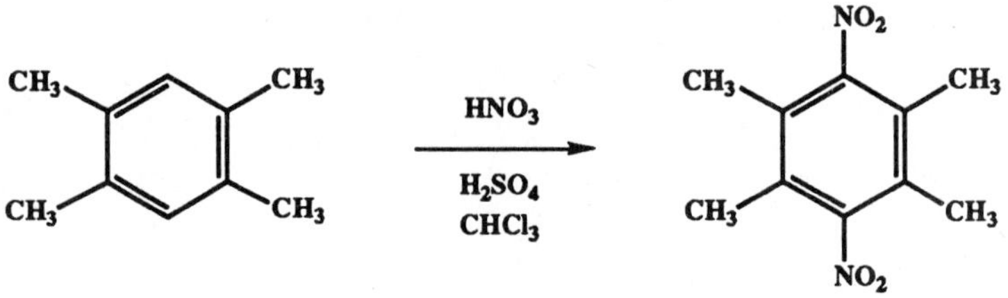

CH_3 CH_3 CH_3 CH_3 — HNO_3 / H_2SO_4, $CHCl_3$ → NO_2 CH_3 CH_3 CH_3 CH_3 NO_2

***OS*, *CV* 2, 290 (1943)**

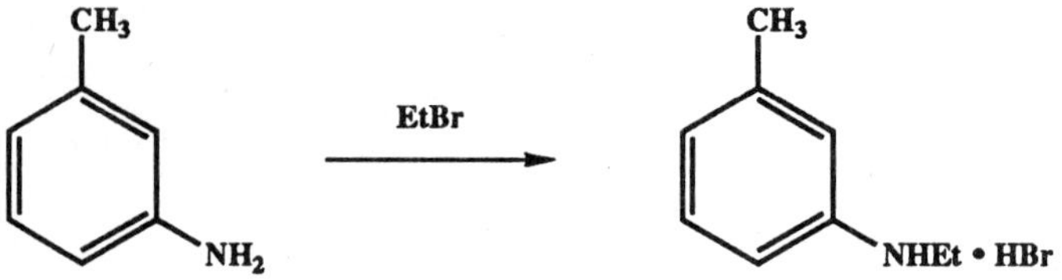

CH_3 NH_2 — EtBr → CH_3 NHEt • HBr

***OS*, *CV* 2, 343 (1943)**

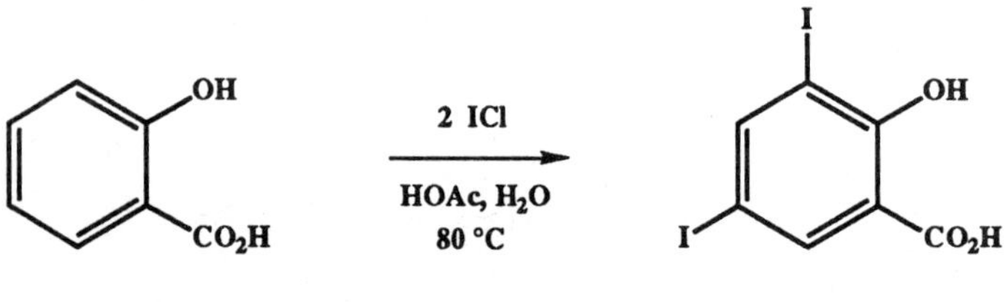

OH CO_2H — 2 ICl / HOAc, H_2O, 80 °C → I OH I CO_2H

***OS*, *CV* 2, 347 (1943)**

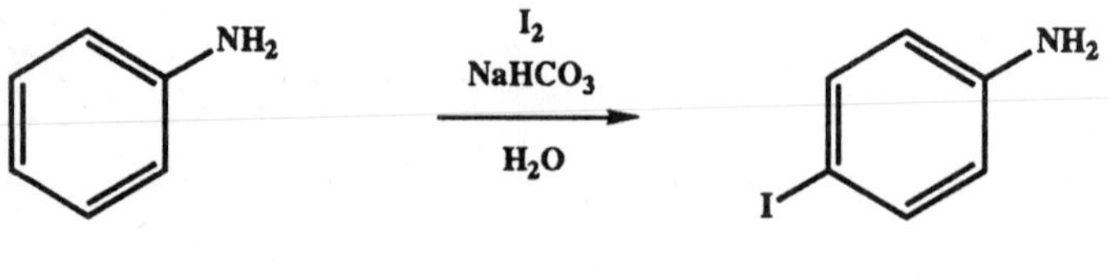

NH_2 — I_2, $NaHCO_3$ / H_2O → NH_2 I

OS, CV 2, 349 (1943)

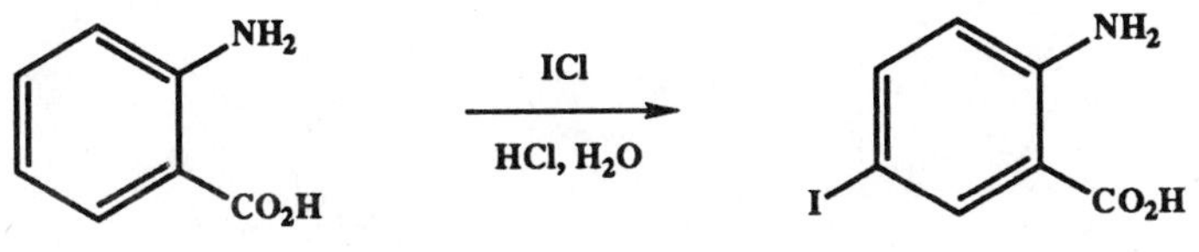

ICl

HCl, H_2O

NH_2 CO_2H I

OS, CV 2, 434 (1943)

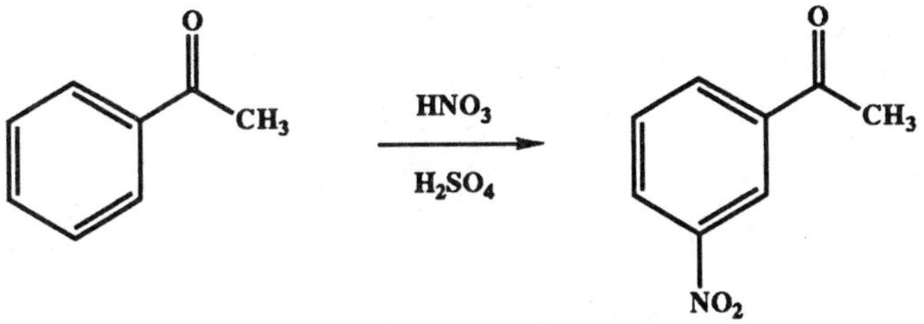

O CH_3

HNO_3

H_2SO_4

NO_2

OS, CV 2, 438 (1943)

H N Ac

HNO_3

H_2SO_4

NO_2

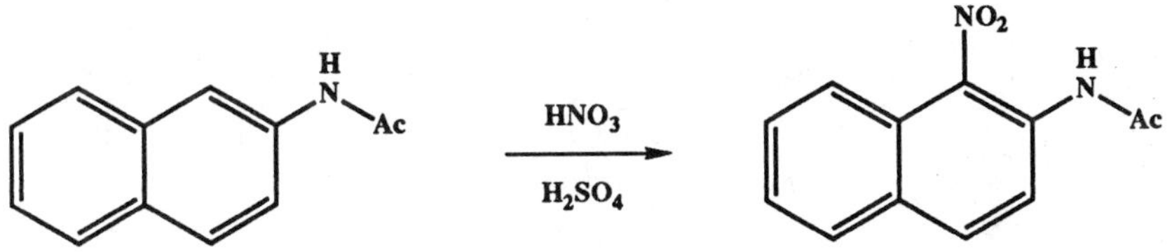

OS, CV 2, 445 (1943)

Cl O_2N

OH

KOH, Δ
Cu (cat.)

OPh O_2N

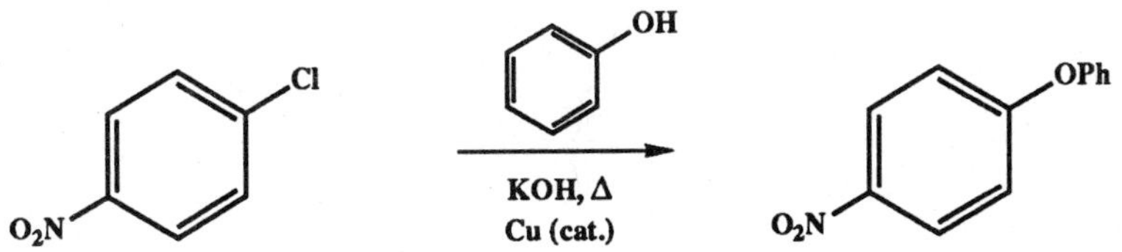

***OS*, *CV* 2, 447 (1943)**

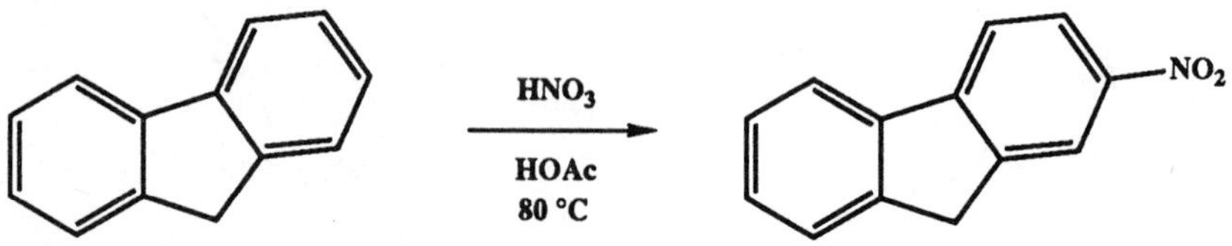

HNO_3, HOAc, 80 °C

-NO_2

***OS*, *CV* 2, 449 (1943)**

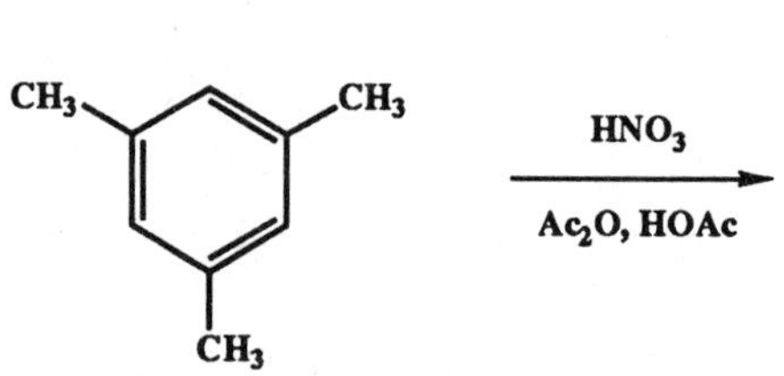

CH_3 CH_3 CH_3

HNO_3, Ac_2O, HOAc

NO_2 CH_3 CH_3 CH_3

***OS*, *CV* 2, 451 (1943)**

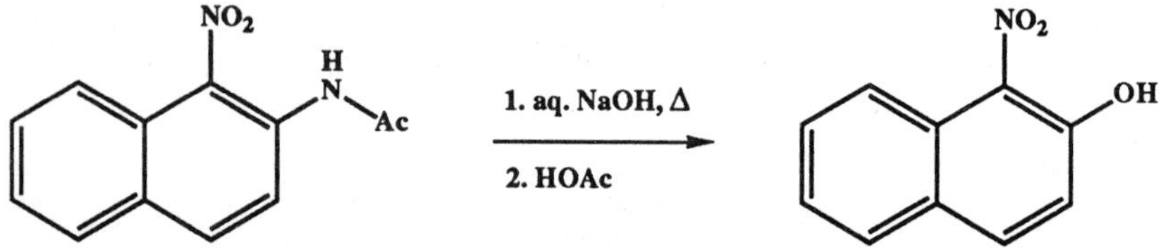

NO_2 H N Ac

1. aq. NaOH, Δ
2. HOAc

NO_2 OH

***OS*, *CV* 2, 453 (1943)**

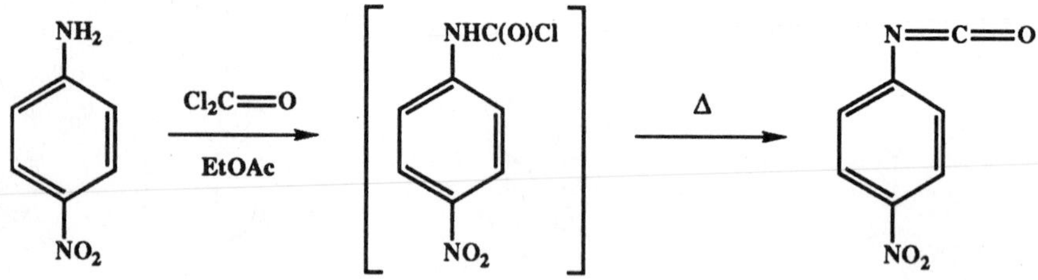

NH_2 NO_2

$Cl_2C{=}O$, EtOAc

[NHC(O)Cl NO_2]

Δ

N=C=O NO_2

OS, CV 2, **459 (1943)**

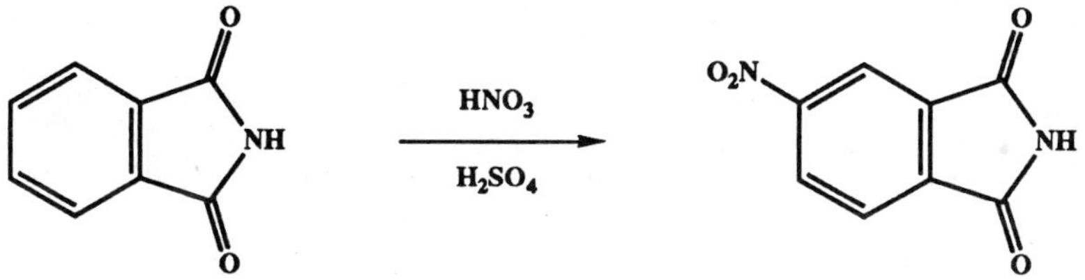

HNO_3 / H_2SO_4; O_2N; NH; O

OS, CV 2, **482 (1943)**

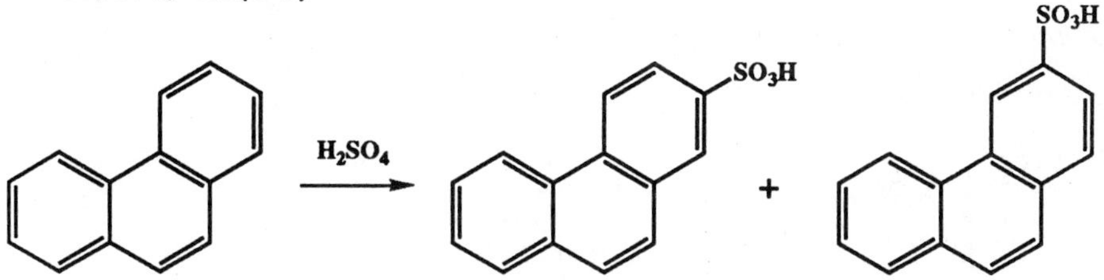

H_2SO_4; SO_3H + SO_3H

OS, CV 2, **539 (1943)**

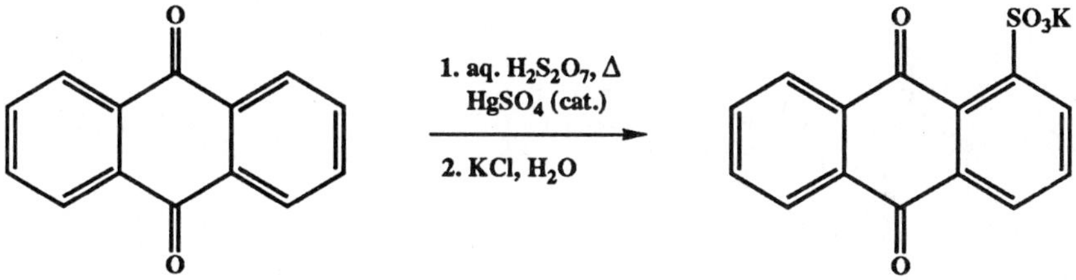

1. aq. $H_2S_2O_7$, Δ, $HgSO_4$ (cat.)
2. KCl, H_2O

O; SO_3K

OS, CV 2, **574 (1943)**

NMe_2 → Br_2, NH_4SCN / HOAc → NMe_2; NCS

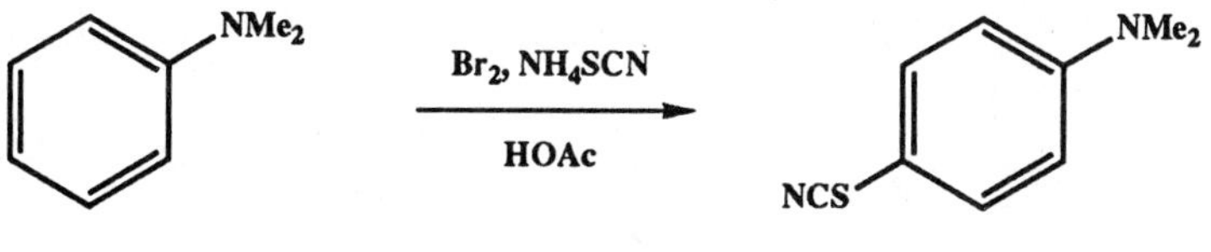

OS, CV 2, 592 (1943)

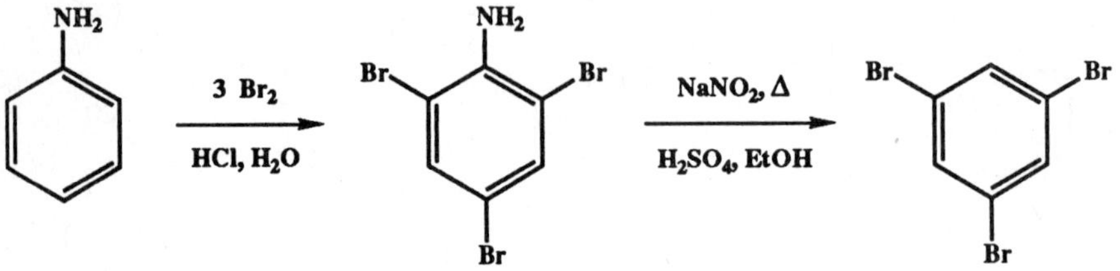

3 Br_2 / HCl, H_2O; $NaNO_2$, Δ / H_2SO_4, EtOH

NH_2 — Br, Br, Br, NH_2 — Br, Br, Br

OS, CV 2, 619 (1943)

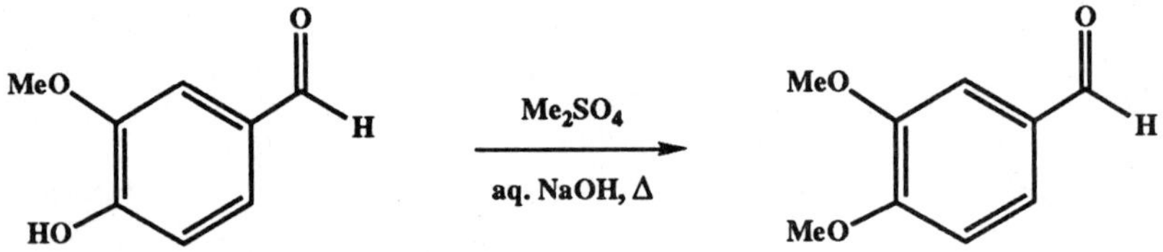

Me_2SO_4 / aq. NaOH, Δ

O, MeO, H, HO — O, MeO, H, MeO

OS, CV 3, 53 (1955)

1. $(NH_4)_2CO_3$ phenol, Δ / 2. NaOH, H_2O

Cl, N — NH_2, N

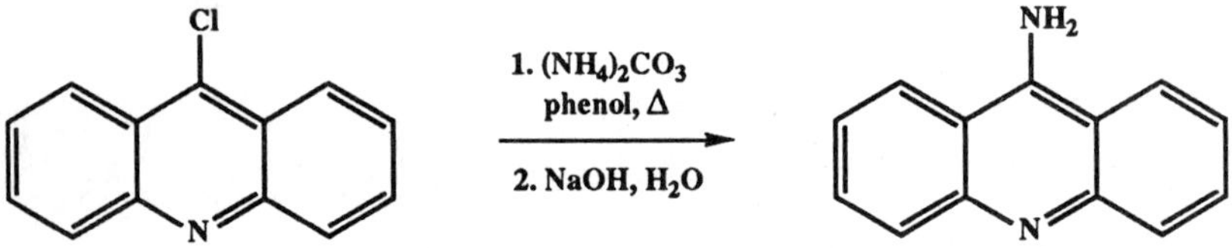

OS, CV 3, 78 (1955)

$ZnCl_2$, NH_4OH / 400 psi, 195 °C

CO_2H, OH — CO_2H, NH_2

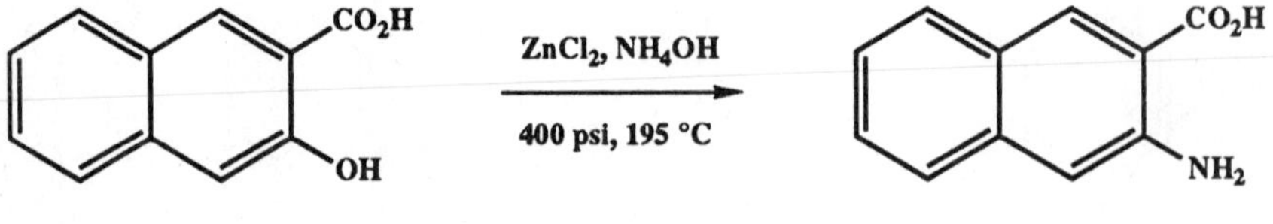

OS, CV 3, 86 (1955)

O_2N–C6H4–Cl → (Na_2S, H_2O, Δ) → H_2N–C6H4–SNa

OS, CV 3, 132 (1955)

2-naphthol → (Br_2, HOAc) → 1,6-dibromo-2-naphthol (Br, OH) → (Sn, Δ) → 6-bromo-2-naphthol (Br, OH)

OS, CV 3, 134 (1955)

phenanthrene → (Br_2, CCl_4, Δ) → 9-bromophenanthrene (Br)

OS, CV 3, 136 (1955)

2-aminopyridine (N, NH_2) → (1. 48% HBr; 2. $NaNO_2$, Br_2, H_2O; 3. NaOH) → 2-bromopyridine (N, Br)

OS, *CV 3*, 138 (1955)

CH_3 CH_3 — Fe, Br_2 / 0 °C → Br CH_3 CH_3

OS, *CV 3*, 140 (1955)

OH NO_2 — *n*-BuBr, K_2CO_3 / acetone, Δ → O-*n*-Bu NO_2

OS, *CV 3*, 239 (1955)

AcHN SO_2Na — Cl O_2N / $(MeOCH_2CH_2)_2O$ → AcHN SO_2 NO_2

OS, *CV 3*, 262 (1955)

SO_2NH_2 NH_2 — 2 HX, 2 H_2O_2 / H_2O → SO_2NH_2 X X NH_2 — H_2SO_4 / H_2O, Δ → X X NH_2

X = Cl, Br

OS, CV 3, 267 (1955)

CO_2Et / OH → 2 SO_2Cl_2 → CO_2Et / Cl / Cl / OH → aq. KOH, Δ then HCl → CO_2H / Cl / Cl / OH

OS, CV 3, 272 (1955)

OH / Cl / N → $POCl_3$, 135 - 140 °C → Cl / Cl / N

OS, CV 3, 288 (1955)

CO_2H → fuming H_2SO_4, Δ → CO_2H / HO_3S / SO_3H → 1. $BaCO_3$, H_2O 2. NaOH, KOH, Δ, then HCl → CO_2H / HO / OH

OS, CV 3, 293 (1955)

O_2N / NO_2 → KCN, aq. MeOH → CN / MeO / NO_2 → KOH, MeOH, Δ → CN / MeO / OMe

OS, *CV 3*, 307 (1955)

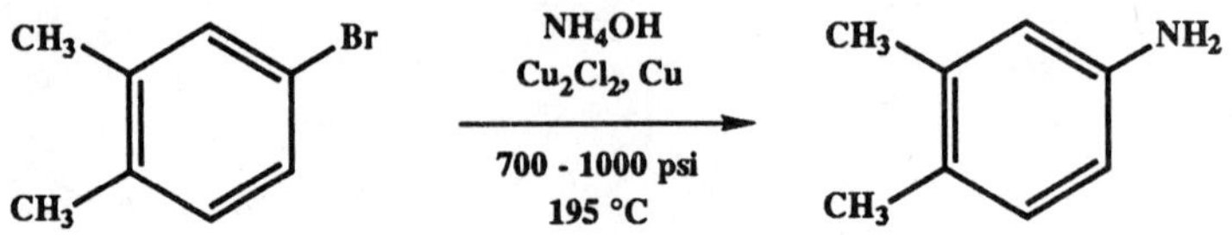
CH3
CH3
Br
NH4OH
Cu2Cl2, Cu
700 - 1000 psi
195 °C
CH3
CH3
NH2

OS, *CV 3*, 337 (1955)

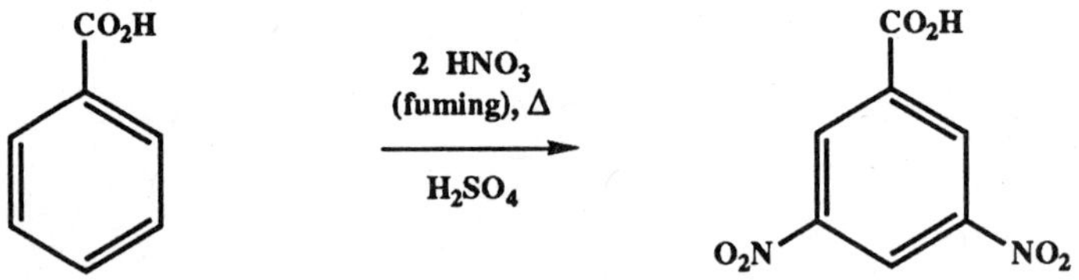
CO2H
2 HNO3
(fuming), Δ
H2SO4
CO2H
O2N
NO2

OS, *CV 3*, 418 (1955)

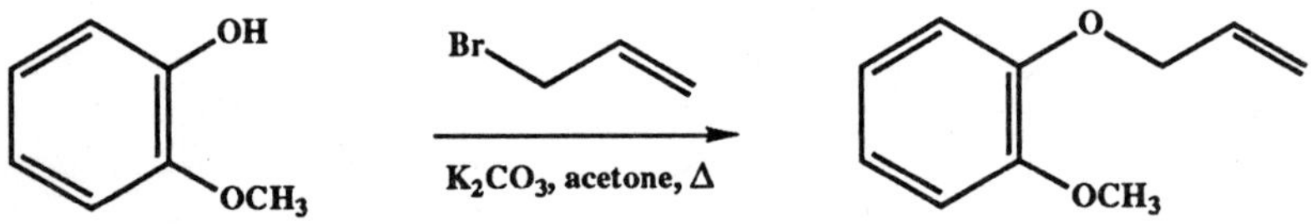
OH
OCH3
Br
K2CO3, acetone, Δ
O
OCH3

OS, *CV 3*, 452 (1955)

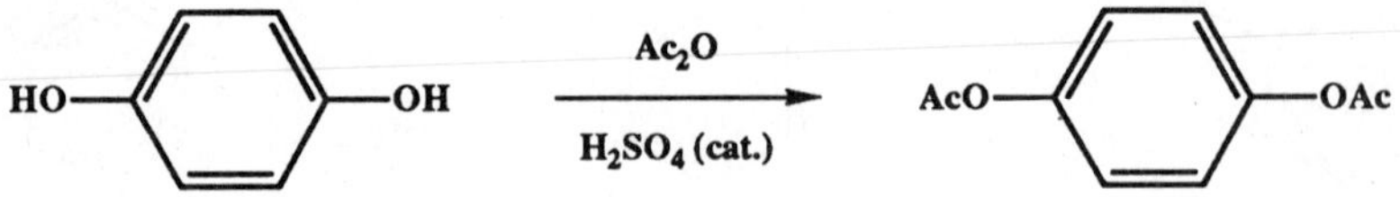
HO
OH
Ac2O
H2SO4 (cat.)
AcO
OAc

OS, CV 3, 564 (1955)

$CH(OCH_3)_2$, NH_2 — 1. aq. H_2SO_4; 2. $NaNO_2$, H_2SO_4; 3. H_2O, Δ → O, H, OH — $(CH_3)_2SO_4$; aq. NaOH; 40 - 50 °C → O, H, OCH_3

OS, CV 3, 566 (1955)

OCH_3, OH — 1. KOH, Δ; 2. Cu (cat.), C_6H_5Br, Δ → OCH_3, O

OS, CV 3, 573 (1955)

O, SO_3Na, O — CH_3NH_2; $NaClO_3$, H_2O; 130 - 135 °C → O, $NHCH_3$, O

OS, CV 3, 575 (1955)

O, $NHCH_3$, O — Br_2, Δ; pyridine → O, $NHCH_3$, O, Br

OS, CV 3, 590 (1955)

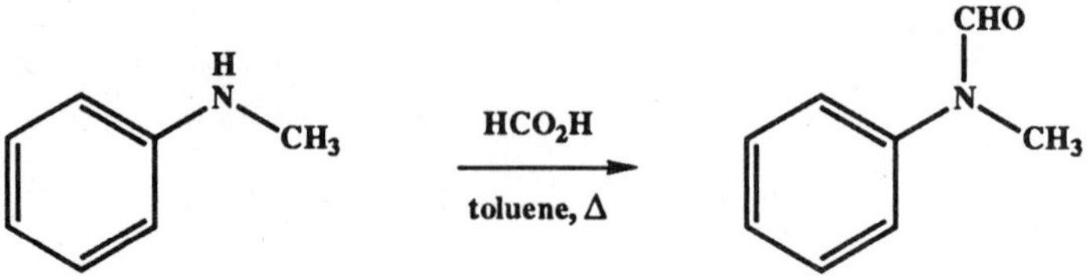
H
N
CH_3
HCO_2H
toluene, Δ
CHO
N
CH_3

OS, CV 3, 644 (1955)

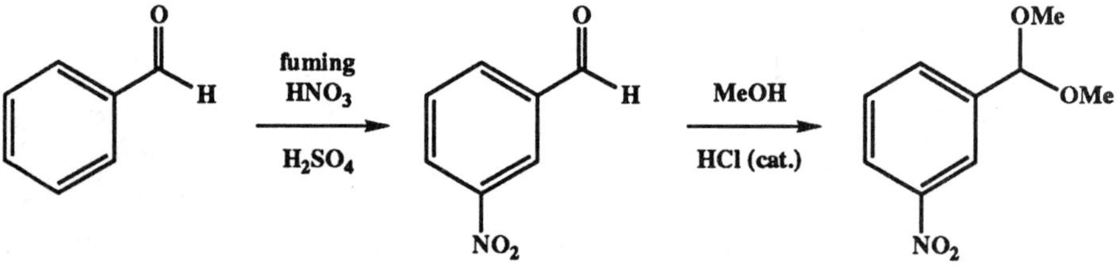
O
H
fuming
HNO_3
H_2SO_4
O
H
NO_2
MeOH
HCl (cat.)
OMe
OMe
NO_2

OS, CV 3, 653 (1955)

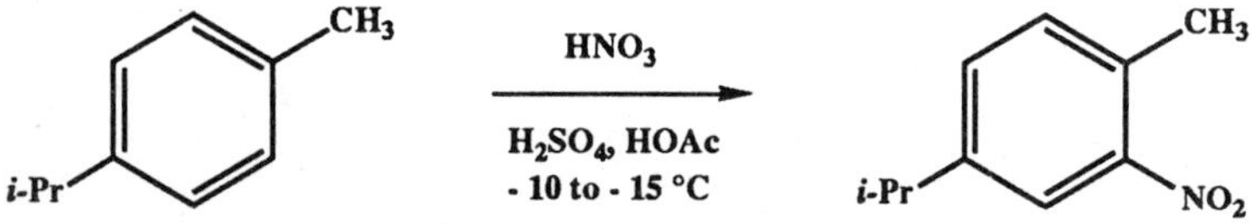
CH_3
i-Pr
HNO_3
H_2SO_4, HOAc
- 10 to - 15 °C
CH_3
i-Pr
NO_2

OS, CV 3, 658 (1955)

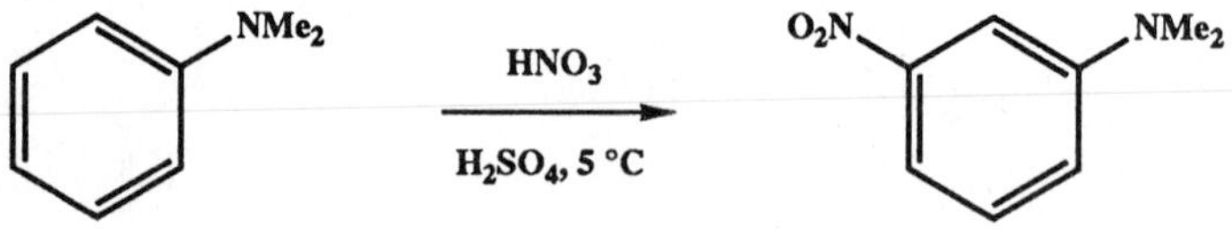
NMe_2
HNO_3
H_2SO_4, 5 °C
O_2N
NMe_2

OS, CV 3, 661 (1955)

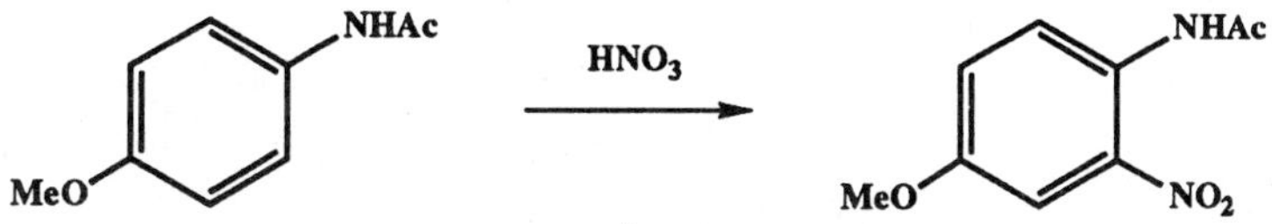

NHAc, MeO — HNO_3 → NHAc, NO_2, MeO

OS, CV 3, 664 (1955)

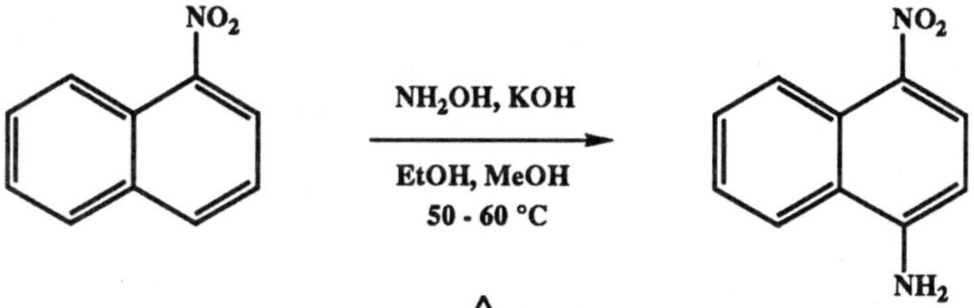

NO_2 — NH_2OH, KOH / EtOH, MeOH / 50 - 60 °C → NO_2, NH_2

OS, CV 3, 667 (1955)

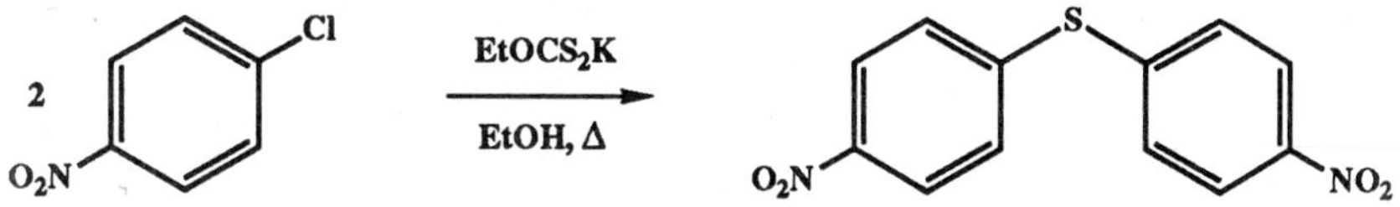

2 Cl, O_2N — $EtOCS_2K$ / EtOH, Δ → S, O_2N, NO_2

OS, CV 3, 753 (1955)

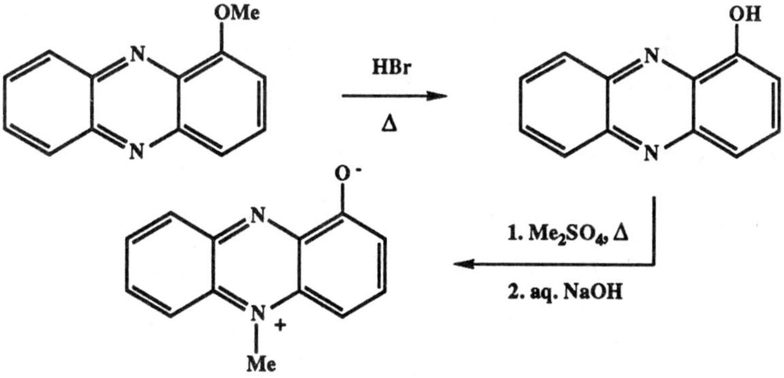

OMe, N, N — HBr / Δ → OH, N, N — 1. Me_2SO_4, Δ / 2. aq. NaOH → O^-, N, N^+, Me

OS, CV 3, 771 (1955)

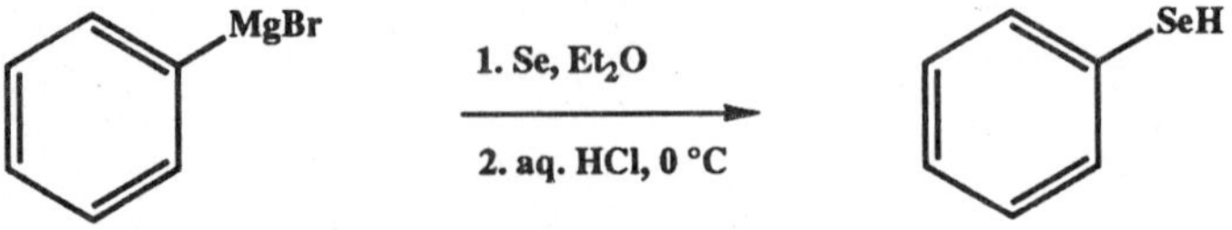

MgBr → SeH

1. Se, Et_2O

2. aq. HCl, 0 °C

OS, CV 3, 796 (1955)

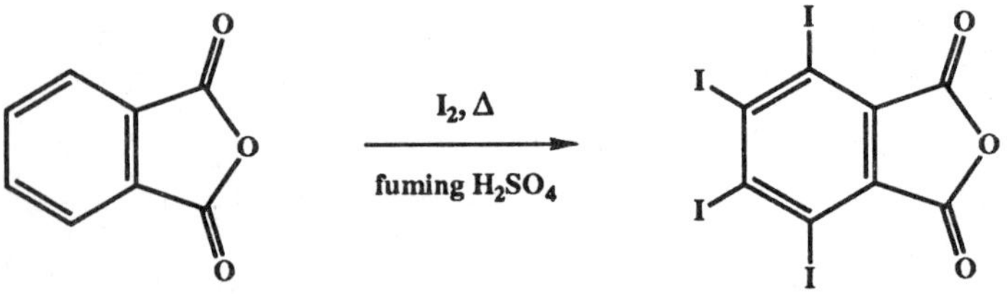

I_2, Δ

fuming H_2SO_4

OS, CV 3, 824 (1955)

CH_3, NH_2 → H_2SO_4, H_2O → [CH_3, NH_3HSO_4] → Δ → HO_3S, CH_3, NH_2

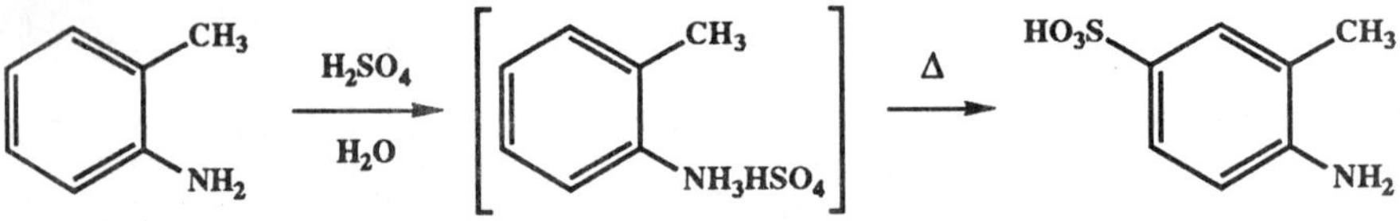

OS, CV 3, 837 (1955)

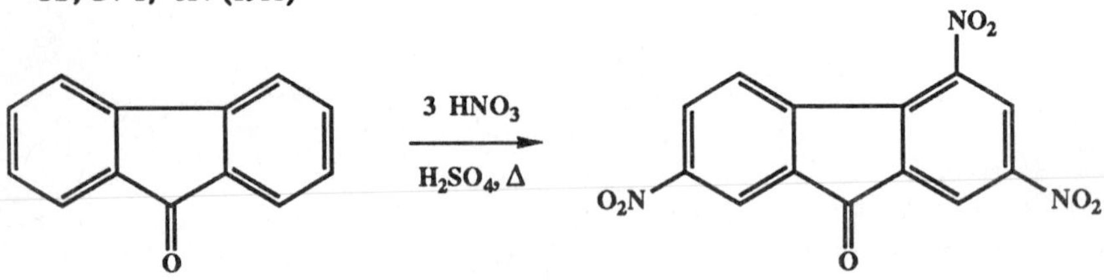

3 HNO_3

H_2SO_4, Δ

NO_2, O_2N, NO_2

OS, *CV* 4, 15 (1963)

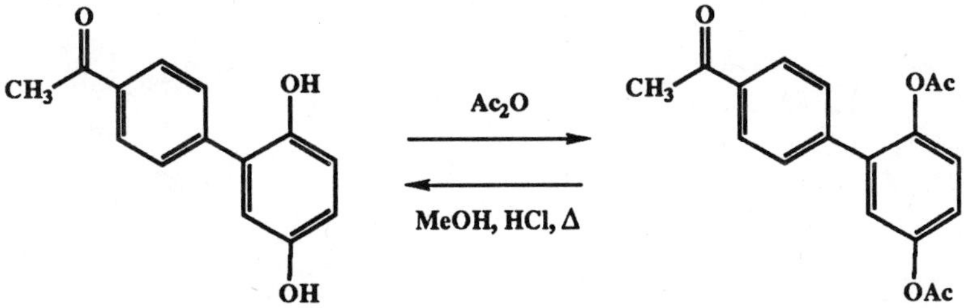
O
CH3
OH
OH
Ac2O
MeOH, HCl, Δ
O
CH3
OAc
OAc

OS, *CV* 4, 34 (1963)

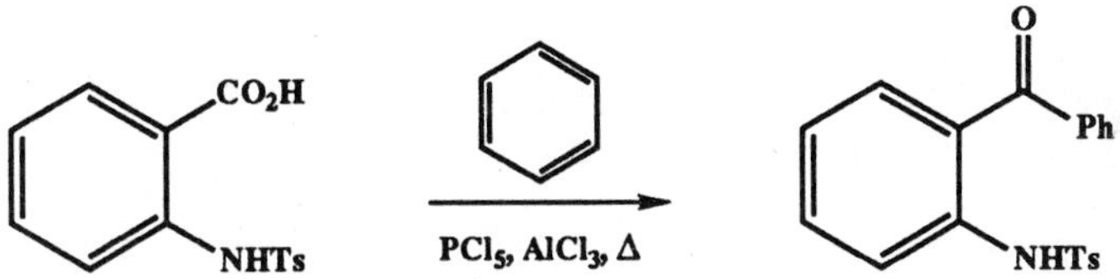
CO2H
NHTs
PCl5, AlCl3, Δ
O
Ph
NHTs

OS, *CV* 4, 42 (1963)

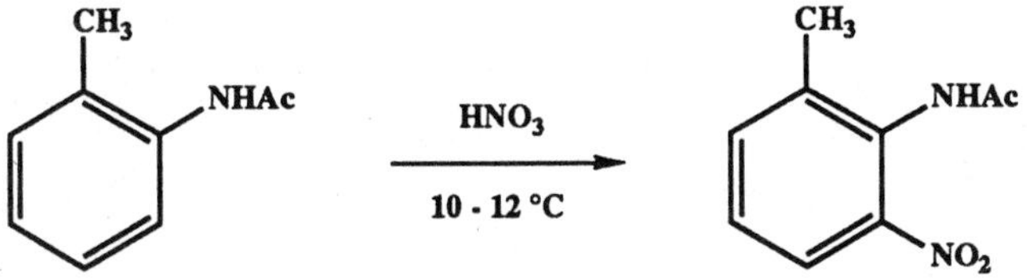
CH3
NHAc
HNO3
10 - 12 °C
CH3
NHAc
NO2

OS, *CV* 4, 49 (1963)

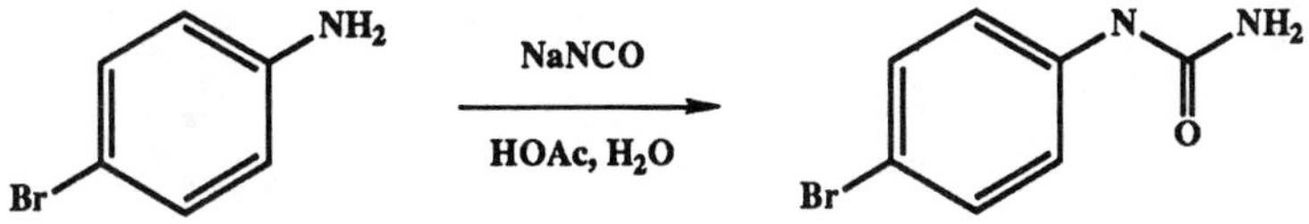
NH2
Br
NaNCO
HOAc, H2O
H
N
NH2
O
Br

OS, CV 4, 52 (1963)

EtO–C6H4–NH_2 • HCl → (O=C(H_2N)(NH_2); HOAc, HCl, H_2O, Δ) → EtO–C6H4–NH–C(O)–NH_2

OS, CV 4, 68 (1963)

3 PhMgBr → (1. 3 $B(OMe)_3$, Et_2O, - 60 °C; 2. H_2O, H_2SO_4, 0 °C) → 3 $PhB(OH)_2$ → (110 °C) → (PhBO)3 (triphenylboroxine)

OS, CV 4, 91 (1963)

2-aminopyridine (NH_2) → ($PhCH_2OH$; aq. KOH, Δ) → 2-($NHCH_2Ph$)pyridine

OS, CV 4, 114 (1963)

CH_3–C6H4–NO_2 → (Fe, Br_2; 80 °C) → 2-Br-1-CH_3-4-NO_2-benzene

OS, CV 4, **180 (1963)**

OS, CV 4, **247 (1963)**

OS, CV 4, **256 (1963)**

OS, CV 4, **283 (1963)**

OS, CV 4, 336 (1963)

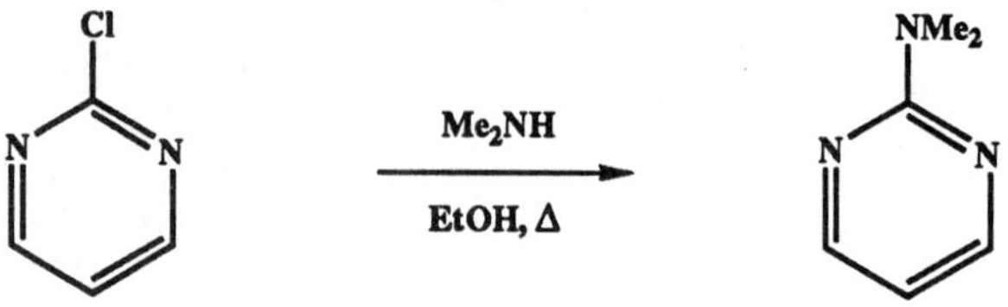
Cl
N
N
Me2NH
EtOH, Δ
NMe2

OS, CV 4, 364 (1963)

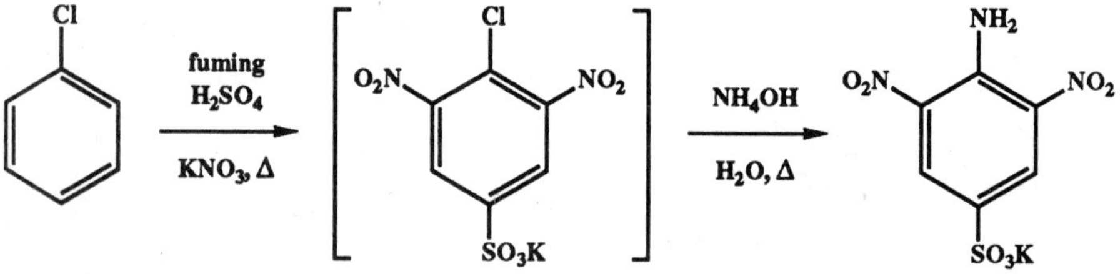
Cl
fuming
H2SO4
KNO3, Δ
O2N
NO2
SO3K
NH4OH
H2O, Δ
NH2

OS, CV 4, 380 (1963)

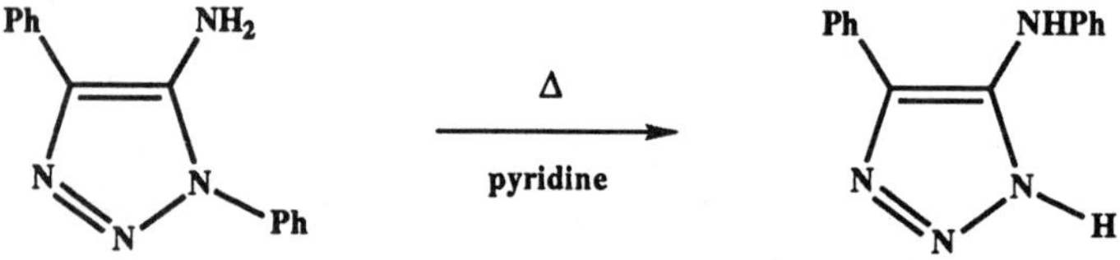
Ph
NH2
N
Ph
Δ
pyridine
NHPh
H

OS, CV 4, 420 (1963)

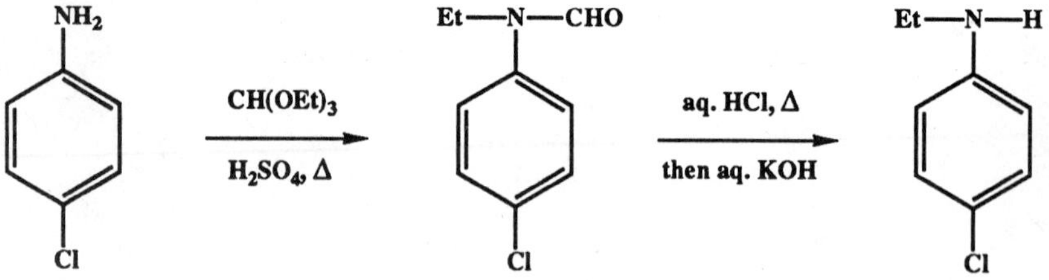
NH2
Cl
CH(OEt)3
H2SO4, Δ
Et—N—CHO
aq. HCl, Δ
then aq. KOH
Et—N—H

OS, *CV* 4, 547 (1963)

MeO, MeO (1,2-dimethoxybenzene) → I_2, CF_3CO_2Ag / $CHCl_3$ → MeO, MeO, I (4-iodo-1,2-dimethoxybenzene)

OS, *CV* 4, 654 (1963)

CH_3, +N, O- (3-methylpyridine N-oxide) → fuming HNO_3 / H_2SO_4, 100 °C → NO_2, CH_3, +N, O-

OS, *CV* 4, 711 (1963)

anthracene → HNO_3, HCl / HOAc → NO_2, Cl → NaOH / H_2O → NO_2

OS, *CV* 4, 722 (1963)

CH═CH—CHO → HNO_3 / Ac_2O, HOAc → CH═CH—CHO, NO_2

OS, *CV* 4, 735 (1963)

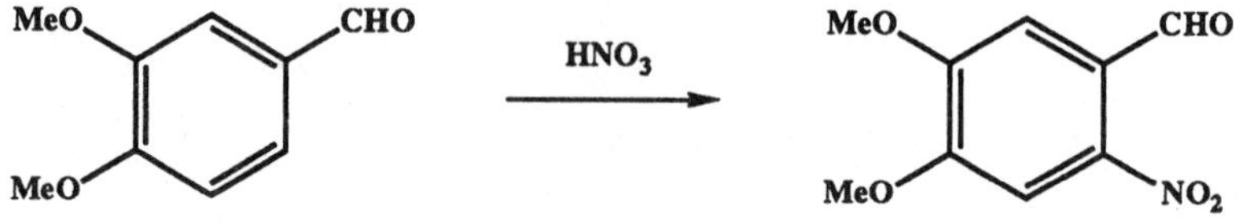

MeO, MeO, CHO — HNO_3 → MeO, MeO, CHO, NO_2

OS, *CV* 4, 784 (1963)

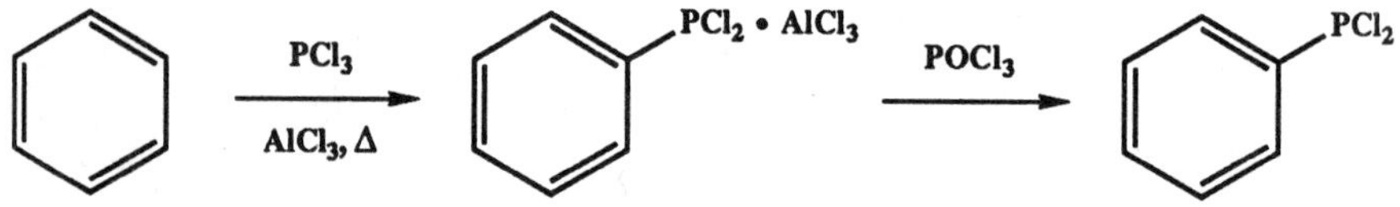

PCl_3 / $AlCl_3$, Δ → $PCl_2 \cdot AlCl_3$ — $POCl_3$ → PCl_2

OS, *CV* 4, 836 (1963)

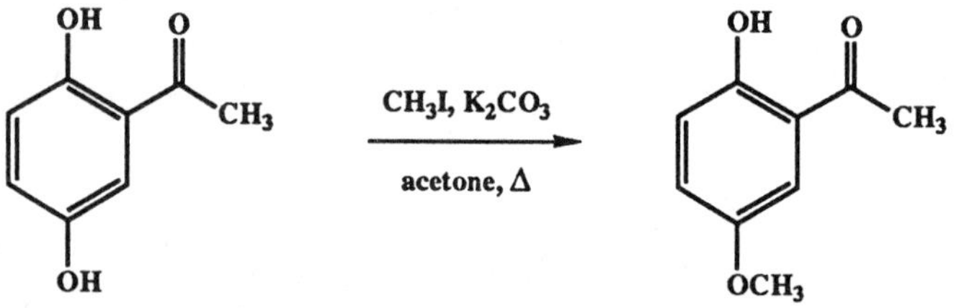

OH, O, CH_3, OH — CH_3I, K_2CO_3 / acetone, Δ → OH, O, CH_3, OCH_3

OS, *CV* 4, 872 (1963)

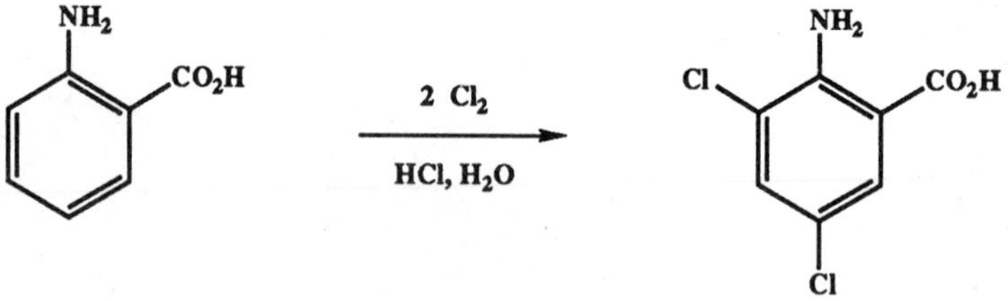

NH_2, CO_2H — 2 Cl_2 / HCl, H_2O → NH_2, Cl, CO_2H, Cl

OS, *CV 4*, **910 (1963)**

3 C_6H_5Cl —(Na, $AsCl_3$; benzene, Δ)→ $(C_6H_5)_3As$ —(H_2O_2; acetone)→ $(C_6H_5)_3AsO$ —(1. C_6H_5MgBr, Δ; 2. HCl, Δ)→ $(C_6H_5)_4AsCl$ • HCl

OS, *CV 4*, **947 (1963)**

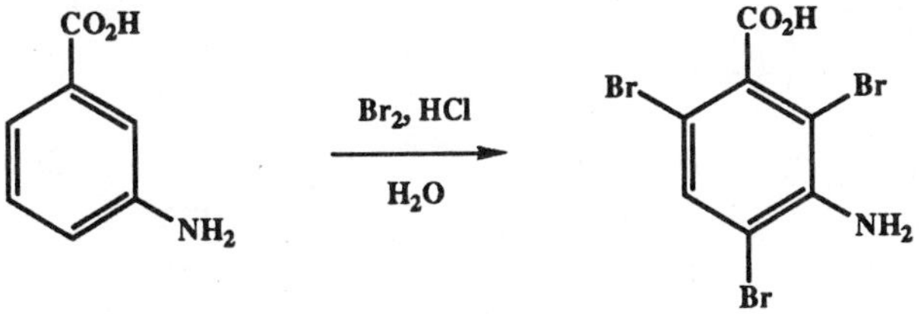

CO₂H / NH₂ —(Br_2, HCl; H_2O)→ Br, Br, CO₂H, NH₂, Br

OS, *CV 5*, **107 (1973)**

Br / Br —(2 $CuS(CH_2)_3CH_3$; quinoline, pyridine, Δ)→ $S(CH_2)_3CH_3$ / $S(CH_2)_3CH_3$

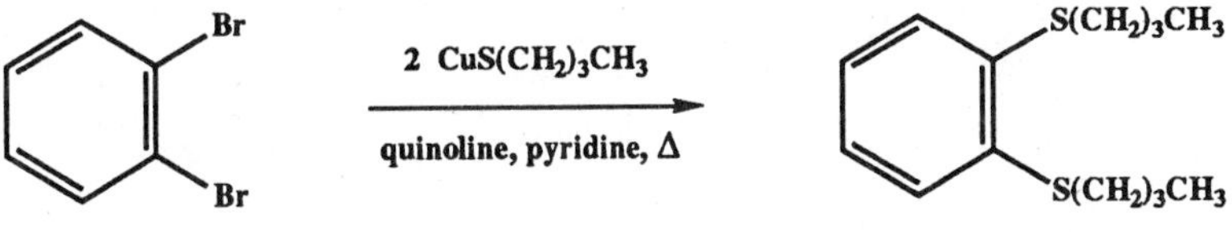

OS, *CV 5*, **117 (1973)**

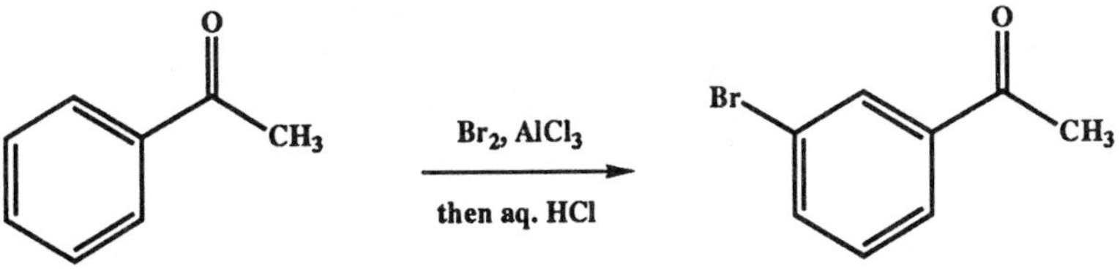

O / CH_3 —(Br_2, $AlCl_3$; then aq. HCl)→ Br / O / CH_3

OS, CV 5, 142 (1973)

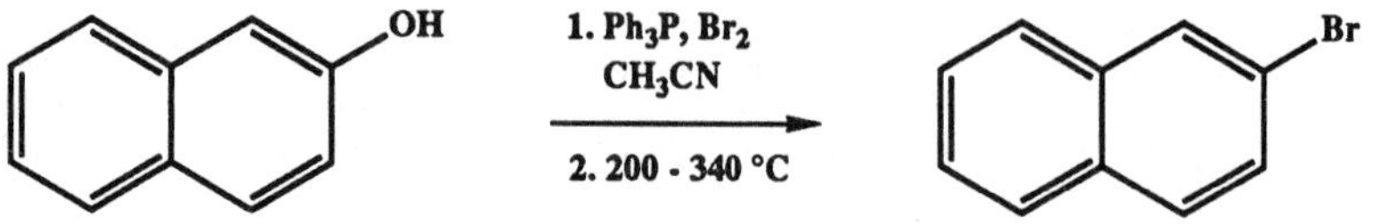

OH

1. Ph_3P, Br_2
CH_3CN
2. 200 - 340 °C

Br

OS, CV 5, 147 (1973)

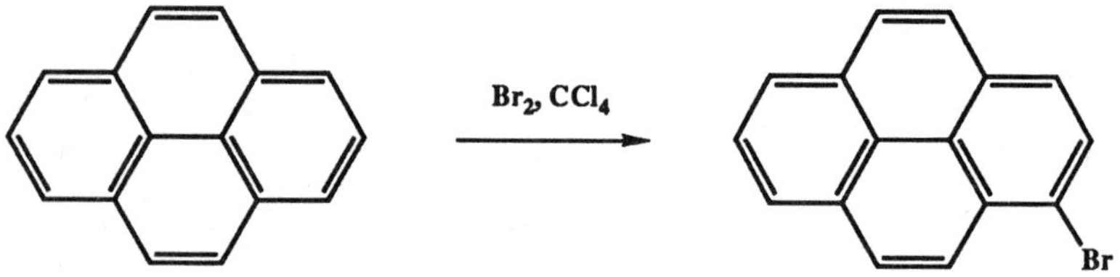

Br_2, CCl_4

Br

OS, CV 5, 206 (1973)

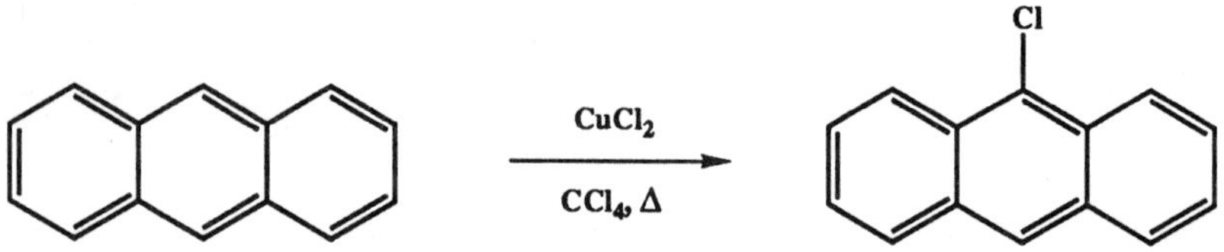

Cl

$CuCl_2$
CCl_4, Δ

OS, CV 5, 223 (1973)

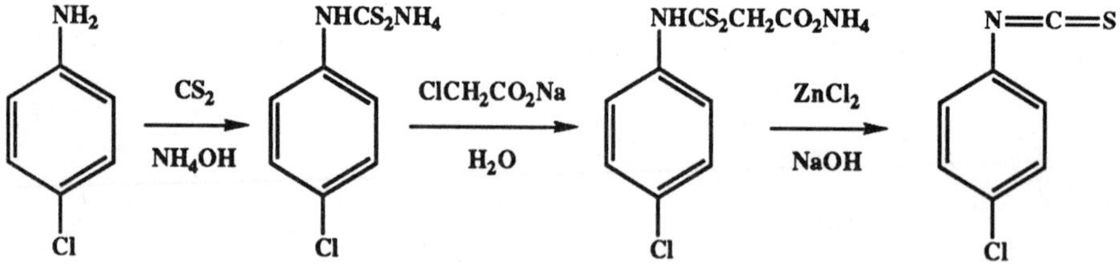

NH_2 / Cl

CS_2
NH_4OH

$NHCS_2NH_4$ / Cl

$ClCH_2CO_2Na$
H_2O

$NHCS_2CH_2CO_2NH_4$ / Cl

$ZnCl_2$
NaOH

N=C=S / Cl

OS, CV 5, 346 (1973)

Br_2 / HOAc; HNO_3 / H_2SO_4

Br, NO_2, N, NH_2

OS, CV 5, 396 (1973)

O, H, SF_3, Δ, CHF_2, S, F

OS, CV 5, 412 (1973)

MeO, OMe, BBr_3 / CH_2Cl_2, - 80 °C, HO, OH

OS, CV 5, 474 (1973)

Cl, O_2N, NO_2, $PhCH_2SH$ / pyridine / MeOH, Δ, SCH_2Ph, SO_2Cl_2 / ethylene chloride, SCl

OS, CV 5, **478 (1973)**

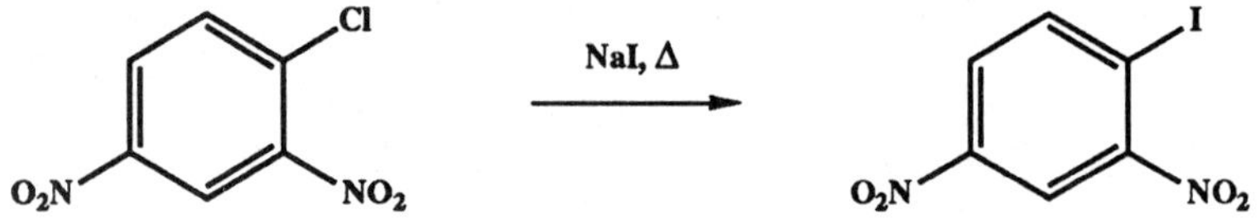
Cl
NaI, Δ
I
O2N
NO2
O2N
NO2

OS, CV 5, **480 (1973)**

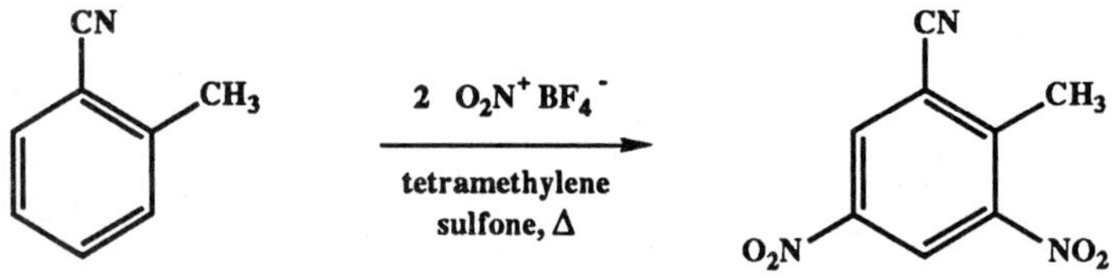
CN
CH3
2 O2N+ BF4-
tetramethylene
sulfone, Δ
CN
CH3
O2N
NO2

OS, CV 5, **496 (1973)**

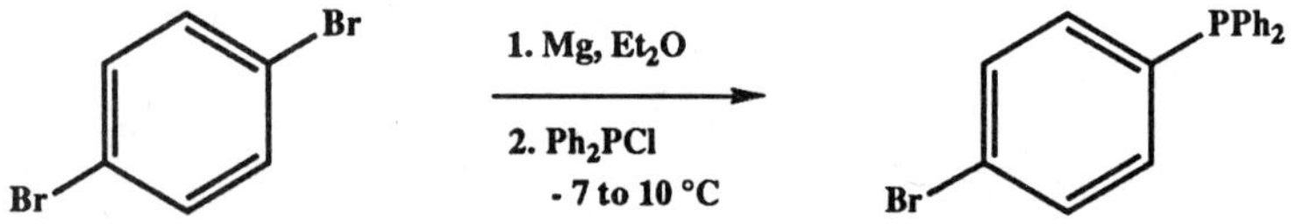
Br
Br
1. Mg, Et2O
2. Ph2PCl
- 7 to 10 °C
PPh2
Br

OS, CV 5, **552 (1973)**

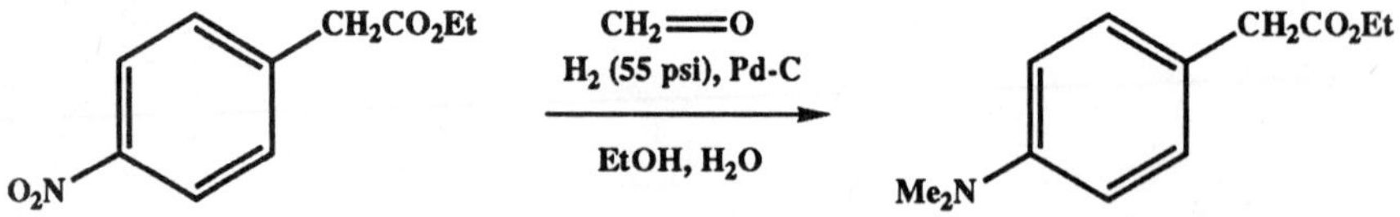
CH2CO2Et
O2N
CH2=O
H2 (55 psi), Pd-C
EtOH, H2O
CH2CO2Et
Me2N

***OS*, *CV 5*, 632 (1973)**

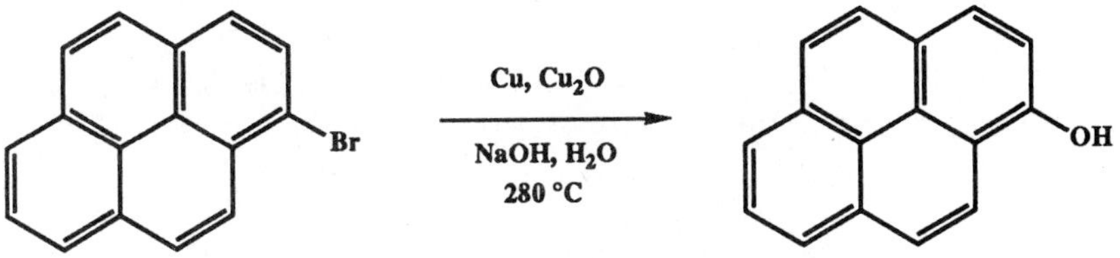

Br

Cu, Cu_2O

NaOH, H_2O
280 °C

OH

***OS*, *CV 5*, 816 (1973)**

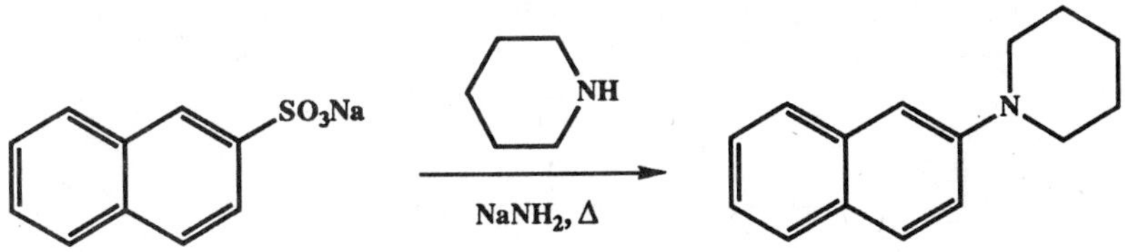

SO_3Na

NH

$NaNH_2$, Δ

N

***OS*, *CV 5*, 829 (1973)**

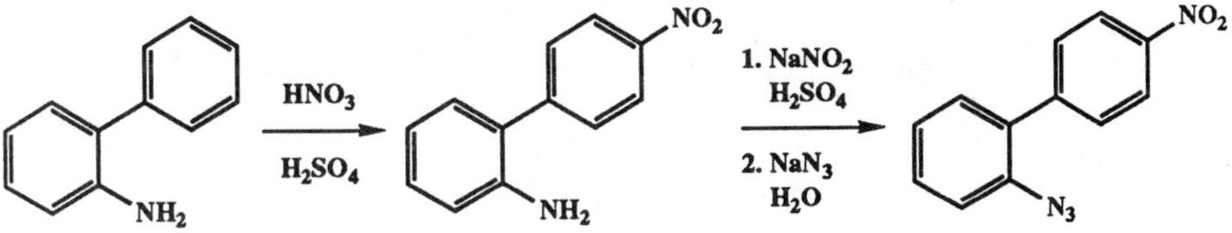

NH_2

HNO_3

H_2SO_4

NO_2

NH_2

1. $NaNO_2$
H_2SO_4

2. NaN_3
H_2O

NO_2

N_3

***OS*, *CV 5*, 918 (1973)**

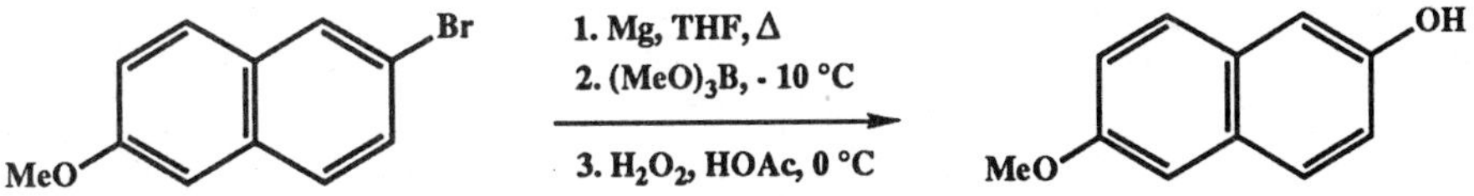

MeO

Br

1. Mg, THF, Δ
2. $(MeO)_3B$, - 10 °C

3. H_2O_2, HOAc, 0 °C

MeO

OH

OS, *CV* 5, 924 (1973)

t-BuO—O–C(=O)–Ph —[PhMgBr; Et_2O, then HCl]→ t-BuO—Ph

OS, *CV* 5, 926 (1973)

t-BuOK + PhBr —[t-BuOH; DMSO, Δ]→ t-BuO—Ph

OS, *CV* 5, 977 (1973)

2 pyridine —[1. $SOCl_2$; 2. EtOH, 0 °C]→ N-(4-pyridyl)pyridinium chloride hydrochloride (+N– ... +NH, Cl^-, Cl^-) —[1. Na_2SO_3, H_2O, Δ; 2. HCl]→ pyridine-4-sulfonic acid (SO_3H)

OS, *CV* 5, 989 (1973)

ethyl isonicotinate (CO_2Et) —[$BrCH_2CO_2Et$; EtOH, Δ]→ N-(CH_2CO_2Et)-4-(CO_2Et)pyridinium Br^- —[1. H_2, Pd-C (100 atm, 90 °C); 2. aq. K_2CO_3, $CHCl_3$, 0 °C]→ N-(CH_2CO_2Et)-4-(CO_2Et)piperidine

OS, CV 5, 1018 (1973)

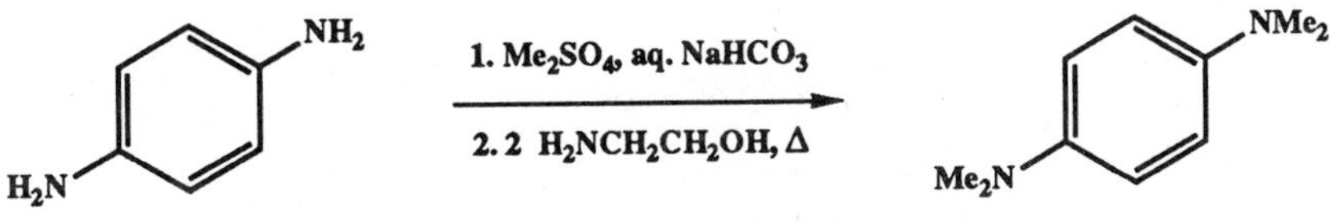

H_2N–C_6H_4–NH_2 → 1. Me_2SO_4, aq. $NaHCO_3$; 2. 2 $H_2NCH_2CH_2OH$, Δ → Me_2N–C_6H_4–NMe_2

OS, CV 5, 1029 (1973)

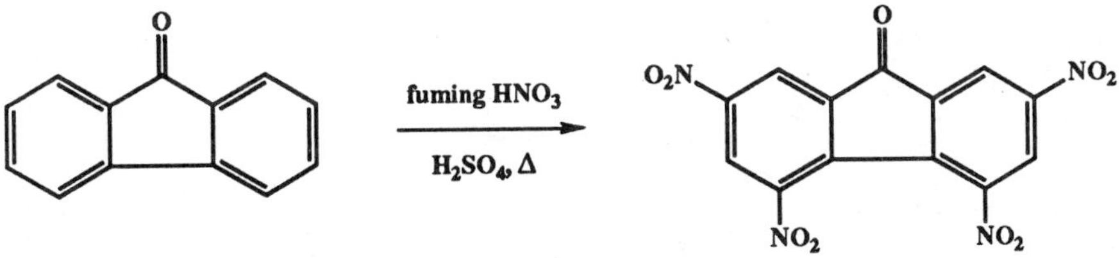

fuming HNO_3, H_2SO_4, Δ

O_2N, NO_2, NO_2, NO_2

OS, CV 5, 1067 (1973)

Cl, Cl → 2 KNO_3, H_2SO_4, 130 °C → Cl, Cl, O_2N, NO_2 → NH_3, $(CH_2OH)_2$, 140 °C → H_2N, NH_2, O_2N, NO_2

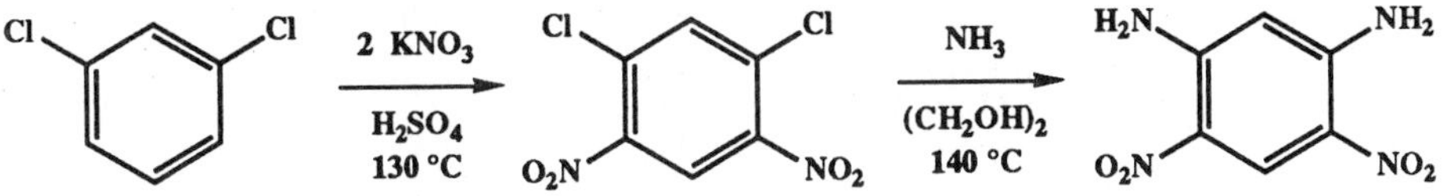

OS, CV 5, 1074 (1973)

NH_2 → O, Cl_3C, CCl_3; hexane, 65 - 70 °C → H, N, CCl_3, O

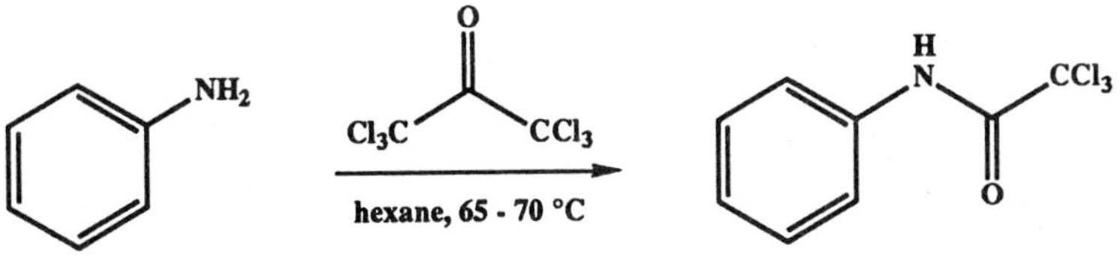

OS, CV 5, 1085 (1973)

NH_2 / CF_3 → 1. $(MeO)_3P{=}O$, Δ; 2. aq. NaOH → NMe_2 / CF_3

OS, CV 5, 1116 (1973)

$$3\ (C_6H_5)_2Hg + 2\ Al \xrightarrow{\text{xylene, }\Delta} 2\ (C_6H_5)_3Al + 3\ Hg$$

OS, CV 6, 150 (1988)

OH → N—N, N, Cl, N, Ph; K_2CO_3, acetone → O, N—N, N, N, Ph

OS, CV 6, 181 (1988)

O, Br, Br, Br, Br + NMe_2 / CF_3 → CH_2Cl_2 → NMe_2 / CF_3 / Br + OH, Br, Br, Br

OS, CV 6, **451 (1988)**

I

(EtO)$_2$PO$^-$ Na$^+$

hν, NH$_3$

O

P(OEt)$_2$

OS, CV 6, **468 (1988)**

MeO

TeCl$_3$

OMe

160 °C

MeO

TeCl$_2$

OMe

OS, CV 6, **533 (1988)**

Br

1. Mg, Et$_2$O

2. Se, Δ

SeMgBr

Br$_2$

SeCl

Cl$_2$, hexane

40 - 50 °C

Se—Se

***OS*, *CV* 6, 700 (1988)**

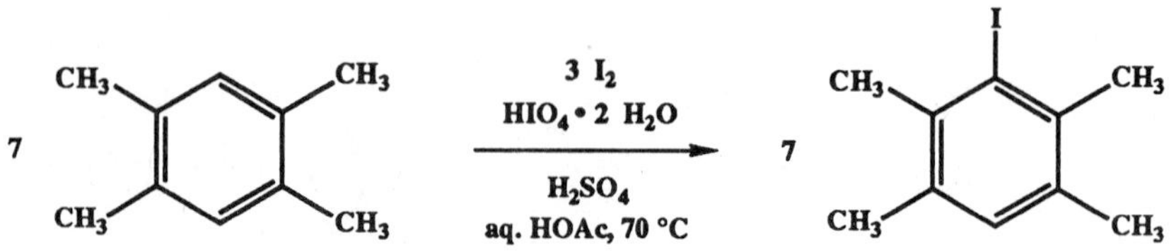

7 1,2,4,5-tetramethylbenzene → ($3\ I_2$, $HIO_4 \cdot 2\ H_2O$, H_2SO_4, aq. HOAc, 70 °C) → 7 3-iodo-1,2,4,5-tetramethylbenzene

***OS*, *CV* 6, 709 (1988)**

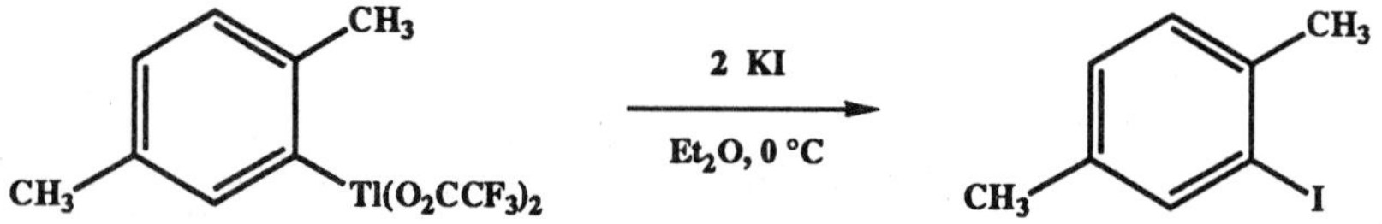

2,5-dimethylphenyl-$Tl(O_2CCF_3)_2$ → (2 KI, Et_2O, 0 °C) → 2-iodo-1,4-dimethylbenzene

***OS*, *CV* 6, 824 (1988)**

2-naphthol (OH) → (Me_2N–C(=S)–Cl, KOH, 10 °C) → 2-naphthyl–O–C(=S)–NMe_2

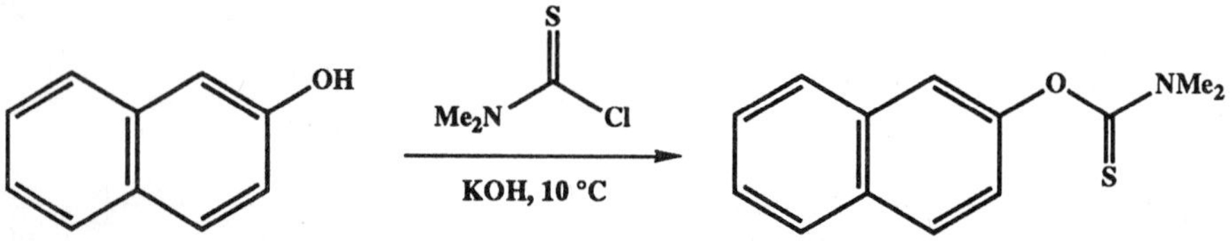

***OS*, *CV* 6, 859 (1988)**

5-methylresorcinol (HO, OH, CH_3) → (Me_2SO_4, K_2CO_3) → 3,5-dimethoxytoluene (MeO, OMe, CH_3)

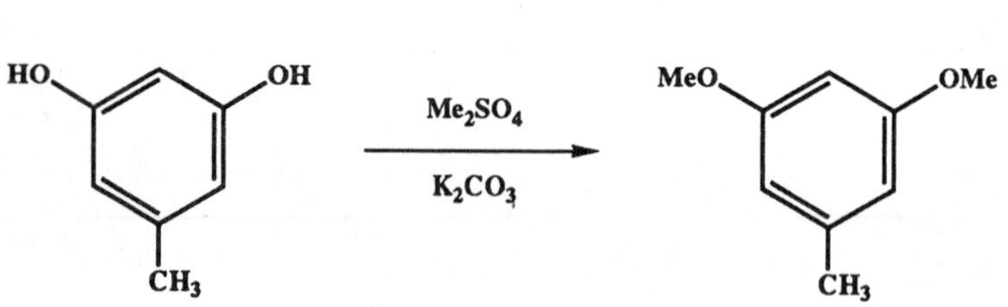

OS, *CV 6*, 875 (1988)

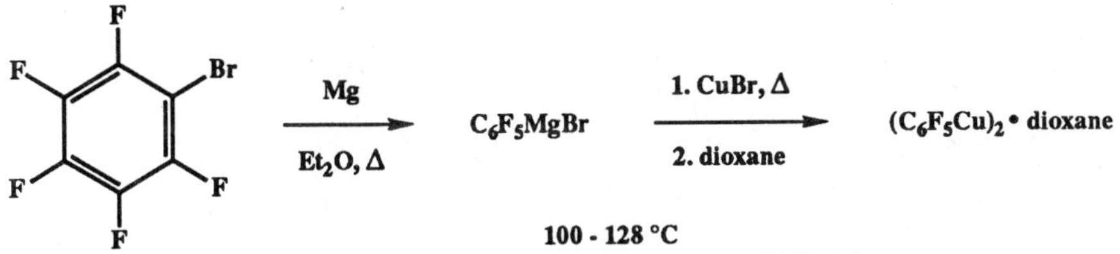

Bromopentafluorobenzene $\xrightarrow[Et_2O,\ \Delta]{Mg}$ C_6F_5MgBr $\xrightarrow[2.\ dioxane]{1.\ CuBr,\ \Delta}$ $(C_6F_5Cu)_2 \cdot$ dioxane

$(C_6F_5Cu)_2 \cdot$ dioxane $\xrightarrow[10^{-3}\ mm\ Hg]{100 - 128\ °C}$ $(C_6F_5Cu)_4$

OS, *CV 7*, 435 (1990); *61*, 35 (1983)

BrCN $\xrightarrow[Et_3N,\ CCl_4,\ -5\ to\ +10\ °C]{C_6H_5OH}$ C_6H_5OCN

OS, *67*, 1 (1988)

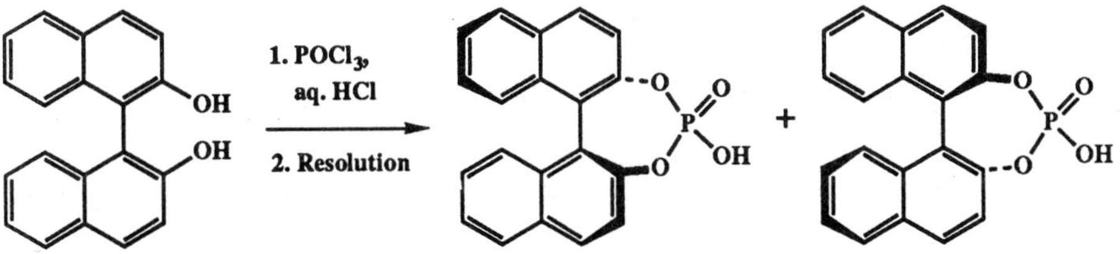

1,1'-Bi-2-naphthol $\xrightarrow[2.\ Resolution]{1.\ POCl_3,\ aq.\ HCl}$ binaphthyl phosphoric acid (P(=O)OH) (*S*) - (+) + (*R*) - (-)

OS, *67*, 20 (1988)

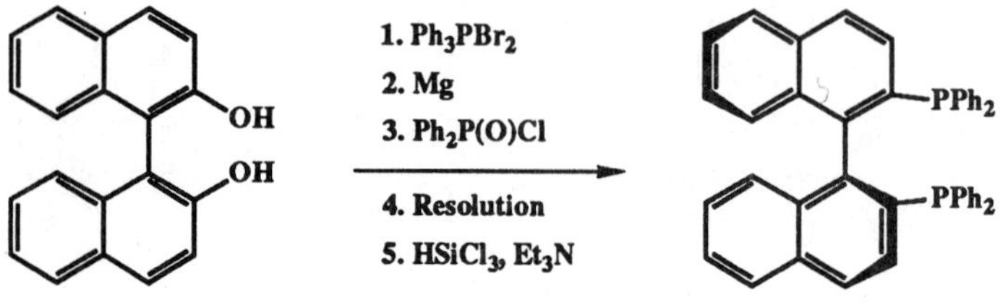

1. Ph_3PBr_2
2. Mg
3. $Ph_2P(O)Cl$
4. Resolution
5. $HSiCl_3$, Et_3N

OH OH

PPh₂ PPh₂

(*S*) - (-) - BINAP + (*R*) - (+) - BINAP

OS, *67*, 222 (1988)

OMe Cl + O N Cl

80% H_2SO_4

OMe Cl Cl

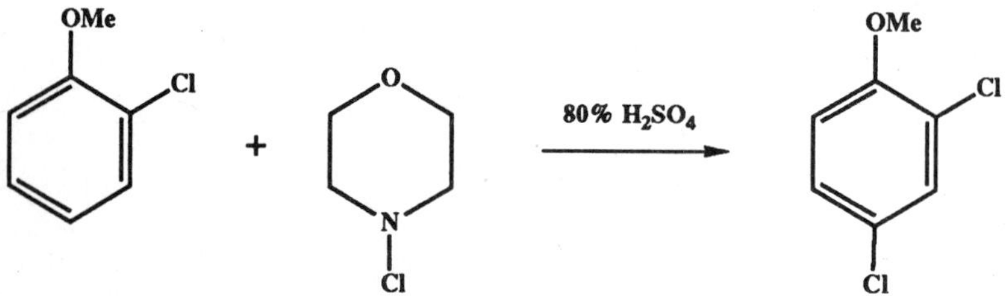

OS, CV 1, 21 (1941)

CH_3CHO —(NaCN, aq. NH_4Cl, Et_2O)→ $CH_3CH(NH_2)CN$ —(1. HCl; 2. $Pb(OH)_2$)→ $CH_3CH(NH_2)CO_2H$

OS, CV 1, 131 (1941)

$HOCH_2CH_2CH_2CN$ —(2 HBr; H_2O, Δ)→ $BrCH_2CH_2CH_2CO_2H$

OS, CV 1, 270 (1941)

$C_6H_5CH_2CN$ —(H_2SO_4, H_2O; EtOH, Δ)→ $C_6H_5CH_2CO_2Et$

OS, CV 1, 289 (1941)

$NC(CH_2)_3CN$ —(HCl, H_2O; Δ)→ $HO_2C(CH_2)_3CO_2H$

OS, CV 1, 298 (1941)

$(H_2NCH_2CN) \bullet H_2SO_4$ —($Ba(OH)_2$; Δ)→ $(H_2NCH_2CO_2)_2Ba$ —(H_2SO_4)→ 2 H_2N—CH_2CO_2H

OS, *CV 1*, 321 (1941)

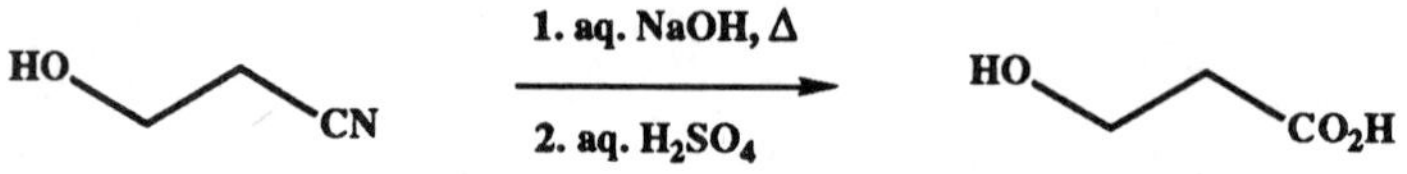

HO–CH2CH2–CN → (1. aq. NaOH, Δ; 2. aq. H_2SO_4) → HO–CH2CH2–CO_2H

OS, *CV 1*, 336 (1941)

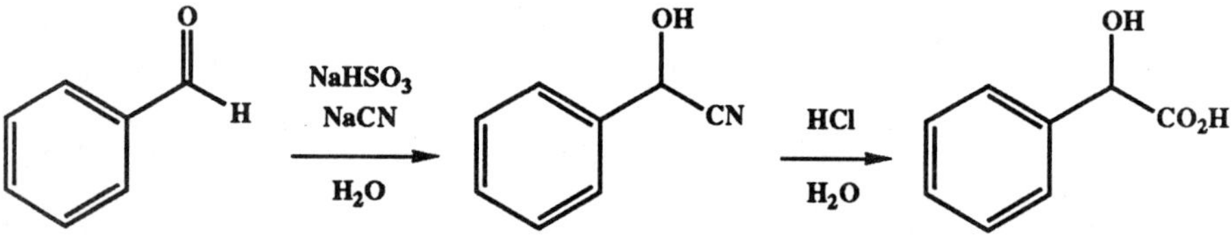

PhCHO → ($NaHSO_3$, NaCN, H_2O) → PhCH(OH)CN → (HCl, H_2O) → PhCH(OH)CO_2H

OS, *CV 1*, 406 (1941)

O_2N–C6H4–CH2CN → (H_2SO_4, H_2O, Δ) → O_2N–C6H4–CH2CO_2H

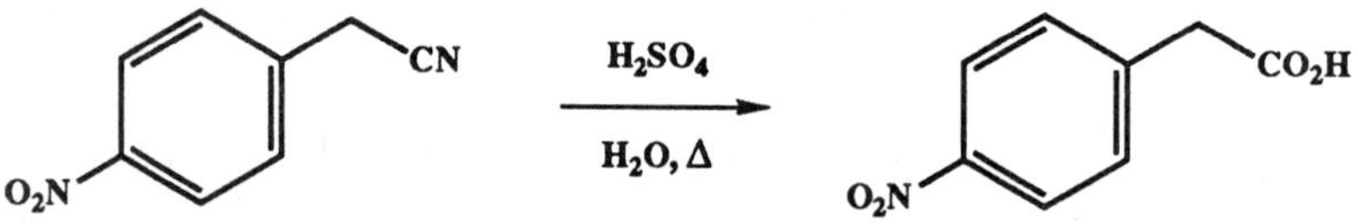

OS, *CV 1*, 436 (1941)

PhCH2CN → (H_2SO_4, H_2O, Δ) → PhCH2CO_2H

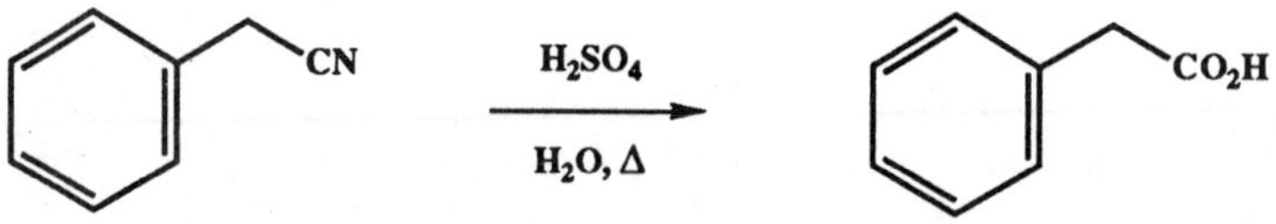

OS, CV 2, 25 (1943)

Cl—$(CH_2)_3CN$

NK

N—$(CH_2)_3CN$

H_2SO_4

H_2O

H_2N CO_2H

OS, CV 2, 29 (1943)

HO CN CH_3 CH_3

1. NH_3, MeOH
2. HBr, H_2O, Δ
3. pyridine, MeOH

H_2N CO_2H CH_3 CH_3

OS, CV 2, 44 (1943)

MeO MeO C≡N

KOH, H_2O_2

H_2O

MeO MeO C(=O) NH_2

OS, CV 2, 284 (1943)

O Me C≡N Ph

HCl

EtOH
- 10 °C

O NH Me OEt Ph

aq. H_2SO_4

Et_2O

O O Me OEt Ph

OS, CV 2, 292 (1943)

$n\text{-}C_{12}H_{25}$—Br → (KCN; EtOH, Δ) → $n\text{-}C_{12}H_{25}$—CN → (1. aq. KOH, Δ; 2. EtOH, HCl) → $n\text{-}C_{12}H_{25}$—CO_2Et

OS, CV 2, 310 (1943)

NC, N, N, CN, N, CN → (EtOH, Δ; HCl, H_2O) → H_2N, O, OEt

OS, CV 2, 376 (1943)

Cl, CO_2H → (1. Na_2CO_3; 2. NaCN) → CN, CO_2Na → (aq. NaOH) → CO_2Na, CO_2Na → (aq. $CaCl_2$) → CO_2^-, Ca^{+2}, CO_2^- → (HCl) → CO_2H, CO_2H

OS, CV 2, 586 (1943)

CH_3, CN → (H_2O_2; aq. NaOH, EtOH, 50 °C) → CH_3, NH_2, O

OS, CV 2, 588 (1943)

2-methylbenzonitrile (o-tolunitrile; ring substituents CH_3 and CN) → o-toluic acid (ring substituents CH_3 and CO_2H)

Reagents: H_2SO_4; H_2O, Δ

OS, CV 3, 34 (1955)

H_2N–CH_2CH_2–CN → H_2N–CH_2CH_2–CO_2H

Reagents: 1. $Ba(OH)_2$, H_2O, 90 °C; 2. CO_2, H_2O, 90 °C

OS, CV 3, 84 (1955)

PhC(O)H → H_2N–CH(Ph)–CN → H_2N–CH(Ph)–CO_2H

Reagents, step 1: NaCN, NH_4Cl; H_2O, MeOH

Reagents, step 2: 1. aq. HCl, Δ; 2. NH_4OH

OS, CV 3, 88 (1955)

PhC(O)Me → H_2N–C(Ph)(Me)–CN → H_2N–C(Ph)(Me)–CO_2H

Reagents, step 1: NaCN, NH_4Cl; H_2O, EtOH

Reagents, step 2: 1. aq. HCl, Δ; 2. pyridine

OS, CV 3, 114 (1955)

Benzoyl cyanide (PhC(O)CN) → benzoylformic acid (PhC(O)CO_2H)

Reagents: HCl, H_2O

OS, *CV* 3, 221 (1955)

Cl, CN; NaOH, Δ; CN; H_2SO_4, H_2O; CO_2H

OS, *CV* 3, 557 (1955)

CH_2Cl, CH_3, CH_3, CH_3; NaCN, Δ, aq. EtOH; CH_2CN, CH_3, CH_3, CH_3; H_2SO_4, H_2O, Δ; CH_2CO_2H, CH_3, CH_3, CH_3

OS, *CV* 3, 560 (1955)

CH_3, CH_3, OH, CN; 98% H_2SO_4, Δ; O, CH_3, NH_2

OS, *CV* 3, 609 (1955)

CH_3, N, CN; H_2S, NH_3, EtOH; S, CH_3, N, C, NH_2

OS, CV 3, 615 (1955)

CH_3—CH=CH—CO_2Et → (1. NaCN, Δ, H_2O, EtOH; 2. $Ba(OH)_2$, Δ then HNO_3)

CH_3 / CO_2H / CO_2H

OS, CV 3, 851 (1955)

CN → (HCl, H_2O, Δ) CO_2H

OS, CV 4, 58 (1963)

HO, CN, Ph, CH_3 → (H_2O, HCl) HO, O, C—NH_2, Ph, CH_3 → (1. NaOH; 2. HCl) HO, O, C—OH, Ph, CH_3

OS, CV 4, 93 (1963)

Me, Et, $PhCH_2$, CN → (1. KOH, Δ, ethylene glycol; 2. HCl) Me, Et, $PhCH_2$, CO_2H

OS, CV 4, 760 (1963)

Ph, CN → (HCl, H_2O) Ph, O, NH_2

OS, CV 4, 790 (1963)

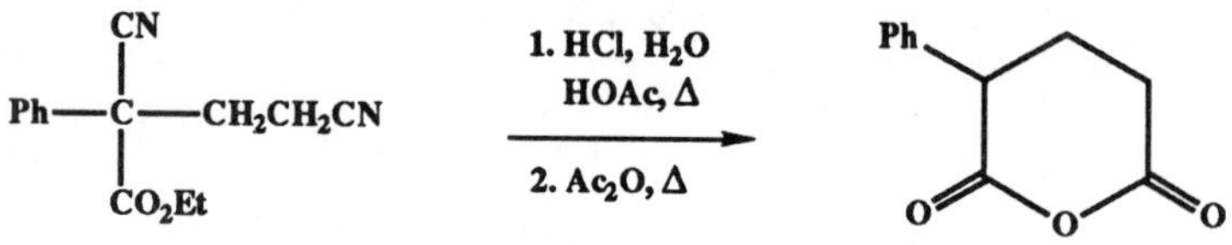
CN
Ph—C—CH2CH2CN
CO2Et
1. HCl, H2O
HOAc, Δ
2. Ac2O, Δ
Ph
O
O
O

OS, CV 4, 804 (1963)

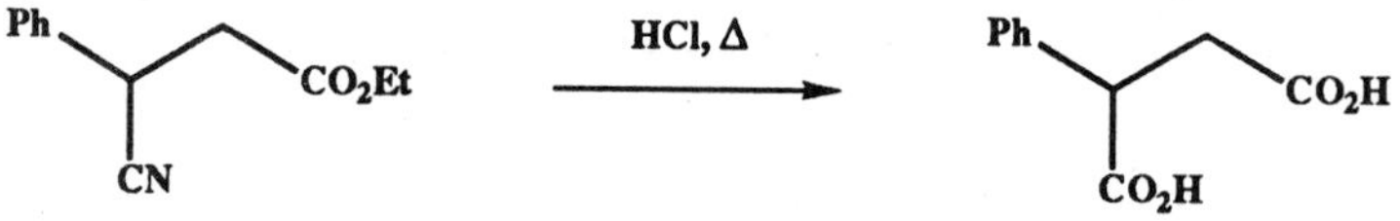
Ph
CO2Et
CN
HCl, Δ
Ph
CO2H
CO2H

OS, CV 5, 73 (1973)

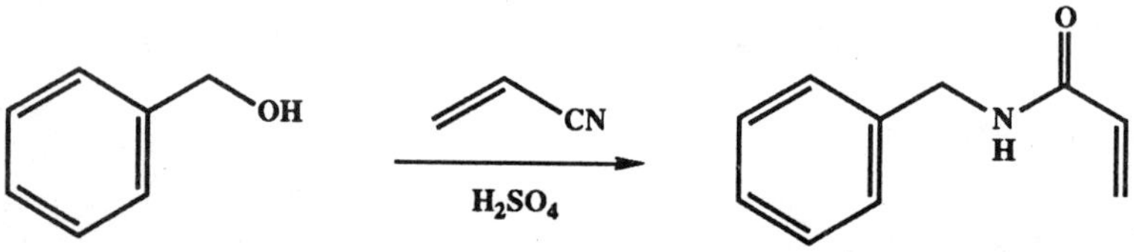
OH
CN
H2SO4
O
N
H

OS, CV 5, 501 (1973)

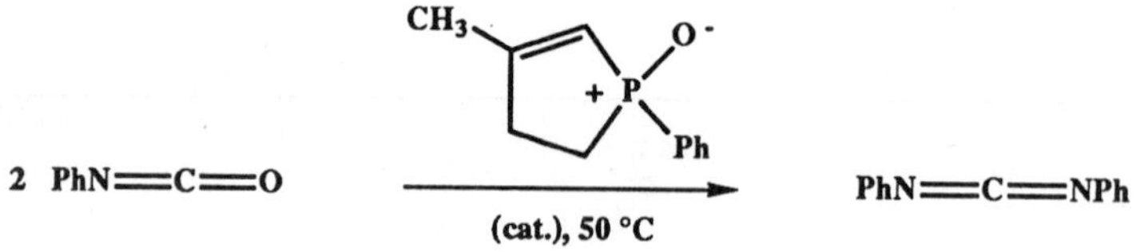
CH3
O-
+P
Ph
2 PhN=C=O
(cat.), 50 °C
PhN=C=NPh

OS, *CV 5*, 504 (1973)

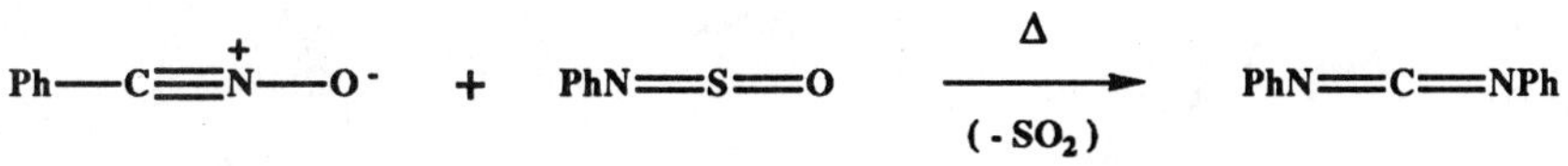

$$Ph-C\equiv\overset{+}{N}-O^{-} + PhN{=}S{=}O \xrightarrow[(-SO_2)]{\Delta} PhN{=}C{=}NPh$$

OS, *65*, 61 (1987)

$$n\text{-Bu-Cl} \xrightarrow[THF]{Mg} n\text{-Bu-MgCl} \xrightarrow{H-\equiv-H} H-\equiv-MgCl \xrightarrow{Me_3SiCl} H-\equiv-SiMe_3$$

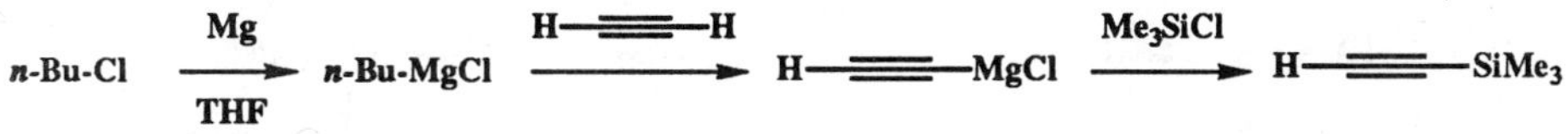

OS, CV 1, 3 (1941)

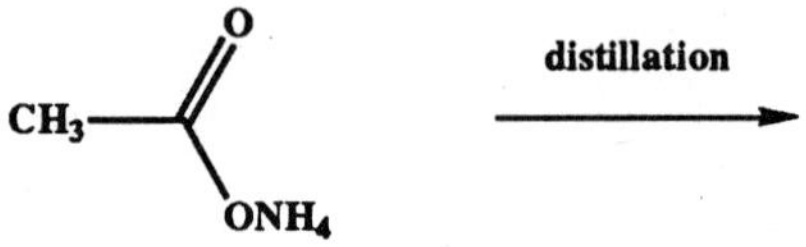

$CH_3C(=O)ONH_4$ —distillation→ $CH_3C(=O)NH_2$

OS, CV 1, 5 (1941)

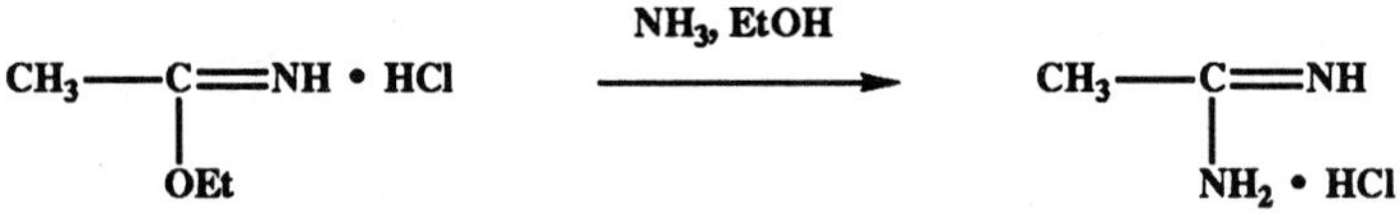

CH_3—C(OEt)=NH • HCl —NH_3, EtOH→ CH_3—C(NH_2)=NH • HCl

OS, CV 1, 12 (1941)

PhCH(OH)CO_2H —Ac—Cl→ PhCH(OAc)CO_2H —$SOCl_2$→ PhCH(OAc)COCl

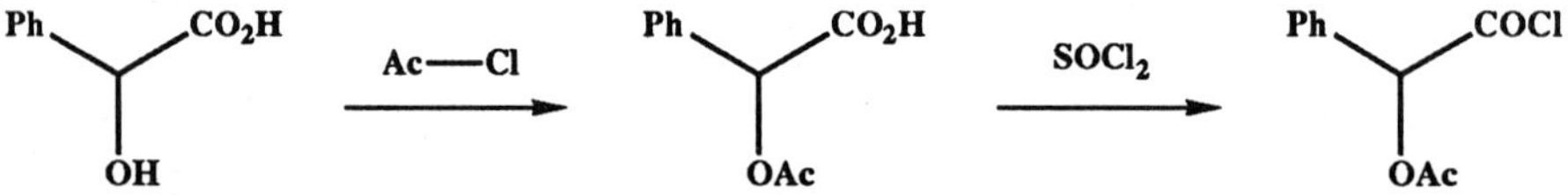

OS, CV 1, 14 (1941)

Saccharin (SO_2, NH, O) —H_2O, HCl, Δ→ 2-(SO_3NH_4)benzoic acid (CO_2H)

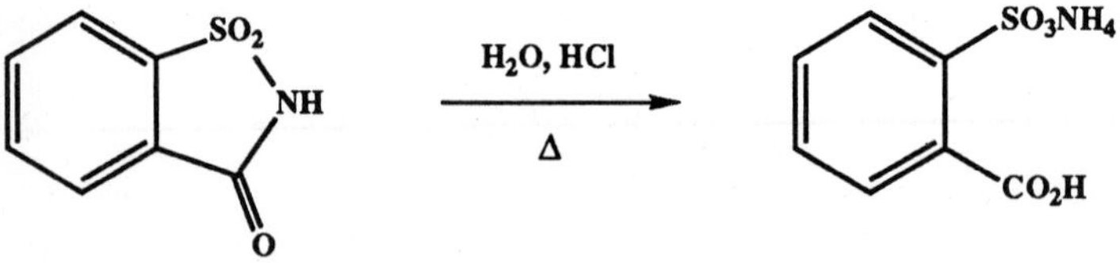

OS, CV 1, 21 (1941)

CH_3CHO —(NaCN, aq. NH_4Cl, Et_2O)→ $CH_3CH(NH_2)CN$ —(1. HCl; 2. $Pb(OH)_2$)→ $CH_3CH(NH_2)CO_2H$

OS, CV 1, 80 (1941)

$PhCHO$ —($PhNH_2$)→ $PhCH{=}NPh$

OS, CV 1, 91 (1941)

$PhCO_2H$ —($(CH_3CO)_2O$, H_3PO_4 (cat.), Δ)→ $PhC(O)OC(O)CH_3$ —(Ph—CO_2H)→ $PhC(O)OC(O)Ph$

OS, CV 1, 138 (1941)

6 *n*-BuOH —(2 $Na_2Cr_2O_7$, 8 H_2SO_4, H_2O)→ 3 $CH_3CH_2CH_2C(O)O$—*n*-Bu

OS, CV 1, 147 (1941)

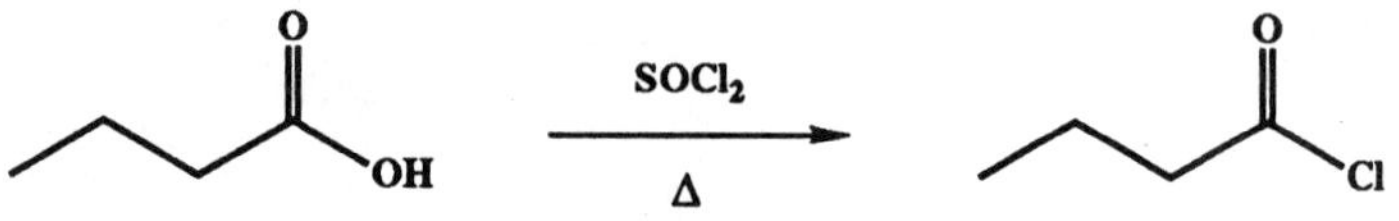

O

OH

$SOCl_2$

Δ

O

Cl

OS, CV 1, 153 (1941)

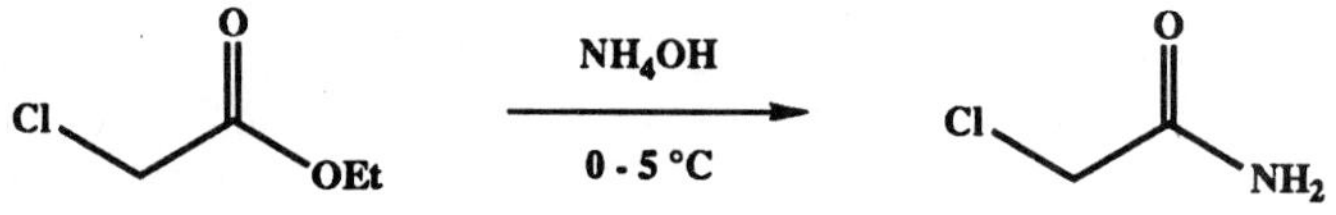

O

Cl

OEt

NH_4OH

0 - 5 °C

O

Cl

NH_2

OS, CV 1, 165 (1941)

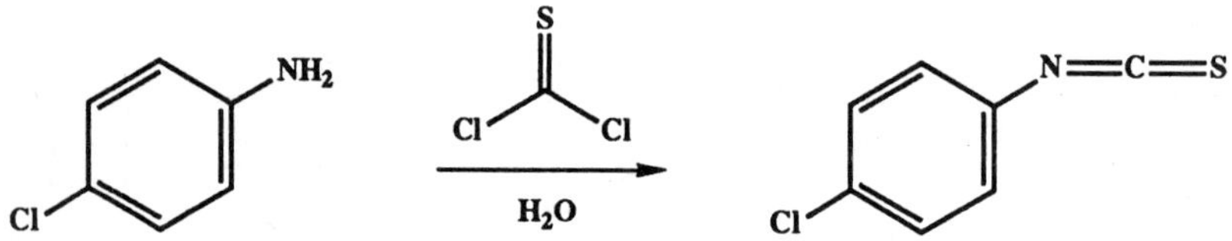

NH_2

Cl

S

Cl

Cl

H_2O

N=C=S

Cl

OS, CV 1, 179 (1941)

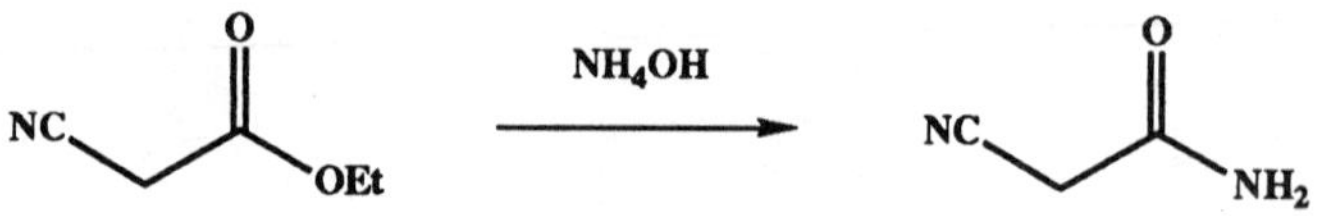

O

NC

OEt

NH_4OH

O

NC

NH_2

OS, CV 1, 205 (1941)

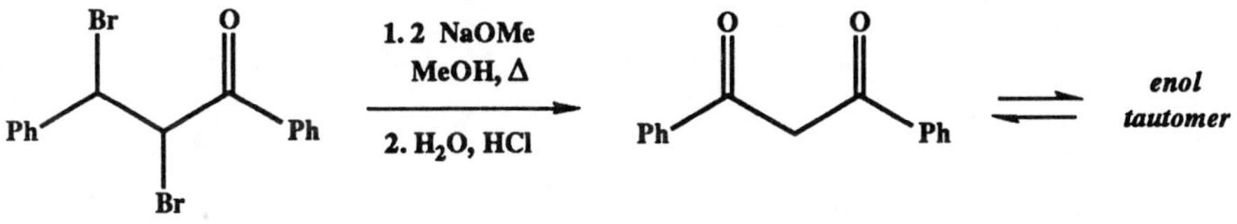

Br, Br, O, Ph, Ph

1. 2 NaOMe
MeOH, Δ

2. H_2O, HCl

O, O, Ph, Ph ⇌ *enol tautomer*

OS, CV 1, 237 (1941)

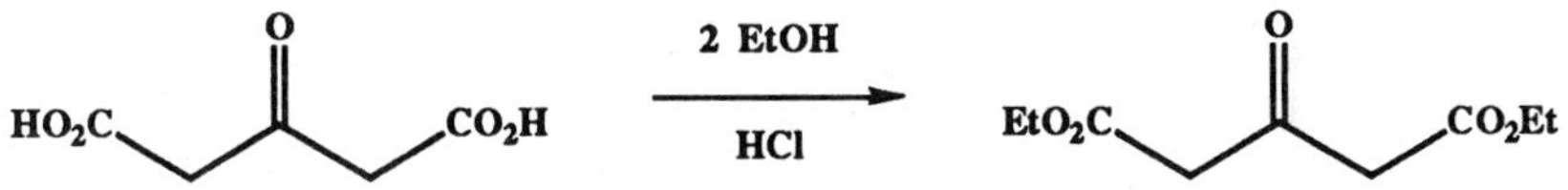

HO_2C, O, CO_2H

2 EtOH

HCl

EtO_2C, O, CO_2Et

OS, CV 1, 241 (1941)

O, Ph, OH, O

H_2SO_4

EtOH

O, Ph, OEt, O

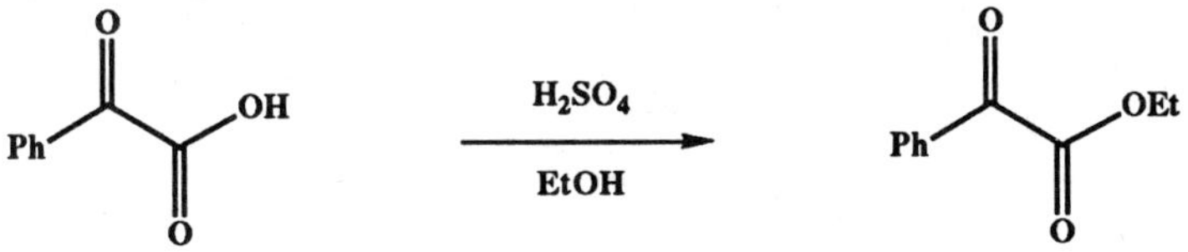

OS, CV 1, 246 (1941)

Br, O, OH

EtOH, H^+

CCl_4, Δ

Br, O, OEt

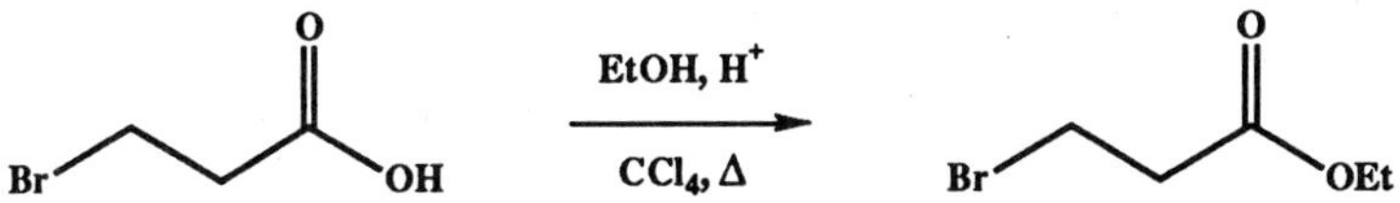

OS, CV 1, 254 (1941)

$ClCH_2CO_2Na$ —[NaCN; H_2O, Δ]→ $NCCH_2CO_2Na$ —[H_2SO_4 (cat.); EtOH, Δ]→ $NCCH_2CO_2Et$

OS, CV 1, 261; 263 (1941)

$HO_2C\text{-}CO_2H$ —[2 EtOH; CCl_4, Δ]→ $EtO_2C\text{-}CO_2Et$

OS, CV 1, 302 (1941)

$H_2N\text{-}C(=NH)\text{-}NH\text{-}C{\equiv}N$ —[NH_4NO_3; 160 °C]→ $H_2N\text{-}C(=NH)\text{-}NH\text{-}C(=NH)\text{-}NH_2 \cdot HNO_3$ —[NH_4NO_3]→ 2 $H_2N\text{-}C(=NH)\text{-}NH_2 \cdot HNO_3$

OS, CV 1, 379 (1941)

Glyceryl tri($C_{13}H_{27}CO_2$) ester —[1. aq. NaOH, Δ; 2. aq. HCl]→ 3 $C_{13}H_{27}CO_2H$

OS, CV 1, 394 (1941)

4-O_2N-C_6H_4-COOH $\xrightarrow{PCl_5,\ \Delta}$ 4-O_2N-C_6H_4-COCl

OS, CV 1, 450 (1941)

PhNH-C(=O)-NH_2 $\xrightarrow[\Delta]{H_2NNH_2 \cdot H_2O}$ PhNH-C(=O)-NH-NH_2

OS, CV 1, 451 (1941)

PhCH=C(CO_2H)(CN) $\xrightarrow[HCl]{EtOH,\ \Delta}$ PhCH=C(CO_2Et)(CN)

OS, CV 1, 453 (1941)

Ph—NH_2 • HCl $\xrightarrow[H_2O]{H_2N-C(=O)-NH_2}$ PhNH-C(=O)-NH_2 $\xrightarrow{aniline}$ PhNH-C(=O)-NHPh

OS, CV 2, 1 (1943)

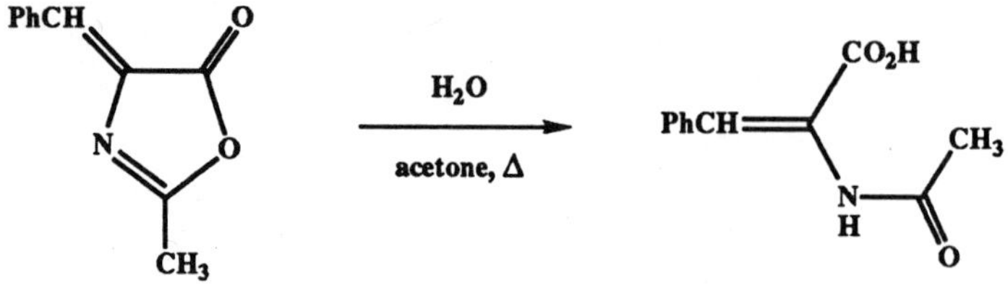

PhCH, O, N, O, CH_3 — H_2O / acetone, Δ → PhCH=, CO_2H, CH_3, N, H, O

OS, CV 2, 24 (1943)

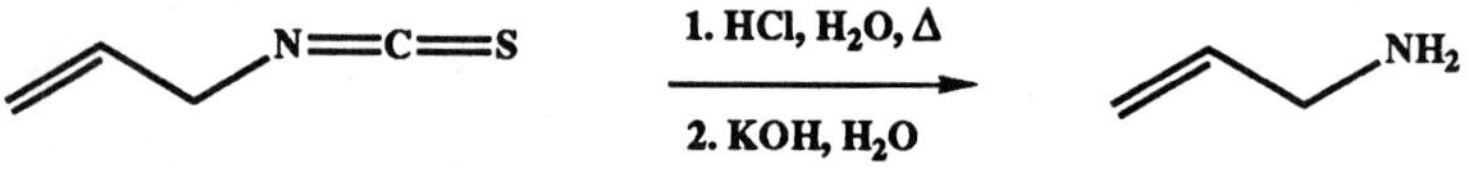

N=C=S — 1. HCl, H_2O, Δ / 2. KOH, H_2O → NH_2

OS, CV 2, 28 (1943)

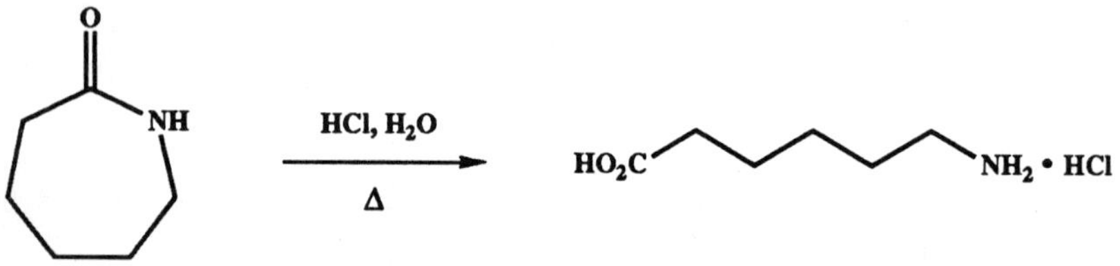

O, NH — HCl, H_2O / Δ → HO_2C, NH_2 • HCl

OS, CV 2, 67 (1943)

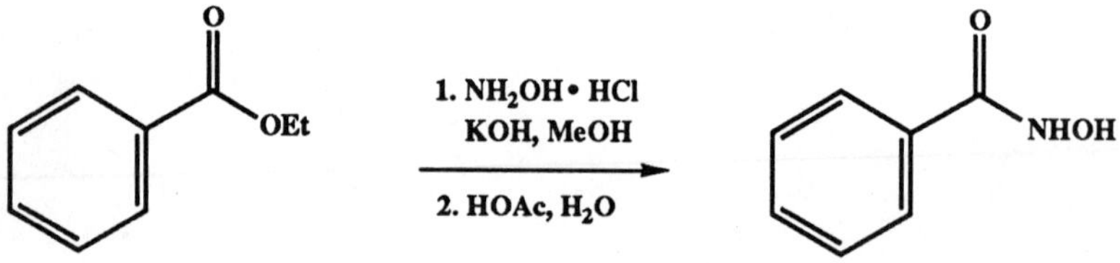

O, OEt — 1. NH_2OH • HCl KOH, MeOH / 2. HOAc, H_2O → O, NHOH

OS, CV 2, 70 (1943)

Benzophenone → benzophenone oxime ($NH_2OH \cdot HCl$; NaOH, EtOH, Δ)

OS, CV 2, 76 (1943)

Caprolactam (NaOH, H_2O) → H_2N—$(CH_2)_5$—CO_2H (PhCOCl) → PhC(=O)—NH—$(CH_2)_5$—CO_2H

OS, CV 2, 137 (1943)

Acrolein (EtOH, HCl) → Cl—CH_2CH_2CH(OEt)(OEt)

OS, CV 2, 204 (1943)

$CH_3C(=O)C(=NOH)CH_3$ (NaO_3SNHOH; H_2O, Δ) → $CH_3C(=NOH)C(=NOH)CH_3$

OS, *CV 2*, 208 (1943)

2 PhC(O)Cl → (H_2NNH_2; NaOH, H_2O) → PhC(O)NHNHC(O)Ph → (Me_2SO_4, NaOH, Δ) → PhC(O)N(Me)N(Me)C(O)Ph → (HCl; H_2O, Δ) → MeNHNHMe

OS, *CV 2*, 260 (1943)

ClCH$_2$C(O)OH → (1. Na, EtOH; 2. HCl) → EtOCH$_2$C(O)OH → (EtOH; HCl) → EtOCH$_2$C(O)OEt

OS, *CV 2*, 264 (1943)

HO_2C(CH$_2$)$_4$$CO_2H$ → (EtOH, H_2SO_4; toluene, Δ) → EtO_2C(CH$_2$)$_4$$CO_2Et$

OS, *CV 2*, 276 (1943)

$HO_2C—(CH_2)_8—CO_2H$ (and diester) → (EtOH, HCl; $(n\text{-}Bu)_2O$, Δ) → $EtO_2C—(CH_2)_8—CO_2H$

OS, CV 2, 278 (1943)

$ClCO_2Et$ + CH_3NH_2 → ($NaOH$; Et_2O, H_2O) → CH_3NHCO_2Et

OS, CV 2, 292 (1943)

n-$C_{12}H_{25}$—Br → (KCN; EtOH, Δ) → n-$C_{12}H_{25}$—CN → (1. aq. KOH, Δ; 2. EtOH, HCl) → n-$C_{12}H_{25}$—CO_2Et

OS, CV 2, 313 (1943)

heptanal → ($NH_2OH \cdot HCl$, Na_2CO_3; H_2O) → heptanal oxime (NOH)

OS, CV 2, 333 (1943)

4-(3,4-dimethoxybenzylidene)-2-phenyloxazol-5-one (MeO, OMe, CH=, N, O, Ph) → ($NaOH$; H_2O, Δ) → 3,4-dimethoxyphenylpyruvic acid (MeO, OMe, CO_2H)

OS, *CV* 2, 411 (1943)

$$2\ H_2N-C(=S)-NH_2 \xrightarrow[H_2O,\ \Delta]{(CH_3)_2SO_4} (HN{=}C(SCH_3){-}NH_2)_2 \cdot H_2SO_4$$

OS, *CV* 2, 414 (1943)

$$HO_2C-CO_2H \xrightarrow[H_2SO_4]{2\ MeOH} MeO_2C-CO_2Me$$

OS, *CV* 2, 422 (1943)

$$CH_3C(=O)CH_2C(=O)OEt \xrightarrow[\text{cat. HCl, EtOH}]{H_2N-C(=O)-NH_2} H_2N-C(=O)-NH-C(CH_3){=}CH-C(=O)OEt$$

OS, *CV* 2, 457 (1943)

$$\text{4-}O_2N\text{-phthalimide} \xrightarrow[2.\ HNO_3]{1.\ \text{aq. NaOH},\ \Delta} \text{4-}O_2N\text{-}C_6H_3(CO_2H)_2$$

OS, CV 2, 503 (1943)

Ph(C=O)CH$_3$ —[NH_4OCHO, Δ]→ [PhCH(CH$_3$)NHCHO] —[HCl, H_2O]→ PhCH(NH$_2$)CH$_3$

dl form

OS, CV 2, 522 (1943)

HCl • HN=C(CH$_3$)– ; HO, OH, OH

$CH_3C{\equiv}N$; $ZnCl_2$, HCl ; Et_2O ; H_2O ; Δ

O=C(CH$_3$)– ; HO, OH, OH

OS, CV 2, 528 (1943)

PCl_5 ; Δ ; Cl, Cl ; $AlCl_3$; Δ ; O, Cl, Cl

OS, CV 2, 622 (1943)

MeO, MeO, H ; 1. $NH_2OH \cdot HCl$ EtOH, H_2O ; 2. Ac_2O, Δ ; MeO, MeO, $C{\equiv}N$

OS, CV 3, 10 (1955)

NH_2 + O ; O ; Δ ; benzene ; H ; N ; O ; O

OS, CV 3, 20 (1955)

NOH ; Ph ; CH_3 ; O ; H_2SO_4 ; H_2O, Δ ; O ; Ph ; CH_3 ; O

OS, CV 3, 28 (1955)

O ; C_6H_{13} ; Cl + O ; C_6H_{13} ; OH ; pyridine ; benzene ; O ; O ; C_6H_{13} ; O ; C_6H_{13}

2 ; O ; Cl ; Cl ; pyridine, Δ ; then aq. HCl ; O ; O ; O ; Cl ; Cl

OS, CV 3, 33 (1955)

O ; OMe ; HCO_2H ; H_2SO_4, 90 °C ; O ; OH

OS, CV 3, 42 (1955)

2 (uric acid) → 2 (alloxan hydrate); $KClO_3$ / HCl, H_2O

OS, CV 3, 46 (1955)

$CH_3CH(OH)CO_2H$ → $CH_3CH(OH)CO_2CH_2CH{=}CH_2$; allyl alcohol (OH) / H_2SO_4, benzene, Δ

OS, CV 3, 66 (1955)

$Et_2C{=}O$ → NaCN, NH_4Cl, NH_4OH, H_2O / MeOH, 60 °C → H_2N–C(Et)$_2$–CN → 1. H_2O, HCl then Δ / 2. $Pb(OH)_2$ → H_2N–C(Et)$_2$–CO_2H

OS, CV 3, 108 (1955)

$PhCOCH_2CO_2Et$ → $PhCOCH_2CONHPh$; Ph—NH_2 / xylene, Δ

OS, CV 3, 116 (1955)

2 PhCOCl —(KSH / EtOH)→ 2 PhCOSK

OS, CV 3, 127 (1955)

$CH_3(CH_2)_4CH_2CHO$ —(KOAc / Ac_2O, Δ)→ $CH_3(CH_2)_4CH{=}CH(OAc)$ —(Br_2 / CCl_4)→ $CH_3(CH_2)_4CHBrCHBr(OAc)$ —(MeOH)→ $CH_3(CH_2)_4CHBrCH(OMe)_2$ —(HCl / H_2O, Δ)→ $CH_3(CH_2)_4CHBrCHO$

OS, CV 3, 141 (1955)

t-Bu—OH + Ac_2O —($ZnCl_2$ / Δ)→ *t*-Bu—OAc

OS, CV 3, 142 (1955)

t-Bu—OH + AcCl —($PhNMe_2$ / Et_2O, Δ)→ *t*-Bu—OAc

OS, *CV 3*, 144 (1955)

t-Bu—OH + AcCl —(Mg / Et_2O)→ *t*-Bu—OAc

OS, *CV 3*, 146 (1955)

$CH_2=CH-CO_2Me$ —(*p*-TsOH, *n*-BuOH, Δ / hydroquinone)→ $CH_2=CH-CO_2$-*n*-Bu

OS, *CV 3*, 154 (1955)

t-Bu–NH–C(=O)–NH_2 —(aq. NaOH, Δ / ethylene glycol)→ *t*-Bu—NH_2

OS, *CV 3*, 164 (1955)

2 $CH_3(CH_2)_4$–C(=O)–OH —($CH_2=C=O$)→ $CH_3(CH_2)_4$–C(=O)–O–C(=O)–$(CH_2)_4CH_3$

OS, CV 3, 167 (1955)

Ph–CH$_2$–O–C(=O)–Cl → (NH_4OH) Ph–CH$_2$–O–C(=O)–NH_2

Ph–CH$_2$–O–C(=O)–Cl → (NaOH, $H_2NCH_2CO_2H$) Ph–CH$_2$–O–C(=O)–NH–CH$_2$–CO_2H

OS, CV 3, 169 (1955)

succinic anhydride → (CH_3OH, Δ) HO_2C–CH$_2$CH$_2$–C(=O)OCH$_3$ → ($SOCl_2$) Cl–C(=O)–CH$_2$CH$_2$–C(=O)OCH$_3$

OS, CV 3, 187 (1955)

Ph–C(=O)–Cl + tetrahydrofuran → ($ZnCl_2$) Ph–C(=O)–O–$(CH_2)_4$–Cl

OS, CV 3, 194 (1955)

2 4-methyl-2-hydroxyquinoline → ($POCl_3$, 80 - 85 °C) 2 4-methyl-2-chloroquinoline

OS, CV 3, 270 (1955)

p-ClPh, p-ClPh, Cl, Cl, Cl — aq. KOH, diethylene glycol, Δ → [p-ClPh, p-ClPh, Cl, Cl] — 1. 3 KOH, 2. H_2SO_4 → p-ClPh, p-ClPh, O, OH

OS, CV 3, 272 (1955)

Cl, NH_2 + EtO, CO_2Et, CO_2Et — Δ → Cl, N, H, EtO_2C, CO_2Et

OS, CV 3, 275 (1955)

$$Et_2NH + CH_2{=}O \xrightarrow[NaCN,\ H_2O]{NaHSO_3} Et_2N{-}CH_2{-}CN$$

OS, CV 3, 360 (1955)

2 $PhNHNH_2$ — CS_2, Et_2O → PhNHNH, S, $PhNHNH_2 \cdot$ HS — Δ → PhNHNH, PhNHNH, S

OS, *CV* 3, 371 (1955)

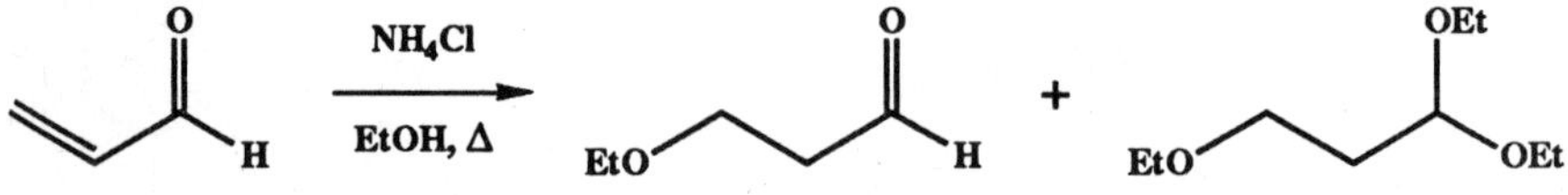

NH_4Cl; EtOH, Δ

H; EtO; H; +; OEt; EtO; OEt

OS, *CV* 3, 374 (1955)

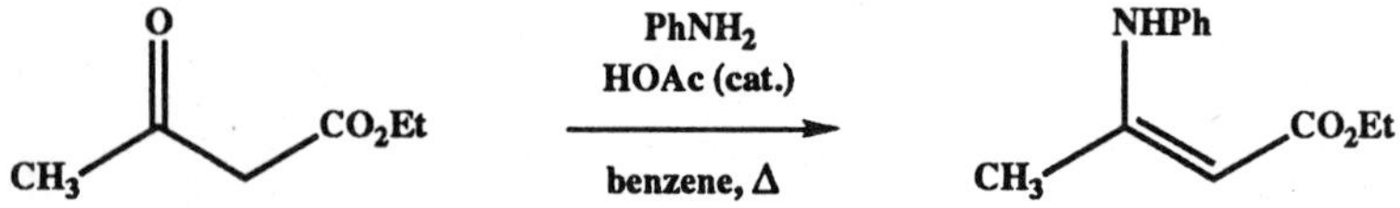

$PhNH_2$
HOAc (cat.)
benzene, Δ

CH_3; CO_2Et; NHPh; CH_3; CO_2Et

OS, *CV* 3, 375 (1955)

$H_2N—NH_2$ → (2 Cl—CO_2Et; aq. Na_2CO_3, EtOH) → $EtO_2C—NH—NH—CO_2Et$

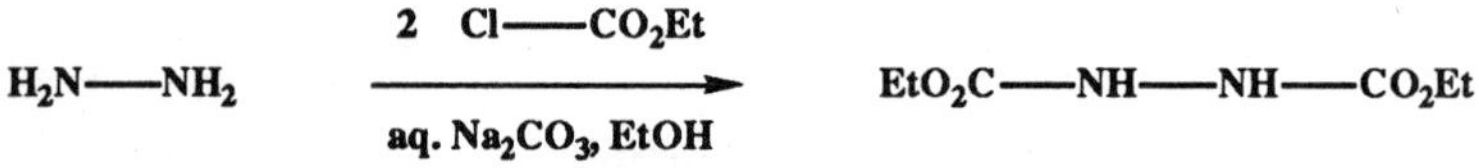

OS, *CV* 3, 381 (1955)

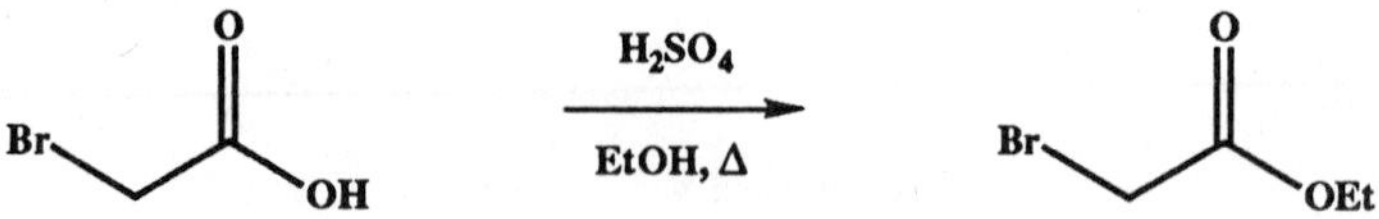

H_2SO_4
EtOH, Δ

Br; OH; Br; OEt

OS, *CV 3*, 422 (1955)

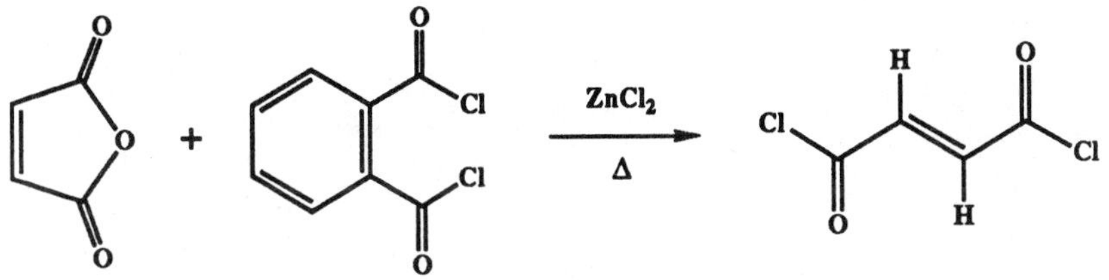
O
O
O
O
Cl
Cl
O
+
ZnCl$_2$
Δ
H
O
Cl
Cl
O
H

OS, *CV 3*, 440 (1955)

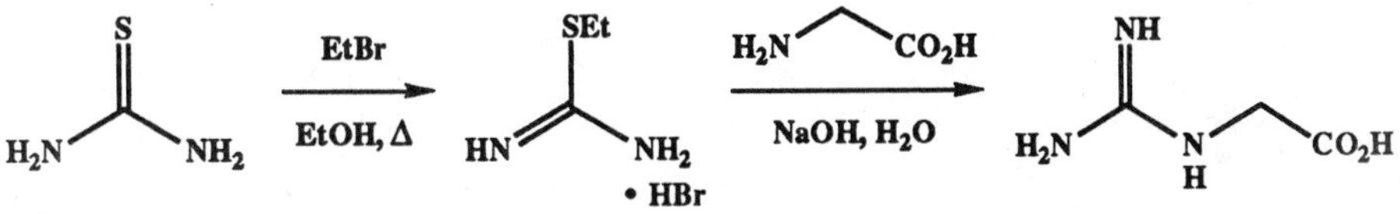
S
H$_2$N
NH$_2$
EtBr
EtOH, Δ
SEt
HN
NH$_2$
• HBr
H$_2$N
CO$_2$H
NaOH, H$_2$O
NH
H$_2$N
N
H
CO$_2$H

OS, *CV 3*, 465 (1955)

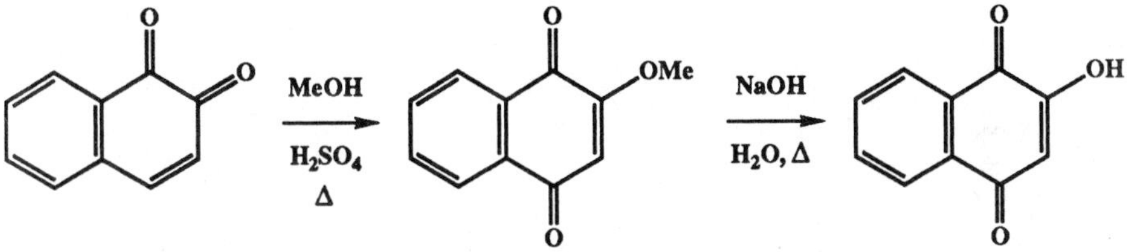
O
O
MeOH
H$_2$SO$_4$
Δ
O
OMe
O
NaOH
H$_2$O, Δ
O
OH
O

OS, *CV 3*, 470 (1955)

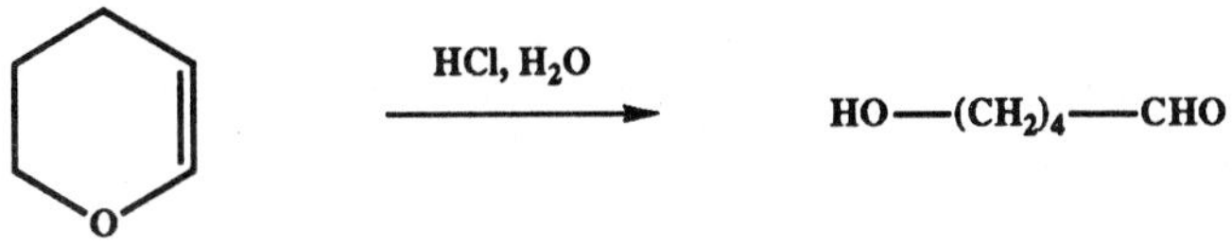
O
HCl, H$_2$O
HO—(CH$_2$)$_4$—CHO

OS, CV 3, 475 (1955)

POCl$_3$, pyridine

OS, CV 3, 490 (1955)

i-Pr–COOH $\xrightarrow{SOCl_2}$ *i*-Pr–COCl $\xrightarrow{NH_4OH}$ *i*-Pr–CONH$_2$

OS, CV 3, 493 (1955)

i-Pr–CONH$_2$ $\xrightarrow{P_2O_5,\ \Delta}$ *i*-Pr—C≡N

OS, CV 3, 501 (1955)

CH_3COCH_3 + $H_2NCH_2CH_2OH$ $\xrightarrow[\text{EtOH}]{H_2\text{ (25 psi)},\ PtO_2}$ $(CH_3)_2CH$–NH–CH_2CH_2OH

OS, CV 3, 516 (1955)

$CH_3CH(OH)CO_2Et$ —(liq. NH_3)→ $CH_3CH(OH)CONH_2$

OS, CV 3, 526 (1955)

$CH_3(CH_2)_4CHBrCHBrCH_2CHBrCHBr(CH_2)_7CO_2H$ —(Zn, HCl, EtOH, Δ)→ $CH_3(CH_2)_4CH{=}CHCH_2CH{=}CH(CH_2)_7CO_2Et$ —(NaOH, EtOH)→ $CH_3(CH_2)_4CH{=}CHCH_2CH{=}CH(CH_2)_7CO_2H$

OS, CV 3, 531 (1955)

$CH_3CH_2CHBrCHBrCH_2CHBrCHBrCH_2CHBrCHBr(CH_2)_7CO_2H$ —(Zn, HCl, EtOH, Δ)→ $CH_3CH_2CH{=}CHCH_2CH{=}CHCH_2CH{=}CH(CH_2)_7CO_2Et$ —(NaOH, EtOH)→ $CH_3CH_2CH{=}CHCH_2CH{=}CHCH_2CH{=}CH(CH_2)_7CO_2H$

OS, *CV* 3, 535 (1955)

$POCl_3$, Δ

ethylene dichloride

OS, *CV* 3, 536 (1955)

H_2SO_4, - 10 °C

liq. NH_3

OS, *CV* 3, 547 (1955)

$SOCl_2$

Δ

l - isomer

l - isomer

OS, *CV* 3, 549 (1955)

$Zn(CN)_2$

HCl, $AlCl_3$

$C_2H_2Cl_4$

HCl

H_2O

OS, CV 3, 555 (1955)

Br; CH_3; CH_3; CH_3 → 1. Mg, Et_2O, Δ; 2. CO_2; 3. H_3O^+ → HO, O; CH_3; CH_3; CH_3 → $SOCl_2$, Δ → Cl, O; CH_3; CH_3; CH_3

OS, CV 3, 584 (1955)

$MeO_2C—(CH_2)_8$—C(=O)NH_2 → P_2O_5, $C_2H_2Cl_4$, 145 °C → $MeO_2C—(CH_2)_8—C≡N$

OS, CV 3, 591 (1955)

NC; CH_3; O; NH_2; CN; O; NH_2 → HCl, H_2O, Δ → HO_2C; CH_3; CO_2H

OS, CV 3, 599 (1955)

$CH_3—NH_2$ → CS_2, aq. NaOH → [CH_3–NH–C(=S)–SNa] → $ClCO_2Et$ → [CH_3–NH–C(=S)–SCO_2Et] → $CH_3—N=C=S$

OS, *CV* 3, 610 (1955)

$CH_3C(O)CO_2H$ → (CH_3OH, *p*-TsOH, benzene, Δ) → $CH_3C(O)CO_2CH_3$

OH

OCH$_3$

OS, *CV* 3, 613 (1955)

HO$_2$C$(CH_2)_8$CO$_2$Me → (SOCl$_2$, Δ) → ClOC$(CH_2)_8$CO$_2$Me → (NH_4OH, 8 °C) → $H_2NOC(CH_2)_8CO_2Me$

OS, *CV* 3, 619 (1955)

Phthalic anhydride → (1. H_2O_2, NaOH, - 10 °C; 2. H_2SO_4, H_2O, - 10 °C) → 2-(O—OH carbonyl)benzoic acid (CO_2H)

OS, *CV* 3, 623 (1955)

Adipic acid (CO_2H, CO_2H) → (SOCl$_2$, Δ) → COCl, COCl → (1. Br$_2$; 2. EtOH) → Br, CO_2Et, CO_2Et, Br

OS, *CV* 3, 646 (1955)

O_2N — CO_2H — p-$TsNH_2$, PCl_5, Δ — O_2N — C≡N

OS, *CV* 3, 649 (1955)

2 O_2N — O — Cl — Na_2O_2, toluene, 0 °C — O_2N — O — O — O — O — NO_2

OS, *CV* 3, 690 (1955)

CH_2OH, OH, OH, H, CHO, HO, H, OH — NH_2OH, Na, MeOH, 65 °C — CH_2OH, OH, NOH, OH, H, CH, HO, H, OH — Ac_2O, NaOAc, HOAc, Δ — CH_2OAc, OAc, N, OAc, H, C, AcO, H, OAc

d - glucose

d - isomer

OS, *CV* 3, 712 (1955)

N=N — CO_2H — $SOCl_2$, Δ — N=N — COCl

***OS, CV 3,* 714 (1955)**

CH=CH—CO_2H

1. $SOCl_2$, Δ

2. phenol, Δ

CH=CH—CO_2Ph

***OS, CV 3,* 717 (1955)**

O

Ph CH_3

H_2, Ra (Ni)

(3500-5000 psi)

NH_3, 150 °C

NH_2

Ph CH_3

***OS, CV 3,* 735 (1955)**

O

Ph N=C=S

$PhNH_2$

acetone

O S

Ph N H N H Ph

NaOH

H_2O, Δ

S

Ph N H NH_2

***OS, CV 3,* 765 (1955)**

CO_2Ph

OH

CH_3

NH_2

$C_6H_3Cl_3$, Δ

CH_3

O

N H

OH

OS, CV 3, 768 (1955)

$HO_2C(CH_2)_8CO_2H$ → ($H_2NC(O)NH_2$, 160 °C) → [$H_2NOC(CH_2)_8CONH_2$] → (220 - 340 °C) → $NC(CH_2)_8CN$ + $NC(CH_2)_8CO_2H$

OS, CV 3, 774 (1955)

$MeOCH_2CH(Br)CO_2Me$ → (aq. NaOH, then H_2SO_4) → $MeOCH_2CH(Br)CO_2H$

OS, CV 3, 827 (1955)

3-$CH_3C_6H_4NH_2$ → (PhCHO) → 3-$CH_3C_6H_4N{=}CHPh$ → (H_2 (1000 psi), Ra (Ni), Et_2O) → 3-$CH_3C_6H_4NHCH_2Ph$

OS, CV 3, 846 (1955)

n-$C_{11}H_{23}C(O)Cl$ → (NaN_3, H_2O, acetone) → [n-$C_{11}H_{23}C(O)N_3$] → (Δ, benzene) → n-$C_{11}H_{23}$—N=C=O

OS, CV 4, 39 (1963)

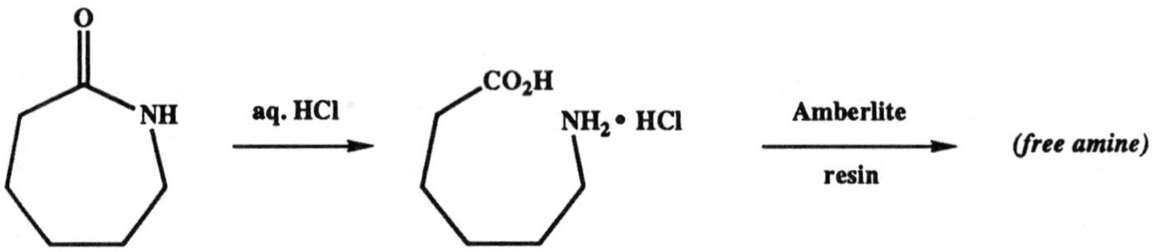

aq. HCl; CO_2H; $NH_2 \cdot HCl$; Amberlite resin; *(free amine)*

OS, CV 4, 80 (1963)

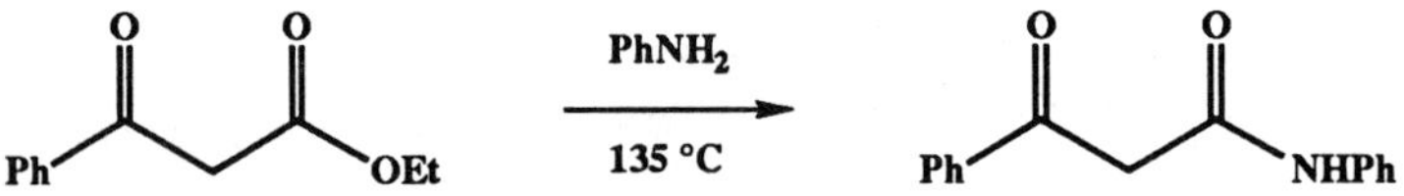

Ph, OEt; $PhNH_2$, 135 °C; Ph, NHPh

OS, CV 4, 101 (1963)

2 $CH_2{=}O$ → HCl, $ClSO_3H$, H_2O → $ClCH_2OCH_2Cl$

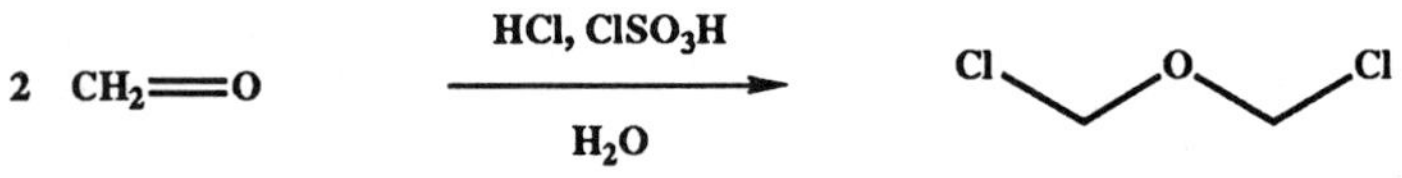

OS, CV 4, 144 (1963)

$ClCH_2C(O)NH_2$ → P_2O_5, Δ, trimethylbenzene → $ClCH_2C{\equiv}N$

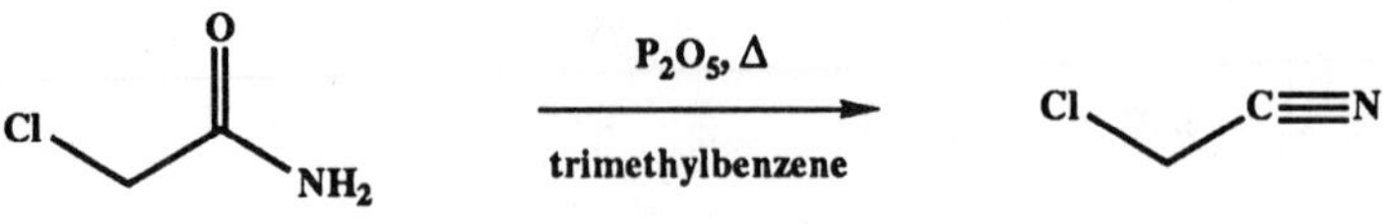

OS, *CV* 4, 154 (1963)

$Me_3\overset{+}{N}CH_2COOH\ Cl^-$ —[$SOCl_2$, 70 °C]→ $Me_3\overset{+}{N}CH_2COCl\ Cl^-$

OS, *CV* 4, 166 (1963)

Nicotinamide N-oxide ($CONH_2$, N^+–O^-) —[PCl_5, $POCl_3$, Δ]→ 2-chloro-3-cyanopyridine (CN, Cl, N)

OS, *CV* 4, 169 (1963)

$PhCH(OH)CO_2Et$ —[$SOCl_2$]→ $PhCH(Cl)CO_2Et$ —[HOAc, HCl, Δ]→ $PhCH(Cl)CO_2H$

OS, *CV* 4, 172 (1963)

2-Cl-C_6H_4-NH-C(=S)-NH_2 —[1. $Pb(OAc)_2$, aq. KOH, Δ; 2. HOAc]→ 2-Cl-C_6H_4-NH-CN

OS, *CV* 4, 178 (1963)

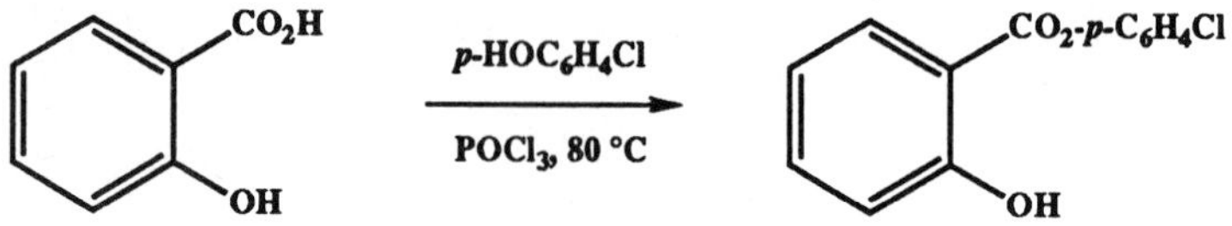

Salicylic acid (CO_2H, OH) → (CO_2-*p*-C_6H_4Cl, OH)

p-HOC_6H_4Cl

$POCl_3$, 80 °C

OS, *CV* 4, 229 (1963)

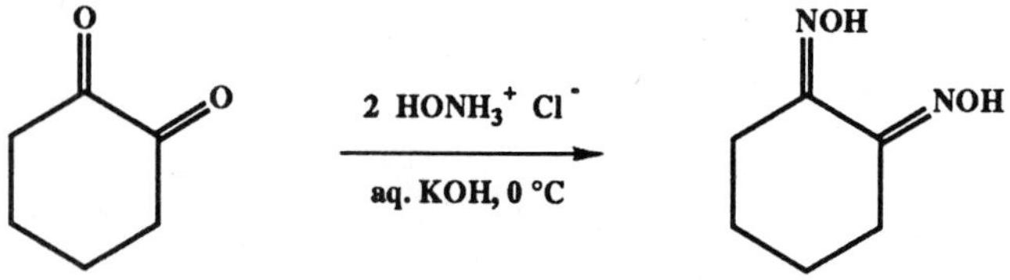

1,2-Cyclohexanedione (O, O) → (NOH, NOH)

2 $HONH_3^+$ Cl^-

aq. KOH, 0 °C

OS, *CV* 4, 242 (1963)

HO, OH, CO_2H, CO_2H → AcO, OAc, O, O, O

Ac_2O, Δ

H_2SO_4

d - isomer

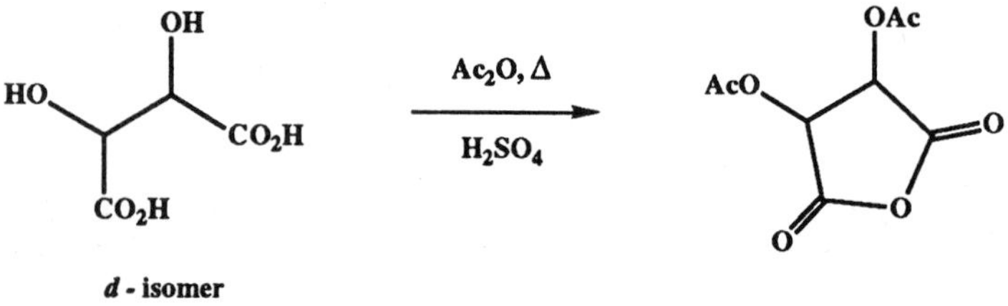

OS, *CV* 4, 261 (1963)

CO_2H / CO_2H + 2 $CH_2{=}C(CH_3)_2$ ⇌ CO_2-*t*-Bu / CO_2-*t*-Bu

H_2SO_4

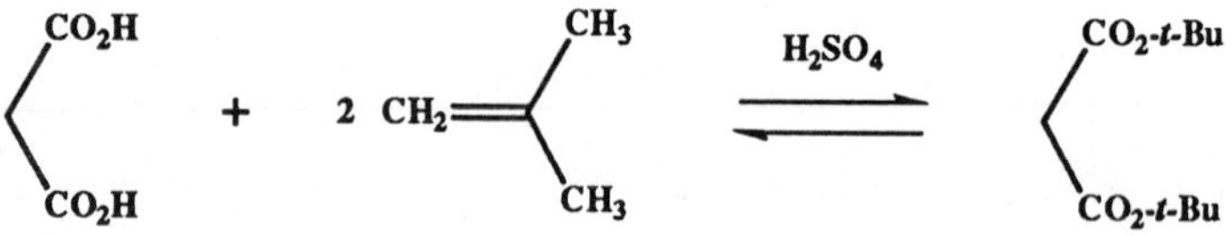

OS, *CV 4*, 263 (1963)

$HO_2C-CH_2-CO_2H$ —(2 $SOCl_2$, 50 °C)→ $ClOC-CH_2-COCl$ —(2 *t*-BuOH, $PhNMe_2$, Δ)→ $t\text{-}BuO_2C-CH_2-CO_2\text{-}t\text{-}Bu$

OS, *CV 4*, 274 (1963)

2 cyclohexanone —(2 NaCN, H_2O; $H_2NNH_2 \cdot H_2SO_4$)→ 1,2-bis(1-cyanocyclohexyl)hydrazine (CN, NC, N—N, H H)

OS, *CV 4*, 276 (1963)

$(NC)_2C{=}C(CN)_2$ + $HOCH_2CH_2OH$ —(urea, 75 °C)→ $(NC)_2C{=}C$(1,3-dioxolan-2-ylidene)

OS, *CV 4*, 302 (1963)

2-furyl-$CH{=}CHCO_2H$ —(2 EtOH, HCl, Δ)→ $EtO_2CCH_2CH_2COCH_2CH_2CO_2Et$

OS, *CV* 4, 304 (1963)

O O O

p-TsOH

EtOH, Δ

CO_2Et

CO_2Et

OS, *CV* 4, 307 (1963)

S S

Et_2N S—S NEt_2

Cl_2

50 - 75 °C

2

S

Et_2N Cl

OS, *CV* 4, 317 (1963)

$CH_3(CH_2)_7$—CH—CH—$(CH_2)_7CO_2H$

OH OCHO

NaOH, Δ

then HCl

$CH_3(CH_2)_7$—CH—CH—$(CH_2)_7CO_2H$

OH OH

OS, *CV* 4, 329 (1963)

HO_2C—C≡C—CO_2K

MeOH

H_2SO_4

MeO_2C—C≡C—CO_2Me

OS, *CV* 4, 339 (1963)

O

OH

$SOCl_2$

Δ

O

Cl

Me_2NH

benzene

O

NMe_2

OS, *CV 4*, 348 (1963)

$CH_3CH(CH_3)CO_2H$ → (P, Br_2; 100 °C) → $(CH_3)_2CBrCOBr$ → (Zn, Δ; EtOAc) → $(CH_3)_2C{=}C{=}O$

OS, *CV 4*, 383 (1963)

PhC(O)NHPh → (PCl_5; Δ) → PhC(Cl)=NPh → ($PhNH_2$, Δ; pyridine) → PhC(NHPh)=NPh

OS, *CV 4*, 390 (1963)

$HO_2C(CH_2)_3CO_2H$ → (phenol, $POCl_3$; benzene, Δ) → $PhO_2C(CH_2)_3CO_2Ph$

OS, *CV 4*, 411 (1963)

H_2N—NH_2 → (2 $ClCO_2Et$, Na_2CO_3; H_2O, EtOH) → EtO_2C—NH—NH—CO_2Et

OS, *CV 4*, 417 (1963)

CO_2Et–CH_2–CO_2Et → (1. KOH, EtOH; 2. aq. HCl, 5 - 10 °C) → CO_2Et–CH_2–CO_2H ⇌ (isobutylene, H_2SO_4) CO_2Et–CH_2–CO_2-*t*-Bu

OS, *CV 4*, 427 (1963)

Cl_2CH–CO_2H → (1. 3 NaOEt, EtOH, Δ; 2. HCl, EtOH) → $(EtO)_2CH$–CO_2Et

OS, *CV 4*, 436 (1963)

$CH_3CH_2CH_2CH_2CH(Et)CONH_2$ → ($SOCl_2$; benzene, 80 °C) → $CH_3CH_2CH_2CH_2CH(Et)C{\equiv}N$

OS, *CV 4*, 441 (1963)

NC, O, Et, NH, Me, NC, O → (H_2SO_4, H_2O; Δ) → Et, Me, CO_2H, CO_2H

OS, *CV 4*, 444 (1963)

n-Bu, Et, CO_2Et, Me → 1. aq. KOH, EtOH, ethylene glycol; 2. HCl → n-Bu, Et, CO_2H, Me

OS, *CV 4*, 486 (1963)

EtO, O, H, H, O, OEt → 1. NH_4OH, NH_4Cl; 2. P_2O_5, Δ → H, NC, CN, H

OS, *CV 4*, 496 (1963)

C≡N, $(CH_2)_3$, CO_2K → dilute HCl → O, C—NH_2, $(CH_2)_3$, CO_2H → HCl, Δ → CO_2H, $(CH_2)_3$, CO_2H

OS, *CV 4*, 499 (1963)

O → 0.2 *N* HNO_3, Δ → OH, CHO

OS, CV 4, **513 (1963)**

$n\text{-}C_6H_{13}C(O)OH \xrightarrow{\text{urea, }\Delta} n\text{-}C_6H_{13}C(O)NH_2$

OS, CV 4, **554 (1963)**

$CH_2{=}C(CO_2H)CH_2CO_2H \xrightarrow{2\ PCl_5,\ \Delta} CH_2{=}C(COCl)CH_2COCl$

OS, CV 4, **558 (1963)**

$i\text{-Pr}CH_2CH_2C(O)CH{=}CHCl \xrightarrow[\text{MeOH, - 10 °C}]{\text{NaOH}} i\text{-Pr}CH_2CH_2C(O)CH_2CH(OMe)_2$

OS, CV 4, **573 (1963)**

$2\text{-}(OCH_3)C_6H_4CH{=}C(CH_3)NO_2 \xrightarrow[H_2O]{\text{Fe, HCl}} 2\text{-}(OCH_3)C_6H_4CH_2C(O)CH_3$

OS, CV 4, 588 (1963)

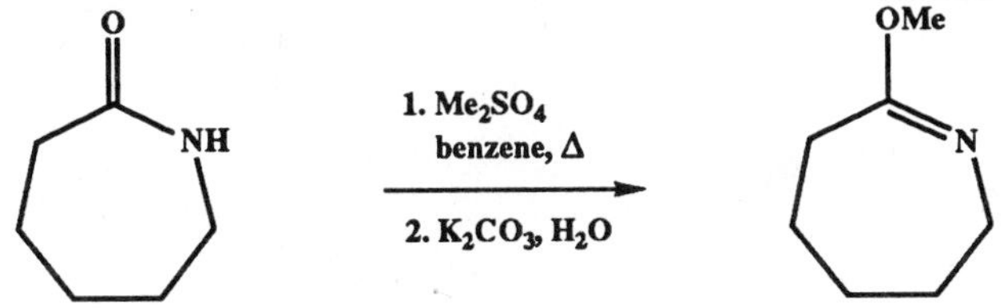

O, NH; 1. Me_2SO_4 benzene, Δ; 2. K_2CO_3, H_2O; OMe, N

OS, CV 4, 590 (1963)

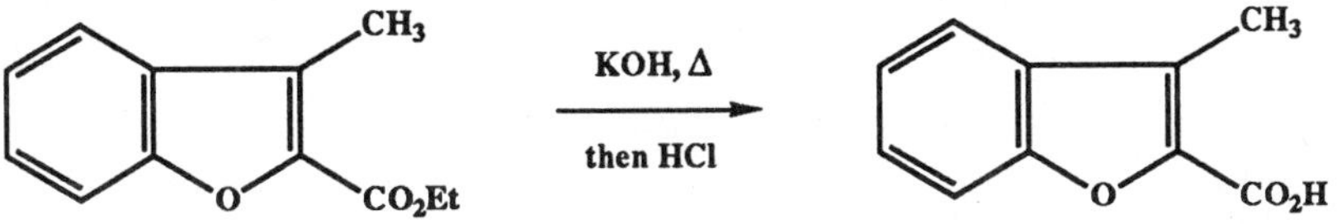

CH_3, O, CO_2Et; KOH, Δ then HCl; CH_3, O, CO_2H

OS, CV 4, 603 (1963)

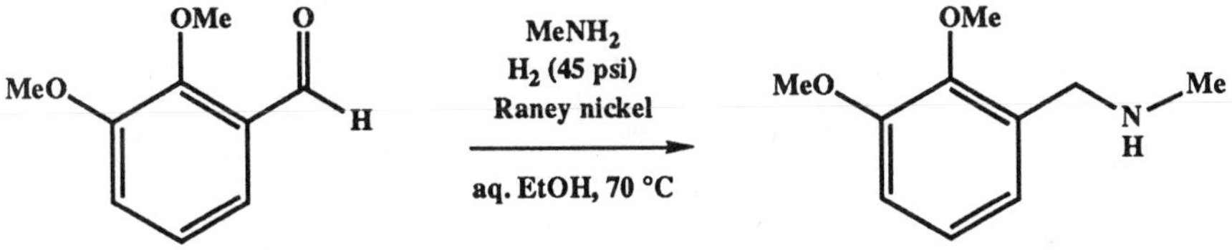

MeO, OMe, O, H; $MeNH_2$ H_2 (45 psi) Raney nickel aq. EtOH, 70 °C; MeO, OMe, N H, Me

OS, CV 4, 608 (1963)

C_9H_{19}, CO_2H, CH_3; 1. Br_2, PBr_3, 90 °C 2. MeOH; C_9H_{19}, CO_2Me, Br, CH_3; quinoline Δ; C_9H_{19}, CO_2Me, CH_3, H; 1. KOH, Δ EtOH 2. H_2SO_4 H_2O; C_9H_{19}, CO_2H, CH_3, H

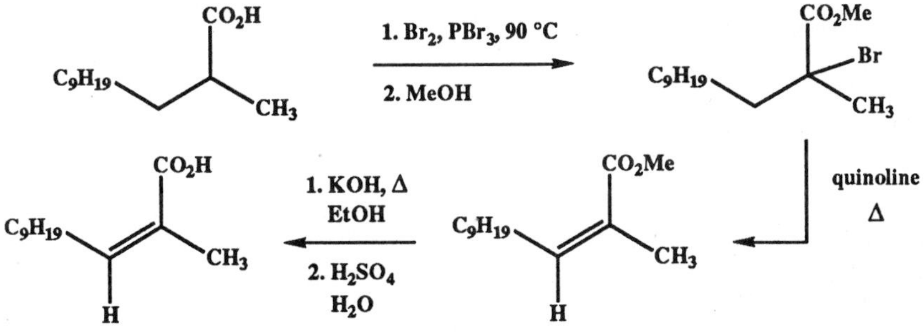

OS, CV 4, 616 (1963)

$C_9H_{19}CH_2CH(CH_3)CO_2H$ → (Br_2, PBr_3; 90 °C) → $C_9H_{19}CH_2CBr(CH_3)COBr$ → (KO-*t*-Bu, Δ) → $C_9H_{19}CH_2C(=CH_2)CO_2$-*t*-Bu → (1. KOH, Δ EtOH; 2. H_2SO_4 H_2O) → $C_9H_{19}CH_2C(=CH_2)CO_2H$

OS, CV 4, 620 (1963)

PhN(H)Me + EtCH(Br)COCl → (benzene) → PhN(Me)C(=O)CH(Br)Et

OS, CV 4, 628 (1963)

methyl 3-methylfuran-2-carboxylate (CH_3, CO_2Me) → (1. aq. NaOH, Δ; 2. HCl) → 3-methylfuran-2-carboxylic acid (CH_3, CO_2H)

OS, CV 4, 635 (1963)

$MeO_2C—(CH_2)_9—CO_2Me$ → (1. $Ba(OH)_2$ aq. MeOH; 2. aq. HCl) → $MeO_2C—(CH_2)_9—CO_2H$

OS, *CV* 4, 664 (1963)

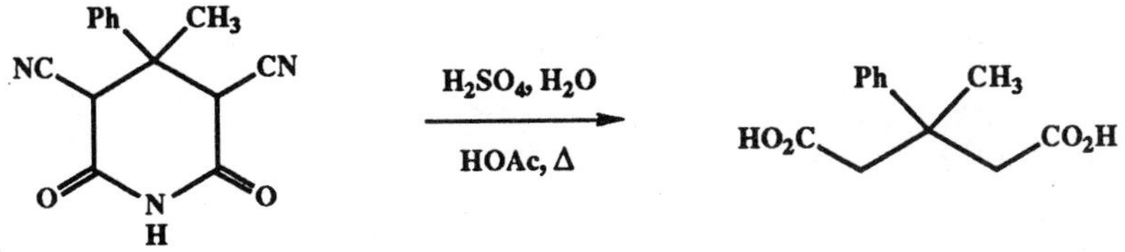

H_2SO_4, H_2O

HOAc, Δ

Ph, CH_3, NC, CN, O, N, H, O → Ph, CH_3, HO_2C, CO_2H

OS, *CV* 4, 667 (1963)

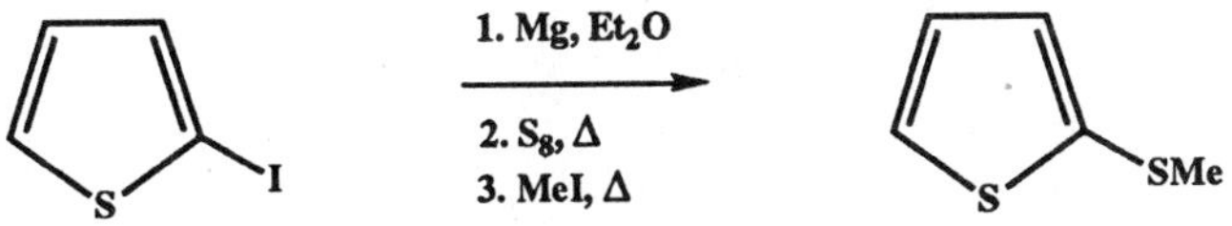

1. Mg, Et_2O

2. S_8, Δ

3. MeI, Δ

S, I → S, SMe

OS, *CV* 4, 706 (1963)

P_2O_5, Δ

O, NH_2, N → CN, N

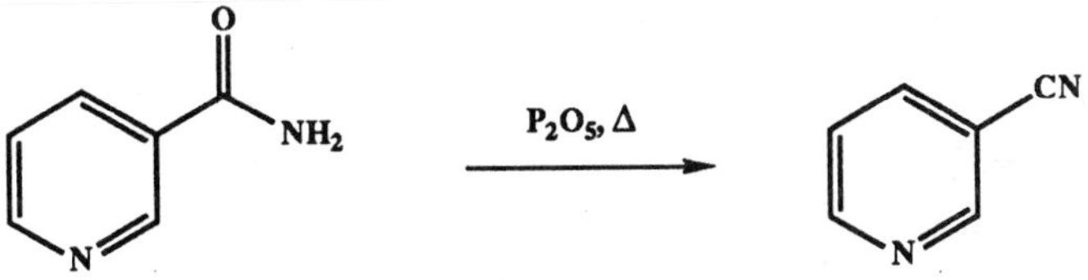

OS, *CV* 4, 715 (1963)

$SOCl_2$

Δ

NaN_3

H_2O

O, OH, NO_2 → O, Cl, NO_2 → O, N_3, NO_2

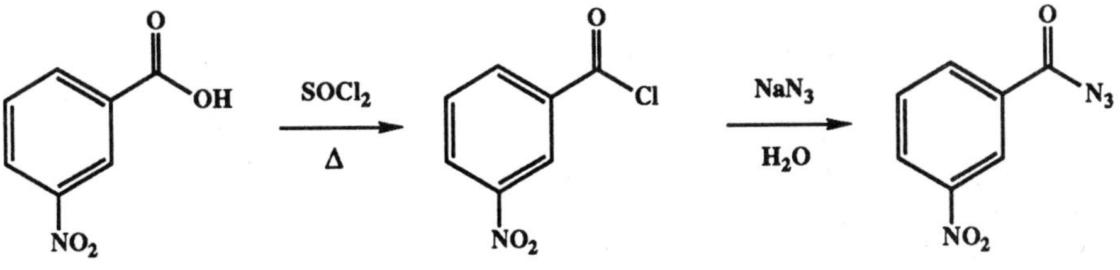

OS, CV 4, 739 (1963)

$n\text{-}C_{17}H_{33}COOH \xrightarrow{SOCl_2,\ \Delta} n\text{-}C_{17}H_{33}COCl$

OS, CV 4, 766 (1963)

Ph—CH=CH—CH$_2$-(succinic anhydride) $\xrightarrow[2.\ aq.\ H_2SO_4]{1.\ aq.\ Na_2CO_3,\ \Delta}$ Ph—CH=CH—CH$_2$—CH(CO_2H)—CH$_2$—CO_2H

OS, CV 4, 819 (1963)

$EtO_2C\text{-}(CH_2)_4\text{-}CO_2Et \xrightarrow[EtOH,\ \Delta]{H_2NNH_2} H_2NNHCO\text{-}(CH_2)_4\text{-}CONHNH_2$

OS, CV 4, 844 (1963)

$HO_2C(Br)C=C(Br)CHO \xrightarrow[EtOH,\ H_2O]{NaNO_2} Na^+\ {}^-C(CHO)_2NO_2 \cdot H_2O$

OS, CV 4, 903 (1963)

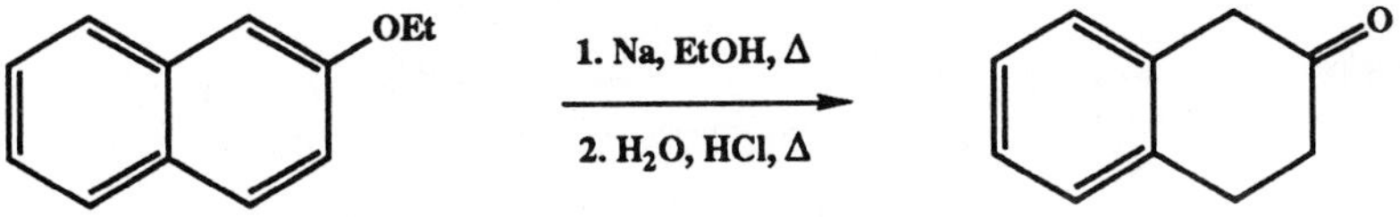

OEt

1. Na, EtOH, Δ

2. H_2O, HCl, Δ

O

OS, CV 4, 924 (1963)

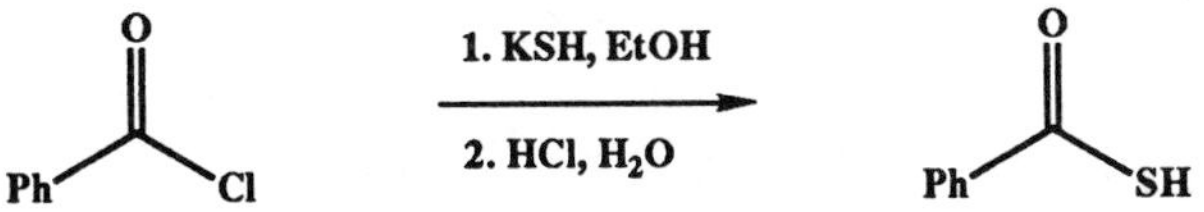

O

Ph Cl

1. KSH, EtOH

2. HCl, H_2O

O

Ph SH

OS, CV 4, 927 (1963)

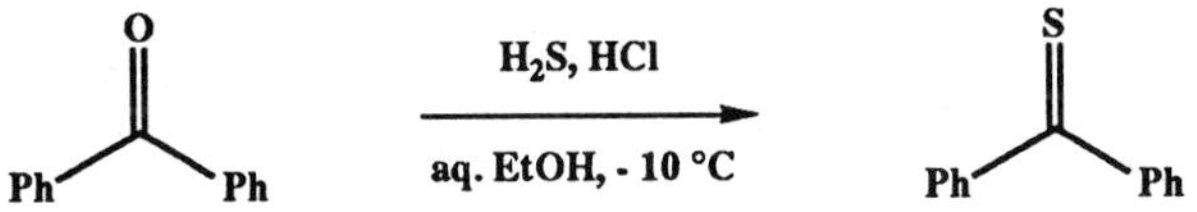

O

Ph Ph

H_2S, HCl

aq. EtOH, - 10 °C

S

Ph Ph

OS, CV 4, 928 (1963)

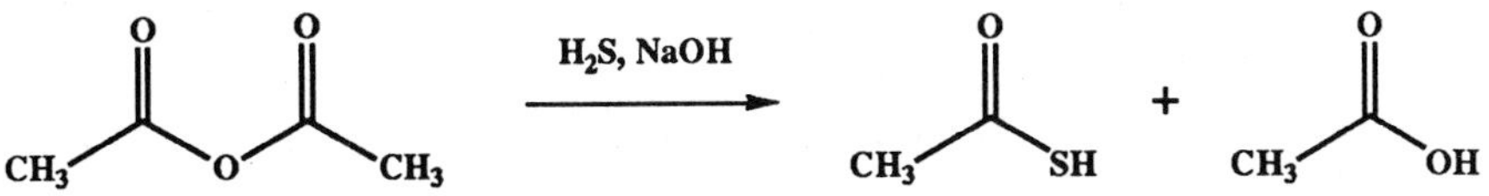

O O

CH_3 O CH_3

H_2S, NaOH

O

CH_3 SH

\+

O

CH_3 OH

OS, CV 4, 977 (1963)

$CH_3(CH_2)_{10}$—CO_2H → (vinyl OAc; $Hg(OAc)_2$, H_2SO_4, Δ) → $CH_3(CH_2)_{10}$—CO_2—(vinyl)

OS, CV 5, 66 (1973)

PhC(O)Cl → (HF) → PhC(O)F

OS, CV 5, 111 (1973)

2 $C_6H_5NMe_2$ → (ClC(O)C(O)Cl; $AlCl_3$, CS_2) → Me_2N–C_6H_4–C(O)C(O)–C_6H_4–NMe_2

OS, CV 5, 155 (1973)

diketene → (NaOAc; *t*-BuOH) → CH_3C(O)CH_2C(O)O-*t*-Bu

OS, CV 5, 162 (1973)

t-BuOH —($NaCNO, CF_3CO_2H$ / benzene)→ *t*-BuO–C(=O)–NH_2

OS, CV 5, 166 (1973)

CH_3S–C(=O)–Cl —(*t*-BuOH, Δ / $CHCl_3$, pyr.)→ CH_3S–C(=O)–O-*t*-Bu —(H_2NNH_2, Δ)→ H_2NNH—CO_2-*t*-Bu

OS, CV 5, 168 (1973)

PhO–C(=O)–Cl —(*t*-BuOH / CH_2Cl_2, quinoline)→ PhO–C(=O)–O-*t*-Bu —(H_2NNH_2, Δ)→ H_2NNH—CO_2-*t*-Bu

OS, CV 5, 171 (1973)

$NCCH_2CO_2H$ —(PCl_5 / Et_2O)→ $NCCH_2COCl$ —(*t*-BuOH / $PhNMe_2$, Δ)→ $NCCH_2CO_2$-*t*-Bu

OS, CV 5, **258 (1973)**

p-TsNH—NH$_2$ → [(HO)$_2$CHCO$_2$H; HCl, H$_2$O] → p-TsNH—N=CHCO$_2$H → [SOCl$_2$, Δ; benzene] → p-TsNH—N=CHCOCl → [trans - CH$_3$CH=CHCH$_2$OH; Et$_3$N, CH$_2$Cl$_2$, 0 °C] → trans-CH$_3$CH=CHCH$_2$O–C(=O)–CHN$_2$

OS, CV 5, **292 (1973)**

cyclohexanone → [2 CH$_2$=CHCH$_2$OH; Me$_2$C(OMe)$_2$; p-TsOH, benzene, Δ] → 1,1-bis(allyloxy)cyclohexane

OS, CV 5, **365 (1973)**

HC(=O)OCH$_3$ → [PCl$_5$; POCl$_3$] → Cl$_2$CH–OCH$_3$

OS, CV 5, **387 (1973)**

Cl$_3$C–C(=O)–NEt$_2$ → [(n-Bu)$_3$P, Et$_2$O; 85 - 95 °C] → Cl$_2$C=C(Cl)–NEt$_2$

OS, *CV 5*, 459 (1973)

p-TsOH

MeOH, Δ

CO_2Me

CO_2Me

OS, *CV 5*, 539 (1973)

O

OH

EtOH, p-TsOH

benzene, Δ

O

OEt

OS, *CV 5*, 545 (1973)

O

O

HBr, EtOH

0 °C

Br

CO_2Et

OS, *CV 5*, 612 (1973)

O

NH

S

1. aq. KOH, Δ

2. aq. HCl

CH_2CO_2H

CO_2H

OS, CV 5, 627 (1973)

1. aq. NaOH, Δ
2. aq. HCl

HO, CH=, NH, HN, O, O → HO, CO_2H, O

OS, CV 5, 642 (1973)

MgBr, S → $PhCO_2O$-*t*-Bu, Et_2O, 0 °C → S, O-*t*-Bu → *p*-TsOH, Δ → S, OH ⇌ *keto form*

OS, CV 5, 645 (1973)

O, H_2N, OEt + $NH_2OH \cdot HCl$ → 1. aq. NaOH, 2. HCl → O, H_2N, N, H, OH

OS, CV 5, 656 (1973)

CHO, N, H → $(NH_4)_2HPO_4$, HOAc, nitropropane, Δ → CN, N, H

OS, *CV 5*, 684 (1973)

$$CH_2{=}CCl_2 + 2\ CH_3OCH_2CH_2OH \xrightarrow[150\ °C]{Na,\ xylene} CH_2{=}C(OCH_2CH_2OCH_3)_2$$

OS, *CV 5*, 762 (1973)

n-Bu, CH_3, CO_2-*sec*-Bu → 1. aq. KOH, EtOH, Δ; 2. aq. HCl → *n*-Bu, CH_3, CO_2H

OS, *CV 5*, 780 (1973)

N-CH_3, O → P_2S_5, CS_2, Δ → N-CH_3, S → 1. CH_3I, Et_2O; 2. *t*-BuOK, Et_2O, Δ → N-CH_3, SCH_3

OS, *CV 5*, 808 (1973)

O → HN, O (morpholine); *p*-TsOH, toluene, Δ → N, O

OS, CV 5, 822 (1973)

2 (pyridine-3-carboxylic acid, CO_2H) —Et_3N, Cl_2CO / benzene→ nicotinic anhydride

OS, CV 5, 887 (1973)

d - gluconolactone —Ac_2O / $ZnCl_2$→ *d* - tetraacetate —Ac_2O / $ZnCl_2$→ *d* - pentaacetate

OS, CV 5, 966 (1973)

Ph-NH-C(=S)-NH-C(OMe)=NH —HCl / EtOH, Δ→ Ph-NH-C(=S)-NH-C(=O)-NH_2

OS, CV 5, 1005 (1973)

$Ph—CH=CH_2$ —2 PCl_5 / benzene→ [$Ph—CH=CH\overset{+}{P}Cl_3$ PCl_6^-] —2 SO_2→ $Ph—CH=CHPOCl_2$

OS, CV 5, 1031 (1973)

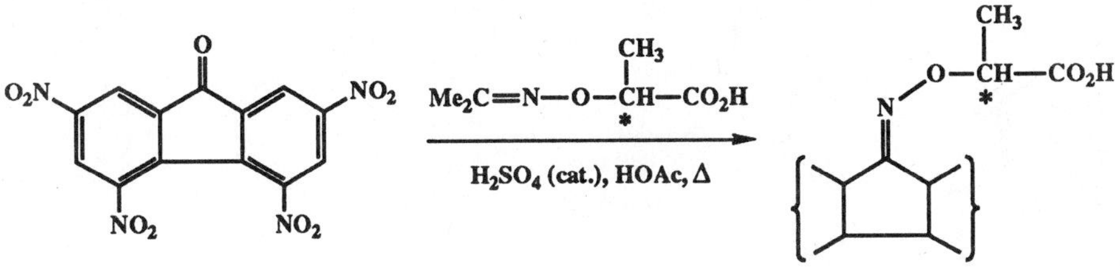

$Me_2C{=}N{-}O{-}CH(CH_3){-}CO_2H$

H_2SO_4 (cat.), HOAc, Δ

OS, CV 5, 1051 (1973)

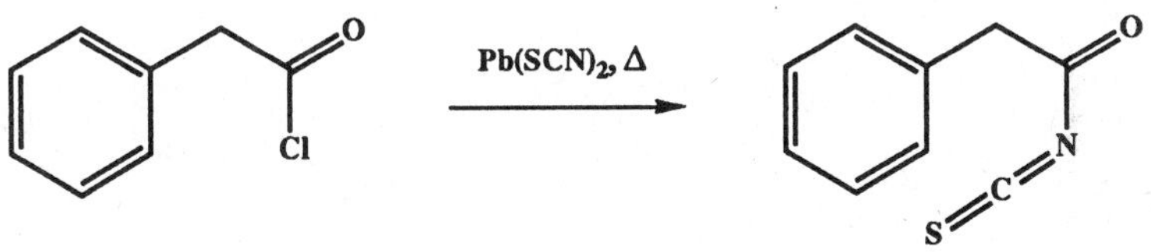

$Pb(SCN)_2$, Δ

OS, CV 5, 1074 (1973)

$Cl_3C{-}CO{-}CCl_3$

hexane, 65 - 70 °C

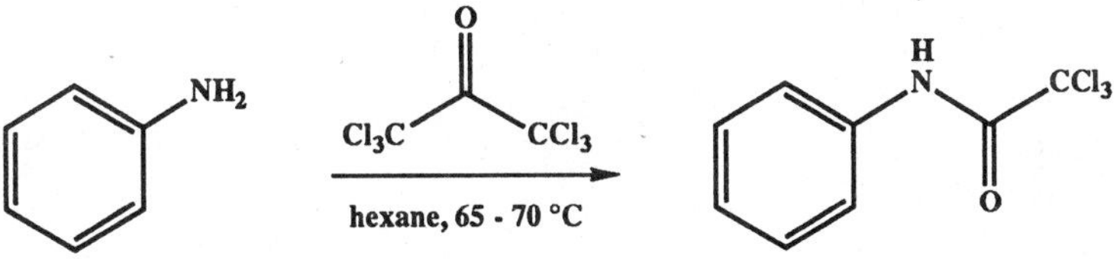

OS, CV 5, 1082 (1973)

$CH_3(CH_2)_5{-}CO_2H$ → $CH_3(CH_2)_5{-}CF_3$

SF_4, 100 - 130 °C

(0.5 - 1 mm Hg)

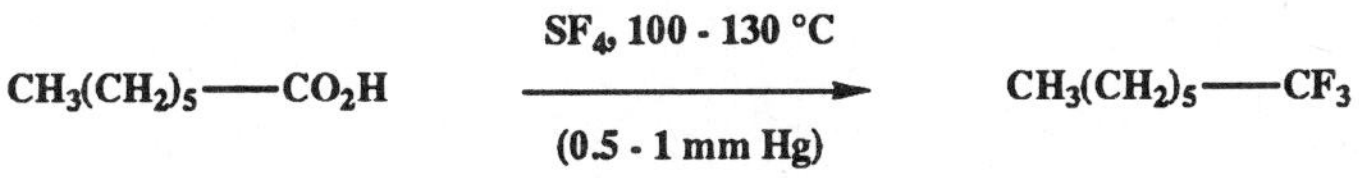

OS, CV 6, 1 (1988)

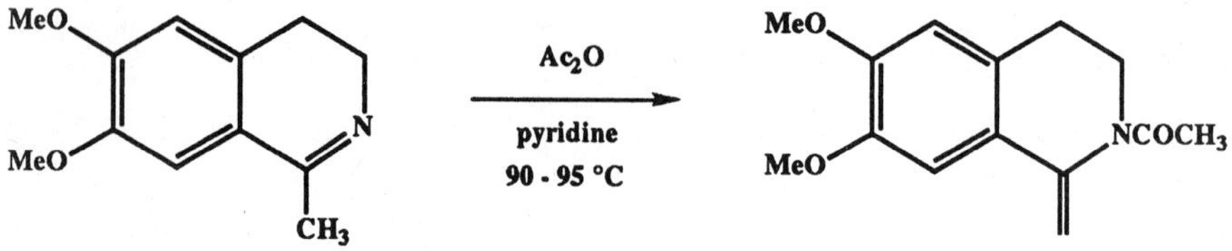

MeO, MeO, N, CH_3 → Ac_2O, pyridine, 90 - 95 °C → MeO, MeO, $NCOCH_3$

OS, CV 6, 8 (1988)

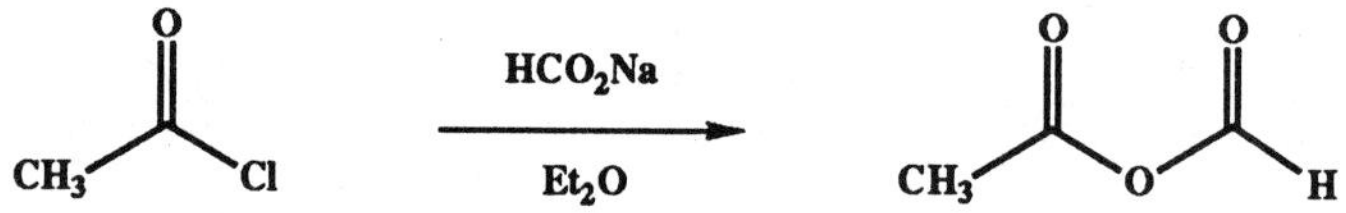

O, CH_3, Cl → HCO_2Na, Et_2O → O, O, CH_3, O, H

OS, CV 6, 10 (1988)

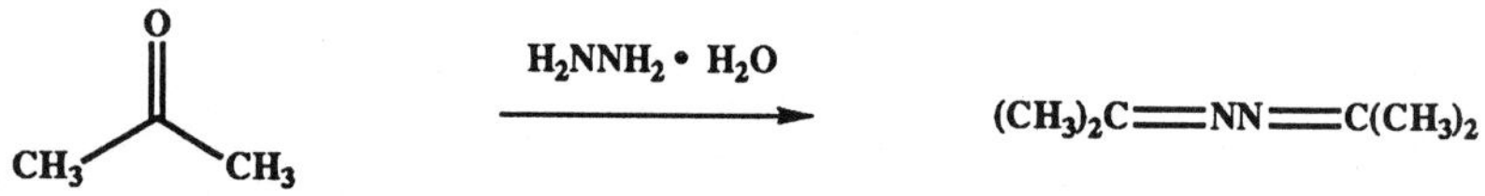

O, CH_3, CH_3 → $H_2NNH_2 \bullet H_2O$ → $(CH_3)_2C{=}NN{=}C(CH_3)_2$

OS, CV 6, 12 (1988)

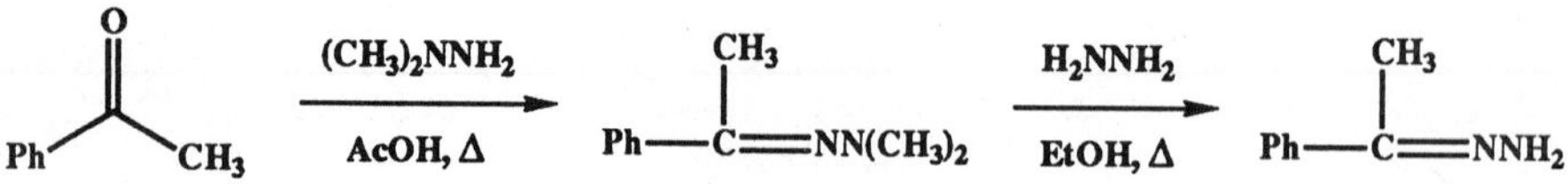

O, Ph, CH_3 → $(CH_3)_2NNH_2$, AcOH, Δ → CH_3, Ph—C=$NN(CH_3)_2$ → H_2NNH_2, EtOH, Δ → CH_3, Ph—C=NNH_2

OS, *CV 6*, 62 (1988)

p-Ts-NHNH$_2$

MeOH, Δ

OH

H

H

H

p-Ts-NHN

OS, *CV 6*, 95 (1988)

CO_2H

1. $ClCO_2Et$, (i-Pr)$_2$NEt, acetone, 0 °C

2. NaN_3, H_2O, 0 °C

O

N_3

OS, *CV 6*, 121 (1988)

O

CH_3

Ac_2O, $HClO_4$

CCl_4

OAc

CH_3

OS, *CV 6*, 190 (1988)

$CH_3CH_2CH_2CH_2CH_2$—CO_2H

1. $SOCl_2$, CCl_4, 65 °C

2. N-bromosuccinimide, HBr, 85 °C

$CH_3CH_2CH_2CH_2CH$—COCl (Br on CH)

OS, CV 6, 199 (1988)

OH, N, Ph, CN; Cl_2CO, $PhNMe_2$, benzene, dioxane, 5 - 6 °C → OCOCl, N, Ph, CN; t-BuOH, pyridine, 5 - 10 °C → OCO_2-t-Bu, N, Ph, CN

OS, CV 6, 207 (1988)

O, t-BuO, OK; $(EtO)_2P(O)Cl$ → O, O, t-BuO, O, $P(OEt)_2$; KN_3 → O, t-BuO, N_3

OS, CV 6, 210 (1988)

O, t-Bu, t-Bu, O; 1. Cl_2, HOAc 2. Et_2NH, Et_2O 3. Cl_2, HOAc 4. Et_2NH, Et_2O → O, Cl, t-Bu, t-Bu, Cl, O; NaN_3, MeOH → O, N_3, t-Bu, t-Bu, N_3, O

OS, CV 6, 259 (1988)

O, Cl, CH_3; t-BuOLi → O, O-t-Bu, CH_3

OS, *CV* 6, 263 (1988)

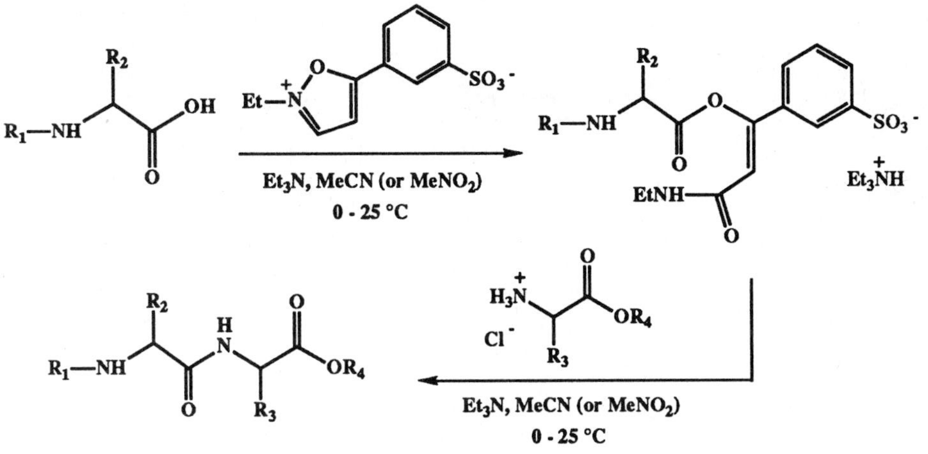

R_2, R_1—NH, OH, O; Et—$\overset{+}{N}$, O, SO_3^-

Et_3N, MeCN (or $MeNO_2$)
0 - 25 °C

R_2, R_1—NH, O, O, SO_3^-, EtNH, O, $Et_3\overset{+}{N}H$

$H_3\overset{+}{N}$, Cl^-, O, OR_4, R_3

Et_3N, MeCN (or $MeNO_2$)
0 - 25 °C

R_2, R_1—NH, H, N, O, OR_4, O, R_3

OS, *CV* 6, 276 (1988)

O, Cl, Cl

1. H_2O_2, aq. NaOH
$MgSO_4$, dioxane

2. aq. H_2SO_4

O, O—OH, Cl

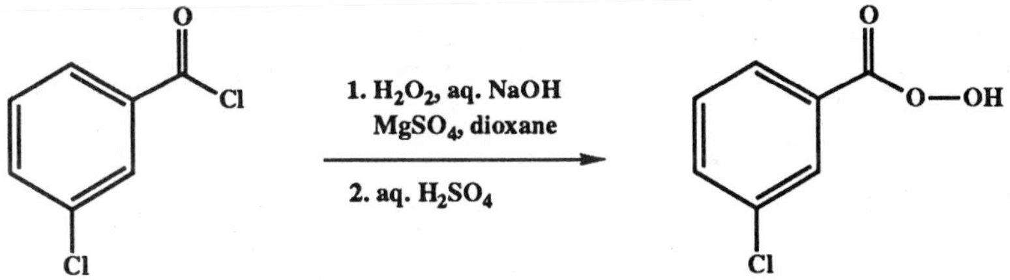

OS, *CV* 6, 282 (1988)

O, NMe_2

Cl_2CO

CH_2Cl_2
0 - 20 °C

Cl, $\overset{+}{N}Me_2$ Cl^-

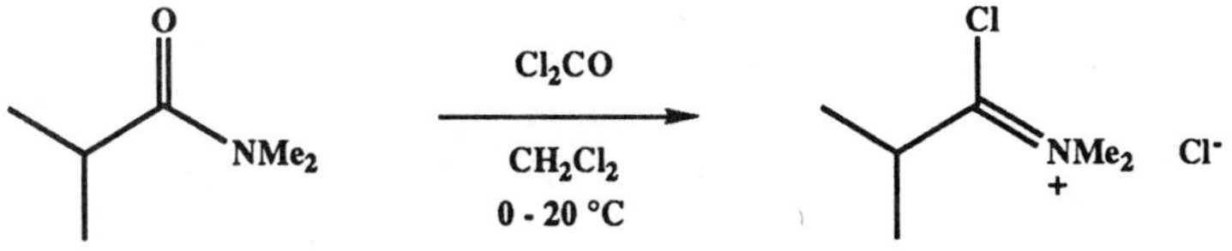

OS, *CV 6*, 293 (1988)

C_8H_{17} H H O

p-TsNHNH$_2$

EtOH, Δ

C_8H_{17} H H p-TsNHN

OS, *CV 6*, 304 (1988)

CH═CH—CO_2H

$ClSO_2NCO$

Δ

O CH═CH—C $NHSO_2Cl$

OS, *CV 6*, 312 (1988)

O Cl

HO OSO_2CH_3 CH_3 CH_3

pyridine, 0 °C

O OSO_2CH_3 O CH_3 CH_3

erythro

$NaBH_4$

pyridine

115 °C

CH_3 O O CH_3

OS, CV 6, 361 (1988)

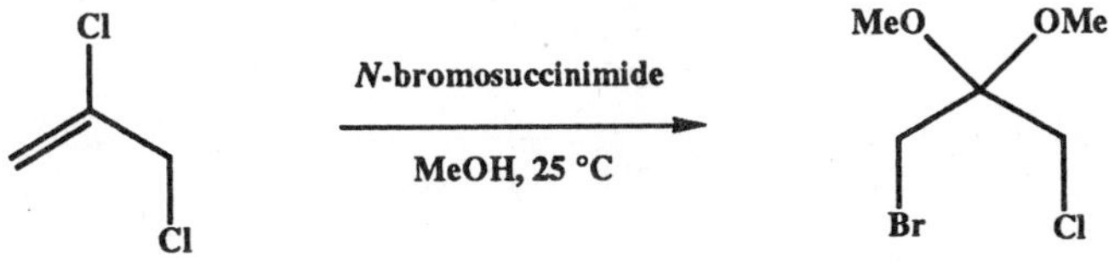

Cl ... Cl → *N*-bromosuccinimide / MeOH, 25 °C → MeO, OMe, Br, Cl

OS, CV 6, 418 (1988)

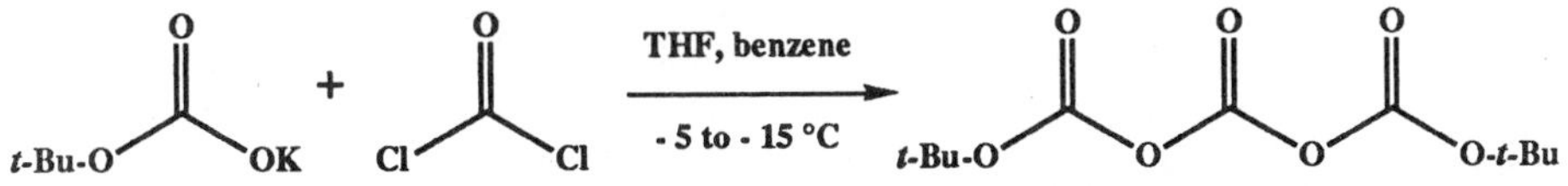

t-Bu-O(C=O)OK + Cl(C=O)Cl → THF, benzene / - 5 to - 15 °C → *t*-Bu-O(C=O)O(C=O)O(C=O)O-*t*-Bu

OS, CV 6, 448 (1988)

$(EtO)_2P(O)—CH_2CHO$ → H_2N-cyclohexyl / MeOH → $(EtO)_2P(O)—CH{=}CH—NH$-cyclohexyl

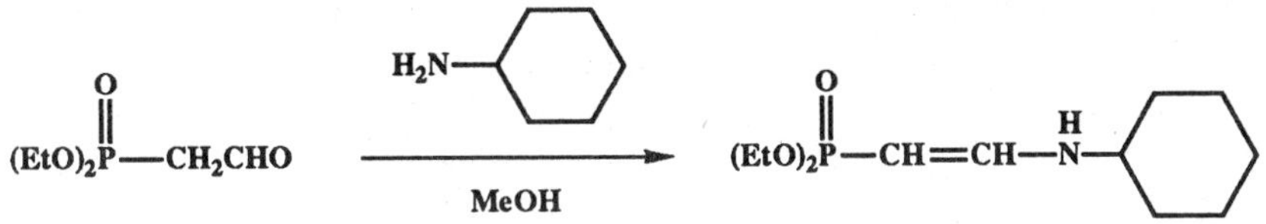

OS, CV 6, 492 (1988)

cyclohexyl-C(O)OMe → Me_2AlNMe_2 / benzene, hexane, Δ → cyclohexyl-C(O)NMe_2

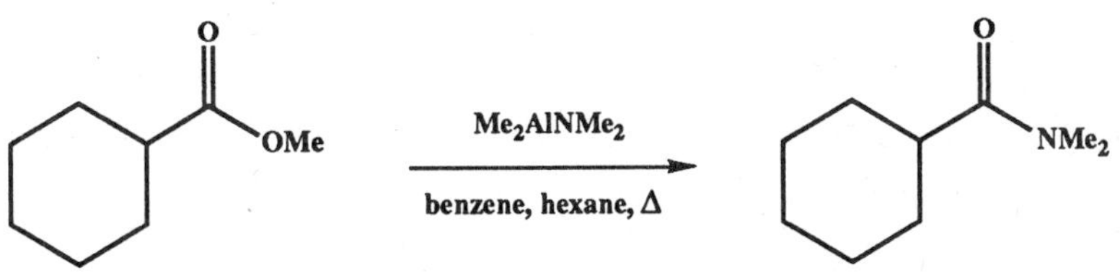

OS, CV 6, **496 (1988)**

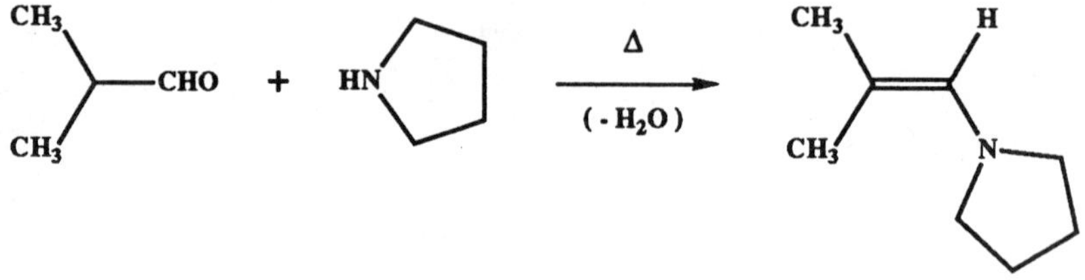

CH_3, CH_3, CHO + HN ; Δ ; (- H_2O) ; CH_3, CH_3, H, N

OS, CV 6, **499 (1988)**

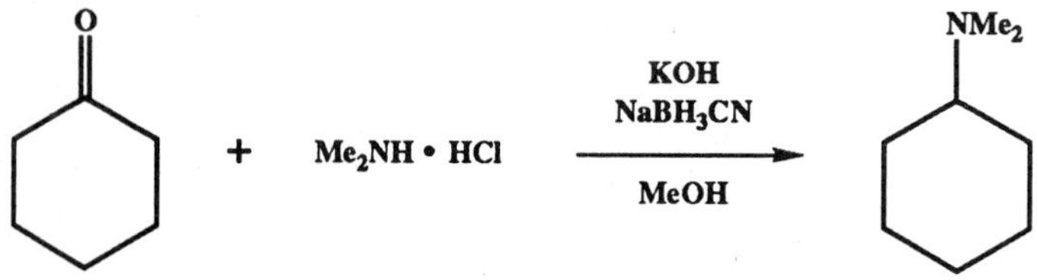

O ; + $Me_2NH \cdot HCl$; KOH, $NaBH_3CN$; MeOH ; NMe_2

OS, CV 6, **505 (1988)**

EtO_2C, O, CO_2Et ; 1. PCl_5, 25 °C ; 2. H_2SO_4, MeOH, Δ ; MeO_2C, Cl, CO_2Me

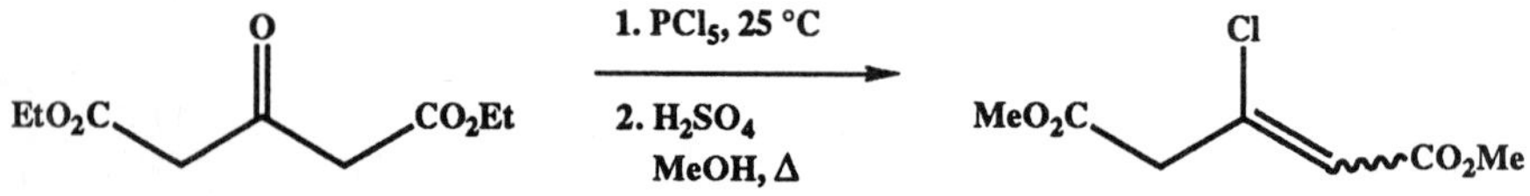

OS, CV 6, **549 (1988)**

Ph_2CH, O, OH ; $SOCl_2$; benzene, Δ ; Ph_2CH, O, Cl

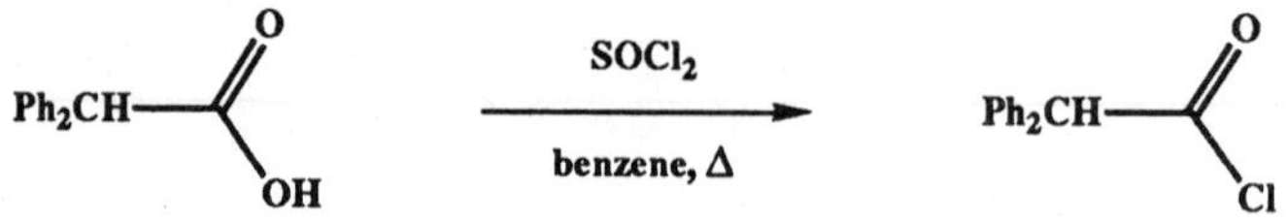

OS, CV 6, 558 (1988)

2 (2-thiophenethiol, SH) + 2 (2-bromothiophene, Br) → Cu_2O ($-Cu_2Br_2$) ($-H_2O$) → 2 (di-2-thienyl sulfide)

OS, CV 6, 576 (1988)

(4-acetoxybenzoic acid: AcO, OH) → $R_3O^+\ BF_4^-$, $(i\text{-}Pr)_2NEt$, CH_2Cl_2, 25 °C → (AcO, OR)

R = Me or Et

OS, CV 6, 618 (1988)

(2-pyrrolyl trichloromethyl ketone: N H, CCl_3, O) → NaOEt, EtOH → (N H, OEt, O)

OS, CV 6, 640 (1988)

(pyridine–SO_3: N+, SO_3) → 1. aq. NaOH, - 20 to - 5 °C; 2. aq. NaOH, 55 to 60 °C → NaO–CH=CH–CH=CH–CH=O (H)

***OS*, *CV* 6, 679 (1988)**

H_2O_2, NaOH
MeOH, 0 °C

Ph
Ph
N—NH_2
benzene, 25 °C

Ph
Ph
N
N
O

***OS*, *CV* 6, 692 (1988)**

Ph
O
CH_3

1. NaH, DME, 20 °C
2. Ac_2O, 5 - 30 °C

Ph
OAc
H
CH_3

***OS*, *CV* 6, 704 (1988)**

S
S
CH_2Ph
N

MeI
DMF
75 - 80 °C

I^-
S
S
CH_2Ph
N^+
Me

I
CH_2Ph

***OS*, *CV* 6, 715 (1988)**

$H_3\overset{+}{N}CH_2CH_2$
Cl^-
O
OH
+
O
Cl
$OCCl_3$

dioxane
55 - 60 °C

O=C=N—CH_2CH_2
O
Cl

OS, CV 6, 751 (1988)

t-Bu—CH_2—C(CH_3)$_2$—N=C(D)—CH(CH_3)CH_2CH_3 —(oxalic acid; H_2O, Δ)→ O=C(D)—CH(CH_3)CH_2CH_3

OS, CV 6, 757 (1988)

$(CH_3)_2CHCOCH_3$ —($(CF_3SO_2)_2O$, pyridine; pentane, - 78 to 25 °C)→ $(CH_3)_2C=C(OSO_2CF_3)CH_3$ + $(CH_3)_2CHC(OSO_2CF_3)=CH_2$

OS, CV 6, 818 (1988)

$PhCOCH_3$ + CH_3NH_2 —($TiCl_4$; Et_2O, hexane, - 30 °C)→ $PhC(=NCH_3)CH_3$

OS, CV 6, 910 (1988)

1-Ph-cyclopentane-1-CO_2H —($ClCO_2Et$; Et_3N, 0 °C, acetone)→ [1-Ph-cyclopentane-1-C(O)OC(O)OEt] —(NaN_3; H_2O)→ 1-Ph-cyclopentane-1-C(O)N_3

OS, *CV 6*, 913 (1988)

Ph CO$_2$Et (cyclopropane) → 1. NaOH, H_2O; 2. HCl → Ph CO$_2$H (cyclopropane)

OS, *CV 6*, 1004 (1988)

$PhNH_2$ → Cl_3C–CO–CF_3, DMSO → PhNH–CO–CF_3

HO–C$_6$H$_4$–CH$_2$–CH(NH$_2$)–CO$_2$H (*l* - (-) - tyrosine) → Cl_3C–CO–CF_3, DMSO → HO–C$_6$H$_4$–CH$_2$–CH(NH–CO–CF$_3$)–CO$_2$H (*l* isomer)

OS, *CV 6*, 1033 (1988)

CH_2=CH–Br → 1. Mg, THF, Δ; 2. Me_3SiCl, Δ → CH_2=CH–SiMe$_3$ → 1. Br_2, - 78 °C; 2. Et_2NH, Δ → CH_2=C(Br)–SiMe$_3$

OS, *CV 7*, 467 (1990); *61*, 1 (1983)

$$CH_3\text{—}\overset{O}{\overset{\|}{C}}\text{—}\overset{O}{\overset{\|}{C}}\text{—}OH + H\text{—}C(Cl)_2\text{—}OMe \xrightarrow{25 - 50\ °C} CH_3\text{—}\overset{O}{\overset{\|}{C}}\text{—}\overset{O}{\overset{\|}{C}}\text{—}Cl + H\text{—}\overset{O}{\overset{\|}{C}}\text{—}OMe$$

OS, *CV* 7, 27 (1990); *61*, 14 (1983)

PhCHO + H_2N-(5-aminotetrazole) → (1. Et_3N, MeOH, 50 °C; 2. H_2 (500 psi), Pd / C) → $PhCH_2NH$-(tetrazol-5-yl)

OS, *CV* 7, 87 (1990); *61*, 48 (1983)

cyclohexyl-C(=O)S-*t*-Bu → (*t*-BuOH, $Hg(O_2CCF_3)_2$, MeCN, 25 °C) → cyclohexyl-C(=O)O-*t*-Bu

OS, *CV* 7, 210 (1990); *61*, 56 (1983)

$(HOCH_2)_2C(CO_2Et)_2$ → (48% HBr, 85 - 90 °C) → $CH_2=C(CO_2H)CH_2Br$ → (EtOH, H^+, reflux) → $CH_2=C(CO_2Et)CH_2Br$

OS, *CV* 7, 302 (1990); *61*, 71 (1983)

quinoline + $CSCl_2$ → ($CaCO_3$, CH_2Cl_2, H_2O) → 2-(N=C=S)-C$_6$H$_4$-CH=CH-CHO

OS, *CV* 7, 81 (1990); *61*, 134 (1983)

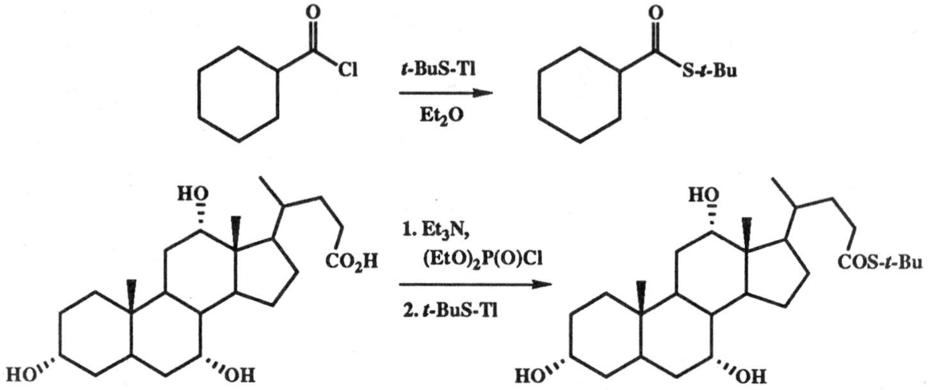

t-BuS-Tl

Et_2O

HO

CO_2H

HO

OH

1. Et_3N, $(EtO)_2P(O)Cl$

2. *t*-BuS-Tl

HO

COS-*t*-Bu

HO

OH

OS, *CV* 7, 77 (1990); *61*, 141 (1983)

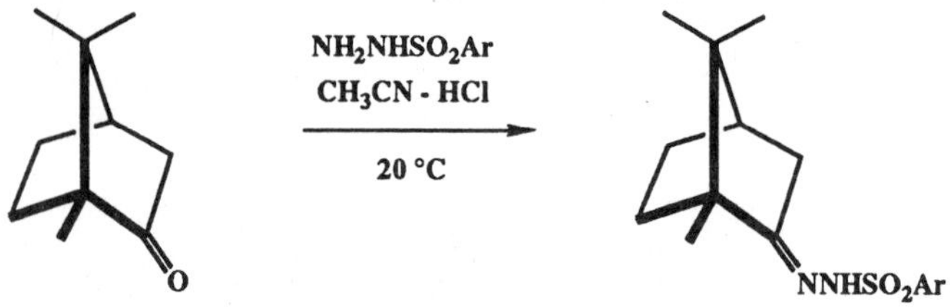

NH_2NHSO_2Ar

CH_3CN - HCl

20 °C

O

$NNHSO_2Ar$

OS, *CV* 7, 351 (1990); *62*, 14 (1984)

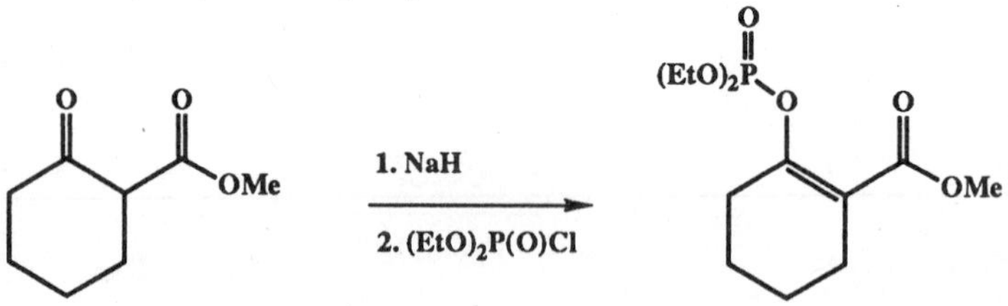

O O

OMe

1. NaH

2. $(EtO)_2P(O)Cl$

O

$(EtO)_2P$ O O

OMe

OS, *CV* 7, 372 (1990); *62*, 158 (1984)

MeO OMe N Me O S S S S P P N Me S

OS, *CV* 7, 30 (1990); *63*, 175 (1985)

Cl H R_1 CbzNH—CH—C—O C=C O N NMe O R_2 H_2N—CH—CO_2Me R_1 O R_2 CbzNH—CH—C—NH—CH—CO_2Me

OS, *CV* 7, 93 (1990); *63*, 183 (1985)

H CO_2H EtO_2C H + *t*-BuOH $C_6H_{11}N$=C=NC_6H_{11} N NMe_2 H CO_2t-Bu EtO_2C H

OS, *CV* 7, 164 (1990); *64*, 175 (1986)

I Ph O O Na_2CO_3 MeOH Ph O CO_2Me

OS, *CV* 7, 495 (1990); *64*, 196 (1986)

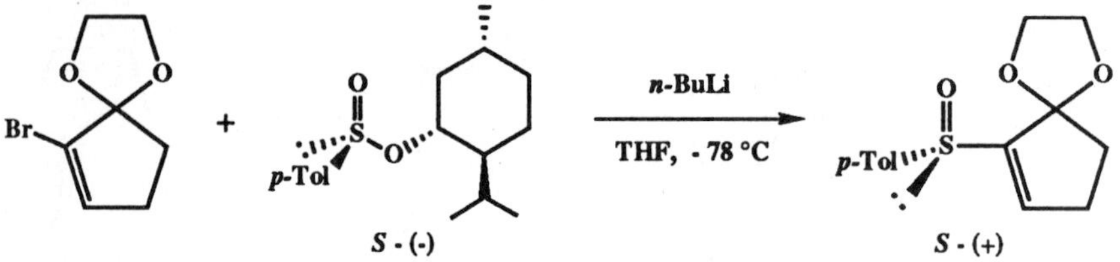

Br · O · O · + · O · S · O · *p*-Tol · *S* - (-) · *n*-BuLi · THF, - 78 °C · O · *p*-Tol · S · O · O · *S* - (+)

OS, *CV* 7, 438 (1990); *64*, 207 (1986)

PhCHO —(NH_2NHTs)→ PhCHNNHTs —(1. NaOMe; 2. Δ, vacuum)→ $PhCHN_2$

OS, *CV* 7, 506 (1990); *64*, 217 (1986)

CF_3CO_2H + CF_3SO_3H —(P_2O_5)→ $CF_3CO_2SO_2CF_3$

OS, *65*, 230 (1987)

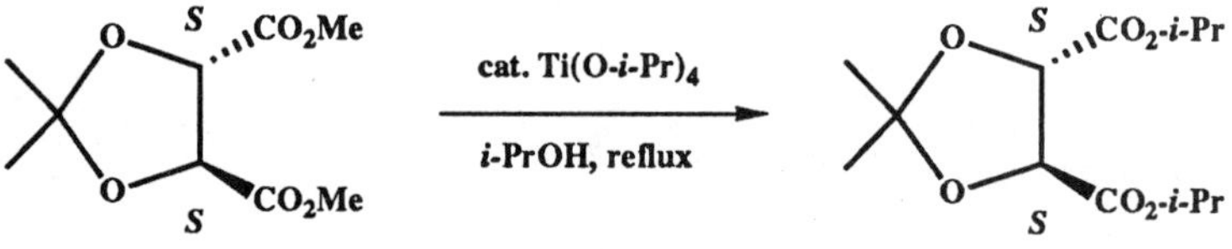

O · *S* · CO_2Me · O · CO_2Me · *S* · cat. Ti(O-*i*-Pr)$_4$ · *i*-PrOH, reflux · O · *S* · CO_2-*i*-Pr · O · CO_2-*i*-Pr · *S*

OS, *66*, 37 (1987)

CO_2Me · O · KOH · MeOH, H_2O · CO_2H · O

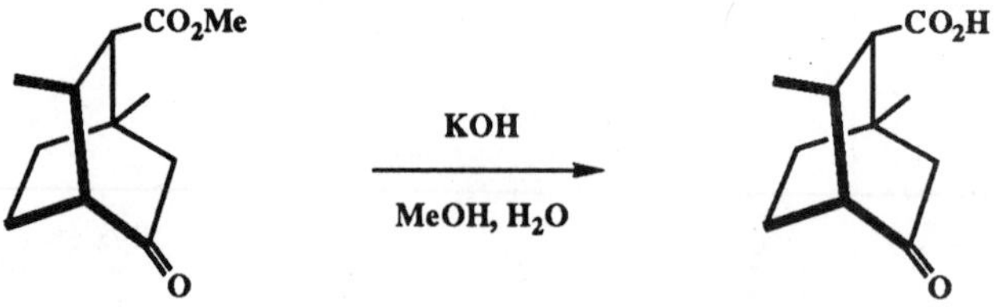

OS, *66*, 108 (1987)

Me_3CS ... CO_2Me ... OH ... CF_3CO_2Ag ... MeO_2C ... O ... Bu_4NF ... HO ... O

OS, *66*, 132 (1987)

CO_2H ... $ClCO_2Et$ / N-methylmorpholine ... CO_2CO_2Et ... NH_3 ... $CONH_2$

OS, *66*, 173 (1987)

C–$CH(CO_2Et)_2$ ... H ... $POCl_3$ ... Cl ... C=$C(CO_2Et)_2$ ... H

OS, *67*, 69 (1988)

BocNH—CH(*S*)—$CO_2H \cdot H_2O$ (CH_2CHMe_2) ... $ClCO_2Me$, MeNHOMe ... BocNH—CH(*S*)—CON(Me)—OMe (CH_2CHMe_2)

OS, *68*, 116 (1989)

O ... $(CF_3SO_2)_2O$, 4-Me-2,6-di-*t*-Bu-pyridine (Me, *t*-Bu, N, *t*-Bu) ... CH_2Cl_2 ... OSO_2CF_3

OS, *CV* 2, 399 (1943)

$$CH_3{-}OH \xrightarrow{P,\ I_2} CH_3{-}I$$

OS, *CV* 2, 404 (1943)

$$(CH_3)_2SO_4 \xrightarrow[\text{aq. } CaCO_3]{KI,\ \Delta} CH_3{-}I$$

OS, *CV* 2, 412 (1943)

$$CH_3{-}OH \xrightarrow[H_2SO_4]{HNO_3} CH_3{-}ONO_2$$

OS, *CV* 5, 797 (1973)

$$3,5\text{-}(O_2N)_2C_6H_3CO_2H \xrightarrow[Et_2O]{p\text{-}CH_3C_6H_4N{=}N{-}NH{-}CH_3} 3,5\text{-}(O_2N)_2C_6H_3CO_2CH_3$$

OS, *CV* 7, 346 (1990); *62*, 101 (1984)

$$CH_3Cl \ + \ Li\ (1\%\ Na) \xrightarrow[25\ °C]{Et_2O} CH_3Li \ + \ LiCl$$

OS, CV 1, 25 (1941)

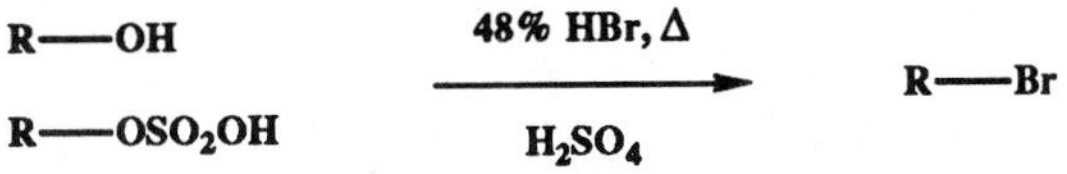

R—OH

R—OSO_2OH

48% HBr, Δ

H_2SO_4

R—Br

R = allyl, *i*-amyl, butyl, dodecyl, ethyl, octyl, $Br(CH_2)_3$

OS, CV 1, 117 (1941)

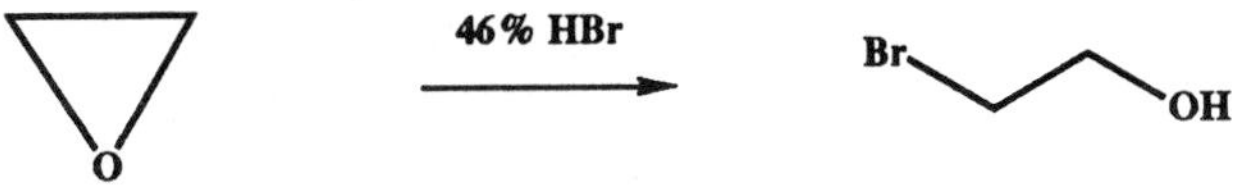

O

46% HBr

Br

OH

OS, CV 1, 119 (1941)

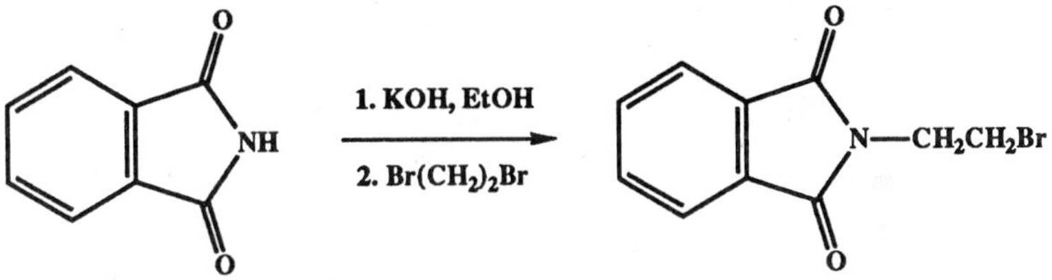

O

NH

O

1. KOH, EtOH

2. $Br(CH_2)_2Br$

O

N—CH_2CH_2Br

O

OS, CV 1, 131 (1941)

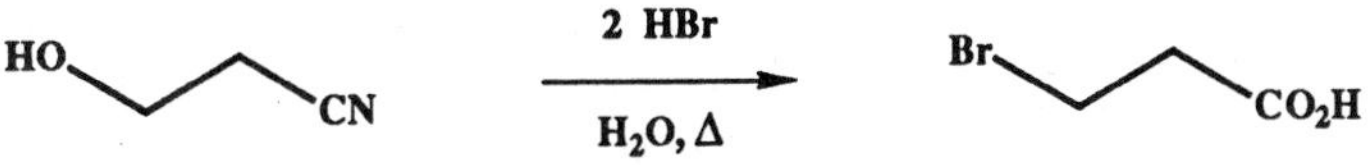

HO

CN

2 HBr

H_2O, Δ

Br

CO_2H

***OS*, *CV 1*, 142 (1941)**

OH → Cl

$ZnCl_2$, HCl

Δ

***OS*, *CV 1*, 145 (1941)**

OH → OMs

NaOH

p-$CH_3C_6H_4SO_2Cl$

***OS*, *CV 1*, 203 (1941)**

2 Br → N—CN

Na_2NCN

EtOH, Δ

***OS*, *CV 1*, 292 (1941)**

HO OH OH → Cl OH Cl

2 HCl, HOAc

100 - 110 °C

***OS*, *CV 1*, 294 (1941)**

HO OH OH → Cl OH OH

HCl, HOAc

105 - 110 °C

OS, *CV 1*, 296 (1941)

Cl–CH_2–CH(OH)–CH_2–OH → (NaOPh; EtOH, Δ) → PhO–CH_2–CH(OH)–CH_2–OH

◇

OS, *CV 1*, 300 (1941)

Cl—CH_2CO_2H → (NH_4OH) → H_2N—CH_2CO_2H

◇

OS, *CV 1*, 377 (1941)

CH_2=O + CH_3OH + HCl → $ClCH_2OCH_3$

◇

OS, *CV 1*, 401 (1941)

$ClCH_2CO_2Na$ → (1. aq. $NaNO_2$; 2. 80 °C) → CH_3NO_2

◇

OS, *CV 1*, 428 (1941)

N-benzoylpiperidine (N–C(=O)Ph) → (PBr_3, Br_2; Δ) → Br—$(CH_2)_5$—Br + PhCN

***OS*, *CV 1*, 435 (1941)**

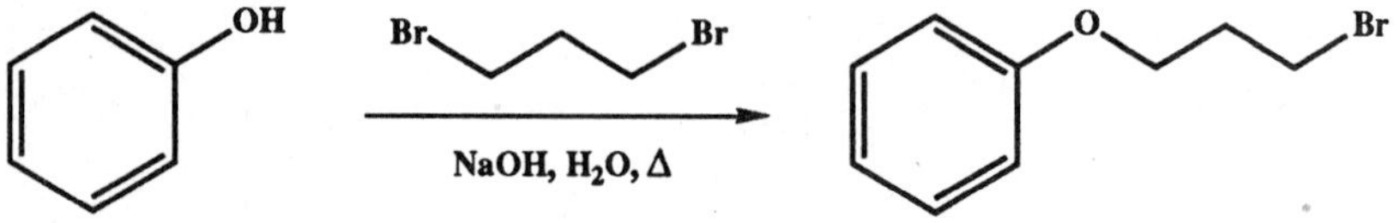

OH Br Br O Br

NaOH, H_2O, Δ

***OS*, *CV 1*, 533 (1941)**

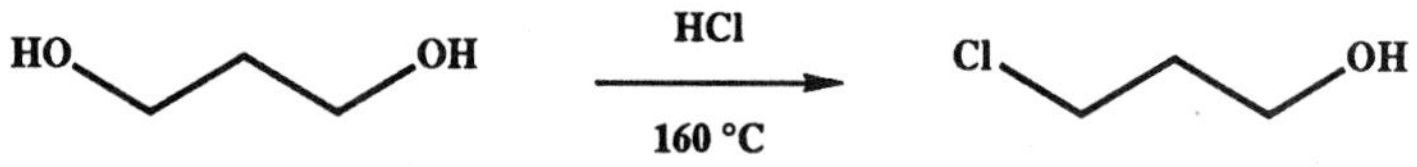

HO OH Cl OH

HCl

160 °C

***OS*, *CV 2*, 5 (1943)**

O Me Br O Me O H O O Me OH

HCO_2K

MeOH

Δ

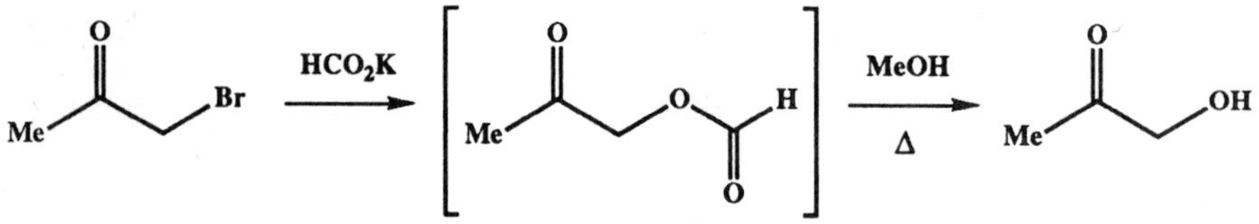

***OS*, *CV 2*, 25 (1943)**

O NK O O N—$(CH_2)_3CN$ O

Cl—$(CH_2)_3CN$

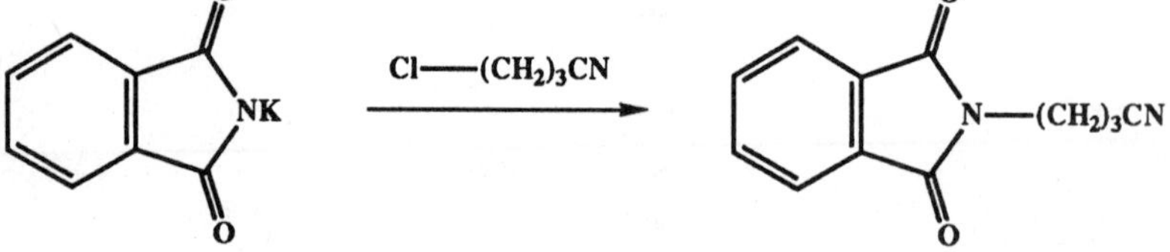

OS, *CV* 2, 83 (1943)

Phthalimide (NH) → N—CH$_2$Ph ; PhCH$_2$Cl, K$_2$CO$_3$, 190 °C

OS, *CV* 2, 85 (1943)

$(CH_3)_3N$ + $ClCH_2CO_2Et$ + H_2NNH_2 →(EtOH) $(CH_3)_3\overset{+}{N}CH_2C(O)NHNH_2$ Cl^-

OS, *CV* 2, 91 (1943)

$HOCH_2CH_2NH_2$ →(HBr, Δ) $BrCH_2CH_2\overset{+}{N}H_3$ Br^-

OS, *CV* 2, 108 (1943)

n-BuOH →(HONO; H_2SO_4, 0 °C) *n*-BuONO

OS, *CV* 2, 112 (1943)

2 *n*-BuOH + $SOCl_2$ →(Δ) $(n\text{-Bu})_2SO_3$

OS, CV 2, **136 (1943)**

$CH_3SCH_2CH_2OH$ —($SOCl_2$, $CHCl_3$)→ $CH_3SCH_2CH_2Cl$

OS, CV 2, **183 (1943)**

$ClCH_2CH_2OH$ —(Et_2NH, Δ; NaOH, H_2O)→ $Et_2NCH_2CH_2OH$

OS, CV 2, **184 (1943)**

CH_3CH_2—I + CH_3CH_2—Br —(Zn-Cu, Δ)→ CH_3CH_2—Zn—CH_2CH_3

OS, CV 2, **246 (1943)**

n-$C_{12}H_{25}$—OH —(HBr, 100 - 120 °C)→ *n*-$C_{12}H_{25}$—Br

OS, CV 2, **260 (1943)**

$ClCH_2CO_2H$ —(1. Na, EtOH; 2. HCl)→ $EtOCH_2CO_2H$ —(EtOH, HCl)→ $EtOCH_2CO_2Et$

OS, *CV* 2, 308 (1943)

HO, OH, OH — P, Br_2 → Br, Br, OH

OS, *CV* 2, 322 (1943)

$C_{16}H_{33}$—OH — P, I_2; 150 °C → $C_{16}H_{33}$—I

OS, *CV* 2, 328 (1943)

O, OH, Cl — NH_4OH; H_2O → O, OH, NH_2 — 1. Cl, O, Ph; NaOH, H_2O; 2. HCl → O, OH, HN, Ph, O

OS, *CV* 2, 345 (1943)

Cl, OH — CH_3SNa; EtOH, Δ → CH_3, S, OH

OS, *CV* 2, 358 (1943)

CH_3, CH_3, OH — PBr_3 → CH_3, CH_3, Br

OS, CV 2, **387 (1943)**

HO–CH_2–CN $\xrightarrow{(CH_3)_2SO_4}$ CH_3O–CH_2–CN

OS, CV 2, **397 (1943)**

CH_3NH_2 + 2 Cl—CH_2CO_2H $\xrightarrow[H_2O]{NaOH}$ $CH_3N(CH_2CO_2H)_2$

OS, CV 2, **476 (1943)**

$C(CH_2OH)_4$ $\xrightarrow[180\ °C]{4\ PBr_3}$ $C(CH_2Br)_4$ $\xrightarrow[MEK\ or\ acetone]{4\ NaI}$ $C(CH_2I)_4$

OS, CV 2, **547 (1943)**

2 $CH_3CH_2CH_2Br$ $\xrightarrow[EtOH,\ \Delta]{Na_2S}$ $CH_3CH_2CH_2$—S—$CH_2CH_2CH_3$

OS, CV 2, **558 (1943)**

$BrCH_2CH_2CH_2Br$ $\xrightarrow[aq.\ EtOH,\ \Delta]{Na_2SO_3}$ $BrCH_2CH_2CH_2SO_3Na$

***OS*, *CV* 2, 563 (1943)**

Br ... SO_3Na —(NH_4OH)→ H_2N ... SO_3H

***OS*, *CV* 2, 564 (1943)**

Br ... NH_3Br —(Na_2SO_3; H_2O, Δ)→ H_2N ... SO_3H

***OS*, *CV* 2, 571 (1943)**

O (tetrahydrofuran) —(HCl, Δ)→ Cl ... OH

***OS*, *CV* 2, 576 (1943)**

2 HO ... Cl —(Na_2S, H_2O; Δ)→ HO ... S ... OH

***OS*, *CV* 3, 50 (1955)**

Cl ... OEt, OEt —(NH_3, MeOH; 140 °C)→ H_2N ... OEt, OEt

OS, CV 3, 172 (1955)

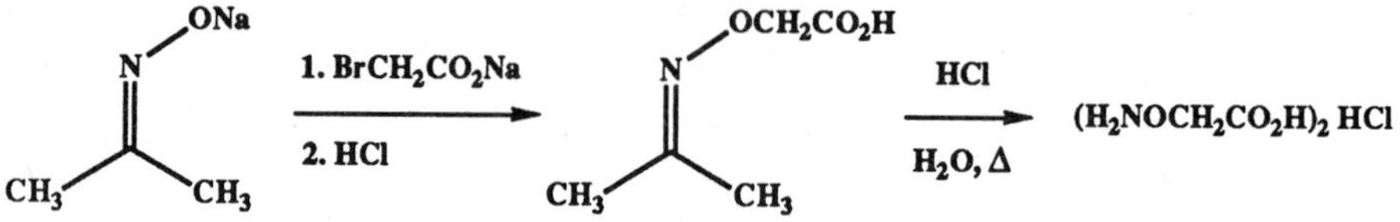
ONa
N
CH3
CH3
1. BrCH2CO2Na
2. HCl
OCH2CO2H
N
CH3
CH3
HCl
H2O, Δ
(H2NOCH2CO2H)2 HCl

OS, CV 3, 203 (1955)

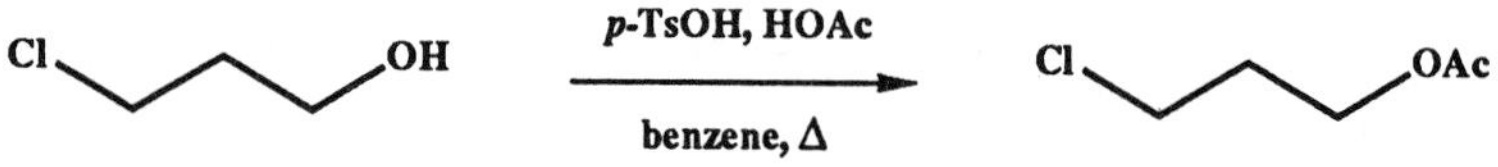
Cl
OH
p-TsOH, HOAc
benzene, Δ
Cl
OAc

OS, CV 3, 227 (1955)

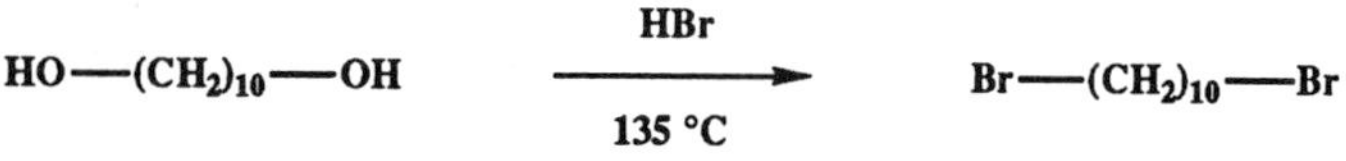
HO—(CH2)10—OH
HBr
135 °C
Br—(CH2)10—Br

OS, CV 3, 254 (1955)

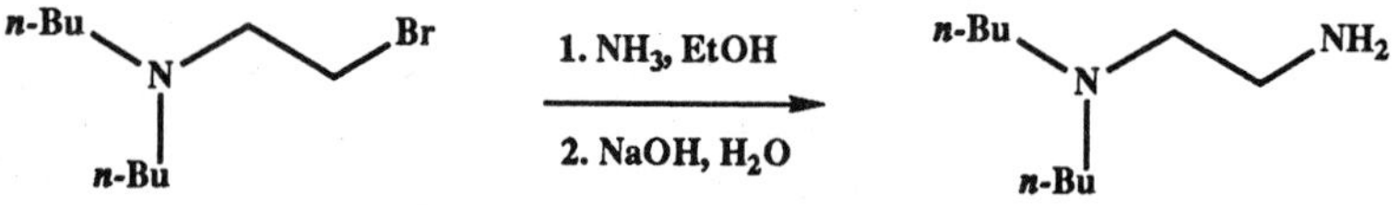
n-Bu
N
n-Bu
Br
1. NH3, EtOH
2. NaOH, H2O
n-Bu
N
n-Bu
NH2

OS, CV 3, 256 (1955)

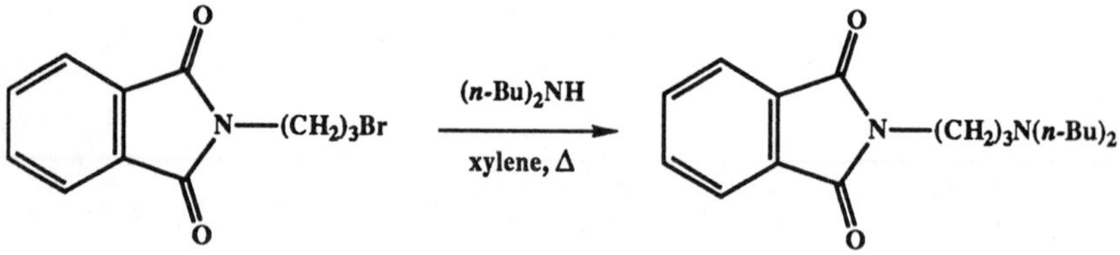
O
N—(CH2)3Br
O
(n-Bu)2NH
xylene, Δ
O
N—(CH2)3N(n-Bu)2
O

OS, CV 3, 363 (1955)

$$n\text{-}C_{12}H_{25}\text{—Br} + S{=}C(NH_2)_2 \xrightarrow[\Delta]{EtOH} n\text{-}C_{12}H_{25}\text{—S—}C({=}NH)NH_2$$

$$2\ n\text{-}C_{12}H_{25}\text{—S—}C({=}NH)NH_2 \xrightarrow[H_2O,\ \Delta]{NaOH} 2\ n\text{-}C_{12}H_{25}\text{—SH} + NC\text{—}NH\text{—}C({=}NH)NH_2$$

OS, CV 3, 366 (1955)

$$CH_3(CH_2)_{10}\text{—}CH_2OH \xrightarrow[\text{pyridine}]{p\text{-TsCl}} CH_3(CH_2)_{10}\text{—}CH_2OTs$$

OS, CV 3, 370 (1955)

$$EtO\text{—}CH_2CH_2\text{—}OH \xrightarrow{PBr_3} EtO\text{—}CH_2CH_2\text{—}Br$$

OS, CV 3, 446 (1955)

$$HO\text{—}(CH_2)_6\text{—}OH \xrightarrow[\text{toluene},\ \Delta]{HCl,\ H_2O} HO\text{—}(CH_2)_6\text{—}Cl$$

OS, CV 3, 650 (1955)

$$p\text{-}O_2N\text{—}C_6H_4\text{—}CH_2Cl \xrightarrow[HOAc,\ \Delta]{NaOAc} p\text{-}O_2N\text{—}C_6H_4\text{—}CH_2OAc$$

***OS*, *CV* 3, 652 (1955)**

NaOH, Δ

aq. MeOH

OAc
O_2N
OH
O_2N

***OS*, *CV* 3, 692 (1955)**

Br_2, Δ

H_2O, SO_2

O
Br
Br

***OS*, *CV* 3, 698 (1955)**

$SOCl_2$

pyridine

O
OH
O
Cl

***OS*, *CV* 3, 793 (1955)**

PBr_3, pyridine

benzene, - 5 °C

O
OH
O
Br

***OS*, *CV* 3, 833 (1955)**

$ZnCl_2$

Ac_2O, Δ

H_2SO_4

H_2O, Δ

O
OH
OAc
OAc
OAc
OH
OH
OH

OS, CV 4, **62 (1963)**

$$HO_2C—(CH_2)_7—CO_2H \xrightarrow[\text{silica gel}]{NH_3,\ 500\ °C} NC—(CH_2)_7—CN$$

OS, CV 4, **84 (1963)**

$$PhC(=O)OCH_2CH_2Cl \xrightarrow[\text{3. AgCl, } H_2O]{\text{1. NaI, MEK, }\Delta\ \ \text{2. } Me_3N\text{, acetone}} PhC(=O)OCH_2CH_2\overset{+}{N}Me_3\ Cl^-$$

OS, CV 4, **98 (1963)**

$$PhCH_2Cl \xrightarrow[EtOH]{Me_3N} PhCH_2\overset{+}{N}Me_3\ Cl^- \xrightarrow[EtOH]{NaOEt} PhCH_2\overset{+}{N}Me_3\ EtO^-$$

OS, CV 4, **106 (1963)**

$$\text{Phthalimido-}NCH_2CH_2OH \xrightarrow{PBr_3,\ \Delta} \text{Phthalimido-}NCH_2CH_2Br$$

OS, CV 4, **128 (1963)**

$$CH_3—C(Cl)=CH—CH_2Cl \xrightarrow[H_2O,\ \Delta]{Na_2CO_3} CH_3—C(Cl)=CH—CH_2OH$$

OS, CV 4, 266 (1963)

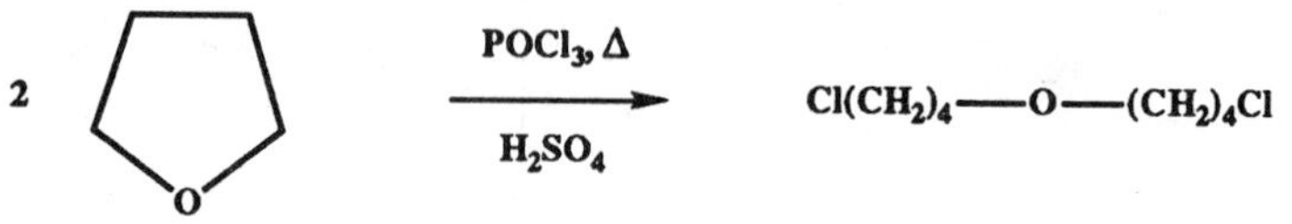

2 (tetrahydrofuran) —[$POCl_3$, Δ / H_2SO_4]→ $Cl(CH_2)_4—O—(CH_2)_4Cl$

OS, CV 4, 295 (1963)

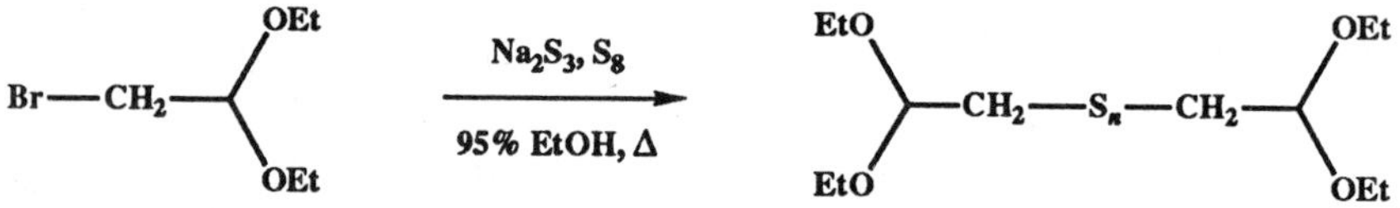

$Br—CH_2—CH(OEt)_2$ —[Na_2S_3, S_8 / 95% EtOH, Δ]→ $(EtO)_2CH—CH_2—S_n—CH_2—CH(OEt)_2$

OS, CV 4, 321 (1963)

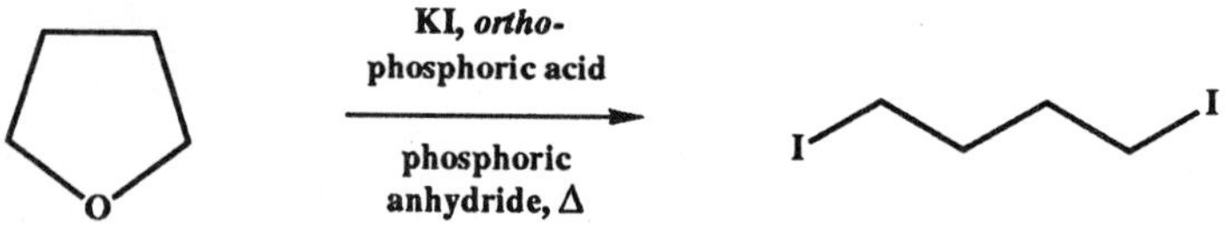

(tetrahydrofuran) —[KI, *ortho*-phosphoric acid / phosphoric anhydride, Δ]→ I—$(CH_2)_4$—I

OS, CV 4, 323 (1963)

HO—$(CH_2)_6$—OH —[KI, *ortho*-phosphoric acid / phosphoric anhydride, Δ]→ I—$(CH_2)_6$—I

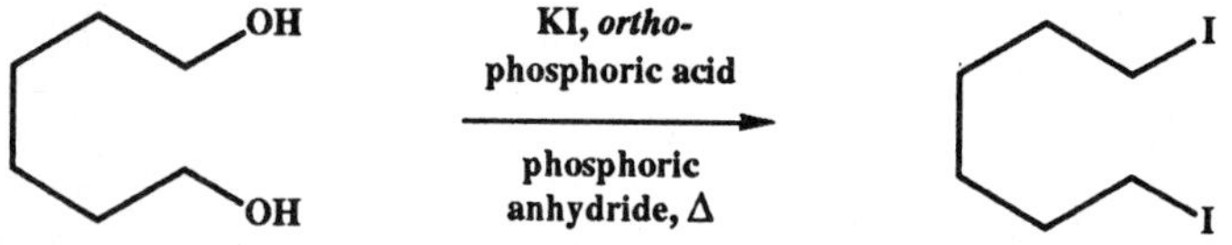

OS, CV 4, 333 (1963)

$Me_2N—CH_2CH_2—OH$ —[$SOCl_2$ / 0 °C]→ $Me_2N—CH_2CH_2—Cl$ • HCl

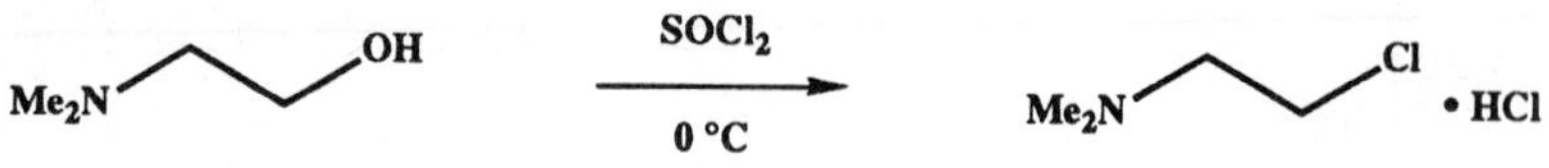

OS, CV 4, **368 (1963)**

I–$(CH_2)_5$–I → ($AgNO_2$, Et_2O) → O_2N–$(CH_2)_5$–NO_2

OS, CV 4, **401 (1963)**

$BrCH_2CH_2Br$ → (2 H_2N–C(=S)–NH_2, hot EtOH) → CH_2CH_2(S–C(=NH)–NH_2)$_2$ • 2 HBr → (1. aq. KOH, Δ; 2. aq. H_2SO_4) → $HSCH_2CH_2SH$

OS, CV 4, **438 (1963)**

Et—I → (AgCN) → Et—N≡C • AgI → (KCN, H_2O) → Et—N≡C

OS, CV 4, **491 (1963)**

2-furylmethyl–OH → (H_2N–C(=S)–NH_2, HCl, H_2O) → 2-furylmethyl–S–C(=NH)–NH_2 • HCl → (NaOH, H_2O, Δ) → 2-furylmethyl–SH

OS, CV 4, **525 (1963)**

CH_3–$(CH_2)_5$–Br → (KF, Δ, ethylene glycol) → CH_3–$(CH_2)_5$–F

***OS, CV 4,* 529 (1963)**

$O((CH_2)_4Cl)_2$ —[2 Na_2SO_3; H_2O, Δ]→ $O((CH_2)_4SO_3Na)_2$

***OS, CV 4,* 576 (1963)**

MeO-C_6H_4-CH_2OH —[HCl]→ MeO-C_6H_4-CH_2Cl

***OS, CV 4,* 582 (1963)**

CH_2NMe_2 / CH_3 (benzene) —[EtBr; EtOH, Δ]→ $CH_2\overset{+}{N}Me_2Et$ Br^- / CH_3 —[1. NaOAc, AcOH, Δ; 2. NaOH, Δ]→ CH_2OH / CH_3

***OS, CV 4,* 597 (1963)**

CH_3-C(=O)- (lactone, O, O) —[HCl; H_2O, Δ]→ CH_3-C(=O)-$CH_2CH_2CH_2$-Cl

***OS*, *CV* 4**, **681 (1963)**

$C(CH_2OH)_4$ —(48% HBr; HOAc, Δ)→ $(HOCH_2)_3C(CH_2Br)$

HOCH2, CH2OH, HOCH2, CH2OH → HOCH2, CH2Br, HOCH2, CH2OH

***OS*, *CV* 4**, **724 (1963)**

$CH_3—(CH_2)_7—Br$ —($AgNO_2$; Et_2O)→ $CH_3—(CH_2)_7—NO_2$

***OS*, *CV* 4**, **753 (1963)**

$C(CH_2OH)_4$ —(4 $PhSO_2Cl$; pyridine)→ $C(CH_2SO_3Ph)_4$ —(4 NaBr; diethylene glycol, Δ)→ $C(CH_2Br)_4$

***OS*, *CV* 4**, **810 (1963)**

phthalide + potassium phthalimide (NK) —(1. DMF, Δ; 2. HOAc, H_2O)→ 2-(phthalimidomethyl)benzoic acid (CO_2H)

OS, *CV 4*, 967 (1963)

KS, KS, S; 1. 2 $ClCH_2CO_2K$; 2. HCl; HO_2CCH_2S, HO_2CCH_2S, S

OS, *CV 5*, 121 (1973)

CH_2, Br, Br; 1. N, N, N, N, Δ; 2. HCl, aq. EtOH; 3. NaOH, H_2O; CH_2, NH_2, Br

OS, *CV 5*, 124 (1973)

CH_2, Br, Br; $EtNH_2$, H_2O; CH_2, NHEt, Br

OS, *CV 5*, 249 (1973)

CH=CH—CH_2OH; Ph_3PBr_2; CH_3CN; CH=CH—CH_2Br

OS, *CV 5*, 545 (1973)

HBr, EtOH; 0 °C

Br ... CO_2Et

OS, *CV 5*, 580 (1973)

CH_3—$(CH_2)_4$—C(F)(H)—CH_2Br → (NaOAc, NaI; DMF, Δ) → CH_3—$(CH_2)_4$—C(F)(H)—CH_2OAc

OS, *CV 5*, 586 (1973)

Cl ... CO_2-*t*-Bu → (NaN_3, Δ; aq. acetone) → N_3 ... CO_2-*t*-Bu → (H_2, Pd-C; MeOH) → H_2N ... CO_2-*t*-Bu

OS, *CV 5*, 621 (1973)

$-CH_2\overset{+}{N}Me_3$ I^- → (NaOH; H_2O, Δ) → $-CH_2OH$

Fe Fe

OS, CV 5, 1046 (1973)

Ph—CCl_3 → (1. H_2S, KOH, EtOH, Δ; 2. $ClCH_2CO_2Na$, H_2O, Δ; 3. HCl, H_2O, 0 °C) → Ph–C(=S)–SCH_2CO_2H

OS, CV 5, 1141 (1973)

n-Bu—Cl → (Mg powder, Δ; methylcyclohexane) → *n*-Bu—MgCl *(unsolvated)*

OS, CV 5, 1145 (1973)

PhO–$CH_2CH_2CH_2$–Br → (Ph_3P; phenol, 90 °C) → PhO–$CH_2CH_2CH_2$–PPh_3^+ Br^- → (EtOAc, Δ) → CH_2=CH–PPh_3^+ Br^-

OS, CV 6, 5 (1988)

$CH_3C(=O)$—$NHCH_2OH$ → ($HSCH_2$—$CH(NH_3^+Cl^-)CO_2H$; aq. HCl, 25 °C) → $CH_3C(=O)$—$NHCH_2SCH_2$—$CH(NH_3^+Cl^-)CO_2H$

l - (-) isomer

OS, CV 6, **75 (1988)**

$HO(CH_2)_3N(CH_2CH_2CO_2Et)_2$ + $SOCl_2$ ⟶ $Cl(CH_2)_3N(CH_2CH_2CO_2Et)_2$

OS, CV 6, **101 (1988)**

$PhCH_2OH$ —(HCHO, HCl, 25 °C)⟶ $PhCH_2OCH_2Cl$

OS, CV 6, **187 (1988)**

1,1-dibromo-2,2-diphenylcyclopropane (Ph, Ph, Br, Br) —(AgOAc, HOAc; 100 - 120 °C)⟶ $Ph_2C{=}C(Br)CH_2OAc$

OS, CV 6, **273 (1988)**

$CH_3C{\equiv}CH$ —(1. $NaNH_2$, liq. NH_3; 2. $Br(CH_2)_3Cl$)⟶ $CH_3C{\equiv}C(CH_2)_3Cl$

OS, CV 6, **324 (1988)**

$HC{\equiv}CCH_2CH_2OH$ —($(CF_3SO_2)_2O$; Na_2CO_3; - 55 to 0 °C)⟶ $HC{\equiv}CCH_2CH_2OSO_2CF_3$

OS, CV 6, 404 (1988)

$PhCH_2$—Cl → (Na_2S, Δ; aq. EtOH) → $PhCH_2$—S—CH_2Ph → (1. Br_2, CCl_4, light, Δ; 2. *m*-CPBA, Et_2O) → PhCH(Br)—SO_2—CH(Br)Ph

OS, CV 6, 432 (1988)

CH_2N_2 → (NaOD, D_2O; THF, Et_2O, 0 °C) → CD_2N_2

OS, CV 6, 448 (1988)

Br—$CH_2CH(OEt)_2$ → ($(EtO)_3P$; Δ) → $(EtO)_2P(=O)$—$CH_2CH(OEt)_2$

OS, CV 6, 482 (1988)

cis-1,2-cyclohexanedimethanol (CH_2OH, CH_2OH; H, H) → ($MeSO_2Cl$; pyridine, - 5 to 0 °C) → bis-mesylate (CH_2OSO_2Me, CH_2OSO_2Me; H, H)

OS, CV 6, 586 (1988)

(E)-alkenol (H, CH_3, OH) → (*p*-TsCl, pyridine; 0 - 5 °C) → tosylate (H, CH_3, OTs)

OS, *CV 6*, **634 (1988)**

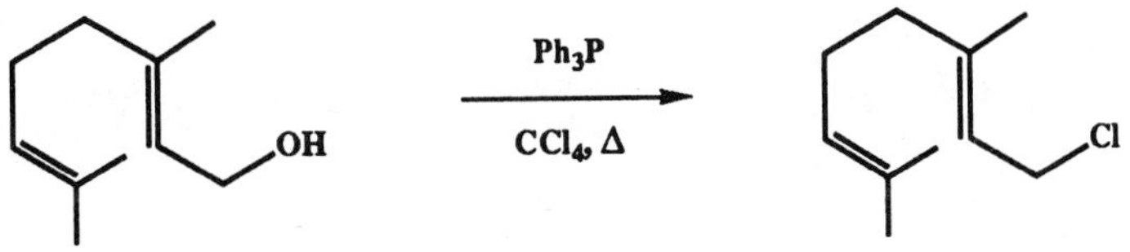

OH → (Ph_3P; CCl_4, Δ) → Cl

OS, CV 6, 638 (1988)

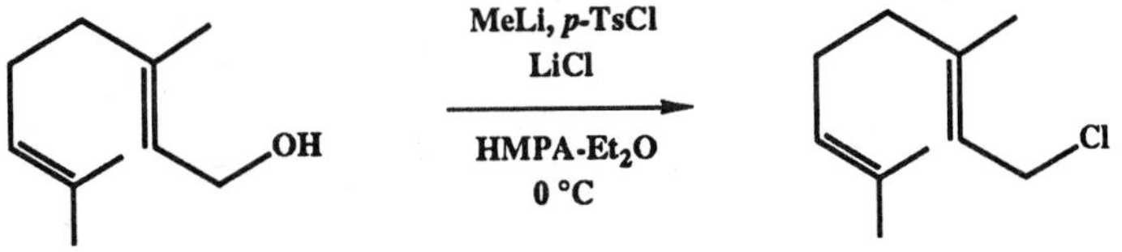

OH → (MeLi, *p*-TsCl, LiCl; HMPA-Et_2O, 0 °C) → Cl

OS, CV 6, 683 (1988)

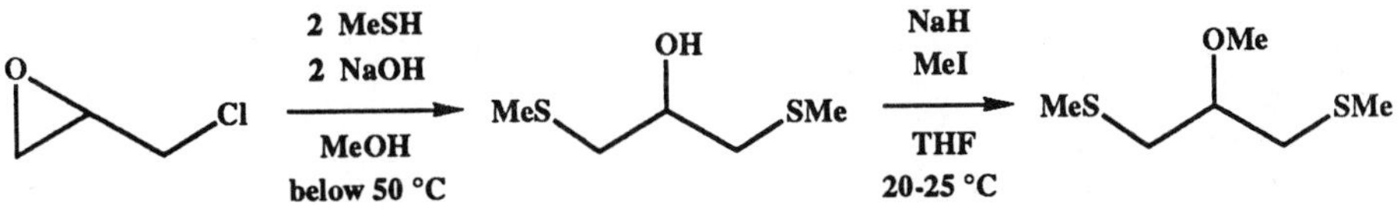

O, Cl → (2 MeSH, 2 NaOH; MeOH, below 50 °C) → MeS, OH, SMe → (NaH, MeI; THF, 20-25 °C) → MeS, OMe, SMe

OS, CV 6, 704 (1988)

Br + HS, S, N → (Et_3N; THF, Δ) → S, S, N

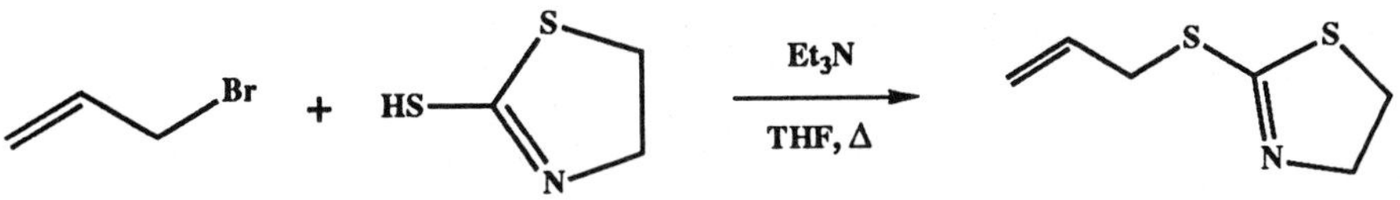

OS, CV 6, **781 (1988)**

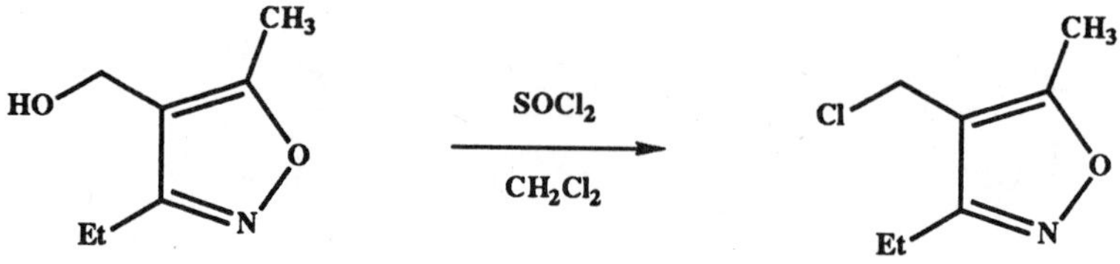

$SOCl_2$

CH_2Cl_2

OS, CV 6, **788 (1988)**

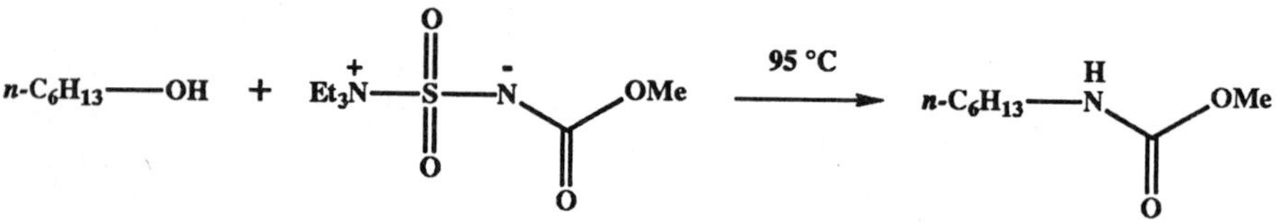

n-C_6H_{13}—OH + $Et_3\overset{+}{N}$—SO_2—$\overset{-}{N}$—C(O)OMe →(95 °C) *n*-C_6H_{13}—NH—C(O)OMe

OS, CV 6, **807 (1988)**

Br—$(CH_2)_5$C(O)—OMe

$Na_2Fe(CO)_4$ • (dioxane)$_{1.5}$

THF (*or N*-methylpyrrolidinone)
25 °C

[$(CH_2)_4$—C(O)—OMe, CH_2, OC—$\overset{-}{Fe}$(CO)(CO)(CO) Na^+]

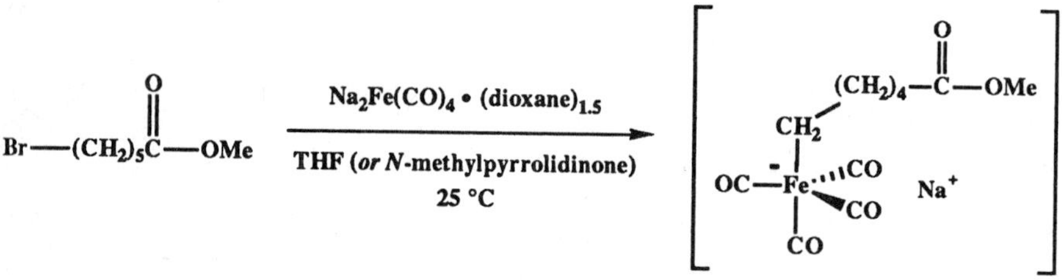

OS, CV 6, **830 (1988)**

t-Bu—CH_2OH →(CH_3I, $(PhO)_3P$) *t*-Bu—CH_2I

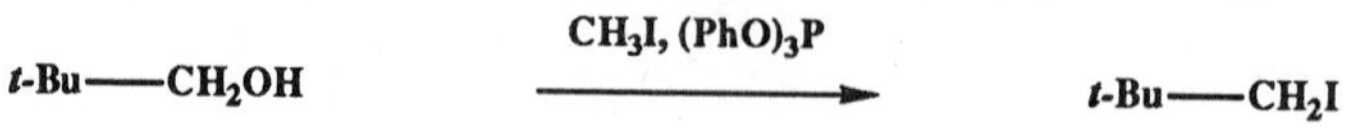

OS, CV 6, **833 (1988)**

$$t\text{-Bu—CH}_2\text{—Br} + \text{PhSNa} \xrightarrow[\text{H}_2\text{O, 70 °C}]{\text{C}_{16}\text{H}_{33}\text{P(Bu)}_3^+\ \text{Br}^-} t\text{-Bu—CH}_2\text{—SPh}$$

OS, CV 6, **835 (1988)**

$$p\text{-O}_2\text{N-C}_6\text{H}_4\text{-CH}_2\text{OH} \xrightarrow[\text{CH}_2\text{Cl}_2\text{, 10 °C}]{\text{Et}_2\text{NSF}_3} p\text{-O}_2\text{N-C}_6\text{H}_4\text{-CH}_2\text{F}$$

OS, CV 6, **951 (1988)**

$$\text{P-C}_6\text{H}_4\text{-CH}_2\text{Cl} \xrightarrow[\text{DMF, 100 °C}]{\text{potassium phthalimide}} \text{P-C}_6\text{H}_4\text{-CH}_2\text{-N(phthalimide)}$$

P = styrene - divinylbenzene copolymer

OS, CV 6, **1016 (1988)**

$$2\ p\text{-CH}_3\text{-C}_6\text{H}_4\text{-SO}_2\text{-SK} \xrightarrow[\text{KI, EtOH, }\Delta;\ n = 2, 3]{\text{Br(CH}_2)_n\text{Br}} p\text{-Me-C}_6\text{H}_4\text{-SO}_2\text{-S—(CH}_2)_n\text{—S-SO}_2\text{-C}_6\text{H}_4\text{-}p\text{-Me}$$

***OS*, *CV 7*, 319 (1990); *61*, 77 (1983)**

$(HOCH_2)_2C(CO_2Et)_2$ —HBr, reflux→ $(BrCH_2)_2CHCO_2H$ —MeOH, $MeSO_3H$, $ClCH_2CH_2Cl$→ $(BrCH_2)_2CHCO_2Me$

***OS*, *CV 7*, 266 (1990); *62*, 58 (1984)**

2-methyl-2-propen-1-ol —1. *n*-BuLi, TMEDA; 2. Me_3SiCl→ Me_3Si–CH_2–C(=CH_2)–CH_2–$OSiMe_3$ —H_2SO_4→ Me_3Si–CH_2–C(=CH_2)–CH_2–OH

***OS*, *CV 7*, 356 (1990); *63*, 140 (1985)**

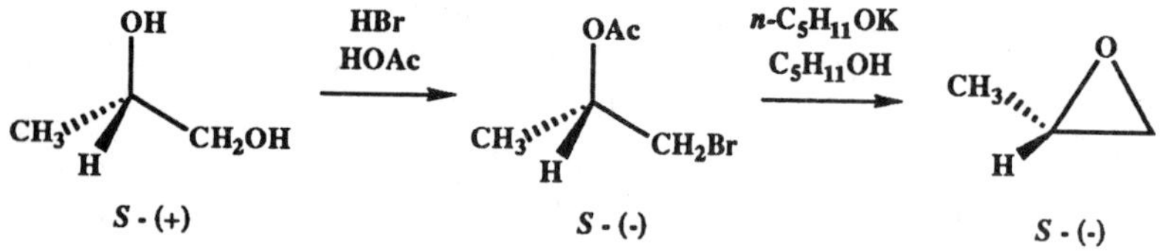

CH_3CH(OH)CH_2OH (*S* - (+)) —HBr, HOAc→ CH_3CH(OAc)CH_2Br (*S* - (-)) —*n*-$C_5H_{11}OK$, $C_5H_{11}OH$→ methyloxirane (*S* - (-))

***OS*, *CV 7*, 453 (1990); *64*, 157 (1986)**

PhSH —1. NaOEt, EtOH; 2. $BrCH_2CH_2Br$→ $PhSCH_2CH_2Br$

***OS*, *65*, 150 (1987)**

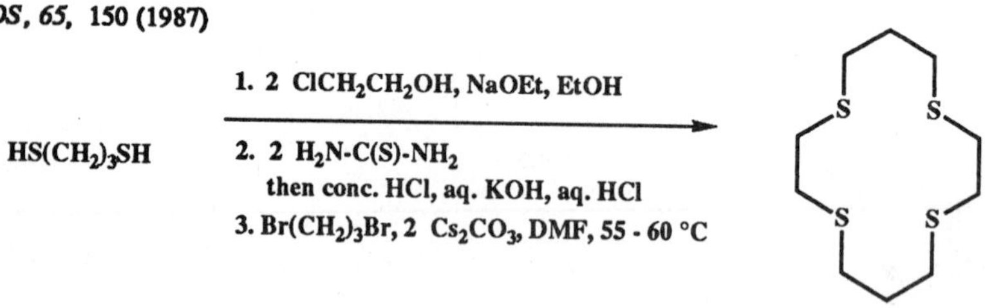

$HS(CH_2)_3SH$ —1. 2 $ClCH_2CH_2OH$, NaOEt, EtOH; 2. 2 H_2N-C(S)-NH_2 then conc. HCl, aq. KOH, aq. HCl; 3. $Br(CH_2)_3Br$, 2 Cs_2CO_3, DMF, 55 - 60 °C→

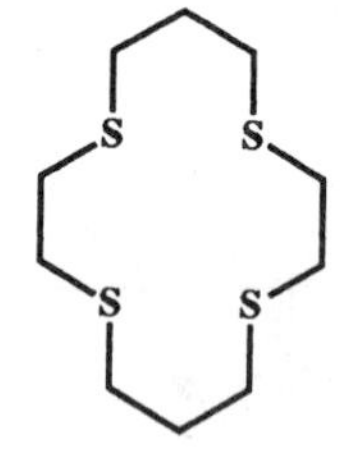

OS, *65*, 243 (1987)

Ph O O HO HO O OMe + NBr O O

$BaCO_3$ / CCl_4 / $C_2H_2Cl_4$

Br PhCOO O HO HO OMe

OS, *66*, 194 (1987)

1. LDA THF, - 75 °C
2. C_2Cl_6 THF, - 50 °C

Cl

$(EtO)_2P(O)H$, *t*-BuOK, DMF

$(EtO)_2P$ O

OS, *66*, 211 (1987)

OH

1. N-chloro-succinimide / Me_2S
2. $(Bu_4N)_3P_2O_7H$
3. Ion exchange

$OP_2O_6(NH_4)_3$

OS, *67*, 105 (1988)

AcO Cl + RR'NH

$Pd(PPh_3)_4$ (4 mol %) THF, 25 °C
or uncatalyzed, MeCN, 80 °C

AcO NRR'

R, R' = Et; R = H, R' = $PhCH_2$

OS, *68*, 8 (1989)

PhSCl → ($NaCH_2NO_2$, EtOH) → $PhSCH_2NO_2$

OS, *68*, 92 (1989)

HOCH_2, CH_2OH, O, O, Me, Me → (NaH, $PhCH_2Br$, THF) → $PhCH_2OCH_2$, CH_2OCH_2Ph, O, O, Me, Me

l isomer

OS, *68*, 188 (1989)

Me_3Si, OH → (1. *p*-TsCl; 2. *p*-MeO-C_6H_4-NH_2) → Me_3Si, H, N, OMe

OS, *68*, 227 (1989)

Cl, O, O, Cl → (1. $PhCH_2NH_2$; 2. NaI, MeCN, Na_2CO_3; I, O, O, I; reflux, 18 hr) → $PhCH_2$—N, O, O, O, O, N—CH_2Ph

OS, CV 1, 12 (1941)

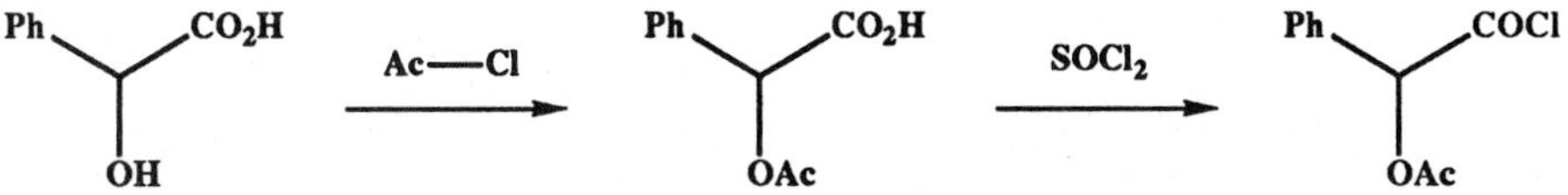

Ph, CO_2H, OH —(Ac—Cl)→ Ph, CO_2H, OAc —($SOCl_2$)→ Ph, COCl, OAc

OS, CV 1, 23 (1941)

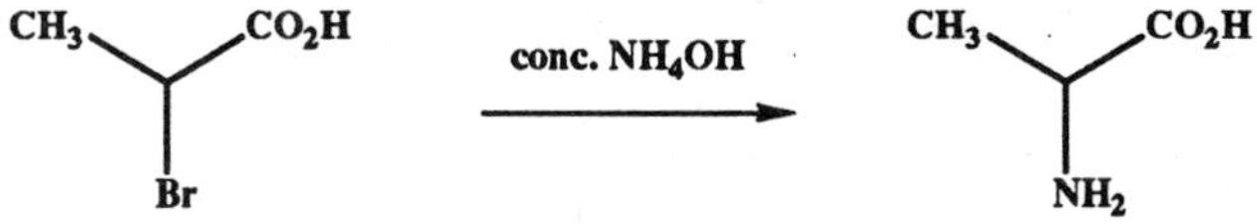

CH_3, CO_2H, Br —(conc. NH_4OH)→ CH_3, CO_2H, NH_2

OS, CV 1, 48 (1941)

CO_2H, Br —(conc. NH_4OH)→ CO_2H, NH_2

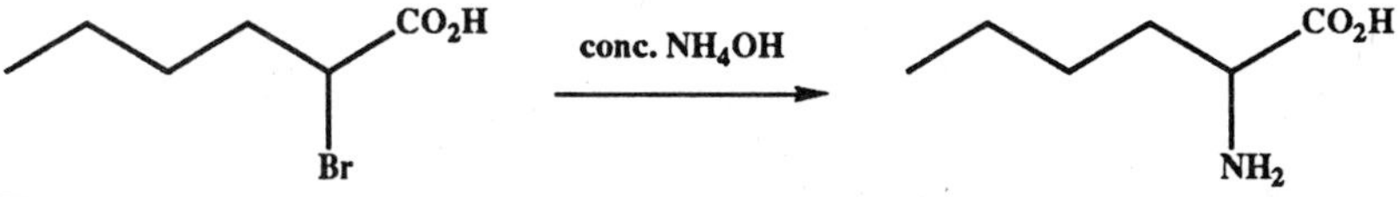

OS, CV 1, 258 (1941)

$CHCl_3$ —(Na, EtOH)→ $CH(OEt)_3$

OS, CV 1, 271 (1941)

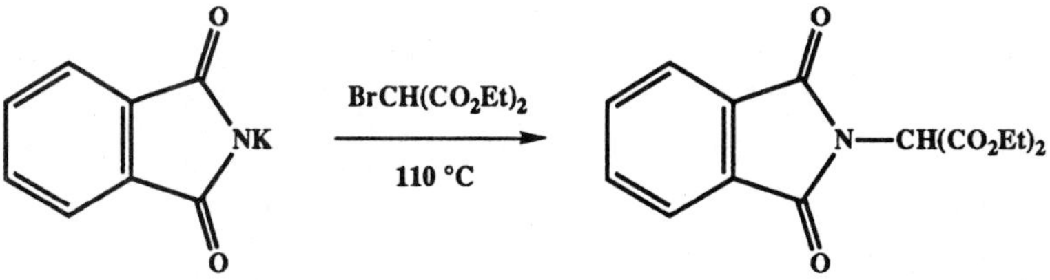

O, NK, O —($BrCH(CO_2Et)_2$, 110 °C)→ O, N—$CH(CO_2Et)_2$, O

OS, CV 1, 364 (1941)

CH_2OH O OH H OH HO H OH — CH_3OH, HCl, Δ → CH_2OH O OH H OCH_3 HO H OH

d - glucose

OS, CV 2, 69 (1943)

OH Ph Ph O — Ac_2O; HOAc, H_2SO_4 → OAc Ph Ph O

OS, CV 2, 89 (1943)

CH_3 Br — Br_2, Δ; light → $CHBr_2$ Br — H_2O, Δ; $CaCO_3$ → O H Br

OS, CV 2, 122 (1943)

CH_2OAc O H H O CH_2OAc O OAc H AcO AcO OAc H OAc OAc — NaOMe; MeOH → CH_2OH O H H O CH_2OH O OH H HO HO OH H OH OH

α - cellobiose octaacetate

OS, *CV* 2, 133 (1943)

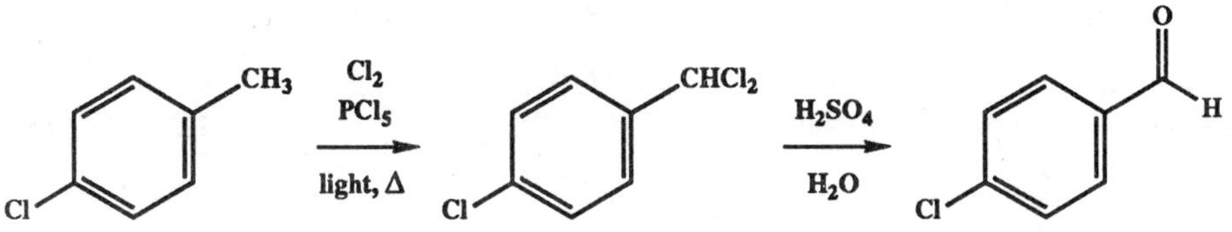

Cl_2, PCl_5, light, Δ; H_2SO_4, H_2O

CH3, Cl, CHCl2, O, H

OS, *CV* 2, 159 (1943)

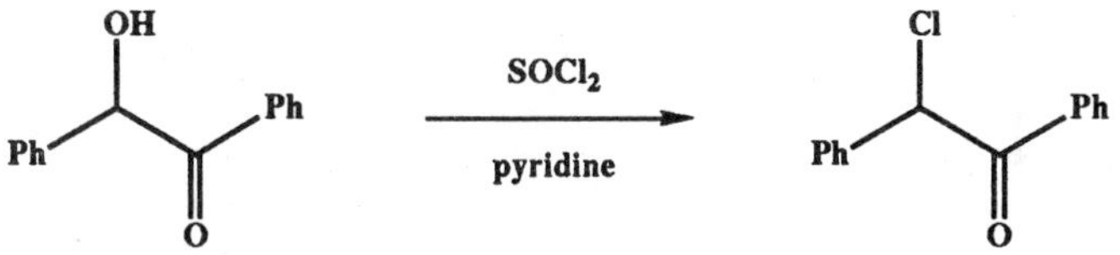

$SOCl_2$, pyridine

OH, Ph, O, Cl

OS, *CV* 2, 365 (1943)

i-PrOH, H_2SO_4, benzene, Δ

CH3, O, OH, O-*i*-Pr

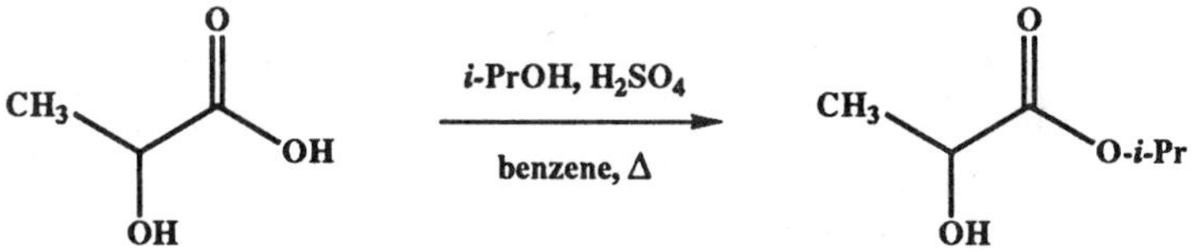

OS, *CV* 2, 366 (1943)

NaSCN, Δ, EtOH, H_2O

CH3, Br, SCN

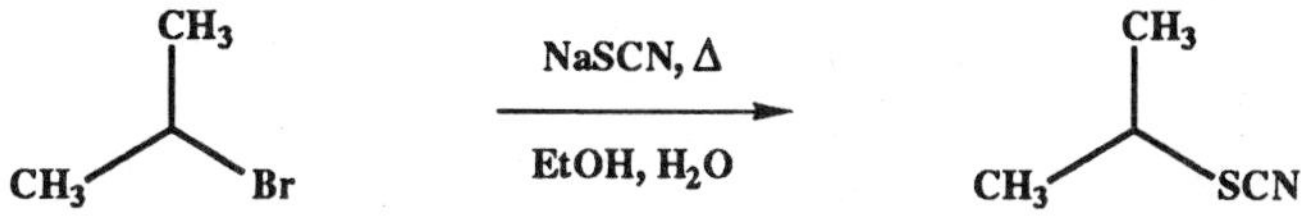

OS, CV 2, 374 (1943)

O
Ph
N—$(CH_2)_4$—$CHCO_2H$
H
Br

NH_4OH

O
Ph
N—$(CH_2)_4$—$CHCO_2H$
H
NH_2

aq. HCl, Δ

HCl • H_2N—$(CH_2)_4$—$CHCO_2H$
NH_2 • HCl

dl - lysine di-HCl

pyridine
EtOH, Δ

HCl • H_2N—$(CH_2)_4$—$CHCO_2H$
NH_2

dl - lysine HCl

OS, CV 3, 11 (1955)

CH_2OH
HO
HO
O
HO
OH

d - glucose

1. Ac_2O
2. HBr

CH_2OAc
AcO
AcO
O
AcO
Br

α - *d* - isomer

OS, CV 3, 101 (1955)

CH_2OAc
OAc
OAc
H
CN
AcO
H
OAc

d - isomer

NaOMe, MeOH
- 12 °C

CH_2OH
OH
OH
O
HO
H
H

d - arabinose

OS, *CV* 3, 123 (1955)

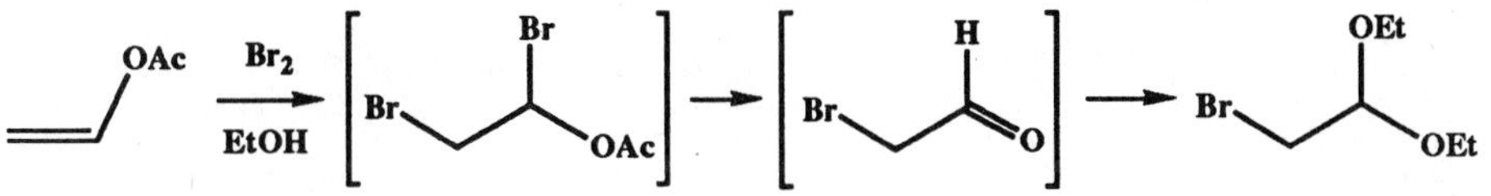

OAc; Br_2, EtOH; Br, Br, OAc; Br, H, O; Br, OEt, OEt

OS, *CV* 3, 127 (1955)

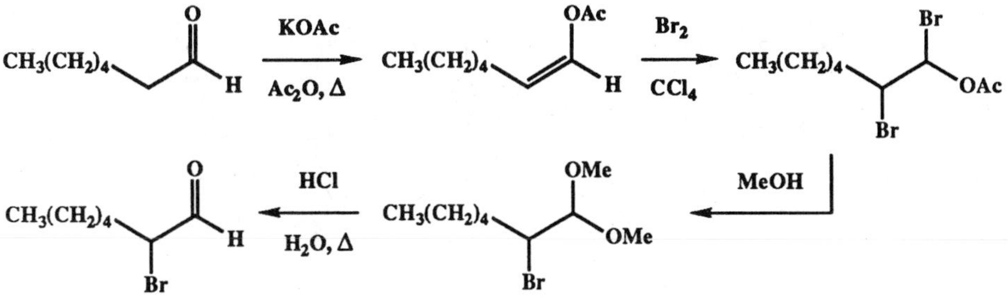

$CH_3(CH_2)_4$, O, H; KOAc, Ac_2O, Δ; $CH_3(CH_2)_4$, OAc, H; Br_2, CCl_4; $CH_3(CH_2)_4$, Br, OAc, Br; MeOH; $CH_3(CH_2)_4$, OMe, OMe, Br; HCl, H_2O, Δ; $CH_3(CH_2)_4$, O, H, Br

OS, *CV* 3, 432 (1955)

CH_2OH, O, OH, H, OH, HO, H, OH

d - glucose

1. Ph_3CCl pyridine, Δ
2. Ac_2O

$CH_2O{-}CPh_3$, O, OAc, OAc, H, AcO, H, H, OAc

HBr, HOAc

CH_2OH, O, OAc, OAc, H, AcO, H, H, OAc

β - *d* - glucose 1,2,3,4-tetraacetate

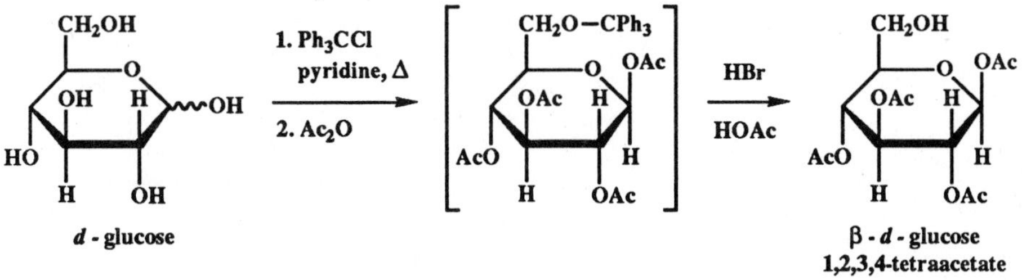

OS, CV 3, 434 (1955)

CH_2OAc, O, H, OAc, H, AcO, Br, H, OAc

Ag_2CO_3; H_2O, acetone

CH_2OAc, O, OH, OAc, H, AcO, H, H, OAc

β - *d* - glucose
2,3,4,6-tetraacetate

OS, CV 3, 471 (1955)

OH, HO_2C, CO_2H, OH

fuming HNO_3; H_2SO_4

ONO_2, HO_2C, CO_2H, ONO_2

OS, CV 3, 495 (1955)

Me, Br, Et

$CH_2(CO_2Et)_2$; Na, EtOH, Δ

Me, $CH(CO_2Et)_2$, Et

1. aq. KOH, Δ, then HCl
2. Br_2, Et_2O
3. Δ

Me, CO_2H, Et, Br

NH_4OH, Δ

Me, CO_2H, Et, NH_2

dl - isoleucine

OS, *CV* 3, 523 (1955)

i-Pr $CH_2CH_2CO_2H$ → (Br_2; PCl_3, Δ) → i-Pr $CH_2CH(Br)CO_2H$ → (NH_4OH) → i-Pr $CH_2CH(NH_2)CO_2H$

OS, *CV* 3, 538 (1955)

$PhCOCH_3$ → (Cl_2; HOAc) → $PhCOCHCl_2$ → (1. aq. NaOH, 65 °C; 2. aq. HCl) → $PhCH(OH)CO_2H$

OS, *CV* 3, 544 (1955)

l - menthol + $ClCH_2CO_2H$ → (1. Na, toluene, Δ; 2. HCl, H_2O) → menthyl-OCH_2CO_2H (*l* - isomer)

OS, *CV* 3, 705 (1955)

$PhCH_2CH(Br)CO_2H$ → (NH_4OH) → $PhCH_2CH(NH_2)CO_2H$

dl - phenylalanine

OS, *CV* 3, 737 (1955)

Br_2, 140 °C; H_2O, Δ

O, O, Br, CO_2H, CHO

OS, *CV* 3, 774 (1955)

MeO, CO_2H, Br

1. NH_4OH, Δ
2. HBr, Δ
3. NH_4OH

HO, CO_2H, NH_2

dl - serine

OS, *CV* 3, 788 (1955)

CH_3, CH_3

Br_2, light, Δ

$CHBr_2$, $CHBr_2$

95% H_2SO_4, 70 - 110 °C

CHO, CHO

OS, *CV* 3, 800 (1955)

CH_2OH, O, OH, H, OH, HO, H, OH

d - glucose

Me_2SO_4, aq. NaOH, CCl_4, Δ

CH_2OMe, O, OMe, H, OMe, MeO, H, OMe

HCl, H_2O, Δ

CH_2OMe, O, OMe, H, OH, MeO, H, OMe

2,3,4,6-tetramethyl-*d* - glucose

OS, CV 3, 813 (1955)

OCH_3, CO_2H, CH_3, Br — 1. NH_4OH heat; 2. HCO_2H, Ac_2O → OCH_3, CO_2H, CH_3, NHCHO — 1. HBr, Δ; 2. NH_4OH, EtOH → OH, CO_2H, CH_3, NH_2

dl - threonine

OS, CV 3, 833 (1955)

OH, O — $ZnCl_2$; Ac_2O, Δ → OAc, OAc, OAc — H_2SO_4; H_2O, Δ → OH, OH, OH

OS, CV 3, 848 (1955)

CH_3, CO_2H, CH_3 — Br_2; PCl_3, Δ → CH_3, CO_2H, CH_3, Br — NH_4OH → CH_3, CO_2H, CH_3, NH_2

dl - valine

OS, CV 4, 25 (1963)

H, N, O, OH, OH, O, N, H, O + O, H, N, HO, O, N, H, O — H_2O → H, N, O, O, N, H, O, HO, O, O, H, N, O, N, H, O

OS, CV 4, 72 (1963)

Ph, Ph, OH → (HO–CH₂CH₂–Cl; cat. H^+, benzene, Δ) → Ph, Ph, O, Cl

OS, CV 4, 110 (1963)

O, CH_3, Br → (Br_2; HOAc) → O, $CHBr_2$, Br → (1. aq. NaOH, 5 °C; 2. HCl) → HO, CO_2H, Br

OS, CV 4, 169 (1963)

OH, Ph, CO_2Et → ($SOCl_2$) → Cl, Ph, CO_2Et → (HOAc; HCl, Δ) → Cl, Ph, CO_2H

OS, CV 4, 427 (1963)

O, Cl, H, Cl, OH → (1. 3 NaOEt, EtOH, Δ; 2. HCl, EtOH) → O, EtO, H, OEt, OEt

***OS*, *CV* 4,** **454 (1963)**

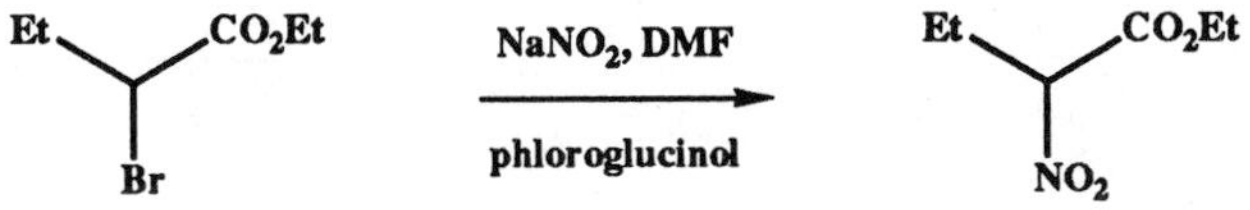

$NaNO_2$, DMF

phloroglucinol

***OS*, *CV* 4,** **464 (1963)**

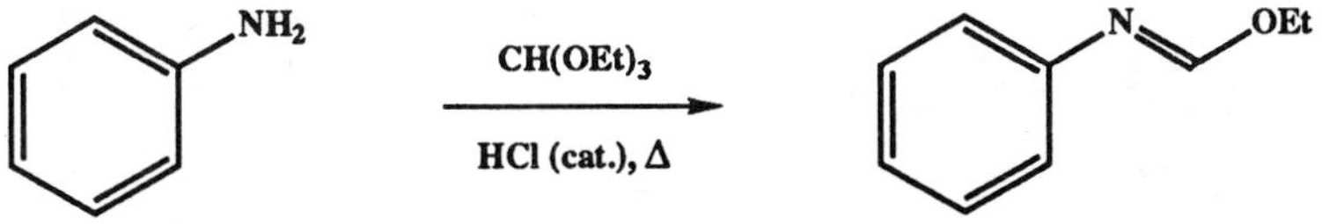

$CH(OEt)_3$

HCl (cat.), Δ

***OS*, *CV* 4,** **466 (1963)**

Δ

benzene

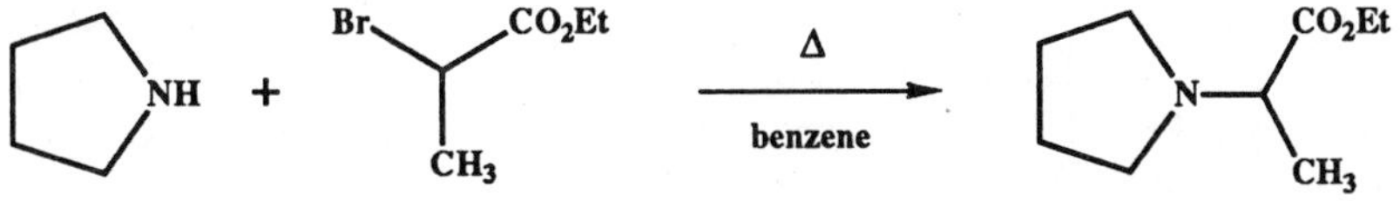

***OS*, *CV* 4,** **590 (1963)**

SO_2Cl_2

0 - 5 °C

NaOPh

benzene, Δ

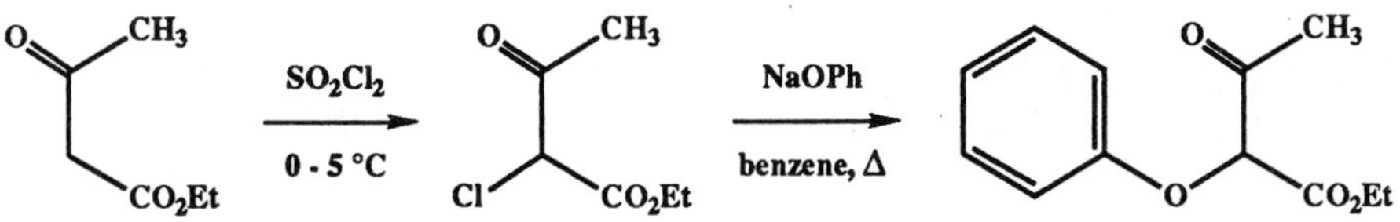

OS, CV 4, 807 (1963)

CH_3 / CH_3 —(4 Br_2; light, Δ)→ $CHBr_2$ / $CHBr_2$ —($(CO_2K)_2$; aq. EtOH, Δ)→ CHO / CHO

OS, CV 5, 1 (1973)

CH_2OH, O, H, OH, H, HO, OH, H, NHAc —(acetyl chloride)→ CH_2OAc, O, H, OAc, H, AcO, Cl, H, NHAc

d - isomer   *d* - isomer

OS, CV 5, 245 (1973)

C_8H_{17}, HO, H, H, H —(CH_2N_2, HBF_4; CH_2Cl_2, Et_2O)→ C_8H_{17}, CH_3O, H, H, H

OS, CV 5, 266 (1973)

NOH • HCl, Cl —(MeOH; Et_3N)→ NOH, OMe

OS, *CV 5*, 414 (1973)

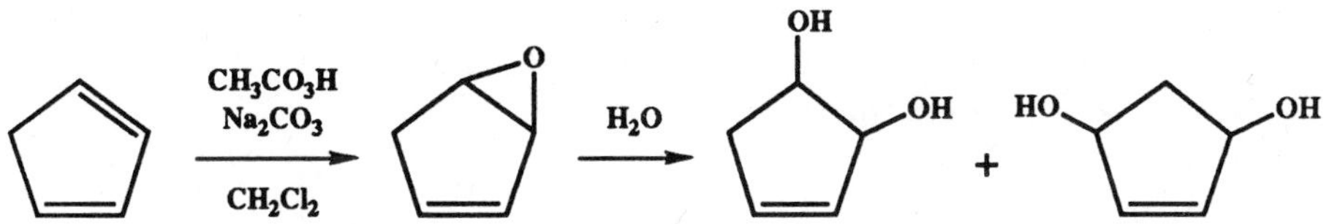

CH_3CO_3H, Na_2CO_3, CH_2Cl_2; H_2O

OH, OH; HO, OH

OS, *CV 5*, 582 (1973)

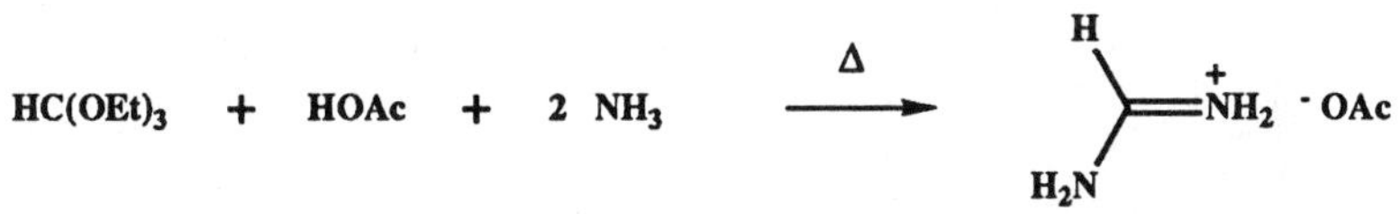

$HC(OEt)_3$ + HOAc + 2 NH_3 $\xrightarrow{\Delta}$ H, H_2N, $\overset{+}{N}H_2$ $^-$OAc

OS, *CV 5*, 591 (1973)

HO, HO, HO, CO_2H $\xrightarrow[H_2SO_4\ (cat.)]{Ac_2O,\ \Delta}$ AcO, AcO, AcO, CO_2H

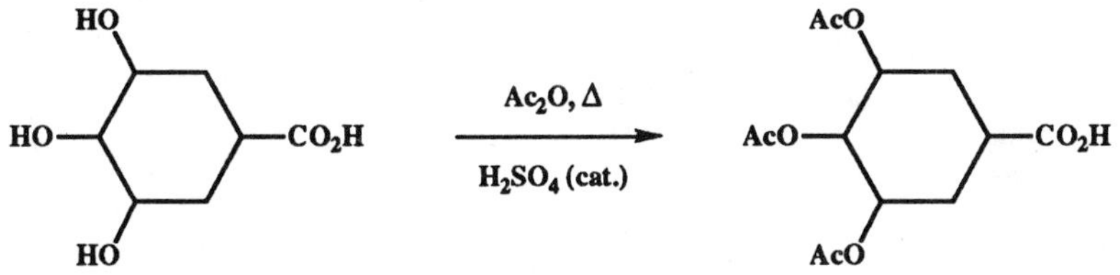

OS, *CV 5*, 608 (1973)

OH $\xrightarrow[HBr\ (cat.)]{PBr_3}$ C_6H_9Br (mixture of bromohexadienes)

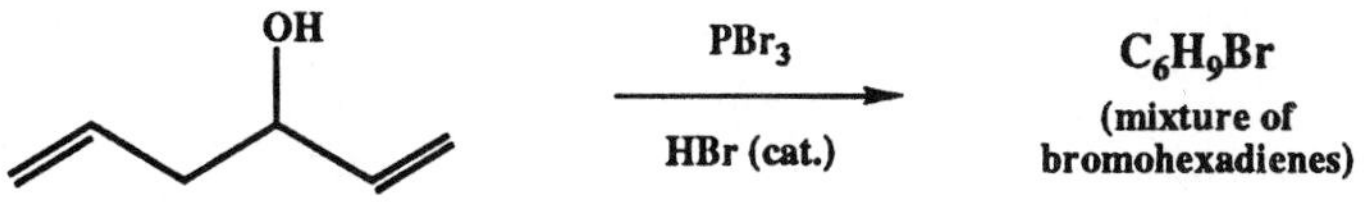

OS, *CV 5*, 859 (1973)

H, OCH_2CH_2Cl — 1. *n*-BuLi, Et_2O / 2. aq. $NaHCO_3$ → H, OH

ca. 6 : 1 *exo* : *endo* — *ca.* 8 : 1 *exo* : *endo*

OS, *CV 5*, 863 (1973)

OAc — NaOMe / MeOH, Δ → OH

OS, *CV 5*, 887 (1973)

CH_2OH, O, OH, H, O, HO, H, OH — Ac_2O / $ZnCl_2$ → CH_2OAc, OH, OAc, H, CO_2H, AcO, H, OAc — Ac_2O / $ZnCl_2$ → CH_2OAc, OAc, OAc, H, CO_2H, AcO, H, OAc

d - gluconolactone — *d* - tetraacetate — *d* - pentaacetate

OS, *CV 6*, 235 (1988)

i-Pr-Br — $Na_2S_2O_3 \cdot 5\ H_2O$ / MeOH, H_2O, Δ → *i*-Pr—S—SO_3Na — *sec*-BuSNa / H_2O, 0 - 5 °C → *i*-Pr—S—S—*sec*-Bu

***OS, CV 6,* 298 (1988)**

$CH_2=CHOEt$; $Hg(OAc)_2$, Δ

C8H17 H H HO OEt Hg(OAc)2, Δ C8H17 H H O

***OS, CV 6,* 403 (1988)**

$PhCH_2CO_2H$ → (Br_2, PCl_3; benzene, Δ) → $Ph\text{—}CH(Br)CO_2H$ → (Na_2S; aq. Na_2CO_3) → $Ph\text{—}CH(CO_2H)\text{—}S\text{—}CH(CO_2H)\text{—}Ph$

***OS, CV 6,* 620 (1988)**

$HC(OEt)_3$ → (H_2S, H_2SO_4; HOAc, hydroquinone) → $H\text{—}C(=S)OEt$

***OS, CV 6,* 683 (1988)**

Epichlorohydrin (O, Cl) → (2 MeSH, 2 NaOH; MeOH, below 50 °C) → $MeSCH_2CH(OH)CH_2SMe$ → (NaH, MeI; THF, 20-25 °C) → $MeSCH_2CH(OMe)CH_2SMe$

OS, CV 6, 776 (1988)

n-Pr–C(=O)SH + $CH_3CH(Br)C(=O)CH_3$ → (Et_3N, Et_2O) n-Pr–C(=O)–S–CH(CH_3)–C(=O)CH_3

OS, CV 6, 830 (1988)

Cyclohexanol (–OH) → $[(PhO)_3PCH_3]$ I → Cyclohexyl iodide (–I)

OS, CV 7, 501 (1990); *62,* 48 (1984)

1. Phthalimide, $EtO_2CN=NCO_2Et$, Ph_3P
2. N_2H_4

OH → NH_2

OS, CV 7, 470 (1990); *63,* 192 (1985)

R; OH; CO_2H; R - (+)

1. $(C_5H_4NS)_2$, Ph_3P benzene, 20 °C
2. $AgClO_4$, MeCN, reflux

R; O; O; R - (+)

OS, *65*, 68 (1987)

t-BuOH, K_2CO_3, reflux

1. PCl_5, hexane
2. liq. NH_3

t-BuO, Cl, t-BuO, Cl

OS, *66*, 151 (1987)

CO_2H, S, H_2N, H, R

$NaNO_2$, HCl

CO_2H, S, Cl, H, R

R = Me, *i*-Pr, *i*-Bu, (*S*) - *sec*-Bu

OS, *66*, 160 (1987)

CO_2H, Cl, H, R

LAH, Et_2O

CH_2OH, Cl, H, R

KOH, H_2O

H_2C, O, H, R

S isomers

R = Me, *i*-Pr, *i*-Bu, (*S*) - *sec*-Bu

R isomers

OS, *68*, 210 (1989)

OH, $CH_3(CH_2)_4$

DMAP, 15 min
Et_2O, 25 °C

O, O, O, $CH_3(CH_2)_4$

***OS*, *CV 1*, 95 (1941)**

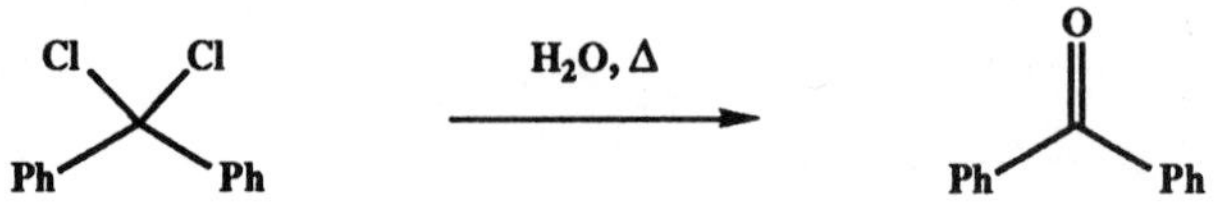
Cl
Cl
Ph
Ph
H_2O, Δ
O
Ph
Ph

***OS*, *CV 1*, 144 (1941)**

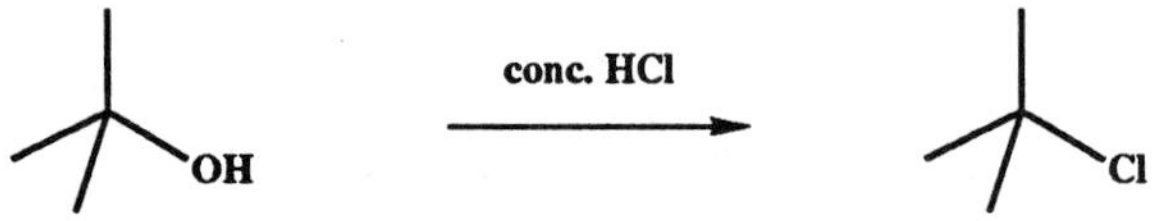
OH
conc. HCl
Cl

***OS*, *CV 2*, 29 (1943)**

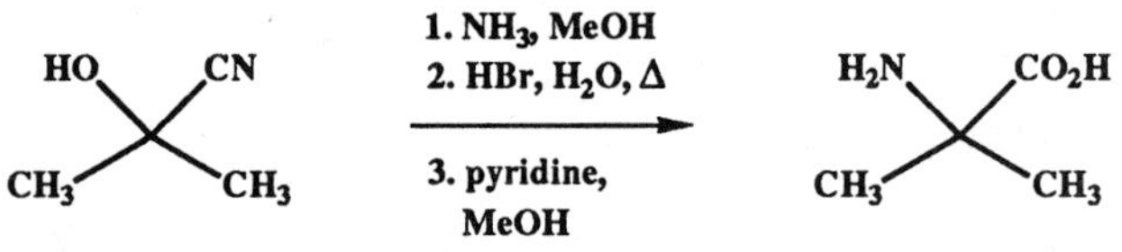
HO
CN
CH_3
CH_3
1. NH_3, MeOH
2. HBr, H_2O, Δ
3. pyridine,
MeOH
H_2N
CO_2H
CH_3
CH_3

***OS*, *CV 2*, 244 (1943)**

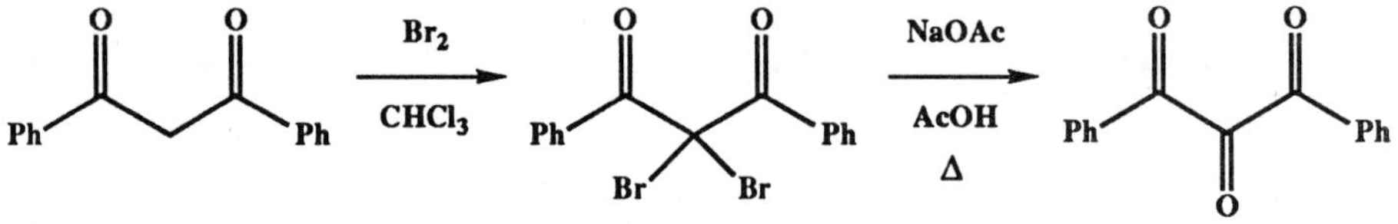
O
O
Ph
Ph
Br_2
$CHCl_3$
O
O
Ph
Ph
Br
Br
NaOAc
AcOH
Δ
O
O
Ph
Ph
O

***OS*, *CV 2*, 573 (1943)**

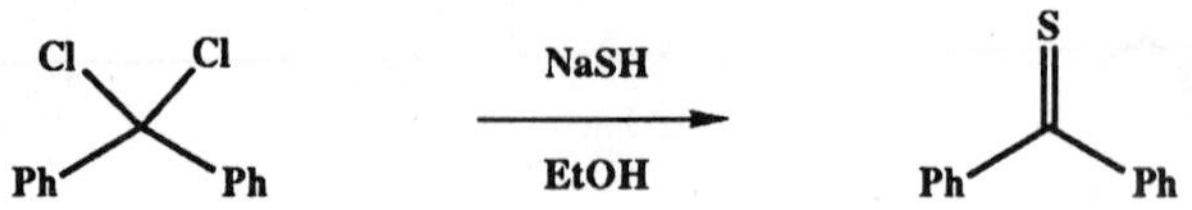
Cl
Cl
Ph
Ph
NaSH
EtOH
S
Ph
Ph

***OS*, *CV 2*, 607 (1943)**

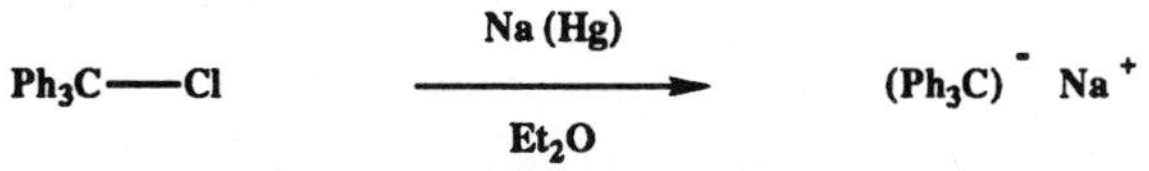

$$Ph_3C{-}Cl \xrightarrow[Et_2O]{Na(Hg)} (Ph_3C)^- \; Na^+$$

***OS*, *CV 3*, 42 (1955)**

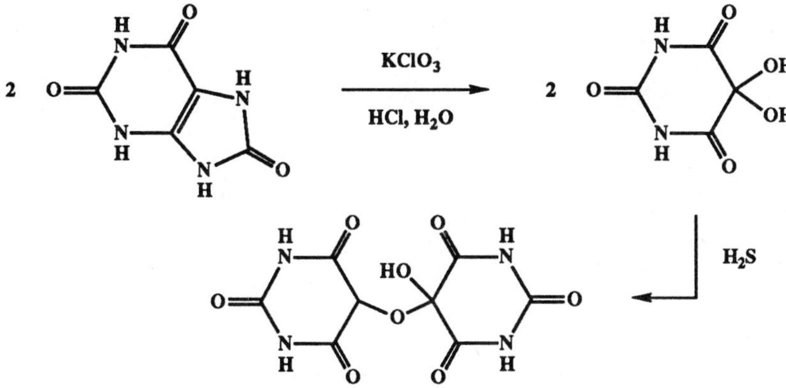

KClO$_3$; HCl, H_2O; H_2S

***OS*, *CV 3*, 841 (1955)**

$$Ph_3C{-}OH \xrightarrow[\text{benzene, } \Delta]{CH_3COCl} Ph_3C{-}Cl$$

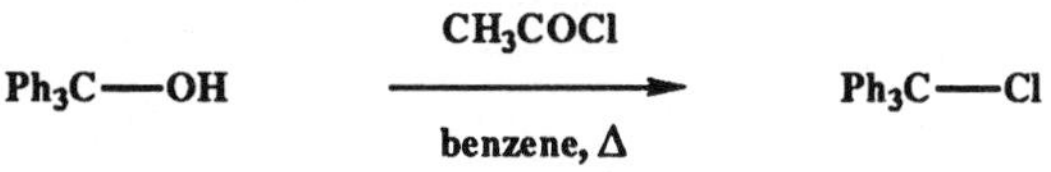

***OS*, *CV 4*, 25 (1963)**

H_2O

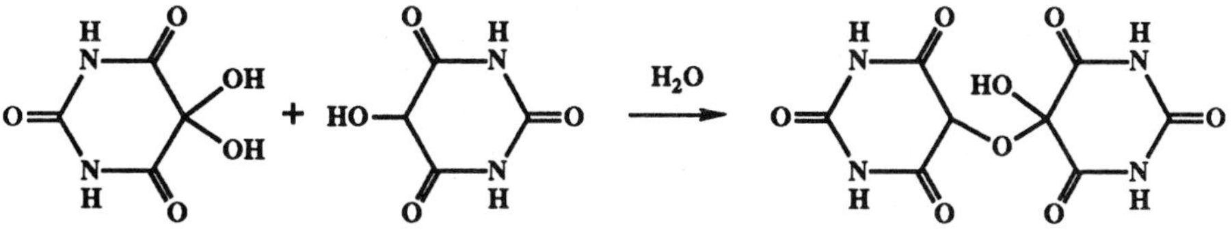

OS, CV 4, **423 (1963)**

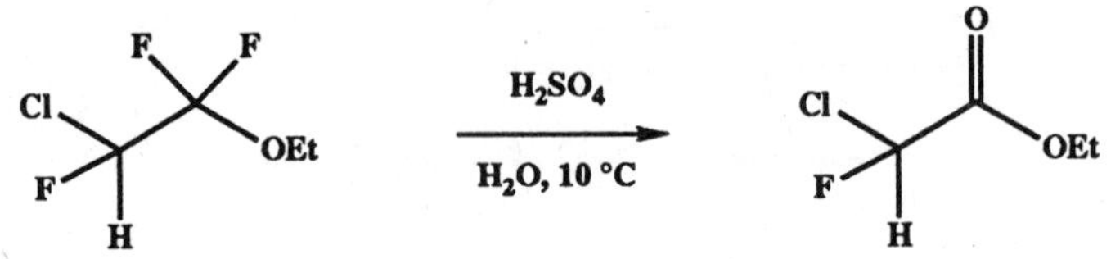
F
F
Cl
OEt
F
H
H_2SO_4
H_2O, 10 °C
O
Cl
OEt
F
H

OS, CV 4, **457 (1963)**

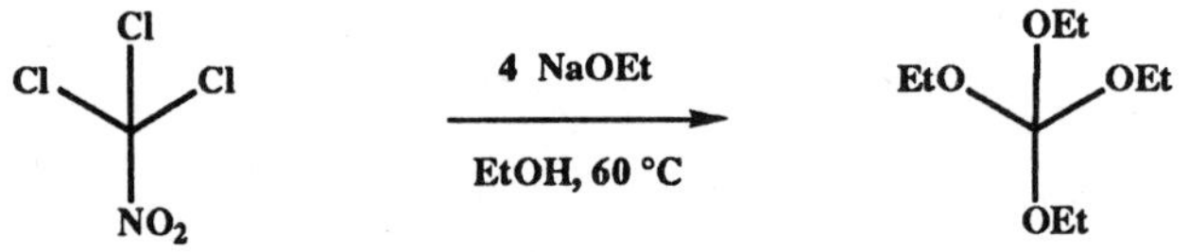
Cl
Cl
Cl
NO_2
4 NaOEt
EtOH, 60 °C
OEt
EtO
OEt
OEt

OS, CV 5, **5 (1973)**

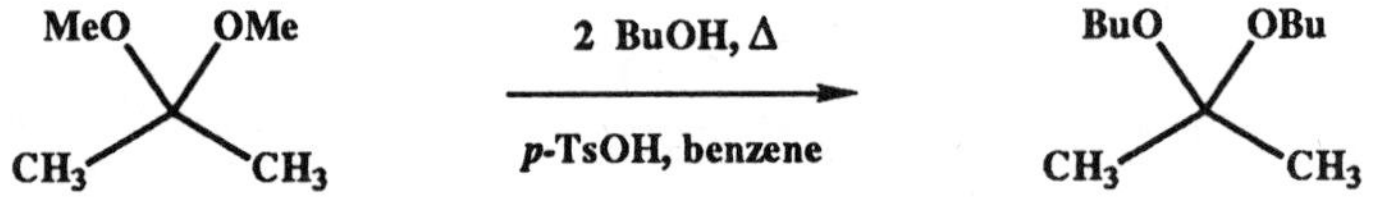
MeO
OMe
CH_3
CH_3
2 BuOH, Δ
p-TsOH, benzene
BuO
OBu
CH_3
CH_3

OS, CV 5, **355 (1973)**

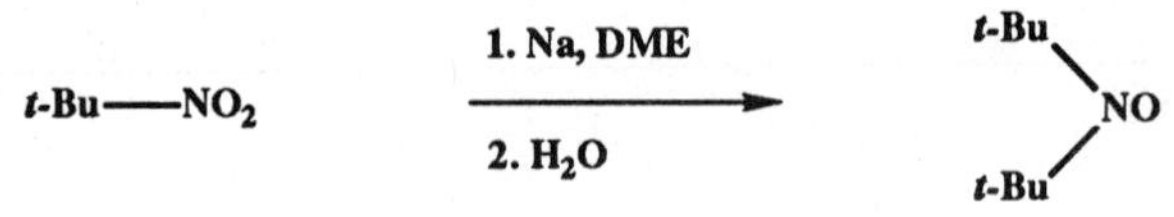
t-Bu—NO_2
1. Na, DME
2. H_2O
t-Bu
NO
t-Bu

OS, *CV 5*, 424 (1973)

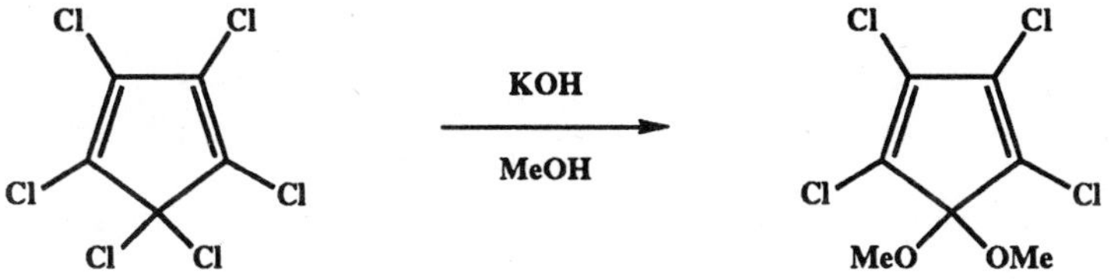

KOH

MeOH

OS, *CV 5*, 471 (1973)

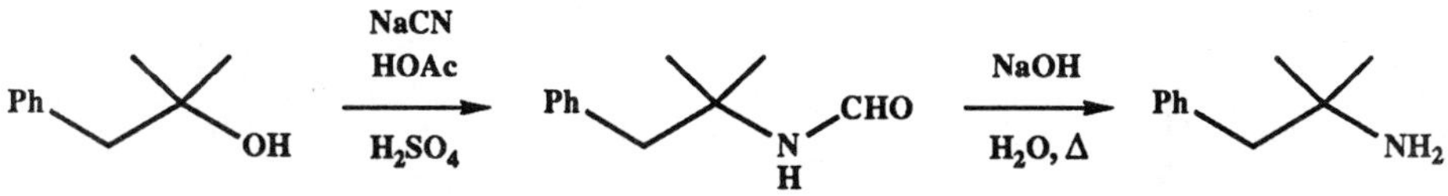

NaCN
HOAc
H_2SO_4

Ph, OH, CHO, N, H

NaOH
H_2O, Δ

NH_2

OS, *CV 6*, 628 (1988)

OH

N • HF(HF)$_x$

F

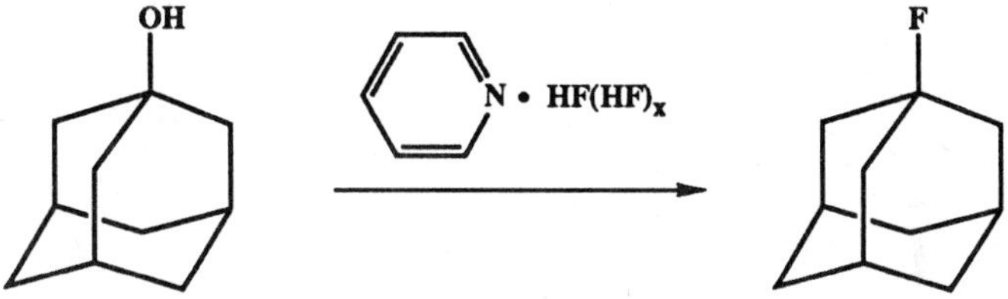

OS, *CV 7*, 433 (1990); *60*, 104 (1981)

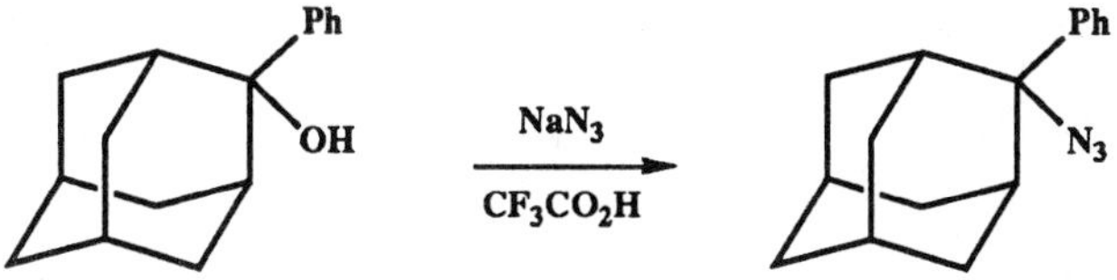

Ph, OH

NaN_3

CF_3CO_2H

Ph, N_3

OS, CV 1, 36 (1941)

R—OH $\xrightarrow{PBr_3, \Delta}$ R—Br

R = ethyl, *i*-Pr, *n*-Pr, *n*-Bu, *sec*-Bu, *t*-Bu

OS, CV 1, 327 (1941)

NH_2

OH

Cl_3C OH

NH_2OH

HCl, H_2O, Δ

H
N
O

HON H

OS, CV 3, 260 (1955)

OH

Cl_3C OH

NH_4OH

KCN, Et_2O

O

HCl_2C NH_2

OS, CV 5, 909 (1973)

NH_2

CH_3

1. *t*-BuOCl
benzene, 5 °C

2. NaOMe, MeOH, Δ
3. HCl, H_2O, Δ

O

NH_3Cl

C - C Bond Formation

OS, CV 1, **95 (1941)**

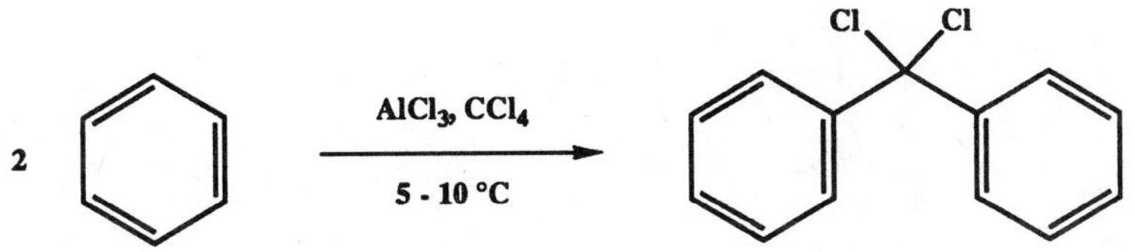

2

$AlCl_3$, CCl_4

5 - 10 °C

Cl Cl

OS, CV 1, **109 (1941)**

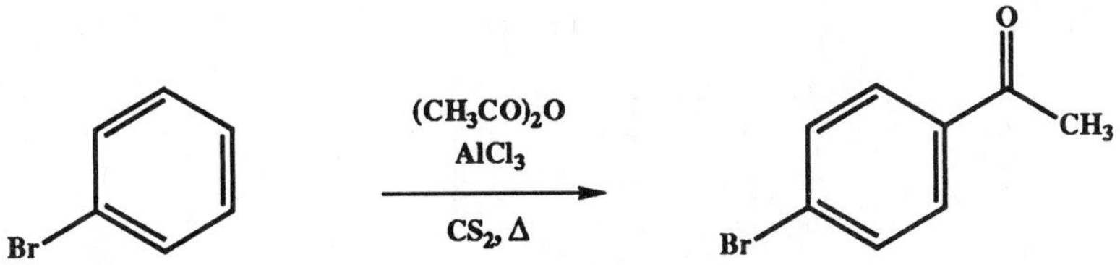

Br

$(CH_3CO)_2O$

$AlCl_3$

CS_2, Δ

O

CH_3

Br

OS, CV 1, **113 (1941)**

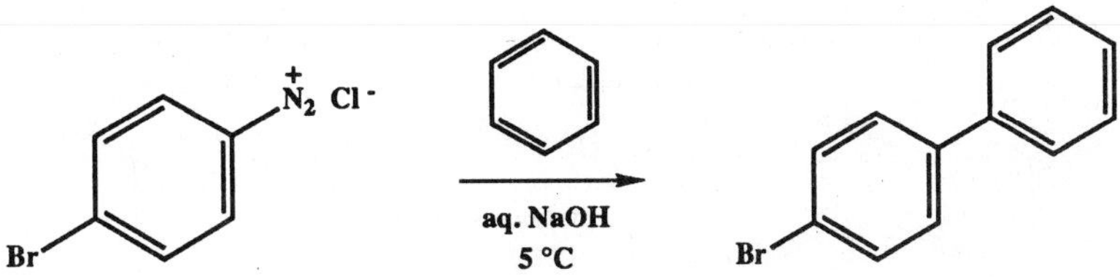

$\overset{+}{N}_2$ Cl^-

Br

aq. NaOH

5 °C

Br

OS, CV 1, **214 (1941)**

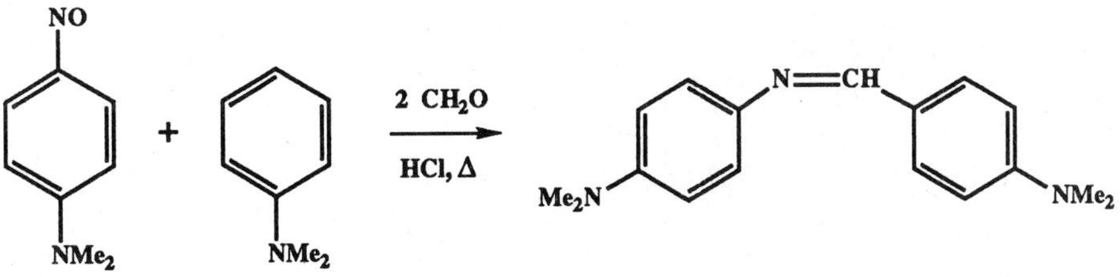

NO

NMe_2

+

NMe_2

2 CH_2O

HCl, Δ

N═CH

Me_2N

NMe_2

OS, CV 1, 217 (1941)

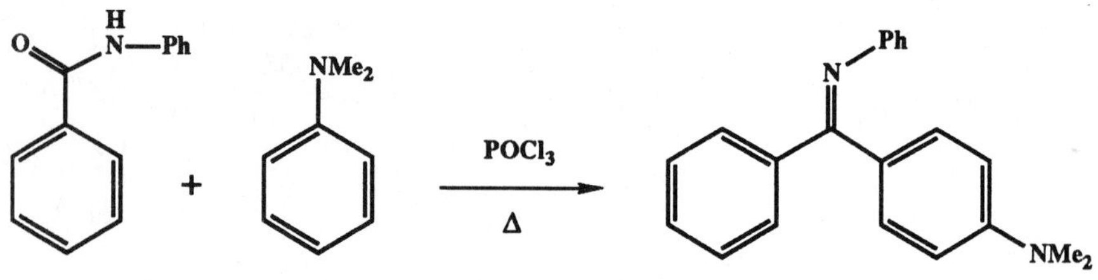

$POCl_3$, Δ

NMe$_2$, N—Ph, Ph

OS, CV 1, 229 (1941)

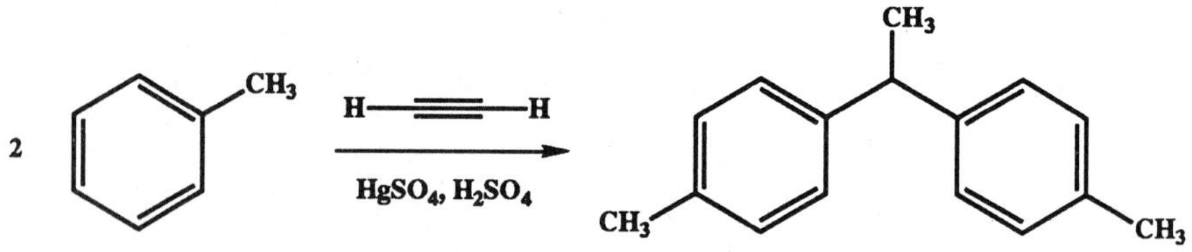

2 CH_3

H—≡—H

$HgSO_4$, H_2SO_4

OS, CV 1, 381 (1941)

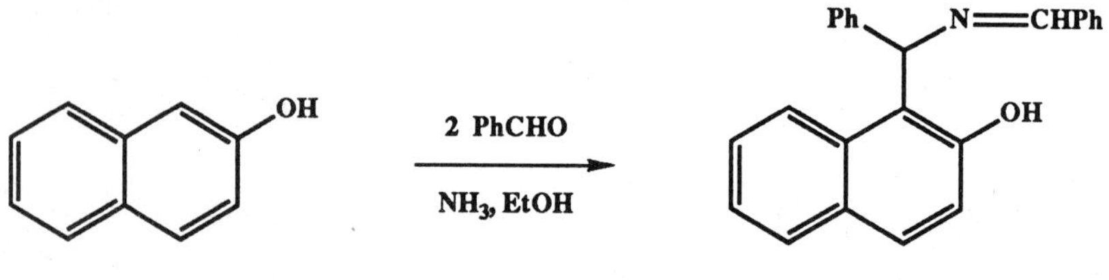

OH

2 PhCHO

NH_3, EtOH

Ph, N=CHPh, OH

OS, CV 1, 514 (1941)

NH_2, R_1, R_2

1. 2 HCl, $NaNO_2$

2. CuCN

CN, R_1, R_2

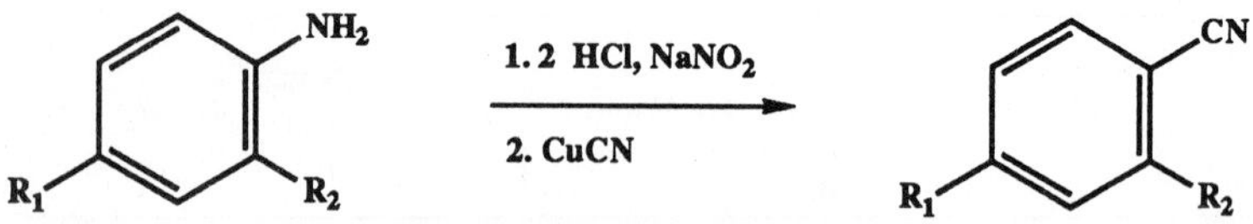

R_1 = Me, R_2 = H

R_1 = H, R_2 = Me

OS, CV 1, **517 (1941)**

CH_3

$AlCl_3$

CO_2H CH_3

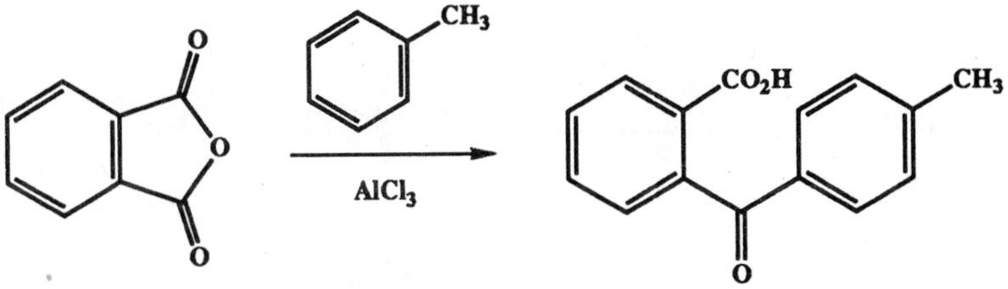

OS, CV 1, **548 (1941)**

$$3\ C_6H_6 \xrightarrow[CCl_4]{AlCl_3} (C_6H_5)_3CCl \bullet AlCl_3$$

OS, CV 2, **3 (1943)**

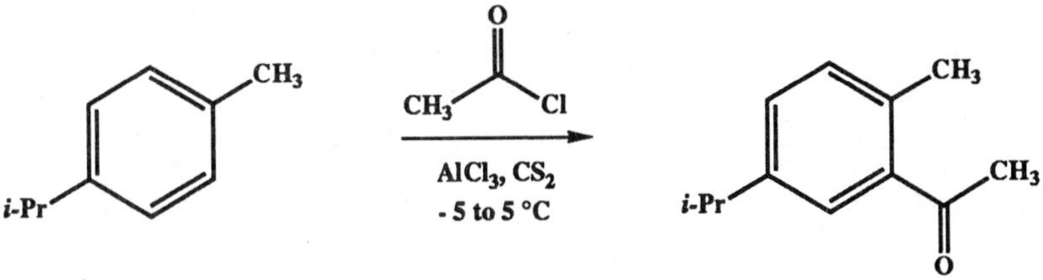

CH_3

i-Pr

CH_3 Cl

$AlCl_3$, CS_2
- 5 to 5 °C

CH_3 CH_3

i-Pr

OS, CV 2, **81 (1943)**

$AlCl_3$, Δ

CO_2H

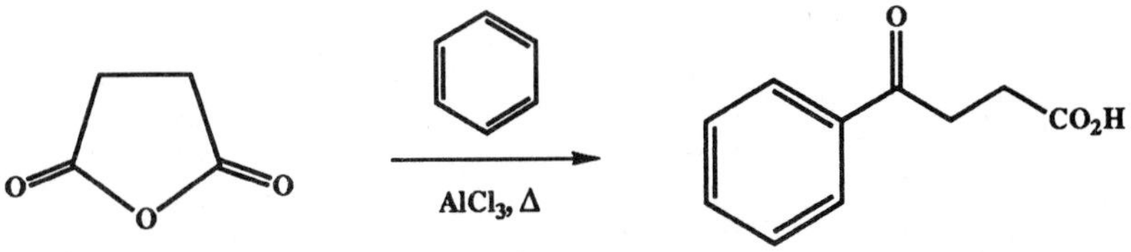

***OS*, *CV* 2, 151 (1943)**

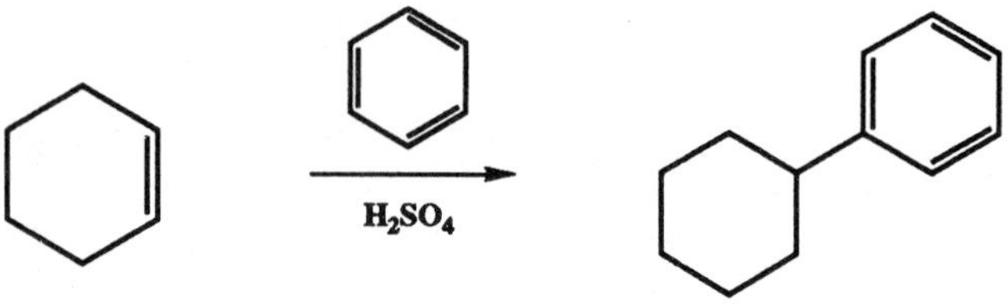

H_2SO_4

***OS*, *CV* 2, 156 (1943)**

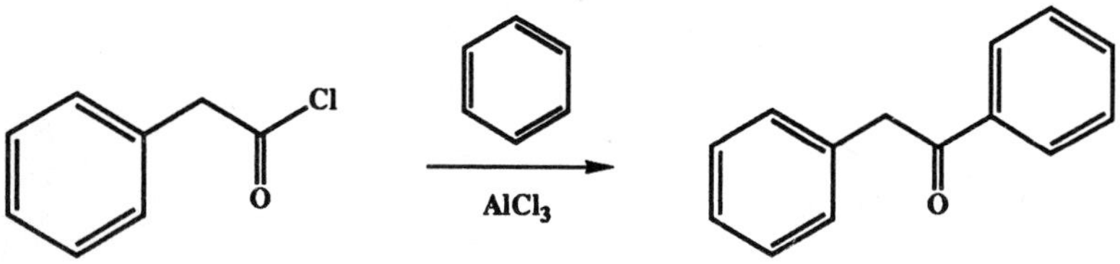

Cl

O

$AlCl_3$

O

***OS*, *CV* 2, 169 (1943)**

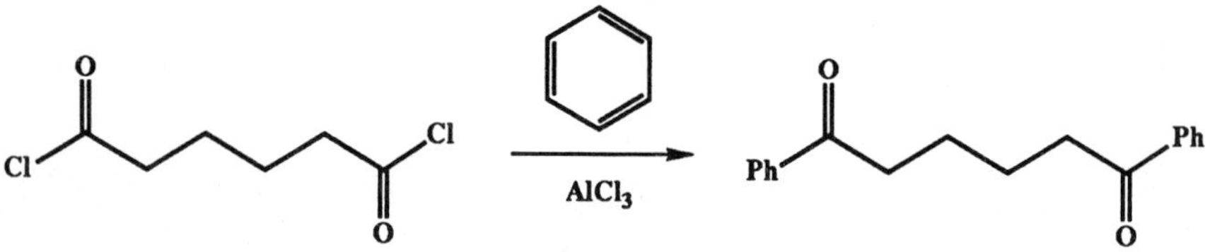

O

Cl

Cl

O

$AlCl_3$

O

Ph

Ph

O

***OS*, *CV* 2, 232 (1943)**

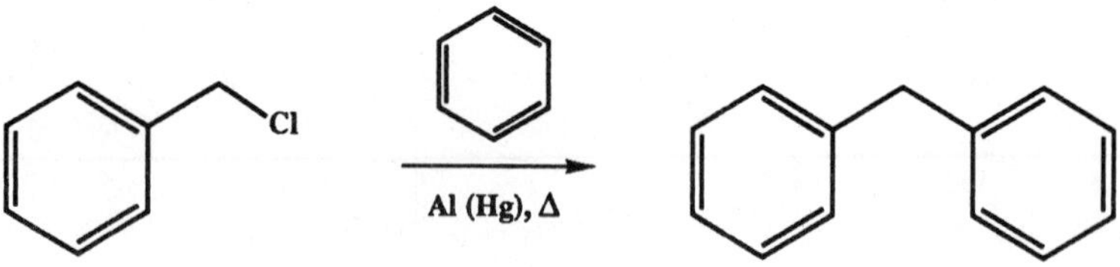

Cl

Al (Hg), Δ

OS, *CV* 2, 236 (1943)

Ph—CH=CH—C(=O)—Ph → (benzene, $AlCl_3$) → Ph–CH(Ph)–CH₂–C(=O)–Ph

OS, *CV* 2, 248 (1943)

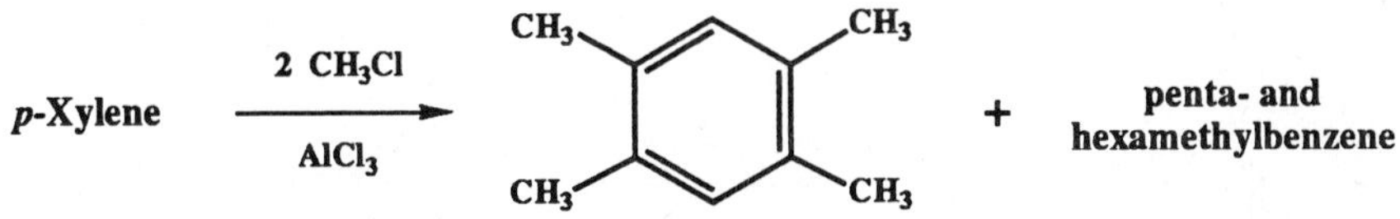

p-Xylene → (2 CH_3Cl, $AlCl_3$) → 1,2,4,5-tetramethylbenzene (CH_3 ×4) + penta- and hexamethylbenzene

OS, *CV* 2, 282 (1943)

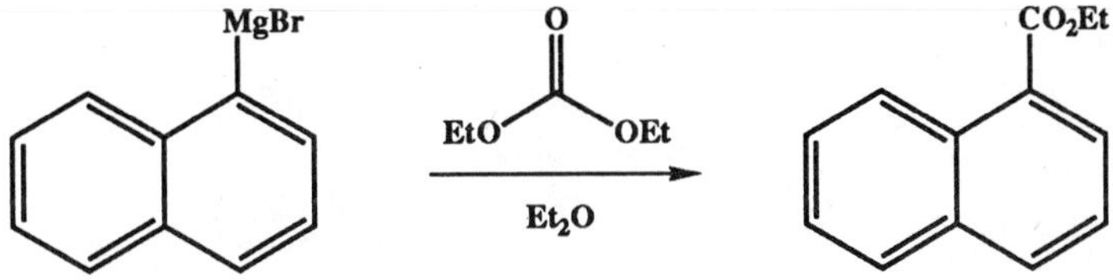

1-naphthyl-MgBr → (EtO–C(=O)–OEt, Et_2O) → 1-naphthyl-CO_2Et

OS, *CV* 2, 304 (1943)

1,2,3-trihydroxybenzene (OH, OH, OH) → ($(CH_3CO)_2O$, $ZnCl_2$, HOAc, 140 - 145 °C) → 2,3,4-trihydroxyacetophenone (OH, OH, OH, $COCH_3$)

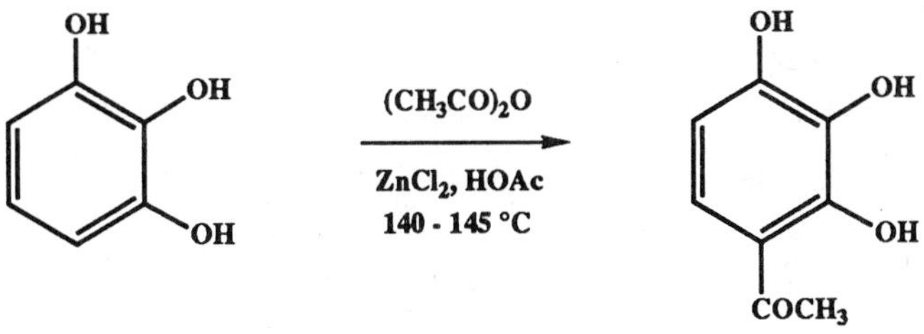

OS, CV 2, **360 (1943)**

Br, CH_3, CH_3, CH_3

1. Mg, Et_2O

2. $(CH_3)_2SO_4$

CH_3, CH_3, CH_3, CH_3

OS, CV 2, **425 (1943)**

Br

1. Mg, Et_2O

2. CO_2, - 7 °C

3. H_2SO_4

CO_2H

OS, CV 2, **517 (1943)**

N

PhLi

Et_2O, toluene

110 °C

N, Ph

OS, CV 2, **522 (1943)**

HO, OH, OH

$CH_3C{\equiv}N$

$ZnCl_2$, HCl

Et_2O

HCl • HN, CH_3

HO, OH, OH

H_2O

Δ

O, CH_3

HO, OH, OH

OS, *CV* 2, 543 (1943)

$AlCl_3$

CS_2, Δ

OH

R_1

R_2

R_1 = H
R_2 = C(O)CH_2CH_3

R_1 = C(O)CH_2CH_3
R_2 = H

OS, *CV* 2, 557 (1943)

HO OH

CO_2, $KHCO_3$

H_2O, Δ
then HCl

CO_2H

HO OH

OS, *CV* 2, 583 (1943)

CH_3

CO, HCl

$AlCl_3$

CH_3

H

O

OS, *CV* 3, 6 (1955)

$AlCl_3$

nitrobenzene
0 °C

O CO_2H

OS, *CV* 3, 23 (1955)

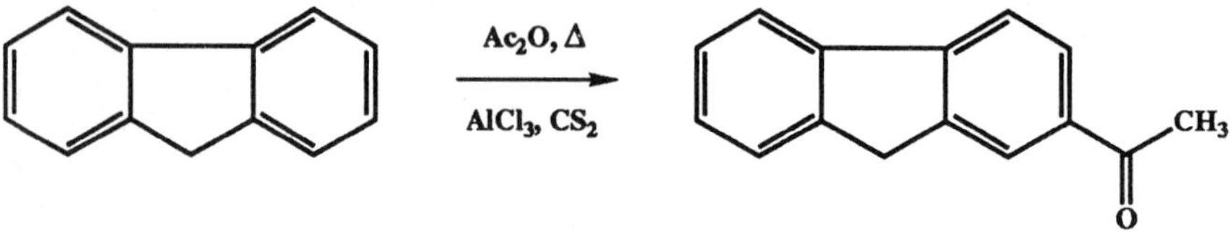

Ac_2O, Δ

$AlCl_3$, CS_2

CH_3

O

OS, *CV* 3, 98 (1955)

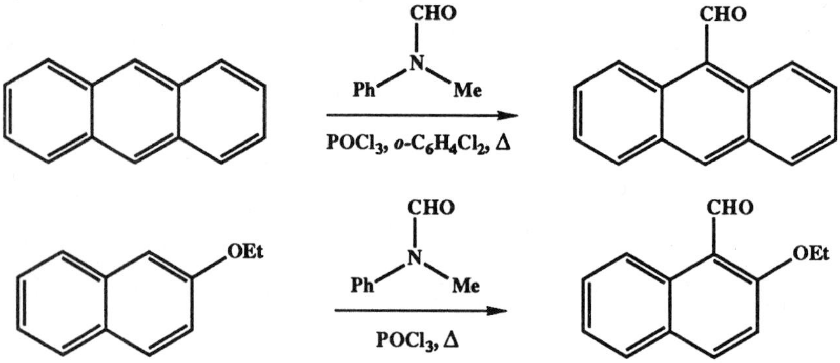

CHO

N

Ph Me

$POCl_3$, o-$C_6H_4Cl_2$, Δ

CHO

OEt

CHO

N

Ph Me

$POCl_3$, Δ

CHO

OEt

OS, *CV* 3, 109 (1955)

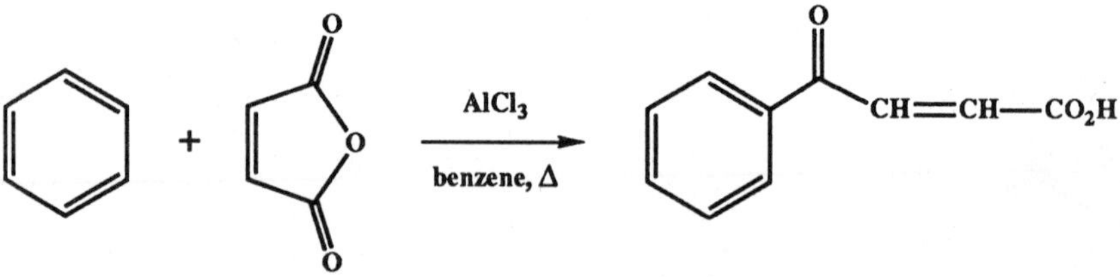

O

+

O

O

$AlCl_3$

benzene, Δ

O

CH=CH—CO_2H

OS, *CV* 3, 157 (1955)

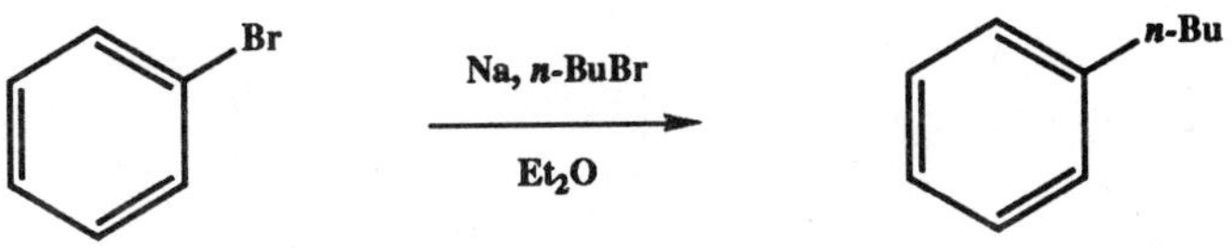

Na, *n*-BuBr

Et_2O

Br

n-Bu

OS, *CV* 3, 183 (1955)

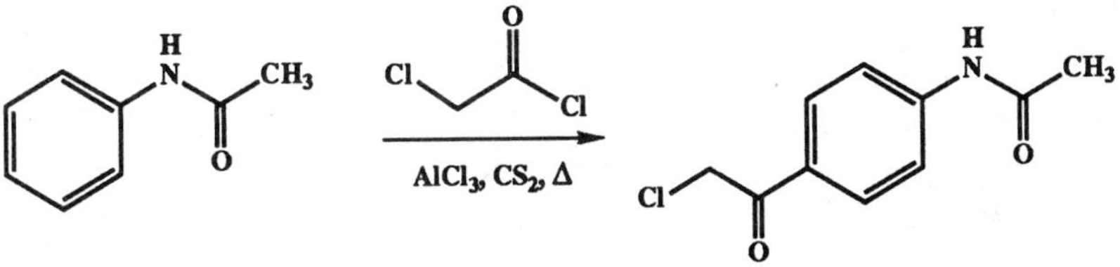

H N CH_3 O

Cl O Cl

$AlCl_3$, CS_2, Δ

Cl O H N CH_3 O

OS, *CV* 3, 195 (1955)

CH_2O, HCl
H_3PO_4, HOAc

80 - 85 °C

CH_2Cl

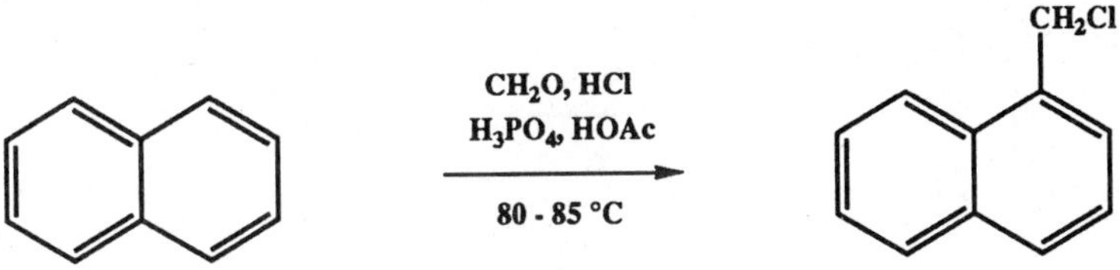

OS, *CV* 3, 200 (1955)

Br Cl

Mg

Et_2O, Δ

MgBr Cl

1. CH_3CHO, Δ
2. aq. NH_4Cl

OH CH_3 Cl

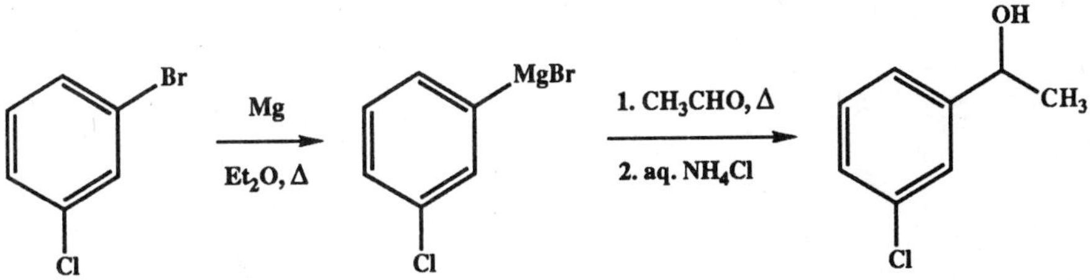

***OS, CV 3*, 212 (1955)**

CuCN, 260 °C

***OS, CV 3*, 248 (1955)**

2 (benzene), $AlCl_3$, 55 °C

***OS, CV 3*, 280 (1955)**

$AlCl_3$, Δ

***OS, CV 3*, 281 (1955)**

$AlCl_3$, Δ

OS, CV 3, **293 (1955)**

O_2N, NO_2 —KCN / aq. MeOH→ MeO, CN, NO_2 —KOH / MeOH, Δ→ MeO, CN, OMe

OS, CV 3, **326 (1955)**

CH_3, CH_3 + O, EtO_2C, CO_2Et —$SnCl_4$→ CH_3, CH_3, CO_2Et, CO_2Et, OH

OS, CV 3, **343 (1955)**

Br, Ph, CH_3, O —AlCl₃, Δ→ Ph, Ph, CH_3, O

OS, CV 3, **347 (1955)**

Br, Ph, CN —$AlCl_3$, Δ→ Ph, Ph, CN

OS, *CV* 3, 353 (1955)

O O CHPh

1. $PhMgBr$
Et_2O, benzene

2. aq. H_2SO_4, Δ

Ph Ph O

OS, *CV* 3, 410 (1955)

2 N + 2 Ac_2O

Zn dust
heat

Zn
HOAc

CH_2CH_3 N

OS, *CV* 3, 418 (1955)

O OCH_3

Δ

OH OCH_3

OS, *CV* 3, 463 (1955)

OH

NaOH, $CHCl_3$
EtOH, Δ

CHO OH

OS, *CV* 3, 468 (1955)

OH O_2N

$CH_2(OMe)_2$
HCl, H_2SO_4, 70 °C

OH Cl O_2N

OS, *CV* 3, 549 (1955)

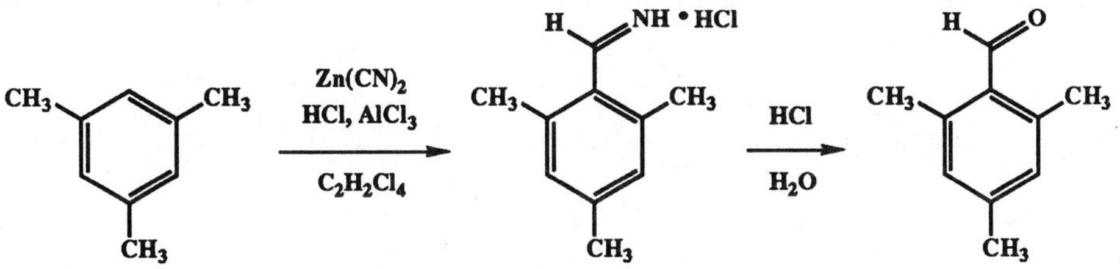

$Zn(CN)_2$, HCl, $AlCl_3$, $C_2H_2Cl_4$; HCl, H_2O

CH_3 CH_3 CH_3 — H NH • HCl — H O

OS, *CV* 3, 553 (1955)

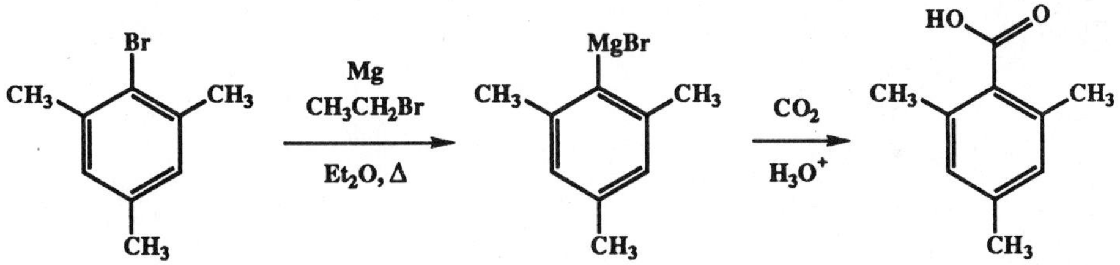

Br — Mg, CH_3CH_2Br, Et_2O, Δ — MgBr — CO_2, H_3O^+ — HO O

CH_3 CH_3 CH_3

OS, *CV* 3, 555 (1955)

Br — 1. Mg, Et_2O, Δ; 2. CO_2; 3. H_3O^+ — HO O — $SOCl_2$, Δ — Cl O

CH_3 CH_3 CH_3

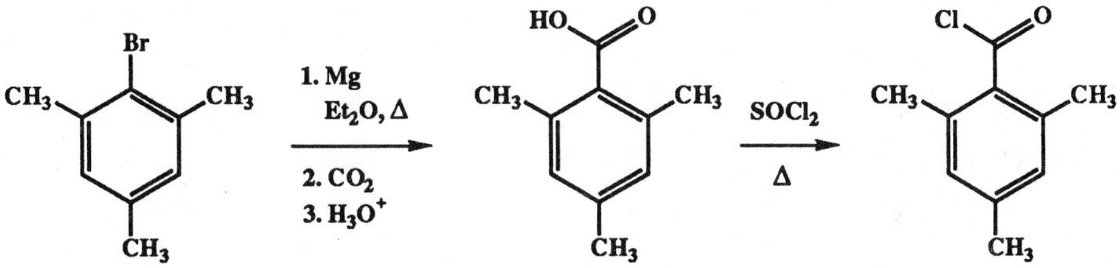

OS, *CV* 3, 557 (1955)

CH_3 CH_3 CH_3 — CH_2═O, HCl, 55 °C — CH_2Cl, CH_3 CH_3 CH_3

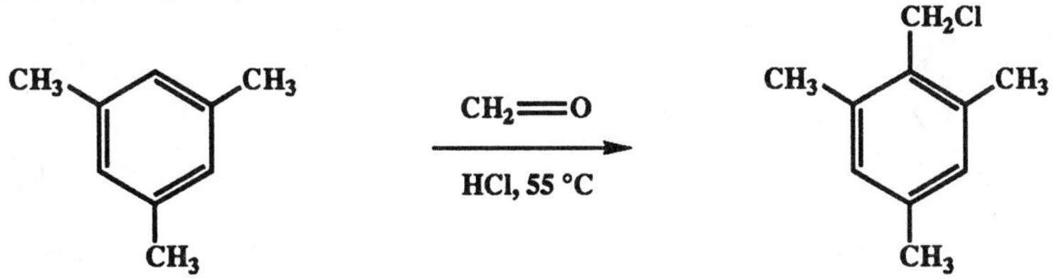

OS, CV 3, 562 (1955)

PhMgBr + MeO–CH₂–C≡N →(Et_2O) [Ph–C(=NMgBr)–CH₂–OMe] →(H_2O, H_2SO_4) Ph–C(=O)–CH₂–OMe

OS, CV 3, 631 (1955)

1-Bromonaphthalene →(CuCN, Δ; pyridine) 1-Naphthalenecarbonitrile (CN)

OS, CV 3, 701 (1955)

9-Bromophenanthrene →(1. Mg, Et_2O, Δ; 2. $HC(OEt)_3$, Δ) 9-Phenanthrene–$CH(OEt)_2$

OS, CV 3, 761 (1955)

Resorcinol (HO, OH) →($ZnCl_2$; HOAc, Δ) 2,4-Dihydroxyacetophenone (HO, OH, C(=O)CH_3)

OS, *CV* 3, 842 (1955)

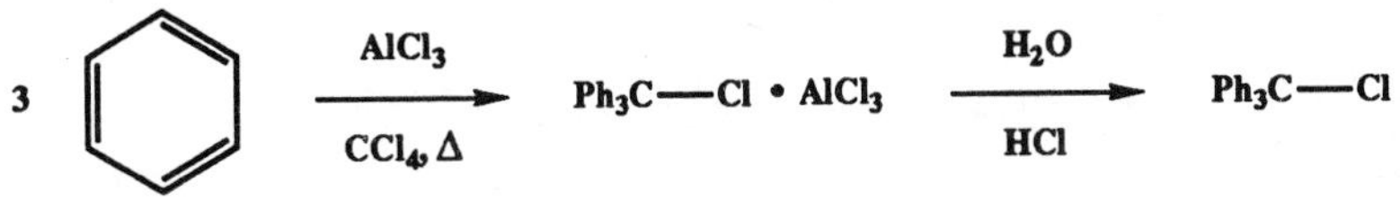

3 → ($AlCl_3$ / CCl_4, Δ) → $Ph_3C—Cl \cdot AlCl_3$ → (H_2O / HCl) → $Ph_3C—Cl$

OS, *CV* 4, 8 (1963)

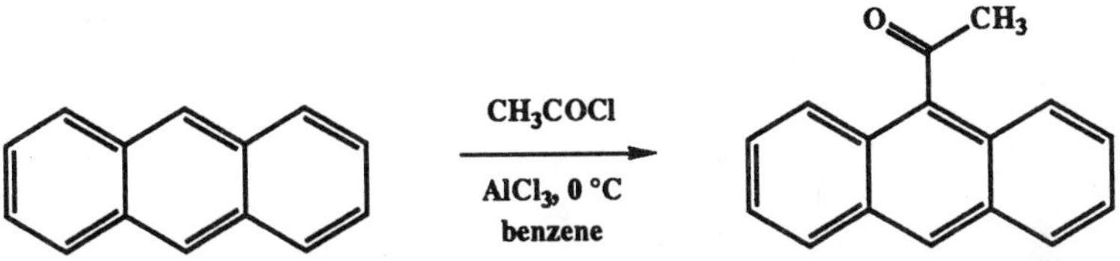

→ (CH_3COCl / $AlCl_3$, 0 °C, benzene) → O, CH_3

OS, *CV* 4, 15 (1963)

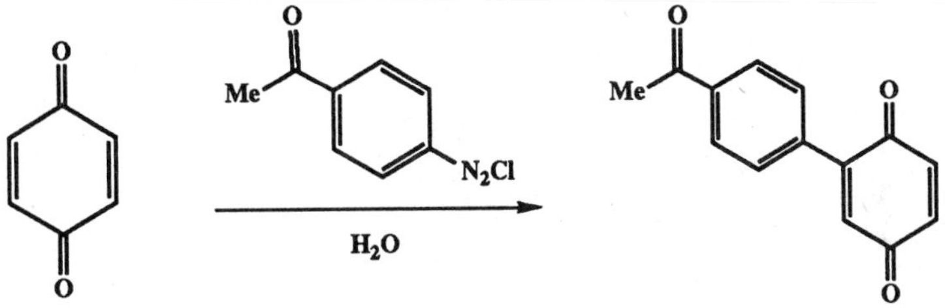

O, O → (Me, O, N_2Cl / H_2O) → Me, O, O, O

OS, *CV* 4, 34 (1963)

CO_2H, NHTs → (PCl_5, $AlCl_3$, Δ) → O, Ph, NHTs

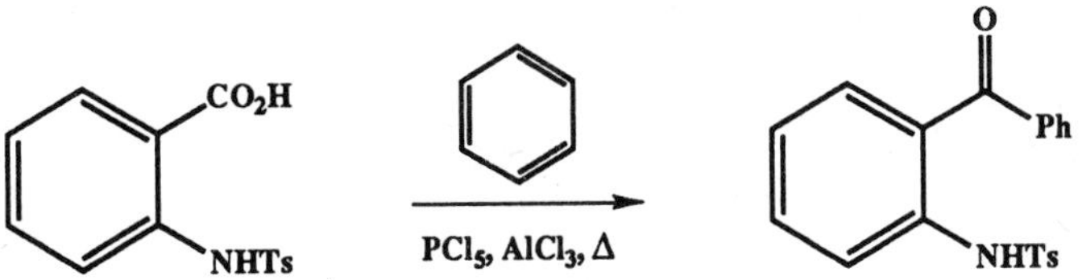

OS, CV 4, 47 (1963)

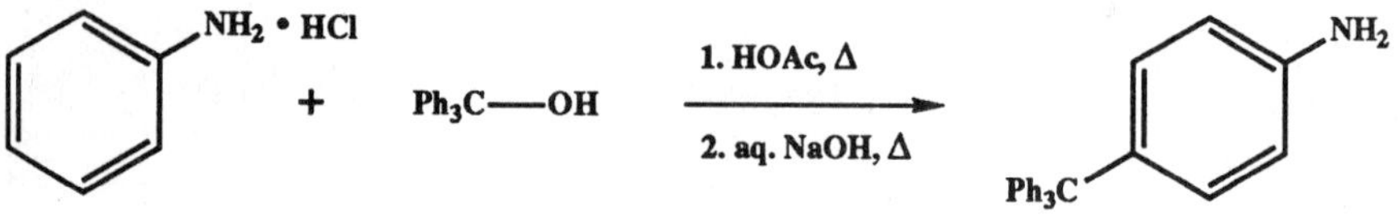

$NH_2 \cdot HCl$ + Ph_3C—OH

1. HOAc, Δ

2. aq. NaOH, Δ

NH_2

Ph_3C

OS, CV 4, 88 (1963)

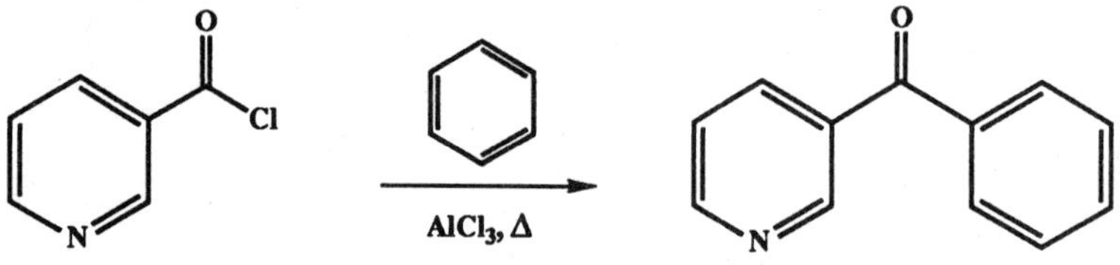

O

Cl

N

$AlCl_3$, Δ

O

N

OS, CV 4, 114 (1963)

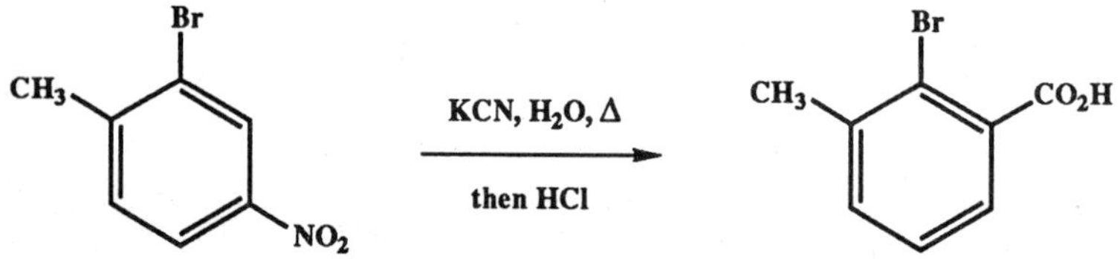

Br

CH_3

NO_2

KCN, H_2O, Δ

then HCl

Br

CH_3

CO_2H

OS, CV 4, 331 (1963)

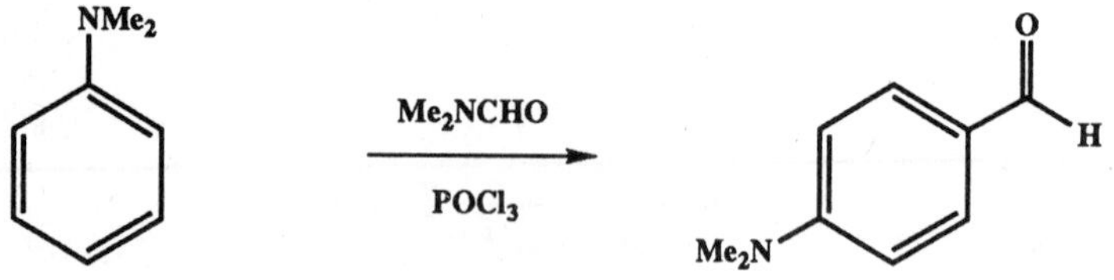

NMe_2

Me_2NCHO

$POCl_3$

O

H

Me_2N

OS, *CV* 4, 520 (1963)

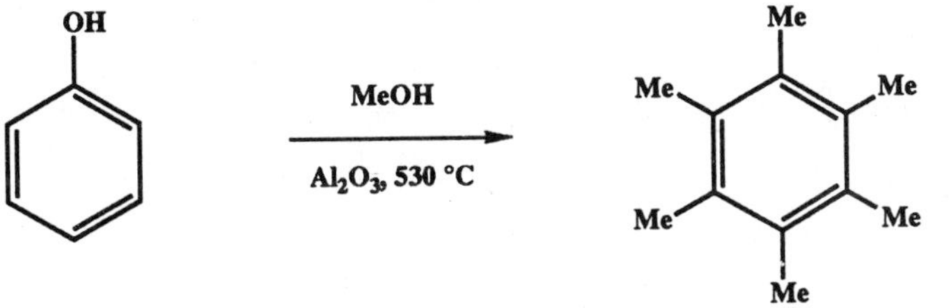

OH

MeOH

Al_2O_3, 530 °C

Me Me Me Me Me Me

OS, *CV* 4, 702 (1963)

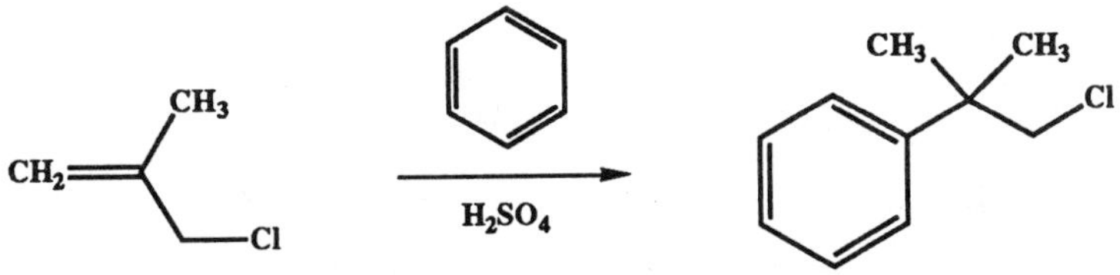

CH_3

CH_2=

Cl

H_2SO_4

CH_3 CH_3 Cl

OS, *CV* 4, 718 (1963)

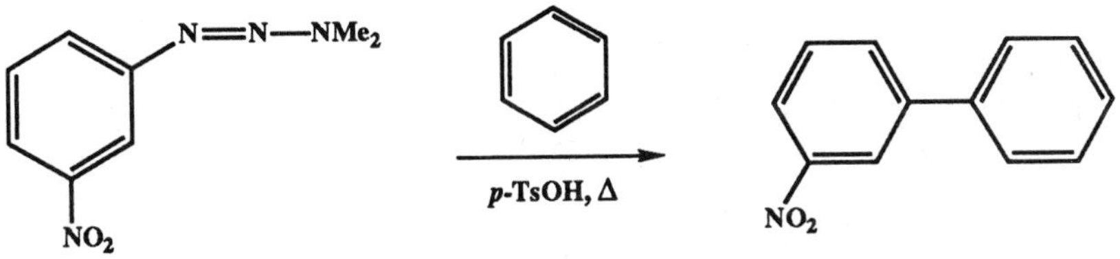

N=N—NMe_2

NO_2

p-TsOH, Δ

NO_2

OS, *CV* 4, 727 (1963)

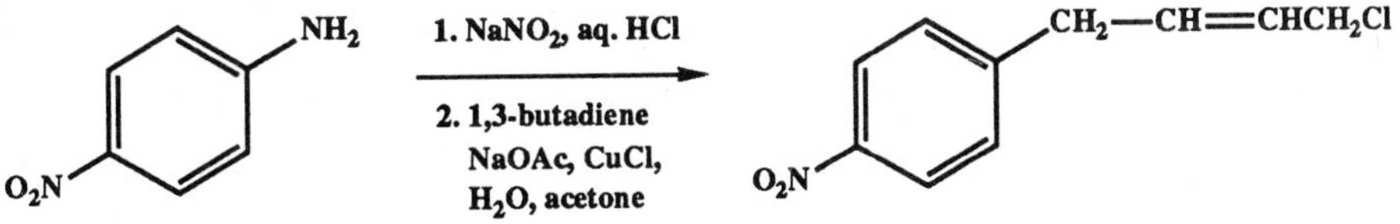

NH_2

O_2N

1. $NaNO_2$, aq. HCl

2. 1,3-butadiene
NaOAc, CuCl,
H_2O, acetone

CH_2—CH=CHCH_2Cl

O_2N

OS, *CV* 4, 866 (1963)

MeO, OH, OMe; 1. hexamethylenetetramine, Δ; 2. H_2SO_4, H_2O → MeO, OH, OMe, CHO

OS, *CV* 4, 953 (1963)

NMe_2 + NC, CN, NC, CN; DMF → (*p*-Me_2N)C_6H_4, CN, NC, CN

OS, *CV* 4, 960 (1963)

Br, Ph, CO_2H, Br; 2 benzene; $AlBr_3$, Δ → Ph, Ph, CO_2H, Ph

OS, *CV* 5, 49 (1973)

CH_3, CH_3, CH_3; Cl, Cl, OMe; $TiCl_4$, CH_2Cl_2 → CH_3, CH_3, Cl, CH_3, OMe

OS, CV 5, 51 (1973)

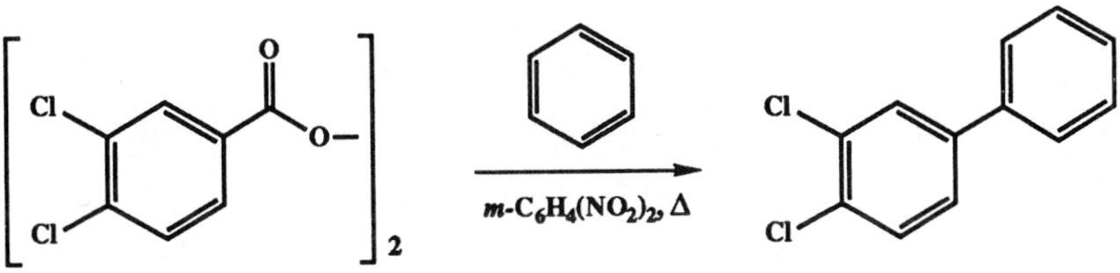

Cl Cl O O 2

m-$C_6H_4(NO_2)_2$, Δ

Cl Cl

OS, CV 5, 111 (1973)

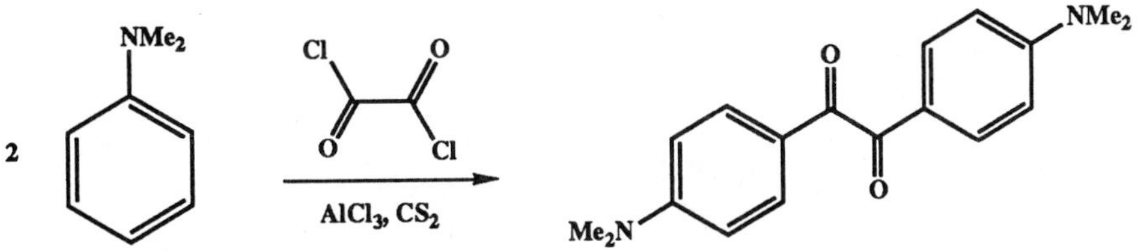

2 NMe_2

Cl O O Cl

$AlCl_3$, CS_2

O O Me_2N NMe_2

OS, CV 5, 130 (1973)

CCl_3 OH

Br

H_2SO_4, SO_3

CCl_3 Br

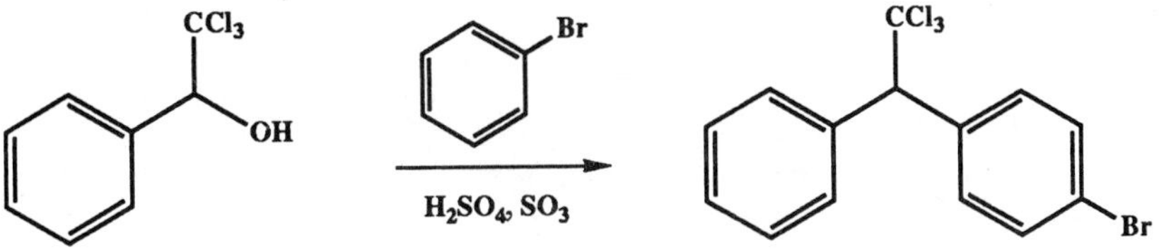

OS, CV 5, 139 (1973)

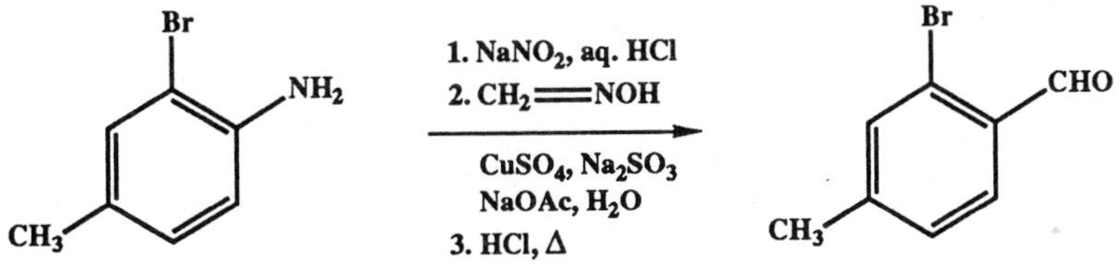

Br NH_2 CH_3

1. $NaNO_2$, aq. HCl
2. CH_2=NOH

$CuSO_4$, Na_2SO_3
NaOAc, H_2O
3. HCl, Δ

Br CHO CH_3

OS, CV 5, 269 (1973)

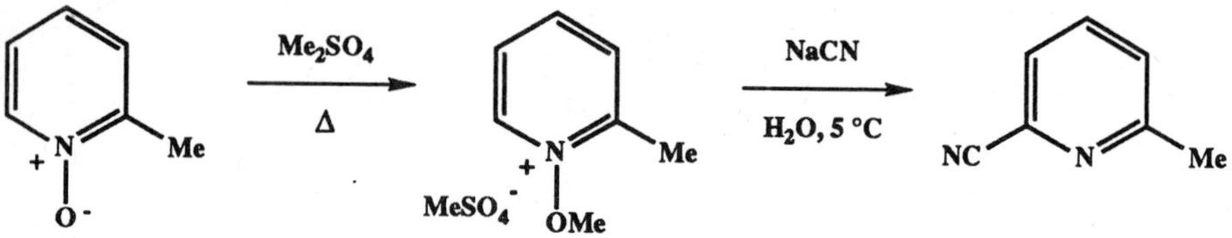

Me_2SO_4, Δ

MeSO$_4^-$ OMe

NaCN

H_2O, 5 °C

NC N Me

OS, CV 5, 422 (1973)

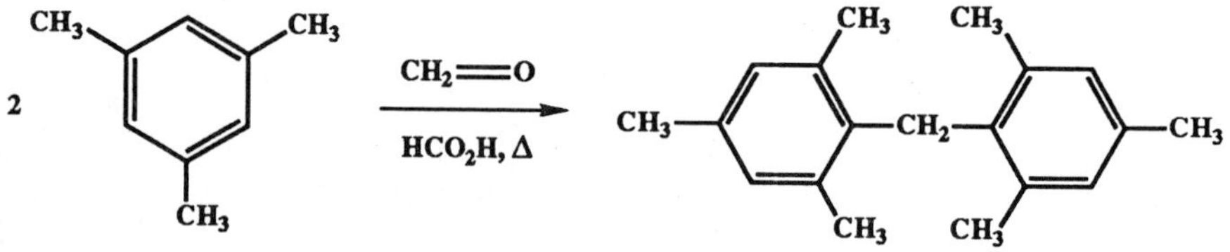

2 CH_3 CH_3 CH_3

CH_2=O

HCO_2H, Δ

CH_3 CH_3 CH_3 CH_2 CH_3 CH_3 CH_3

OS, CV 5, 434 (1973)

Fe

1. $Me_2NCH_2NMe_2$
H_3PO_4, HOAc, Δ

2. MeI, MeOH, Δ

$CH_2\overset{+}{N}Me_3$ I^-

Fe

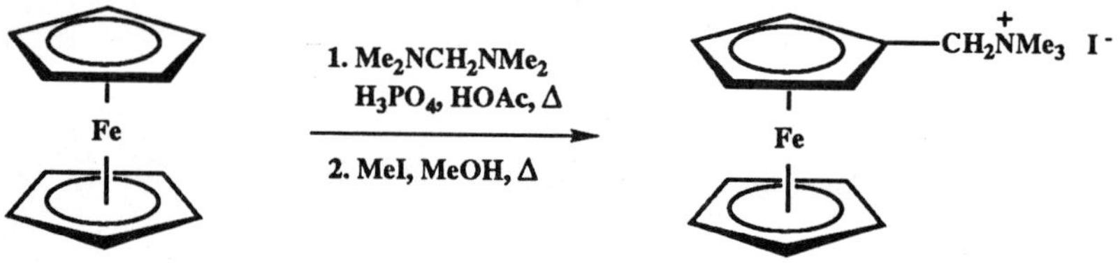

OS, CV 5, 520 (1973)

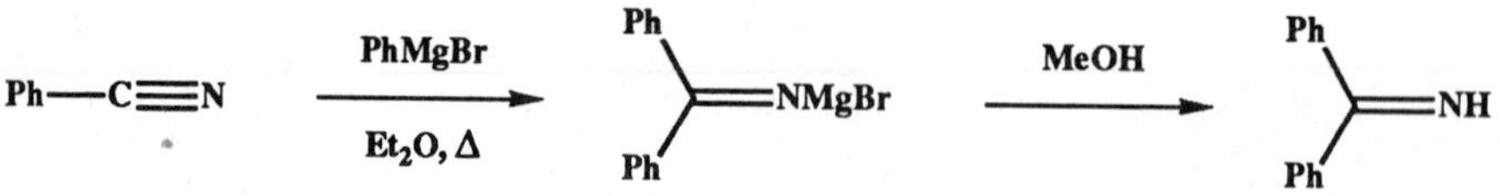

Ph—C≡N

PhMgBr

Et_2O, Δ

Ph Ph =NMgBr

MeOH

Ph Ph =NH

OS, *CV 5*, 706 (1973)

CH_3 CH_3 CH_3

O O Cl Cl

$AlCl_3$, CS_2, Δ
then aq. HCl

CO_2H CH_3 CH_3 CH_3

OS, *CV 5*, 890 (1973)

Cl Cl Cl Cl Cl Cl

Mg, benzene
ethylene bromide

MgCl Cl Cl Cl Cl Cl

1. CO_2
2. aq. HCl

CO_2H Cl Cl Cl Cl Cl

OS, *CV 5*, 1058 (1973)

O Cl Cl

p-MeC_6H_4MgBr
Et_2O

[*p*-MeC_6H_4 OMgBr Cl Cl]

1. EtMgBr, $FeCl_3$
2. NH_4Cl, HCl

p-MeC_6H_4 OH

OS, *CV 6*, 21 (1988)

p-AcO-C_6H_4—$\overset{+}{N}$≡N Br^-

CH_2=$CHCO_2H$
CuBr, HBr

p-AcO-$C_6H_4CH_2$—$CHCO_2H$
Br

***OS*, *CV* 6, 34 (1988)**

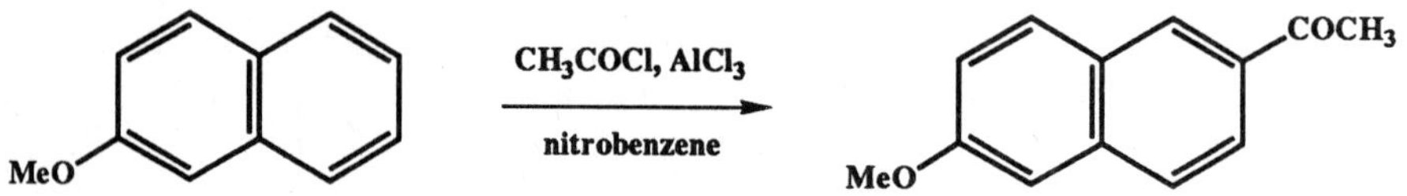
MeO
CH_3COCl, $AlCl_3$
nitrobenzene
$COCH_3$
MeO

***OS*, *CV* 6, 36 (1988)**

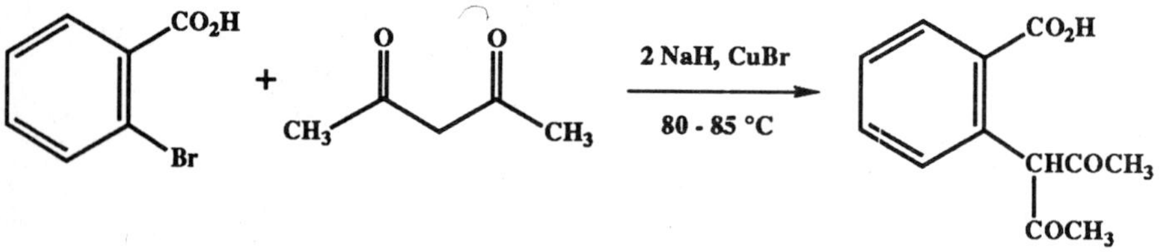
CO_2H
Br
+
O
O
CH_3
CH_3
2 NaH, CuBr
80 - 85 °C
CO_2H
$CHCOCH_3$
$COCH_3$

***OS*, *CV* 6, 64 (1988)**

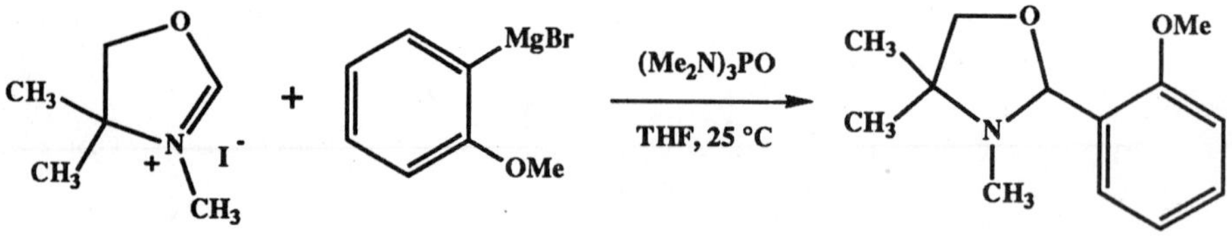
O
CH_3
CH_3
N
+
I⁻
CH_3
+
MgBr
OMe
$(Me_2N)_3PO$
THF, 25 °C
CH_3
CH_3
O
N
CH_3
OMe

OS, CV 6, 115 (1988)

KCN, PhCOCl

CH_2Cl_2, H_2O

NCOPh

NC H

NCOPh

NC H

NaH

$PhCH_2Cl$

DMF

- 10 °C

NCOPh

NC CH_2Ph

NaOH, Δ

aq. EtOH

N

CH_2Ph

OS, CV 6, 407 (1988)

Cl

Cl

+ 2 *n*-BuMgBr

1. $NiCl_2$(dppp), cat.

Et_2O, 0 °C

2. Δ, then aq. HCl

n-Bu

n-Bu

(dppp) = $Ph_2P(CH_2)_3PPh_2$

OS, CV 6, 465 (1988)

MeO OMe

$ClSO_2N{=}C{=}O$

CH_2Cl_2

MeO OMe

$NHSO_2Cl$

O

OS, CV 6, 478 (1988)

NMe$_2$ / CH$_3$ → (n-BuLi, Me$_2$NCH$_2$CH$_2$NMe$_2$, hexane, 25 °C) → NMe$_2$ / Li / CH$_3$ → (1. Ph$_2$C=O, Et$_2$O, 25 °C; 2. HOAc) → NMe$_2$ / OH / CPh$_2$ / CH$_3$

OS, CV 6, 581 (1988)

CO$_2$Et / NH$_2$ → (1. t-BuOCl, CH$_2$Cl$_2$, CH$_3$CN, - 40 °C; 2. CH$_3$SCH$_3$, - 30 to - 40 °C) → CO$_2$Et / HN–S$^+$(CH$_3$)$_2$ Cl$^-$ → (Et$_3$N, Δ) → CO$_2$Et / NH$_2$ / SCH$_3$

OS, CV 6, 625 (1988)

Fe + O / Cl / Cl → (AlCl$_3$, CH$_2$Cl$_2$, 0 - 25 °C) → Fe / O / Cl

OS, CV 6, 722 (1988)

CH$_3$—Ni(Br)(Br)Ni—CH$_3$ + 2 Br → (DMF, 25 - 60 °C) → 2

OS, *CV 6*, 815 (1988)

Ph—HgOAc + $CH_2=C(CH_3)CH_2OH$ → [$Pd(OAc)_2$; CH_3CN, Et_2O] → [Ph–CH_2–C(CH_3)(PdOAc)–CH_2OH] → Ph–CH_2–CH(CH_3)–CHO

OS, *CV 6*, 873 (1988)

C_6F_6 + $NCCH_2CO_2Et$ → [K_2CO_3; DMF; 110 - 120 °C] → $C_6F_5CH(CN)CO_2Et$

OS, *CV 6*, 928 (1988)

$CH_3COCH_2COCH_3$ → [1. $NaNH_2$, NH_3; 2. Ph_2ICl; 3. aq. HCl] → $PhCH_2COCH_2COCH_3$

OS, *CV 7*, 162 (1990); *60*, 49 (1981)

bis(1,3-diphenylimidazolidin-2-ylidene) + 3-(NHCOMe)phenol → [chlorobenzene; 100 °C] → 2,4-bis(1,3-diphenylimidazolidin-2-yl)-5-(NHCOMe)phenol

OS, *CV* 7, 420 (1990); *61*, 8 (1983)

$C_5H_{11}C_6H_5 + COCl_2 \xrightarrow[CH_2Cl_2]{AlCl_3} p\text{-}C_5H_{11}C_6H_4COCl$

OS, *CV* 7, 361 (1990); *61*, 82 (1983)

$PhI + CH_2{=}C(CH_3)CH_2OH + Et_3N \xrightarrow{Pd(OAc)_2} PhCH_2CH(CH_3)CHO + Et_3NH^+ I^-$

OS, *CV* 7, 229 (1990); *62*, 24 (1984)

Anisole (OMe) → 1. $Pb(OAc)_4$, 2. AcOH → p-MeO-C₆H₄-$Pb(OAc)_3$ → 2-oxocyclohexane-CO_2Et, pyridine → 2-(4-methoxyphenyl)-2-CO_2Et-cyclohexanone (O, OMe, CO_2Et)

OS, *CV* 7, 172 (1990); *62*, 39 (1984)

$C_6H_5NMe_2$ → 1. *n*-BuLi; 2. (Z)-PhCH=CHBr (H, H, Ph, Br), $PdCl_2$, Ph_3P, MeLi → (Z)-2-(2-phenylethenyl)-N,N-dimethylaniline (H, H, Ph, NMe_2)

OS, *CV 7*, 105 (1990); *62*, 67 (1984)

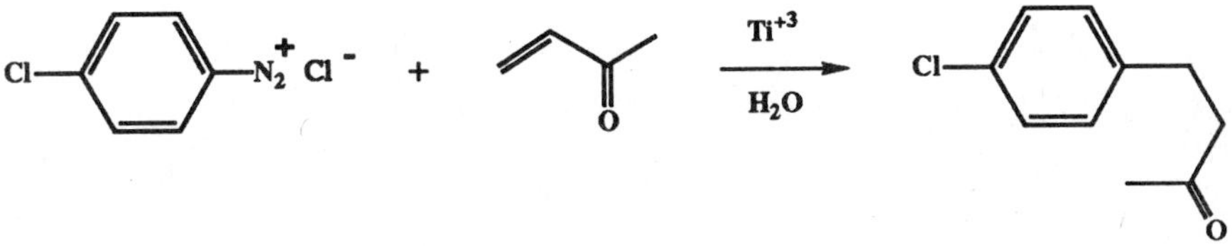

Cl– C_6H_4 –N_2^+ Cl^- + (methyl vinyl ketone) → (Ti^{+3}, H_2O) → 4-(4-chlorophenyl)-2-butanone

OS, *65*, 203 (1987)

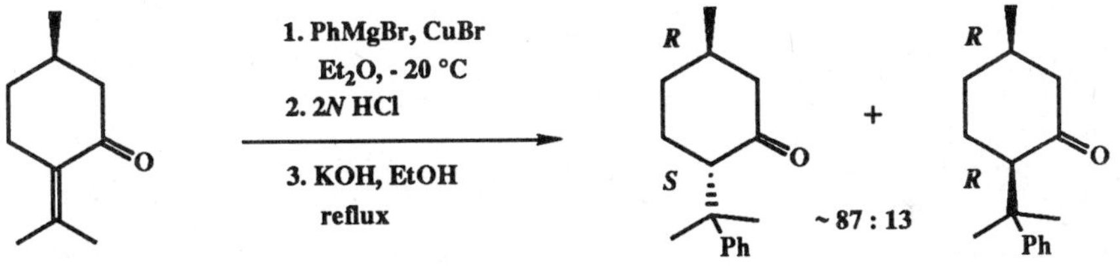

R - (+)

1. PhMgBr, CuBr
   Et_2O, - 20 °C
2. 2*N* HCl
3. KOH, EtOH
   reflux

R, *S* (Ph) + *R*, *R* (Ph)

~ 87 : 13

OS, *66*, 67 (1987)

CH_3, I → 1. *t*-BuLi; 2. $ZnCl_2$ → CH_3, ZnCl → (Br–C_6H_4–NO_2, $Pd(Ph_3P)_4$) → CH_3, NO_2 biaryl

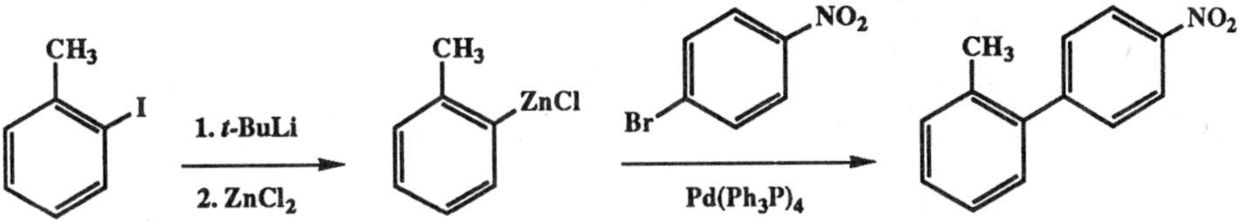

***OS*, *CV 1*, 94 (1941)**

2 PhCHO —NaCN / EtOH, Δ→ PhCH(OH)C(O)Ph

***OS*, *CV 1*, 199 (1941)**

2 CH_3COCH_3 —$Ba(OH)_2$ / Δ→ $(CH_3)_2C(OH)CH_2COCH_3$

***OS*, *CV 1*, 235 (1941)**

2 CH_3CO_2Et —1. Na, Δ; 2. HOAc→ $CH_3COCH_2CO_2Et$

***OS*, *CV 1*, 238 (1941)**

EtO₂C–CO₂Et + CH_3COCH_3 —NaOEt, EtOH→ $CH_3COCH_2COCO_2Et$

OS, CV 1, 290 (1941)

$CH_2{=}O$ + 2 $CH_2(CO_2Et)_2$ —(Et_2NH, Δ)→ $(EtO_2C)_2CH{-}CH_2{-}CH(CO_2Et)_2$

OS, CV 1, 425 (1941)

4 $CH_2{=}O$ + CH_3CHO —($Ca(OH)_2$, H_2O)→ $C(CH_2OH)_4$

OS, CV 2, 126 (1943)

2 $(CO_2Et)_2$ + CH_3COCH_3 —(NaOEt, EtOH)→ $EtO_2C{-}CO{-}CH_2{-}CO{-}CH_2{-}CO{-}CO_2Et$

OS, CV 2, 194 (1943)

$PhCH_2CH_2CH_2CO_2Et$ —($EtO_2C{-}CO_2Et$; NaOEt, EtOH, Et_2O)→ $PhCH_2CH_2CH(CO_2Et)COCO_2Et$

OS, CV 2, 272 (1943)

O

CH_3 OEt

\+ $EtO_2C—CO_2Et$

1. Na, EtOH, Et_2O
2. HOAc, H_2O

O

CH_3 OEt

O CO_2Et

OS, CV 2, 287 (1943)

CN

1. Na, EtOH
2. $EtO_2C—CO_2Et$
then HCl

CN

CO_2Et

O

OS, CV 2, 288 (1943)

O

Ph OEt

\+ $EtO_2C—CO_2Et$

1. Na, EtOH
2. H_2SO_4, H_2O

O

Ph OEt

O CO_2Et

OS, CV 2, 487 (1943)

CN

O

CH_3 OEt

Na, EtOH, Δ

CN

CH_3

O

OS, CV 2, 531 (1943)

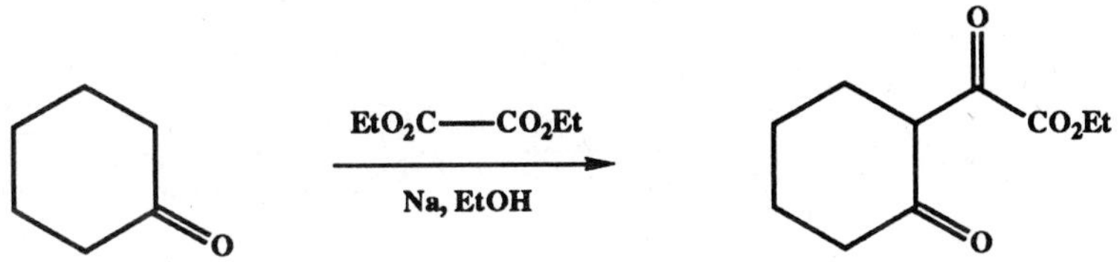

Cyclohexanone + $EtO_2C—CO_2Et$ → (Na, EtOH) → 2-oxocyclohexyl glyoxylate ethyl ester (CO_2Et)

OS, CV 3, 16 (1955)

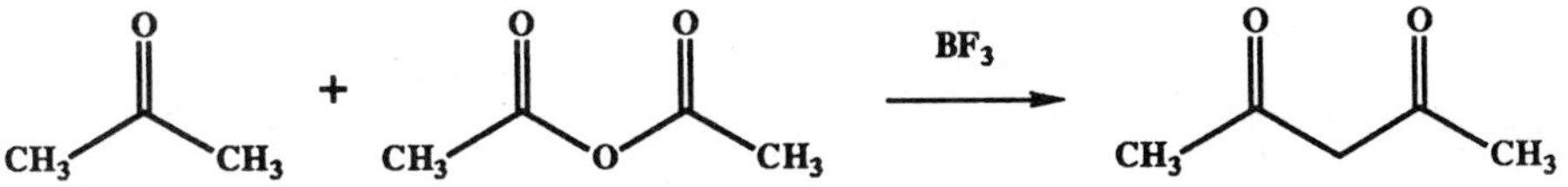

CH_3COCH_3 + $CH_3C(O)OC(O)CH_3$ → (BF_3) → $CH_3COCH_2COCH_3$

OS, CV 3, 17 (1955)

CH_3COCH_3 + CH_3CO_2Et → (1. Na, EtOH, Et_2O; 2. H_2SO_4) → $CH_3COCH_2COCH_3$

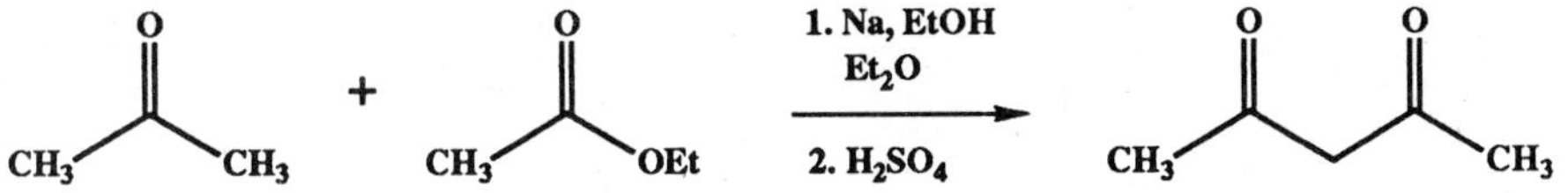

OS, CV 3, 251 (1955)

$PhCO_2Et$ + CH_3COPh → (1. NaOEt, Δ; 2. aq. H_2SO_4) → $PhCOCH_2COPh$

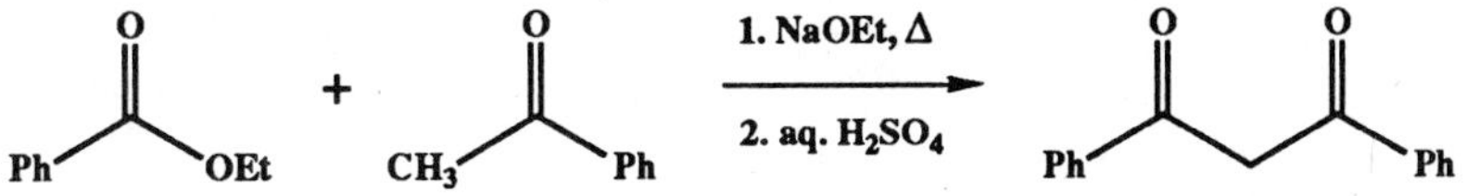

OS, CV 3, 291 (1955)

O; CH_3; *i*-Pr

1. Na, NH_3
2. *i*-Pr-CH_2CO_2Et, Δ
3. HCl, H_2O

O; O; *i*-Pr; *i*-Pr

OS, CV 3, 305 (1955)

O; Ph; CH_3

CH_2=O
$(CH_3)_2NH_2Cl$
HCl, EtOH, Δ

O; Ph; $N(CH_3)_2$

OS, CV 3, 379 (1955)

O; CH_3; CO_2Et

1. Na, EtOH
2. $PhCO_2Et$, Δ
3. H_2SO_4, H_2O

O; Ph; CO_2Et

OS, CV 3, 387 (1955)

OH; CH_3; O

1. $CH_3CH_2CO_2Et$
Na powder, Δ
2. HOAc, H_2O

OH; CH_3; O; O

***OS*, *CV* 3, 510 (1955)**

$EtO_2C-CH_2CH_2-CO_2Et$ → (EtO_2C-CO_2Et; KOEt, Et_2O, Δ) → $EtO_2C-CH_2-CH(CO_2Et)-C(=O)-CO_2Et$

***OS*, *CV* 4, 141 (1963)**

$C_{16}H_{33}-CH_2-CO_2Et$ → (1. EtO_2C-CO_2Et, NaOEt, EtOH, 60 °C; 2. HOAc, H_2O) → $C_{16}H_{33}-CH(CO_2Et)-C(=O)-CO_2Et$

***OS*, *CV* 4, 174 (1963)**

4-$Cl-C_6H_4-CH_2-CN$ → ($Ph-CH_2-CO_2Et$; Na, EtOH, Δ then HCl, H_2O) → 4-$Cl-C_6H_4-CH(CN)-C(=O)-CH_2-Ph$

***OS*, *CV* 4, 210 (1963)**

$CH_3-C(=O)-CH_3$ → (HCO_2Et, NaOMe, Et_2O) → [$CH_3-C(=O)-CH=CH-ONa$] → ($H_2N-C(=O)-CH_2-CN$; piperidine acetate, Δ, then HOAc) → 6-methyl-3-cyano-2(1H)-pyridinone (CH_3, N-H, =O, CN)

OS, CV 4, 278 (1963)

2 [γ-butyrolactone] → 1. NaOMe, MeOH, Δ; 2. HCl, Δ → [$Cl(CH_2)_3$C(=O)$(CH_2)_3Cl$] → aq. NaOH, Δ → [dicyclopropyl ketone]

OS, CV 4, 285 (1963)

PhC(=O)OC(=O)OEt + $(EtO_2C)_2$CHMgOEt → Et_2O, then aq. H_2SO_4 → PhC(=O)CH$(CO_2Et)_2$

OS, CV 4, 461 (1963)

$PhCH_2CN$ → 1. $(EtO)_2CO$, NaOEt, toluene, Δ; 2. HOAc, H_2O → PhCH(CN)CO_2Et

OS, CV 4, 479 (1963)

2-(PhCO$_2$)C$_6$H$_4$C(=O)CH_3 → KOH, pyridine → 2-(HO)C$_6$H$_4$C(=O)CH_2C(=O)Ph

***OS*, *CV 4*, 907 (1963)**

5 CH_2=O

CaO, H_2O

$HOCH_2$ CH_2OH $HOCH_2$ CH_2OH

***OS*, *CV 5*, 27 (1973)**

H_2N OH

2 Ac_2O

pyridine, Δ

Ac H N CH_3

***OS*, *CV 5*, 187 (1973)**

CH_3

NaOMe
EtOCHO

Et_2O

CH_3 Na^+ - CHO

1. KNH_2, NH_3
2. *n*-BuBr, Et_2O

3. NaOH, H_2O

CH_3 *n*-Bu

***OS*, *CV 5*, 198 (1973)**

NaH
$EtOCO_2Et$

benzene, Δ

CO_2Et

OS, CV 5, **381 (1973)**

EtO_2C–CH_2–CO_2Et → (2 CH_2=O; aq. $KHCO_3$) → $(EtO_2C)_2C(CH_2OH)_2$

OS, CV 5, **687 (1973)**

$EtO_2CCH_2CH_2CO_2Et$ + EtO_2C–CO_2Et → (Na, EtOH; Et_2O, toluene) → EtO_2C–CO–CH(CO_2Et)–CH_2–CO_2Et

OS, CV 5, **718 (1973)**

PhCOCH$_2$COCH$_3$ → (MeO_2C—C_6H_4-*p*-OMe; NaH, DME, Δ, then HCl) → PhCOCH$_2$COCH$_2$CO–C_6H_4-*p*-OMe

OS, CV 5, **833 (1973)**

CH_3NO_2 + CH_2=O → (1. KOH, MeOH; 2. H_2SO_4) → $HOCH_2CH_2NO_2$

OS, CV 5, 937 (1973)

Ph–C(=O)–OEt → 1. $CH_3S(=O)–CH_2K$; 2. HCl, H_2O → Ph–C(=O)–CH(OH)–SCH_3 → $Cu(OAc)_2$, $CHCl_3$ → Ph–C(=O)–C(=O)–H

OS, CV 6, 611 (1988)

2,6-dimethylpyridine (CH_3, N, CH_3) → 1. KNH_2, NH_3; 2. $(EtO)_2CO$, NH_3; 3. NH_4Cl → 6-methyl-2-pyridyl–CH_2CO_2Et

OS, CV 6, 692 (1988)

Ph(H)C=C(OAc)(CH_3) → 1. 2 CH_3Li, DME; 2. $ZnCl_2$, Et_2O; 3. *n*-PrCHO, 0 - 10 °C; NH_4Cl, H_2O, 0 °C → CH_3–C(=O)–CH(Ph)–CH(OH)–*n*-Pr (R^*, R^*)

threo

OS, CV 6, 901 (1988)

cyclohexyl–N=$CHCH_3$ → 1. (*i*-Pr)$_2$NLi, Et_2O; 2. benzophenone → C_6H_{11}–N=CH–CH_2–C(OH)Ph_2 (N····H–O)

OS, *CV* 7, 386 (1990); *60*, 92 (1981)

$CH_3COCH_2CH_2OCH_2Ph$ + $^{13}CH_3CO_2H$ → (LDA, THF, 0 - 5 °C) → (THF, 0 - 25 °C) → $CH_3C(OH)(^{13}CH_2CO_2H)CH_2CH_2OCH_2Ph$

OS, *CV* 7, 213 (1990); *61*, 5 (1983)

$HO_2CCH_2CO_2Et$ → (1. *n*-BuLi, THF; 2. $Me_2CHCOCl$) → $Me_2CHCOCH_2CO_2Et$

OS, *CV* 7, 351 (1990); *62*, 14 (1984)

cyclohexanone → ($MeOCO_2Me$, KH, NaH) → 2-(CO_2Me)cyclohexanone → 2-[$OP(O)(OEt)_2$]cyclohexene-1-CO_2Me → (Me_2CuLi) → 2-Me-cyclohexene-1-CO_2Me

OS, *CV* 7, 381 (1990); *63*, 79 (1985)

$CH_3CH_2CH(CN)OCH(CH_3)OCH_2CH_3$ → (1. LDA, THF, - 78 °C; 2. acetone; 3. aq. H_2SO_4; 4. aq. NaOH) → $CH_3CH_2COC(OH)(CH_3)_2$

***OS*, *CV* 7, 185 (1990); *63*, 89 (1985)**

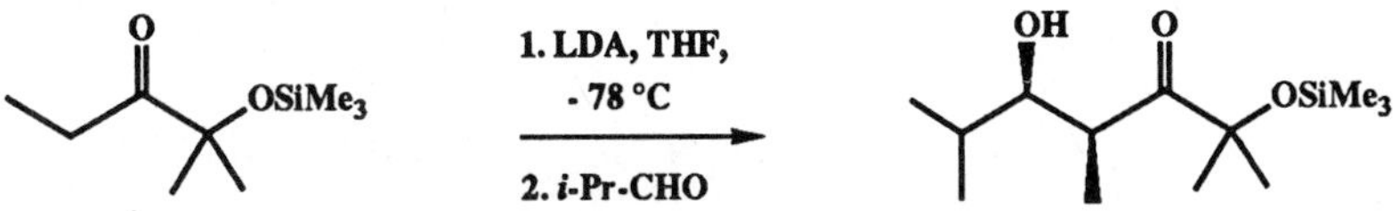

O, OSiMe$_3$

1. LDA, THF, - 78 °C
2. *i*-Pr-CHO

OH, O, OSiMe$_3$

***OS*, *CV* 7, 190 (1990); *63*, 99 (1985)**

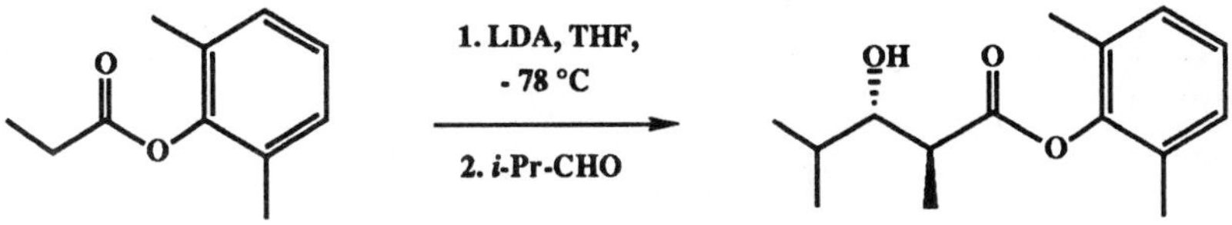

O, O

1. LDA, THF, - 78 °C
2. *i*-Pr-CHO

OH, O, O

***OS*, *65*, 6 (1987)**

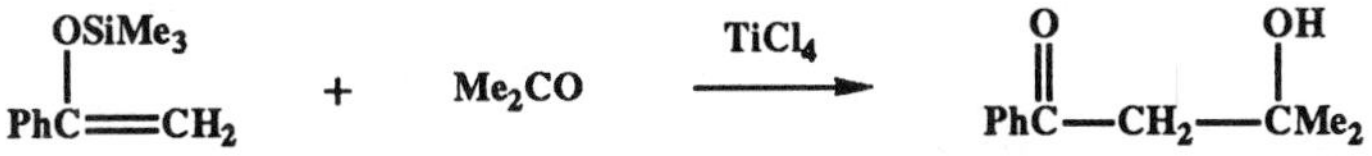

$PhC(OSiMe_3){=}CH_2$ + Me_2CO →($TiCl_4$) $PhC(O){-}CH_2{-}C(OH)Me_2$

***OS*, *68*, 64 (1989)**

CH$_2$, MeO$_2$C, H + CH_3CH_2CHO → N, N (5 mol %) → CH$_2$, MeO$_2$C, OH

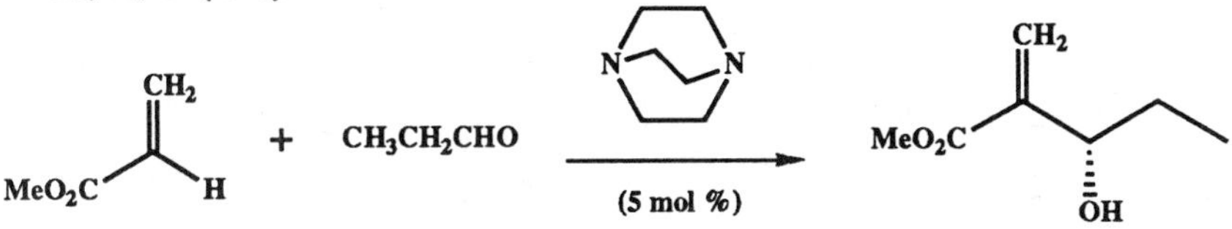

***OS*, *68*, 83 (1989)**

CH$_2$Ph, O, *S*, N, CH$_3$, O, O

1. Bu_2BOTf, Et_3N
2. PhCHO

CH$_2$Ph, O, OH, *S*, N, Ph, O, CH$_3$, O

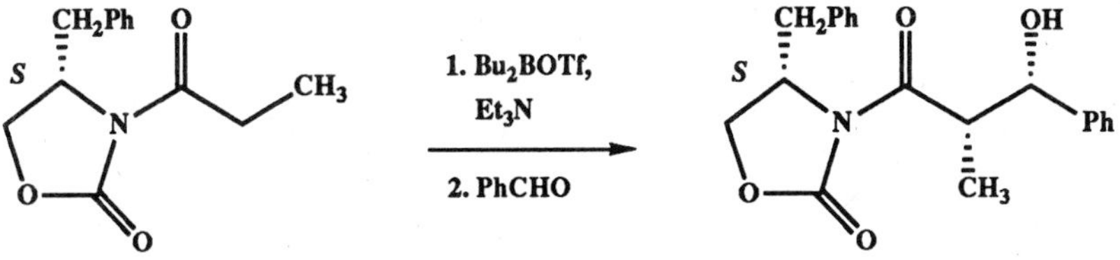

C - C BOND FORMATION - COUPLING

OS, *CV 1*, 222 (1941)

2 (2-aminobenzoic acid: NH_2, CO_2H) → 1. 2 $NaNO_2$, aq. HCl; 2. 2 Cu^+, Δ → (biphenyl-2,2'-dicarboxylic acid: CO_2H, CO_2H)

OS, *CV 1*, 459 (1941)

2 $CH_3C(O)CH_3$ → Mg, $HgCl_2$, benzene → [$(CH_3)_2C(O)–C(O)(CH_3)_2$ Mg] → H_2O, Δ → $(CH_3)_2C(OH)–C(OH)(CH_3)_2$

OS, *CV 2*, 71 (1943)

2 (benzophenone) → *i*-PrOH, sunlight, HOAc (cat.) → $Ph_2C(OH)–C(OH)Ph_2$

OS, *CV 2*, 114 (1943)

2 *n*-PrC(O)OEt → 4 Na, xylene, Et_2O → [*n*-Pr(NaO)C=C(ONa)*n*-Pr] → H_2SO_4, H_2O → *n*-PrC(O)CH(OH)*n*-Pr

C - C BOND FORMATION - COUPLING

OS, *CV* 2, 273 (1943)

2 Br–CH(CO_2Et)$_2$ —(Na_2CO_3, Δ)→ $(EtO_2C)_2C=C(CO_2Et)_2$

OS, *CV* 3, 121 (1955)

2 $CH_2=CHCH_2Cl$ —(Mg, Et_2O)→ $CH_2=CHCH_2CH_2CH=CH_2$

OS, *CV* 3, 339 (1955)

2 1-chloro-2-nitrobenzene —(Cu bronze, sand, 220 °C)→ 2,2'-dinitrobiphenyl

OS, *CV* 3, 401 (1955)

2 $EtO_2C—(CH_2)_8—CO_2K$ —(electrolysis, (Pt cathode))→ $EtO_2C—(CH_2)_{16}—CO_2Et$

OS, *CV* 4, 273 (1963)

1,1'-azobis(cyclohexanecarbonitrile) (CN, NC, N=N) —(Δ, toluene)→ 1,1'-bi(cyclohexanecarbonitrile) (CN, NC)

OS, *CV 4*, 367 (1963)

2 O_2N–C_6H_4–CH_3 —[air (O_2); aq. KOH, MeOH]→ O_2N–C_6H_4–CH_2CH_2–C_6H_4–NO_2

OS, *CV 4*, 372 (1963)

2 Et—CH(NO_2)—Cl —[1. KOH, H_2O, 0 - 10 °C; 2. HCl]→ Et—C(NO_2)=C(NO_2)—Et

OS, *CV 4*, 392 (1963)

PhCHO + $PhCH_2CN$ —[NaCN; aq. MeOH, Δ]→ PhCH(CN)–CH(CN)Ph

OS, *CV 4*, 869 (1963)

2 $CH_3COCH_2COCH_3$ —[1. NaOH, aq. MeOH; 2. I_2, Et_2O]→ $(CH_3CO)_2CH$=$CH(COCH_3)_2$

OS, *CV 4*, 872 (1963)

2 (2-amino-3,5-dichlorobenzoic acid) → 1. NaOH, H_2O; 2. $NaNO_2$, aq. HCl; 3. "Cu^+", NH_4OH → 6,6'-dicarboxy-2,2',4,4'-tetrachlorobiphenyl (Cl, Cl, Cl, Cl, CO_2H, CO_2H)

OS, *CV 4*, 877 (1963)

4 $CH_2(CN)_2$ → 8 Br_2, KBr; H_2O, Δ → KBr $[Br_2C(CN)_2]_4$ → Cu powder; benzene, Δ → 2 $(NC)_2C{=}C(CN)_2$

OS, *CV 4*, 914 (1963)

2 Ph_2CCl_2 → Cu powder; benzene, Δ → $Ph_2C{=}CPh_2$

OS, *CV 5*, 102 (1973)

2 pyridine → Raney Ni W7-J; Δ → 2,2'-bipyridine

OS, *CV 5*, 344 (1973)

TsO^- H_3N^+–CH(CN)$_2$ → NaCN; H_2O, 0 °C → (NC)(H_2N)C=C(CN)(NH_2)

OS, CV 5, 445 (1973)

O_2N–C(CH$_3$)$_2$–CH$_2$CH$_2$–CO_2Me → (1. KOH, H_2O, Δ; 2. electricity (steel cathode)) → O_2N–C(CH$_3$)$_2$–(CH$_2$)$_4$–C(CH$_3$)$_2$–NO_2

OS, CV 5, 463 (1973)

$$2\ MeO_2C{-}(CH_2)_8{-}CO_2H \xrightarrow[\text{2. electricity (Pt cathode)}]{\text{1. Na, MeOH}} MeO_2C{-}(CH_2)_{16}{-}CO_2Me$$

OS, CV 5, 517 (1973)

$$2\ Ph{-}C{\equiv}C{-}H \xrightarrow[\text{pyridine-MeOH}]{Cu(OAc)_2,\ \Delta} Ph{-}C{\equiv}C{-}C{\equiv}C{-}Ph$$

OS, CV 5, 883 (1973)

2 p-$CH_3C_6H_4CH_2\overset{+}{N}Me_3\ Br^-$ → (1. Ag_2O, H_2O; 2. toluene, Δ) → [2.2]paracyclophane

OS, CV 5, 1026 (1973)

$$2\ t\text{-BuOH} \xrightarrow[H_2SO_4,\ H_2O]{H_2O_2,\ FeSO_4} 2\left[HO-C(CH_3)_2-CH_2\cdot\right] \longrightarrow HO-C(CH_3)_2-CH_2CH_2-C(CH_3)_2-OH$$

OS, CV 6, 468 (1988)

$$(4\text{-MeOC}_6H_4)_2TeCl_2 \xrightarrow[(CH_3OCH_2CH_2)_2O]{\text{Raney nickel, }\Delta} 4\text{-MeOC}_6H_4-C_6H_4\text{-4-OMe}$$

OS, CV 6, 488 (1988)

$$2\ 4\text{-CH}_3C_6H_4MgBr + 2\ TlBr \xrightarrow[THF,\ \Delta]{\text{benzene}} 4\text{-CH}_3C_6H_4-C_6H_4\text{-4-CH}_3$$

OS, CV 7, 181 (1990); *60,* 1 (1981)

$$2\ MeO_2C(CH_2)_4COOH \xrightarrow[NaOMe,\ MeOH]{\text{electricity}} MeO_2C(CH_2)_8CO_2Me + 2\ CO_2$$

***OS, CV* 7, 485 (1990); *60,* 41 (1981)**

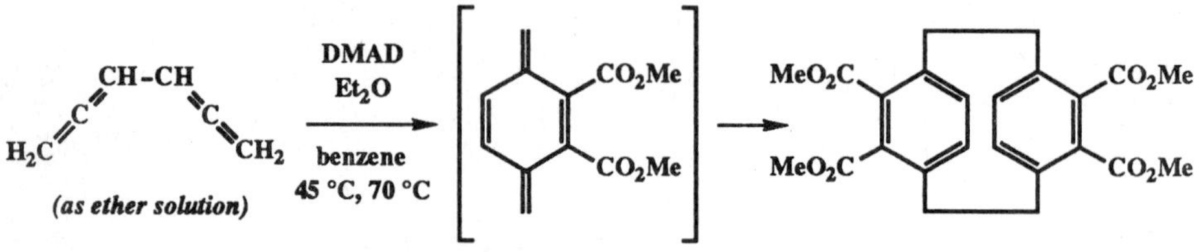

H$_2$C=C=CH-CH=C=CH$_2$ (as ether solution) → DMAD, Et$_2$O / benzene, 45 °C, 70 °C → [CO$_2$Me, CO$_2$Me] → MeO$_2$C, MeO$_2$C, CO$_2$Me, CO$_2$Me

***OS, CV* 7, 479 (1990); *60,* 58 (1981)**

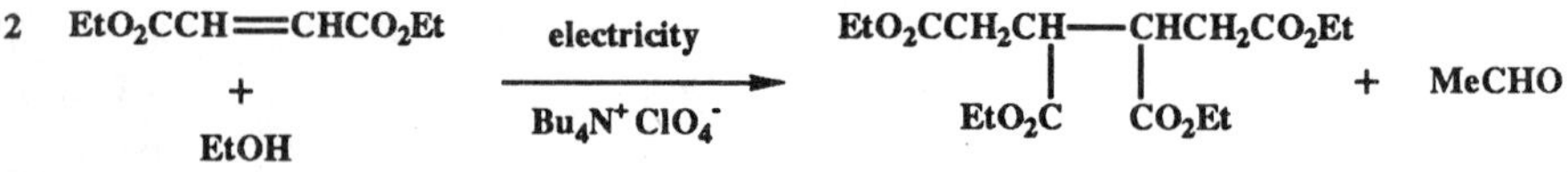

$$2\ EtO_2CCH{=}CHCO_2Et + EtOH \xrightarrow[Bu_4N^+ ClO_4^-]{electricity} EtO_2CCH_2CH(CO_2Et){-}CH(CO_2Et)CH_2CO_2Et + MeCHO$$

***OS, CV* 7, 482 (1990); *60,* 78 (1981)**

$$2\ (MeO_2C)_2CH_2 \xrightarrow[MeOH,\ NaI]{electricity} (MeO_2C)_2CHCH(CO_2Me)_2 + H_2$$

***OS, CV* 7, 1 (1990); *60,* 113 (1981)**

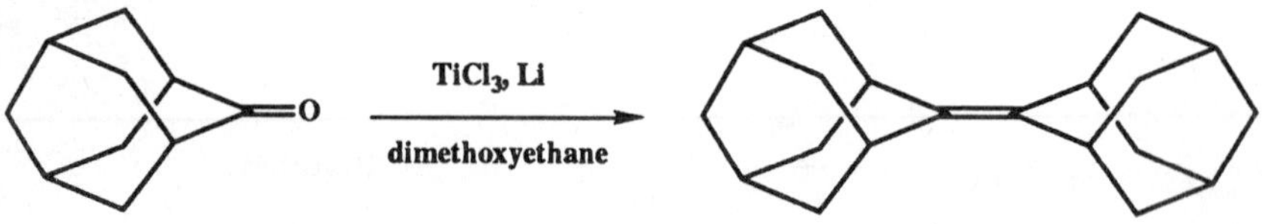

=O → TiCl$_3$, Li / dimethoxyethane →

C - C BOND FORMATION - COUPLING

OS, *65*, 52 (1987)

$Me_3Si-C\equiv C-H$ → (O_2, CuCl; $Me_2NCH_2CH_2NMe_2$) → $Me_3Si-C\equiv C-C\equiv C-SiMe_3$

OS, *65*, 108 (1987)

$N-C_6H_{11}$, H, MeO, MeO, X

A (X = Br)

1. *n*-BuLi, THF, - 78 °C
2. CuI • $P(OEt)_3$, - 78 °C
3. A (X = I), - 78 °C
4. 25 °C, 18 hr

O, O, HC, CH, MeO, OMe, MeO, OMe

OS, *68*, 198 (1989)

NaH / THF → Na^+ − → I_2 / CuBr • Me_2S

C - C BOND FORMATION - CYANATION

***OS*, *CV 1*, 21 (1941)**

CH_3CHO → (NaCN, aq. NH_4Cl, Et_2O) → $CH_3CH(NH_2)CN$ → (1. HCl; 2. $Pb(OH)_2$) → $CH_3CH(NH_2)CO_2H$

***OS*, *CV 1*, 46 (1941)**

$CH_2{=}CHCH_2Br$ → (CuCN, Δ) → $CH_2{=}CHCH_2CN$

***OS*, *CV 1*, 107 (1941)**

$C_6H_5CH_2Cl$ → (aq. NaCN, EtOH, Δ) → $C_6H_5CH_2CN$

***OS*, *CV 1*, 156 (1941)**

$Cl(CH_2)_3Br$ → (KCN, Δ, aq. EtOH) → $Cl(CH_2)_3CN$

***OS*, *CV 1*, 254 (1941)**

$ClCH_2CO_2Na$ → (NaCN, H_2O, Δ) → $NCCH_2CO_2Na$ → (H_2SO_4 (cat.), EtOH, Δ) → $NCCH_2CO_2Et$

OS, CV 1, 256 (1941)

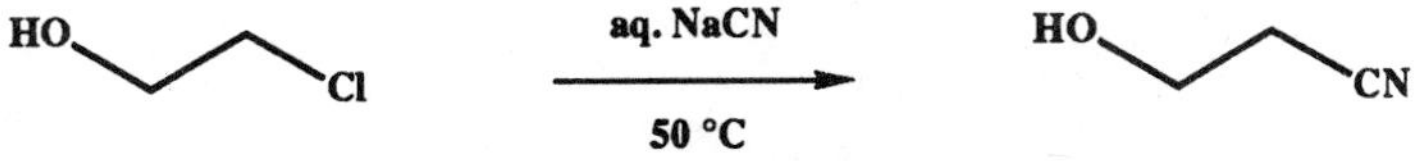

HO–CH_2CH_2–Cl → (aq. NaCN, 50 °C) → HO–CH_2CH_2–CN

OS, CV 1, 336 (1941)

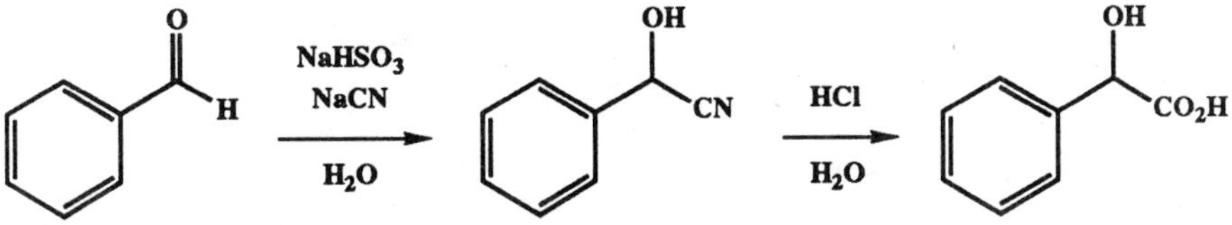

PhCHO → ($NaHSO_3$, NaCN, H_2O) → PhCH(OH)CN → (HCl, H_2O) → PhCH(OH)CO_2H

OS, CV 1, 355 (1941)

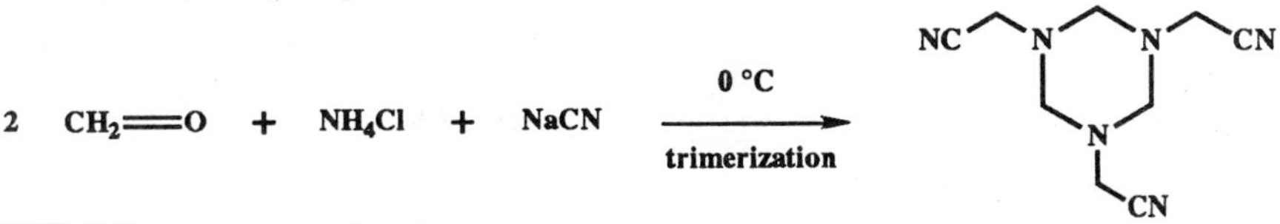

2 CH_2=O + NH_4Cl + NaCN → (0 °C, trimerization) → 1,3,5-tris(cyanomethyl)hexahydro-1,3,5-triazine (NC, N, N, CN, N, CN)

OS, CV 1, 451 (1941)

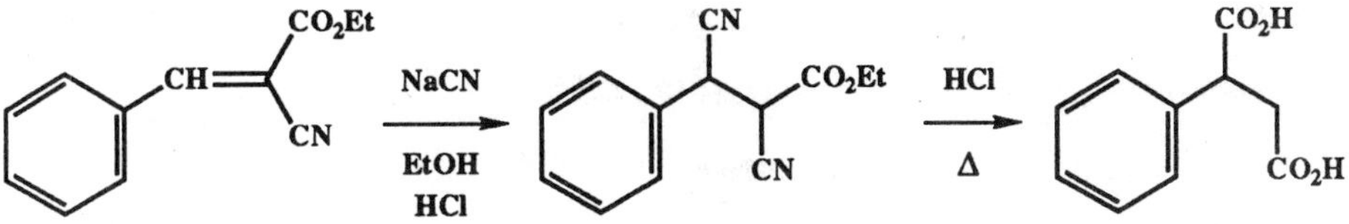

PhCH=C(CO_2Et)(CN) → (NaCN, EtOH, HCl) → PhCH(CN)CH(CN)CO_2Et → (HCl, Δ) → PhCH(CO_2H)CH_2CO_2H

OS, CV 1, 514 (1941)

NH_2, R_1, R_2 → (1. 2 HCl, $NaNO_2$; 2. CuCN; R_1 = Me, R_2 = H; R_1 = H, R_2 = Me) → CN, R_1, R_2

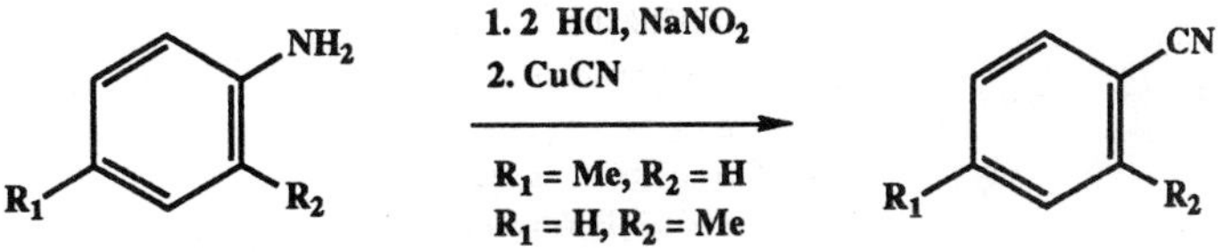

***OS, CV 1,* 536 (1941)**

Br(CH₂)₄Br → NC(CH₂)₄CN

$Br(CH_2)_4Br \xrightarrow[\text{aq. EtOH, } \Delta]{\text{NaCN}} NC(CH_2)_4CN$

***OS, CV 2,* 7 (1943)**

$CH_3COCH_3 \xrightarrow[H_2SO_4,\ H_2O]{\text{NaCN}} (CH_3)_2C(OH)CN$

***OS, CV 2,* 29 (1943)**

$CH_3COCH_3 \xrightarrow[Et_2O,\ H_2O]{NaCN,\ NH_4Cl} (CH_3)_2C(OH)CN$

***OS, CV 2,* 292 (1943)**

$n\text{-}C_{12}H_{25}\text{—}Br \xrightarrow[\text{EtOH, } \Delta]{\text{KCN}} n\text{-}C_{12}H_{25}\text{—}CN \xrightarrow[\text{2. EtOH, HCl}]{\text{1. aq. KOH, } \Delta} n\text{-}C_{12}H_{25}\text{—}CO_2Et$

***OS, CV 2,* 376 (1943)**

$ClCH_2CO_2H \xrightarrow{\text{1. } Na_2CO_3;\ \text{2. NaCN}} NCCH_2CO_2Na \xrightarrow{\text{aq. NaOH}} CH_2(CO_2Na)_2 \xrightarrow{\text{aq. } CaCl_2} CH_2(CO_2^-)_2\ Ca^{+2} \xrightarrow{\text{HCl}} CH_2(CO_2H)_2$

OS, CV 2, 498 (1943)

$$PhCH=CH-C(=O)-Ph \xrightarrow[EtOH,\ H_2O]{KCN,\ HOAc} PhCH(CN)-CH_2-C(=O)-Ph$$

OS, CV 3, 66 (1955)

$$Et_2C=O \xrightarrow[MeOH,\ 60\ °C]{NaCN,\ NH_4Cl,\ NH_4OH,\ H_2O} Et_2C(NH_2)CN \xrightarrow[2.\ Pb(OH)_2]{1.\ H_2O,\ HCl\ then\ \Delta} Et_2C(NH_2)CO_2H$$

OS, CV 3, 84 (1955)

$$PhCHO \xrightarrow[H_2O,\ MeOH]{NaCN,\ NH_4Cl} PhCH(NH_2)CN \xrightarrow[2.\ NH_4OH]{1.\ aq.\ HCl,\ \Delta} PhCH(NH_2)CO_2H$$

OS, CV 3, 88 (1955)

$$PhC(=O)Me \xrightarrow[H_2O,\ EtOH]{NaCN,\ NH_4Cl} PhC(Me)(NH_2)CN \xrightarrow[2.\ pyridine]{1.\ aq.\ HCl,\ \Delta} PhC(Me)(NH_2)CO_2H$$

OS, *CV* 3, 112 (1955)

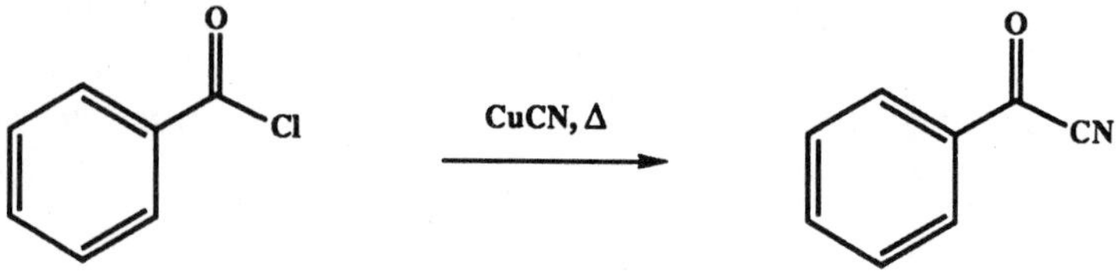

O, Cl → CuCN, Δ → O, CN

OS, *CV* 3, 174 (1955)

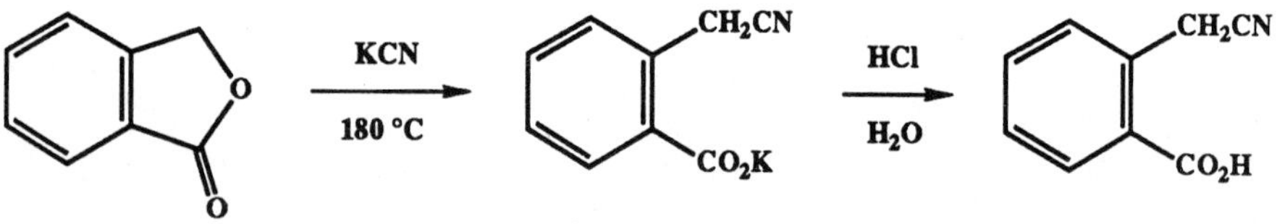

O, O → KCN, 180 °C → CH_2CN, CO_2K → HCl, H_2O → CH_2CN, CO_2H

OS, *CV* 3, 212 (1955)

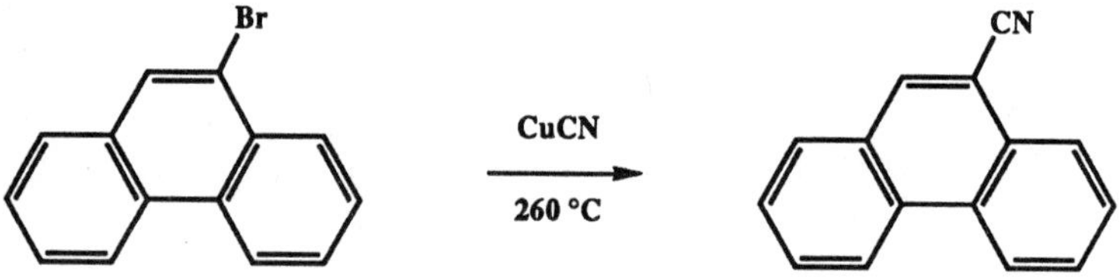

Br → CuCN, 260 °C → CN

OS, *CV* 3, 275 (1955)

$$Et_2NH + CH_2{=}O \xrightarrow[NaCN,\ H_2O]{NaHSO_3} Et_2N{-}CH_2{-}CN$$

OS, CV 3, 293 (1955)

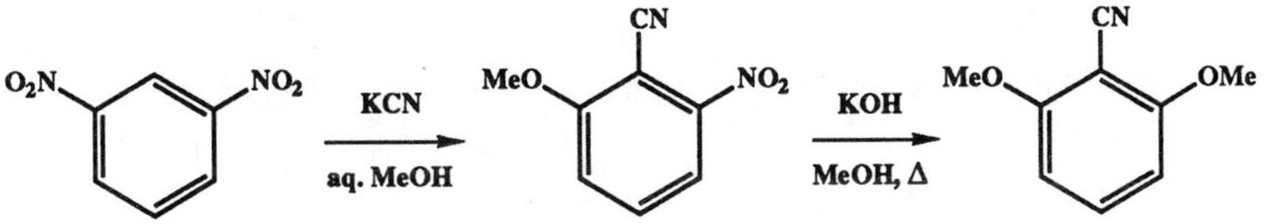

O_2N, NO_2 → KCN, aq. MeOH → MeO, CN, NO_2 → KOH, MeOH, Δ → MeO, CN, OMe

OS, CV 3, 372 (1955)

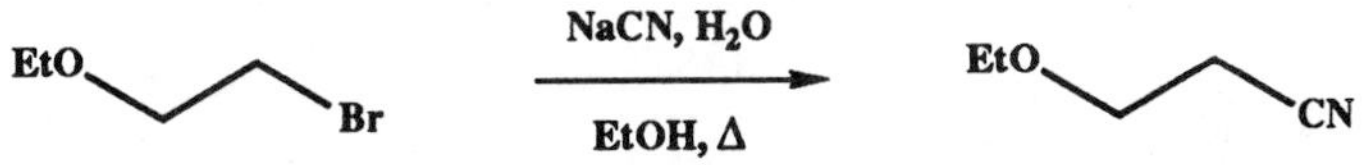

EtO, Br → $NaCN, H_2O$, EtOH, Δ → EtO, CN

OS, CV 3, 436 (1955)

$CH_2{=}O$ → KCN, H_2O, then H_2SO_4 → $HO—CH_2—CN$

OS, CV 3, 549 (1955)

CH_3, CH_3, CH_3 → $Zn(CN)_2$, $HCl, AlCl_3$, $C_2H_2Cl_4$ → H, NH • HCl, CH_3, CH_3, CH_3 → HCl, H_2O → H, O, CH_3, CH_3, CH_3

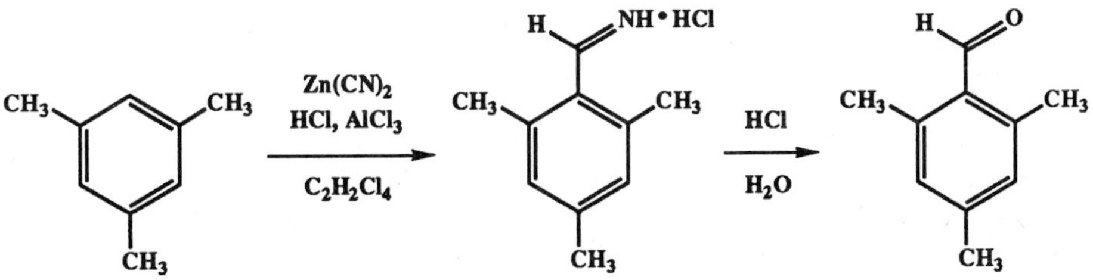

OS, CV 3, **557 (1955)**

CH_2Cl, CH_3, CH_3, CH_3 $\xrightarrow[\text{aq. EtOH}]{\text{NaCN, }\Delta}$ CH_2CN, CH_3, CH_3, CH_3

OS, CV 3, **615 (1955)**

$CH_3—CH{=}CH—CO_2Et$ $\xrightarrow[\text{2. Ba(OH)}_2\text{, }\Delta\text{ then HNO}_3]{\text{1. NaCN, }\Delta\text{ H}_2\text{O, EtOH}}$ CH_3, CO_2H, CO_2H

OS, CV 3, **631 (1955)**

Br $\xrightarrow[\text{pyridine}]{\text{CuCN, }\Delta}$ CN

OS, CV 4, **58 (1963)**

O, Ph, CH_3 $\xrightarrow[\text{HCl, Et}_2\text{O}]{\text{NaCN, H}_2\text{O}}$ HO, CN, Ph, CH_3

OS, *CV* 4, 274 (1963)

2 [cyclohexanone] —(2 NaCN, H_2O; $H_2NNH_2 \cdot H_2SO_4$)→ [1,1'-hydrazodicyclohexanecarbonitrile: CN, NC, N—N, H H]

OS, *CV* 4, 393 (1963)

Ph—CH=C(Ph)(CN) —(KCN, H_2O; Et_2O, MeOH, Δ)→ Ph, CN, Ph, CN

OS, *CV* 4, 496 (1963)

[γ-butyrolactone, O, O] —(KCN; 190 - 195 °C)→ KO_2C~~~CN

OS, *CV* 4, 576 (1963)

MeO-C6H4-CH2Cl —(NaCN, Δ; NaI, acetone)→ MeO-C6H4-CH2CN

OS, CV 4, 641 (1963)

PhCOCl

KCN, H_2O

CN O Ph

PhLi, CH_3I

Et_2O, dioxane

- 10 °C

Ph

CH_3 CN O

OS, CV 4, 804 (1963)

CO_2Et

PhCH

CO_2Et

KCN, aq. EtOH

then HCl

Ph

CO_2Et

CN

OS, CV 5, 239 (1973)

F F

Cl F

1. NaCN, H_2O

CH_3CN, 75 °C

2. H_2SO_4, H_2O

F F

F

CO_2H

Cl

OS, CV 5, 269 (1973)

Me

O^-

Me_2SO_4

Δ

Me

$MeSO_4^-$ OMe

NaCN

H_2O, 5 °C

NC Me

OS, CV 5, 437 (1973)

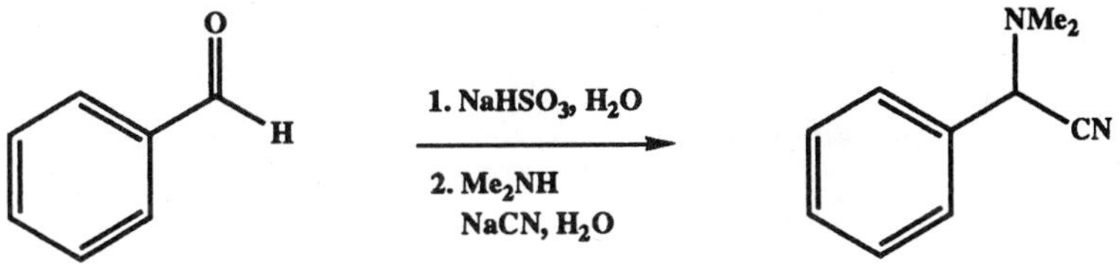

O ; H ; 1. $NaHSO_3$, H_2O ; 2. Me_2NH ; $NaCN$, H_2O ; NMe_2 ; CN

OS, CV 5, 578 (1973)

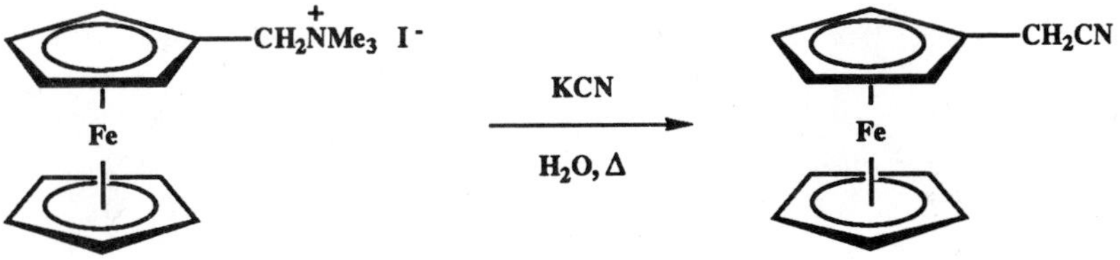

$\overset{+}{C}H_2NMe_3$ I^- ; Fe ; KCN ; H_2O, Δ ; CH_2CN ; Fe

OS, CV 5, 614 (1973)

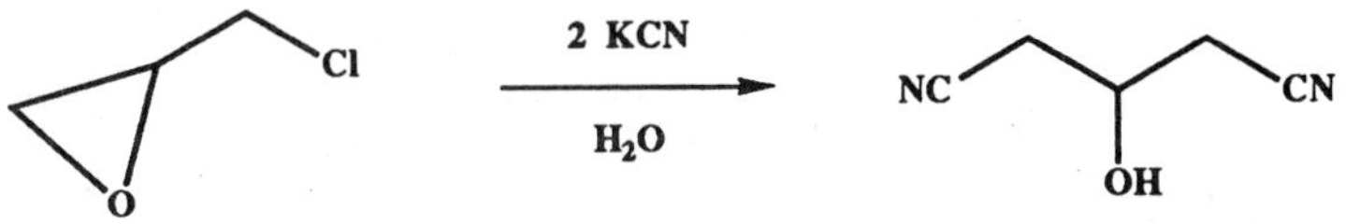

Cl ; O ; 2 KCN ; H_2O ; NC ; CN ; OH

OS, CV 6, 14 (1988)

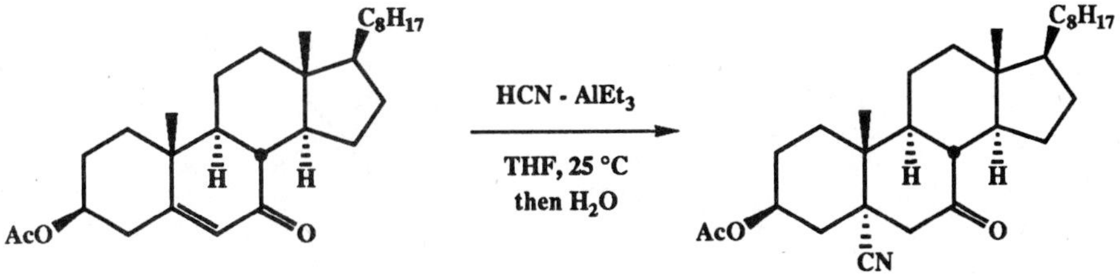

C_8H_{17} ; AcO ; H ; H ; O ; HCN - $AlEt_3$; THF, 25 °C ; then H_2O ; C_8H_{17} ; AcO ; H ; H ; O ; CN

OS, *CV 6*, 41 (1988)

+ p-Ts-$CH_2N{\equiv}C$ → (t-BuOK, EtOH; DME, 5 - 35 °C) → C≡N

OS, *CV 6*, 115 (1988)

N → (KCN, PhCOCl; CH_2Cl_2, H_2O) → N, Ph, NC, H, O

OS, *CV 6*, 307 (1988)

O, MeO → (Et_2AlCN; benzene, toluene, - 15 °C then H_3O^+) → HO, CN, MeO

OS, *CV 6*, 334 (1988)

O → (1. H_2NNHCO_2Me MeOH, AcOH, Δ; 2. HCN, 0 °C) → NC, H H, N—N—CO_2Me

OS, *CV 7*, 20 (1990); *60*, 14 (1981)

$$Ph_2C{=}O \xrightarrow[\text{ZnI}_2,\ \text{CH}_2\text{Cl}_2]{\text{Me}_3\text{SiCN}} Ph_2C(OSiMe_3)(CN) \xrightarrow[\text{THF}]{3\,N\ \text{HCl}} Ph_2C(OH)(CN)$$

OS, *CV 7*, 517 (1990); *60*, 126 (1981)

$$\text{1,4-benzoquinone} + Me_3SiCN \xrightarrow[\text{CCl}_4]{\text{18-crown-6, KCN}} \text{4-(OSiMe}_3\text{)-4-(CN)-cyclohexa-2,5-dienone}$$

OS, *CV 7*, 521 (1990); *62*, 196 (1984)

$$MeO{-}C_6H_4{-}CHO \xrightarrow[\text{KCN},\ \text{Zn(CN)}_2]{\text{Me}_3\text{SiCl}} MeO{-}C_6H_4{-}CH(OSiMe_3)(CN)$$

OS, CV 1, 54 (1941)

3 (salicylic acid: OH, CO_2H) → $CH_2{=}O$, 4 $NaNO_2$; 4 H_2SO_4 (dil. NH_4OH) → HO, CO_2H, CO_2H, =O, HO, CO_2H *(as ammonium salt)*

OS, CV 1, 77 (1941)

PhCHO (O, H) → $CH_3C(O)CH_3$, aq. NaOH; then HCl → $CH{=}CH{-}C(=O){-}CH_3$

OS, CV 1, 78 (1941)

PhCHO (O, H) → $CH_3C(O)Ph$, aq. NaOH; EtOH → $CH{=}CH{-}C(=O){-}Ph$

OS, CV 1, 81 (1941)

PhCHO (O, H) → $CH_3C(O)$*t*-Bu, aq. NaOH; EtOH → $CH{=}CH{-}C(=O){-}$*t*-Bu

OS, CV 1, **181 (1941)**

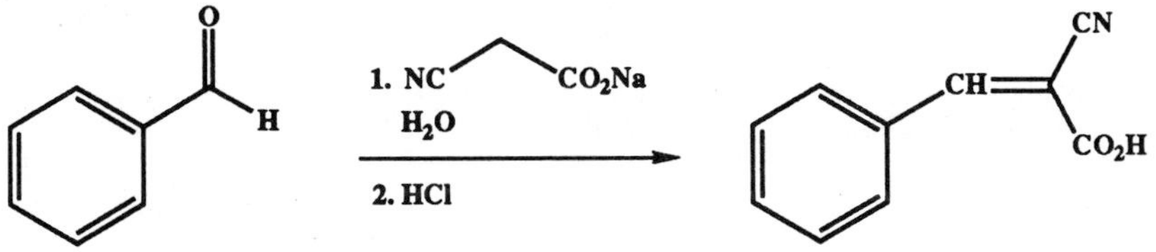

1. NC–CH_2–CO_2Na, H_2O; 2. HCl

CH=C(CN)–CO_2H

OS, CV 1, **252 (1941)**

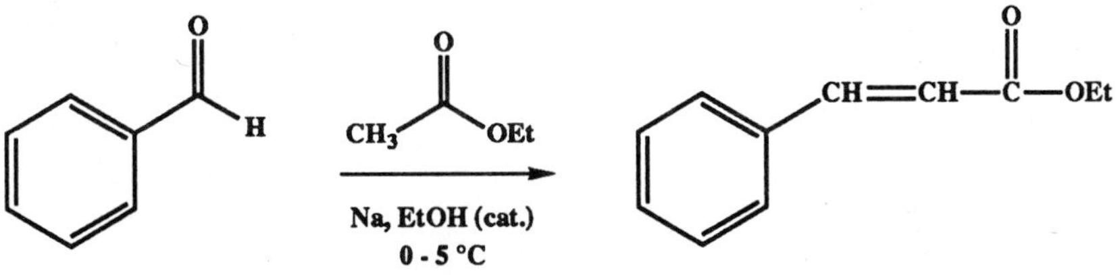

$CH_3C(O)OEt$; Na, EtOH (cat.), 0 - 5 °C

CH=CH–C(O)–OEt

OS, CV 1, **283 (1941)**

CHO

$CH_3C(O)CH_3$; NaOH, H_2O then H_2SO_4

CH=CH–C(O)–CH_3

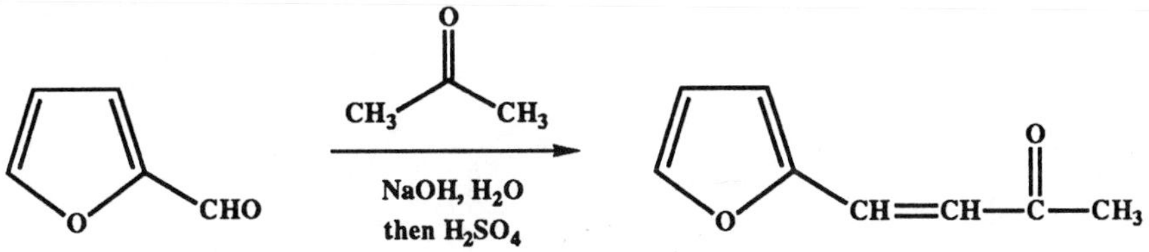

OS, CV 1, **398 (1941)**

NO_2

NaOAc, Ac_2O; 180 °C

CH=CH–CO_2H

NO_2

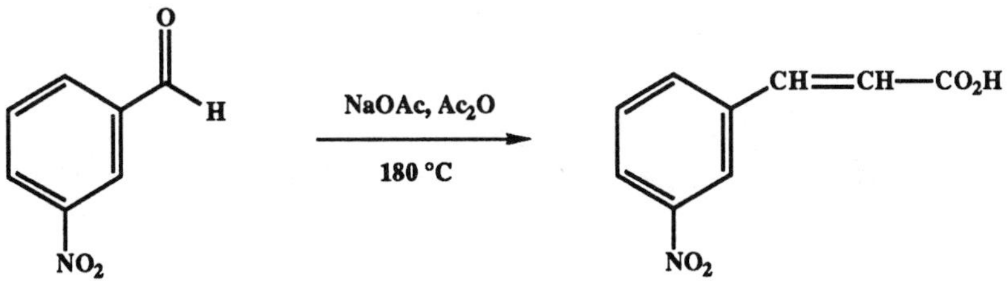

OS, CV 1, **413 (1941)**

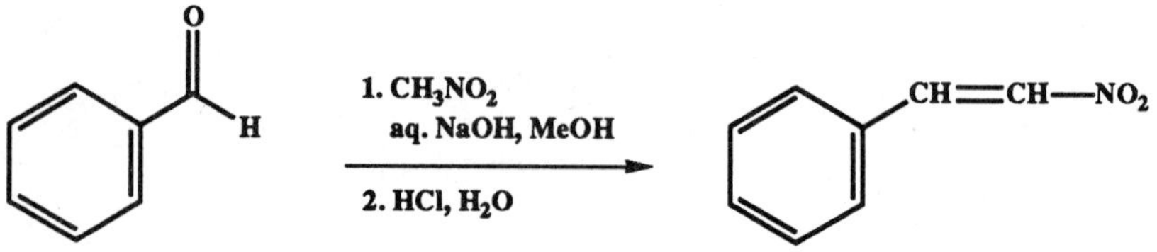

1. CH_3NO_2, aq. NaOH, MeOH
2. HCl, H_2O

CH=CH—NO_2

OS, CV 2, **61 (1943)**

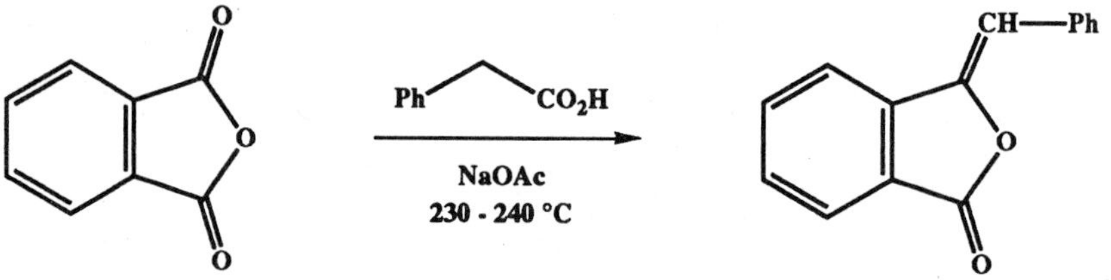

Ph CO_2H

NaOAc
230 - 240 °C

CH—Ph

OS, CV 2, **167 (1943)**

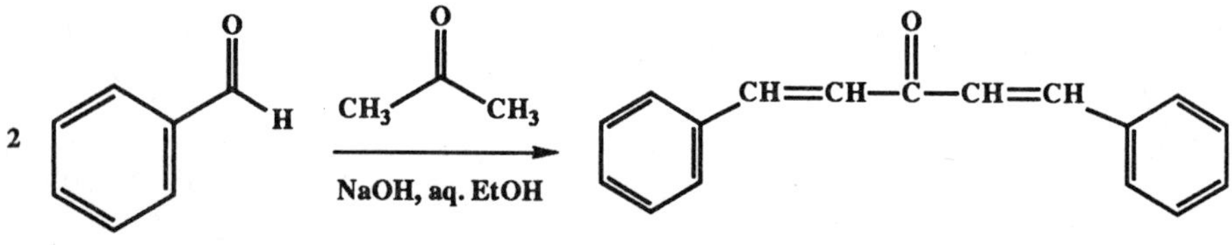

2

CH_3 CH_3

NaOH, aq. EtOH

CH=CH—C—CH=CH

OS, CV 2, **229 (1943)**

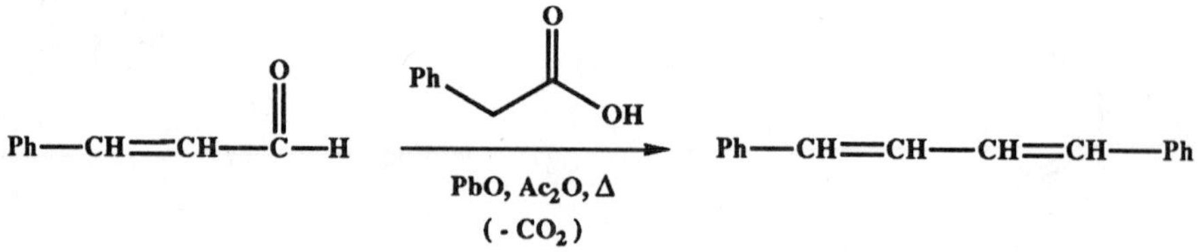

Ph—CH=CH—C—H

Ph OH

PbO, Ac_2O, Δ
(- CO_2)

Ph—CH=CH—CH=CH—Ph

OS, CV 3, 39 (1955)

H N O O= N H O → Ph(C=O)H, H_2O, Δ → H N O O= N H O =CH—Ph

OS, CV 3, 367 (1955)

2 Ph(C=O)CH_3 → Al(O-*t*-Bu)$_3$, xylene, Δ → Ph, CH_3, =CH, O, Ph

OS, CV 3, 377 (1955)

Ph(C=O)H + CO_2Et, CO_2Et → $PhCO_2H$ (cat.), piperidine, benzene, Δ → PhCH= CO_2Et, CO_2Et

OS, CV 3, 395 (1955)

HC(OEt)$_3$ + CO_2Et, CO_2Et → Ac_2O, $ZnCl_2$, Δ → EtOCH= CO_2Et, CO_2Et

OS, *CV 3*, 399 (1955)

$CH_3CH_2C(=O)CH_2CH_3$ + $NCCH_2CO_2Et$ —(NH_4OAc, Δ; HOAc, benzene)→ $(CH_3CH_2)_2C{=}C(CN)CO_2Et$

OS, *CV 3*, 425 (1955)

2-furaldehyde + $CH_2(CO_2H)_2$ —(Δ; pyridine)→ 2-furyl–CH=CHCO$_2$H

OS, *CV 3*, 426 (1955)

2-furaldehyde —(1. KOAc, Ac_2O, Δ; 2. HCl, H_2O)→ 2-furyl–CH=CHCO$_2$H

OS, *CV 3*, 586 (1955)

OH, MeO, CHO + O, NH, N, Me, NH —(155 °C)→ HO, MeO, O, H, N, NH, N, Me

OS, CV 3, 715 (1955)

PhCHO + $PhCH_2CN$ —(NaOEt, EtOH)→ PhCH=C(Ph)CN

OS, CV 3, 747 (1955)

CHO —(acetone, NaOEt, EtOH, - 5 °C)→ CH=CHCOCH$_3$

OS, CV 3, 783 (1955)

$CH_3CH{=}CH{-}CHO$ —($HO_2C\text{-}CH_2\text{-}CO_2H$, pyridine, Δ)→ $CH_3CH{=}CH{-}CH{=}CHCO_2H$

OS, CV 4, 93 (1963)

Me(Et)C=O —(EtO_2CCH_2CN, β - alanine, Δ, HOAc, benzene)→ Me(Et)C=C(CO_2Et)(CN) —(1. $PhCH_2MgCl$; 2. aq. H_2SO_4)→ Me(Et)($PhCH_2$)C–CH(CO_2Et)(CN)

OS, *CV 4*, **132 (1963)**

Ph, Ph C=O + CO_2Et–CH_2CH_2–CO_2Et → (1. KO-*t*-Bu, Δ; 2. HCl, H_2O) → $Ph_2C=C(CO_2Et)CH_2CO_2Et$

OS, *CV 4*, **234 (1963)**

Cyclohexanone + $NCCH_2CO_2H$ → (NH_4OAc; benzene, Δ) → cyclohexylidene-C(CO_2H)(CN)

OS, *CV 4*, **293 (1963)**

$(CH_3CHO)_3$ → (3 $CH_2(CO_2Et)_2$; Ac_2O, Δ) → 3 $CH_3CH{=}C(CO_2Et)_2$

OS, *CV 4*, **327 (1963)**

2,3-(MeO)$_2C_6H_3$CHO → ($HO_2C\text{–}CH_2\text{–}CO_2H$, piperidine; pyridine, Δ, then aq. HCl) → 2,3-(MeO)$_2C_6H_3$–CH=CH–CO_2H

OS, CV 4, **387 (1963)**

Ph–CO–Ph → ($PhCH_2CN$; Na, NH_3; Et_2O, Δ) → $Ph_2C{=}C(Ph)CN$

Ph, Ph, O, Ph, CN, Na, NH_3, Et_2O, Δ, Ph, Ph, Ph, CN

OS, CV 4, **408 (1963)**

OMe, H, MeO, O, O, CO_2Et, CH_3, piperidine, HOAc, benzene, Δ, OMe, O, CH_3, MeO, CH=, CO_2Et

OS, CV 4, **463 (1963)**

CH_3, O, Ph, +, CN, CO_2Et, NH_4OAc, Δ, HOAc, benzene, CH_3, CN, Ph, CO_2Et

OS, CV 4, **515 (1963)**

O, n-C_7H_{15}—N(H), NH_2, NC, CN, $CH(OEt)_3$, Δ, O, n-C_7H_{15}—N(H), N(H)—CH=, CN, CN

OS, CV 4, 536 (1963)

HCO_2Et
NaOEt *or* NaH
EtOH, Et_2O
O
CHOH

OS, CV 4, 573 (1963)

H
O
OCH_3
$CH_3CH_2NO_2$
n-$BuNH_2$
toluene, Δ
CH_3
CH=
NO_2
OCH_3

OS, CV 4, 730 (1963)

CHO
NO_2
$PhCH_2CO_2H$
Et_3N, Ac_2O, Δ
H
Ph
CO_2H
NO_2

OS, CV 4, 731 (1963)

CHO
NO_2
$CH_2(CO_2H)_2$
EtOH, C_5H_5N, Δ
CH=CHCO$_2$H
NO_2
Cu powder
quinoline, Δ
CH=CH$_2$
NO_2

C - C Bond Formation - Olefination

OS, CV 4, 771 (1963)

CH=CH—CHO → 1. CH_3MgBr, Et_2O, 10 °C; 2. 30% H_2SO_4

H

H

OS, CV 4, 777 (1963)

Ph—CHO + Ph—CH_2CO_2H → Et_3N, Ac_2O, Δ

H CO_2H

Ph Ph

cis

OS, CV 5, 361 (1973)

O

H

Me_2N

Ph_3P=CCl_2

heptane, 10 °C

CH=CCl_2

Me_2N

OS, CV 5, 390 (1973)

O

H

Cl—CF_2—CO_2Na

Ph_3P

diglyme, Δ

CF_2

H

OS, *CV 5*, 431 (1973)

$MeO-CH=N^+Me_2\ MeOSO_3^-$ + cyclopentadienide Na^+ —(THF)→ 6-(dimethylaminomethylene)fulvene ($C_5H_4=CH-NMe_2$)

OS, *CV 5*, 499 (1973)

$PhCH{=}CHCH_2\overset{+}{P}Ph_3\ Cl^-$ + PhCHO —(LiOEt, EtOH)→ $PhCH{=}CH{-}CH{=}CHPh$ (all-trans)

OS, *CV 5*, 509 (1973)

$Ph_2C{=}O$ + $(EtO)_2P(O){-}CH_2CO_2Et$ —(NaH, DME, Δ)→ $Ph_2C{=}CH{-}CO_2Et$

OS, *CV 5*, 547 (1973)

cyclohexanone + $(EtO)_2P(O){-}CH_2CO_2Et$ —(NaH, benzene)→ cyclohexylidene$=CH{-}CO_2Et$

OS, CV 5, 567 (1973)

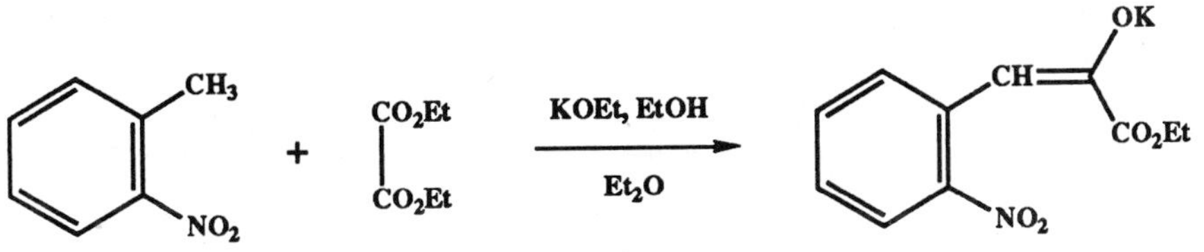

CH_3 NO_2 + CO_2Et CO_2Et — KOEt, EtOH / Et_2O → $CH=$ OK CO_2Et NO_2

OS, CV 5, 585 (1973)

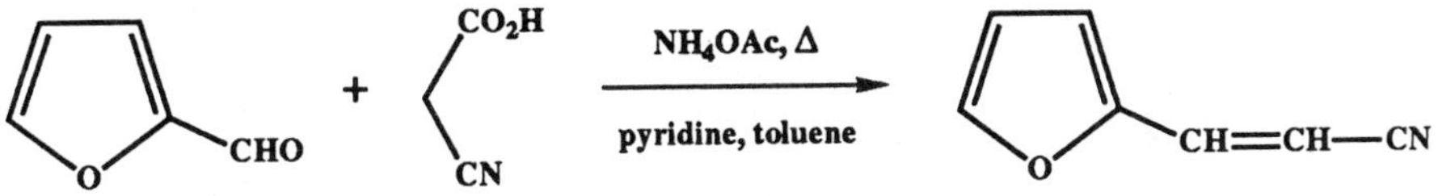

CHO + CO_2H CN — NH_4OAc, Δ / pyridine, toluene → CH=CH—CN

OS, CV 5, 627 (1973)

HO CHO + O NH HN O — piperidine / Δ → HO CH= O NH HN O

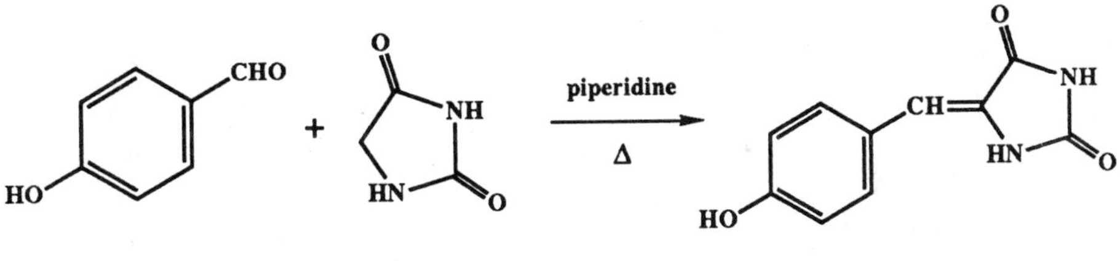

OS, CV 5, 751 (1973)

O — $Ph_3P=CH_2$ / Et_2O, Δ → CH_2

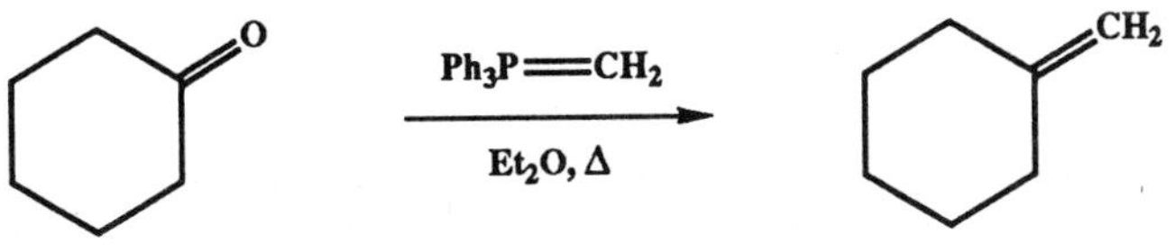

OS, CV 5, 949 (1973)

Ph–CO–CF_3 → ($Cl—CF_2—CO_2Na$, Ph_3P; diglyme, 140 °C) → Ph–C(=CF_2)–CF_3

OS, CV 5, 985 (1973)

2 PhCH=CH–CHO + $Ph_3\overset{+}{P}CH_2$–C_6H_4–$CH_2\overset{+}{P}Ph_3$ 2 Cl^- → (LiOEt, EtOH)

OS, CV 6, 95 (1988)

CH_2=CH–CHO + $CH_2(CO_2H)_2$ → (pyridine) → CH_2=CH–CH=CH–CO_2H

OS, CV 6, 358 (1988)

cyclohexanone + $(EtO)_2\overset{O}{\overset{\|}{P}}-CH{=}CH-\overset{H}{N}-C_6H_{11}$ → (NaH, THF) → $C_6H_{10}{=}CH-CH{=}N-C_6H_{11}$

OS, *CV 6*, 442 (1988)

CO_2Et ZnCl$_2$ Ac$_2$O, Δ CO_2Et 1. CH_3MgI CuCl, Et_2O 2. H_2O, H_2SO_4 CO_2Et

OS, *CV 7*, 332 (1990); *60*, 88 (1981)

+ $(CH_2O)_n$ CF_3COO^- $CH_3\overset{+}{N}H_2Ph$ THF, reflux

OS, *CV 7*, 108 (1990); *62*, 179 (1984)

CH_3CN KOH CN

PhCHO CH_3CN KOH Ph CN

OS, *CV 7*, 232 (1990); *62*, 202 (1984)

Cl + $Ph_3P{=}C(H)CO_2Et$ Et_3N Me(H)C=C=C(H)CO_2Et

OS, *CV* 7, 359 (1990); *63*, 198 (1985)

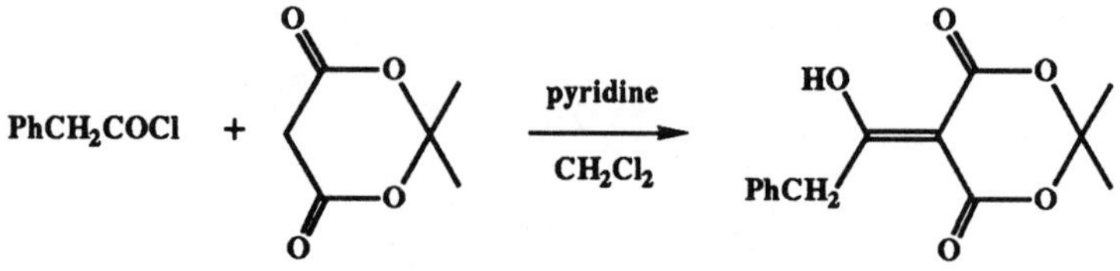

$PhCH_2COCl$ + (2,2-dimethyl-1,3-dioxane-4,6-dione) → (pyridine, CH_2Cl_2) → HO, $PhCH_2$ (5-(1-hydroxy-2-phenylethylidene)-2,2-dimethyl-1,3-dioxane-4,6-dione)

OS, *CV* 7, 34 (1990); *63*, 214 (1985)

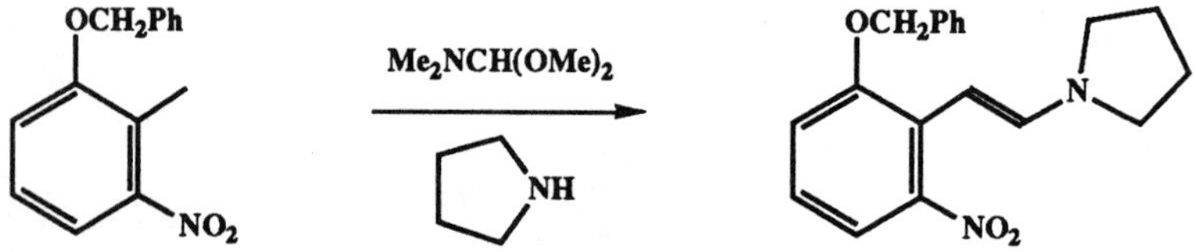

OCH_2Ph, NO_2 → ($Me_2NCH(OMe)_2$, pyrrolidine NH) → OCH_2Ph, N, NO_2

OS, *CV* 7, 142 (1990); *64*, 63 (1986)

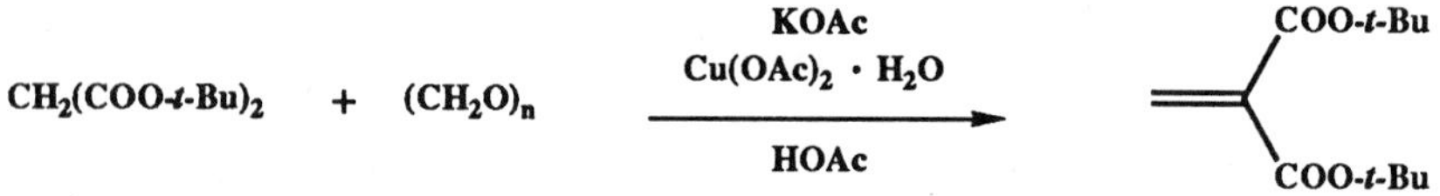

$CH_2(COO\text{-}t\text{-}Bu)_2$ + $(CH_2O)_n$ → (KOAc, $Cu(OAc)_2 \cdot H_2O$, HOAc) → $COO\text{-}t\text{-}Bu$, $COO\text{-}t\text{-}Bu$

OS, *CV* 7, 258 (1990); *64*, 164 (1986)

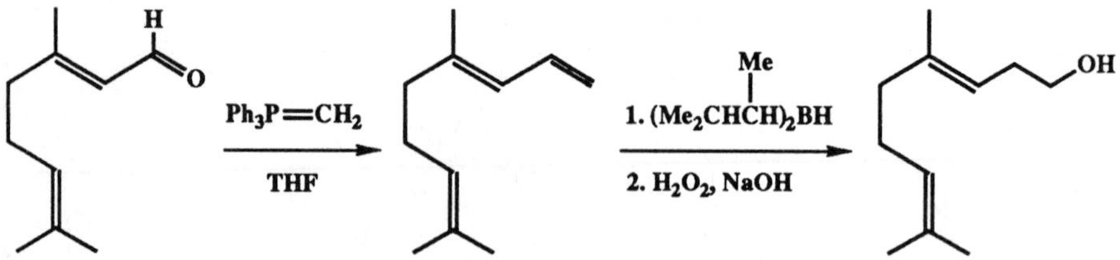

H, O → ($Ph_3P{=}CH_2$, THF) → diene → (1. $(Me_2CHCH(Me))_2BH$; 2. H_2O_2, NaOH) → OH

OS, CV 7, 476 (1990); *64*, 189 (1986)

1. *t*-BuOK, THF
2. CS_2
3. MeI

OS, 65, 17 (1987)

$SnCl_4$

OS, 65, 81 (1987)

Zn - CH_2Br_2 - $TiCl_4$

THF, CH_2Cl_2

r.t.

OS, 65, 119 (1987)

$(EtO)_2PCH_2N{=}CHPh$ → Ph(Me)C=CH—N=CHPh

1. *n*-BuLi, THF, - 78 °C
2. PhCOMe, - 78 °C --> Δ

1. *n*-BuLi, THF, - 78 °C
2. $CH_2{=}CHCH_2Br$, - 78 °C --> r.t.
3. H_3O^+, r.t.

OS, *66*, 220 (1987)

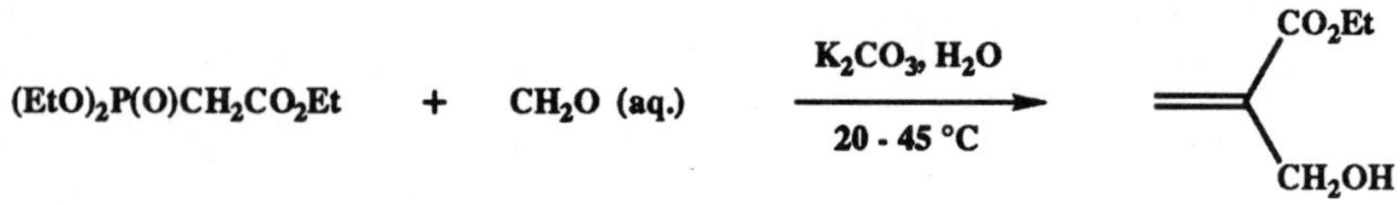

$(EtO)_2P(O)CH_2CO_2Et$ + CH_2O (aq.) → K_2CO_3, H_2O / 20 - 45 °C → CO_2Et, CH_2OH

OS, *67*, 170 (1988)

N, OMe + O, O, O, O → 1. $Ni(acac)_2$, $CHCl_3$ / 2. Na, EtOH, Δ → N H, H, CO_2Et

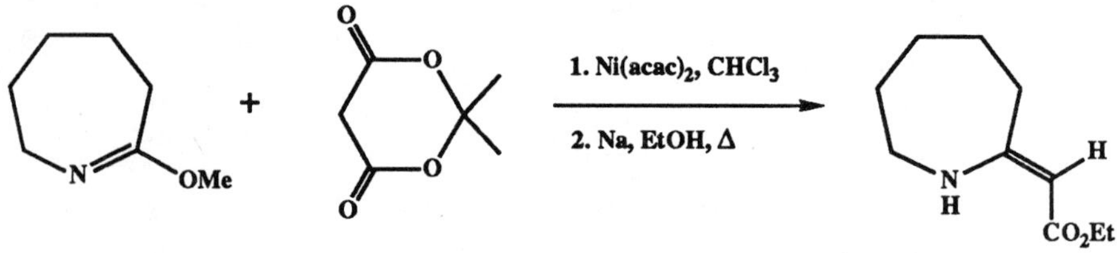

OS, CV 1, **186 (1941)**

Cyclohexyl-MgBr + 2,3-dibromopropene (Br–CH$_2$–C(Br)=CH$_2$) $\xrightarrow[\Delta]{Et_2O}$ cyclohexyl–CH$_2$–C(Br)=CH$_2$

OS, CV 1, **188 (1941)**

Cyclohexyl-MgCl $\xrightarrow[2.\ H_2O]{1.\ (CH_2O)_x,\ Et_2O}$ cyclohexyl–CH$_2$OH

OS, CV 1, **226 (1941)**

EtO–C(=O)–CH$_3$ $\xrightarrow[2.\ NH_4Cl]{1.\ 2\ PhMgBr,\ Et_2O}$ Ph$_2$C(OH)–CH$_3$

OS, CV 1, **248 (1941)**

CH$_3$–C(=O)–CH$_2$–C(=O)–OEt $\xrightarrow[NaOEt,\ EtOH]{n\text{-}BuBr,\ \Delta}$ CH$_3$–C(=O)–CH(Bu)–C(=O)–OEt

***OS*, *CV 1*, 250 (1941)**

EtO(C=O)CH$_2$(C=O)OEt $\xrightarrow[\text{NaOEt, EtOH}]{n\text{-BuBr, }\Delta}$ EtO(C=O)CH(Bu)(C=O)OEt

***OS*, *CV 1*, 272 (1941)**

$EtO_2C—CH{=}CH—CO_2Et \xrightarrow[\text{Na, EtOH, }\Delta]{CH_2(CO_2Et)_2} (EtO_2C)_2CH—CH(CO_2Et)—CH_2—CO_2Et$

***OS*, *CV 1*, 306 (1941)**

$CH_3(CH_2)_3MgBr$ + ethylene oxide $\xrightarrow{Et_2O}$ $CH_3(CH_2)_5OH$

***OS*, *CV 1*, 361 (1941)**

sec-butyl-MgCl $\xrightarrow[\text{2. aq. } H_2SO_4]{\text{1. } CO_2\text{, } Et_2O\text{, -5 to - 12 °C}}$ 2-methylbutanoic acid (CO_2H)

OS, CV 1, 471 (1941)

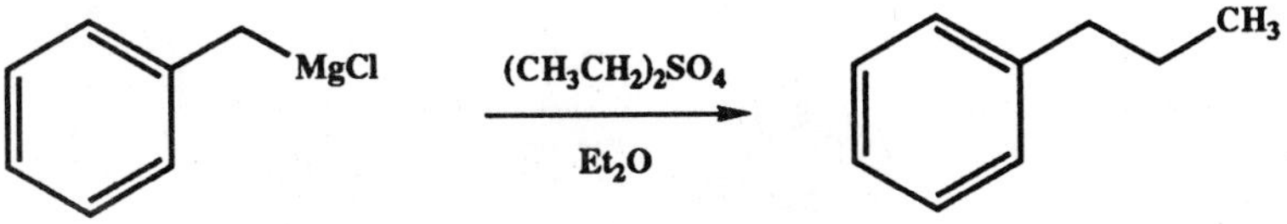

$C_6H_5CH_2MgCl$ → [$(CH_3CH_2)_2SO_4$, Et_2O] → $C_6H_5CH_2CH_2CH_3$

OS, CV 1, 524 (1941)

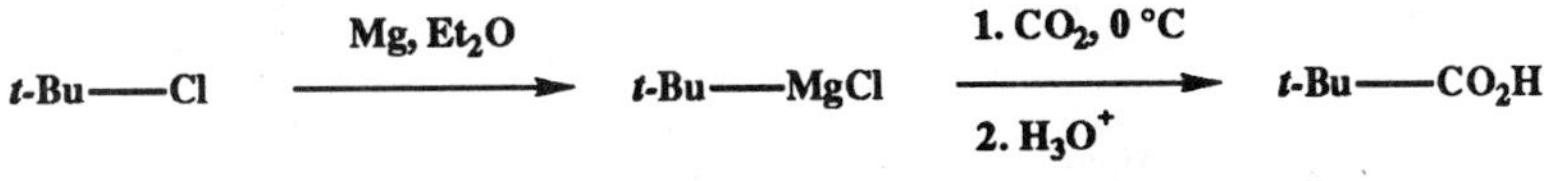

t-Bu—Cl → [Mg, Et_2O] → *t*-Bu—MgCl → [1. CO_2, 0 °C; 2. H_3O^+] → *t*-Bu—CO_2H

OS, CV 2, 8 (1943)

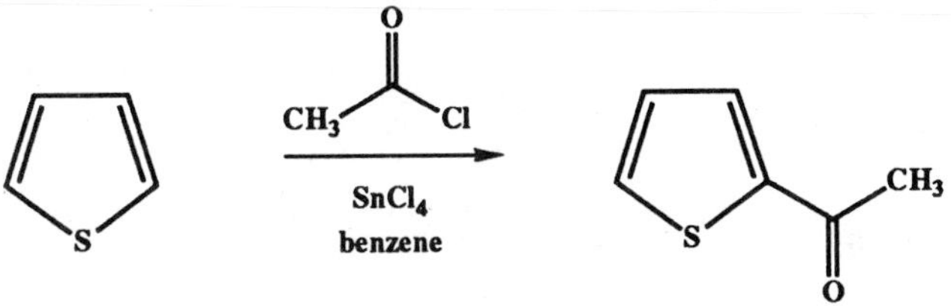

thiophene → [CH_3COCl; $SnCl_4$, benzene] → 2-acetylthiophene (CH_3, O, S)

OS, CV 2, 47 (1943)

$C_6H_5CH_2MgCl$ + 2 p-Me$C_6H_4$$SO_3$-$C_4H_9$ → [Et_2O] → $C_6H_5CH_2C_4H_9$ + $(p\text{-}MeC_6H_4SO_3)_2Mg$ + C_4H_9Cl

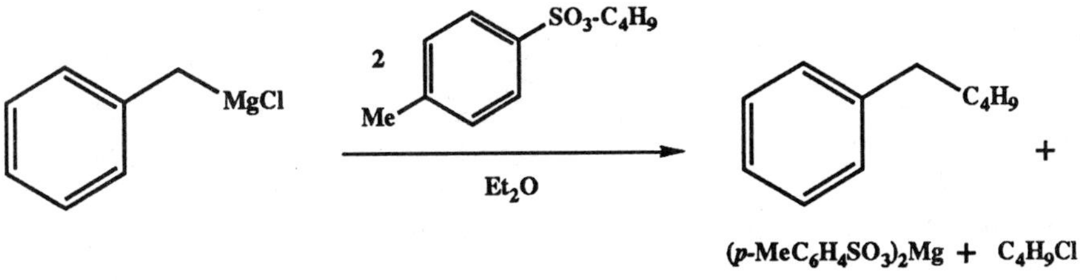

***OS, CV 2,* 179 (1943)**

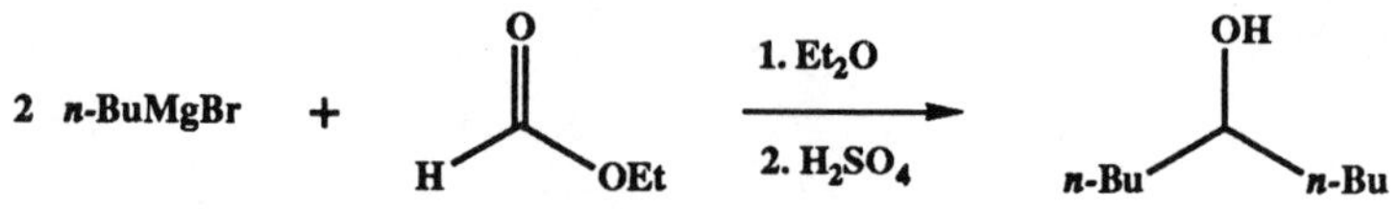

2 *n*-BuMgBr + H–C(=O)–OEt → (1. Et_2O; 2. H_2SO_4) *n*-Bu–CH(OH)–*n*-Bu

***OS, CV 2,* 198 (1943)**

2,4-dimethylpyrrole (Me, Me, H, N–H) → (EtMgBr; Et_2O, Δ) [N-MgBr pyrrole ⇌ 2-MgBr pyrrole (N–H)] → ($ClCO_2Et$; Et_2O, Δ) Me, Me, CO_2Et pyrrole (N–H)

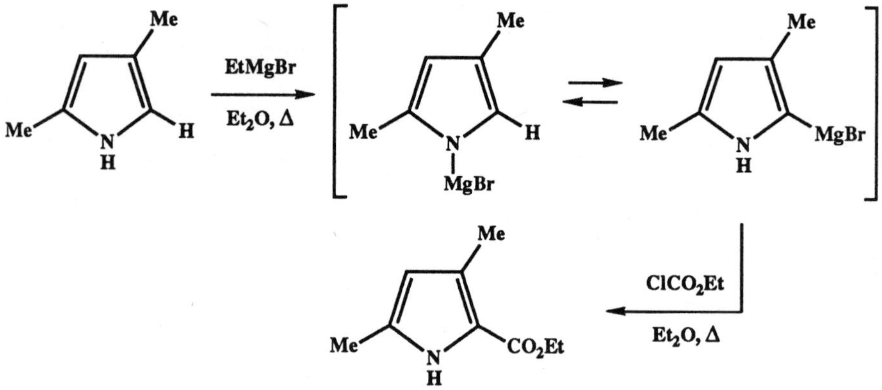

***OS, CV 2,* 262 (1943)**

CH_3–C(=O)–CH_2–C(=O)–OEt → (1. Na, EtOH; 2. $ClCH_2CO_2Et$, Δ) CH_3–C(=O)–CH(CH_2CO_2Et)–C(=O)–OEt

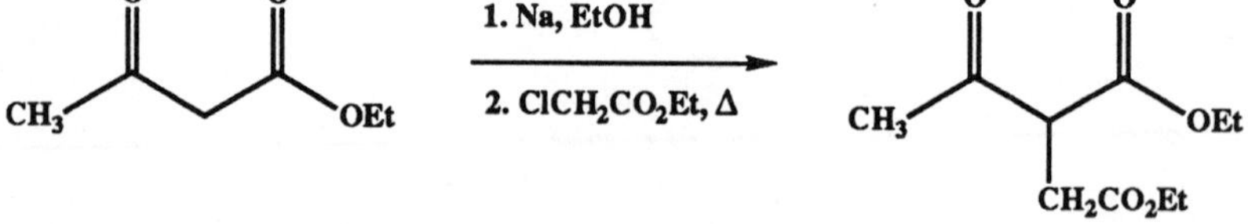

OS, CV 2, 266 (1943)

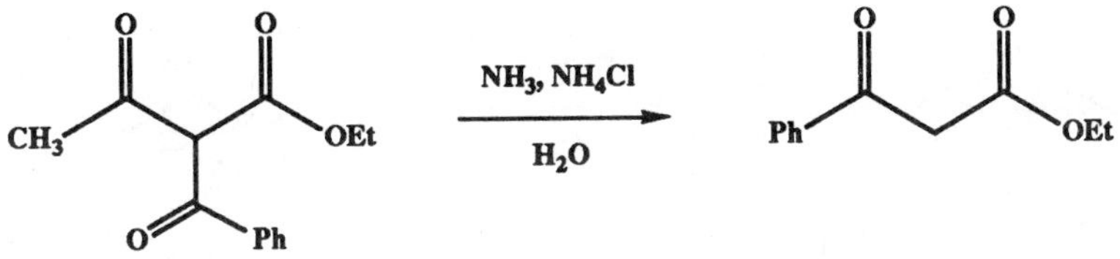

O O
CH_3 OEt
O Ph

NH_3, NH_4Cl

H_2O

O O
Ph OEt

OS, CV 2, 268 (1943)

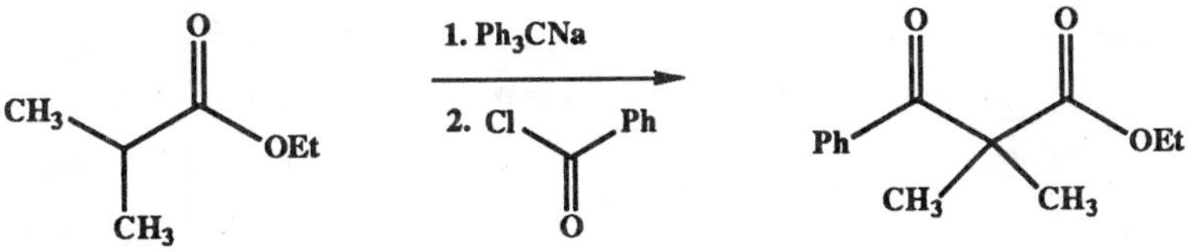

O
CH_3 OEt
CH_3

1. Ph_3CNa

2. Cl Ph
O

O O
Ph OEt
CH_3 CH_3

OS, CV 2, 279 (1943)

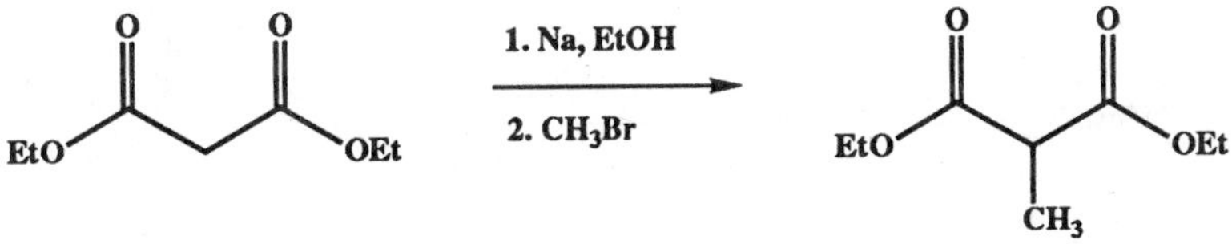

O O
EtO OEt

1. Na, EtOH

2. CH_3Br

O O
EtO OEt
CH_3

OS, CV 2, 312 (1943)

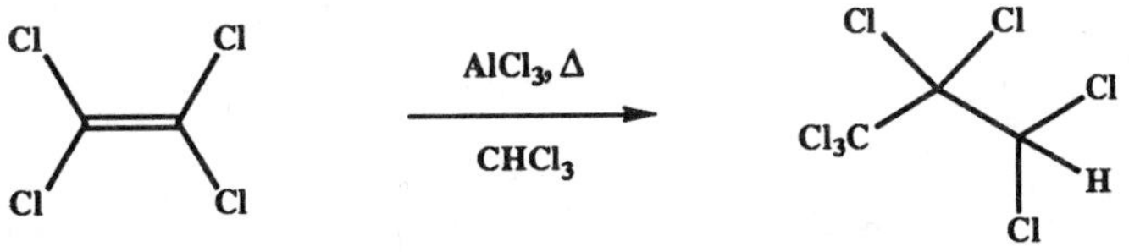

Cl Cl
Cl Cl

$AlCl_3$, Δ

$CHCl_3$

Cl Cl
Cl
Cl_3C
H
Cl

***OS*, *CV* 2, 323 (1943)**

MgBr (n-pentylmagnesium bromide) → $CH(OEt)_3$, Et_2O, Δ → hexanal diethyl acetal (OEt, OEt)

***OS*, *CV* 2, 384 (1943)**

N-phthalimido-$\bar{C}(CO_2Et)_2$ Na^+ → $ClCH_2CH_2SCH_3$, 165 °C → N-phthalimido-$C(CO_2Et)_2CH_2CH_2SCH_3$

***OS*, *CV* 2, 389 (1943)**

$PhCH_2CO_2H$ + CH_3CO_2H → ThO_2 (cat.), 430-450 °C → $PhCH_2C(O)CH_3$

***OS*, *CV* 2, 406 (1943)**

$HC(O)CH_3$ → 1. *i*-PrMgBr, Et_2O; 2. H_2O → *i*-Pr–CH(OH)–CH_3

OS, CV 2, 474 (1943)

EtO, O, O, OEt; $C_7H_{15}Br$, Δ; Na, *n*-BuOH; EtO, O, O, OEt, C_7H_{15}

OS, CV 2, 520 (1943)

S; Cl, O, Ph; $AlCl_3$, CS_2; S, Ph, O

OS, CV 2, 594 (1943)

CO_2Et, CO_2Et; Mg, EtOH; CCl_4, Δ; EtOMg, CO_2Et, CO_2Et; 1. $ClCO_2Et$, Et_2O; 2. HOAc; EtO_2C, CO_2Et, CO_2Et

OS, CV 2, 596 (1943)

CO_2Me, CO_2Me; 1. Na, xylene; 2. $ClCO_2Me$, Δ; MeO_2C, CO_2Me, CO_2Me

OS, CV 2, **602 (1943)**

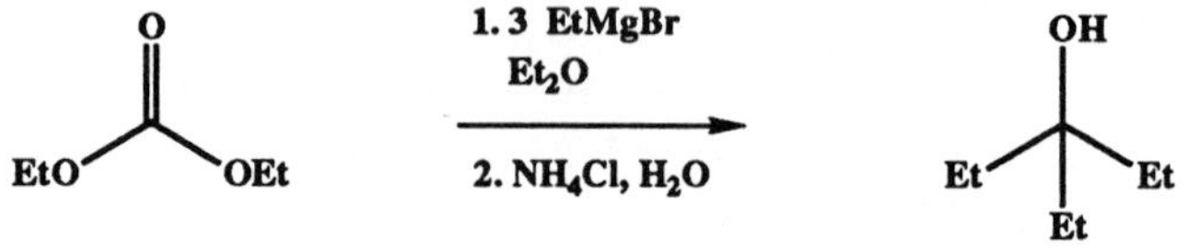

O

EtO OEt

1. 3 EtMgBr
Et_2O

2. NH_4Cl, H_2O

OH

Et Et

Et

OS, CV 2, **606 (1943)**

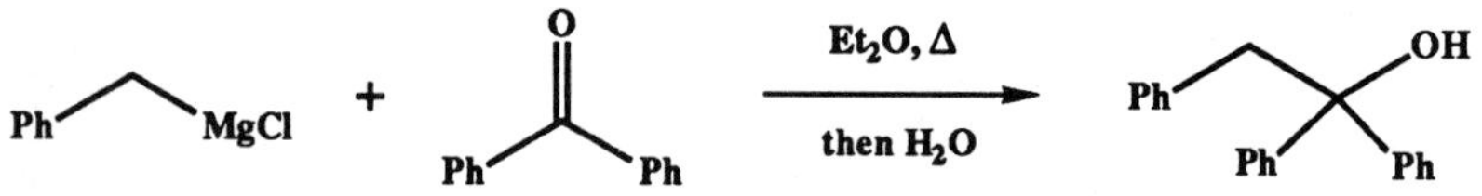

Ph MgCl +

O

Ph Ph

Et_2O, Δ

then H_2O

Ph OH

Ph Ph

OS, CV 3, **14 (1955)**

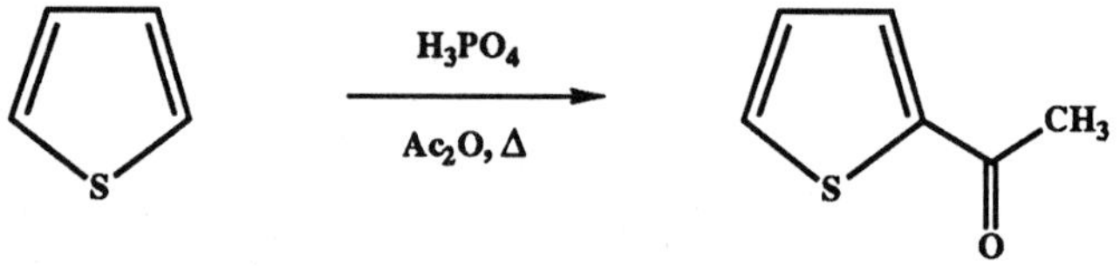

S

H_3PO_4

Ac_2O, Δ

S CH_3

O

OS, CV 3, **26 (1955)**

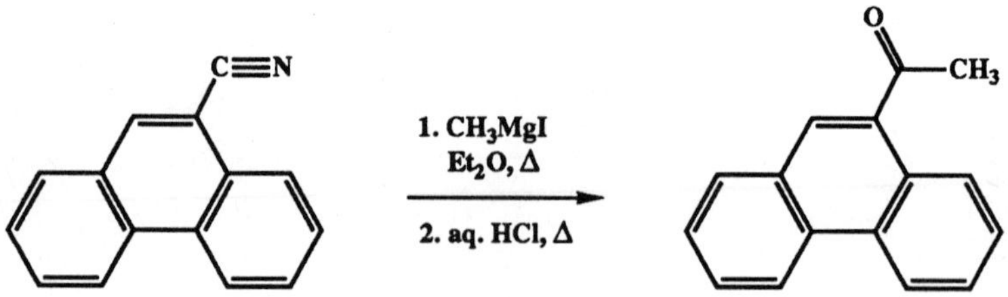

C≡N

1. CH_3MgI
Et_2O, Δ

2. aq. HCl, Δ

O

CH_3

OS, CV 3, **44 (1955)**

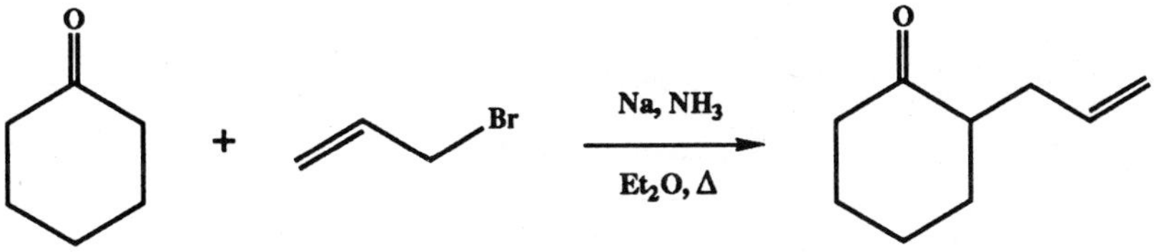

Na, NH_3

Et_2O, Δ

OS, CV 3, **119 (1955)**

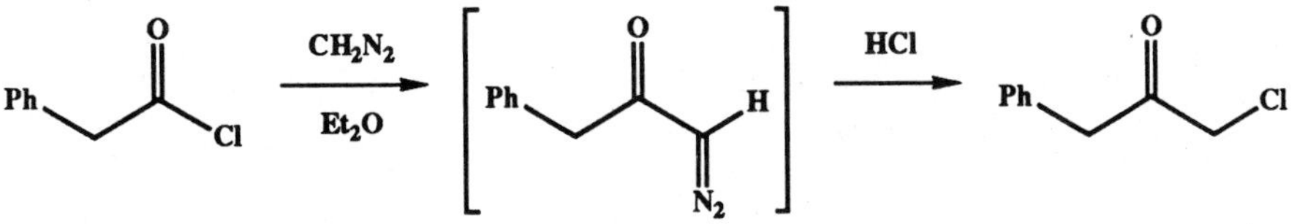

CH_2N_2

Et_2O

HCl

OS, CV 3, **197 (1955)**

$CH_2{=}O$

HCl, 0 °C

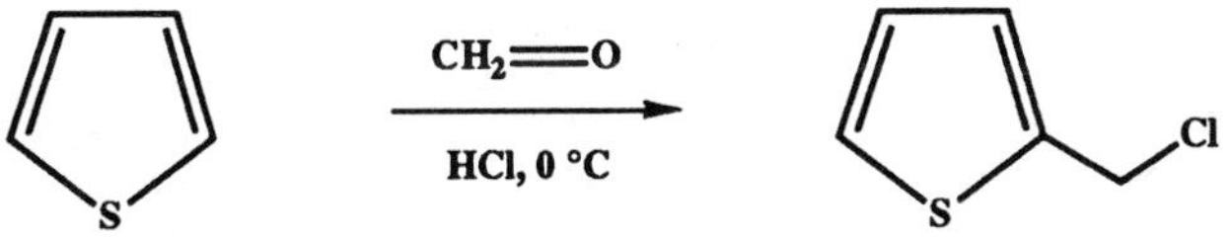

OS, CV 3, **219 (1955)**

1. Na, NH_3

2. $C_6H_{11}Br$
toluene, Δ

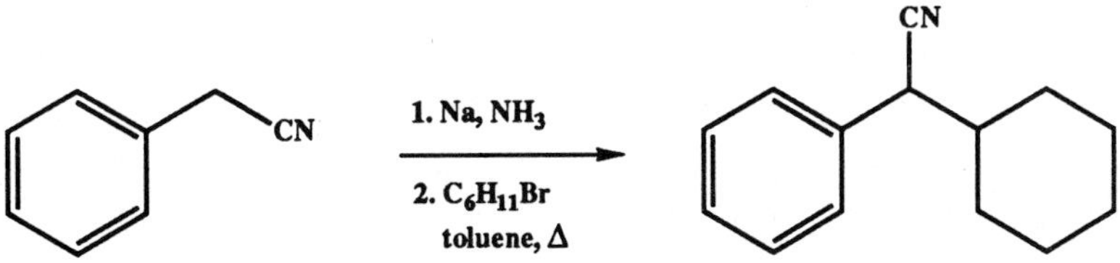

OS, CV 3, 237 (1955)

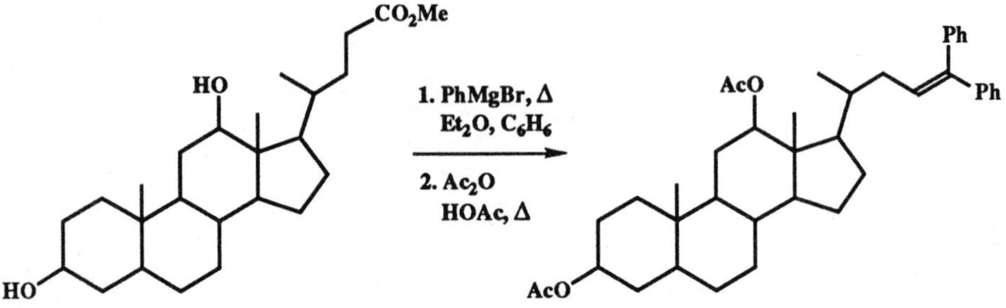

1. PhMgBr, Δ
   Et_2O, C_6H_6

2. Ac_2O
   HOAc, Δ

HO, HO, CO_2Me → AcO, AcO, Ph, Ph

OS, CV 3, 286 (1955)

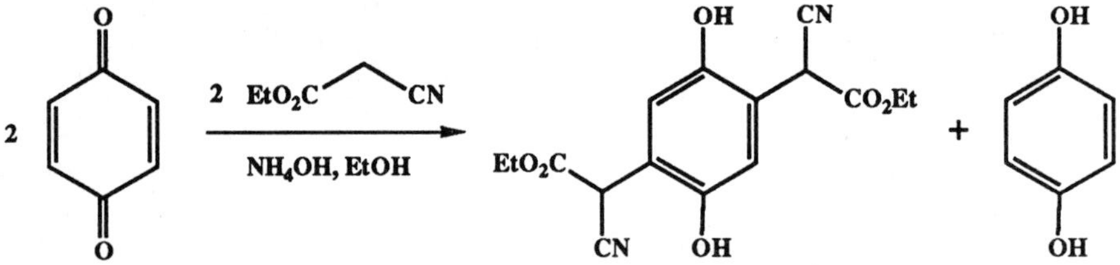

2 (quinone) + 2 EtO_2C–CH_2–CN, NH_4OH, EtOH → OH, CN, CO_2Et, EtO_2C, CN, OH + OH, OH

OS, CV 3, 320 (1955)

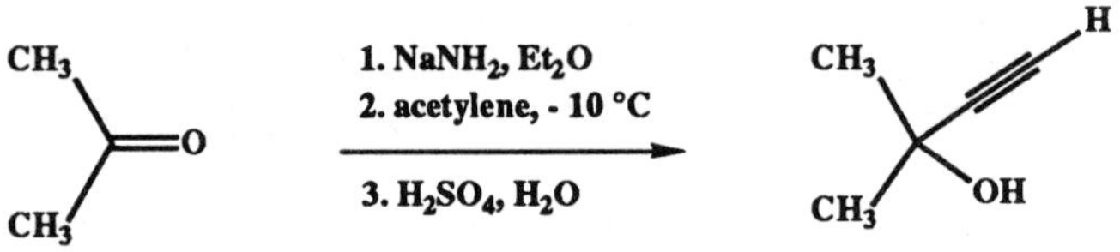

CH_3, CH_3, O

1. $NaNH_2$, Et_2O
2. acetylene, - 10 °C
3. H_2SO_4, H_2O

→ CH_3, CH_3, OH, H

OS, CV 3, 385 (1955)

n-C_3H_7–CHO (O, H) + CO_2Et, CN

H_2 (15 - 30 psi)
Pd-C
piperidine
HOAc

→ n-C_3H_7, CO_2Et, CN

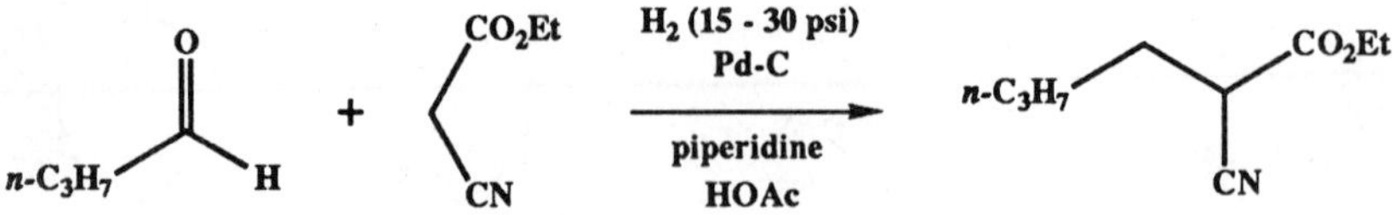

***OS*, *CV* 3, 390 (1955)**

$CH_3COCH_2CO_2Et$ + CH_3COCl → (Mg, C_6H_6, Δ) → $CH_3COCH(COCH_3)CO_2Et$

***OS*, *CV* 3, 397 (1955)**

$(CH_3CH_2)_2C{=}C(CN)CO_2Et$ → (1. Na, EtOH, - 5 °C; 2. CH_3I, Δ) → $CH_3CH{=}C(CH_2CH_3)C(CN)(CH_3)CO_2Et$

***OS*, *CV* 3, 405 (1955)**

$CH_3COCH_2CO_2Et$ → (*i*-PrOH, BF_3) → $CH_3COCH(i\text{-}Pr)CO_2Et$

***OS*, *CV* 3, 408 (1955)**

PhCHO + $BrCH_2CO_2Et$ → (1. Zn, Et_2O, benzene, Δ; 2. H_2SO_4, H_2O) → $PhCH(OH)CH_2CO_2Et$

OS, CV 3, 413 (1955)

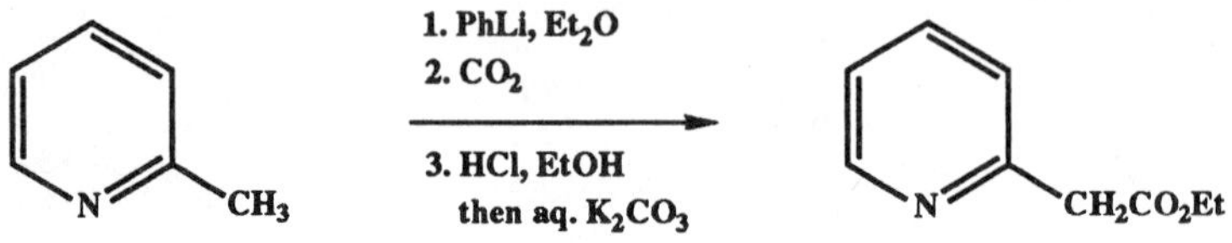

1. PhLi, Et_2O
2. CO_2
3. HCl, EtOH then aq. K_2CO_3

N CH₃ → N CH_2CO_2Et

OS, CV 3, 416 (1955)

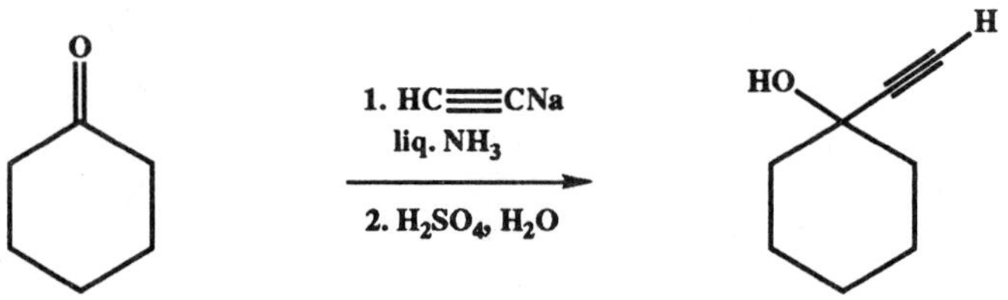

O

1. HC≡CNa liq. NH_3
2. H_2SO_4, H_2O

HO H

OS, CV 3, 495 (1955)

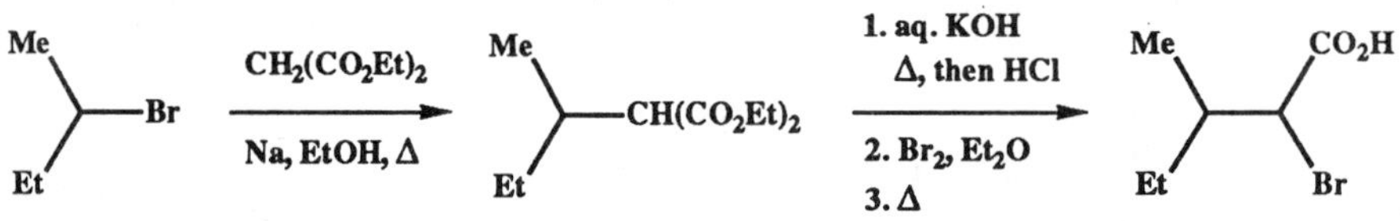

Me Br Et

$CH_2(CO_2Et)_2$
Na, EtOH, Δ

Me $CH(CO_2Et)_2$ Et

1. aq. KOH Δ, then HCl
2. Br_2, Et_2O
3. Δ

Me CO_2H Et Br

OS, CV 3, 591 (1955)

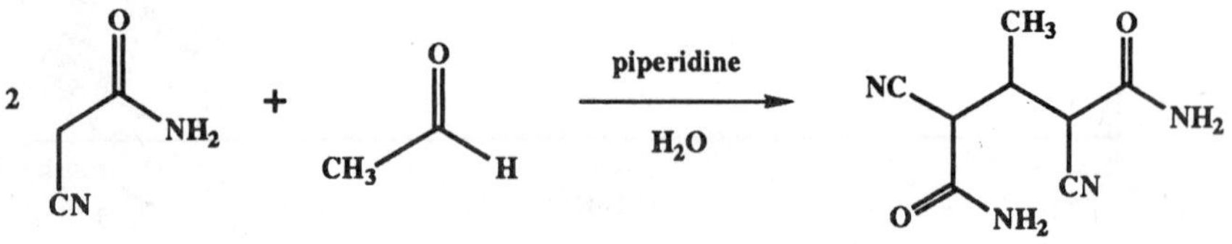

2 O NH_2 CN + O CH_3 H

piperidine
H_2O

CH_3 O NC NH_2 CN O NH_2

OS, CV 3, 601 (1955)

$CH_3CH(CH_3)CH_2CH_2Br$ → [1. Mg, Et_2O, Δ; 2. $CdCl_2$, Et_2O, Δ; 3. $ClC(O)CH_2CH_2CO_2Me$] → $CH_3CH(CH_3)CH_2CH_2C(O)CH_2CH_2CO_2Me$

OS, CV 3, 696 (1955)

$CH_3CH{=}CH{-}C(O){-}H$ → [1. CH_3MgCl, Et_2O; 2. NH_4Cl, H_2O] → $CH_3CH{=}CH{-}C(OH)(CH_3){-}H$

OS, CV 3, 705 (1955)

$PhCH_2Cl$ + $CH_2(CO_2Et)_2$ → [Na, EtOH, Δ] → $PhCH_2CH(CO_2Et)_2$

OS, CV 3, 727 (1955)

$PhC(O)CH_3$ + $ClCH_2CO_2Et$ → [$NaNH_2$, benzene] → epoxide: $C(CH_3)(Ph){-}O{-}CH(CO_2Et)$

OS, CV 3, 729 (1955)

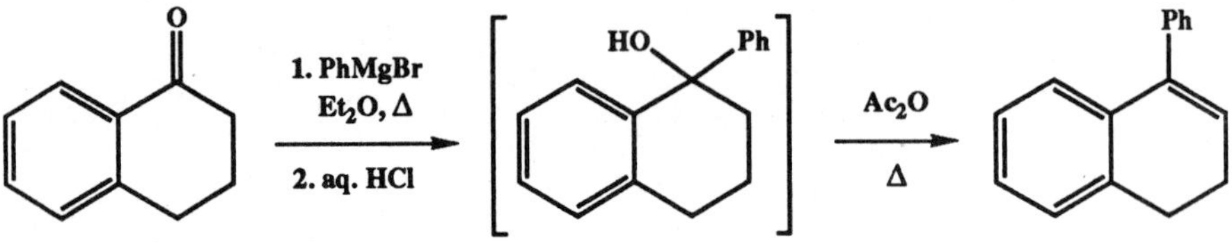

1. PhMgBr
Et_2O, Δ
2. aq. HCl

HO Ph

Ac_2O
Δ

Ph

OS, CV 3, 757 (1955)

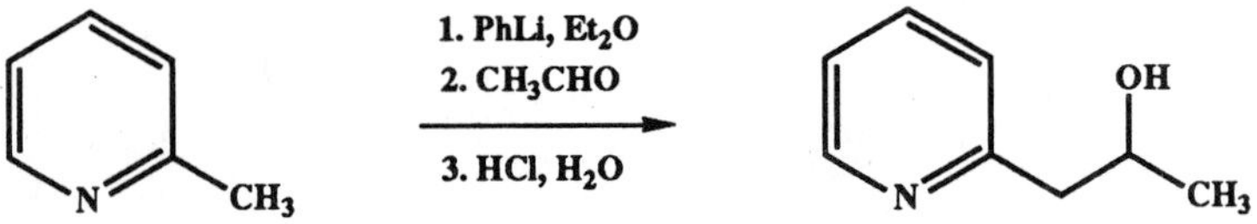

N CH_3

1. PhLi, Et_2O
2. CH_3CHO
3. HCl, H_2O

N OH CH_3

OS, CV 3, 831 (1955)

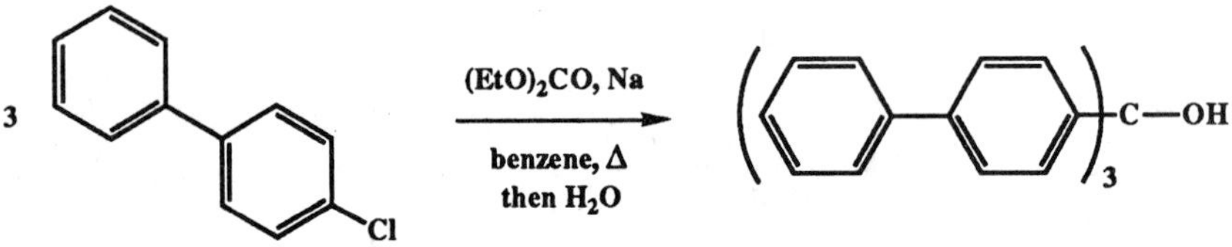

3 Cl

$(EtO)_2CO$, Na
benzene, Δ
then H_2O

C—OH
3

OS, CV 3, 839 (1955)

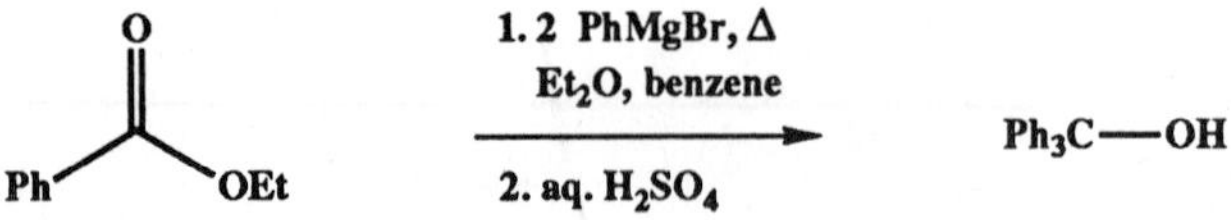

O
Ph OEt

1. 2 PhMgBr, Δ
Et_2O, benzene
2. aq. H_2SO_4

Ph_3C—OH

C - C BOND FORMATION - GENERAL

OS, *CV* 4, 5 (1963)

$CH_3CH(NH_2)CO_2H$ —[$(CH_3CO)_2O$ / pyridine, Δ]→ $CH_3CH(NHCOCH_3)COCH_3$

OS, *CV* 4, 55 (1963)

[Na⁺ phthalimido-$\bar{C}(CO_2Et)_2$] —[$ClCH_2CO_2Et$ / Δ]→ phthalimido-$C(CO_2Et)_2CH_2CO_2Et$

OS, *CV* 4, 93 (1963)

Me(Et)C=O —[EtO_2CCH_2CN / β - alanine, Δ / HOAc, benzene]→ Me(Et)C=C(CO_2Et)(CN) —[1. $PhCH_2MgCl$ / 2. aq. H_2SO_4]→ Me(Et)($PhCH_2$)C–CH(CO_2Et)(CN)

OS, *CV* 4, 117 (1963)

n-Bu—Br —[H—≡—H / Na, NH_3]→ *n*-Bu—≡—H

OS, CV 4, **120 (1963)**

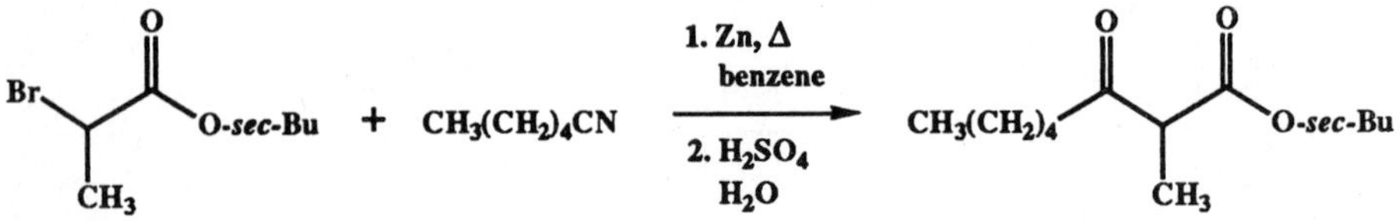

Br, O, O-*sec*-Bu, CH_3 + $CH_3(CH_2)_4CN$ → 1. Zn, Δ benzene; 2. H_2SO_4 H_2O → $CH_3(CH_2)_4$, O, O, O-*sec*-Bu, CH_3

OS, CV 4, **221 (1963)**

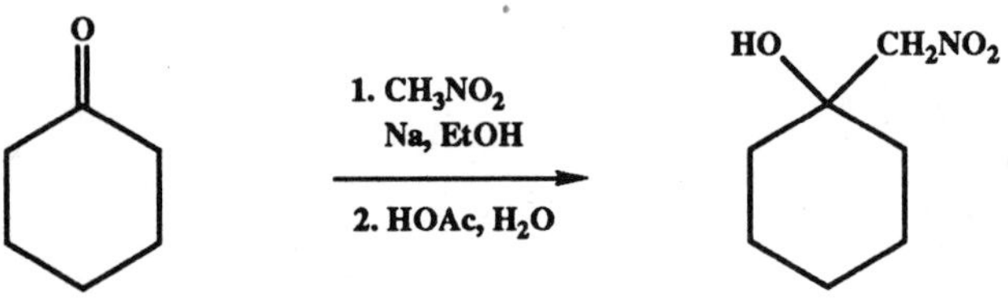

O → 1. CH_3NO_2 Na, EtOH; 2. HOAc, H_2O → HO, CH_2NO_2

OS, CV 4, **281 (1963)**

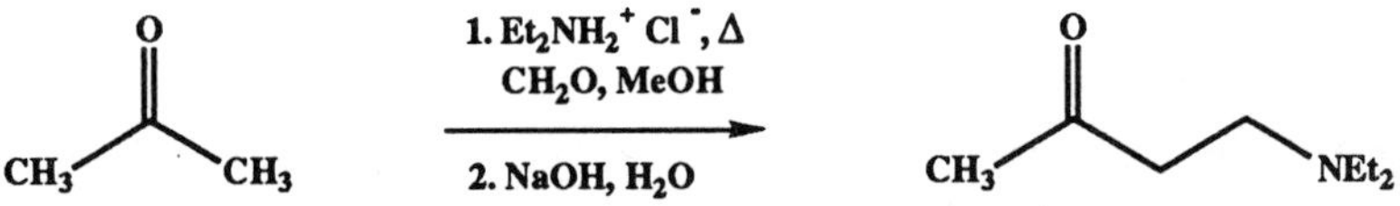

O, CH_3, CH_3 → 1. $Et_2NH_2^+$ Cl^-, Δ CH_2O, MeOH; 2. NaOH, H_2O → O, CH_3, NEt_2

OS, CV 4, **291 (1963)**

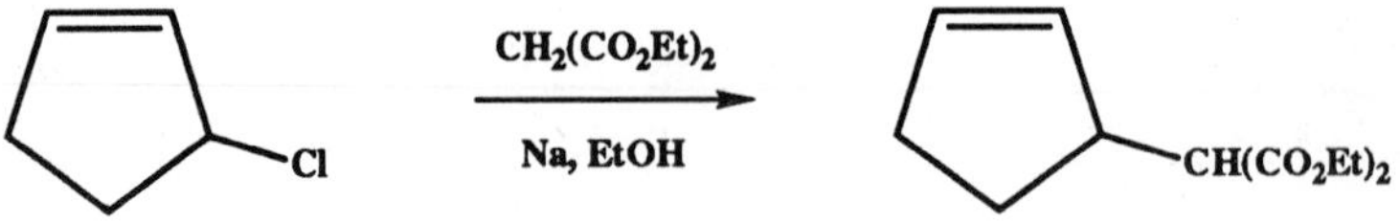

Cl → $CH_2(CO_2Et)_2$; Na, EtOH → $CH(CO_2Et)_2$

C - C BOND FORMATION - GENERAL

***OS*, *CV* 4, 415 (1963)**

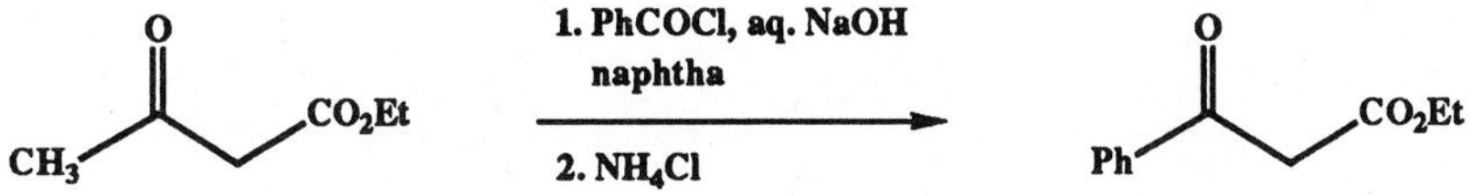

$CH_3COCH_2CO_2Et$ → 1. PhCOCl, aq. NaOH, naphtha; 2. NH_4Cl → $PhCOCH_2CO_2Et$

***OS*, *CV* 4, 444 (1963)**

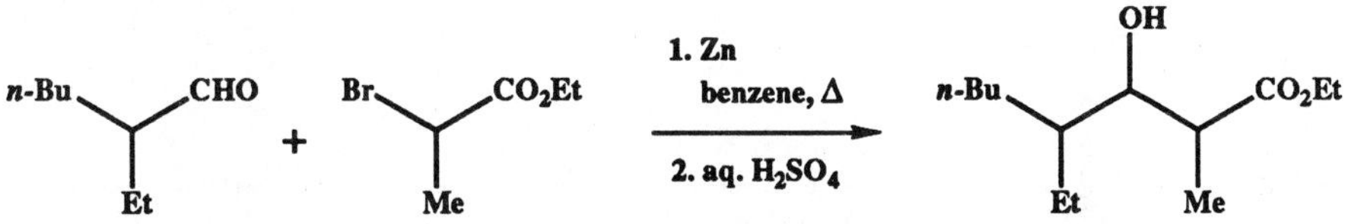

n-Bu–CH(Et)–CHO + Br–CH(Me)–CO_2Et → 1. Zn, benzene, Δ; 2. aq. H_2SO_4 → *n*-Bu–CH(Et)–CH(OH)–CH(Me)–CO_2Et

***OS*, *CV* 4, 459 (1963)**

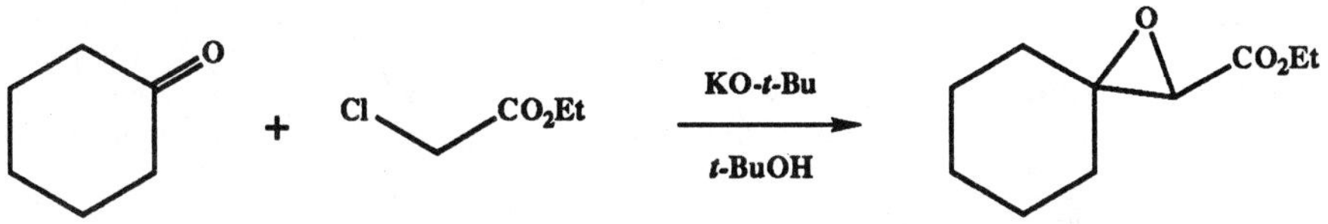

cyclohexanone + $ClCH_2CO_2Et$ → KO-*t*-Bu, *t*-BuOH → spiro epoxide (cyclohexane–oxirane)–CO_2Et

***OS*, *CV* 4, 471 (1963)**

2 cyclohexanone → 1. acetylene, KOH, CaC_2, benzene; 2. aq. HCl, 0 °C → 1,1'-ethynylenebis(cyclohexanol) (OH, HO)

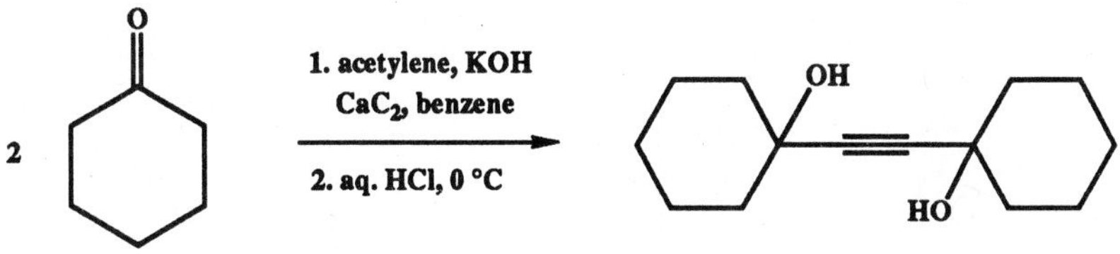

OS, CV 4, 539 (1963)

DMF, $POCl_3$

NaOH, H_2O

CHO

N H

OS, CV 4, 555 (1963)

2 $MeO_2C-(CH_2)_4-C(=O)-Cl$

1. Et_3N, benzene

2. aq. KOH, Δ then HCl

$HO_2C-(CH_2)_4-C(=O)-(CH_2)_4-CO_2H$

OS, CV 4, 560 (1963)

2 $C_{10}H_{21}$ Cl O

2 Et_3N

Et_2O

$C_{10}H_{21}$ O C O $C_{10}H_{21}$

OS, CV 4, 601 (1963)

$n-C_5H_{11}$ O O

1. 2 CH_3MgBr Et_2O, Δ

2. HCl, H_2O benzene

$n-C_5H_{11}$ OH CH_3 OH CH_3

***OS*, *CV* 4, 605 (1963)**

Ph—CH=N—Me → 1. $PhCH_2MgCl$, Δ, Et_2O, benzene; 2. HCl, H_2O, then HCl (g) → Ph–CH(CH_2Ph)–NH–Me • HCl

***OS*, *CV* 4, 623 (1963)**

Na, MeOH; 220 °C

Me

***OS*, *CV* 4, 626 (1963)**

CH_3 … O → CH_2=O, Me_2NH; HOAc, H_2O → CH_3 … O … NMe_2

***OS*, *CV* 4, 641 (1963)**

N → PhCOCl; KCN, H_2O → N, Ph, CN, O

PhLi, CH_3I; Et_2O, dioxane, - 10 °C → N, Ph, CH_3, CN, O

***OS, CV 4*, 649 (1963)**

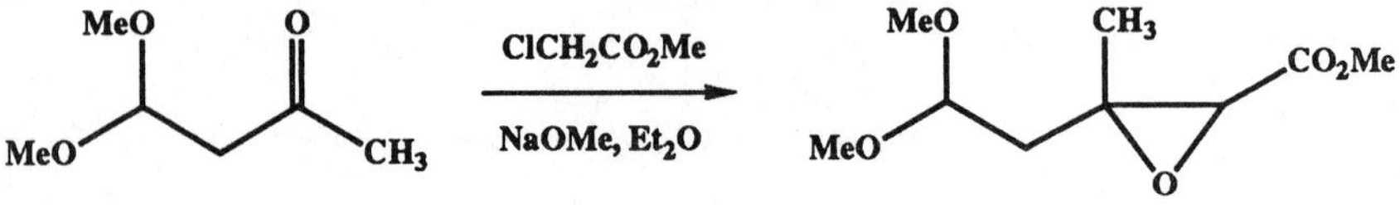

MeO, MeO, O, CH_3 → $ClCH_2CO_2Me$ / NaOMe, Et_2O → MeO, MeO, CH_3, CO_2Me, O

***OS, CV 4*, 708 (1963)**

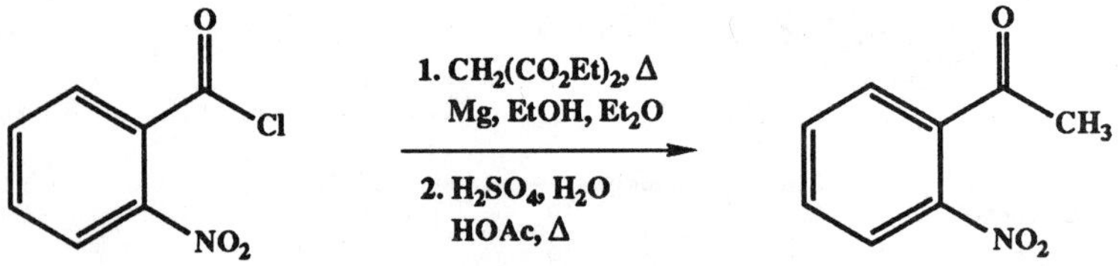

O, Cl, NO_2 → 1. $CH_2(CO_2Et)_2$, Δ Mg, EtOH, Et_2O / 2. H_2SO_4, H_2O HOAc, Δ → O, CH_3, NO_2

***OS, CV 4*, 748 (1963)**

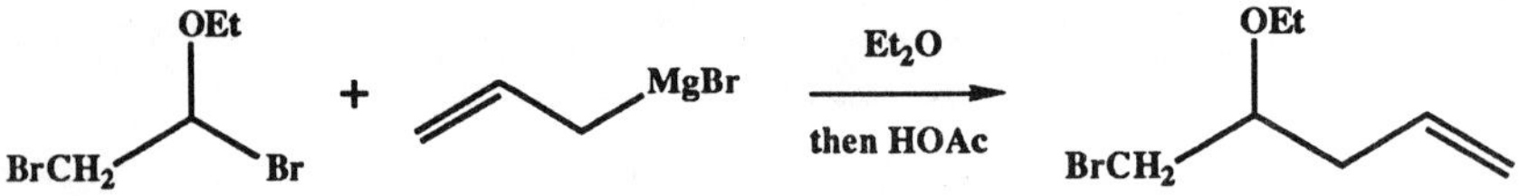

OEt, $BrCH_2$, Br + MgBr → Et_2O / then HOAc → OEt, $BrCH_2$

***OS, CV 4*, 766 (1963)**

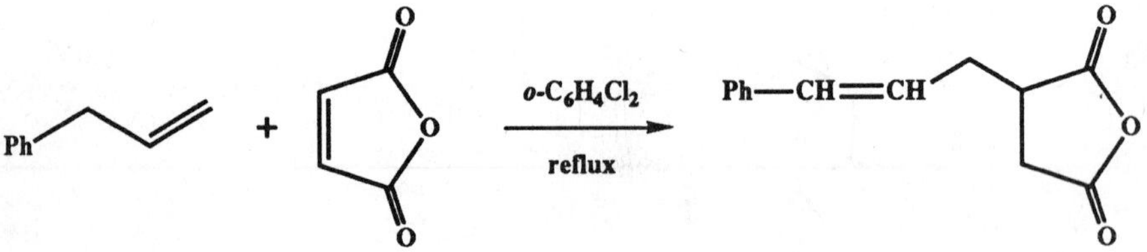

Ph + O, O, O → o-$C_6H_4Cl_2$ / reflux → Ph—CH=CH, O, O, O

OS, CV 4, 776 (1963)

$$\mathrm{Ph{-}C(CN)(H){-}CO_2Et} \xrightarrow[t\text{-BuOH, KOH, MeOH}]{\mathrm{CH_2{=}CHCN}} \mathrm{Ph{-}C(CN)(CH_2CH_2CN){-}CO_2Et}$$

OS, CV 4, 792 (1963)

$$\mathrm{C_6H_5{-}CH{=}CH{-}CHO} \xrightarrow[\text{2. } NH_4Cl,\ H_2O]{\text{1. } HC{\equiv}CMgBr,\ THF} \mathrm{C_6H_5{-}CH{=}CH{-}CH(OH){-}C{\equiv}CH}$$

OS, CV 4, 801 (1963)

$$\mathrm{Ph{-}C{\equiv}C{-}H} \xrightarrow[ZnI_2,\ \Delta]{HC(OEt)_3} \mathrm{Ph{-}C{\equiv}C{-}CH(OEt)_2}$$

OS, CV 4, 831 (1963)

$$\text{pyrrole (N–H)} \xrightarrow[\text{ethylene dichloride}]{Me_2NCHO,\ POCl_3} \left[\text{2-pyrrolyl–CH(OPOCl}_2\text{)(NMe}_2\text{)}\right] \xrightarrow[H_2O,\ \Delta]{NaOAc} \text{pyrrole-2-CHO}$$

***OS*, *CV* 4, 854 (1963)**

$$2\ C_{17}H_{35}{-}CO_2H \xrightarrow{MgO} (C_{17}H_{35}CO_2)_2Mg \xrightarrow[(-CO_2)]{340\ °C} C_{17}H_{35}C(=O)C_{17}H_{35}$$

***OS*, *CV* 4, 915 (1963)**

Thiophene $\xrightarrow[POCl_3]{PhN(Me)CHO}$ 2-thiophenecarboxaldehyde (CHO)

***OS*, *CV* 4, 962 (1963)**

$$Ph_2CH{-}CN \xrightarrow[\text{2. } PhCH_2Cl,\ Et_2O]{\text{1. K, } NH_3} Ph_2C(CN)CH_2Ph$$

***OS*, *CV* 4, 980 (1963)**

Thiophene $\xrightarrow[HCl]{CH_3CHO}$ 2-(1-chloroethyl)thiophene ($CH(Cl)CH_3$)

OS, CV 5, 20 (1973)

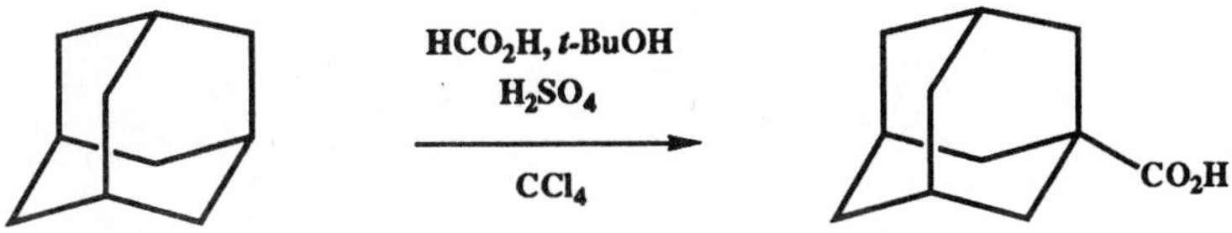

HCO$_2$H, *t*-BuOH
H$_2$SO$_4$
CCl$_4$
CO$_2$H

OS, CV 5, 25 (1973)

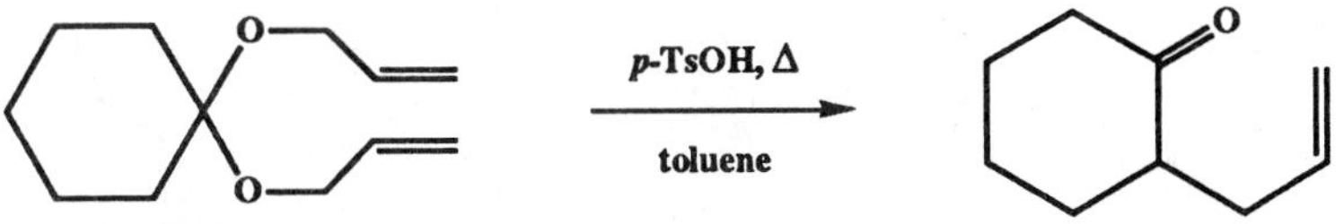

O
O
p-TsOH, Δ
toluene
O

OS, CV 5, 46 (1973)

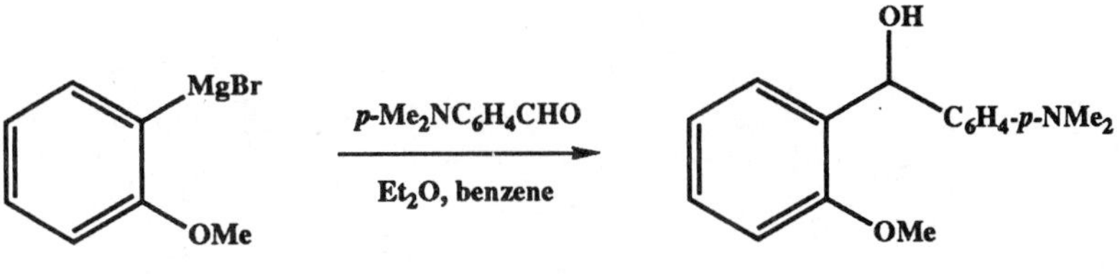

MgBr
OMe
p-Me$_2$NC$_6$H$_4$CHO
Et$_2$O, benzene
OH
C$_6$H$_4$-*p*-NMe$_2$
OMe

OS, CV 5, 76 (1973)

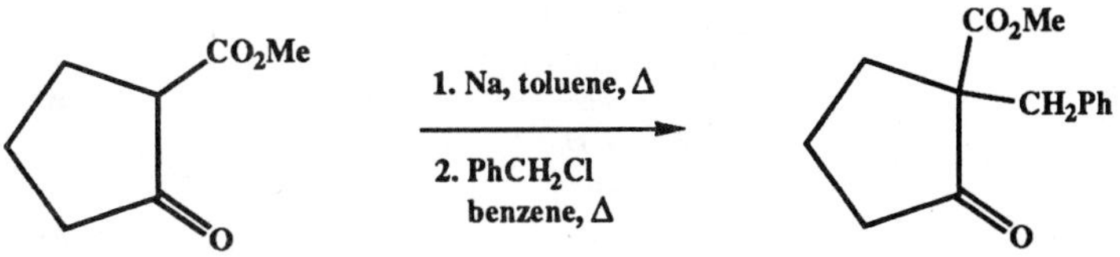

CO$_2$Me
O
1. Na, toluene, Δ
2. PhCH$_2$Cl
benzene, Δ
CO$_2$Me
CH$_2$Ph
O

***OS*, *CV 5*, 187 (1973)**

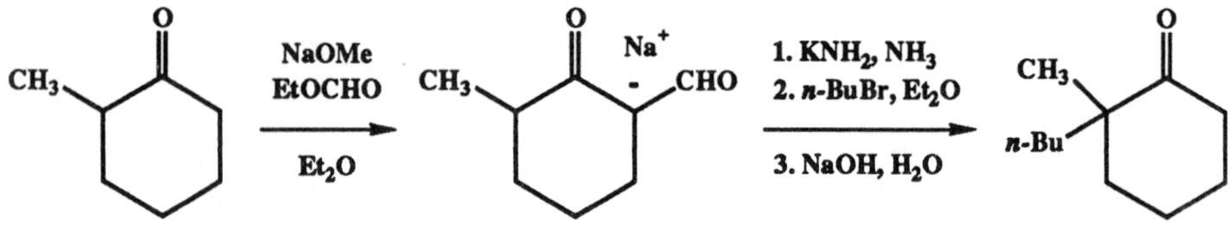

NaOMe, EtOCHO, Et_2O

1. KNH_2, NH_3
2. *n*-BuBr, Et_2O
3. NaOH, H_2O

***OS*, *CV 5*, 215 (1973)**

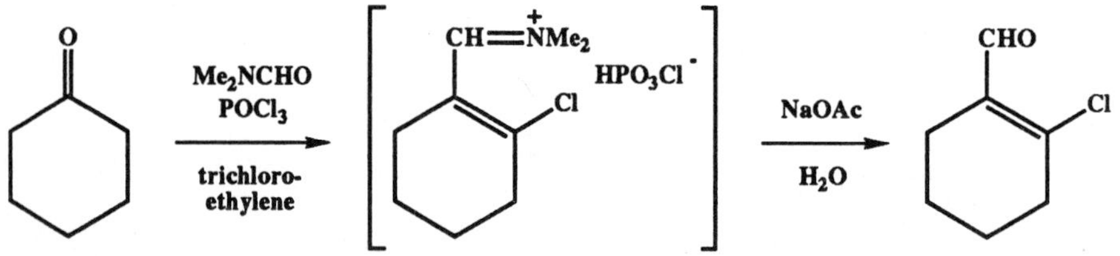

Me_2NCHO, $POCl_3$, trichloroethylene

NaOAc, H_2O

***OS*, *CV 5*, 384 (1973)**

$ClCO(CH_2)_3CO_2Et$

Na, Et_2O

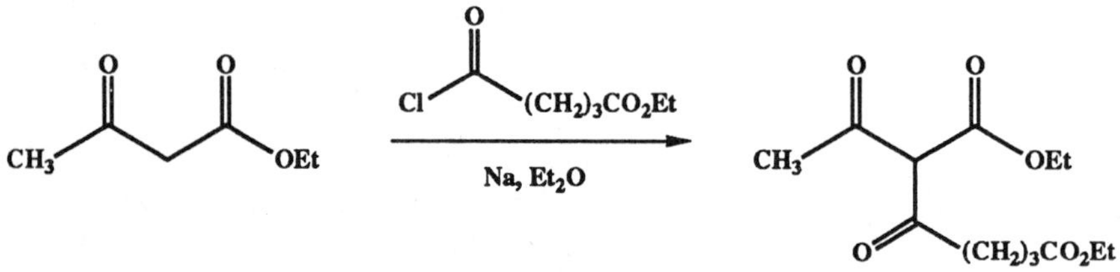

***OS*, *CV 5*, 439 (1973)**

1. $Mg(OMe)_2$, CO_2, DMF, Δ
2. MeOH, HCl, -5 to 5 °C

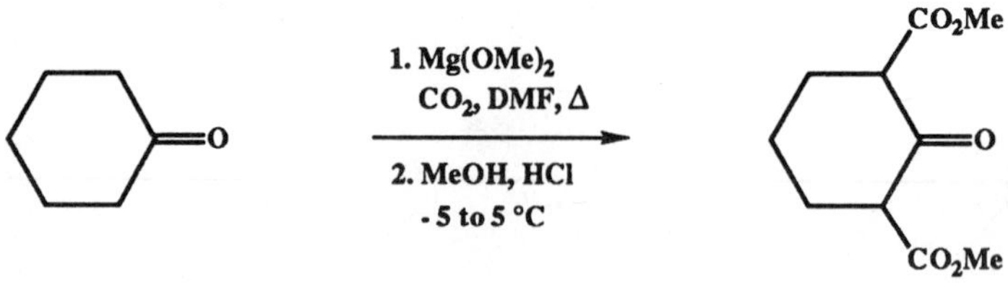

OS, *CV 5*, 452 (1973)

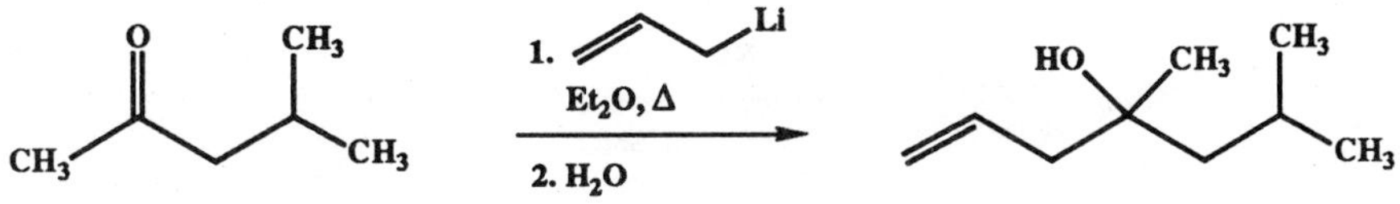

1. allyllithium, Et_2O, Δ
2. H_2O

OS, *CV 5*, 523 (1973)

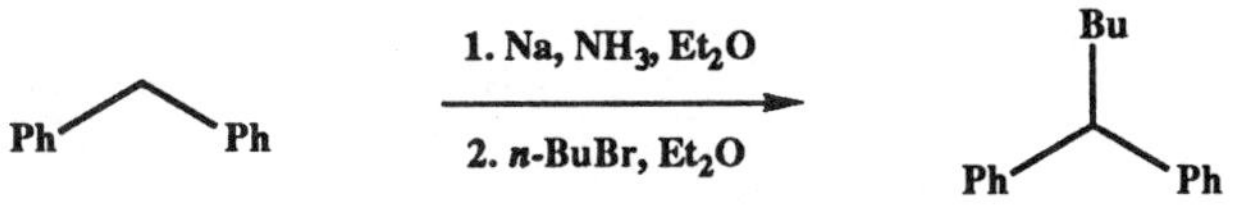

1. Na, NH_3, Et_2O
2. *n*-BuBr, Et_2O

OS, *CV 5*, 526 (1973)

1. 2 Na, NH_3, Et_2O
2. $PhCH_2Cl$, Et_2O then HCl

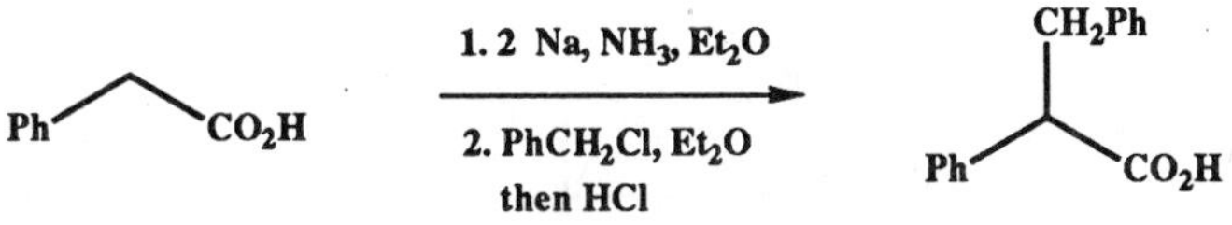

OS, *CV 5*, 533 (1973)

1. $ClCO(CH_2)_8COCl$, Et_3N, $CHCl_3$, 35 °C
2. HCl, H_2O, Δ

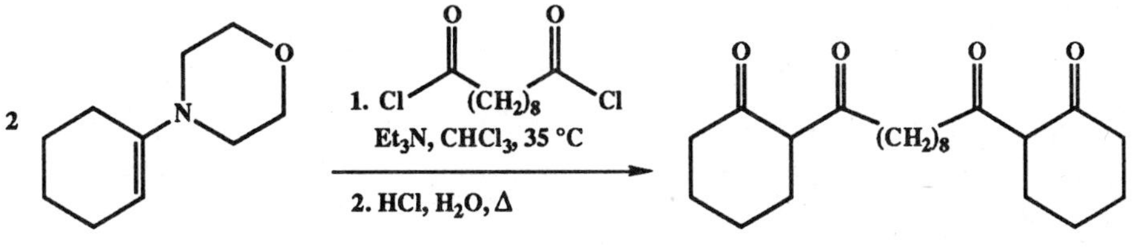

C - C BOND FORMATION - GENERAL

***OS*, *CV 5*, 559 (1973)**

Ph–CH$_2$–CO$_2$Et → (1. Na, NH$_3$; 2. PhCH$_2$CH$_2$Br, Et$_2$O) → Ph–CH(CH$_2$CH$_2$Ph)–CO$_2$Et

***OS*, *CV 5*, 564 (1973)**

Ph–C(=O)–Ph + CH$_3$CO$_2$Et → (1. Li, NH$_3$; 2. NH$_4$Cl) → Ph$_2$C(OH)CH$_2$CO$_2$Et

***OS*, *CV 5*, 572 (1973)**

Et(CH$_3$)C=O + NC–CH$_2$–CO$_2$Et → (KCN, EtOH, HOAc) → [NC–C(Et)(CH$_3$)–CH(CO$_2$Et)–CN] → (aq. HCl, Δ) → HO$_2$C–C(Et)(CH$_3$)–CH$_2$CO$_2$H

***OS*, *CV 5*, 589 (1973)**

2 CH$_3$CH$_2$CH$_2$CO$_2$H → (Fe, Δ) → [(CH$_3$CH$_2$CH$_2$CO$_2$)$_2$Fe] → CH$_3$CH$_2$CH$_2$–C(=O)–CH$_2$CH$_2$CH$_3$

OS, *CV 5*, **608 (1973)**

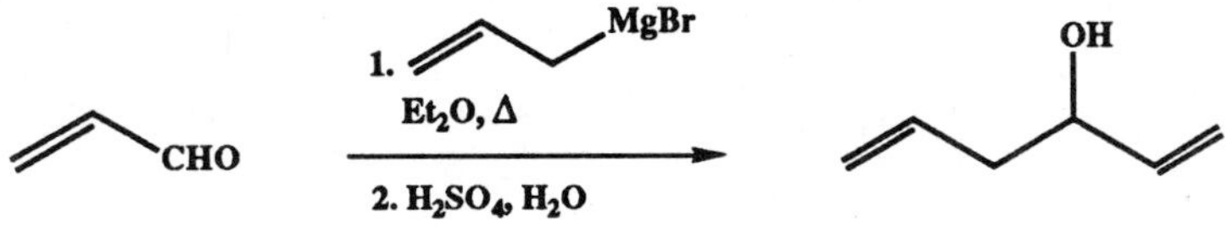

CHO

1. MgBr

Et_2O, Δ

2. H_2SO_4, H_2O

OH

OS, *CV 5*, **654 (1973)**

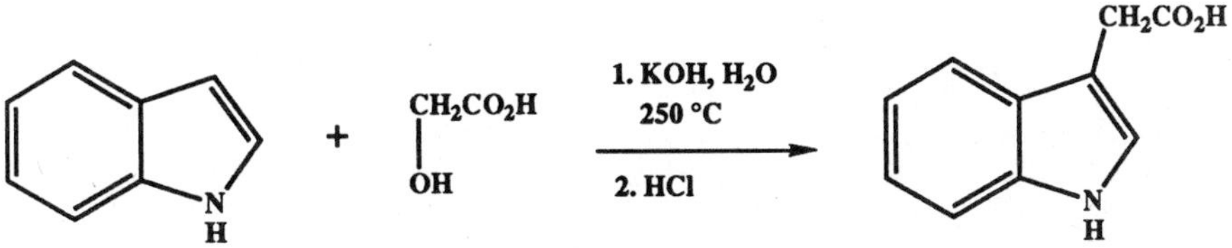

N H + CH_2CO_2H OH

1. KOH, H_2O 250 °C

2. HCl

CH_2CO_2H

N H

OS, *CV 5*, **739 (1973)**

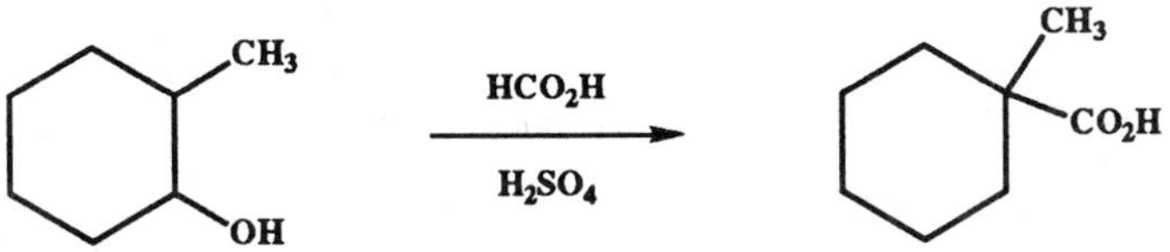

CH_3 OH

HCO_2H

H_2SO_4

CH_3 CO_2H

OS, *CV 5*, **743 (1973)**

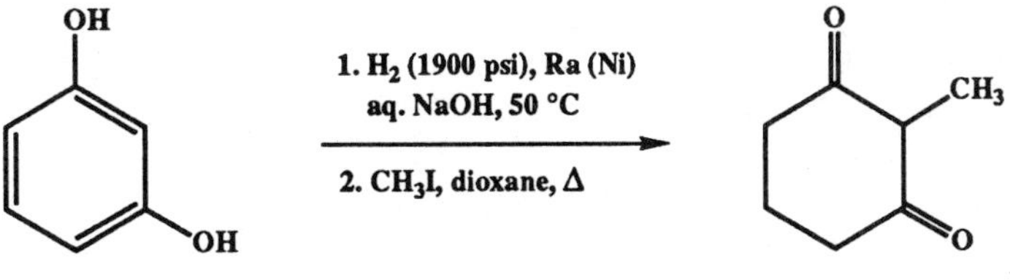

OH OH

1. H_2 (1900 psi), Ra (Ni) aq. NaOH, 50 °C

2. CH_3I, dioxane, Δ

O CH_3 O

***OS, CV 5,* 755 (1973)**

$CH_2{=}S^{+}(O^{-})(CH_3)-CH_3$ + cyclohexanone → DMSO, 60 °C → 1-oxaspiro[2.5]octane

***OS, CV 5,* 762 (1973)**

$CH_3CH{=}CHCO_2$-*sec*-Bu → 1. *n*-BuMgBr, Et_2O, Δ; 2. HCl, H_2O → *n*-Bu–CH(CH_3)–CH_2–CO_2-*sec*-Bu

***OS, CV 5,* 767 (1973)**

$CH_3COCH_2COCH_3$ → $CH_2{=}C(CH_3)CH_2Cl$, K_2CO_3, EtOH, Δ → [$CH_3COCH(CH_2C(CH_3){=}CH_2)COCH_3$] → $CH_3COCH_2CH_2C(CH_3){=}CH_2$

***OS, CV 5,* 775 (1973)**

Cyclohexanecarboxylic acid (OH) → 1. LiH, DME, Δ; 2. CH_3Li, Et_2O; 3. aq. HCl → cyclohexyl methyl ketone (CH_3)

OS, CV 5, 785 (1973)

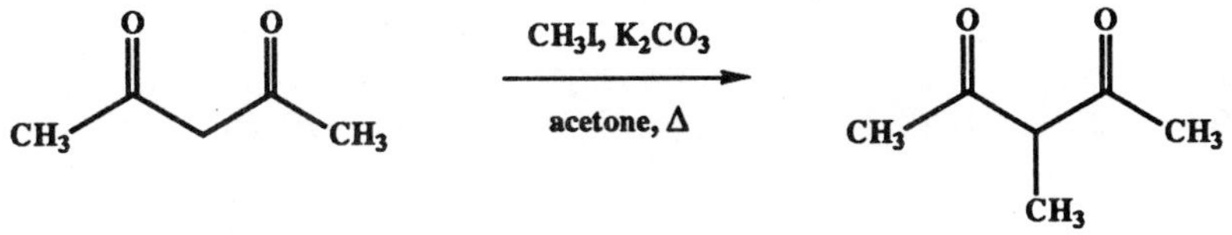

CH_3I, K_2CO_3

acetone, Δ

CH₃ labels: CH_3 CH_3 CH_3 CH_3 CH_3

OS, CV 5, 848 (1973)

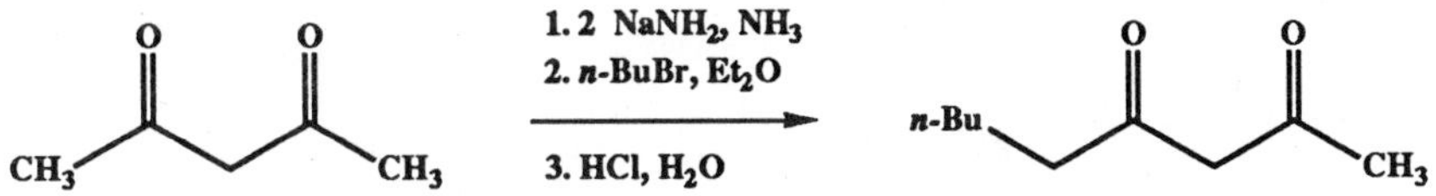

1. 2 $NaNH_2$, NH_3
2. *n*-BuBr, Et_2O
3. HCl, H_2O

n-Bu

OS, CV 5, 893 (1973)

$(EtO)_3P$

pet. ether, 0 °C

Cl Cl Cl Cl Cl Cl → Cl Cl Et Cl Cl Cl

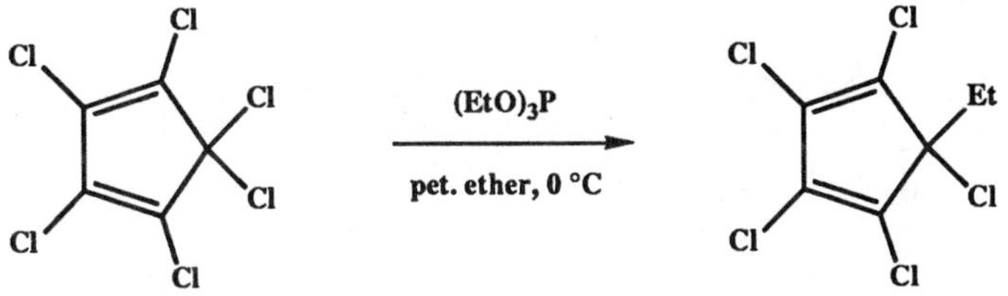

OS, CV 5, 1013 (1973)

NC CN NC CN

$CH_2(CN)_2$

pyridine

H_2O, Δ

$(py\text{-}H)^+$ $^-C(CN)_2$

$Me_4N^+ Cl^-$

H_2O, Δ

Me_4N^+ $^-C(CN)_2$

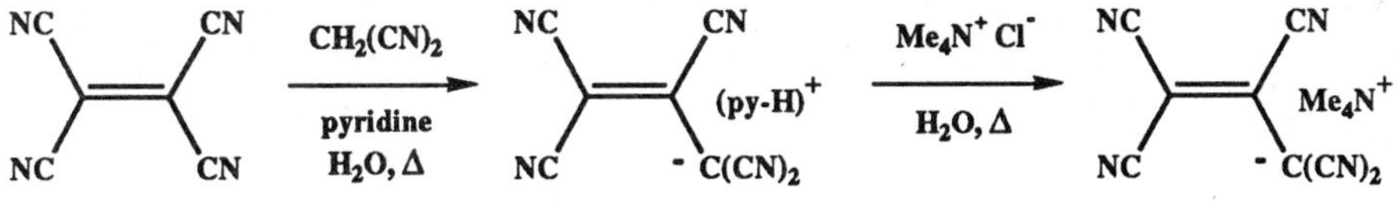

OS, CV 5, **1043 (1973)**

$CH_3-C\equiv C-H$ → (1. Na, NH_3; 2. CO_2, THF, Et_2O; 3. HCl, H_2O) → $CH_3-C\equiv C-CO_2H$

OS, CV 5, **1076 (1973)**

n-C_6H_{13} (alkene) → ($CHCl_3$, $FeCl_3$, benzoin; Et_2NH_4Cl, MeOH, 130 °C) → n-C_6H_{13}–CH(Cl)–CH_2–$CHCl_2$

OS, CV 5, **1092 (1973)**

OH, CH_3, CH_3 → (1. Li, toluene, Δ; 2. CH_3I, 135 °C) → O, CH_3, CH_3, CH_3

OS, CV 6, **51 (1988)**

O, CH_3 → (2 Li, H_2O; NH_3, Et_2O, - 33 °C) → [O^- Li^+, CH_3] → (Br; NH_3, Et_2O, - 33 °C) → O, CH_3

OS, CV 6, **109 (1988)**

Indole + 2-(methylthio)-2-phenyl-1,3-dithiane → ($BF_3 \cdot Et_2O$; $CHCl_3$, 25 °C) 3-(2-phenyl-1,3-dithian-2-yl)indole

OS, CV 6, **121 (1988)**

OS, CV 6, **223 (1988)**

***OS*, *CV* 6, 240 (1988)**

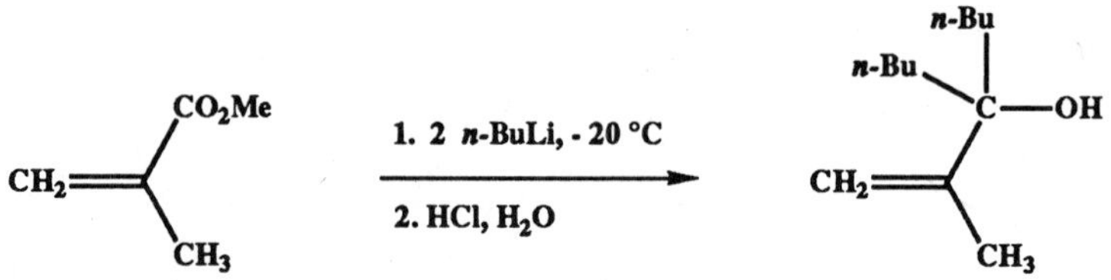

$CH_2=C(CH_3)CO_2Me$ → (1. 2 *n*-BuLi, - 20 °C; 2. HCl, H_2O) → $CH_2=C(CH_3)C(n\text{-}Bu)_2OH$

***OS*, *CV* 6, 242 (1988)**

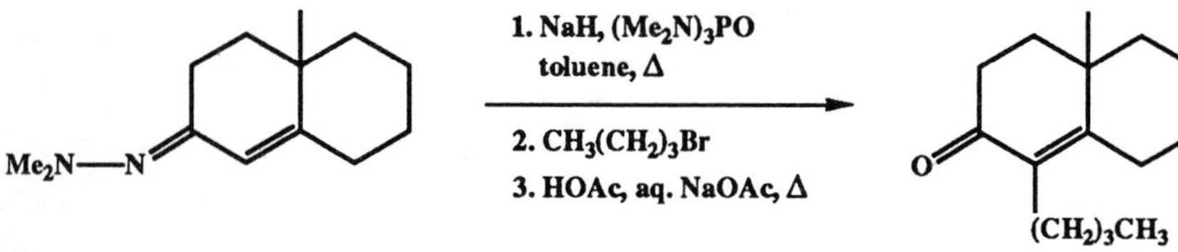

$Me_2N—N=$ (octalin) → (1. NaH, $(Me_2N)_3PO$ toluene, Δ; 2. $CH_3(CH_2)_3Br$; 3. HOAc, aq. NaOAc, Δ) → O= (octalone) with $(CH_2)_3CH_3$

***OS*, *CV* 6, 245 (1988)**

n-BuCH_2COCH_3 → (Ac_2O; *p*-TsOH) → *n*-Bu$CH=C(OAc)CH_3$ → (1. BF_3 • HOAc, Ac_2O; 2. NaOAc, H_2O, Δ) → *n*-Bu$CH(COCH_3)_2$

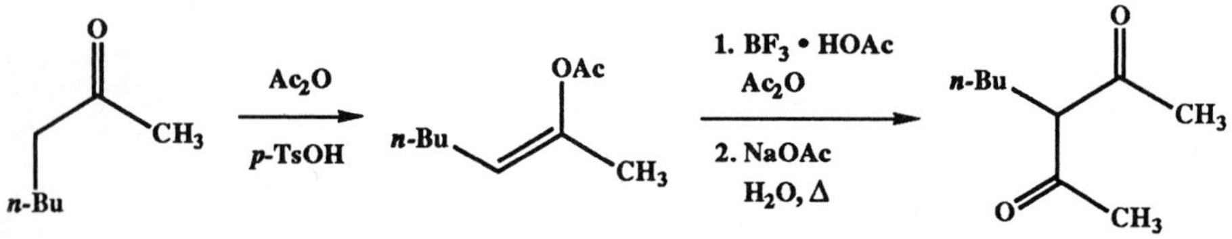

***OS*, *CV* 6, 248 (1988)**

PhS [*t*-Bu] CuLi + ClC(O)Ph → (THF; - 60 to - 65 °C) → *t*-BuC(O)Ph

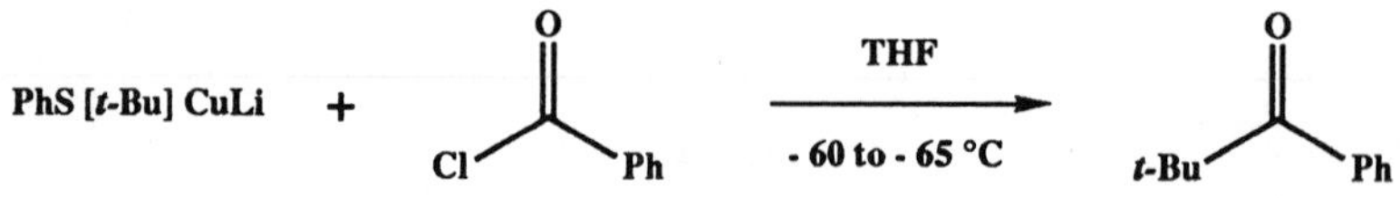

OS, *CV 6*, 273 (1988)

$CH_3-C\equiv C-H$ → (1. $NaNH_2$, liq. NH_3; 2. $Br(CH_2)_3Cl$) → $CH_3-C\equiv C-(CH_2)_3Cl$

OS, *CV 6*, 298 (1988)

C_8H_{17} H H 220 - 225 °C C_8H_{17} H H CHO O

OS, *CV 6*, 316 (1988)

S S 1. *n*-BuLi, THF, - 20 °C 2. $Cl(CH_2)_3Br$, - 75 °C to rt 3. *n*-BuLi, - 75 °C to rt S S

OS, *CV 6*, 338 (1988)

\+ CO + H_2 → (Rh_2O_3, benzene; 60-150 atm., 100 °C) CHO

OS, CV 6, 386 (1988)

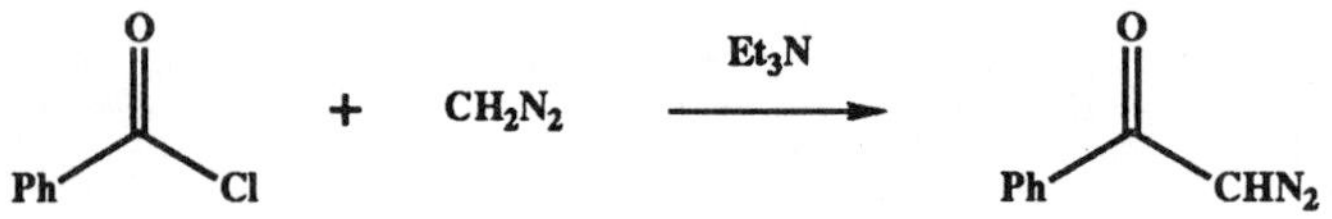

OS, CV 6, 442 (1988)

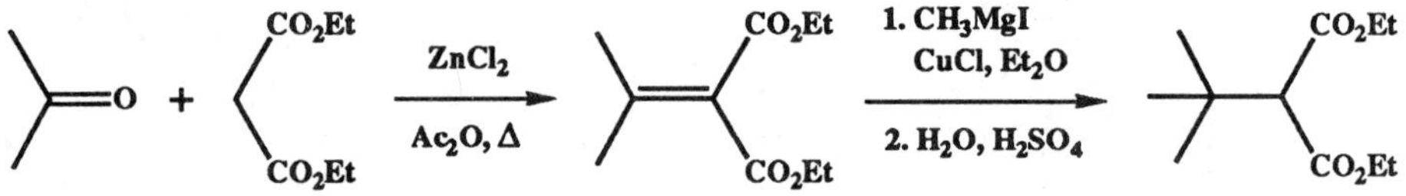

OS, CV 6, 474 (1988)

Me_2NH + $CH_2{=}O$ $\xrightarrow{H_2O,\ 25\ °C}$ $Me_2NCH_2NMe_2$

OS, CV 6, 482 (1988)

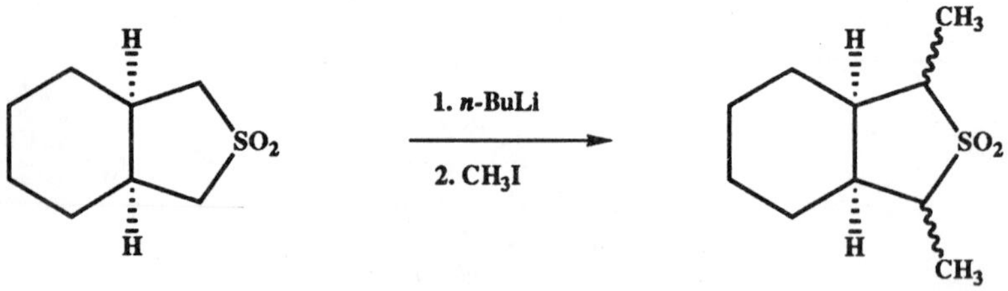

OS, CV 6, 491 (1988)

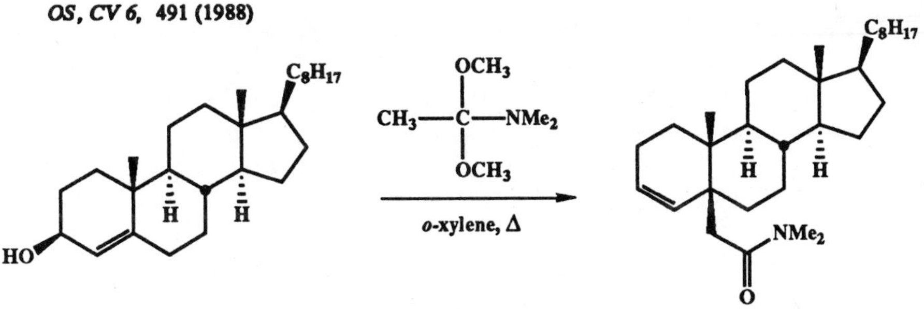

$CH_3-C(OCH_3)_2-NMe_2$

o-xylene, Δ

OS, CV 6, 503 (1988)

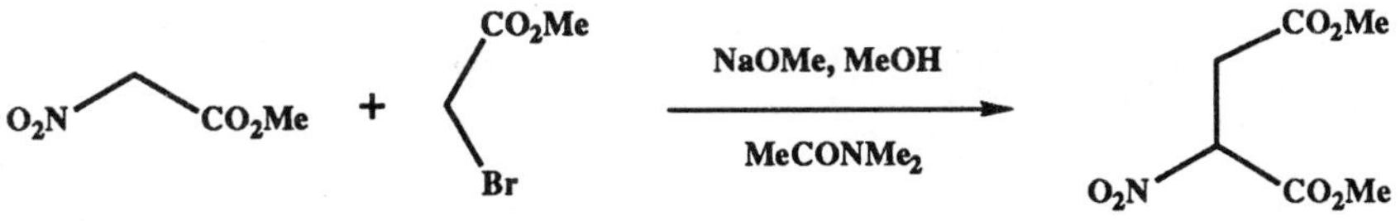

O_2N CO_2Me + CO_2Me Br

NaOMe, MeOH

$MeCONMe_2$

OS, CV 6, 517 (1988)

Ph Br + [CH_3 CH_3 CO_2^- Li^+ Na^+]

THF

CH_3 CH_3 Ph CO_2H

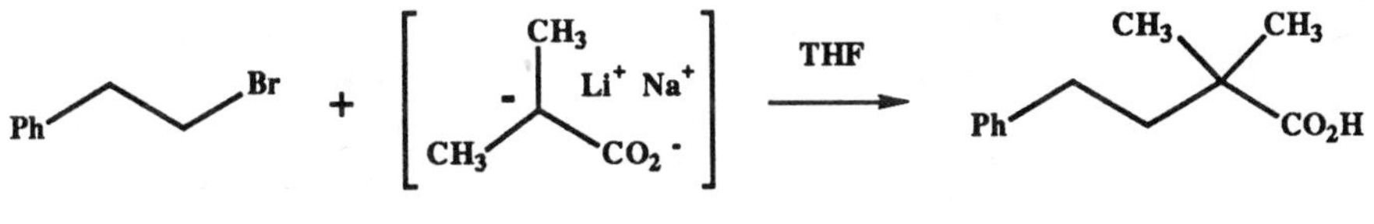

OS, CV 6, 526 (1988)

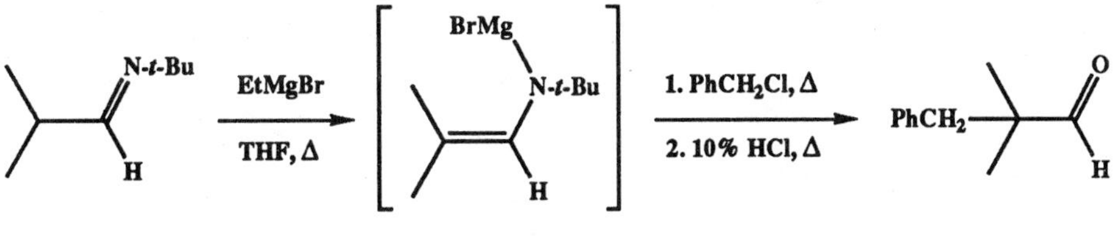

N-*t*-Bu H

EtMgBr

THF, Δ

[BrMg N-*t*-Bu H]

1. $PhCH_2Cl$, Δ

2. 10% HCl, Δ

$PhCH_2$ O H

***OS, CV 6,* 531 (1988)**

Ph—C≡C—Ph → $CH_3S(O)CH_2^-$ Na^+ → 2,3-diphenyl-1,3-butadiene

***OS, CV 6,* 537 (1988)**

$Ph_2C{=}O$ → (CH_3Li, Et_2O, 25 °C) → [$Ph_2C(OLi)CH_3$] → (Li, NH_3 then NH_4Cl) → Ph_2CHCH_3

***OS, CV 6,* 542 (1988)**

Pyrrolidine (N–H) → (1. EtO—N=O, THF, 20 °C; 2. $(i\text{-}Pr)_2NLi$, THF, - 78 °C) → [2-Li-N-NO-pyrrolidine] → (1. $Ph_2C{=}O$, THF, - 78 °C; 2. H_2O; 3. H_2 (200 mm), Ra (Ni), MeOH) → 2-($C(Ph)_2OH$)-pyrrolidine (N–H)

***OS, CV 6,* 564 (1988)**

$Cl—CH_2—CH(OEt)_2$ → ($NaNH_2$, NH_3) → Na^+ $^-C{\equiv}C—OEt$ → (EtBr) → $Et—C{\equiv}C—OEt$

OS, CV 6, 584 (1988)

$CH_3C(OEt)_3$

Δ

C_8H_{17} HO H H CO_2Et

OS, CV 6, 586 (1988)

Na^+ $NC—\bar{C}H—CO_2Et$

DMF, 100 °C

H CH_3 OTs CO_2Et CN

OS, CV 6, 595 (1988)

1. *n*-BuLi, pentane

2. EtBr, 0 - 25 °C

H

OS, CV 6, 598 (1988)

O

$LiCH_2CO_2Et$, THF

then aq. HCl

HO CH_2CO_2Et

OS, CV 6, **606 (1988)**

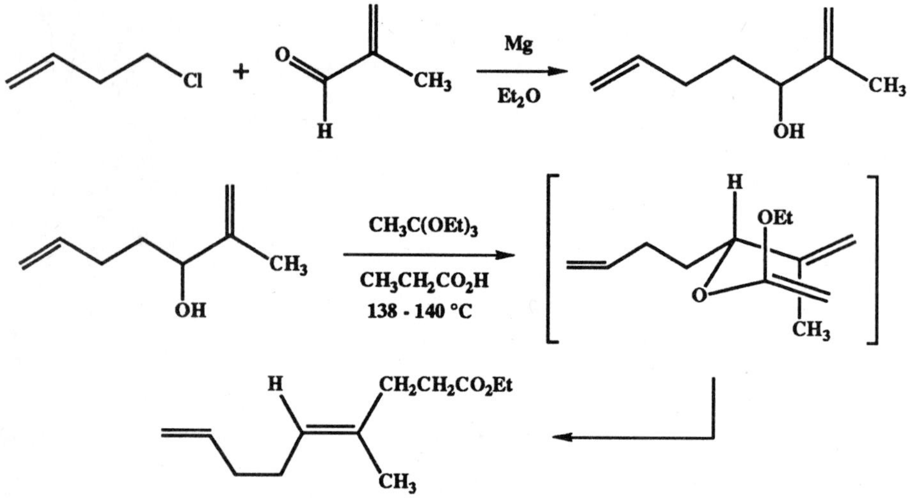

Mg

Et_2O

$CH_3C(OEt)_3$

$CH_3CH_2CO_2H$

138 - 140 °C

OS, CV 6, **613 (1988)**

CH_2N_2, Et_3N

Et_2O, 0 °C

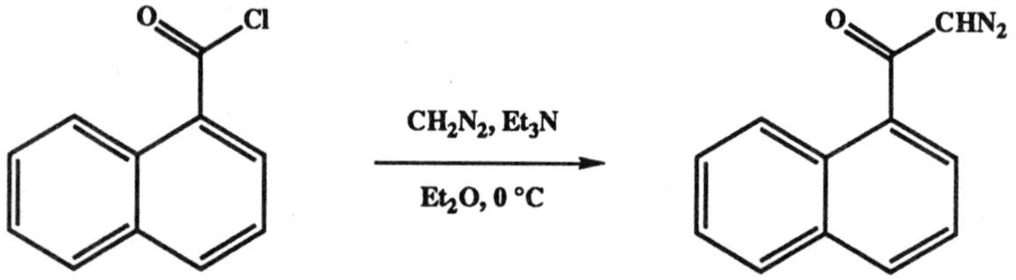

OS, CV 6, **615 (1988)**

benzoyl peroxide

hν, Δ

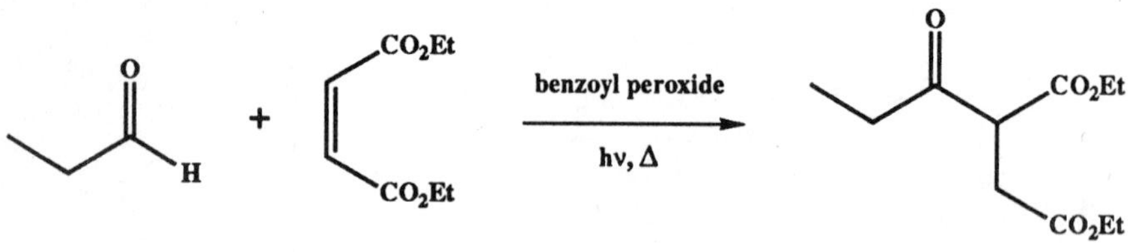

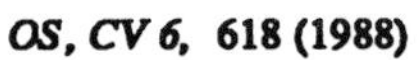
OS, CV 6, 618 (1988)

$Cl_3C(C{=}O)Cl$

CCl3

N H

OS, CV 6, 648 (1988)

NO_2 + O

$(i\text{-}Pr)_2NH$

$CHCl_3$, 60 °C

NO_2

OS, CV 6, 666 (1988)

Me_2CuLi

Et_2O
- 78 °C

OLiCuMe

1. $SiMe_3$ ... O
- 20 °C

2. NH_4Cl, NH_4OH
3. KOH, aq. MeOH reflux

OS, CV 6, 675 (1988)

Cl Cl O

CH_3MgBr

Et_2O, Δ

Cl CH_3 O

OS, CV 6, 683 (1988)

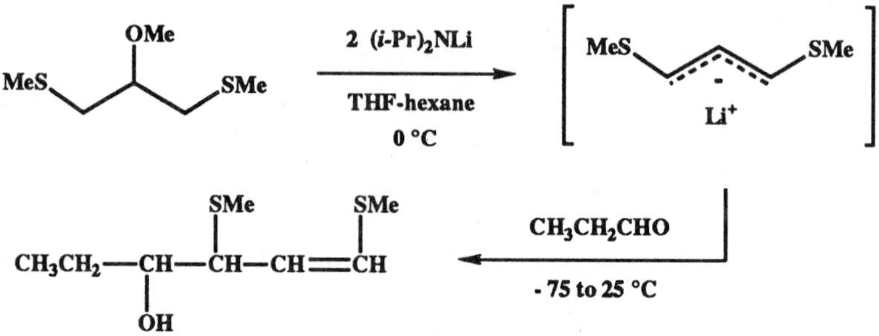

MeS, OMe, SMe → 2 $(i\text{-}Pr)_2NLi$, THF-hexane, 0 °C → [MeS, SMe, Li^+]

CH_3CH_2CHO, - 75 to 25 °C

CH_3CH_2—CH(OH)—CH(SMe)—CH=CH(SMe)

OS, CV 6, 704 (1988)

Br + HS (S, N) → Et_3N, THF, Δ → S, S, N

1. *n*-BuLi, THF, - 55 °C
2. $PhCH_2Br$, - 55 to 0 °C

S, S, CH_2Ph, N

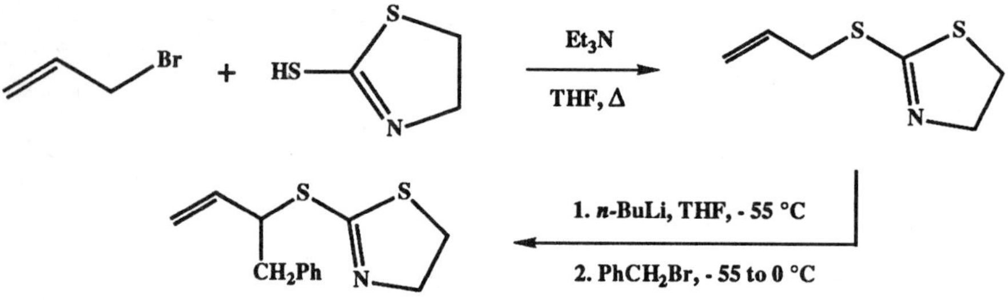

OS, CV 6, 737 (1988)

SPh, SPh → *n*-BuLi, THF, - 20 °C → LiCH(SPh)(SPh)

1. CHO, - 78 °C
2. $(MeO)_2SO_2$, Δ

OMe, SPh, SPh

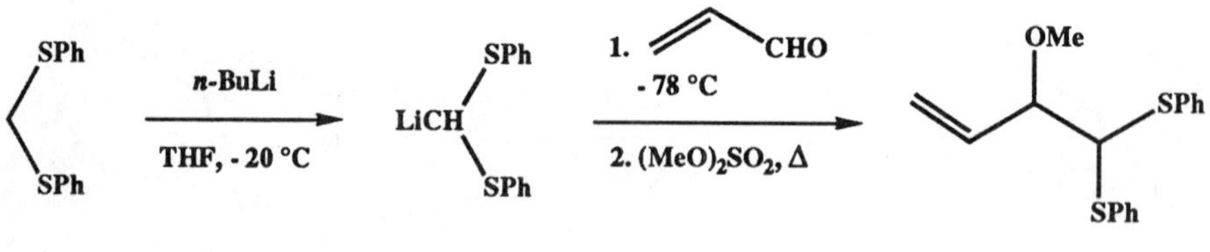

***OS, CV 6,* 751 (1988)**

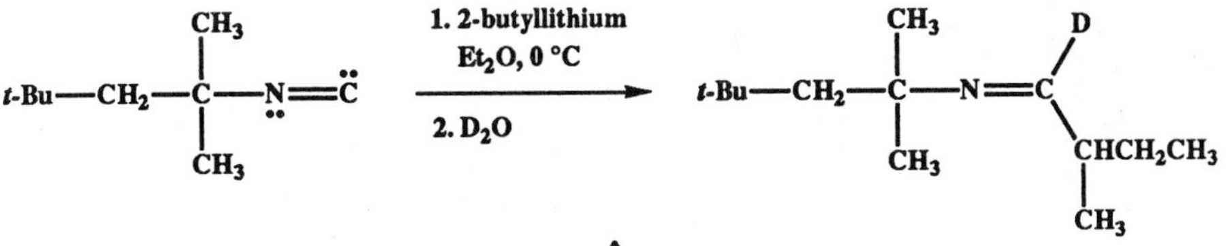

1. 2-butyllithium, Et_2O, 0 °C
2. D_2O

***OS, CV 6,* 762 (1988)**

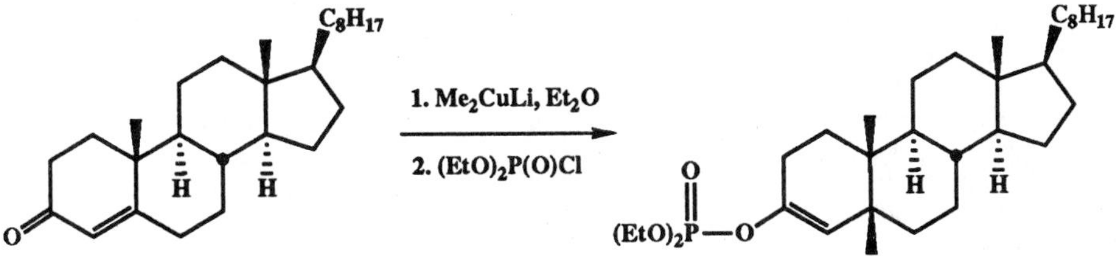

1. Me_2CuLi, Et_2O
2. $(EtO)_2P(O)Cl$

***OS, CV 6,* 776 (1988)**

$C_6H_5P[(CH_2)_3NMe_2]_2$, LiBr
CH_3CN, 70 °C

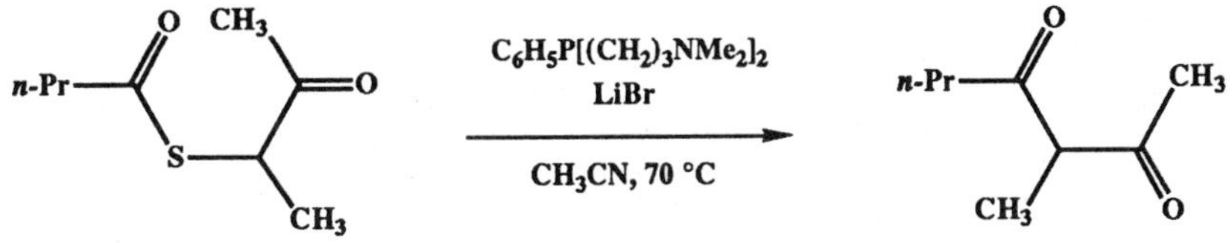

***OS, CV 6,* 781 (1988)**

1. NaH, DMF, benzene
2. aq. HCl, HOAc, Δ

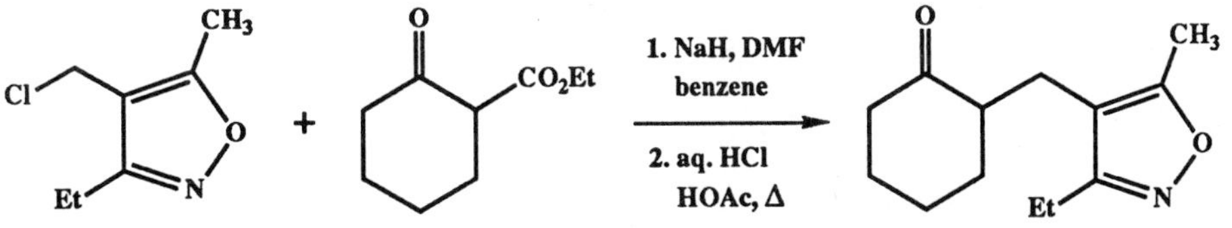

C - C Bond Formation - General

***OS, CV 6,* 786 (1988)**

OH

+

Li

$Me_2N(CH_2)_2NMe_2$

pentane, 25 °C

OLi

Li

H_2O

OH

***OS, CV 6,* 797 (1988)**

2 CH_3NO_2 $\xrightarrow[160\ °C]{2\ KOH}$ $KO_2C—CH{=}NO_2K$ $\xrightarrow[-15\ °C]{H_2SO_4,\ MeOH}$ $MeO_2C—CH_2NO_2$

***OS, CV 6,* 807 (1988)**

$\left[OC—\overset{-}{Fe}(CO)_3—CH_2(CH_2)_4—\overset{O}{\overset{\|}{C}}—OMe \right] Na^+$

1. CO, THF, 25 °C

2. HOAc

$H—\overset{O}{\overset{\|}{C}}—(CH_2)_5—\overset{O}{\overset{\|}{C}}—OMe$

CH_3I, 0 - 25 °C

N-methylpyrrolidinone

$CH_3—\overset{O}{\overset{\|}{C}}—(CH_2)_5—\overset{O}{\overset{\|}{C}}—OMe$

***OS, CV 6,* 818 (1988)**

NCH_3

Ph CH_3

+ $Br(CH_2)_3Cl$

$(i\text{-}Pr)_2NLi$

THF, - 30 °C

Cl

NCH_3

Ph

OS, CV 6, 845 (1988)

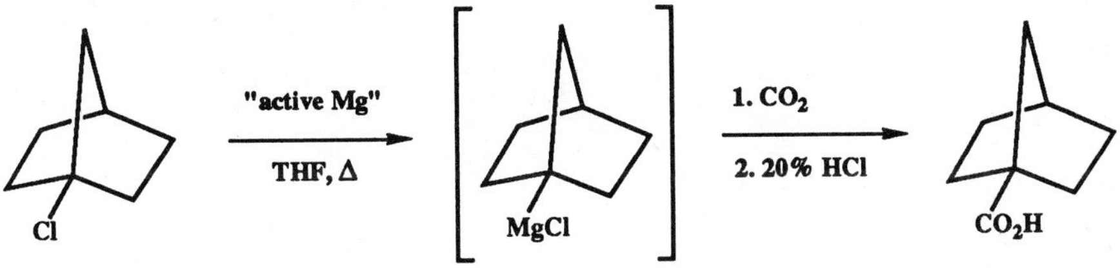

Cl → "active Mg", THF, Δ → [MgCl] → 1. CO_2, 2. 20% HCl → CO_2H

OS, CV 6, 866 (1988)

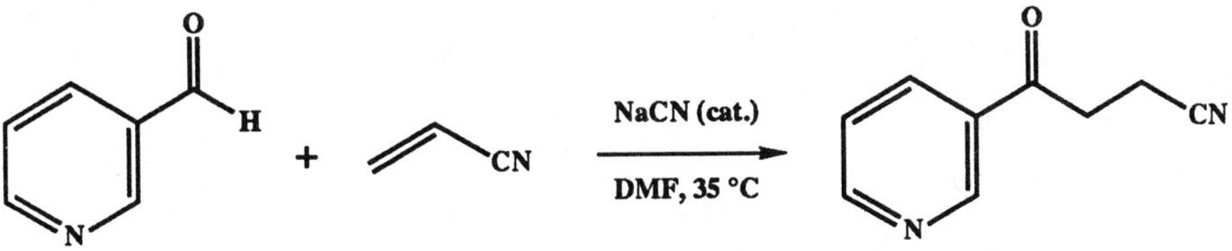

O, H, N + CN → NaCN (cat.), DMF, 35 °C → O, CN, N

OS, CV 6, 869 (1988)

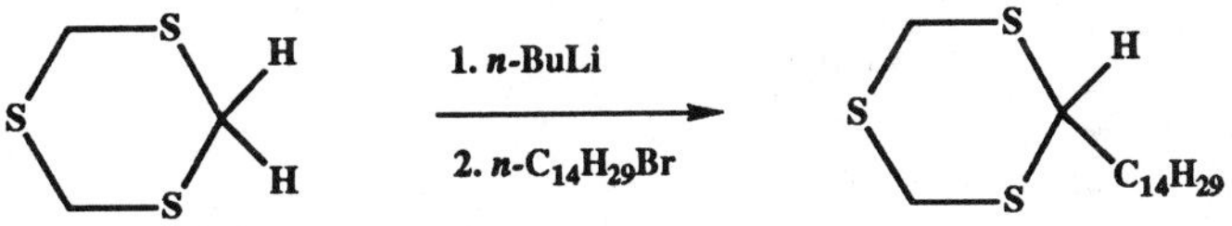

S, S, S, H, H → 1. *n*-BuLi, 2. *n*-$C_{14}H_{29}Br$ → S, S, S, H, $C_{14}H_{29}$

OS, CV 6, 883 (1988)

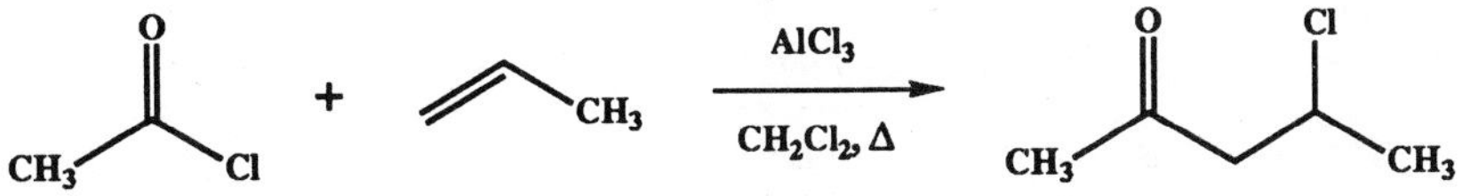

O, CH_3, Cl + CH_3 → $AlCl_3$, CH_2Cl_2, Δ → O, Cl, CH_3, CH_3

OS, CV 6, 890 (1988)

$AgNO_3$, $(NH_4)_2S_2O_8$, H_2O, 60 - 65 °C

OPh

CO_2H

OPh

OS, CV 6, 897 (1988)

+ EtBr

NaOH

$[Et_3NCH_2Ph]^+ Cl^-$

CN

Et

CN

OS, CV 6, 919 (1988)

$(C_6H_{13})_3B$ + N_2 Ph

H_2O, Δ

C_6H_{13} Ph

OS, CV 6, 925 (1988)

Ph—≡—MgBr + BrCH$_2$—≡—H

CuCl, THF, Δ

Ph—≡—CH$_2$—≡—H

***OS*, *CV 6*, 940 (1988)**

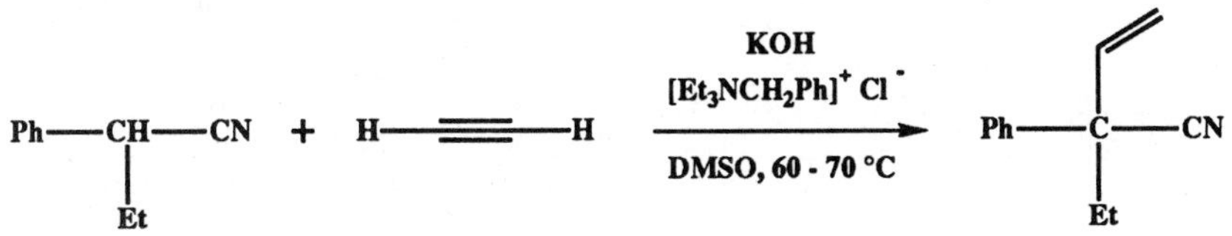

Ph—CH(Et)—CN + H—≡—H → (KOH, $[Et_3NCH_2Ph]^+ Cl^-$, DMSO, 60 - 70 °C) → Ph—C(Et)(CH=CH$_2$)—CN

***OS*, *CV 6*, 991 (1988)**

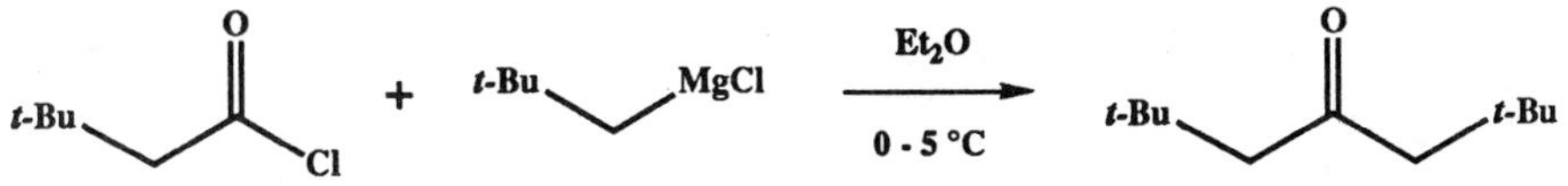

t-BuCH_2COCl + t-BuCH_2MgCl → (Et_2O, 0 - 5 °C) → t-BuCH_2COCH_2t-Bu

***OS*, *CV 6*, 1001 (1988)**

OMe, PF_6^-, $Fe(CO)_3$, + ; + 5,5-dimethylcyclohexane-1,3-dione → (H_2O, EtOH, Δ) → OMe, $Fe(CO)_3$

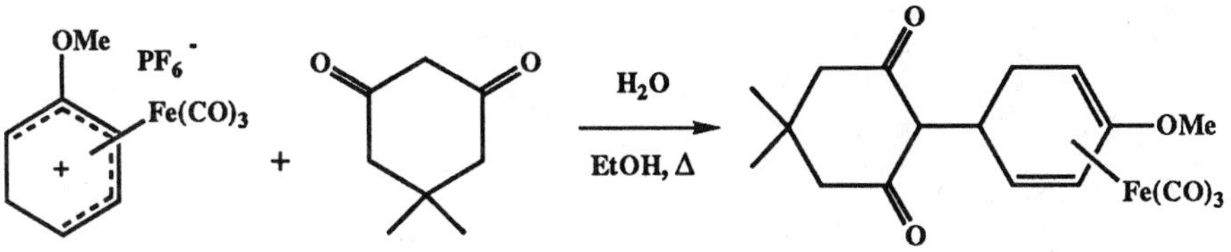

***OS*, *CV 6*, 1033 (1988)**

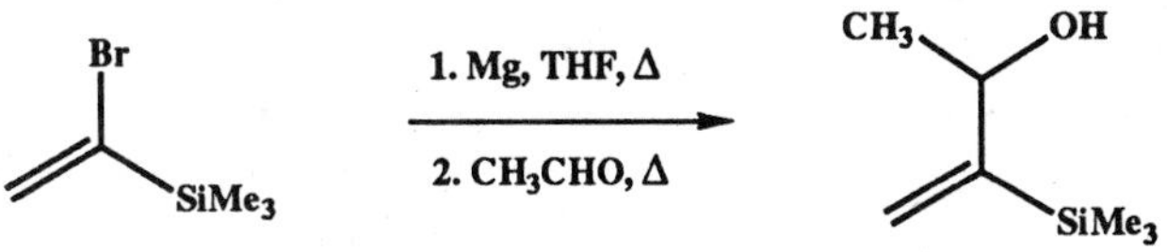

Br, $SiMe_3$ → (1. Mg, THF, Δ; 2. CH_3CHO, Δ) → CH_3, OH, $SiMe_3$

***OS*, *CV* 7, 287 (1990); *60*, 72 (1981)**

2 imidazole + 3 C_6H_5COCl → (3 Et_3N; CH_3CN; *ca.* 20 °C) 1,3-dibenzoyl-2-(1-benzoylimidazol-2-yl)-4-imidazoline (N—Bz, H, N—Bz, N—Bz)

***OS*, *CV* 7, 334 (1990); *60*, 81 (1981)**

H—C≡C—CH₂—O—(tetrahydropyran-2-yl) → (1. EtMgBr; 2. $ClCO_2Me$; 3. Dowex 50 (H^+ form), MeOH) CO_2Me—C≡C—CH_2OH + MeO—(tetrahydropyran-2-yl)

***OS*, *CV* 7, 414 (1990); *60*, 117 (1981)**

1-(OTMS)cyclohexene + 2-nitropropene (CH_3, NO_2) → ($SnCl_4$) 2-(2-oxopropyl)cyclohexanone

***OS*, *CV* 7, 447 (1990); *61*, 42 (1983)**

Ph—C(H)=O (*Re* / *Si* faces) + *n*-BuLi → (Me₂N—CH₂—C(MeO)(H)—C(MeO)(H)—CH₂—NMe₂ "*DDB*"; isopentane, - 140 °C) (*R*)-1-phenyl-1-pentanol (Ph, HO, H) (2 parts) + (*S*)-1-phenyl-1-pentanol (HO, H, Ph) (1 part)

OS, CV 7, 249 (1990); *61,* 59 (1983)

OMe

CO_2H

1. Na, NH_3
2. $C_7H_{15}Br$
3. H^+, H_2O

O

C_7H_{15}

OS, CV 7, 271 (1990); *61,* 65 (1983)

O O

Br

1. *n*-BuLi
2. CH_2O

O O

OH

OS, CV 7, 77 (1990); *61,* 141 (1983)

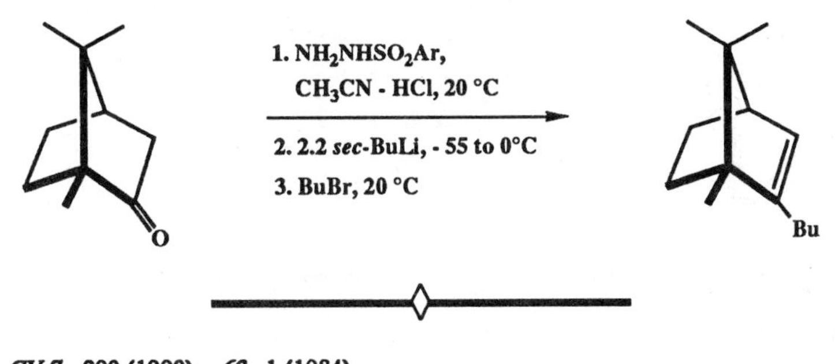

OS, CV 7, 290 (1990); *62,* 1 (1984)

n-BuLi

1. CuBr · Me_2S
2. HC≡CH
3. I_2

H H

n-Bu I

OS, *CV* 7, 351 (1990); *62*, 14 (1984)

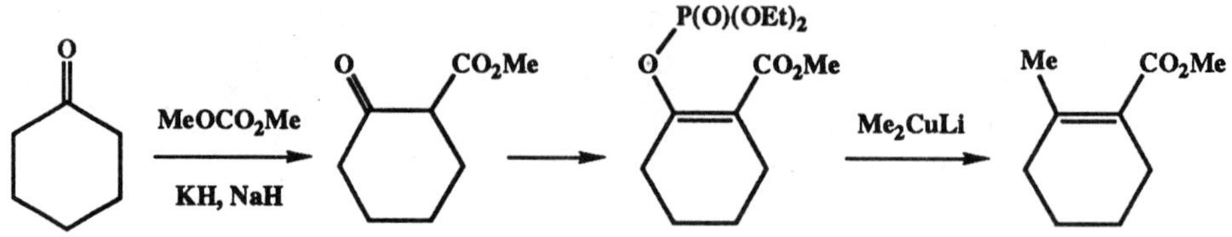

O

MeOCO₂Me

KH, NaH

O CO₂Me

P(O)(OEt)₂

O CO₂Me

Me₂CuLi

Me CO₂Me

OS, *CV* 7, 245 (1990); *62*, 31 (1984)

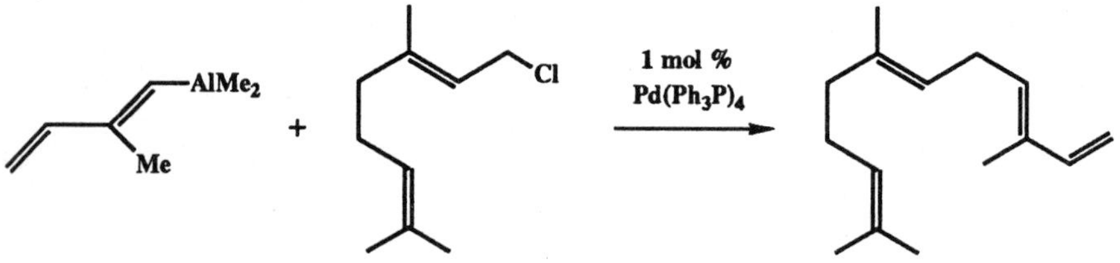

AlMe₂

Me

+

Cl

1 mol %

Pd(Ph₃P)₄

OS, *CV* 7, 501 (1990); *62*, 48 (1984)

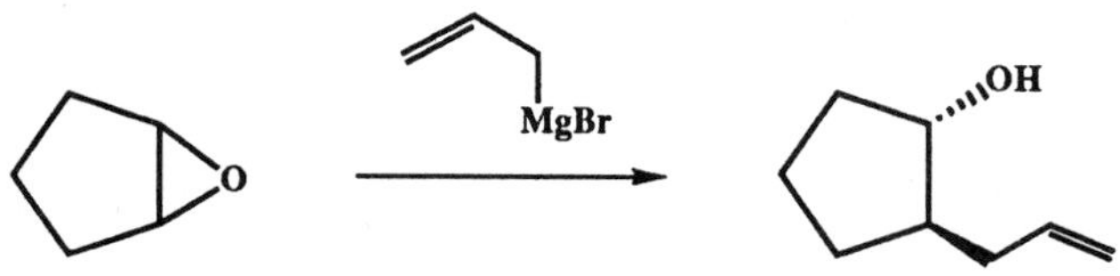

O

MgBr

OH

OS, *CV* 7, 443 (1990); *62*, 86 (1984)

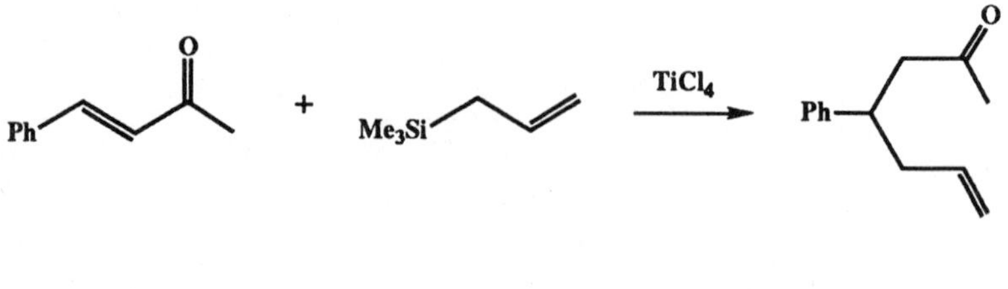

Ph

O

+

Me₃Si

TiCl₄

Ph

O

OS, CV 7, 424 (1990); *62,* 95 (1984)

Me$_3$SiCl

Et$_3$N

OSiMe$_3$

Cl

TiCl$_4$

O

OS, CV 7, 102 (1990); *62,* 111 (1984)

CHO

hν

N

Me

OH

OS, CV 7, 177 (1990); *62,* 125 (1984)

CHO

MgBr

OH

OS, CV 7, 95 (1990); *62,* 170 (1984)

CHO

HO

S

N

CH$_2$Ph

Cl$^-$

Et$_3$N, EtOH

O

OH

OS, *CV* 7, 363 (1990); *63*, 26 (1985)

aq. HOAc

70 °C

HO

O

O

O

O

O

OS, *CV* 7, 368 (1990); *63*, 37 (1985)

aq. HOAc

O

O

O

O

O

O

OS, *CV* 7, 339 (1990); *63*, 44 (1985)

Na

- Na+

$BrCH_2CO_2Me$

CH_2CO_2Me

OS, *CV* 7, 153 (1990); *63*, 109 (1985)

CO_2Et

HO—H

CH_2

CO_2Et

S - (-) isomer

2 LDA

- 75 to - 20 °C

CO_2Et

LiO—H

CH

LiO OEt

Br

CO_2Et

HO—*S*—H

CH_2—*R*—H

CO_2Et

(+) isomer

(92% *u*; 8% *l*)

OS, *CV* 7, 276 (1990); *63*, 203 (1985)

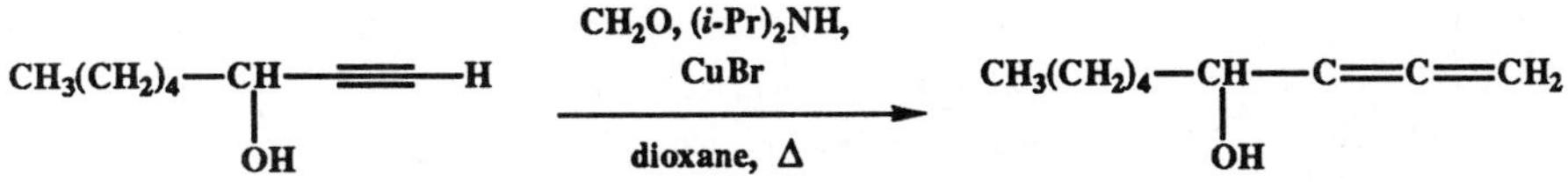

OS, *CV* 7, 236 (1990); *64*, 1 (1986)

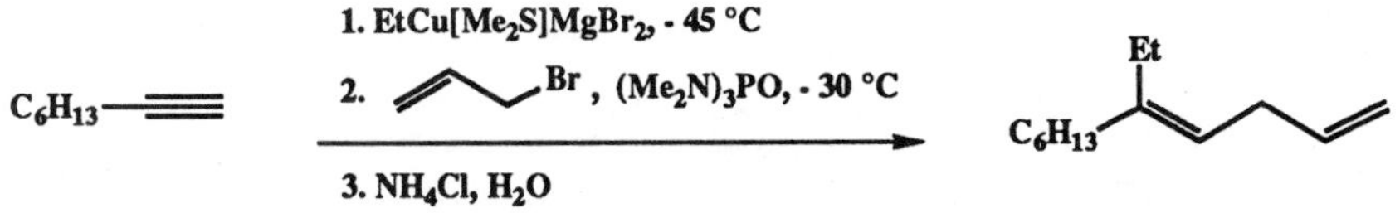

OS, *CV* 7, 456 (1990); *64*, 10 (1986)

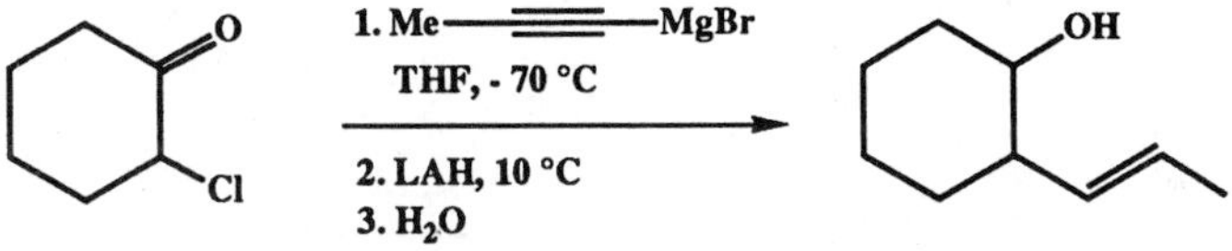

OS, *CV* 7, 208 (1990); *64*, 68 (1986)

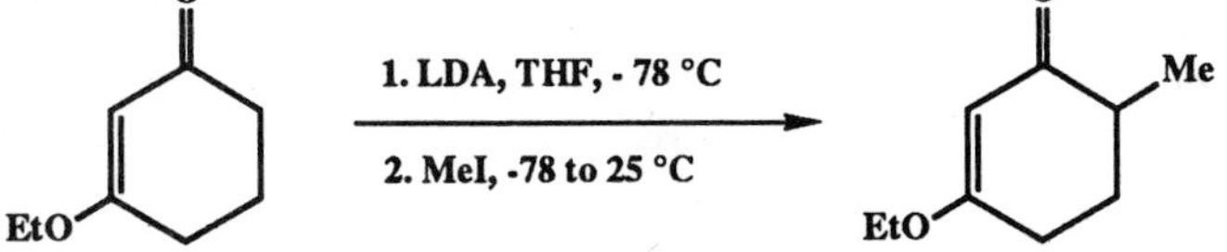

OS, *CV* 7, 241 (1990); *64*, 73 (1986)

1. LDA, $(Me_2N)_3PO$, THF, - 78 °C
2. $Cl_2C=CHCl$, warm to r.t., remove solvents and $(i\text{-}Pr)_2NH$
3. DIBAL, toluene, 0 °C
4. H^+

EtO Me O Me Cl Cl O

OS, *CV* 7, 226 (1990); *64*, 108 (1986)

$$Me-C\equiv C-H \xrightarrow[\text{2. ClCO}_2\text{Et, Et}_2\text{O, 0 °C}]{\text{1. } n\text{-BuLi, Et}_2\text{O, - 78 °C}} Me-C\equiv C-CO_2Et$$

OS, *CV* 7, 451 (1990); *64*, 114 (1986)

$$PhCH_2CH_2Cl \xrightarrow{Mg} PhCH_2CH_2MgCl \xrightarrow{\text{N-CHO (piperidine)}} PhCH_2CH_2CHO$$

OS, *CV* 7, 323 (1990); *64*, 144 (1986)

$$CH_2(CO_2Me)_2 \xrightarrow{KOH} MeO_2CCH_2COOK \xrightarrow[POCl_3]{2\ DMF}$$

CO_2Me H H O OH

C - C Bond Formation - General

OS, *65*, 12 (1987)

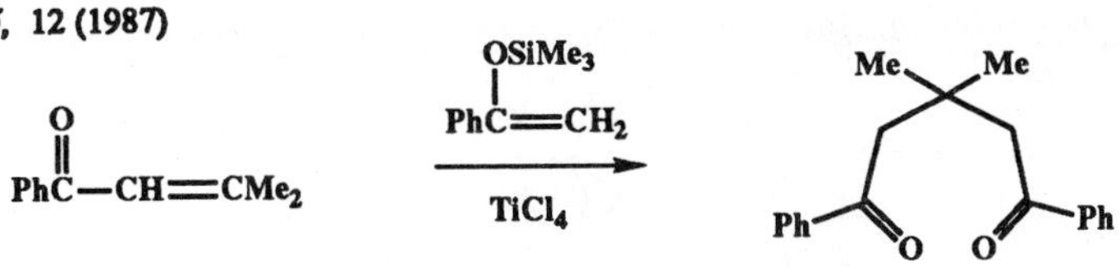

O
PhC—CH═CMe2
OSiMe3
PhC═CH2
TiCl4
Me
Me
Ph
O
O
Ph

OS, *65*, 17 (1987)

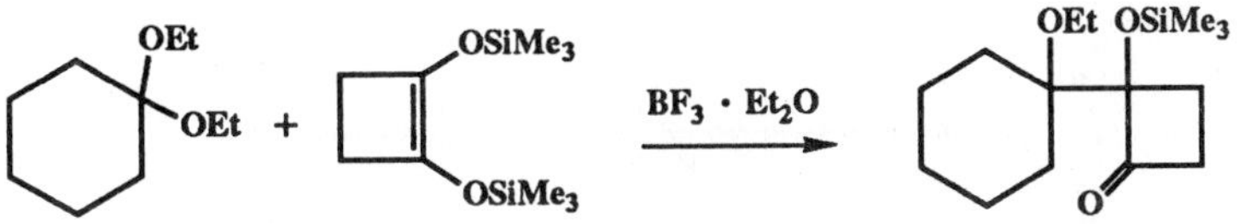

OEt
OEt
+
OSiMe3
OSiMe3
BF3 · Et2O
OEt OSiMe3
O

OS, *65*, 26 (1987)

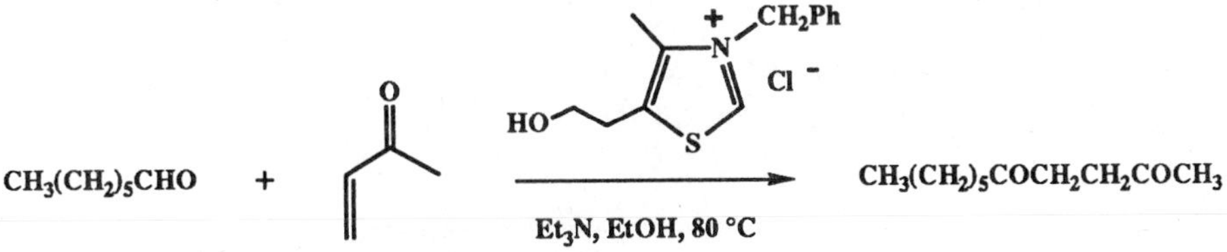

CH3(CH2)5CHO
+
O
+
CH2Ph
N
Cl -
HO
S
Et3N, EtOH, 80 °C
CH3(CH2)5COCH2CH2COCH3

OS, *65*, 42 (1987)

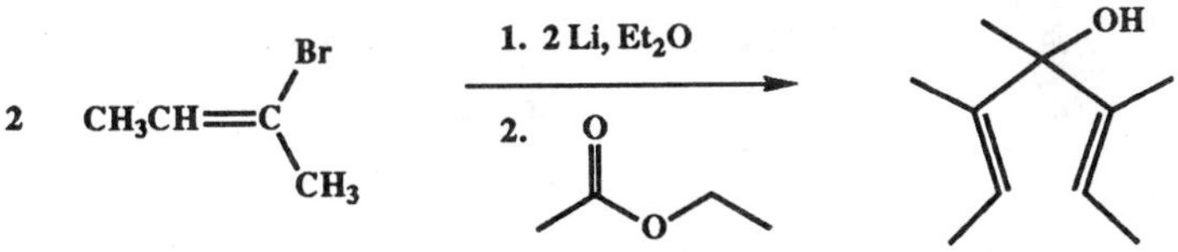

2
CH3CH═C
Br
CH3
1. 2 Li, Et2O
2.
O
O
OH

OS, *65*, 47 (1987)

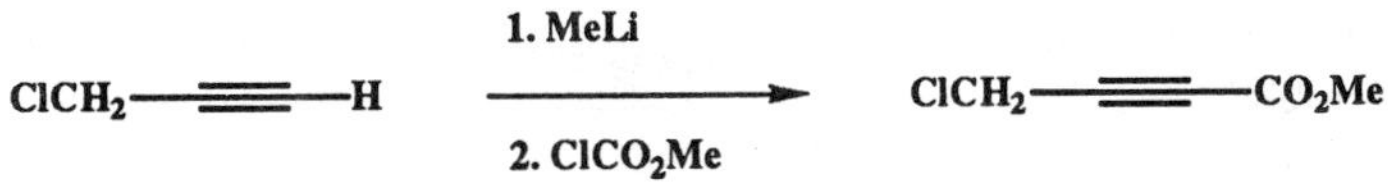

ClCH2
H
1. MeLi
2. ClCO2Me
ClCH2
CO2Me

OS, 65, 119 (1987)

$(EtO)_2PCH_2N{=}CHPh$ → 1. *n*-BuLi, THF, - 78 °C; 2. PhCOMe, - 78 °C --> Δ → Ph(Me)C=CH—N=CHPh

→ 1. *n*-BuLi, THF, - 78 °C; 2. $CH_2{=}CHCH_2Br$, - 78 °C --> r.t.; 3. H_3O^+, r.t. → Ph, CHO, Me

OS, 65, 183 (1987)

Me, O, Me → 1. OMe, N, NH_2 *"SAMP"*; 2. LDA, Et_2O, 0 °C; 3. *n*-PrI, - 110 °C → H, OMe, N, N, Me, Me, *ZSS*, Me

OS, 65, 236 (1987)

OAc, AcO, O, AcO, AcO, Br + CN → Bu_3SnH; Et_2O, hν → OAc, AcO, O, AcO, AcO, CN

OS, 66, 1 (1987)

$MeSO_3CH_2$—≡—$SiMe_3$ → MeMgCl, CuBr, LiBr; THF → $H_2C{=}C{=}C(SiMe_3)Me$

OS, *66*, 43 (1987)

$Zn(CH_2CH_2CO_2Et)_2$ + cyclohexenone → (CuBr · Me_2S; Me_3SiCl) → 1-($OSiMe_3$)-3-($CH_2CH_2CO_2Et$)cyclohexene

OS, *66*, 52 (1987)

2 $CH_2=CHCO_2Me$ → ($Pd(CH_3CN)_4(BF_4)_2$; $LiBF_4$) → MeO_2C–$CH_2CH_2CH=CH$–CO_2Me

OS, *66*, 60 (1987)

C_8H_{17}–C≡C–H → (1. $(i\text{-}Bu)_2AlH$; 2. 1 mol % $Pd(Ph_3P)_4$, $ZnCl_2$, I–CH=CH–C_4H_9) → C_8H_{17}–CH=CH–CH=CH–C_4H_9

OS, *66*, 75 (1987)

$CH_2(CO_2Me)_2$ + $Me_2C{=}CHCH_2Br$ → (NaOMe, MeOH, 0 °C) → $Me_2C{=}CHCH_2CH(CO_2Me)_2$

$Me_2C{=}CHCH_2CH(CO_2Me)_2$ → (HC≡C–CH_2Br; NaH, THF, 0 °C to r.t.) → HC≡C–CH_2–$C(CO_2Me)_2$–CH_2–CH=CMe_2

OS, *66*, 87 (1987)

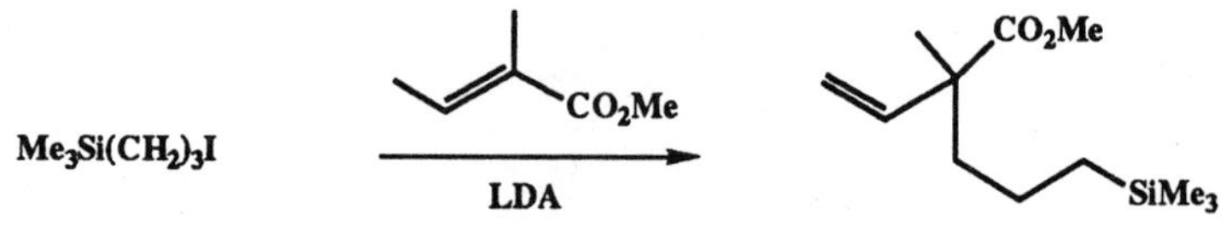

$Me_3Si(CH_2)_3I$ → (CO_2Me ester; LDA) → CO_2Me, $SiMe_3$

◊

OS, *66*, 95 (1987)

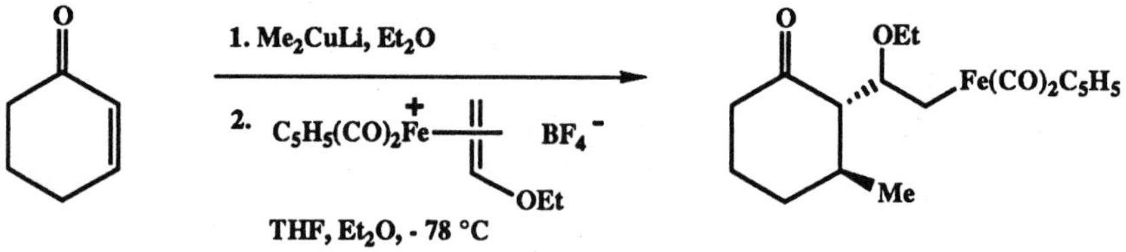

1. Me_2CuLi, Et_2O
2. $C_5H_5(CO)_2Fe^+$ (OEt) BF_4^-
THF, Et_2O, - 78 °C

OEt, $Fe(CO)_2C_5H_5$, Me

◊

OS, *66*, 116 (1987)

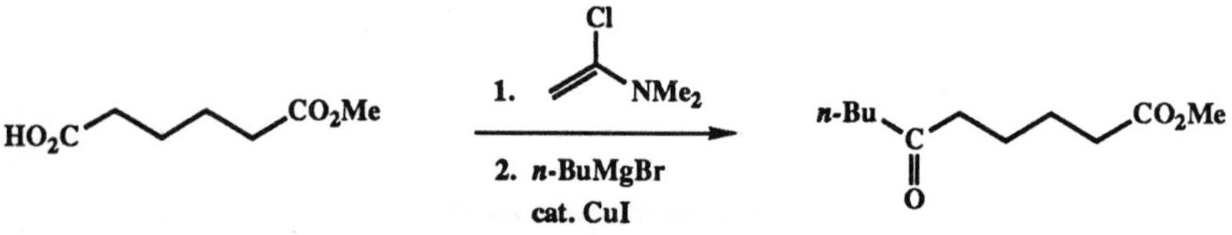

HO_2C ... CO_2Me

1. Cl, NMe_2
2. *n*-BuMgBr cat. CuI

n-Bu C(=O) ... CO_2Me

◊

OS, *67*, 48 (1988)

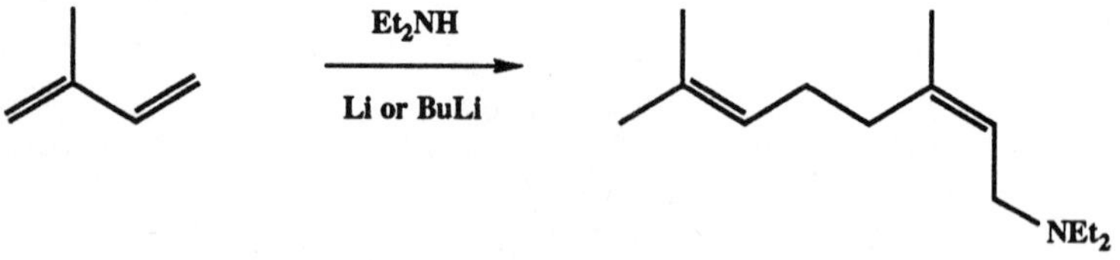

Et_2NH
Li or BuLi

NEt_2

C - C BOND FORMATION - GENERAL

***OS*, *67*, 60 (1988)**

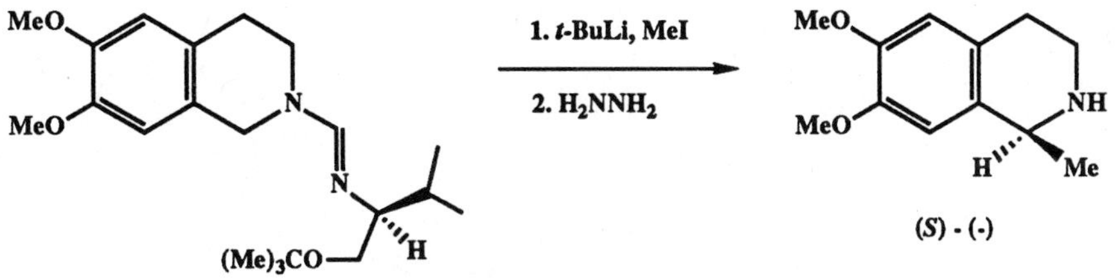

MeO, MeO, N, N, $(Me)_3CO$, H → 1. *t*-BuLi, MeI; 2. H_2NNH_2 → MeO, MeO, NH, H, Me

(*S*) - (-)

***OS*, *67*, 86 (1988)**

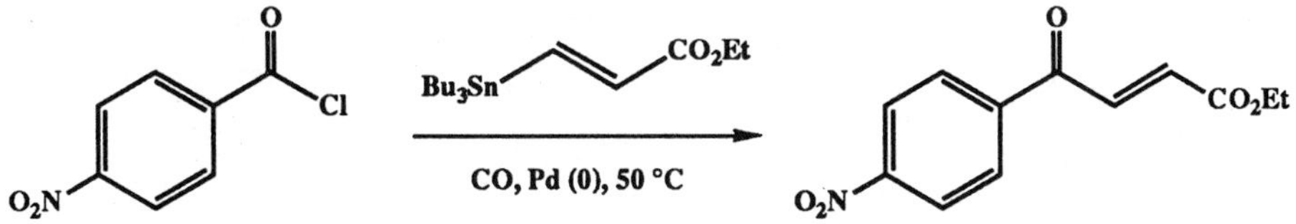

O, Cl, O_2N → Bu_3Sn / CO_2Et; CO, Pd (0), 50 °C → O, CO_2Et, O_2N

***OS*, *67*, 98 (1988)**

CO_2Et, I → Zn(Cu) → [CO_2Et, ZnI] → COCl; Pd (0) → CO_2Et, O

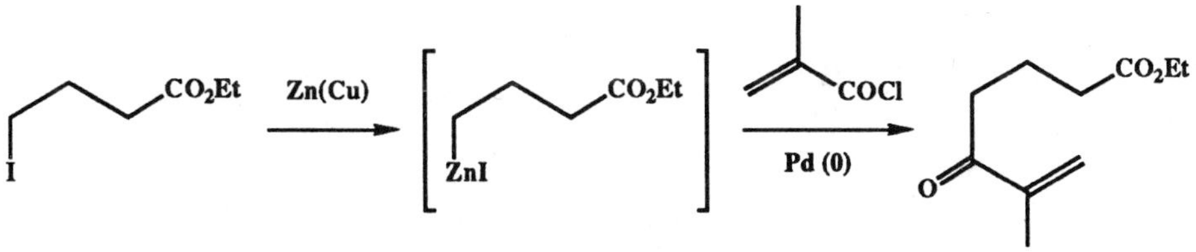

***OS*, *67*, 125 (1988)**

n-C_8H_{17}—CH—CO_2Et (with $SiMePh_2$ on CH) → 1. MeMgBr, THF; 2. MeLi; 3. KO-*t*-Bu → n-$C_8H_{17}CH{=}CMe_2$

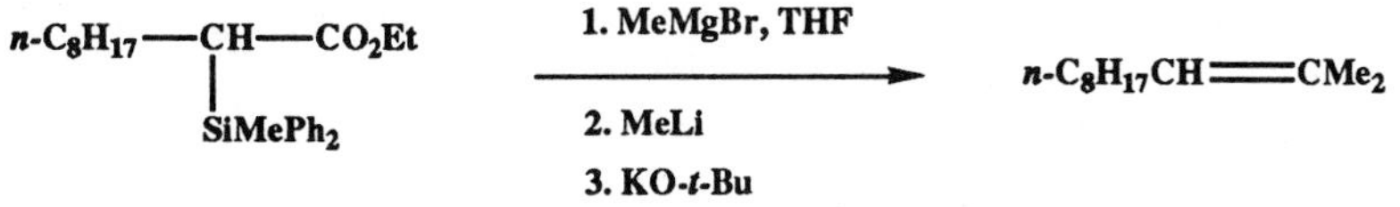

***OS*, *67*, 141 (1988)**

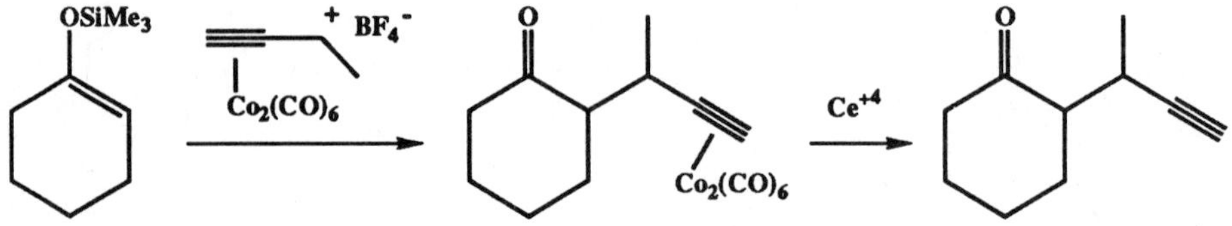

OSiMe$_3$ + $Co_2(CO)_6$-complexed propargyl cation BF_4^- → $Co_2(CO)_6$ complex; Ce^{+4} →

***OS*, *67*, 180 (1988)**

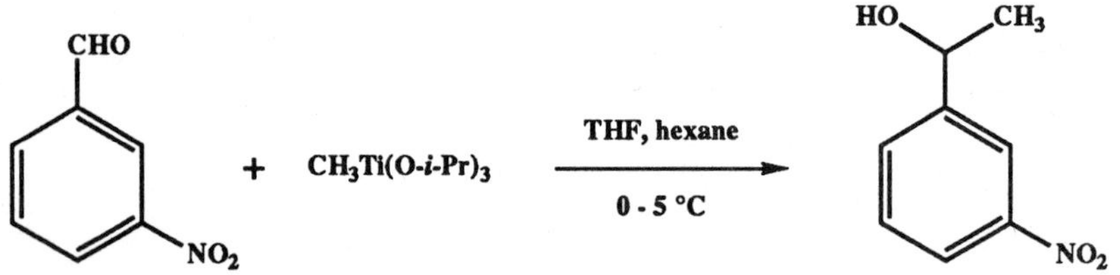

CHO, NO_2 + $CH_3Ti(O\text{-}i\text{-}Pr)_3$ —THF, hexane / 0 - 5 °C→ HO, CH_3, NO_2

***OS*, *67*, 193 (1988)**

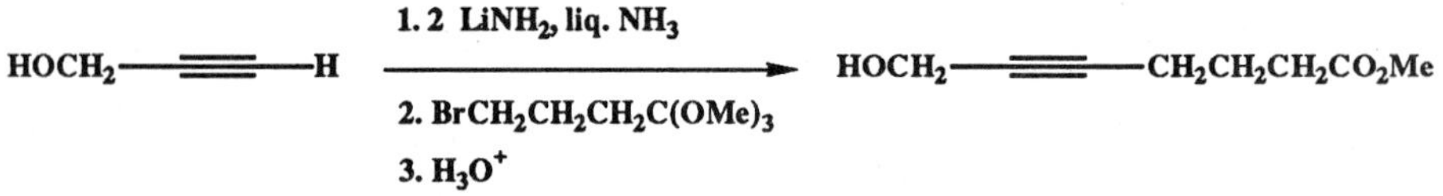

$HOCH_2$—≡—H —1. 2 $LiNH_2$, liq. NH_3 / 2. $BrCH_2CH_2CH_2C(OMe)_3$ / 3. H_3O^+→ $HOCH_2$—≡—$CH_2CH_2CH_2CO_2Me$

***OS*, *67*, 205 (1988)**

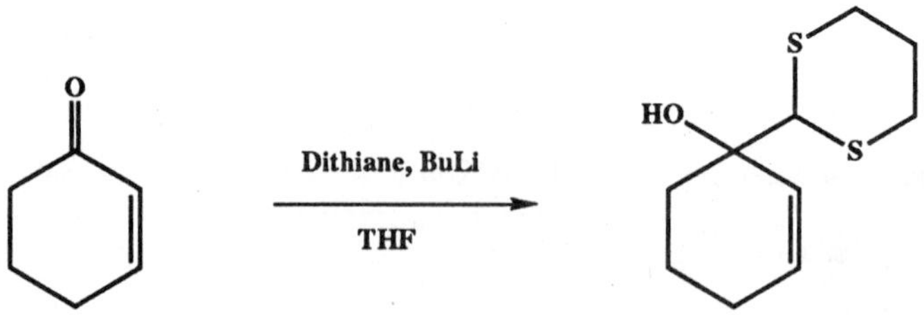

O —Dithiane, BuLi / THF→ HO, S, S

***OS*, *67*, 210 (1988)**

OEt

Br

1. *t*-BuLi, Et_2O, - 78 °C

2. O H

OEt

OH

***OS*, *68*, 14 (1989)**

Li—≡—H

THF, - 78 °C

O

H

OH

(-) isomer

97 : 3 *endo* : *exo*

***OS*, *68*, 56 (1989)**

MeI

Triton B

O O

O O

***OS*, *68*, 116 (1989)**

OSO_2CF_3

+ Me_3Sn

$Pd(Ph_3P)_4$, LiCl

THF

CO

$Pd(Ph_3P)_4$, LiCl

THF

O

***OS*, *68*, 130 (1989)**

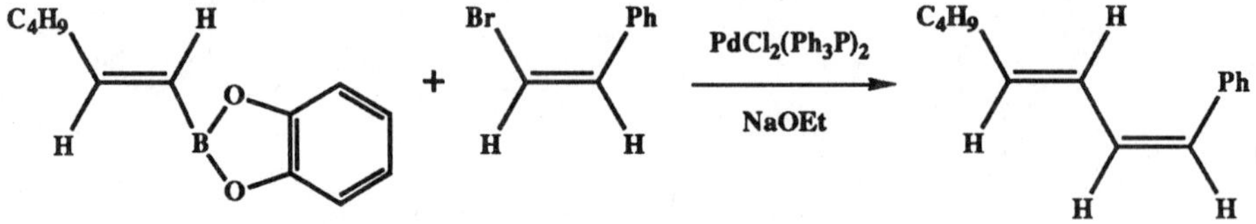

C_4H_9 H H B O O + Br Ph H H

$PdCl_2(Ph_3P)_2$

NaOEt

C_4H_9 H H Ph H H

***OS*, *68*, 162 (1989)**

O $OPO(NMe_2)_2$

1. *n*-BuLi

2. MeI

3. HCOOH

Me O O

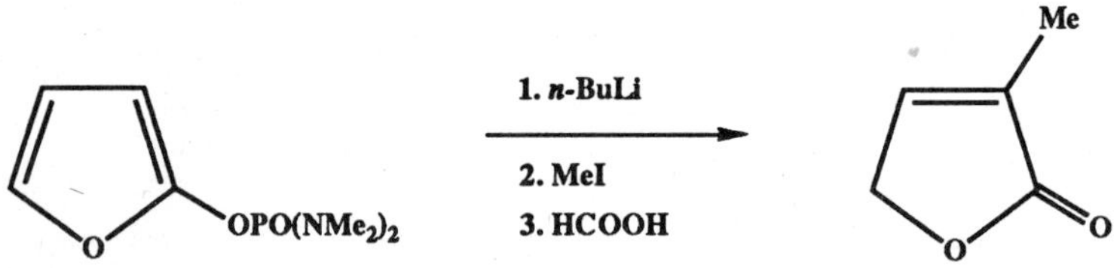

CLEAVAGE

OS, CV 1, 10 (1941)

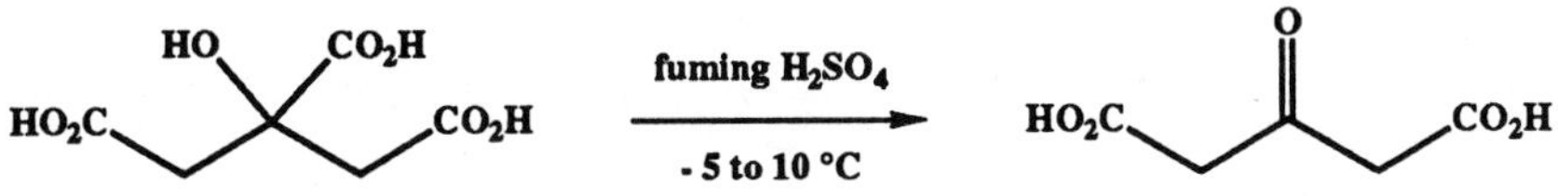
HO
CO_2H
HO_2C
CO_2H
fuming H_2SO_4
- 5 to 10 °C
O
HO_2C
CO_2H

OS, CV 1, 56 (1941)

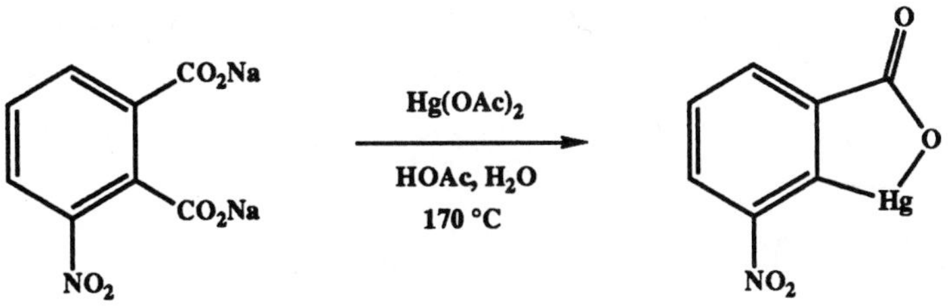
CO_2Na
CO_2Na
NO_2
$Hg(OAc)_2$
HOAc, H_2O
170 °C
O
O
Hg
NO_2

OS, CV 1, 192 (1941)

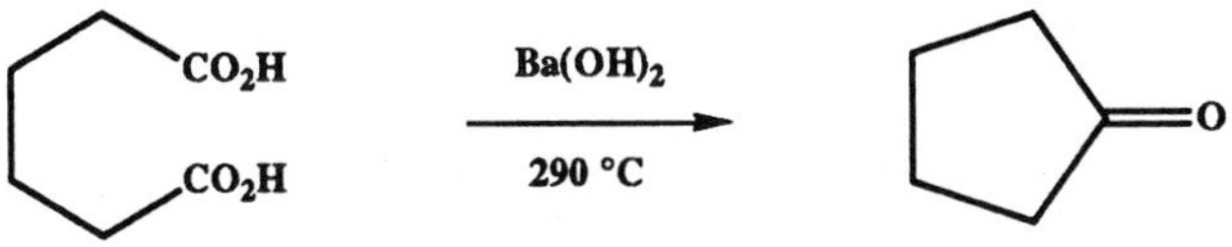
CO_2H
CO_2H
$Ba(OH)_2$
290 °C
O

OS, CV 1, 274 (1941)

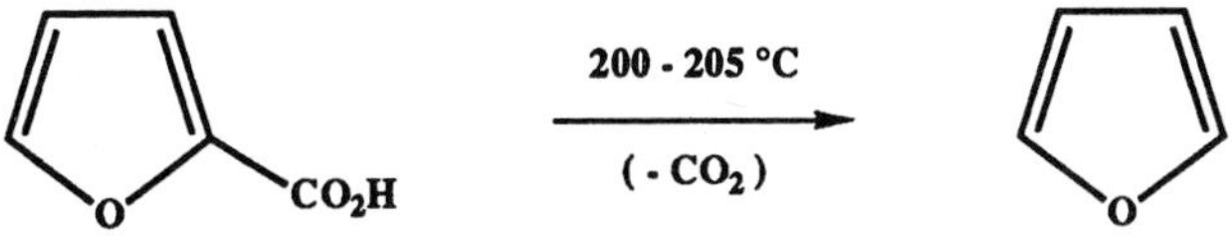
O
CO_2H
200 - 205 °C
(- CO_2)
O

OS, CV 1, **290 (1941)**

EtO_2C, CO_2Et, EtO_2C, CO_2Et → HCl, H_2O, Δ → HO_2C, CO_2H

OS, CV 1, **351 (1941)**

O, O, CH_3, OEt, Bu → 1. NaOH, H_2O; 2. H_2SO_4 → O, CH_3, Bu

OS, CV 1, **401 (1941)**

$ClCH_2CO_2Na$ → 1. aq. $NaNO_2$; 2. 80 °C → CH_3NO_2

OS, CV 1, **440 (1941)**

CH=CH—CO_2H → cat. hydroquinone, Δ → CH=CH_2

OS, CV 1, **451 (1941)**

CO_2Et, CH=, CN → NaCN, EtOH, HCl → CN, CO_2Et, CN → HCl, Δ → CO_2H, CO_2H

OS, CV 1, 455 (1941)

CO_2H H_2N NH_2 NH_2 — H_2O, HCl, Δ → HO OH OH

OS, CV 1, 473 (1941)

OH OH HO_2C CO_2H OH OH — 1. 2 NH_4OH, 2. Δ → N H

OS, CV 1, 475 (1941)

OH HO_2C CO_2H OH — $KHSO_4$, Δ → O CH_3 CO_2H

OS, CV 1, 523 (1941)

CO_2Et EtO_2C CO_2Et EtO_2C — HCl, H_2O, Δ → CO_2H HO_2C CO_2H

***OS*, *CV 1*, 541 (1941)**

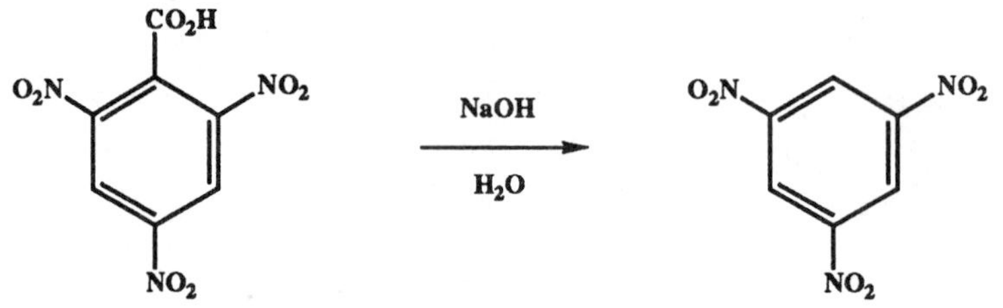

CO_2H O_2N NO_2 NO_2 — NaOH, H_2O → O_2N NO_2 NO_2

***OS*, *CV 2*, 21 (1943)**

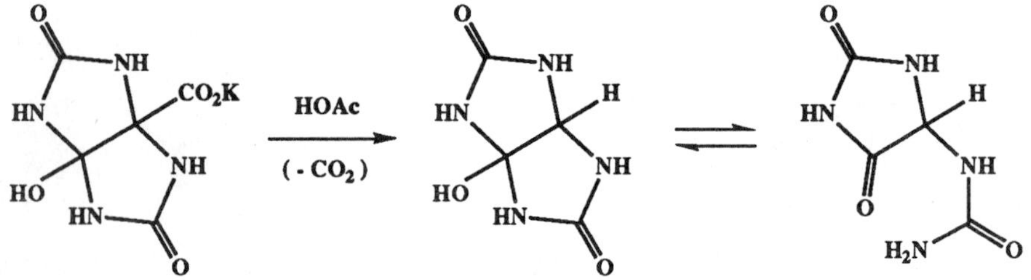

O NH CO_2K HN NH HO HN O — HOAc, (- CO_2) → O NH H HN NH HO HN O ⇌ O NH H HN NH O H_2N O

***OS*, *CV 2*, 61 (1943)**

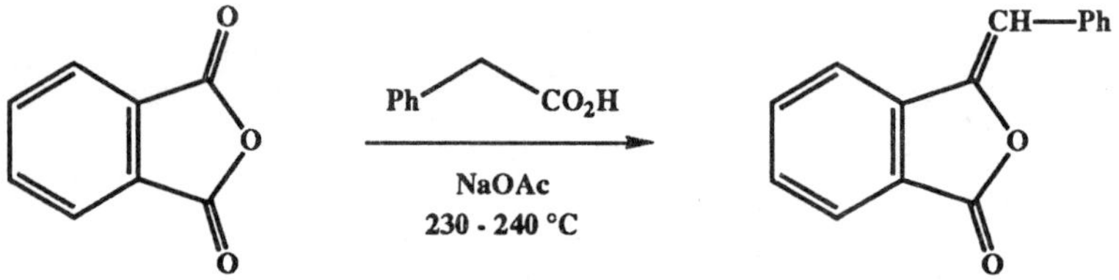

O O O — Ph CO_2H, NaOAc, 230 - 240 °C → CH—Ph O O

***OS*, *CV 2*, 93 (1943)**

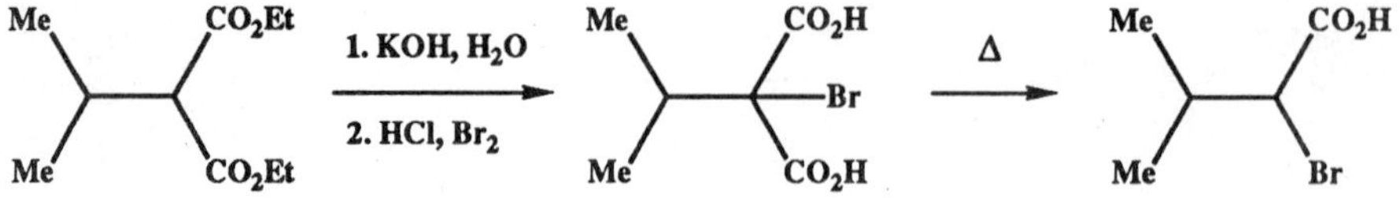

Me CO_2Et Me CO_2Et — 1. KOH, H_2O; 2. HCl, Br_2 → Me CO_2H Br Me CO_2H — Δ → Me CO_2H Me Br

OS, CV 2, **100 (1943)**

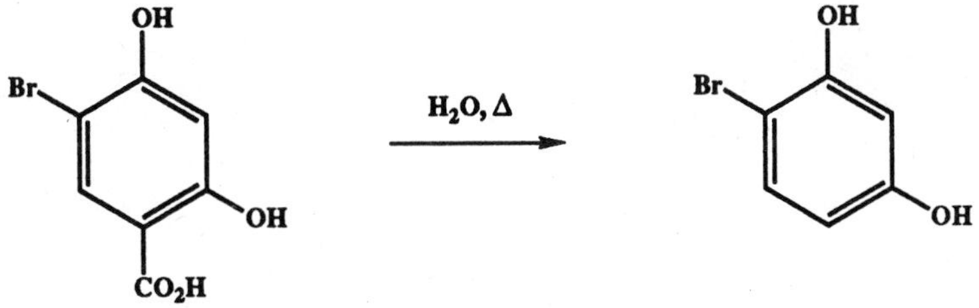

OH Br OH CO_2H — H_2O, Δ → OH Br OH

OS, CV 2, **214 (1943)**

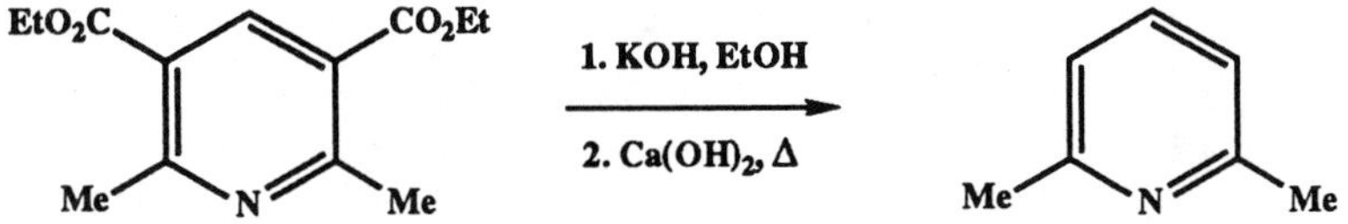

EtO_2C CO_2Et Me N Me — 1. KOH, EtOH; 2. $Ca(OH)_2$, Δ → Me N Me

OS, CV 2, **217 (1943)**

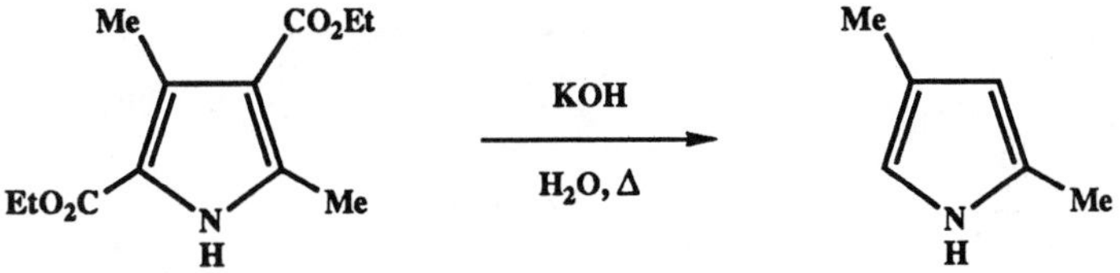

Me CO_2Et EtO_2C N H Me — KOH, H_2O, Δ → Me N H Me

OS, CV 2, **229 (1943)**

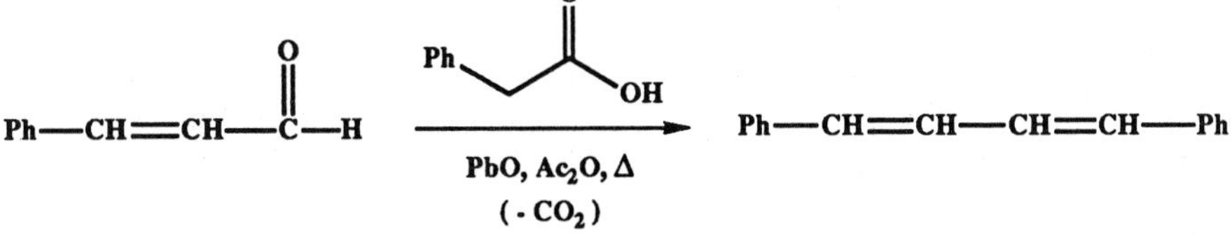

Ph—CH=CH—C(=O)—H — PhCH_2C(=O)OH; PbO, Ac_2O, Δ (- CO_2) → Ph—CH=CH—CH=CH—Ph

***OS*, *CV* 2, 368 (1943)**

HO$_2$C, OH, O, OH, HO, O; Δ; O, O, O

***OS*, *CV* 2, 384 (1943)**

O, N—C(CO$_2$Et)$_2$, CH$_2$CH$_2$SCH$_3$, O

NaOH, EtOH, Δ, then HCl

CO$_2$H, H, N—C(CO$_2$H)$_2$, CH$_2$CH$_2$SCH$_3$, O

HCl, H$_2$O, Δ

CH$_3$SCH$_2$CH$_2$, CO$_2$H, NH$_2$

***dl* - methionine**

***OS*, *CV* 2, 389 (1943)**

PhCH$_2$CO$_2$H + CH$_3$CO$_2$H → ThO$_2$ (cat.), 430-450 °C → Ph, O, CH$_3$

***OS*, *CV* 2, 391 (1943)**

Ph, O, CH$_3$, CN → H$_2$SO$_4$, H$_2$O → Ph, O, CH$_3$

***OS*, *CV* 2, 416 (1943)**

CH_3, CH_3, CO_2Et, CO_2Et —— 1. KOH, H_2O / 2. H_2SO_4, Δ ——> CH_3, CH_3, CO_2H

***OS*, *CV* 2, 474 (1943)**

C_7H_{15}, CO_2Et, CO_2Et —— 1. KOH, H_2O, Δ / 2. HCl ——> C_7H_{15}, CO_2H, CO_2H —— Δ ——> C_7H_{15}, CO_2H

***OS*, *CV* 2, 512 (1943)**

CN —— CH_3ONO_2 / Na, EtOH ——> NO_2Na, CN —— 1. NaOH, H_2O, Δ / 2. HCl ——> NO_2

***OS*, *CV* 3, 213 (1955)**

CO_2Et, CO_2Et —— KOH / then HCl ——> CO_2H, CO_2H —— Δ ——> H, CO_2H

OS, CV 3, 267 (1955)

CO_2H Cl Cl OH — Δ, $PhNMe_2$ → Cl Cl OH

OS, CV 3, 272 (1955)

OH CO_2Et Cl N — 1. aq. NaOH, Δ 2. HCl 3. 250 °C Dowtherm A *or* Ph_2O → OH Cl N

OS, CV 3, 286 (1955)

OH CN CO_2Et EtO_2C CN OH — HCl, H_2O, Δ → OH CO_2H HO_2C OH

OS, CV 3, 317 (1955)

O CH_3 CO_2Et O CH_3 CH_3 CO_2Et — HOAc, H_2SO_4, Δ → O CH_3 CH_3 CO_2Et — 1. NaOH EtOH, Δ 2. H_2SO_4 → O CH_3 CH_3

OS, *CV* 3, 326 (1955)

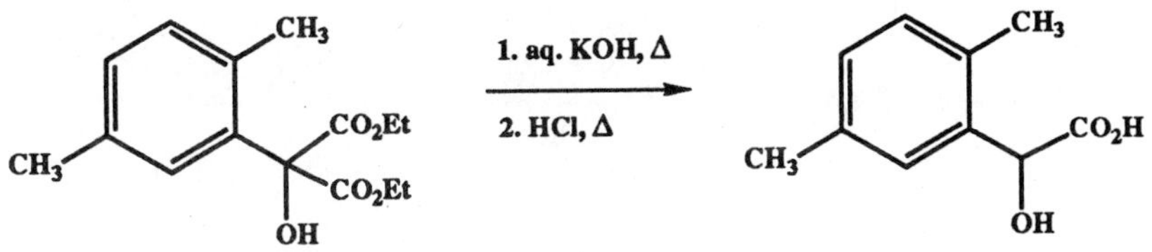

1. aq. KOH, Δ
2. HCl, Δ

OS, *CV* 3, 401 (1955)

$$2\ EtO_2C—(CH_2)_8—CO_2K \xrightarrow[\text{(Pt cathode)}]{\text{electrolysis}} EtO_2C—(CH_2)_{16}—CO_2Et$$

OS, *CV* 3, 425 (1955)

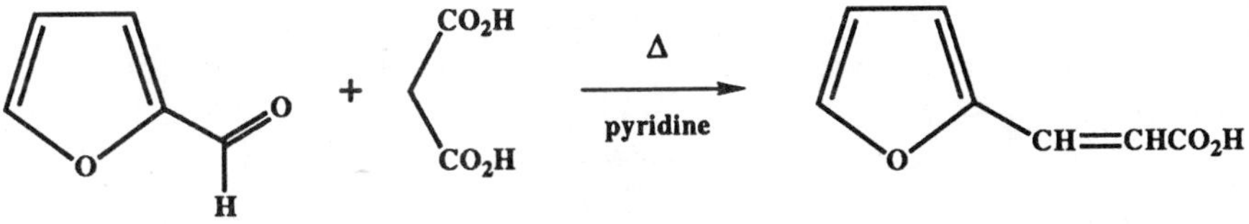

Δ
pyridine

CH=CHCO₂H

OS, *CV* 3, 471 (1955)

CuO, Δ

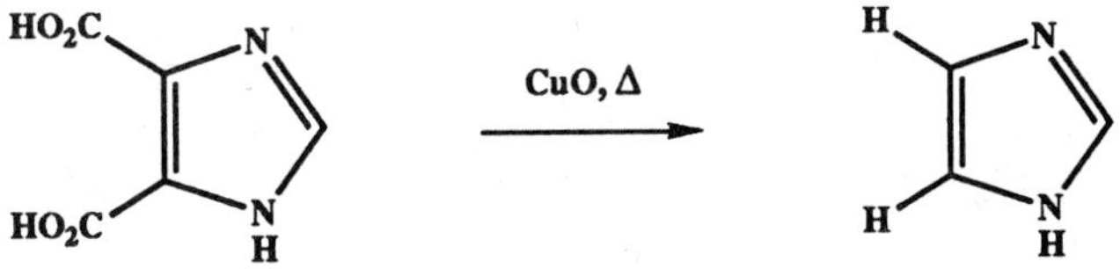

CLEAVAGE - DECARBOXYLATION

***OS, CV 3,* 495 (1955)**

Me, Et, $CH(CO_2Et)_2$ → 1. aq. KOH, Δ, then HCl; 2. Br_2, Et_2O; 3. Δ → Me, Et, CO_2H, Br

***OS, CV 3,* 510 (1955)**

O, EtO_2C, CO_2Et, CO_2Et → H_2O, HCl → O, HO_2C, CO_2H

***OS, CV 3,* 513 (1955)**

O, CH_3, CH_3, EtO_2C, N, H, CH_3 → H_2NNH_2; Na, EtOH, Δ → CH_3, CH_3, H, N, H, CH_3

***OS, CV 3,* 578 (1955)**

MeO_2C, CO_2H → 1. $AgNO_3$, KOH, H_2O; 2. Br_2, CCl_4, Δ → MeO_2C, Br

***OS*, *CV* 3, 591 (1955)**

HCl

H_2O, Δ

***OS*, *CV* 3, 621 (1955)**

Br_2

H_2O, Δ

***OS*, *CV* 3, 637 (1955)**

Δ

H_2O

***OS*, *CV* 3, 705 (1955)**

Δ

***OS, CV 3,* 733 (1955)**

CH3, Ph, H, CO2Et, O — 1. Na, EtOH; 2. aq. HCl, Δ → Ph, O, H, CH3

***OS, CV 3,* 783 (1955)**

$CH_3CH{=}CH{-}CHO$ — $HO_2C\text{-}CH_2\text{-}CO_2H$; pyridine, Δ → $CH_3CH{=}CH{-}CH{=}CHCO_2H$

***OS, CV 4,* 5 (1963)**

CH3, O, OH, NH2 — $(CH_3CO)_2O$; pyridine, Δ → CH3, O, CH3, HN, CH3, O

***OS, CV 4,* 55 (1963)**

O, O, N—C(CO2Et)2, CO2Et — HOAc, HCl; H2O, Δ → HO2C, CO2H, NH2

***dl* - aspartic acid**

OS, CV 4, **93 (1963)**

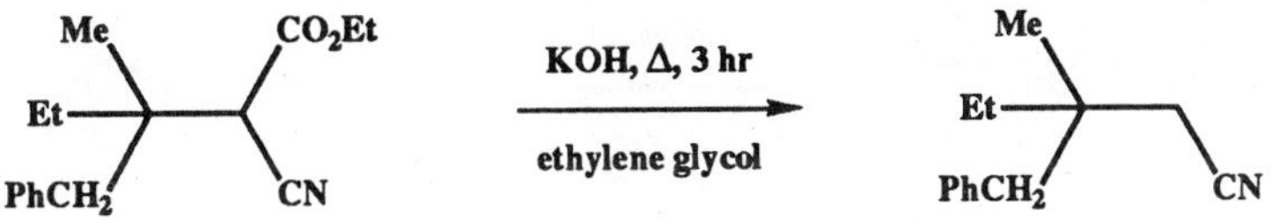

Me, Et, $PhCH_2$ / CO_2Et, CN —(KOH, Δ, 3 hr; ethylene glycol)→ Me, Et, $PhCH_2$ / CN

OS, CV 4, **176 (1963)**

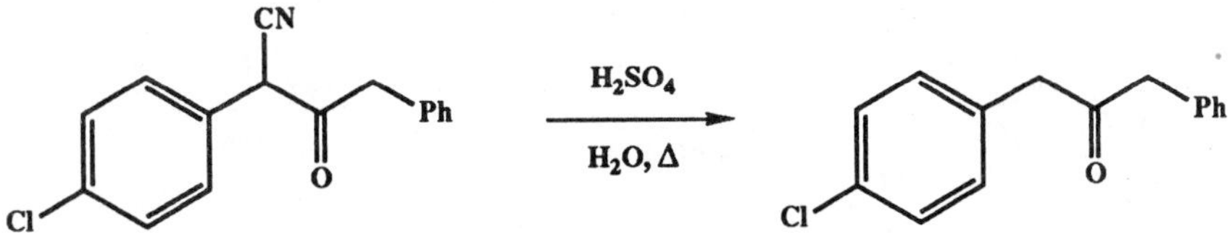

CN, O, Ph, Cl —(H_2SO_4; H_2O, Δ)→ O, Ph, Cl

OS, CV 4, **201 (1963)**

2 OH, HO_2C, CO_2H —(fuming H_2SO_4; Δ)→ CO_2H, O, O

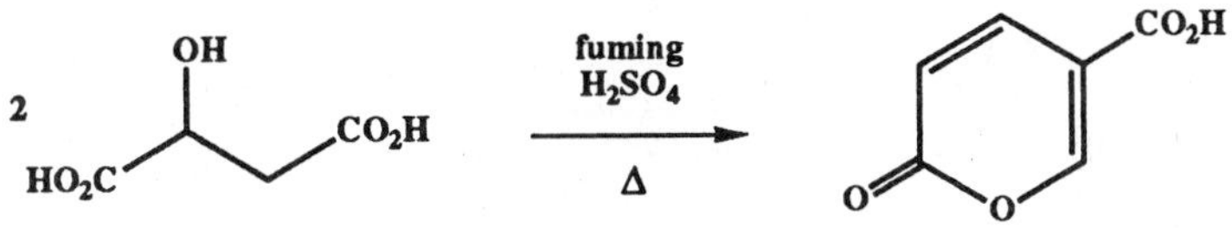

OS, CV 4, **234 (1963)**

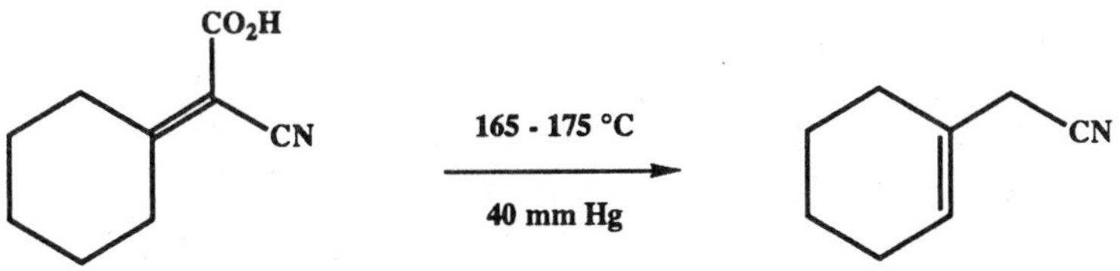

CO_2H, CN —(165 - 175 °C; 40 mm Hg)→ CN

OS, CV 4, 254 (1963)

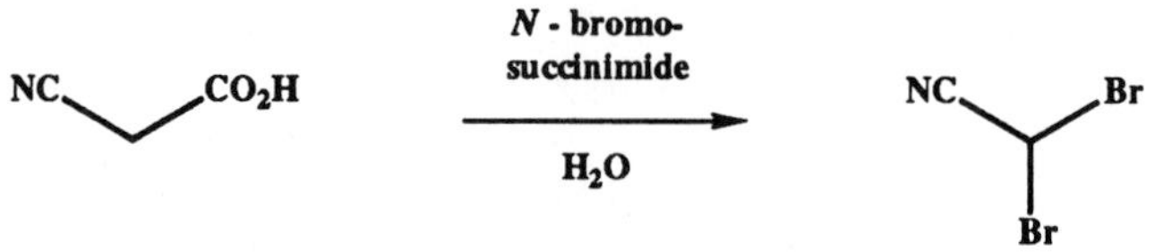
NC
CO2H
N - bromo-
succinimide
H2O
NC
Br
Br

OS, CV 4, 278 (1963)

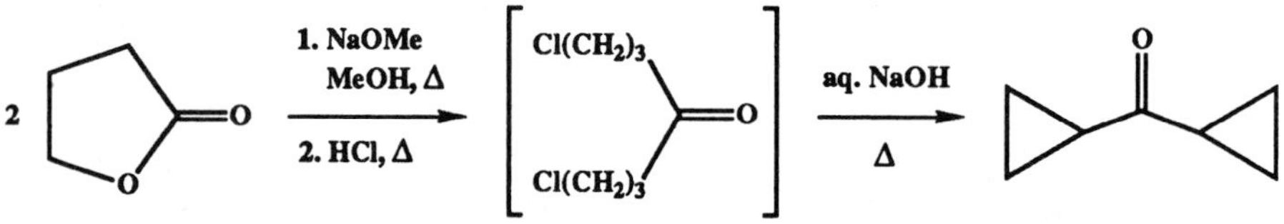
2
O
O
1. NaOMe
MeOH, Δ
2. HCl, Δ
Cl(CH2)3
O
Cl(CH2)3
aq. NaOH
Δ
O

OS, CV 4, 327 (1963)

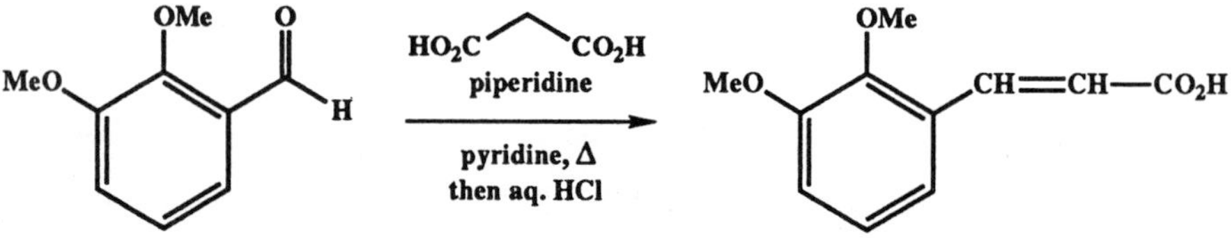
OMe
O
MeO
H
HO2C
CO2H
piperidine
pyridine, Δ
then aq. HCl
OMe
MeO
CH=CH—CO2H

OS, CV 4, 337 (1963)

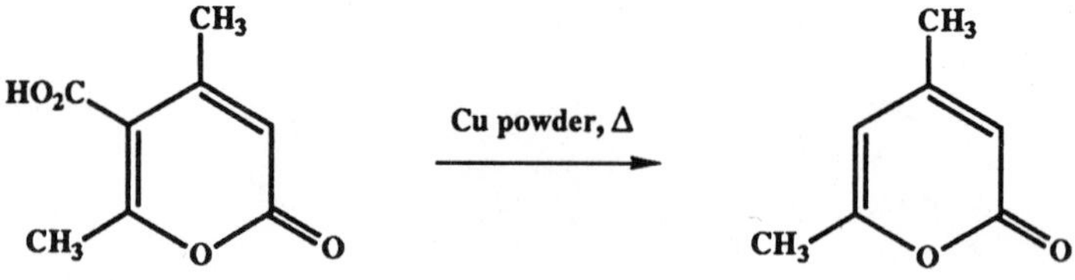
CH3
HO2C
CH3
O
O
Cu powder, Δ
CH3
CH3
O
O

***OS*, *CV* 4, 441 (1963)**

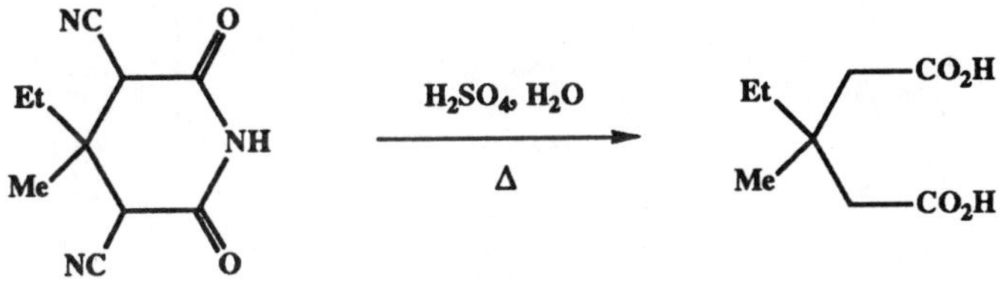

NC, O, Et, NH, Me, NC, O

H_2SO_4, H_2O

Δ

Et, Me, CO_2H, CO_2H

***OS*, *CV* 4, 555 (1963)**

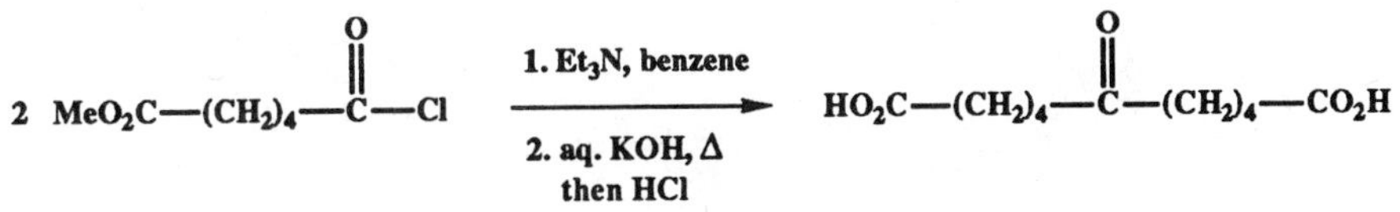

2 $MeO_2C—(CH_2)_4—C(=O)—Cl$

1. Et_3N, benzene
2. aq. KOH, Δ then HCl

$HO_2C—(CH_2)_4—C(=O)—(CH_2)_4—CO_2H$

***OS*, *CV* 4, 560 (1963)**

$C_{10}H_{21}$, O, C=O, $C_{10}H_{21}$

2% H_2SO_4, Δ

or 2% KOH, Δ

$C_{10}H_{21}$, O, $C_{10}H_{21}$

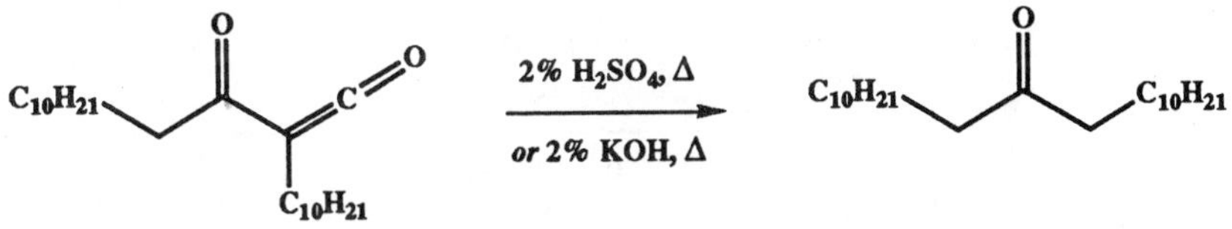

***OS*, *CV* 4, 590 (1963)**

CH_3, O, CO_2H

280 °C

CH_3, O

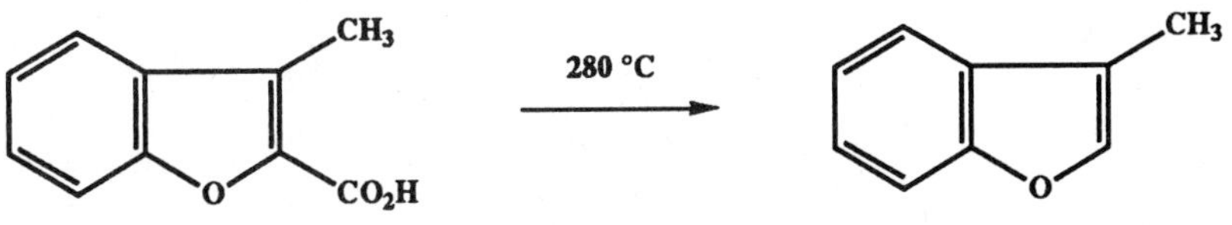

OS, CV 4, 597 (1963)

HCl

H_2O, Δ

CH_3 O CH_3 O O O Cl

OS, CV 4, 628 (1963)

CH_3 CO_2H O

Cu powder

quinoline, Δ

CH_3 O

OS, CV 4, 630 (1963)

CH_3—CH═CH—CO_2Me + EtO_2C CO_2Et

1. Na, EtOH, Δ
2. HCl, H_2O, Δ
3. 180 - 190 °C
(- CO_2)

CH_3 CO_2H CO_2H

OS, CV 4, 633 (1963)

CH_3 O CO_2H

PhN_2^+ Cl^-

aq. NaOAc, 50 °C
(- CO_2)

CH_3 O CH═N—NHPh

OS, CV 4, 641 (1963)

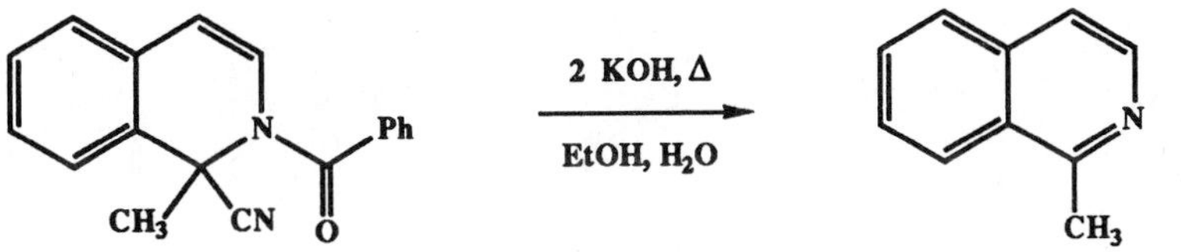

N Ph O CH3 CN

2 KOH, Δ

EtOH, H_2O

N CH_3

OS, CV 4, 664 (1963)

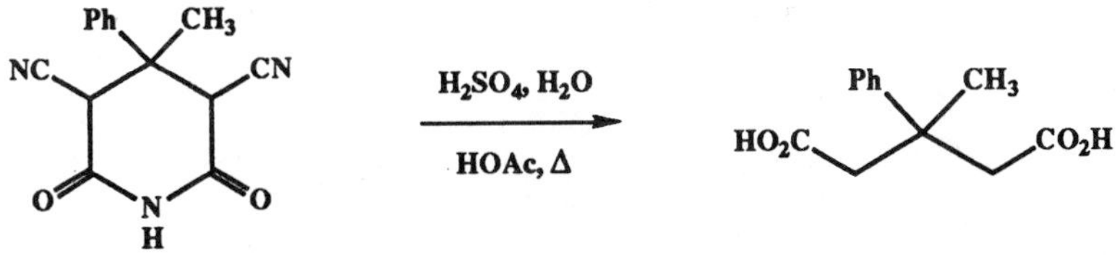

Ph CH_3 NC CN O N H O

H_2SO_4, H_2O

HOAc, Δ

Ph CH_3 HO_2C CO_2H

OS, CV 4, 688 (1963)

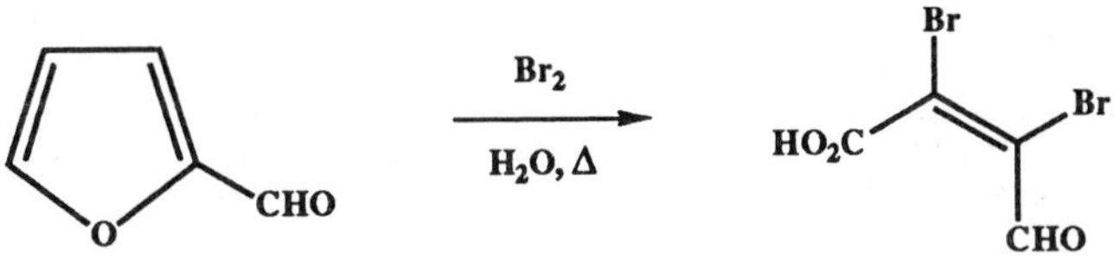

O CHO

Br_2

H_2O, Δ

Br Br HO_2C CHO

OS, CV 4, 708 (1963)

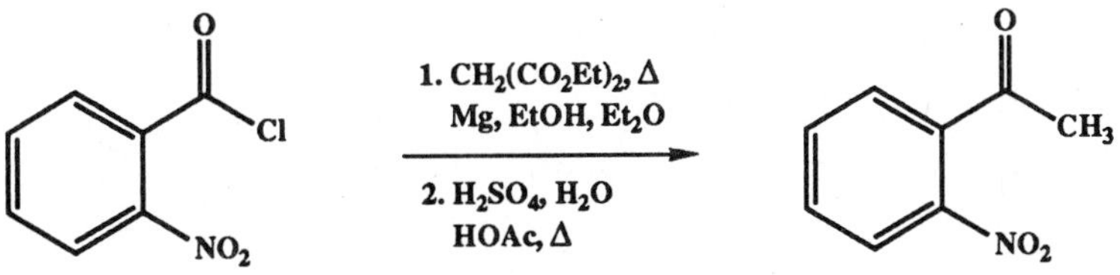

O Cl NO_2

1. $CH_2(CO_2Et)_2$, Δ
   Mg, EtOH, Et_2O

2. H_2SO_4, H_2O
   HOAc, Δ

O CH_3 NO_2

OS, CV 4, 731 (1963)

CHO

$CH_2(CO_2H)_2$

EtOH, C_5H_5N, Δ

NO_2

$CH{=}CHCO_2H$

Cu powder

quinoline, Δ

NO_2

$CH{=}CH_2$

NO_2

OS, CV 4, 790 (1963)

CN

Ph—C—CH_2CH_2CN

CO_2Et

aq. HCl

HOAc, Δ

H

Ph—C—$CH_2CH_2CO_2H$

CO_2H

OS, CV 4, 804 (1963)

CO_2Et

PhCH=

CO_2Et

KCN, aq. EtOH

then HCl

Ph

CO_2Et

CN

OS, CV 4, 816 (1963)

CHO

CHO

+ $MeNH_2$ +

HO_2C—

=O

HO_2C—

aq. NaOH

Na_2HPO_4, Δ

MeN

=O

***OS*, *CV 4*, 844 (1963)**

Br, Br, HO_2C, CHO → ($NaNO_2$; EtOH, H_2O) → $Na^+ \; ^-C(CHO)_2$—NO_2 • H_2O

***OS*, *CV 4*, 854 (1963)**

$$2\ C_{17}H_{35}\text{—}CO_2H \xrightarrow{MgO} (C_{17}H_{35}CO_2)_2Mg \xrightarrow[(-CO_2)]{340\ °C} C_{17}H_{35}C(=O)C_{17}H_{35}$$

***OS*, *CV 4*, 857 (1963)**

H, CO_2H, Ph, Ph (*cis*) → (Cu chromite; quinoline, Δ) → H, H, Ph, Ph (*cis*)

***OS*, *CV 5*, 27 (1973)**

H_2N, O, OH → (2 Ac_2O; pyridine, Δ) → Ac, H, N, O, CH_3

OS, *CV 5*, 51 (1973)

m-$C_6H_4(NO_2)_2$, Δ

OS, *CV 5*, 76 (1973)

CO_2Me CH_2Ph O

LiI • 2 H_2O

2,4,6-collidine, Δ

CH_2Ph O

OS, *CV 5*, 126 (1973)

CO_2H

Br_2, HgO

tetrachloroethane

Br

OS, *CV 5*, 130 (1973)

CCl_3 Br

aq. KOH, Δ

diethylene glycol

Br

OS, CV 5, 187 (1973)

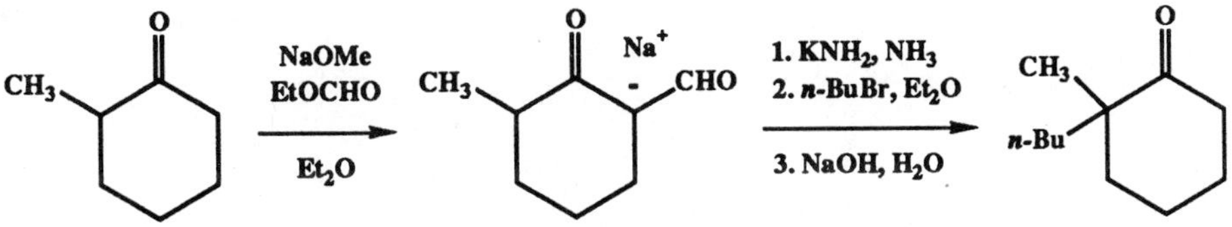

CH_3 — NaOMe, EtOCHO, Et_2O → CH_3, Na^+, CHO → 1. KNH_2, NH_3; 2. *n*-BuBr, Et_2O; 3. NaOH, H_2O → CH_3, *n*-Bu

OS, CV 5, 251 (1973)

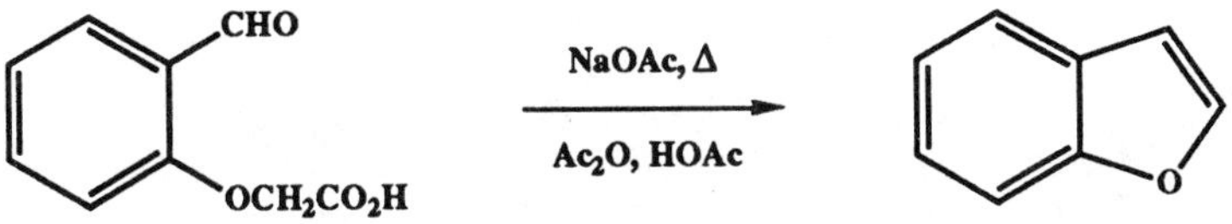

CHO, OCH_2CO_2H — NaOAc, Δ; Ac_2O, HOAc → O

OS, CV 5, 273 (1973)

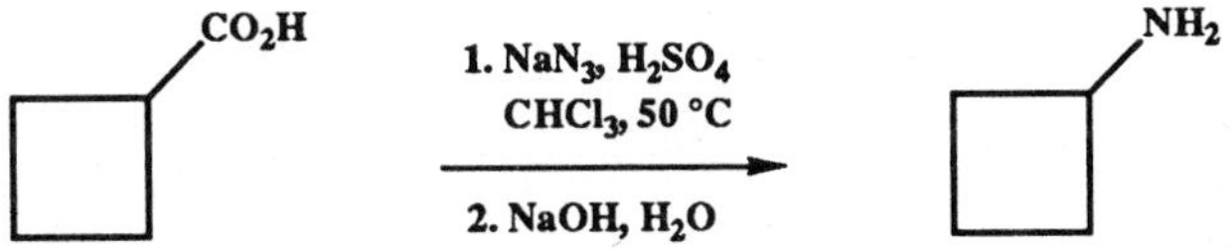

CO_2H — 1. NaN_3, H_2SO_4, $CHCl_3$, 50 °C; 2. NaOH, H_2O → NH_2

OS, CV 5, 277 (1973)

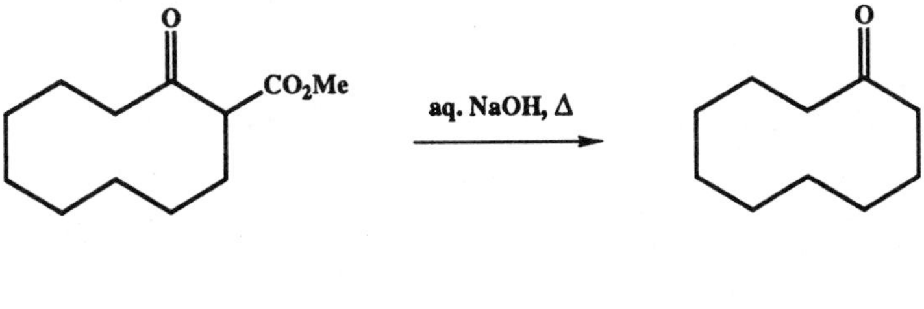

O, CO_2Me — aq. NaOH, Δ → O

OS, CV 5, 288 (1973)

O, CO_2Et, EtO_2C, O → O, O

185 - 195 °C

H_2O

OS, CV 5, 390 (1973)

O, H → CF_2, H

$Cl—CF_2—CO_2Na$

Ph_3P

diglyme, Δ

OS, CV 5, 445 (1973)

O_2N, CO_2Me → O_2N, NO_2

1. KOH, H_2O, Δ

2. electricity (steel cathode)

OS, CV 5, 463 (1973)

2 $MeO_2C—(CH_2)_8—CO_2H$ → $MeO_2C—(CH_2)_{16}—CO_2Me$

1. Na, MeOH

2. electricity (Pt cathode)

OS, CV 5, 572 (1973)

Et, CH_3 C=O + CO_2Et, CN (ethyl cyanoacetate) —KCN / EtOH, HOAc→ [NC–C(Et)(CH_3)–CH(CO_2Et)–CN] —aq. HCl / Δ→ HO_2C–C(Et)(CH_3)–CH_2CO_2H

OS, CV 5, 585 (1973)

Furfural (CHO) + CO_2H, CN —NH_4OAc, Δ / pyridine, toluene→ furyl–CH=CH—CN

OS, CV 5, 589 (1973)

$$2\ CH_3CH_2CH_2CO_2H \xrightarrow[\Delta]{Fe} [(CH_3CH_2CH_2CO_2)_2Fe] \longrightarrow CH_3CH_2CH_2—C(=O)—CH_2CH_2CH_3$$

OS, CV 5, 635 (1973)

3-hydroxyquinoline-4-carboxylic acid (CO_2H, OH, N) —boiling diethyl succinate→ 3-hydroxyquinoline (OH, N)

OS, CV 5, 687 (1973)

EtO$_2$C, O, CO$_2$Et, CO$_2$Et — aq. HCl, Δ → O, HO$_2$C, CO$_2$H

OS, CV 5, 747 (1973)

CH$_3$, O, O, O, O, CO$_2$Et — H$_3$PO$_4$ / H$_2$O, Δ → CH$_3$, O, O, O

OS, CV 5, 982 (1973)

HO$_2$C, O, O — 650 °C → O, O

OS, CV 5, 1037 (1973)

I$^+$, Ph, CO$_2^-$ + Ph, Ph, O, Ph, Ph — C$_6$H$_4$Et$_2$, Δ / (- CO, - CO$_2$) → Ph, Ph, Ph, Ph

***OS, CV 6,* 95 (1988)**

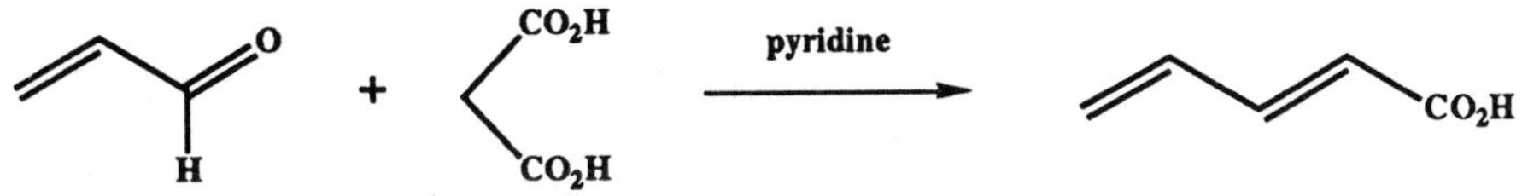

O H + CO_2H CO_2H —pyridine→ CO_2H

***OS, CV 6,* 115 (1988)**

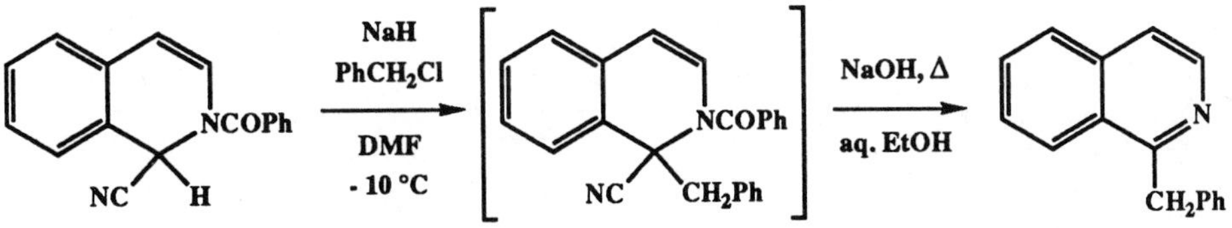

NCOPh NC H —NaH, $PhCH_2Cl$, DMF, - 10 °C→ [NCOPh NC CH_2Ph] —NaOH, Δ, aq. EtOH→ N CH_2Ph

***OS, CV 6,* 179 (1988)**

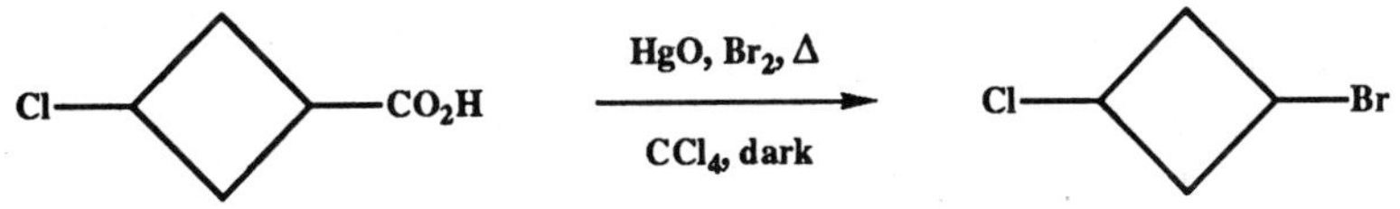

Cl CO_2H —HgO, Br_2, Δ, CCl_4, dark→ Cl Br

***OS, CV 6,* 271 (1988)**

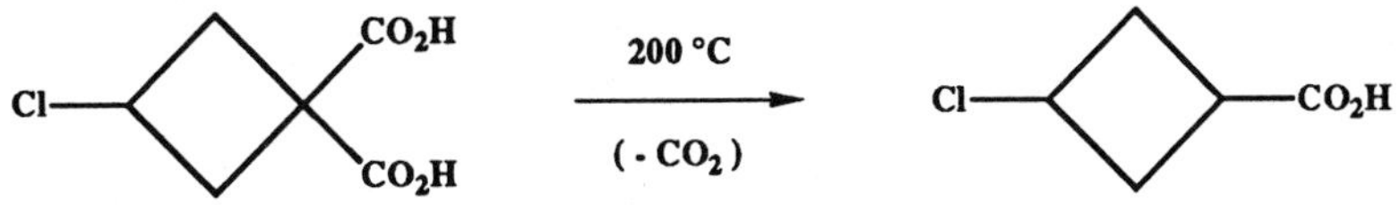

Cl CO_2H CO_2H —200 °C, (- CO_2)→ Cl CO_2H

OS, CV 6, 403 (1988)

Ph—CHCO$_2$H / S / Ph—CHCO$_2$H —(H_2O_2, HOAc)→ Ph—CHCO$_2$H / SO$_2$ / Ph—CHCO$_2$H —(Br_2, KBr, aq. HOAc, light, 80 °C)→ Ph—CHBr / SO$_2$ / Ph—CHBr

OS, CV 6, 418 (1988)

t-Bu-O–C(=O)–O–C(=O)–O–C(=O)–O-t-Bu —(diazabicyclo-octane, CCl_4, 25 °C)→ t-Bu-O–C(=O)–O–C(=O)–O-t-Bu

OS, CV 6, 615 (1988)

O, CO$_2$Et, CO$_2$Et —(boric acid, 150 °C)→ O, CO$_2$Et

OS, CV 6, 781 (1988)

Cl, CH$_3$, O, N, Et + O, CO$_2$Et —(1. NaH, DMF, benzene; 2. aq. HCl, HOAc, Δ)→ O, CH$_3$, O, N, Et

OS, CV 6, 873 (1988)

H_2SO_4, HOAc

H_2O, Δ

OS, CV 6, 890 (1988)

$AgNO_3$

$(NH_4)_2S_2O_8$

H_2O, 60 - 65 °C

OS, CV 6, 932 (1988)

aq. HCl

Δ

OS, CV 6, 965 (1988)

1. aq. HCl

2. aq. KOH

CLEAVAGE - DECARBOXYLATION

OS, *CV 7*, 181 (1990); *60*, 1 (1981)

$$2\ MeO_2C(CH_2)_4COOH \xrightarrow[\text{NaOMe, MeOH}]{\text{electricity}} MeO_2C(CH_2)_8CO_2Me + 2\ CO_2$$

OS, *CV 7*, 213 (1990); *61*, 5 (1983)

$$HO_2CCH_2CO_2Et \xrightarrow[\text{2. } Me_2CHCOCl]{\text{1. } n\text{-BuLi, THF}} Me_2CHCOCH_2CO_2Et$$

OS, *CV 7*, 381 (1990); *63*, 79 (1985)

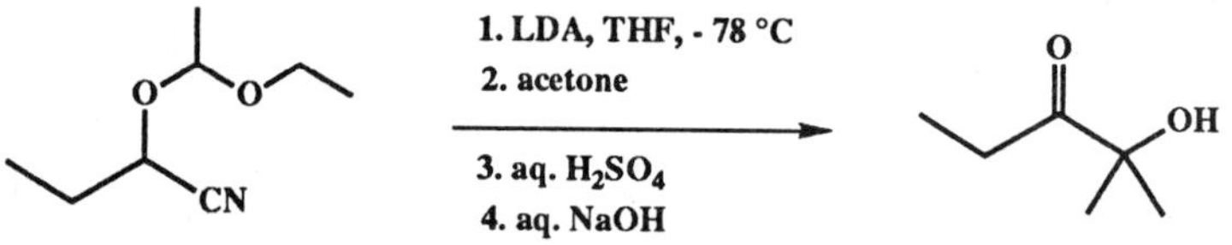

1. LDA, THF, - 78 °C
2. acetone
3. aq. H_2SO_4
4. aq. NaOH

CN, O, O, OH

OS, *CV 7*, 359 (1990); *63*, 198 (1985)

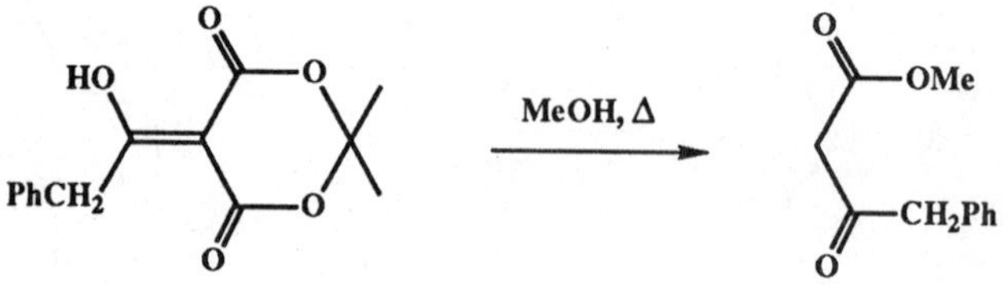

MeOH, Δ

HO, $PhCH_2$, O, O, O, O → O, OMe, CH_2Ph, O

OS, *66*, 29 (1987)

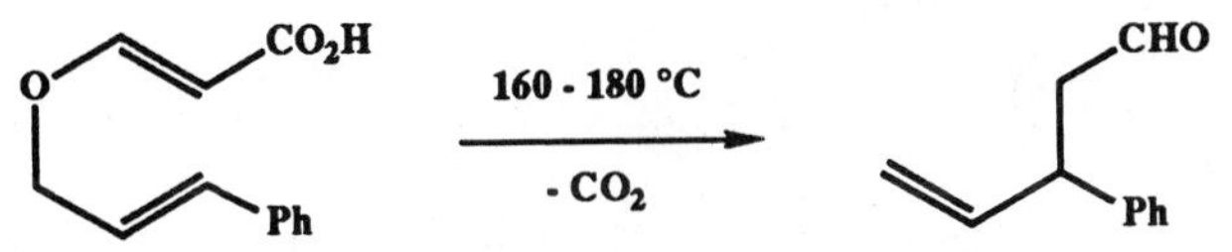

160 - 180 °C

- CO_2

OS, *66*, 173 (1987)

Et_3N

90 °C

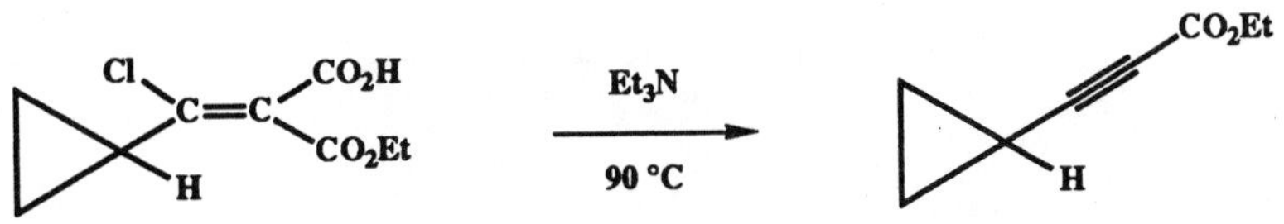

OS, *68*, 210 (1989)

CCl_4

reflux

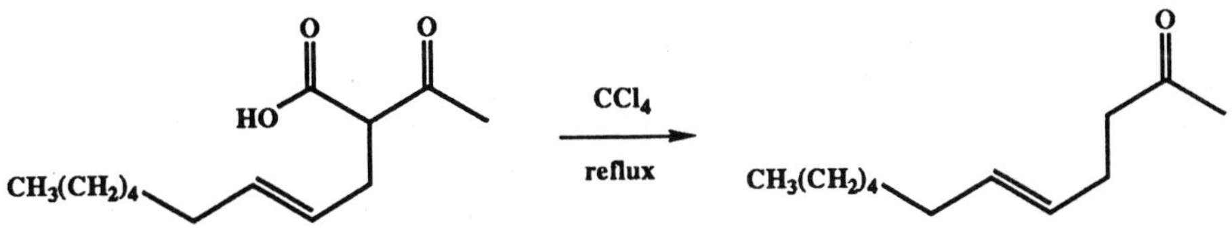

***OS*, *CV 1*, 18 (1941)**

OH (cyclohexanol) → 50% HNO_3, Δ / NH_4VO_3 (cat.) → $HO_2C—(CH_2)_4—CO_2H$

***OS*, *CV 1*, 149 (1941)**

CHO, OH → H_2O_2 / NaOH → OH, OH

***OS*, *CV 1*, 385 (1941)**

N, N, Me → HNO_3, Δ → CO_2H, N

***OS*, *CV 1*, 526 (1941)**

O, *t*-Bu, CH_3 → 1. 4 NaOH, 3 Br_2 / 2. H_2SO_4 → O, *t*-Bu, OH

***OS*, *CV 2*, 44 (1943)**

O, MeO, MeO, NH_2 → NaOCl, NaOH / H_2O → MeO, MeO, NH_2

***OS*, *CV* 2,** 53 (1943)

$CH_3(CH_2)_5CH(OH)CH_2CH{=}CH{-}(CH_2)_7CO_2H$ → 1. $KMnO_4$, KOH, H_2O; 2. H_2SO_4 → $HO_2C(CH_2)_7CO_2H$

***OS*, *CV* 2,** 302 (1943)

Furfural → $NaClO_3$, V_2O_5 (cat.), H_2O, 75 °C → $HO_2C{-}CH{=}CH{-}CO_2H$

***OS*, *CV* 2,** 333 (1943)

3,4-(MeO)$_2C_6H_3CH_2COCO_2H$ → 1. H_2O_2, aq. NaOH; 2. HCl, H_2O → 3,4-(MeO)$_2C_6H_3CH_2CO_2H$

***OS*, *CV* 2,** 428 (1943)

2-Naphthyl-$COCH_3$ → 1. NaOH, Cl_2, H_2O; 2. HCl → 2-Naphthyl-COOH

OS, *CV 2*, 523 (1943)

O
CO_2H
CO_2H
CHO
CO_2H

$KMnO_4$

aq. NaOH, Δ

$NaHSO_3$

HCl, H_2O

OS, *CV 3*, 39 (1955)

H
N
O
O
=CH—Ph
N
H
O

H
N
O
OH
O
OH
N
H
O

CrO_3, HOAc

H_2O, 50 - 60 °C

OS, *CV 3*, 234 (1955)

Ph
AcO
Ph
AcO

O
OH
HO
HO

1. CrO_3
HOAc

2. KOH
3. HCl

OS, *CV 3*, 302 (1955)

O
CH_3
CH_3
CH_3

O
CH_3
OH
CH_3

1. KOCl
aq. dioxane

2. H_2SO_4

OS, CV 3, 420 (1955)

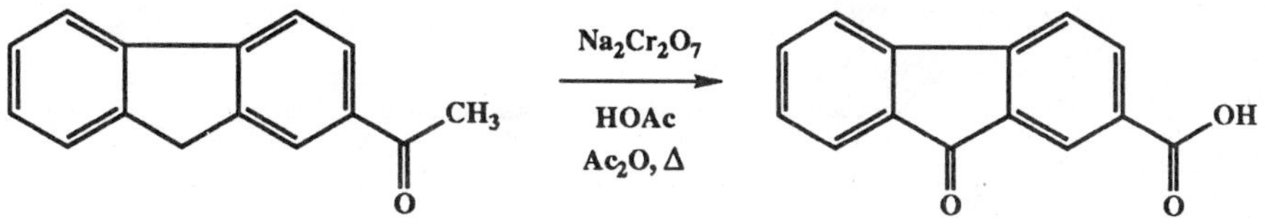

$Na_2Cr_2O_7$

HOAc

Ac_2O, Δ

CH₃ O OH O O

OS, CV 3, 449 (1955)

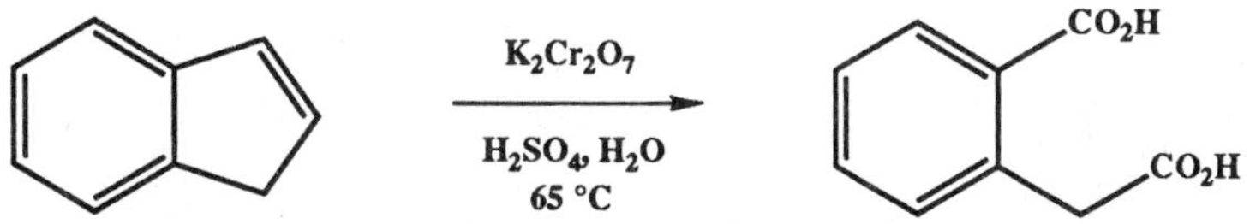

$K_2Cr_2O_7$

H_2SO_4, H_2O

65 °C

CO_2H CO_2H

OS, CV 3, 759 (1955)

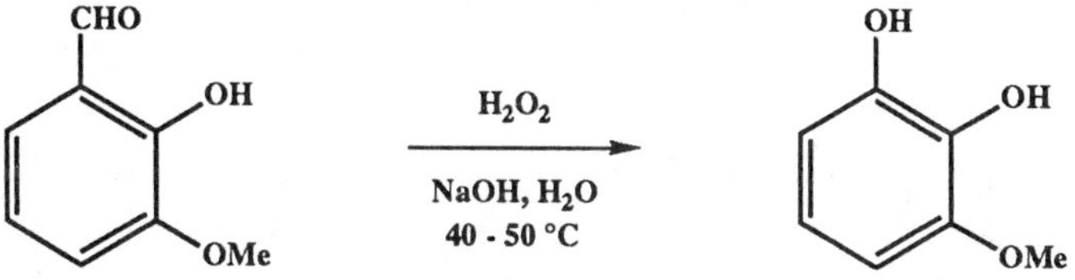

CHO OH OMe

H_2O_2

NaOH, H_2O

40 - 50 °C

OH OH OMe

OS, CV 3, 791 (1955)

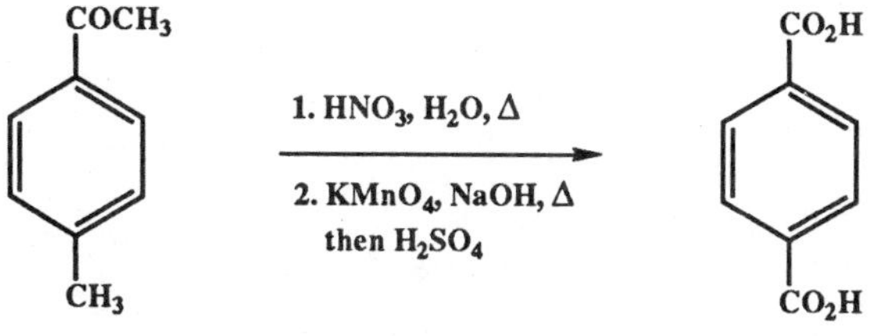

$COCH_3$ CH_3

1. HNO_3, H_2O, Δ

2. $KMnO_4$, NaOH, Δ then H_2SO_4

CO_2H CO_2H

OS, CV 3, 803 (1955)

$$4\ (CH_3CO)_2O + 4\ HNO_3 \longrightarrow C(NO_2)_4 + 7\ CH_3CO_2H + CO_2$$

OS, CV 3, 822 (1955)

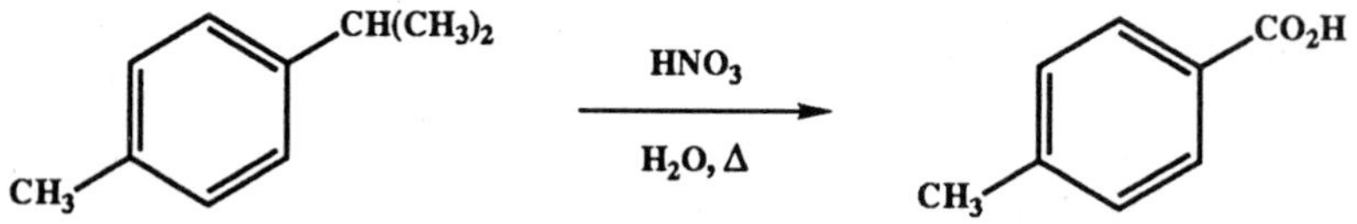

$CH(CH_3)_2$; CH_3 — HNO_3 / H_2O, Δ → CO_2H; CH_3

OS, CV 4, 19 (1963)

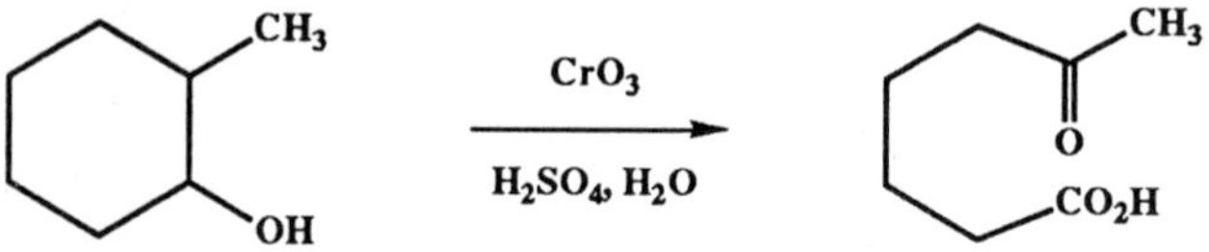

CH_3; OH — CrO_3 / H_2SO_4, H_2O → CH_3; O; CO_2H

OS, CV 4, 124 (1963)

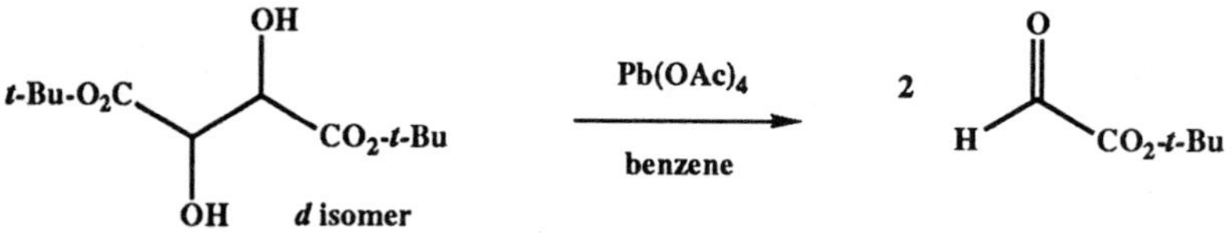

OH; *t*-Bu-O_2C; CO_2-*t*-Bu; OH; *d* isomer — $Pb(OAc)_4$ / benzene → 2 O; H; CO_2-*t*-Bu

OS, CV 4, 136 (1963)

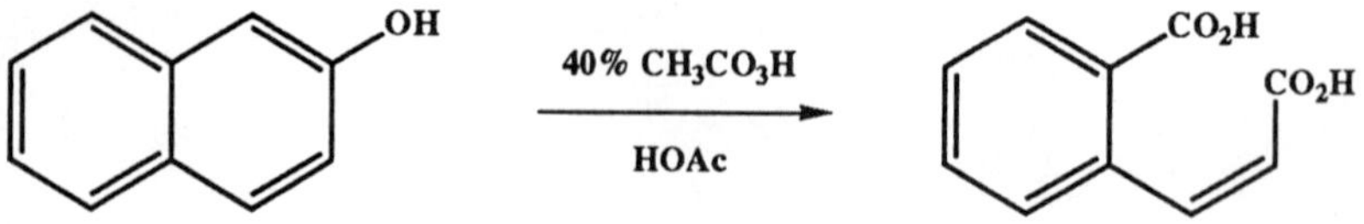

OH — 40% CH_3CO_3H / HOAc → CO_2H; CO_2H

OS, *CV 4*, 345 (1963)

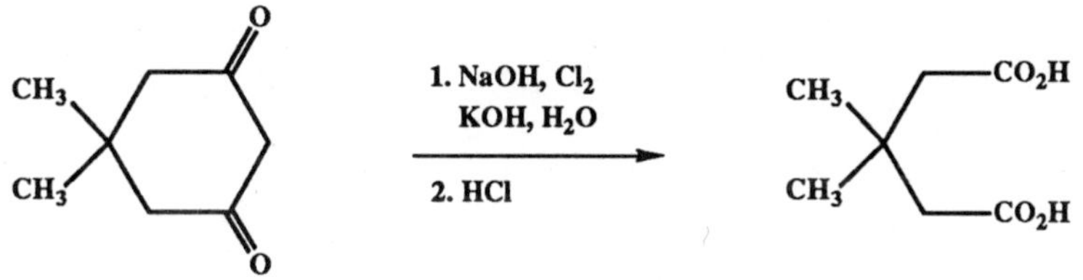

1. NaOH, Cl_2
KOH, H_2O
2. HCl

CH_3 CH_3 CO_2H CO_2H O O

OS, *CV 4*, 484 (1963)

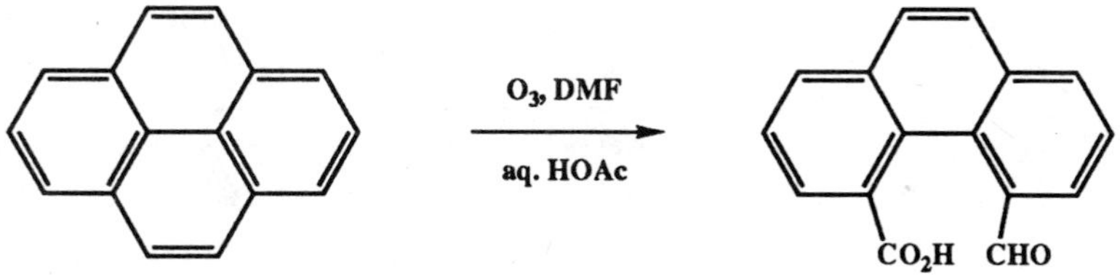

O_3, DMF
aq. HOAc

CO_2H CHO

OS, *CV 4*, 824 (1963)

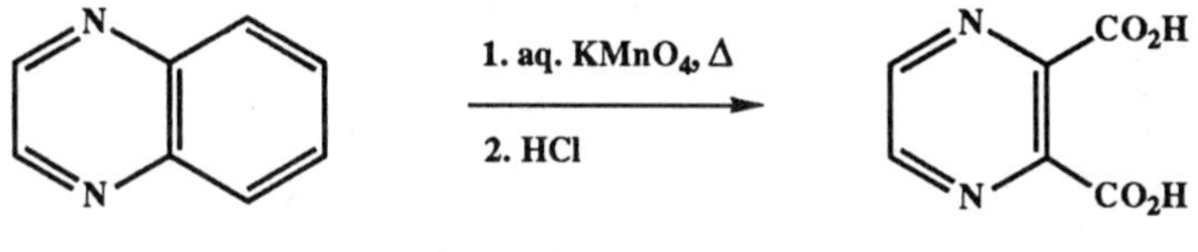

1. aq. $KMnO_4$, Δ
2. HCl

N N CO_2H CO_2H

OS, *CV 5*, 46 (1973)

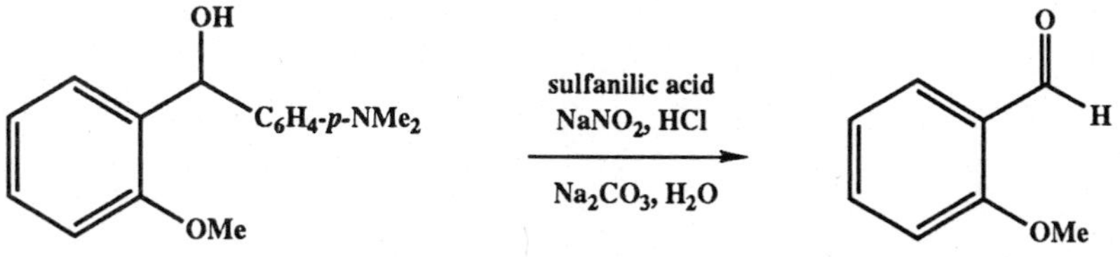

OH C_6H_4-*p*-NMe_2 OMe

sulfanilic acid
$NaNO_2$, HCl
Na_2CO_3, H_2O

O H OMe

OS, CV 5, **393 (1973)**

CH_2—CCl / CF_2=CF → $KMnO_4$, NaOH, then H_2SO_4 → CH_2—CO_2H / CF_2—CO_2H

OS, CV 5, **489 (1973)**

1. O_3, MeOH, - 30 °C
2. NaI *or* KI, HOAc

CHO CHO

OS, CV 5, **493 (1973)**

O_3, MeOH, - 30 °C, then HCl

OMe, O, O, OMe

NaOH, EtOH, Δ

CO_2H, CHO

OS, CV 5, **818 (1973)**

OOH → 70% H_2SO_4, 15 - 25 °C → OH

OS, *CV 6*, 662 (1988)

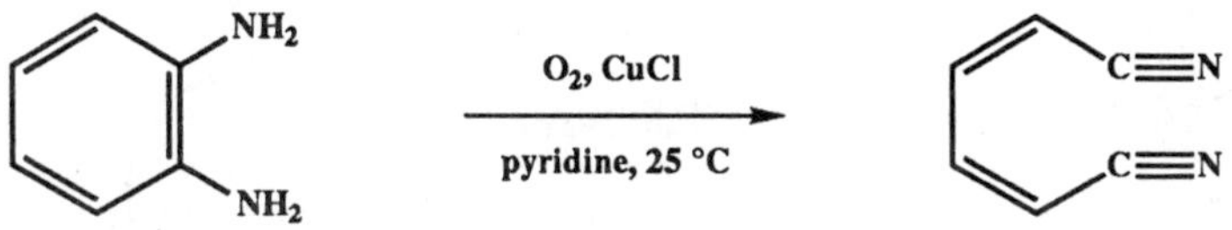

NH_2 / NH_2 —[O_2, CuCl / pyridine, 25 °C]→ C≡N / C≡N

OS, *CV 6*, 690 (1988)

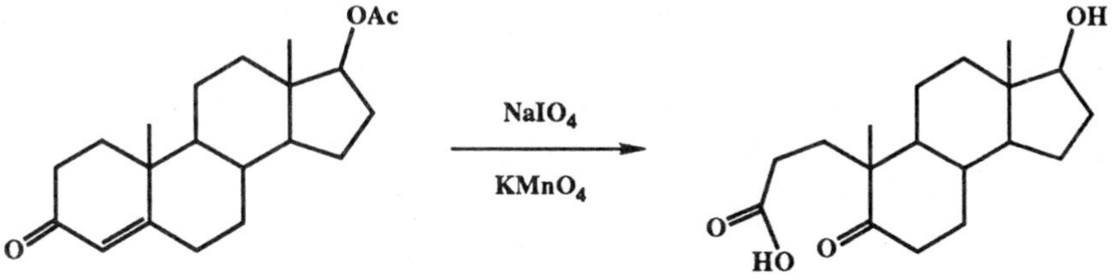

OAc / O —[$NaIO_4$ / $KMnO_4$]→ OH / O / HO / O

OS, *CV 6*, 958 (1988)

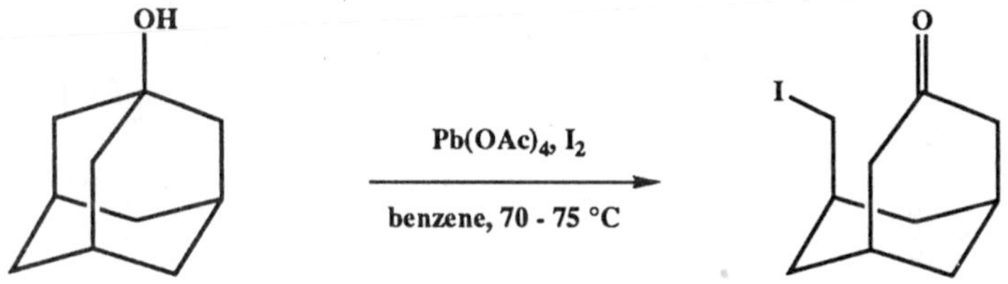

OH —[$Pb(OAc)_4$, I_2 / benzene, 70 - 75 °C]→ O / I

endo

OS, *CV 6*, 976 (1988)

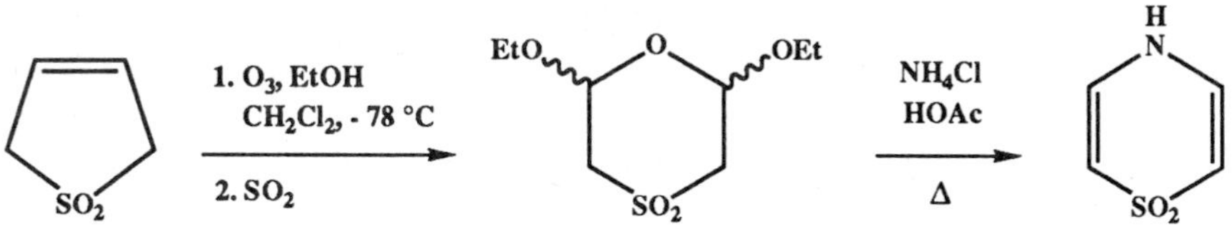

SO_2 —[1. O_3, EtOH, CH_2Cl_2, - 78 °C / 2. SO_2]→ EtO / O / OEt / SO_2 —[NH_4Cl, HOAc / Δ]→ H / N / SO_2

ca. 55 : 45 *cis* : *trans*

OS, *CV 6*, 1024 (1988)

$Ce(NH_4)_2(NO_3)_6$

aq. CH_3CN, Δ

OS, *CV 7*, 397 (1990); *60*, 11 (1981)

$CH_3(CH_2)_{17}CH{=}CH_2$ → $CH_3(CH_2)_{17}CO_2H$

$KMnO_4$, Adogen 464

CH_2Cl_2, H_2O

OS, *CV 7*, 185 (1990); *63*, 89 (1985)

H_5IO_6

THF, 25 °C

OS, *CV 7*, 168 (1990); *64*, 150 (1986)

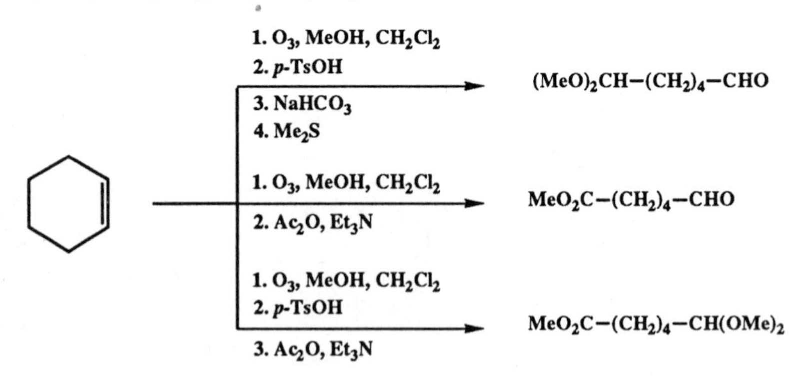

OS, *66*, 180 (1987)

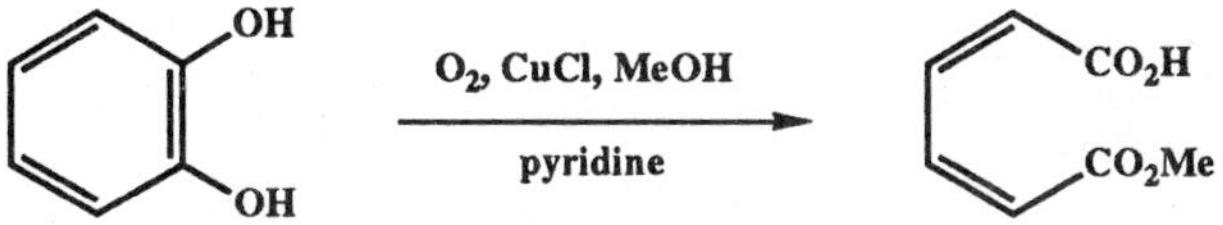

OH
OH

O_2, CuCl, MeOH
pyridine

CO_2H
CO_2Me

OS, *68*, 41 (1989)

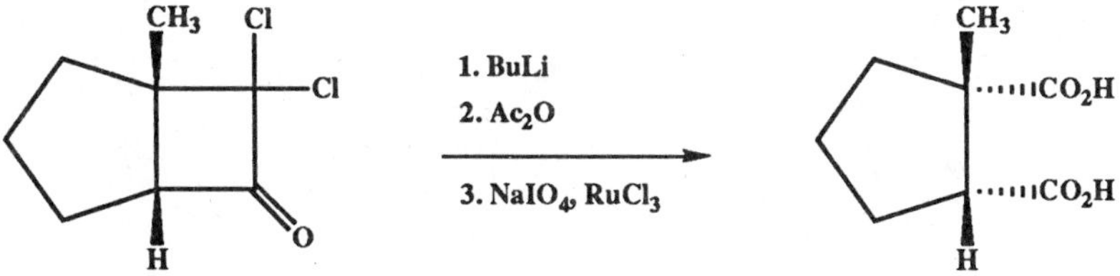

CH_3 Cl
Cl
H
O

1. BuLi
2. Ac_2O
3. $NaIO_4$, $RuCl_3$

CH_3
CO_2H
CO_2H
H

OS, *68*, 162 (1989)

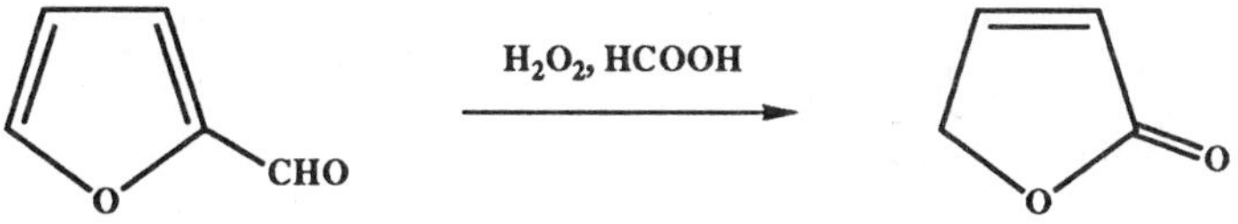

O CHO

H_2O_2, HCOOH

O O

OS, CV 1, **330 (1941)**

$CH_3COCH_3 \xrightarrow[(-CH_4)]{\Delta} [CH_2{=}C{=}O]$

OS, CV 2, **19 (1943)**

succinimide $\xrightarrow[H_2O]{KOH,\ Br_2}$ $H_2N\text{–}CH_2CH_2\text{–}CO_2H$

OS, CV 2, **102 (1943)**

cyclohexene $\xrightarrow{\text{cracking}}$ butadiene + $CH_2{=}CH_2$

OS, CV 2, **266 (1943)**

$MeCOCH(COPh)CO_2Et \xrightarrow[H_2O]{NH_3,\ NH_4Cl} PhCOCH_2CO_2Et$

***OS*, *CV 2*, 279 (1943)**

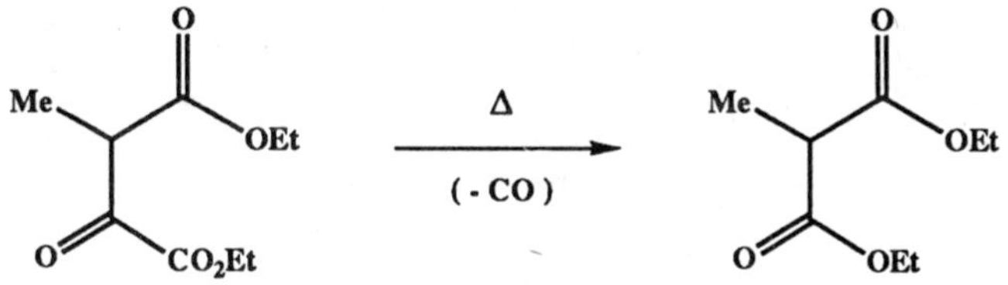

Me, O, OEt, O, CO_2Et —Δ (- CO)→ Me, O, OEt, O, OEt

***OS*, *CV 2*, 288 (1943)**

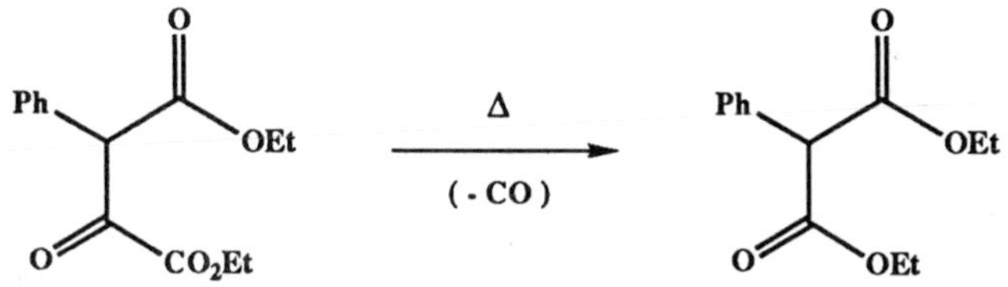

Ph, O, OEt, O, CO_2Et —Δ (- CO)→ Ph, O, OEt, O, OEt

***OS*, *CV 2*, 531 (1943)**

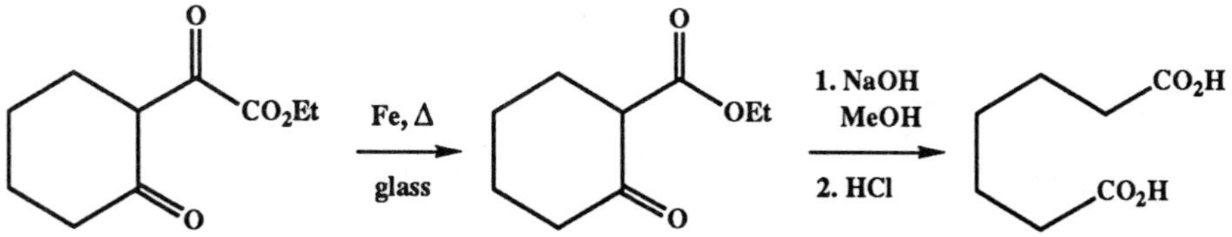

O, CO_2Et, O —Fe, Δ glass→ O, OEt, O —1. NaOH MeOH 2. HCl→ CO_2H, CO_2H

OS, CV 2, 535 (1943)

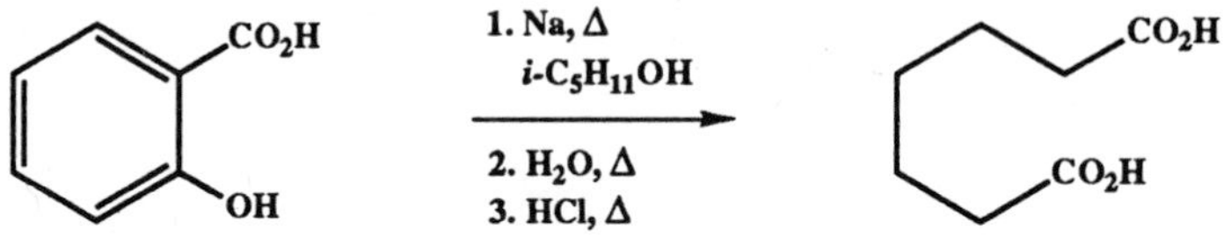

1. Na, Δ, *i*-$C_5H_{11}OH$
2. H_2O, Δ
3. HCl, Δ

OS, CV 3, 101 (1955)

NaOMe, MeOH, - 12 °C

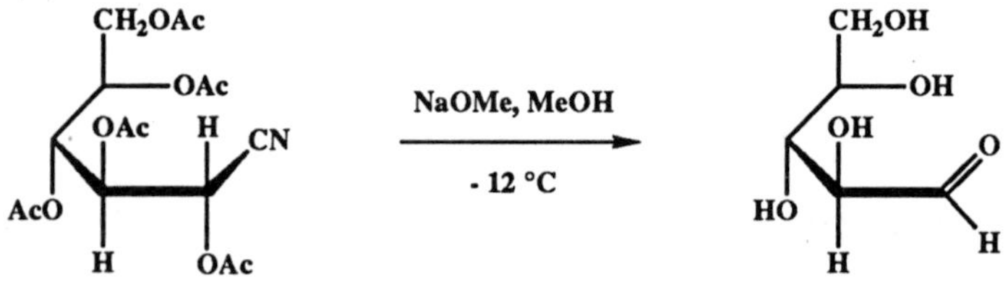

d - isomer → *d* - arabinose

OS, CV 3, 281 (1955)

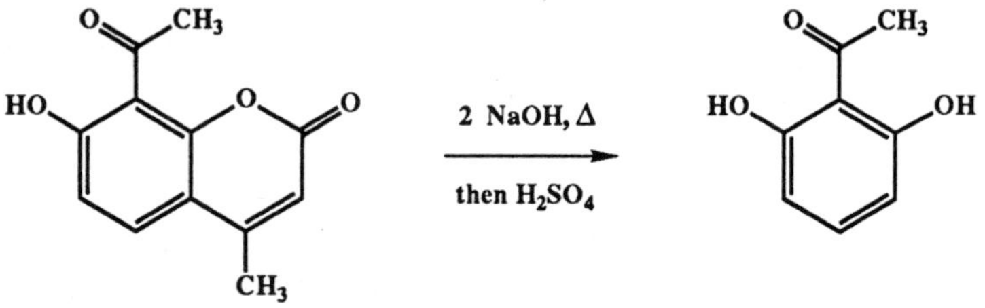

2 NaOH, Δ, then H_2SO_4

OS, CV 3, 329 (1955)

$PhNH_2$, Na, Cu, 230 °C

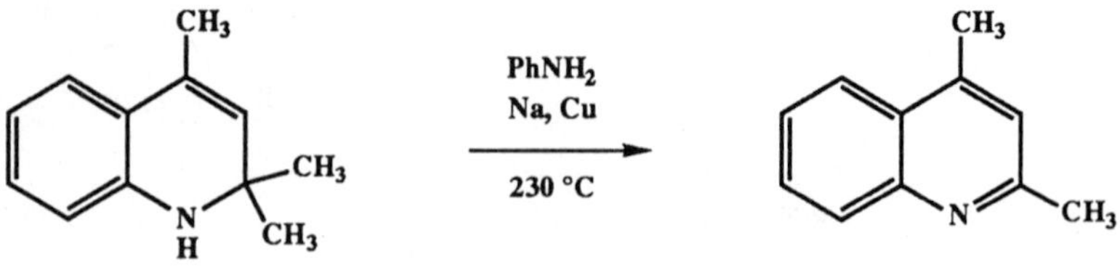

OS, CV 3, 379 (1955)

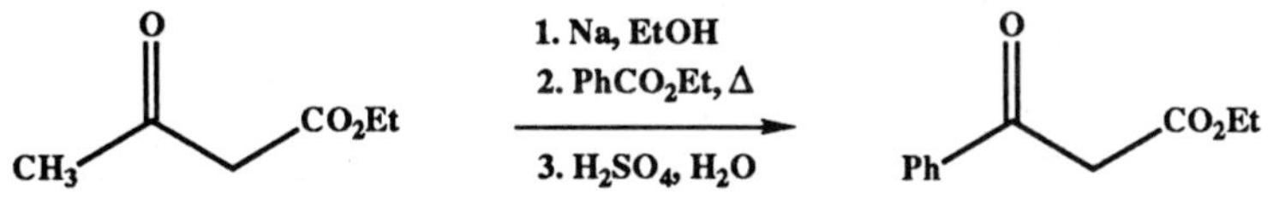

O
CH_3 CO_2Et
1. Na, EtOH
2. $PhCO_2Et$, Δ
3. H_2SO_4, H_2O
O
Ph CO_2Et

OS, CV 3, 807 (1955)

Ph Ph Ph Ph O
O O O
C_6H_5Br, Δ
(- CO)
Ph Ph Ph Ph O O O
Br_2
Δ
Ph Ph Ph Ph O O O

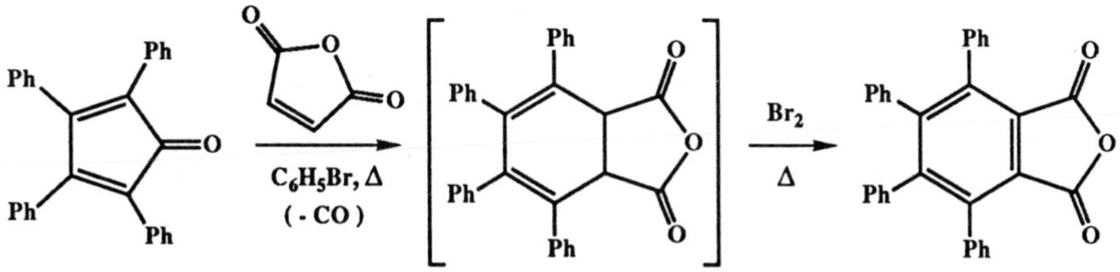

OS, CV 4, 45 (1963)

O
NH_2
N
Br_2, 70 °C
aq. NaOH
NH_2
N

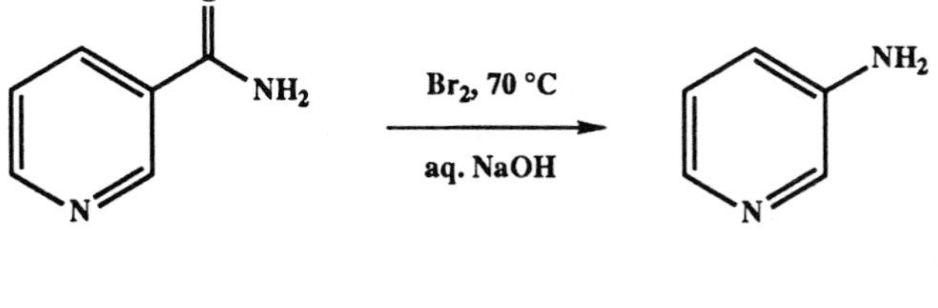

OS, CV 4, 141 (1963)

$C_{16}H_{33}$—CH—CO_2Et (with C(=O)CO_2Et on CH) —Δ, (- CO)→ $C_{16}H_{33}$—CH—CO_2Et (with CO_2Et on CH)

OS, CV 4, 238 (1963)

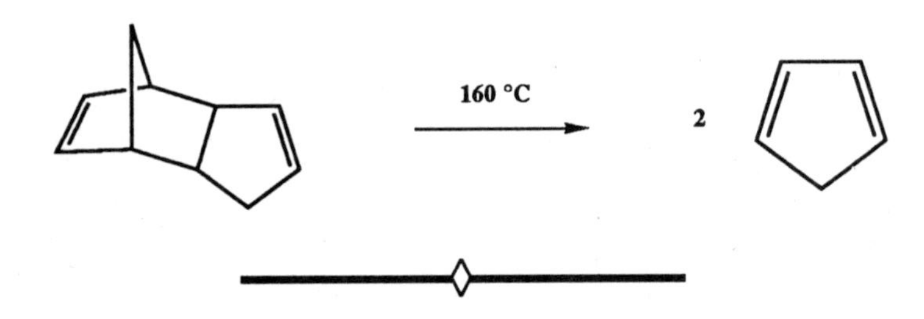

OS, CV 4, 276 (1963)

(NC)$_2$C=C(CN)$_2$ + HO–CH$_2$CH$_2$–OH —urea, 75 °C→ (NC)$_2$C=C(–O–CH$_2$CH$_2$–O–)

OS, CV 4, 415 (1963)

CH_3C(=O)CH_2CO_2Et —1. PhCOCl, aq. NaOH, naphtha; 2. NH_4Cl→ PhC(=O)CH_2CO_2Et

***OS*, *CV 4*, 819 (1963)**

$H_2NNH-C(=O)-(CH_2)_4-C(=O)-NHNH_2$ —($NaNO_2$; HCl, H_2O)→ [$N_3-C(=O)-(CH_2)_4-C(=O)-N_3$] —(Δ, benzene)→ $O=C=N-(CH_2)_4-N=C=O$ —(HCl, H_2O)→ $HCl \cdot H_2N-(CH_2)_4-NH_2 \cdot HCl$

***OS*, *CV 4*, 953 (1963)**

$C_6H_5NMe_2$ + $(NC)_2C=C(CN)_2$ —(DMF)→ $(p\text{-}Me_2N)C_6H_4(NC)C=C(CN)_2$

***OS*, *CV 4*, 957 (1963)**

2-formyl-2,4,4-trimethylcyclopentanone (O, CHO, CH_3, CH_3, CH_3) —(NaOH; H_2O, Et_2O)→ 2,4,4-trimethylcyclopentanone (O, CH_3, CH_3, CH_3)

***OS*, *CV 5*, 179 (1973)**

$CH_3-C(=O)-C(=N_2)-C(=O)-O\text{-}t\text{-}Bu$ —(NaOMe; MeOH, 0 - 5 °C)→ $H-C(=N_2)-C(=O)-O\text{-}t\text{-}Bu$

OS, CV 5, 235 (1973)

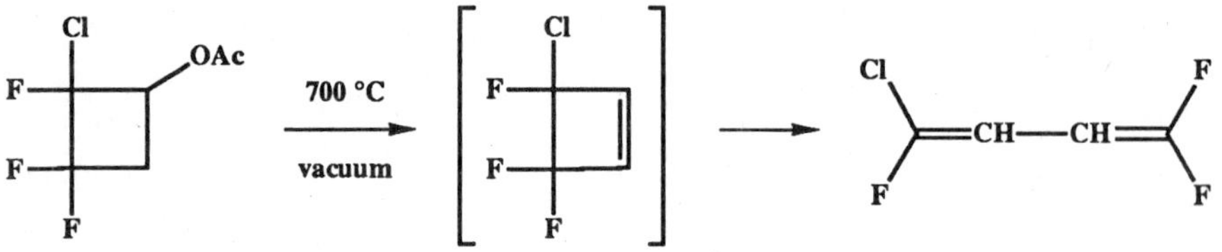

700 °C, vacuum

OS, CV 5, 266 (1973)

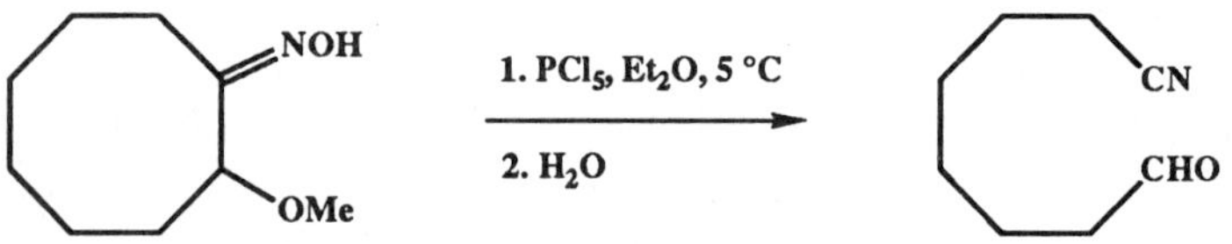

1. PCl_5, Et_2O, 5 °C
2. H_2O

OS, CV 5, 297 (1973)

hν, CH_2Cl_2
(- 2 CO)

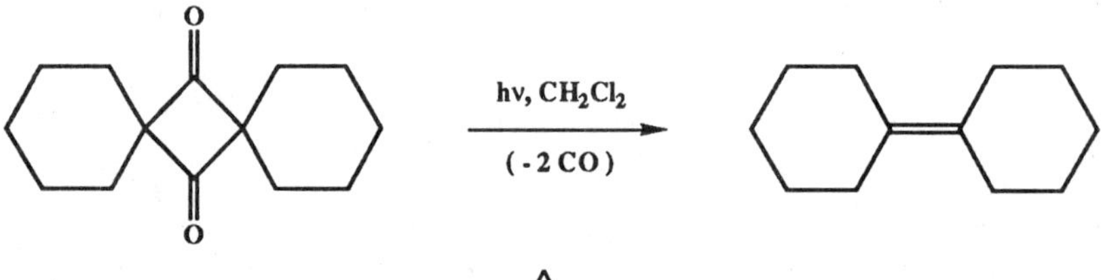

OS, CV 5, 384 (1973)

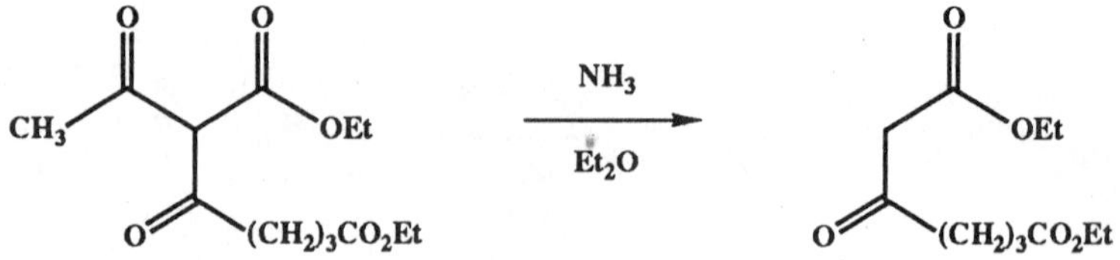

NH_3
Et_2O

OS, *CV 5*, 533 (1973)

NaOH

EtOH, Δ

$NaO_2C(CH_2)_5$ $(CH_2)_8$ $(CH_2)_5CO_2Na$

OS, *CV 5*, 604 (1973)

Δ

(- CO)

OS, *CV 5*, 679 (1973)

550 °C

2 $CH_2{=}C{=}O$

OS, *CV 5*, 706 (1973)

$AlCl_3$, CS_2, Δ

then aq. HCl

OS, CV 5, **734 (1973)**

CO_2Me

CO_2Me

pyrolysis

$CH_2{=}C{=}CH{-}CO_2Me$

+

$CH_2{=}CH{-}CO_2Me$

OS, CV 5, **767 (1973)**

O O

CH_3 CH_3

Cl

K_2CO_3, EtOH, Δ

O O

CH_3 CH_3

O

CH_3

OS, CV 5, **1013 (1973)**

NC CN

NC CN

$CH_2(CN)_2$

pyridine

H_2O, Δ

NC CN

$(py\text{-}H)^+$

NC $^-C(CN)_2$

$Me_4N^+\ Cl^-$

H_2O, Δ

NC CN

Me_4N^+

NC $^-C(CN)_2$

OS, CV 5, **1037 (1973)**

+

I

Ph

CO_2^-

+

Ph

Ph

O

Ph

Ph

$C_6H_4Et_2$, Δ

(- CO, - CO_2)

Ph

Ph

Ph

Ph

***OS*, *CV 5*, 1074 (1973)**

$C_6H_5NH_2$ + Cl_3C–CO–CCl_3 → (hexane, 65 - 70 °C) → C_6H_5NH–CO–CCl_3

***OS*, *CV 6*, 187 (1988)**

1,1-dibromo-2,2-diphenylcyclopropane → (AgOAc, HOAc; 100 - 120 °C) → Ph_2C=C(Br)CH_2OAc

***OS*, *CV 6*, 196 (1988)**

cis-3,4-diacetoxybicyclo[4.2.0]octa-2,4-diene (trans-OAc) + MeO_2C–C≡C–CO_2Me → (Δ, C_6H_6) → [AcO, AcO, CO_2Me, CO_2Me adduct] → AcO–CH=CH–CH=CH–OAc + $C_6H_4(CO_2Me)_2$

***OS*, *CV 6*, 210 (1988)**

2,5-diazido-3,6-di-*t*-butyl-1,4-benzoquinone (N_3, *t*-Bu, O) → (Δ, benzene) → 2 (*t*-Bu)(NC)C=C=O

***OS*, *CV 6*, 268 (1988)**

NC, NC, CN, CN, O

$(n\text{-Bu})_2S$

NC, NC, O

+

n-Bu, n-Bu, S, CN, CN

***OS*, *CV 6*, 368 (1988)**

H, CO_2Me

NaN_3, H_2SO_4

$CHCl_3$, 40 °C

O

***OS*, *CV 6*, 389 (1988)**

O, CHOH

p-TsN_3, Et_3N

CH_2Cl_2, - 5 °C

O, N_2

***OS*, *CV 6*, 422 (1988)**

MeO_2C, CO_2Me, Cl, Cl

Δ

vacuum

Cl, Cl

+

CO_2Me, CO_2Me

OS, CV 6, 590 (1988)

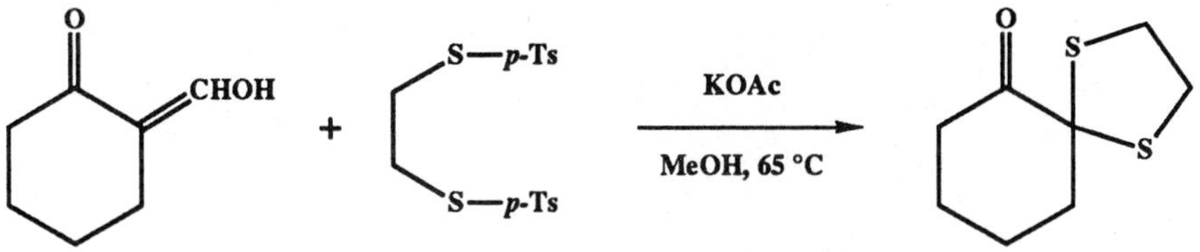
O
CHOH
+
S—p-Ts
S—p-Ts
KOAc
MeOH, 65 °C
O
S
S

OS, CV 6, 618 (1988)

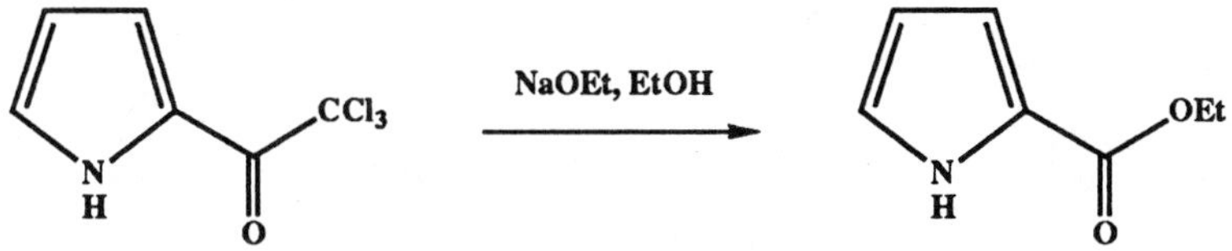
N
H
CCl3
O
NaOEt, EtOH
N
H
OEt
O

OS, CV 6, 625 (1988)

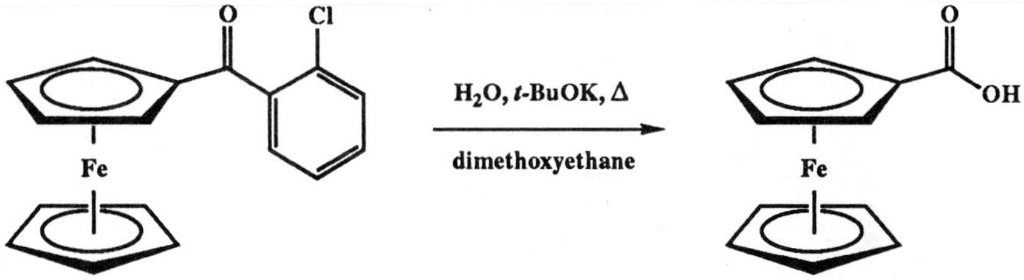
O
Cl
Fe
H2O, t-BuOK, Δ
dimethoxyethane
O
OH
Fe

OS, CV 6, 679 (1988)

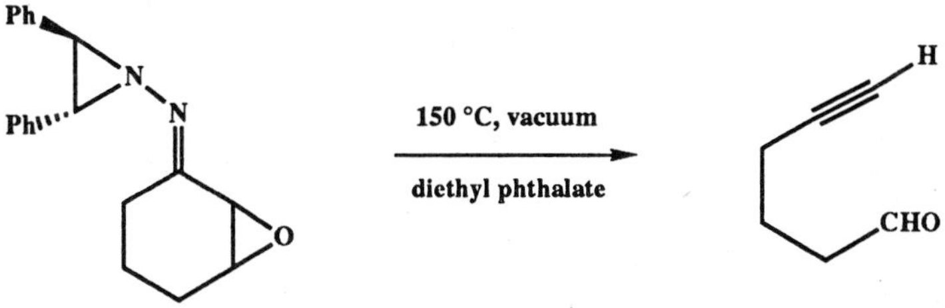
Ph
Ph
N
N
O
150 °C, vacuum
diethyl phthalate
H
CHO

OS, *CV 7*, 12 (1990); *60*, 6 (1981)

Ph, H, Cl, Cl — 2 NaOH / 2 EtOH, Δ → Ph, $(EtO)_2CH$

OS, *CV 7*, 297 (1990); *63*, 127 (1985)

O, HO, O, OH, H, H, OH, OH — *d* isomer — 1. 2 H_2O_2 2 Na_2CO_3 / 2. 4 HCl → HO, OH, H, H, O, O — *d* isomer

OS, *65*, 17 (1987)

OEt, OEt + $OSiMe_3$, $OSiMe_3$ — $SnCl_4$ → $OSiMe_3$, CO_2Et

OS, *66*, 132 (1987)

O, C—NH_2 — 1. $(CF_3CO_2)_2$ PhI / 2. H_2O → $\overset{+}{N}H_3$ Cl^-

PROTECTION / DEPROTECTION

OS, CV 1, 1 (1941)

CH_3CHO → (2 EtOH, $CaCl_2$) → $CH_3CH(OEt)_2$

OS, CV 3, 644 (1955)

PhCHO → (fuming HNO_3, H_2SO_4) → 3-$NO_2C_6H_4CHO$ → (MeOH, HCl (cat.)) → 3-$NO_2C_6H_4CH(OMe)_2$

OS, CV 4, 21 (1963)

$CH_2=CHCHO$ + $HC(OEt)_3$ → (NH_4NO_3, EtOH) → $CH_2=CHCH(OEt)_2$

OS, CV 6, 954 (1988)

$CH_2=CHCHO$ → (1. Br_2, 0 °C; 2. $HC(OEt)_3$, EtOH) → $BrCH_2CHBrCH(OEt)_2$

PROTECTION / DEPROTECTION - [+ P] - CHO ; CO

***OS*, *CV* 7, 521 (1990); *62*, 196 (1984)**

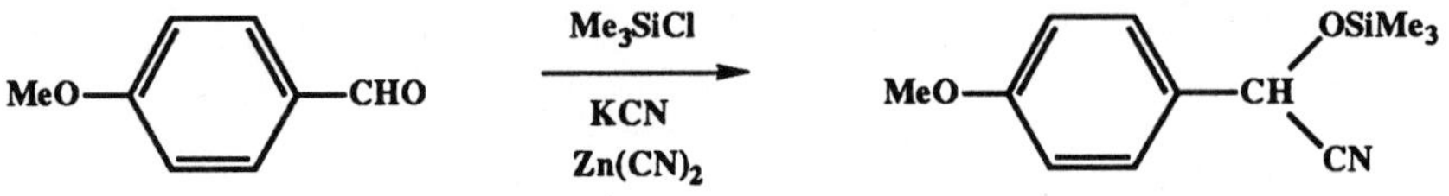
MeO
CHO
Me3SiCl
KCN
Zn(CN)2
MeO
CH
OSiMe3
CN

***OS*, *CV* 7, 59 (1990); *62*, 140 (1984)**

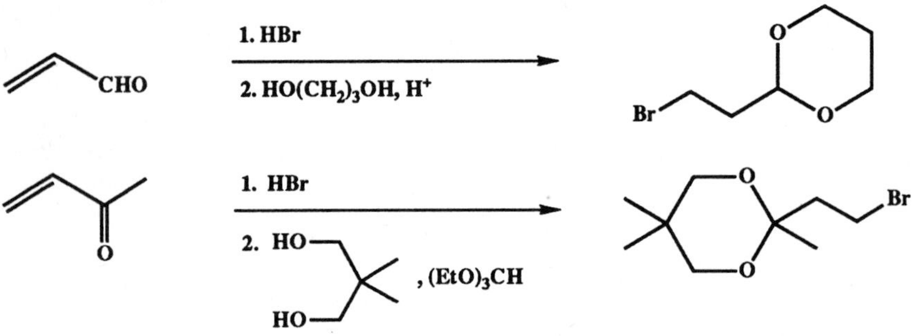
CHO
1. HBr
2. HO(CH2)3OH, H+
Br
O
O
O
1. HBr
2. HO
HO
, (EtO)3CH
O
O
Br

***OS*, *CV* 5, 303 (1973)**

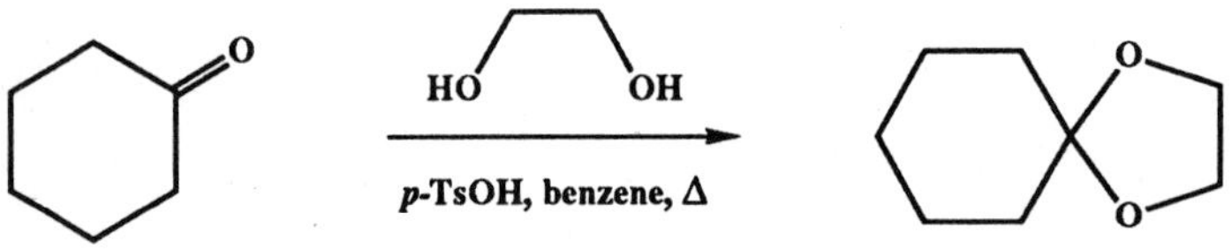
O
HO
OH
p-TsOH, benzene, Δ
O
O

***OS*, *CV* 6, 327 (1988)**

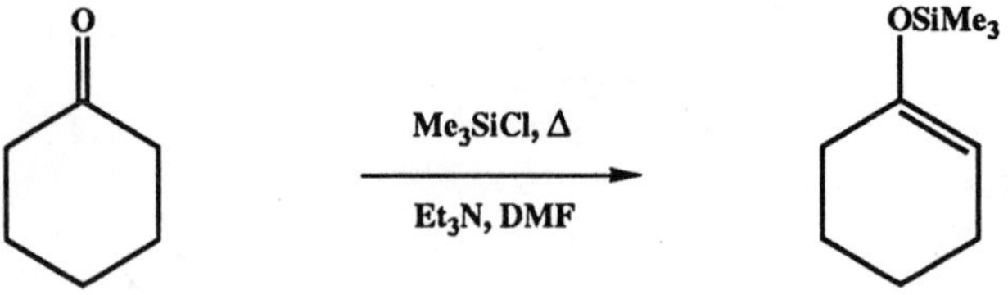
O
Me3SiCl, Δ
Et3N, DMF
OSiMe3

OS, CV 6, 445 (1988)

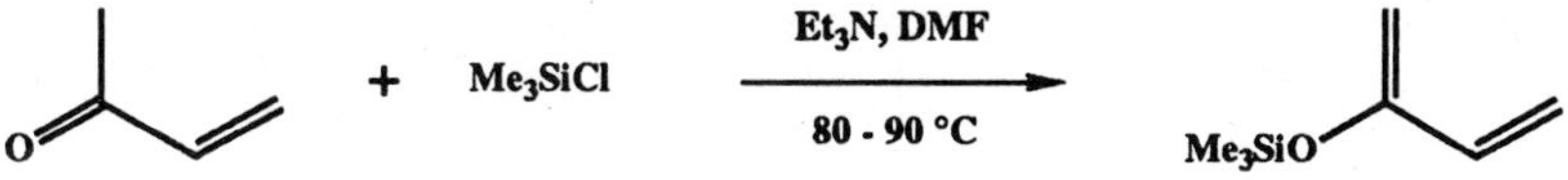
+
Me3SiCl
Et3N, DMF
80 - 90 °C
Me3SiO

OS, CV 7, 517 (1990); *60,* 126 (1981)

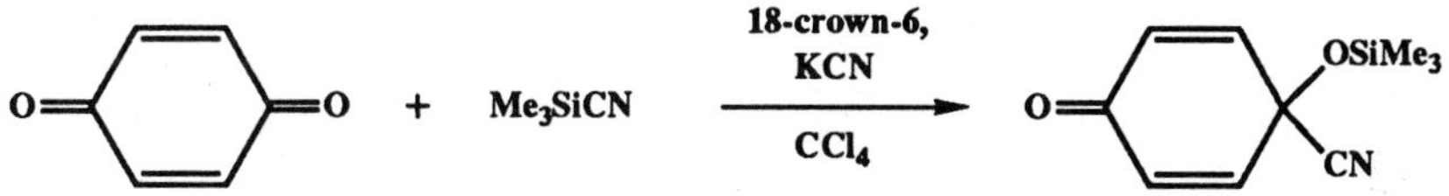
+
Me3SiCN
18-crown-6,
KCN
CCl4
OSiMe3
CN

OS, CV 7, 124 (1990); *61,* 74 (1983)

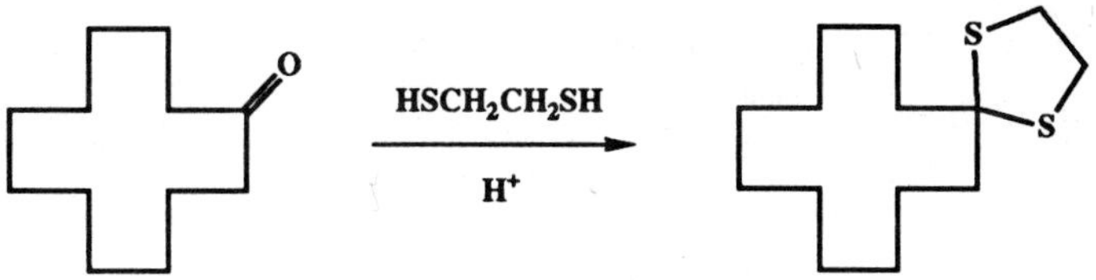
HSCH2CH2SH
H+

OS, CV 7, 512 (1990); *61,* 122 (1983)

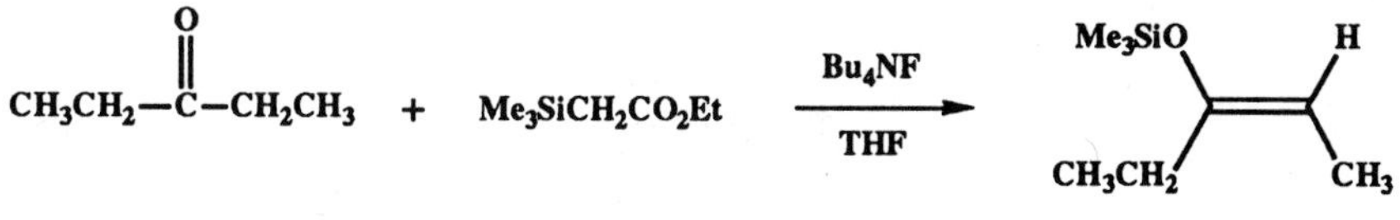
CH3CH2—C—CH2CH3
+
Me3SiCH2CO2Et
Bu4NF
THF
Me3SiO
H
CH3CH2
CH3

OS, *CV 7*, 312 (1990); *61*, 147 (1983)

OMe

Me_3SiCl
Et_3N, $ZnCl_2$
benzene

OMe
Me_3SiO

OS, *CV 7*, 424 (1990); *62*, 95 (1984)

O

Me_3SiCl, Et_3N

$OSiMe_3$

OS, *CV 7*, 241 (1990); *64*, 73 (1986)

Me Cl
Cl
O

1. $HOCH_2CH_2OH$, *p*-TsOH, C_6H_6, reflux; remove solvents, *p*-TsOH
2. 2 equiv. *n*-BuLi, THF, - 78 °C

Me
O
O

OS, *CV 7*, 282 (1990); *64*, 118 (1986)

O

1. LDA, DME, - 15 °C
2. Me_3SiCl

$OSiMe_3$

***OS*, *65*, 1 (1987)**

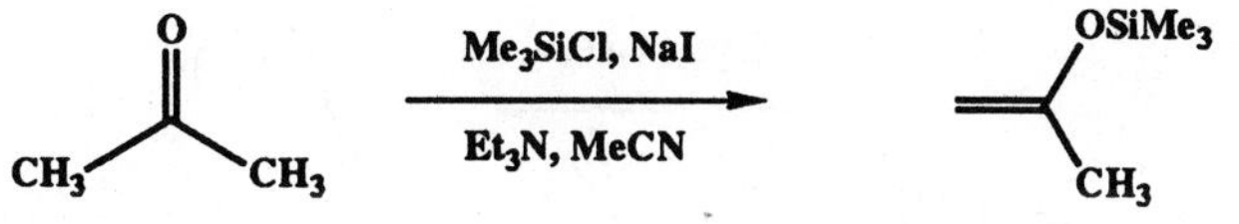

Me_3SiCl, NaI

Et_3N, MeCN

***OS*, *65*, 32 (1987)**

H_2SO_4

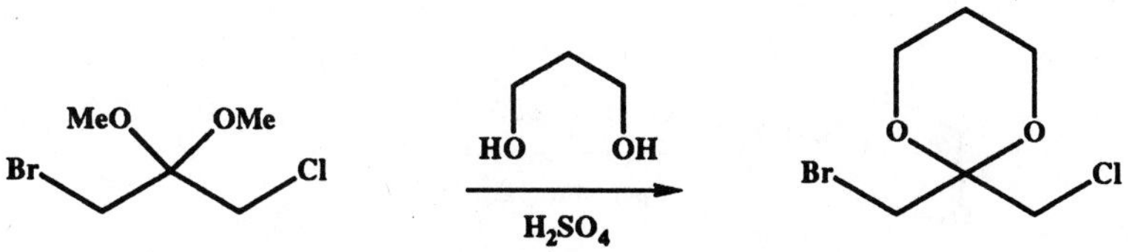

***OS*, *67*, 202 (1988)**

$HC(OCH_3)_3$

CH_3OH, *p*-TsOH

55 °C, 5 hr

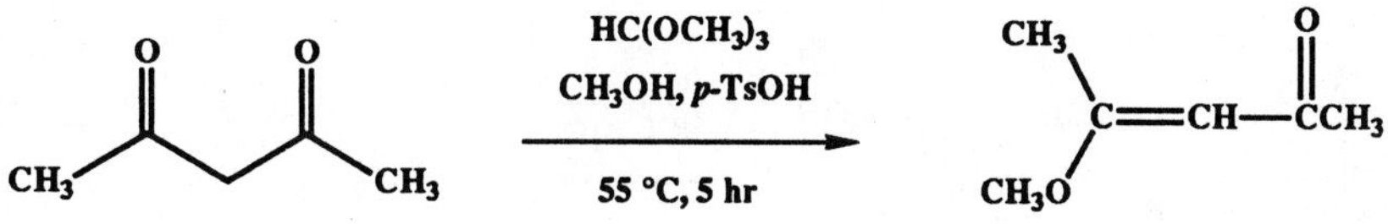

***OS*, *CV 1*, 285 (1941)**

Ac_2O, NaOAc

benzene, Δ

CH_2OH → CH_2OAc

***OS*, *CV 3*, 502 (1955)**

HO OH OH

acetone, *p*-TsOH

pet. ether, Δ

O O OH

***OS*, *CV 4*, 679 (1963)**

O Ph H + $HOCH_2$ CH_2OH $HOCH_2$ CH_2OH

HCl, H_2O

Ph O—CH_2 O—CH_2 CH_2OH CH_2OH

***OS*, *CV 7*, 381 (1990); *63*, 79 (1985)**

O OH

$OSiMe_3$
MeC=$NSiMe_3$

100 °C

O $OSiMe_3$ + O NH_2

OS, *CV* 7, 297 (1990); *63*, 127 (1985)

HO OH H H O O *D* isomer → $Me_2C(OMe)_2$ / *p*-TsOH → O O H H O O *D* isomer

OS, *CV* 7, 34 (1990); *63*, 214 (1985)

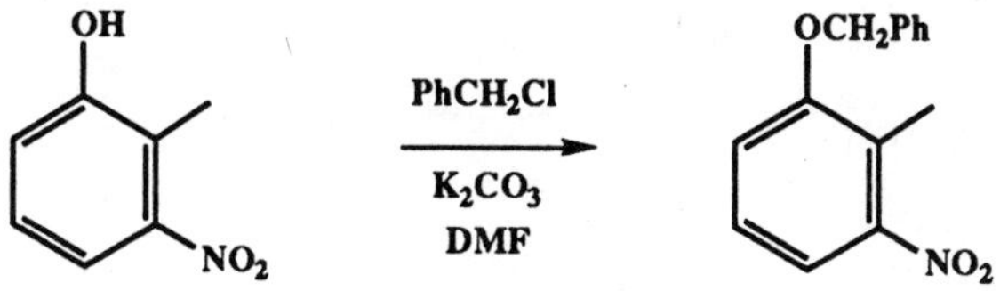

OH NO_2 → $PhCH_2Cl$ / K_2CO_3 DMF → OCH_2Ph NO_2

OS, *CV* 7, 160 (1990); *64*, 80 (1986)

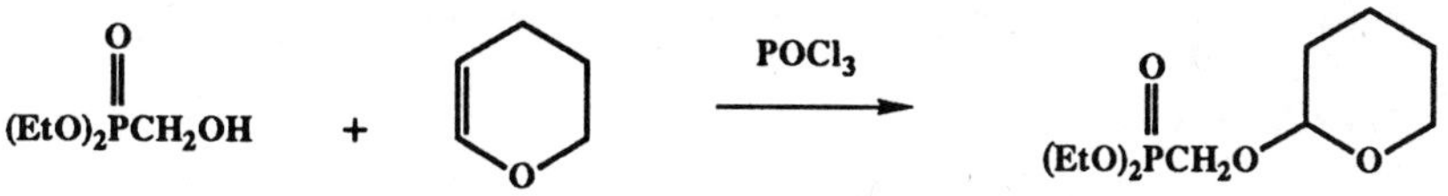

$(EtO)_2P(O)CH_2OH$ + (3,4-dihydro-2H-pyran, O) → $POCl_3$ → $(EtO)_2P(O)CH_2O$–(tetrahydropyran-2-yl, O)

OS, CV 6, 104 (1988)

$PhCH_2Br$, KOH

DMSO, 25 °C

N

H

N

CH_2Ph

OS, CV 6, 106 (1988)

NaH

HMPA

25 °C

N

H

N

Na^+ -

$PhCH_2Cl$

0 - 25 °C

N

CH_2Ph

OS, CV 6, 203 (1988)

N

H

CO_2H

l - proline

NH

Me_2N NMe_2

t-$BuOCO_2Ph$

DMSO

S

N

CO_2H

CO_2-*t*-Bu

OS, CV 6, 1004 (1988)

$PhNH_2$

Cl_3C—$COCF_3$

DMSO

O

PhNH CF_3

CO_2H

NH_2

HO

l - (-) - tyrosine

Cl_3C—$COCF_3$

DMSO

CO_2H

HN CF_3

HO

O

l isomer

OS, *CV 7*, 70 (1990); *63*, 160 (1985)

$PhCH_2CH(NH_2)CO_2H$ → (t-Bu-O_2C—O—CO_2-t-Bu, NaOH, H_2O, t-BuOH, 20 - 40 °C) → $PhCH_2CH(NHCO_2$-t-Bu$)CO_2H$

OS, *CV 7*, 75 (1990); *63*, 171 (1985)

$PhCH_2CH(NH_2)CO_2H$ → (t-Bu-O_2C—O—N=C(CN)Ph) → $PhCH_2CH(NHCO_2$-t-Bu$)CO_2H$

OS, *67*, 187 (1988)

PhNHOH → (AcCl, $NaHCO_3$, CH_2Cl_2, H_2O, 0 °C) → PhN(Ac)OH

OS, CV 7, **411 (1990);** ***60,*** **66 (1981)**

CO_2H

CO_2H

OAc

, H^+

O

O

O

O

OS, 68, **155 (1989)**

HO

(-) - menthol

1. *n*-BuLi, THF, < 20 °C

2.

O

Ph

OMe

THF, 1 hr

1. *n*-BuLi, THF, < 20 °C

2.

CO_2Me

N

, THF, 1 hr

O

Ph

O

(*1R*) - (-) isomer

O

N

O

(*1R*) - (-) isomer

OS, CV 1, **214 (1941)**

$Me_2N-C_6H_4-N=CH-C_6H_4-NMe_2$ → (CH_2O, 50% HOAc) → $H-C(=O)-C_6H_4-NMe_2$

(+ $Me_2NC_6H_4N=CH_2$)

OS, CV 2, **305 (1943)**

$HOCH_2CH(OH)CH(OEt)_2$ → (H_2SO_4, H_2O) → $HOCH_2CH(OH)CHO$

dl - glyceraldehyde

OS, CV 2, **323 (1943)**

$CH_3(CH_2)_4CH(OEt)_2$ → (H_2SO_4, H_2O, Δ) → $CH_3(CH_2)_4CHO$

OS, CV 2, **441 (1943)**

$O_2N-C_6H_4-CH(OAc)_2$ → (H_2SO_4, aq. EtOH, Δ) → $O_2N-C_6H_4-CHO$

OS, CV 3, 564 (1955)

$CH(OCH_3)_2$; NH_2; 1. aq. H_2SO_4; 2. $NaNO_2$, H_2SO_4; 3. H_2O, Δ; O; H; OH; $(CH_3)_2SO_4$; aq. NaOH; 40 - 50 °C; O; H; OCH_3

OS, CV 3, 701 (1955)

$CH(OEt)_2$; aq. H_2SO_4; Δ; CHO

OS, CV 5, 49 (1973)

CH_3; CH_3; Cl; CH_3; OMe; HCl, H_2O; CH_3; CH_3; H; CH_3; O

OS, CV 6, 64 (1988)

CH_3; CH_3; O; N; CH_3; OMe; $(CO_2H)_2$; H_2O, Δ; O; H; OMe

***OS*, *CV 6*, 312 (1988)**

H_3O^+

CHO

***OS*, *CV 6*, 358 (1988)**

=CH—CH=N—

H_3O^+

Δ

a

CHO

b

ca. 85 : 15

a : *b*

***OS*, *CV 6*, 448 (1988)**

$(EtO)_2P(=O)—CH_2CH(OEt)_2$ —(aq. HCl, Δ)→ $(EtO)_2P(=O)—CH_2CHO$

***OS*, *CV 6*, 683 (1988)**

$CH_3CH_2—CH(OH)—CH(SMe)—CH=CH(SMe)$

$HgCl_2$
$CaCO_3$
THF, H_2O
55 °C

$CH_3CH_2—CH(OH)—CH=CH—CHO$

OS, CV 6, 751 (1988)

t-Bu—CH_2—C(CH_3)(CH_3)—N=C(D)—CH(CH_3)CH_2CH_3 → (oxalic acid; H_2O, Δ) → O=C(D)—CH(CH_3)CH_2CH_3

OS, CV 6, 869 (1988)

2-$C_{14}H_{29}$-1,3-dithiane → (HgCl₂, HgO; aq. MeOH, Δ) → $(MeO)_2CH$—$C_{14}H_{29}$ → (p-TsOH; THF, Δ) → OHC—$C_{14}H_{29}$

OS, CV 6, 893 (1988)

Ph-azirine—CH(OMe)₂ → (HOAc, H_2O; dioxane, 90 °C) → Ph-azirine—CHO

OS, CV 6, 901 (1988)

C_6H_{11}—N=CH—CH_2—C(OH)(Ph)₂ → (oxalic acid; H_2O, 100 °C) → OHC—CH=C(Ph)₂

OS, CV 6, 905 (1988)

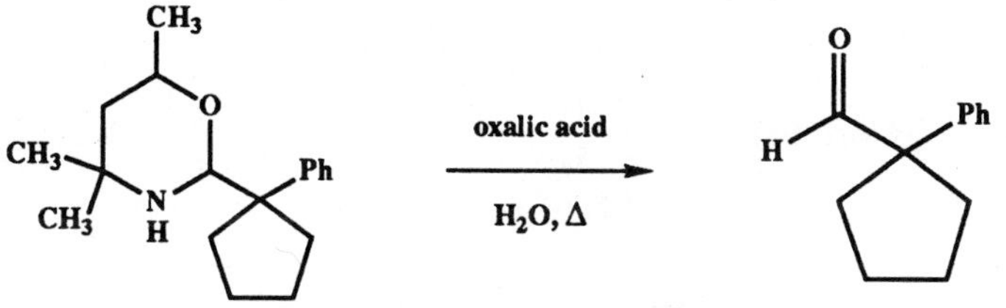

oxalic acid

H_2O, Δ

OS, CV 7, 12 (1990); *60,* 6 (1981)

H_2O, HCO_2H

+ 4 to - 4 °C

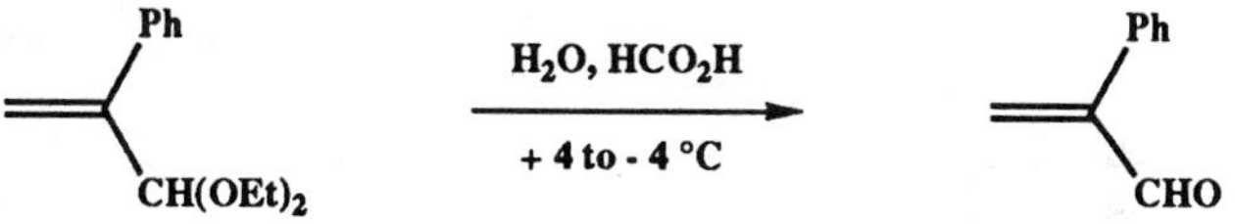

OS, CV 7, 162 (1990); *60,* 49 (1981)

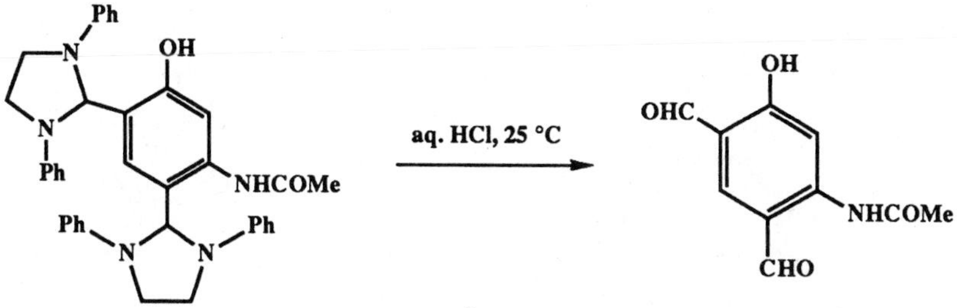

aq. HCl, 25 °C

OS, CV 7, 287 (1990); *60,* 72 (1981)

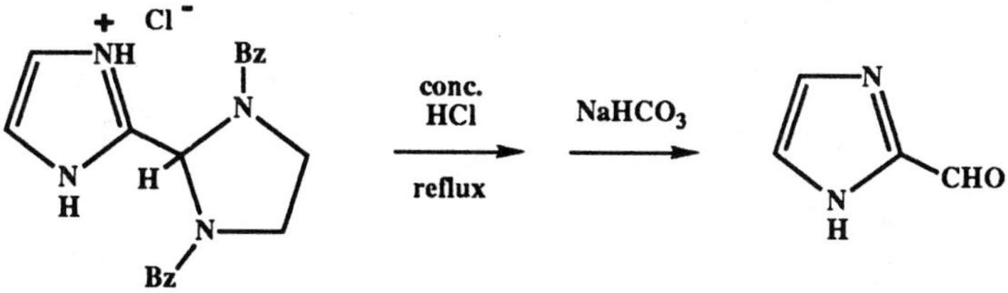

conc. HCl

reflux

$NaHCO_3$

OS, *65*, 119 (1987)

Ph, CH_3 C=CH—N=CHPh

1. *n*-BuLi, THF, - 78 °C
2. CH_2=CHCH$_2$Br, - 78 °C --> r.t.
3. H_3O^+, r.t.

Ph, CHO, CH_3

OS, *67*, 33 (1988)

NEt_2

(*R*) - (-)

H_2SO_4

CHO

(*R*) - (+)

OS, *CV 3*, 541 (1955)

CH_2OH, O, OH, HO, OCH_3, HO, H, H

α - methyl - *d* - mannoside

H_2O, H_2SO_4

Δ

CH_2OH, O, OH, HO, OH, HO, H, H

d - mannose

OS, *67*, 205 (1988)

S, HO, S

HgO, $BF_3 \cdot Et_2O$

H_2O, THF

CHO, OH

OS, *CV 1*, 217 (1941)

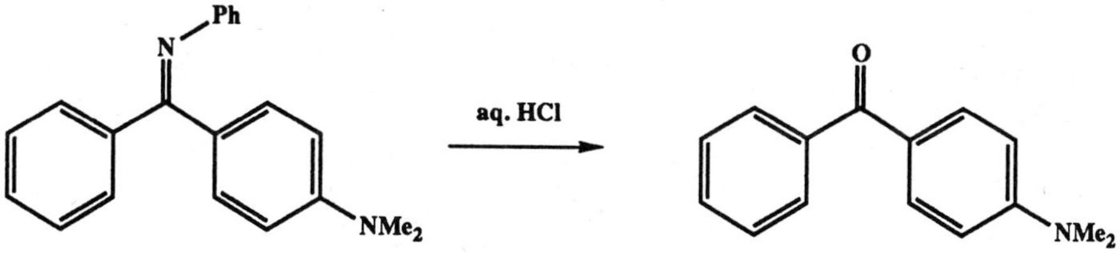

N-phenyl imine of 4-(dimethylamino)benzophenone (N–Ph, NMe$_2$) → aq. HCl → 4-(dimethylamino)benzophenone (O, NMe$_2$)

OS, *CV 2*, 519 (1943)

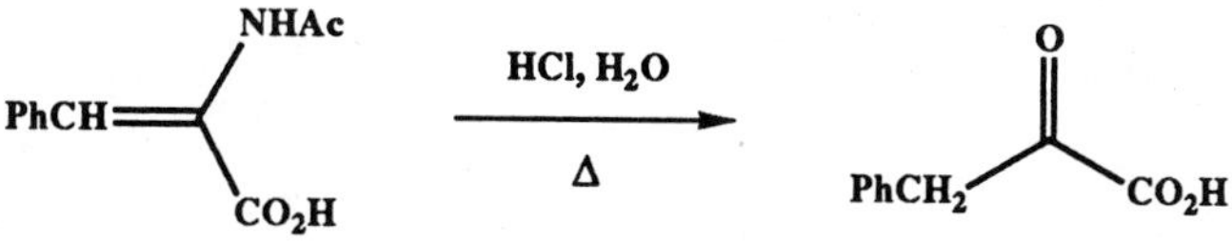

PhCH=C(NHAc)(CO$_2$H) → HCl, H_2O, Δ → $PhCH_2C(O)CO_2H$

OS, *CV 5*, 91 (1973)

7,7-dimethoxynorbornene (MeO, OMe) → H_2SO_4, H_2O → 7-norbornenone (O)

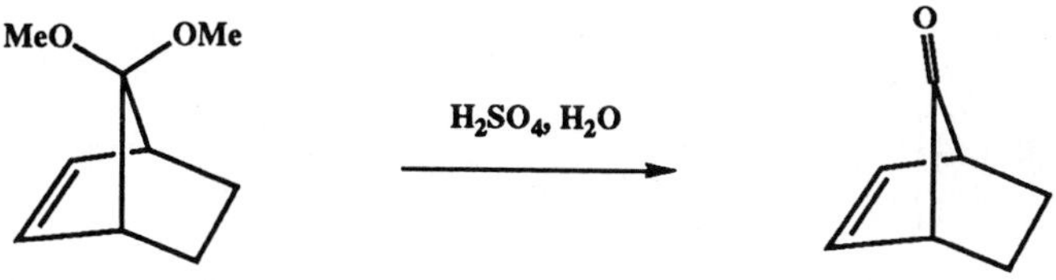

OS, *CV 5*, 294 (1973)

3-ethoxycyclohex-2-en-1-ol (OEt, OH) → H_2SO_4, H_2O → cyclohex-2-enone (O)

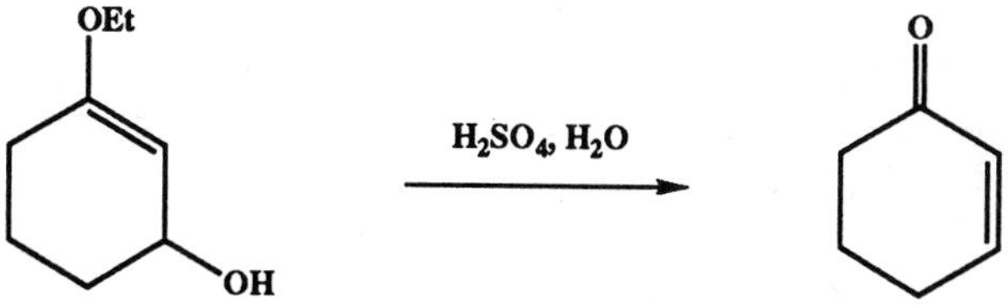

OS, *CV 6*, 109 (1988)

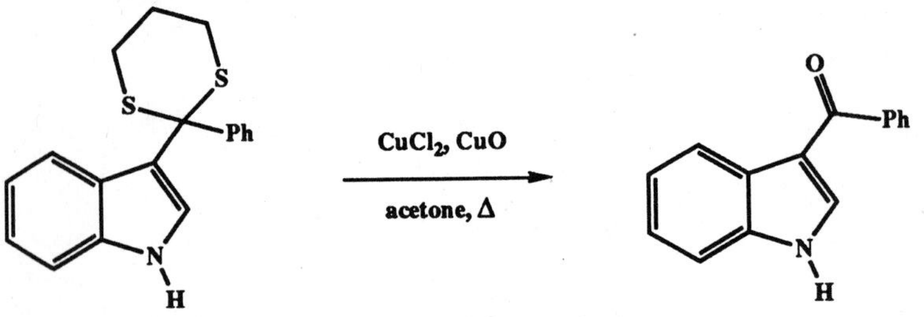

S, S, Ph, N, H

$CuCl_2$, CuO

acetone, Δ

O, Ph, N, H

OS, *CV 6*, 142 (1988)

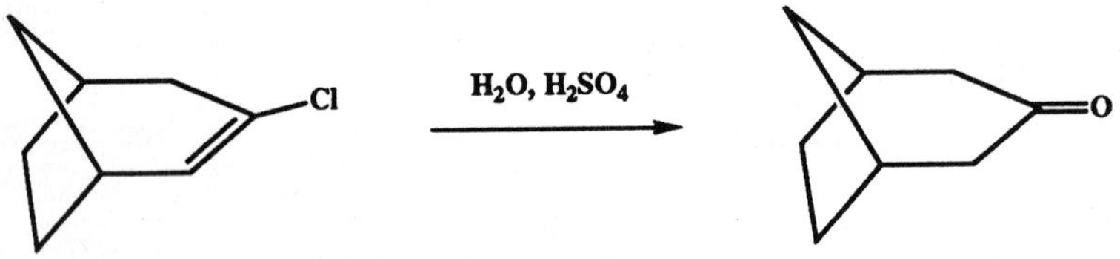

Cl

H_2O, H_2SO_4

O

OS, *CV 6*, 167 (1988)

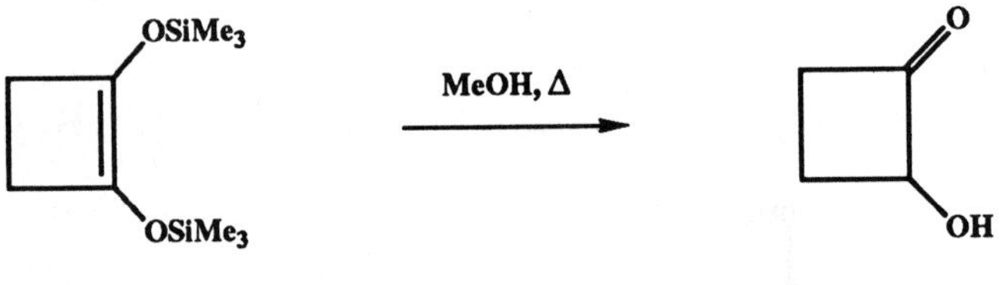

$OSiMe_3$

$OSiMe_3$

MeOH, Δ

O

OH

OS, *CV 6*, 242 (1988)

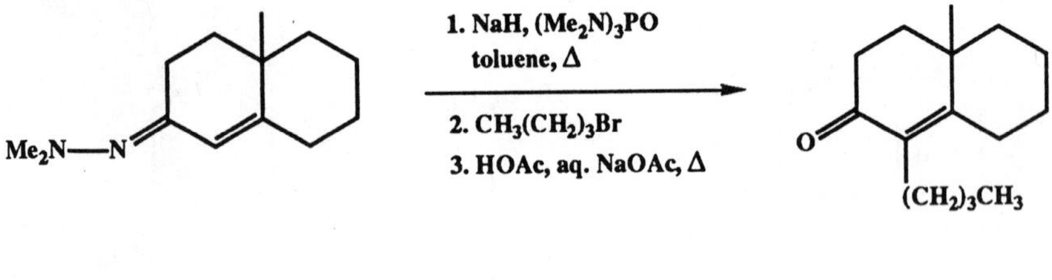

Me_2N—N

1. NaH, $(Me_2N)_3PO$
toluene, Δ

2. $CH_3(CH_2)_3Br$
3. HOAc, aq. NaOAc, Δ

O

$(CH_2)_3CH_3$

OS, CV 6, **316 (1988)**

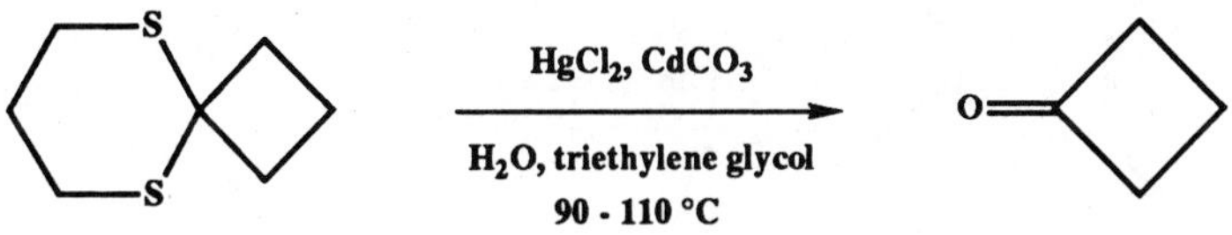

$HgCl_2$, $CdCO_3$

H_2O, triethylene glycol

90 - 110 °C

OS, CV 6, **361 (1988)**

MeO OMe

H_2SO_4

aq. CH_2Cl_2, 0 °C

O

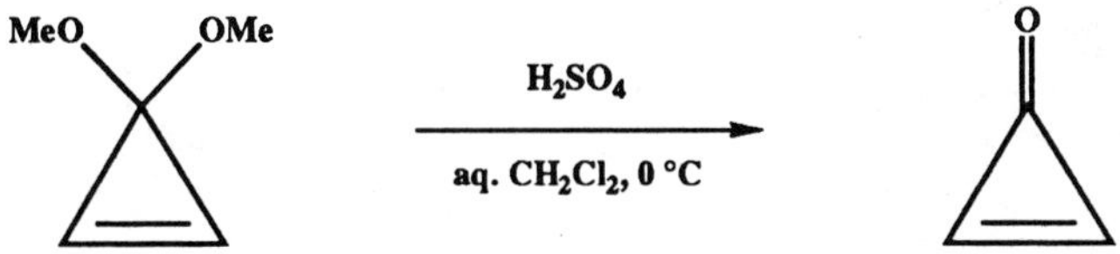

OS, CV 6, **996 (1988)**

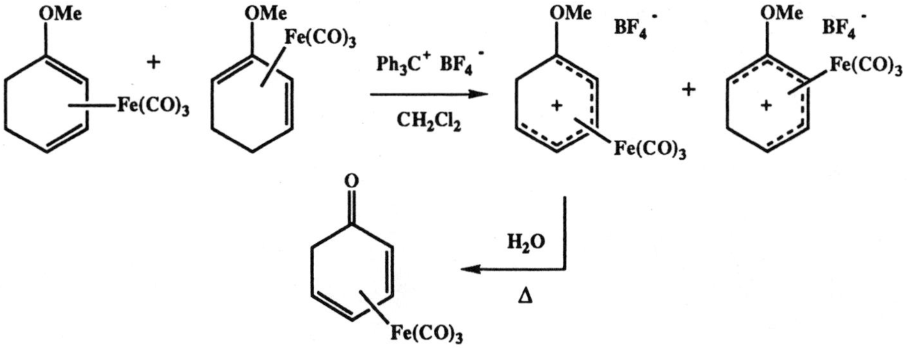

OMe + OMe Fe(CO)3

Fe(CO)3

Ph_3C^+ BF_4^-

CH_2Cl_2

OMe BF_4^- + OMe BF_4^-

Fe(CO)3 Fe(CO)3

O

H_2O

Δ

Fe(CO)3

OS, *CV 7*, 495 (1990); *64*, 196 (1986)

$CuSO_4$, acetone

p-Tol S

S

OS, *65*, 17 (1987)

$OSiMe_3$

H_3O^+

CO_2Et

OS, *65*, 183 (1987)

O_3, CH_2Cl_2
- 78 °C

Me Me

ZSS Me

S

OS, *68*, 25 (1989)

H OMe
H $SiMe_3$

H_3O^+

$CH_3C(=O)—SiMe_3$

OS, CV 1, 150 (1941)

OMe
OH
48% HBr
Δ
OH
OH

OS, CV 2, 549 (1943)

O
O
H
O
PCl_5
Cl
Cl
O
O
H
Cl Cl
H_2O
OH
OH
H
O

OS, CV 3, 586 (1955)

HO
MeO
O OH
Me
N
H
HI, P
Ac_2O, Δ
HO
HO
O OH
Me
N
H

OS, CV 6, 859 (1988)

MeO
OMe
CH_3
EtSNa
DMF, Δ
HO
OMe
CH_3

***OS*, *67*, 13 (1988)**

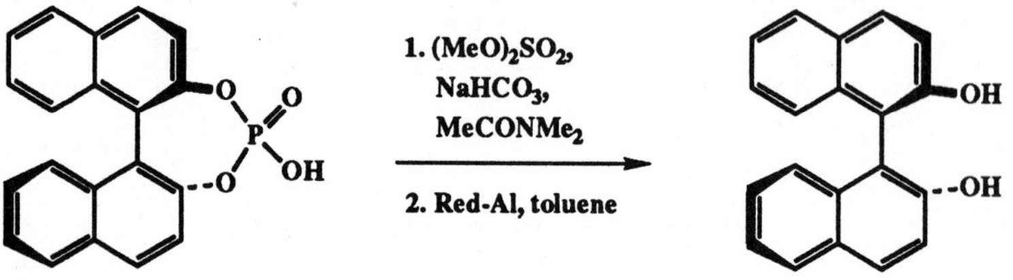
O
O
P
OH
O
1. (MeO)2SO2,
NaHCO3,
MeCONMe2
2. Red-Al, toluene
OH
OH

(*R*) - (-) (*R*) - (+)

***OS*, *CV 6*, 348 (1988)**

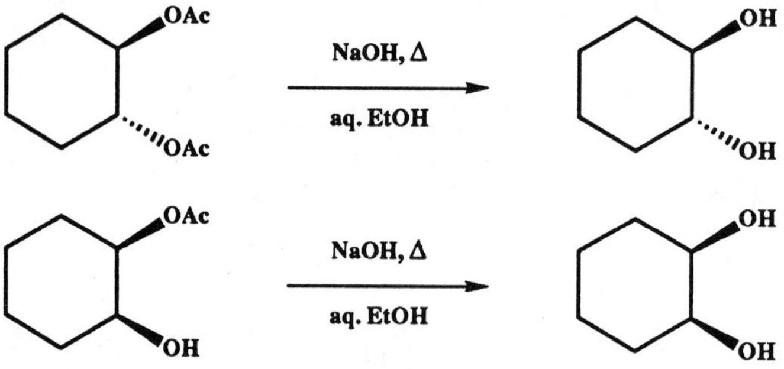
OAc
OAc
NaOH, Δ
aq. EtOH
OH
OH
OAc
OH
NaOH, Δ
aq. EtOH
OH
OH

***OS*, *CV 6*, 353 (1988)**

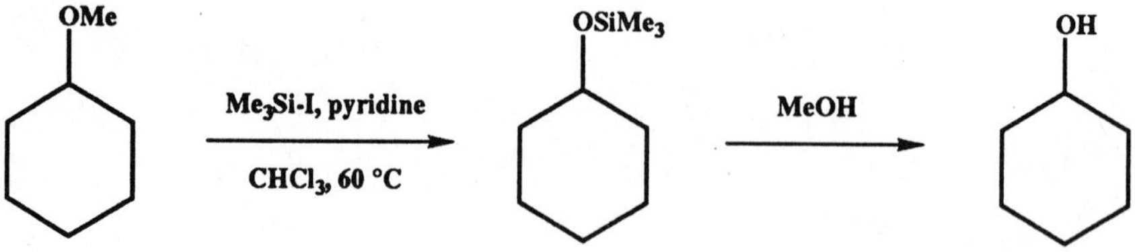
OMe
Me3Si-I, pyridine
CHCl3, 60 °C
OSiMe3
MeOH
OH

OS, CV 6, 1024 (1988)

NaOH

MeOH, H_2O

OAc

OH

CH_3 CH_3 CH_3 CH_3

OS, CV 3, 813 (1955)

OCH_3 CO_2H CH_3 Br

1. NH_4OH heat

2. HCO_2H Ac_2O

OCH_3 CO_2H CH_3 NHCHO

1. HBr, Δ

2. NH_4OH EtOH

OH CO_2H CH_3 NH_2

dl - threonine

OS, CV 7, 4 (1990); *62,* 149 (1984)

H Ac N O O H

KOH, MeOH

NH_2 OH

OS, CV 1, 381 (1941)

Ph N=CHPh OH

1. aq. HCl, Δ

2. aq. KOH

Ph NH_2 OH

OS, CV 2, 374 (1943)

$$\text{PhC(=O)–NH–}(CH_2)_4\text{–CH(NH}_2\text{)CO}_2\text{H} \xrightarrow{\text{aq. HCl, }\Delta} \text{HCl} \cdot \text{H}_2\text{N–}(CH_2)_4\text{–CH(NH}_2 \cdot \text{HCl)CO}_2\text{H}$$

dl - lysine di-HCl

OS, CV 2, 491 (1943)

$$\text{PhCH}_2\text{CH(NHAc)CO}_2\text{H} \xrightarrow[H_2O]{HCl} \text{PhCH}_2\text{CH(NH}_2\text{)CO}_2\text{H}$$

dl - β - phenylalanine

OS, CV 6, 252 (1988)

$$\text{PhCH}_2\text{O–C(=O)–NH–CH(CO}_2\text{H)–(CH}_2)_2\text{SMe} \xrightarrow[\text{Me}_2\text{NCOMe, liq. NH}_3\text{, - 33 °C}]{\text{H}_2\text{ (1 atm), Pd, Et}_3\text{N}} \text{H}_2\text{N–CH(CO}_2\text{H)–(CH}_2)_2\text{SMe}$$

l isomer

l - methionine

$$\text{PhCH}_2\text{O–C(=O)–NH–CH(CH}_2\text{O-}t\text{-Bu)–C(=O)–NH–CH(CH}_2\text{S-}t\text{-Bu)–CO}_2\text{-}t\text{-Bu} \xrightarrow{\text{(same)}} \text{H}_2\text{N–CH(CH}_2\text{O-}t\text{-Bu)–C(=O)–NH–CH(CH}_2\text{S-}t\text{-Bu)–CO}_2\text{-}t\text{-Bu}$$

l, l isomer

OS, *CV 3*, 661 (1955)

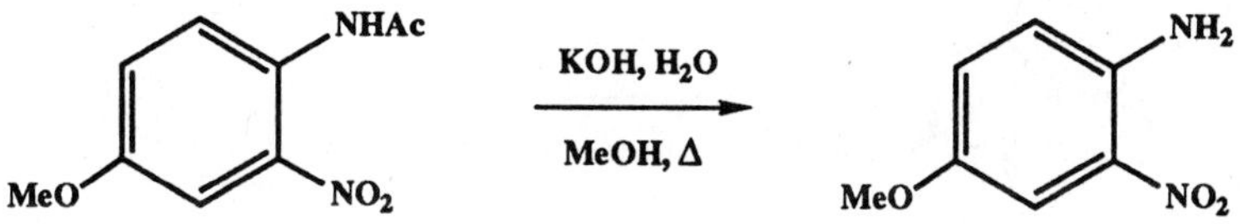

MeO, NHAc, NO_2 → (KOH, H_2O / MeOH, Δ) → MeO, NH_2, NO_2

OS, *CV 4*, 34 (1963)

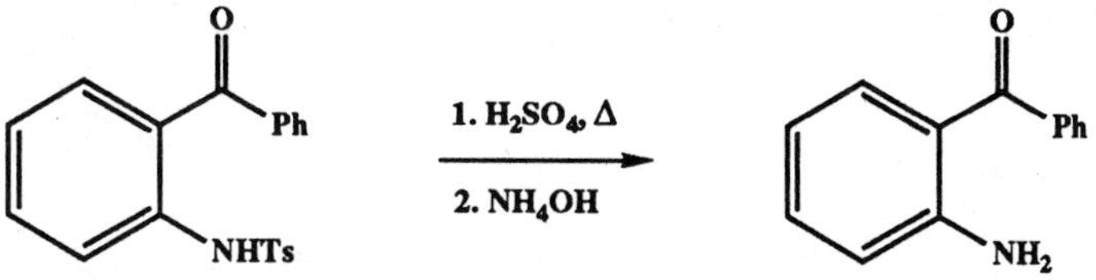

O, Ph, NHTs → (1. H_2SO_4, Δ / 2. NH_4OH) → O, Ph, NH_2

OS, *CV 4*, 42 (1963)

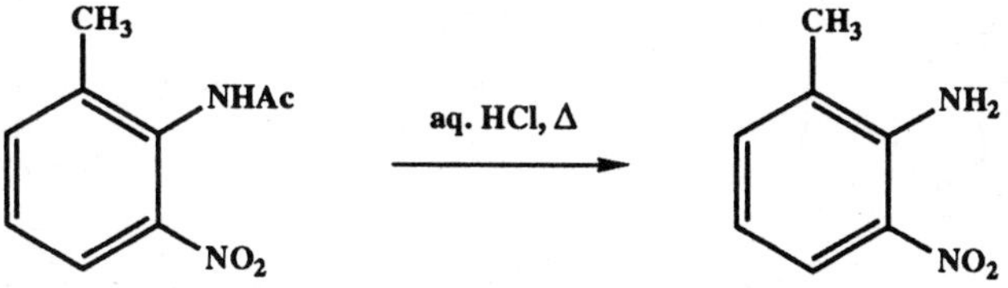

CH_3, NHAc, NO_2 → (aq. HCl, Δ) → CH_3, NH_2, NO_2

OS, *CV 5*, 406 (1973)

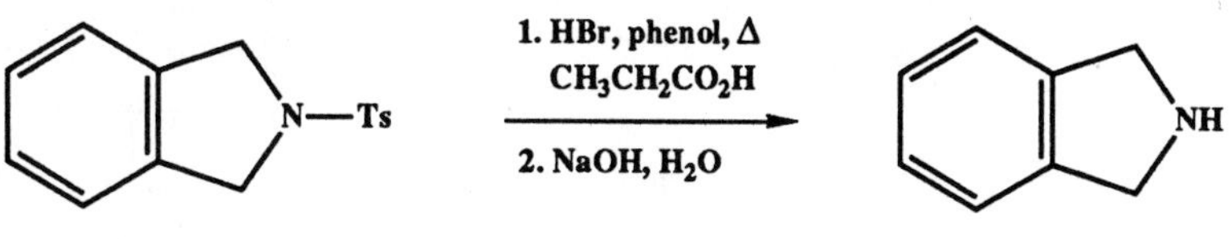

N—Ts → (1. HBr, phenol, Δ, $CH_3CH_2CO_2H$ / 2. NaOH, H_2O) → NH

OS, CV 6, 56 (1988)

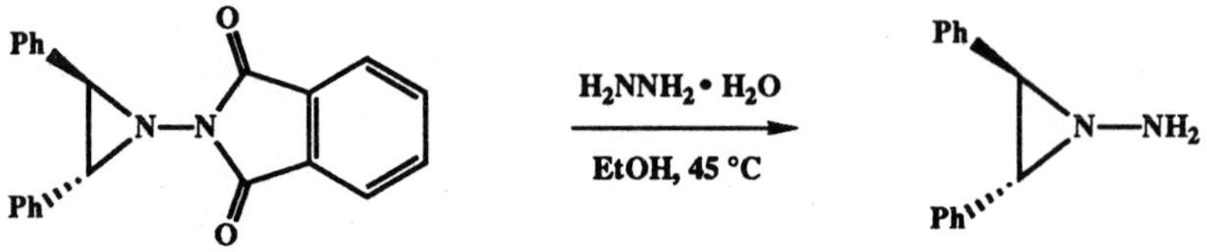

Ph, Ph, N, N, O, O

$H_2NNH_2 \cdot H_2O$

EtOH, 45 °C

Ph, Ph, N—NH_2

OS, CV 6, 507 (1988)

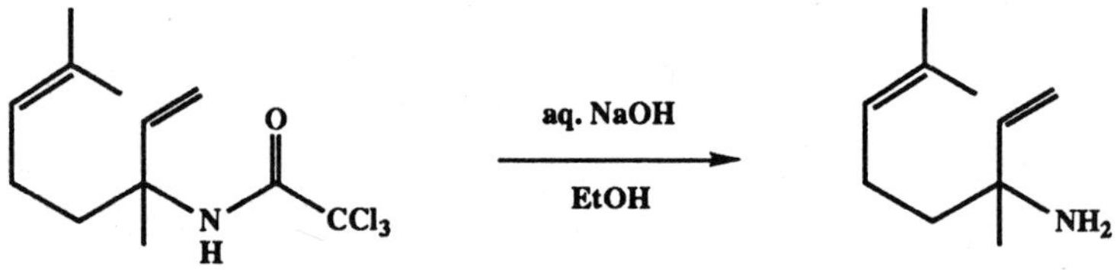

O, N, H, CCl_3

aq. NaOH

EtOH

NH_2

OS, CV 6, 652 (1988)

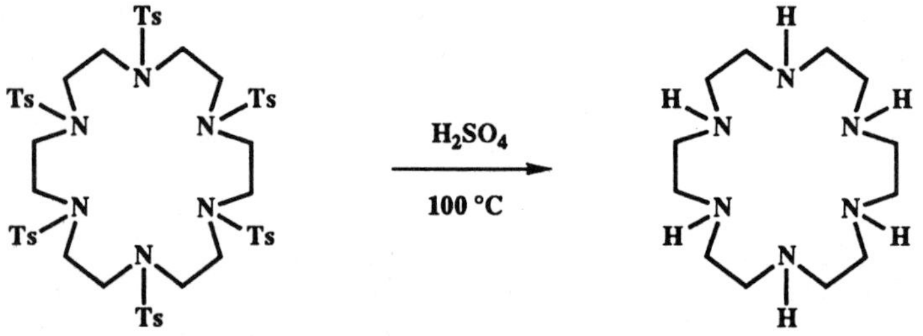

Ts, N, Ts, N, Ts, N, Ts, N, Ts, N, Ts, N

H_2SO_4

100 °C

H, N, H, N, H, N, H, N, H, N, H, N

OS, CV 6, 951 (1988)

P, N, O, O

H_2NNH_2

EtOH, Δ

P, NH_2

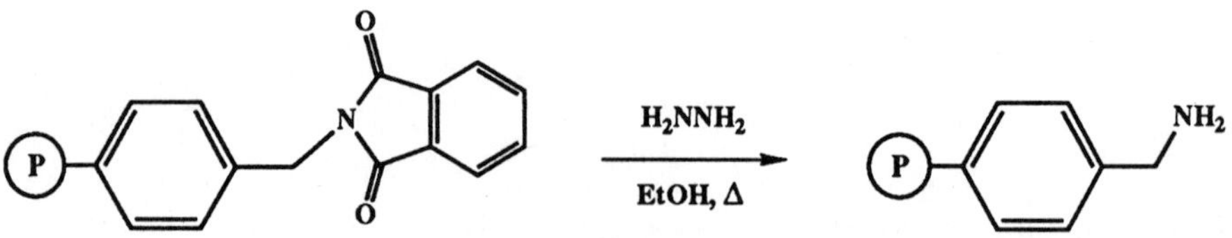

P = styrene - divinylbenzene copolymer

OS, CV 2, 384 (1943)

NaOH
EtOH, Δ
then HCl

N—$C(CO_2Et)_2$
$CH_2CH_2SCH_3$

CO_2H
H
N—$C(CO_2H)_2$
$CH_2CH_2SCH_3$

HCl
H_2O, Δ

$CH_3SCH_2CH_2$ CO_2H NH_2

dl - methionine

OS, CV 4, 55 (1963)

CO_2Et
N—$C(CO_2Et)_2$

HOAc, HCl
H_2O, Δ

HO_2C CO_2H NH_2

dl - aspartic acid

OS, CV 5, 93 (1973)

CCl_3

H_3PO_4, Δ

CO_2H

exo

OS, CV 7, 190 (1990); *63,* 99 (1985)

OH O

1. KOH,
MeOH
2. H_3O^+

OH COOH

OS, 68, 83 (1989)

CH$_2$Ph, S, O, OH, N, Ph, O, CH$_3$, O — H_2O_2, LiOH → HO, O, OH, Ph, CH$_3$, (2S*, 3S*) + CH$_2$Ph, S, NH, O, O

OS, CV 1, 388 (1941)

NH$_2$, NO$_2$, SO$_3$H — aq. H_2SO_4, Δ → NH$_2$, NO$_2$

OS, CV 4, 364 (1963)

NH$_2$, O$_2$N, NO$_2$, SO$_3$K — H_2O, H_2SO_4, Δ → NH$_2$, O$_2$N, NO$_2$

OS, CV 3, 809 (1955)

CH$_3$, NH$_2$ — 1. $NaNO_2$, aq. HCl; 2. EtOCS$_2$K, 40 - 45 °C → CH$_3$, S, S, OEt — 1. KOH, EtOH, Δ; 2. H_2SO_4, H_2O → CH$_3$, SH

OS, *CV* *6*, 824 (1988)

1. KOH, Δ ethylene glycol

2. HCl

SH

OS, *CV* *4*, 641 (1963)

N Ph CH_3 CN O

2 KOH, Δ

EtOH, H_2O

N CH_3

OS, *CV* *6*, 115 (1988)

NCOPh NC H

NaH $PhCH_2Cl$

DMF - 10 °C

NCOPh NC CH_2Ph

NaOH, Δ

aq. EtOH

N CH_2Ph

OS, *CV* *7*, 27 (1990); *61*, 14 (1983)

$PhCH_2NH$— N N N N H

NaOBr, H_2O

CH_2Cl_2, 0 °C

$PhCH_2N{\equiv}C$

OS, CV 3, 731 (1955)

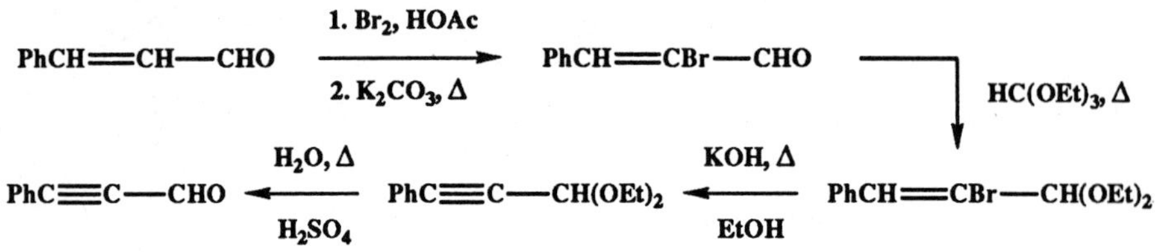

$$PhCH{=}CH{-}CHO \xrightarrow[2.\ K_2CO_3,\ \Delta]{1.\ Br_2,\ HOAc} PhCH{=}CBr{-}CHO \xrightarrow{HC(OEt)_3,\ \Delta} PhCH{=}CBr{-}CH(OEt)_2$$

$$PhCH{=}CBr{-}CH(OEt)_2 \xrightarrow[EtOH]{KOH,\ \Delta} PhC{\equiv}C{-}CH(OEt)_2 \xrightarrow[H_2SO_4]{H_2O,\ \Delta} PhC{\equiv}C{-}CHO$$

OS, CV 6, 567 (1988)

3-methoxy-4-ethoxybenzaldehyde (CH_3O, CH_3CH_2O) $\xrightarrow[benzene,\ \Delta]{HOCH_2CH_2OH,\ p\text{-TsOH}}$ 1,3-dioxolane acetal (CH_3O, CH_3CH_2O, H) $\xrightarrow[2.\ HCl,\ H_2O]{1.\ Ph_2PLi,\ THF,\ 25\ °C}$ 3-hydroxy-4-ethoxybenzaldehyde (HO, CH_3CH_2O)

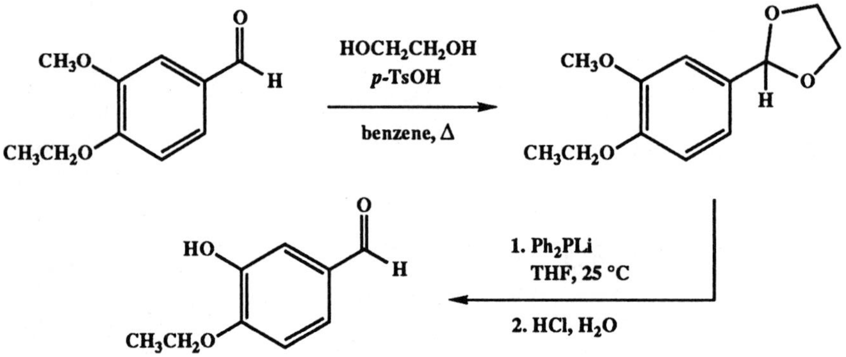

OS, CV 4, 903 (1963)

β-tetralone $\underset{aq.\ Na_2CO_3}{\overset{NaHSO_3,\ aq.\ EtOH}{\rightleftharpoons}}$ bisulfite adduct (SO_3Na, OH)

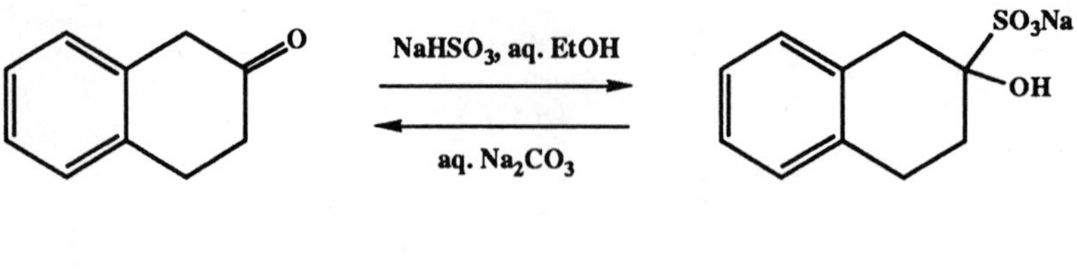

OS, CV 7, 20 (1990); *60,* 14 (1981)

$Ph_2C{=}O$ → (Me_3SiCN; ZnI_2, CH_2Cl_2) → $Ph_2C(OSiMe_3)(CN)$ → (3 *N* HCl; THF) → $Ph_2C(OH)(CN)$

OS, CV 7, 271 (1990); *61,* 65 (1983)

2-bromo-2-cyclopentenone → ($HOCH_2CH_2OH$; *p*-TsOH) → ethylene ketal → (1. *n*-BuLi; 2. CH_2O; 3. $(COOH)_2$, H_2O) → 2-(hydroxymethyl)-2-cyclopentenone

OS, CV 3, 432 (1955)

d - glucose → (1. Ph_3CCl, pyridine, Δ; 2. Ac_2O) → [6-O-trityl tetraacetate, $CH_2O{-}CPh_3$, OAc] → (HBr; HOAc) → β - *d* - glucose 1,2,3,4-tetraacetate

OS, *68*, 92 (1989)

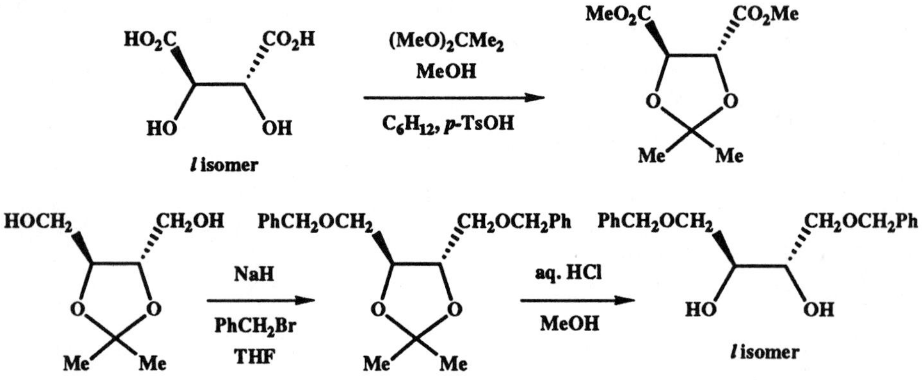

HO_2C CO_2H HO OH *l* isomer → [$(MeO)_2CMe_2$, MeOH, C_6H_{12}, *p*-TsOH] → MeO_2C CO_2Me O O Me Me

$HOCH_2$ CH_2OH O O Me Me → [NaH, $PhCH_2Br$, THF] → $PhCH_2OCH_2$ CH_2OCH_2Ph O O Me Me → [aq. HCl, MeOH] → $PhCH_2OCH_2$ CH_2OCH_2Ph HO OH *l* isomer

OS, *67*, 52 (1988)

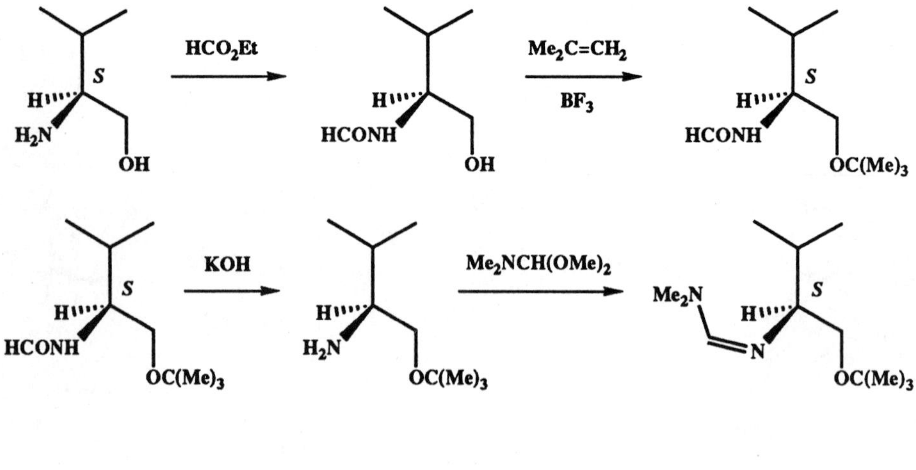

S H H_2N OH → [HCO_2Et] → H HCONH OH → [$Me_2C{=}CH_2$, BF_3] → *S* H HCONH $OC(Me)_3$

S H HCONH $OC(Me)_3$ → [KOH] → H H_2N $OC(Me)_3$ → [$Me_2NCH(OMe)_2$] → Me_2N *S* H N $OC(Me)_3$

OS, CV 1, 111 (1941)

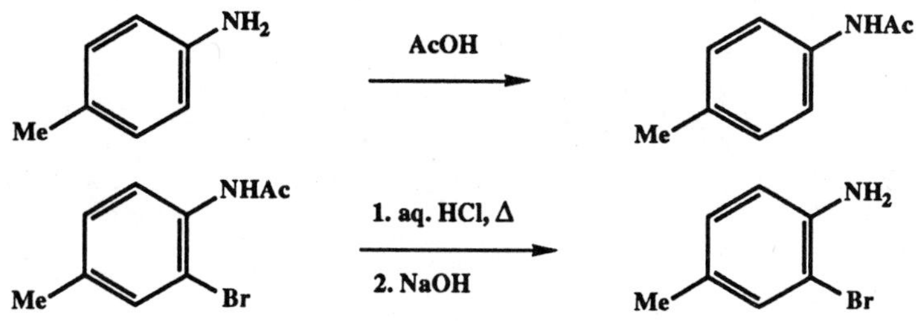

NH_2 — Me — AcOH → NHAc — Me

NHAc — Me — Br — 1. aq. HCl, Δ / 2. NaOH → NH_2 — Me — Br

OS, CV 2, 208 (1943)

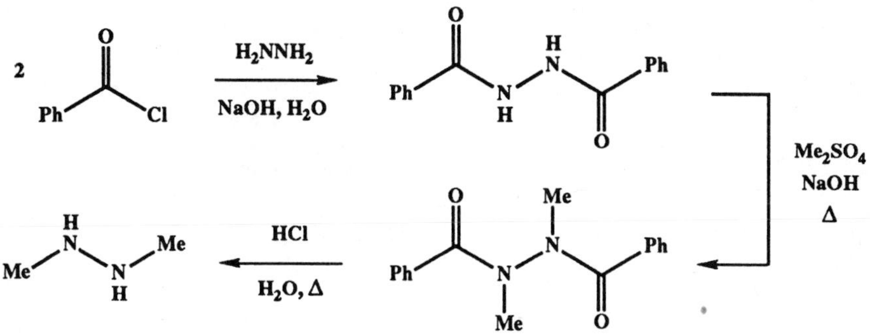

2 Ph–C(O)–Cl — H_2NNH_2 / NaOH, H_2O → PhC(O)NH–NHC(O)Ph — Me_2SO_4 / NaOH / Δ → PhC(O)N(Me)–N(Me)C(O)Ph — HCl / H_2O, Δ → MeNH–NHMe

OS, 67, 60 (1988)

MeO, MeO, NH — Me_2N–CH=N–C(H)(*S*)(i-Pr)–CH$_2$O-*t*-Bu / (+) - camphorsulfonic acid → MeO, MeO, N–CH=N–, *t*-Bu-O, H — 1. *t*-BuLi, MeI / 2. H_2NNH_2 → MeO, MeO, NH, H, Me — (*S*) - (-)

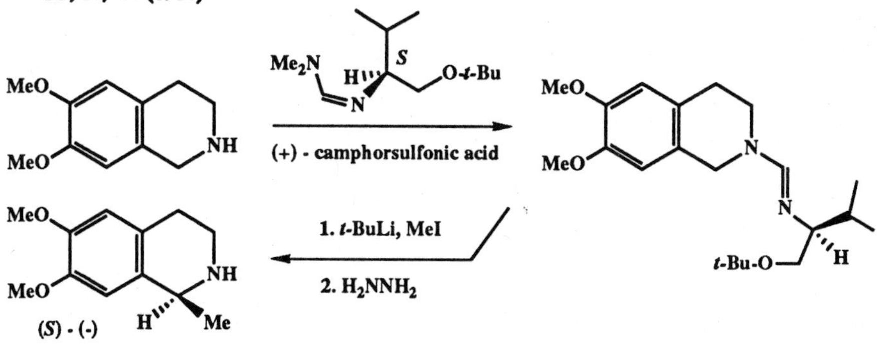

OS, *CV 2*, 97 (1943)

OS, *68*, 162 (1989)

Miscellaneous

OS, CV 1, 67 (1941)

Mesquite Gum → (H_2SO_4, H_2O, Δ) → *l* - arabinose

HO, OH, OH, H, CHO, H, H, OH

OS, CV 1, 194 (1941)

Human hair (keratin) → (1. Soap, H_2O; 2. aq. HCl, Δ; 3. NaOH, NaOAc; 4. aq. HCl) → *l* - cystine

HO_2C, *R*, S—S, *R*, CO_2H, NH_2, NH_2

OS, CV 1, 280 (1941)

Corn cobs ($C_5H_{10}O_5$) → (10% H_2SO_4, NaCl, Δ; (- H_2O)) →

O, CHO

OS, CV 1, 286 (1941)

Wheat Gluten → (1. HCl, Δ; 2. $PhNH_2$, H_2O) → *d* - glutamic acid

HO_2C, *R*, CO_2H, NH_2

OS, CV 1, 314 (1941)

$$NaCN + H_2SO_4 \xrightarrow{H_2O} HCN + NaHSO_4$$

OS, CV 1, 335 (1941)

$$C_{12}H_{22}O_{11} \xrightarrow[H_2O, \Delta]{HCl} 2\ CH_3COCH_2CH_2CO_2H$$

OS, CV 1, 366 (1941)

$$\text{Castor Oil} \xrightarrow[\Delta]{NaOH, H_2O} CH_3(CH_2)_5CH(OH)CH_3 + NaO_2C(CH_2)_8CO_2Na$$

OS, CV 1, 371 (1941)

Vegetable ivory waste $(C_6H_{10}O_5)_n$ $\xrightarrow[CH_3OH, \Delta]{H_2SO_4, HCl}$

CH₂OH, O, OH, HO, OCH₃, HO, H, H

d - α-methyl mannoside

OS, CV 1, 421, 424 (1941)

$HOOC{-}COOH \cdot 2\ H_2O \xrightarrow{\Delta} HOOC{-}COOH$

OS, CV 1, 463 (1941)

$H_2PtCl_6 \xrightarrow[\Delta]{6\ NaNO_3} Pt(NO_3)_4 \xrightarrow{\Delta} PtO_2 \xrightarrow{H_2O} PtO_2 \cdot H_2O$

OS, CV 1, 497 (1941)

$HO_2C{-}CH(OH){-}CH(OH){-}CO_2H$ → $HO_2C{-}CH(OH){-}CH(OH){-}CO_2H$

1. aq. NaOH, Δ
2. HCl
3. $CaCl_2$ then H_2SO_4

d - tartaric acid → *dl* - tartaric acid (+ *meso*)

OS, CV 1, 538 (1941)

Nutmegs → (Et_2O extraction) → $C_{13}H_{27}CO_2CH_2{-}CH(O_2CC_{13}H_{27}){-}CH_2O_2CC_{13}H_{27}$

OS, CV 2, 49 (1943)

Gelatine —hydrolysis→ Amino acids

H_2N–C(=NH)–NH–$(CH_2)_3$–CH(NH_2)–CO_2H ⇌ (PhCHO, aq. NaOH; reverse: HCl, H_2O *(as HCl salt)*) H_2N–C(=NH)–NH–$(CH_2)_3$–CH(N=CHPh)–CO_2H • H_2O

OS, CV 2, 53 (1943)

Castor Oil —KOH, EtOH, Δ→ $CH_3(CH_2)_5$–CH(OH)–CH_2–CH=CH–$(CH_2)_7CO_2H$

OS, CV 2, 120 (1943)

Milk —0.05 *M* HCl, pH = 4.6→ Casein

OS, CV 2, 124 (1943)

Cotton (cellulose) —H_2SO_4, Ac_2O→ α - cellobiose octaacetate

OS, CV 2, 142 (1943)

$Cu(NO_3)_2$, $Ba(NO_3)_2$, $(NH_4)_2CrO_4$ ⟶ Complex chromates $\xrightarrow{\text{ignition}}$ $CuCr_2O_4$, $BaCr_2O_4$

OS, CV 2, 150 (1943)

$$NaCN + Br_2 \xrightarrow{H_2O} BrCN + NaBr$$

OS, CV 2, 258 (1943)

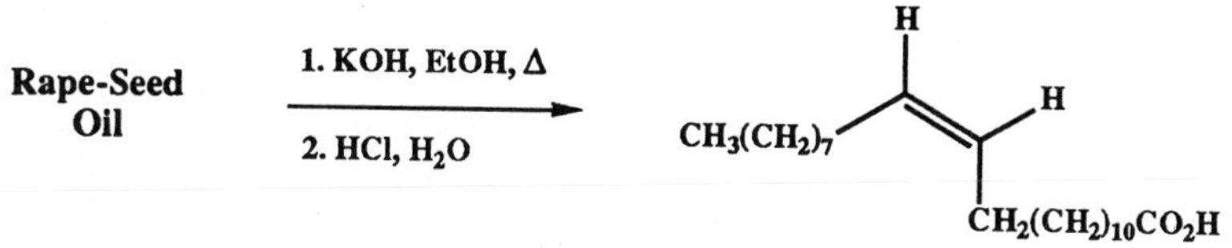
Rape-Seed Oil
1. KOH, EtOH, Δ
2. HCl, H2O
H
H
CH3(CH2)7
CH2(CH2)10CO2H

OS, CV 2, 330 (1943)

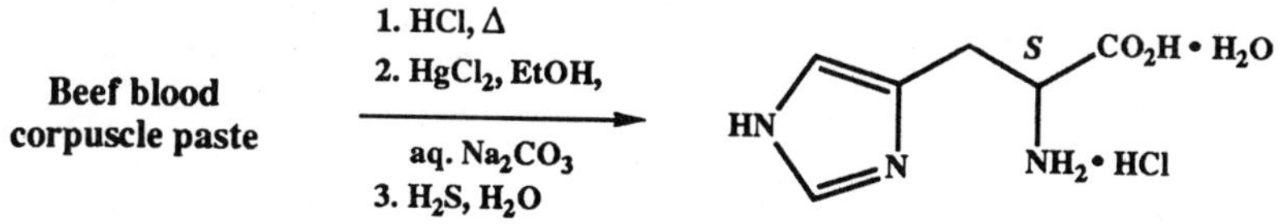
Beef blood corpuscle paste
1. HCl, Δ
2. HgCl2, EtOH,
aq. Na2CO3
3. H2S, H2O
HN
N
S
CO2H • H2O
NH2 • HCl

l - histidine HCl

OS, CV 2, 338 (1943)

$$H_2 + Br_2 \longrightarrow 2\ HBr$$

***OS*, *CV* 2,** **393 (1943)**

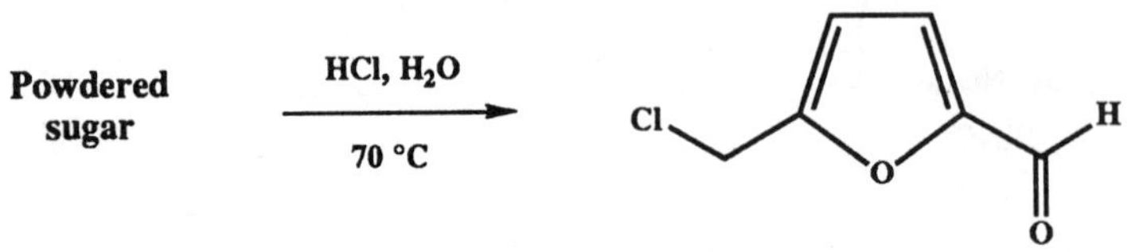

Powdered sugar $\xrightarrow[70\ °C]{HCl,\ H_2O}$ 5-(chloromethyl)furan-2-carbaldehyde (Cl, H, O)

***OS*, *CV* 2,** **506 (1943);** ***CV* 5,** **932 (1973)**

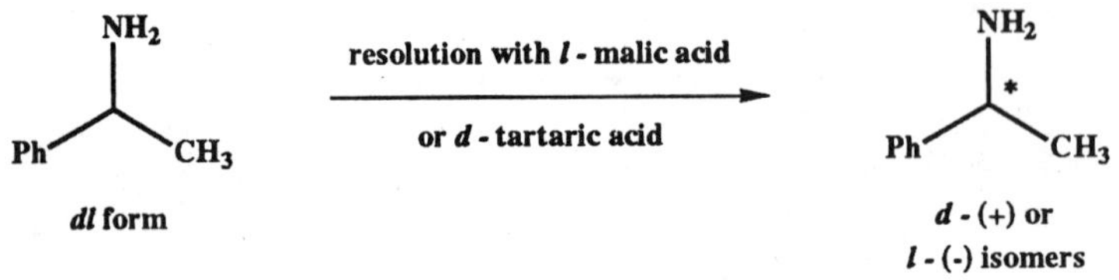

NH_2 / Ph / CH_3 (*dl* form) $\xrightarrow[\text{or } d\text{ - tartaric acid}]{\text{resolution with } l\text{ - malic acid}}$ NH_2 / Ph / CH_3 (*)

d - (+) or
l - (-) isomers

***OS*, *CV* 2,** **551 (1943)**

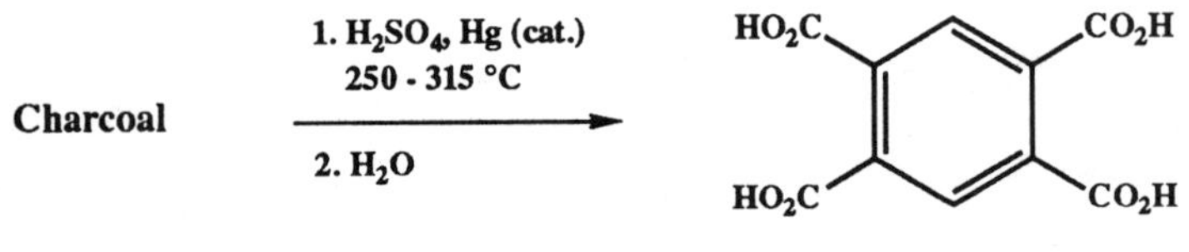

Charcoal $\xrightarrow[2.\ H_2O]{1.\ H_2SO_4,\ Hg\ (cat.)\ 250 - 315\ °C}$ HO_2C, CO_2H, HO_2C, CO_2H (benzenetetracarboxylic acid)

***OS*, *CV* 2,** **555 (1943)**

NH_4SCN $\xrightarrow[2.\ (NH_4)_2Cr_2O_7\ \ 3.\ H_2O]{1.\ 140 - 150\ °C}$ $NH_4\,[Cr(NH_3)_2(SCN)_4] \cdot H_2O$

Reinecke salt

***OS*, *CV* 2,** **612 (1943)**

Casein → 1. Pancreatin, NaF, aq. Na_2CO_3, toluene, 37 °C; 2. H_2SO_4, H_2O; 3. $HgSO_4$ → ***l*-tryptophane** + ***l*-tyrosine**

***OS*, *CV* 3,** **48 (1955)**

$$3\ t\text{-BuOH} + Al \xrightarrow[\text{benzene, } \Delta]{HgCl_2 \text{ (cat.)}} Al(O\text{-}t\text{-Bu})_3$$

***OS*, *CV* 3,** **176, 181 (1955)**

$$NiAl_2 + 6\ NaOH \xrightarrow{H_2O} \text{Raney Nickel W-6} + 2\ Na_3AlO_3 + 3\ H_2$$

$$NiAl_2 + 6\ NaOH \xrightarrow{H_2O} \text{Raney Nickel W-2} + 2\ Na_3AlO_3 + 3\ H_2$$

***OS*, *CV* 3,** **428 (1955)**

Glucose → Emulsin (β-glucosidase), toluene → **β - gentiobiose** → 8 Ac_2O → *(octa-acetate)*

OS, CV 3, 430 (1955)

Crab shells —(6 *N* HCl)→ Chitin —(HCl, Δ)→ *d* - glucosamine

CH$_2$OH, O, OH, H, OH, HO, H, NH$_2$ • HCl

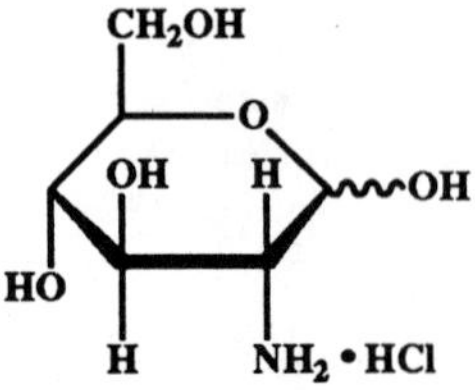

d - glucosamine

OS, CV 3, 442 (1955)

Defibrinated beef blood —(NaCl / HOAc, Δ)→ Hemin $C_{34}H_{32}O_4N_4FeCl$

OS, CV 3, 605 (1955)

Bayberry wax —(H_2SO_4 / MeOH, Δ)→ Methyl myristate $C_{13}H_{27}CO_2Me$ + Methyl palmitate $C_{15}H_{31}CO_2Me$

OS, CV 3, 673 (1955)

Oxygen —(silent electrical discharge)→ Ozone

OS, CV 3, 685 (1955)

$PdCl_2$ + H_2 → →

Pd - $BaSO_4$	(5% Pd)
Pd - C	(5% Pd)
Pd - C	(10% Pd)
$PdCl_2$ - C	(5% Pd)

OS, CV 3, **778 (1955)**

$$2\ Na + 2\ NH_3 \xrightarrow{\Delta} 2\ NaNH_2 + H_2$$

OS, CV 4, **1 (1963)**

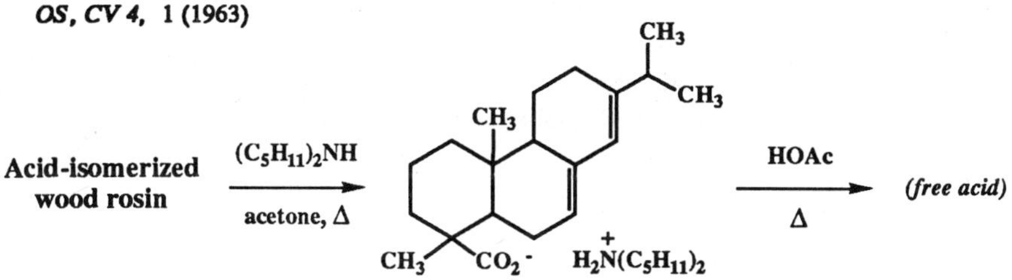

Acid-isomerized wood rosin $\xrightarrow[\text{acetone, }\Delta]{(C_5H_{11})_2NH}$ [structure: CH_3, CH_3, CH_3, CH_3, CO_2^- $H_2\overset{+}{N}(C_5H_{11})_2$] $\xrightarrow[\Delta]{HOAc}$ *(free acid)*

OS, CV 4, **207 (1963)**

$$NaCN + I_2 \xrightarrow[Et_2O]{H_2O} ICN + NaI$$

OS, CV 4, **950 (1963)**

$$CCl_4 + AlCl_3 + PCl_3 \xrightarrow{\Delta} [Cl_3CPCl_3]^+ [AlCl_4]^- \xrightarrow[CH_2Cl_2]{H_2O} Cl_3C{-}P(=O)Cl_2$$

OS, CV 5, **699 (1973)**

Longleaf pine *(Pinus palustris)* oleoresin $\xrightarrow[\text{acetone}]{NH_2C(CH_3)_2CH_2OH}$ [amine salt] $\xrightarrow{10\%\ H_3PO_4}$ [structure: CO_2H]

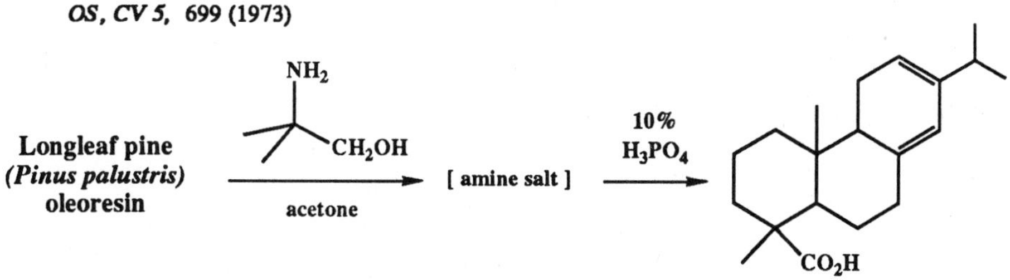

***OS*, *CV 6*, 826 (1988)**

CH_3–CH–NH_2 (1-naphthyl)

racemate

Resolution with (-) - DAG

acetone, reflux

DAG = 2,3 : 4,6-di-*O*-isopropylidene-2-keto-L-gulonic acid hydrate

NH_2, CH_3, C, H

***S* - (-) - isomer**

***OS*, *CV 1*, 418 (1941)**

O, O, O

***n*-C_6H_{13}**

CH_3 OH

$CO_2C_8H_{17}$

CO_2H

Brucine

acetone

(resolution)

$CO_2C_8H_{17}$

CO_2^-

(brucine salt)

$CO_2\overset{*}{C}_8H_{17}$

CO_2-(Brucine)

HCl

EtOH

$CO_2\overset{*}{C}_8H_{17}$

CO_2H

NaOH

H_2O

n-C_6H_{13}

CH_3 * OH

***d* and *l* resolved**